PHYSICAL FOUNDATIONS OF CONTINUUM MECHANICS

A. Ian Murdoch's *Physical Foundations of Continuum Mechanics* will interest engineers, mathematicians, and physicists who study the macroscopic behaviour of solids and fluids or engage in molecular dynamical simulations. In contrast to standard works on the subject, Murdoch's book examines physical assumptions implicit in continuum modelling from a molecular perspective. In so doing, physical interpretations of concepts and fields are clarified by emphasising both their microscopic origin and sensitivity to scales of length and time. Murdoch expertly applies this approach to theories of mixtures, generalised continua, fluid flow through porous media, and systems whose molecular content changes with time. Elements of statistical mechanics are included, for comparison, and two extensive appendices address relevant mathematical concepts and results. This unique and thorough work is an authoritative reference for both students and experts in the field.

A. Ian Murdoch is Professor Emeritus of Mathematics at the University of Strathclyde, Glasgow. His work on continuum mechanics has been widely published in such journals as the *Archive for Rational Mechanics and Analysis*, *Proceedings of the Royal Society*, *Journal of Elasticity*, *International Journal of Engineering Science*, *Continuum Mechanics and Thermodynamics*, and the *Quarterly Journal of Mechanics & Applied Mathematics*. He is the co-editor of two books: *Modelling Macroscopic Phenomena at Liquid Boundaries* and *Modelling Coupled Phenomena in Saturated Porous Materials*, and author of published lecture notes, *Foundations of Continuum Modelling*. Dr Murdoch has taught and lectured at many distinguished mathematics and engineering schools around the world.

Physical Foundations of Continuum Mechanics

A. Ian Murdoch
University of Strathclyde

CAMBRIDGE
UNIVERSITY PRESS

CAMBRIDGE UNIVERSITY PRESS
Cambridge, New York, Melbourne, Madrid, Cape Town,
Singapore, São Paulo, Delhi, Mexico City

Cambridge University Press
32 Avenue of the Americas, New York, NY 10013-2473, USA

www.cambridge.org
Information on this title: www.cambridge.org/9780521765589

First published 2012

Printed in the United States of America

A catalog record for this publication is available from the British Library.

Library of Congress Cataloging in Publication Data
Murdoch, A. I.
Physical foundations of continuum mechanics / A. Ian Murdoch.
pages cm
Includes bibliographical references and index.
ISBN 978-0-521-76558-9
1. Continuum mechanics. 2. Fluid mechanics. I. Title.
QC155.7.M87 2013
531–dc23 2012015692

ISBN 978-0-521-76558-9 Hardback

For Frances, Duncan, and Margaret

Contents

Preface

This work is intended to supplement and complement standard texts on continuum mechanics by drawing attention to physical assumptions implicit in continuum modelling. Particular attention is paid to linking continuum concepts, fields, and relations with underlying molecular behaviour via local averaging in both space and time. The aim is to clarify physical interpretations of concepts and fields and in so doing provide a sound basis for future studies. The contents should be of interest to engineers, mathematicians, and physicists who study macroscopic material behaviour.

The contents are the result of a long-standing study of formal and axiomatic presentations of continuum mechanics. Some of the issues were first addressed in courses delivered under the auspices of CISM[1] (Udine, 1986, 1987), University of Cairo (1994, 1996), and AMAS[2] (Warsaw, 2002; Bydgoszcz, 2003), and other topics treated in published papers. Here the opportunity has been taken to elaborate upon and extend earlier works and to present a unified, more readily accessible treatment of the subject matter.

Given the differing backgrounds of the intended readership, two extensive appendices have been included which develop relevant mathematical concepts and results. In particular, the use of direct (i.e., co-ordinate-free) notation is explained and related to that of Cartesian tensors.

No work exists in isolation: the author is above all indebted to his teachers Mort Gurtin and Walter Noll who introduced him to the mathematical precision and clarity of exposition to be found in modern continuum mechanics. The use of weighting function methodology, central to much of the discussion, and the role of projection operators in statistical mechanics were explained at length to the author by Dick Bedeaux. Appreciation of porous media modelling was gained by interactions with Jozef Kubik and Majid Hassanizadeh. It is also a pleasure to acknowledge the support and encouragement over the years of Mort Gurtin, Peter Chadwick, Harley Cohen, Paolo Podio-Guidugli, Gianpietro del Piero, Angelo Morro, Gérard Maugin, Witold Kosiński, Antonio Romano, Ahmed Ghaleb, David Steigman, and

[1] International Centre for Mechanical Sciences.

[2] Centre of Excellence for Advanced Materials and Structures, Institute of Fundamental Technological Research, Polish Academy of Sciences.

Eliot Fried. Extensive and comprehensive secretarial support for a TeX illiterate was provided in outstanding fashion by Mary McAuley. Finally, I am greatly indebted to my wife Margaret for her patience, support, and encouragement throughout the preparation of this work.

1 Introduction

1.1 Motivation

Material behaviour at length scales greatly in excess of molecular dimensions (i.e., *macroscopic* behaviour) is usually modelled in terms of the continuum viewpoint. From such a perspective the matter associated with any physical system (or *body*) of interest is, at any instant, considered to be distributed continuously throughout some spatial region (deemed to be the region 'occupied' by the system at this instant). Reproducible macroscopic phenomena are modelled in terms of deterministic continuum theories. Such theories have been highly successful, particularly in engineering contexts, and include those of elasticity, fluid dynamics, and plasticity. The totality of such theories constitutes (deterministic) continuum mechanics. The link between actual material behaviour and relevant theory is provided by experimentation/observation. Specifically, it is necessary to relate local experimental measurements to continuum field values. However, the value of any local measurement made upon a physical system is the consequence of a local (both in space *and* time) interaction with this system. Further, local measurement values exhibit erratic features if the scale (in space-time) is sufficiently fine, and such features become increasingly evident with diminishing scale. Said differently, sufficiently sensitive instruments *always* yield measurement values which fluctuate chaotically in both space and time (i.e., these values change perceptibly, in random fashion, with both location and time), and the 'strength' of these fluctuations increases with instrument sensitivity (i.e., with increasingly fine-scale interaction between instrument and system). This *intrinsic* property of material behaviour can only be understood in terms of the essentially discrete nature of matter; that is, it proves necessary to adopt a microscopic viewpoint. Accordingly, such fundamental understanding requires that measurement values be related to local interactions with (or 'samplings' of) fundamental discrete entities (that is, molecules, atoms, or ions) of the system.

While the understanding of small-scale material behaviour requires a microscopic basis, the success of deterministic continuum mechanics might suggest that such considerations are of little relevance to engineering practice. There are two main reasons why this is not the case. Firstly, erratic material behaviour can be manifest at the macroscopic level, as evidenced by turbulent fluid motions. Recourse to stochastic continuum modelling is necessary in such cases. The natures of the fields and balance relations of stochastic continuum mechanics can be fully understood only from the standpoint of microscopic considerations. Secondly, the macroscopic behaviour of any material system ultimately derives from its microscopic constitution, and in certain circumstances microstructural features may persist on a macroscopic scale and must be incorporated into continuum descriptions. (For example, in nematic phases of liquid crystals the co-operative effect of elongated molecules which tend to align with their neighbours is modelled in terms of a *director* field.) Further reasons for exploring the relationship between microscopic and macroscopic aspects of material behaviour are that it enhances the physical interpretation of continuum fields, clarifies basic continuum concepts, elucidates fundamental assumptions implicit in continuum modelling, and thereby improves awareness of the range of applicability of continuum mechanics. Such insight is essential in studies of nanoscale behaviour and in interpreting the results of molecular dynamical simulations.

The preceding remarks serve to motivate attempts to identify continuum field values with local space-time averages of microscopic quantities and to establish the balance relations satisfied by such fields. While these objectives constitute the main part of what follows, their consideration leads to natural implications for the modelling of fluid flow through porous media and for the manner in which observer consensus places restrictions upon constitutive relations. Elements of the probabilistic approach of classical statistical mechanics are outlined for comparison of viewpoints.

1.2 Contents

Basic elements of continuum mechanics are summarised in Chapter 2 for later reference. Included are discussions of the different physical interpretations to be placed on the notion of '*material point*' in solids and fluids, and the special case of rigid bodies.

Attention is drawn in Chapter 3 to conceptual problems associated with the continuum viewpoint. In particular, the manifest dependence of solid boundaries on scale is shown to imply similar sensitivity in mass density. Also discussed are the scale dependence of velocity, the inability to interpret the stress within a rarefied gas (i.e., its pressure) as a force per unit area, and the inappropriateness of deterministic continuum modelling at small length scales.

Local spatial averaging of the masses and momenta of fundamental discrete entities, modelled as point masses, is effected in Chapter 4 in terms of a weighting function w. The continuity equation is established for quite general, suitably normalised choices of w. A simple, physically distinguished choice w_ϵ, with associated length scale ϵ, is defined, and the corresponding boundary of any system of point masses at scale ϵ is thereby delineated. The physical interpretations of volume integrals of the mass and momentum densities appropriate to w_ϵ are obtained. Alternative choices of w are motivated and derived.

In Chapter 5 the velocity field (the ratio of the w-based density of momentum to that of mass) is employed to generate the corresponding motion map. (This is in contrast to the more standard derivation of velocity from a postulated motion map.) After discussing the subatomic origin of molecular interactions, a general *local* form of linear momentum balance is established directly (rather than being obtained in conventional fashion as the localisation of an integral relation) on the basis of assumed pairwise interactions between point masses of quite general nature. (In particular, the interaction between a pair of point masses may depend upon other point masses adjacent to each of this pair.) The balance relation contains an interaction force density $\mathbf{f}_w$. The usual form of balance follows from determination of an interaction stress tensor $\mathbf{T}_w^-$ for which $\operatorname{div}\mathbf{T}_w^- = \mathbf{f}_w$. The corresponding Cauchy stress tensor is $\mathbf{T}_w := \mathbf{T}_w^- - \mathcal{D}_w$, where $\mathcal{D}_w$ is a symmetric tensor of thermal character. (Here *thermal* refers to any quantity which depends upon velocities of individual point masses relative to the local w-scale continuum velocity field values: such relative velocities have random character and are also termed *thermal*.) The non-uniqueness of $\mathbf{T}_w^-$ is explored, and three distinct classes of solutions for pairwise-balanced interactions are examined and compared.

Local forms of energy balance are obtained directly in Chapter 6. If $(eq)_i$ denotes the equation which governs the motion of point mass P_i in an inertial frame, and $\mathbf{v}_i$ denotes the velocity of $\mathbf{v}_i$, then such forms of balance follow by summing relations $(eq)_i \,.\, \mathbf{v}_i\, w$ over all point masses. [Linear momentum balance followed from a similar sum of relations $(eq)_i\, w$.] The distinction between fields of thermal and mechanical character depends upon the presence or otherwise of thermal velocities in their definitions. If interactions are governed by separation-dependent pair potentials, the standard form of balance is obtained in which the internal energy density is the sum of densities of energy of assembly and of heat content (a local density of kinetic energy associated with thermal velocities).

Fine-scale relations are obtained in Chapter 7 by taking suitable moments. Summation of weighted products of masses with displacements of point masses from a given location $\mathbf{x}$ yields a measure $\mathbf{d}_w$ of local inhomogeneity. The time evolution of $\mathbf{d}_w$ gives rise to a relation which expresses moment of mass conservation. Summation of tensorial products of the preceding displacements with $(eq)_i\, w$ yields a generalised local moment of momentum balance. The skew part of this balance constitutes the usual moment of momentum balance: skew tensors can simply be replaced by their equivalent axial vector counterparts. Couple stresses and body couples emerge naturally, together with an internal moment of momentum density. A corresponding fine-scale energy balance is derived, and relative magnitudes of relevant fields are discussed. In contrast with axiomatic approaches, in which (axial vector-valued) moment of momentum balance is considered to determine the symmetry or otherwise of the Cauchy stress tensor, the explicit forms $\mathbf{T}_w$ obtained in Chapter 5 yield this information directly. Moment of momentum balance constitutes an evolution equation for internal moment of momentum, with contributions from $\mathbf{T}_w$, body couple density, and the divergence of the couple stress (a third-order tensor field).

Time averaging is introduced in Chapter 8 with the aim of obtaining field values which reflect local space-time averages: it is such averages that are to be related to local measurements. Time-averaged versions of the continuity equation and balances of momentum and energy are derived. Systems with changing material content are studied in terms of a 'membership' function for the system in question. Global considerations are addressed (with examination of details specific to rocketry and jet propulsion) before the corresponding local forms of balance for mass, momentum and energy are established.

The methodology developed in preceding chapters is applied to mixtures in Chapter 9 which includes the resolution of a paradox associated with incorrect interpretation of the notion of *partial stress* and an introduction to the modelling of reacting constituents.

Fluid flow through porous media is analysed in Chapter 10 at two different scales, one at which pore structure is evident (here the scale-dependent notion of boundary, established in Chapter 3, proves indispensable) and the other at which pores are no longer distinguishable. The small-scale balance of momentum is averaged over so-called representative elementary volumes using an appropriate weighting function. For the case in which incompressible linearly viscous fluid saturates pores, a sequence of relations is obtained, each of which follows from a specific and transparent modelling assumption, culminating in the Brinkman equation and Darcy 'law'.

An alternative averaging procedure is outlined in Chapter 11. This addresses behaviour which is scale-insensitive over a range of length scales (a typical assumption in continuum modelling) and is implemented in terms of so-called ϵ-cells.

Although specific constitutive equations are not discussed (other than for fluid flows in porous media), the definitions of field values in terms of microscopic quantities have implications which are analysed in Chapter 12. These implications are imposed by the fundamental requirement that observers must be able to agree upon the physical interpretations of the fields employed in continuum modelling. Matters are subtle: time averaging must be effected, instant by instant, over the same sets of molecules for all observers if a consensus is to be established. In accomplishing such averaging a crucial role is played by inertial observers. Once field values are established for this class of observers, it is possible to envisage how these values appear to a general observer. The objective natures of time-averaged fields (of mass, momentum, interaction force and external body force densities, together with those of stress and heat flux) then follow. The nature of objectivity in a general scientific context is discussed, and its specific form in deterministic continuum mechanics is characterised in terms of five distinct aspects of consensus. Such consensus mandates restrictions upon response functions. For elastica, these restrictions are those universally accepted. The standard definition of a viscous fluid (as a material for which the stress depends upon the current local values of mass density and velocity gradient *computed in terms of the frame of a general observer*) is shown to simplify to its standard (spin-independent) form if the local measure $\mathbf{d}_w$ of inhomogeneity introduced in Chapter 7 vanishes. However, if the stress depends upon density and velocity gradient with respect to an (any) *inertial* observer, then objectivity does not exclude spin-dependence. Since the physical admissibility (or otherwise) of spin-dependent fluids has been the subject of controversy for forty years, remarks are made which concern the fundamental assumption in classical physics that in principle material behaviour is independent of its observation. Statements which are intended to formalise the consequences of this assumption are not equivalent, and are variously termed *material frame-indifference*, *invariance under superposed rigid body motions*, and *objectivity*. These are listed and compared. Only objectivity emerges as imposing no restriction upon Nature. Further, from the perspective offered by objectivity, there is no requirement that observers should choose the same response function(s) for a given material, restrictions upon response functions which follow from objectivity involve only proper orthogonal tensors, and materials sensitive to spin relative to inertial frames are physically admissible. A personal history of involvement in the controversy is appended.

Chapter 13 examines two approaches to so-called non-local behaviour in the light of previous chapters: namely, the general viewpoint of Edelen, and the *peridynamics* introduced by Silling. Shortcomings in the physical basis of the long-range 'particle–particle' interactions of the latter theory are highlighted, and attention is drawn to the similarity of what is being attempted with the porous medium considerations of Chapter 10.

Elements of classical statistical mechanics are presented in Chapter 14. After introducing the concepts of dynamics in phase space, ensembles, and ensemble averaging in terms of probability density functions, strictly local forms of the continuity equation and linear momentum balance are obtained in the manner of Noll's revision of the pioneering work of Irving and Kirkwood. Two generalisations of this approach, due to Pitteri and to Admal and Tadmor, are discussed. A completely different perspective, due to Zwanzig, is outlined and applied to so-called continuously reproducible behaviour at prescribed scales of length and time. Key features are the selection of an appropriate projection operator coupled with postulates of local equilibrium and dynamic ergodicity. Semigroup formalism leads to a master equation and corresponding Fokker-Planck and fluctuation-dissipation equations. Attention is drawn to the need for a rigorous proof of a semigroup result central to projection operator methodology.

Remarks and suggestions are made in Chapter 15 which concern issues and topics not covered in this volume but which might benefit from the same approach and methodology. These relate to boundaries and interfacial regions, generalised continua, reacting mixtures, configurational forces, electromagnetic phenomena, and irreversibility. The question is raised of whether it might prove possible to derive, motivate, or otherwise gain insight into, the second law of thermodynamics on the basis of scale-dependent, corpuscular, and weighting function considerations.

Two extensive appendices introduce basic mathematical tools, results, and notation. While these will be familiar to many, the intention is to provide a comprehensive, readily accessible source of background material that might be required when studying the main text.

Appendix A is concerned with vectors and linear algebra. Starting from absolute basic, relevant concepts, definitions and results are developed both in direct (basis-free) and Cartesian tensor notation.

The geometry of Euclidean space is discussed in Appendix B, and isometries and homogeneous deformations are defined and characterised. Differentiation of scalar, vector, and linear transformation fields is treated in co-ordinate-free manner and related to equivalent Cartesian tensor formulations. Elements of integration over spatial regions are included, together with statements of divergence theorems and proofs of identities. Generalisations of differential and integral calculus to $\mathbb{R}^n$ are discussed in order to appreciate the phase-space analyses of Chapter 14.

Serious study of any work of this kind requires pencil and paper to hand for checking calculations and results. This is encouraged by the inclusion of many (usually simple and straightforward) exercises. The reader is also prompted on occasion by queries which are intended to help ensure that attention is paid to detail.

2 Some Elements of Continuum Mechanics

2.1 Preamble

In this chapter we address fundamental aspects of continuum modelling in respect of kinematics, mass conservation, balances of linear and rotational momentum, and balance of energy.

After considering the role of mass density in modelling the presence of 'matter', we discuss the manner in which the detailed macroscopic distortion of any material body can be monitored. This is markedly different for solids and fluids, but in both cases it is possible to motivate the notion of material point and thereby establish basic kinematic concepts such as deformation, motion, and velocity. The formal (axiomatic) approach to kinematics is outlined for comparison. Mass conservation is motivated for solids and postulated to hold in general. Dynamical considerations are first addressed for a body as a whole. In addition to tractions on boundaries, the possibility of surface and body couples is considered. Global balances of linear and rotational momentum are postulated and applied to rigid bodies both to emphasise their often-neglected status as a special case of material continua and to develop familiarity with notation, concepts and basic manipulations. Local forms of balance are derived in standard fashion by postulating balances for matter in arbitrary subregions of the region instantaneously occupied by the body, invoking a transport theorem, and then establishing the existence of stress and couple stress tensors and a heat flux vector. It is these local forms of balance that can be *derived* directly from molecular considerations using the weighting function methodology to be introduced in Chapter 4.

2.2 Matter and Its Distribution

Any specific material system of interest (e.g., a rubber tyre, brick, steel girder, liquid in a container, ocean current, atmospheric air, or water in an aquifer) is termed a *body*, B say. The presence of the matter which constitutes B is described in terms of its mass. Specifically, the measure of matter associated with a body is provided by a *mass density function* ρ of position and time which takes non-negative values. The function ρ for a given body has two physical *m*ass *d*ensity interpretations:

M.D.1. The spatial region considered to be occupied by the body at time t, B_t say, is that region in which ρ takes positive values at time t. That is,[1]

$$B_t := \{\mathbf{x} \in \mathcal{E} : \rho(\mathbf{x},t) > 0\}. \tag{2.2.1}$$

M.D.2. The mass, or amount of matter, of B material within any region R at time t is

$$m(R,t) := \int_R \rho(\mathbf{x},t)dV_{\mathbf{x}}. \tag{2.2.2}$$

Remark 2.2.1. In order for (2.2.2) to make sense, ρ must be spatially integrable at all times of interest. It is *assumed* that ρ has continuous partial derivatives with respect to both location $\mathbf{x}$ and time t. Accordingly, at any time of interest, ρ is a continuous function of position and is hence everywhere integrable.

2.3 Motion of Matter: Kinematics and Material Points

As time goes by, a given body B may change position and/or shape. Such time-dependent change is termed a *motion* of the body. To model physical behaviour associated with a motion, it is useful to define the *trajectory* of this body as

$$\mathcal{T}_B := \{(\mathbf{x},t) : \mathbf{x} \in B_t,\ t \in I\}. \tag{2.3.1}$$

Here I denotes the time interval over which the behaviour of B is being modelled. Functions of space and time defined on $\mathcal{T}_B$ are termed *fields*. In particular, ρ is the mass density field.

The detailed prescription of change of position and/or shape of B is modelled in terms of *material points*. Specifically, with each pair $(\mathbf{x},t) \in \mathcal{T}_B$ is associated a material point together with its *velocity* $\mathbf{v}(\mathbf{x},t)$. While the concept of material point is a primitive notion in formal continuum mechanics,[2] in order to link this with observation and experimentation it is necessary to be somewhat specific. (The next subsection contains an outline of the formal, axiomatic approach.)

For a *solid* body (in which any given molecule has near-neighbours which remain so as the body moves and/or changes shape), any group of neighbouring molecules can be 'doped' or, at least in principle, identified in some way. The motion of any such group can be monitored. If the group is localised at point $\hat{\mathbf{x}} \in B_{t_0}$ at time t_0, then at any subsequent time t it will be localised at some point $\mathbf{x} \in B_t$. Formalising this, we write

$$\mathbf{x} = \chi_{t_0}(\hat{\mathbf{x}},t) \tag{2.3.2}$$

and term χ_{t_0} the *motion map corresponding to the situation at time* t_0. Of course, the velocity at time t of that group localised at $\hat{\mathbf{x}}$ at time t_0 will be

[1] Here and henceforth $\mathcal{E}$ will denote Euclidean space; that is, 'space' as we perceive it. Any element $\mathbf{x}$ of $\mathcal{E}$ is a geometrical point. See Appendix B.1.

[2] Cf., e.g., Gurtin [1]. In the general continuum mechanics literature material points are also termed *particles* (cf., e.g., Truesdell & Noll [2] and Chadwick [3]) or, in fluid dynamics, *fluid particles* (cf., e.g., Landau & Lifschitz [4] and Paterson [5]). The term *material point* was introduced by Noll to avoid the common identification of *particle* with *point mass*. The latter has a definite mass, while, as will be seen, a material point has no associated mass but only, at any given time, a motion-dependent mass density.

$\dot{\chi}_{t_0}(\hat{\mathbf{x}},t) := (\partial/\partial t)\{\chi_{t_0}(\hat{\mathbf{x}},t)\}$. Thus the velocity of the doped group located at $\mathbf{x}$ at time t, written as $\mathbf{v}(\mathbf{x},t)$, is precisely $\dot{\chi}_{t_0}(\hat{\mathbf{x}},t)$, via (2.3.2). That is, the *velocity field* $\mathbf{v}$ on $\mathcal{T}_B$ is given by

$$\mathbf{v}(\mathbf{x},t) := \dot{\chi}_{t_0}(\hat{\mathbf{x}},t) \qquad \text{where} \quad \mathbf{x} = \chi_{t_0}(\hat{\mathbf{x}},t). \tag{2.3.3}$$

Similarly, the acceleration field $\mathbf{a}$ on $\mathcal{T}_B$ is given by

$$\mathbf{a}(\mathbf{x},t) := \ddot{\chi}_{t_0}(\hat{\mathbf{x}},t) \qquad \text{where} \quad \mathbf{x} = \chi_{t_0}(\hat{\mathbf{x}},t). \tag{2.3.4}$$

In the case of *liquids* and *gases*, molecules close together at a given time do not remain so but diffuse rapidly.[3] An indication of *gross* molecular motion can be gained by the insertion and observation of small bubbles or suspended particles in liquids and smoke particles or balloons in gases. At any instant, such observations furnish velocity values of bubbles, particles or balloons which would seem, intuitively, to be representative of the instantaneous fluid velocity values at the locations of these 'foreign' objects. The modelling assumption made in fluid dynamics is that for a fluid body B there is a velocity field $\mathbf{v}$ defined on its trajectory $\mathcal{T}_B$. We can visualise an intuitive sense of fluid motion by looking at the situation at some time t_0 and then, on choosing any point $\hat{\mathbf{x}} \in B_{t_0}$, 'follow' the fluid by moving in such a way as always to have the same velocity as the local value of the fluid velocity. If, in such a motion, we arrive at point $\mathbf{x}$ at time t, then we can again write (2.3.2), where, by the foregoing, relation (2.3.3) [and, similarly, relation (2.3.4)] will also be satisfied. Further, with each $\hat{\mathbf{x}} \in B_{t_0}$ we can identify a *hypothetical* 'material point' which is to be regarded as located at $\mathbf{x}$ at time t.

Accordingly, *for both solid and fluid bodies* we have the concept of a motion (corresponding to the situation at some given time) which prescribes the distortion and movement of the relevant body in fine detail. This motion, given by (2.3.2), is related to the associated velocity and acceleration fields by (2.3.3) and (2.3.4).

For any $t \in I$ [see (2.3.1)], the motion map

$$\chi_{t_0}(\cdot,t) : B_{t_0} \longrightarrow B_t \tag{2.3.5}$$

is assumed to be bijective. That is, if $\hat{\mathbf{x}}$ and $\hat{\mathbf{y}}$ are any pair of distinct points in B_{t_0}, then, for any $t \in I$, $\chi_{t_0}(\hat{\mathbf{x}},t)$ and $\chi_{t_0}(\hat{\mathbf{y}},t)$ will not coincide, *and* for each $\mathbf{x} \in B_t$ there exists an $\hat{\mathbf{x}} \in B_{t_0}$ for which (2.3.2) holds.

Point to ponder 1. Consider how one might be led to the bijectivity hypothesis by recalling how a motion can be physically monitored (via doped molecular clusters for solids and immersed entities for fluids).

Point to ponder 2. Note the intrinsic difficulty of monitoring the internal deformation of *solids* and the necessity of remote sensing via a scanning procedure, and how in engineering practice one may only make measurements on the surface of a body (e.g., via attached strain gauges, transducers, or optical monitoring devices).

Point to ponder 3. Note that for fluids the flows of interest can involve very different length scales. For example, the velocity profile of flow down a pipe can only

[3] Typical molecular speeds for fluids (which may be macroscopically motionless) at *s*tandard *t*emperature and *p*ressure (STP) are, on average, of order 10^3 ms^{-1}. Further, individual molecular trajectories are highly erratic, much more so than the Brownian motion of small suspended particles (cf., e.g., Brush [6]).

be monitored at scales smaller than cross-sectional dimensions, while atmospheric wind velocity may be of interest at small scale (motion over an aerofoil), medium scale (motion around a skyscraper), or large scale (weather reporting). Accordingly, the notion of material point would appear to be context/scale-dependent.

Point to ponder 4. The question of scale dependence also arises with solids: consider deformations of small crystalline samples and motions of the Earth (namely terrestrial – solid – tides and seismic waves).

Summary. The notion of material point has been motivated quite differently for solid and fluid phases of matter. In a solid one can, roughly speaking, identify the position of a material point at a given time with the location of a small cluster of neighbouring molecules. The motion of this material point then can be tracked (at least in principle) by monitoring the motion of this cluster since any cluster of near-neighbouring molecules maintains its integrity. On the other hand, for fluids a material point can, loosely speaking, be thought of as a hypothetical immersed object whose motion is governed by the action thereon of fluid molecules with which it interacts/collides. Of course, the particular interacting/colliding molecules in question change rapidly with time. What should be clear is that

the key role played by the notion of material point, whether the body concerned is in solid, liquid, or gaseous state, is that of tracking the macroscopic distortion/flow of the body as time passes.

*2.4[4] The Formal (Axiomatic) Approach to Matter and Material Points

In formal continuum mechanics[5] the notion of material point is *primitive* (i.e., a formal concept which serves as a building block for subsequent development of the subject but is otherwise undefined). A body B is considered to be a set of material points. Any possible physical manifestation of the body is termed a *configuration*. More precisely, a configuration $\boldsymbol{\kappa}$ is a map

$$\boldsymbol{\kappa} : B \longrightarrow \mathcal{E}. \tag{2.4.1}$$

It is assumed that in no configuration can two distinct material points coincide. That is, if $\mathbf{X}, \mathbf{Y} \in B$ are distinct material points, then $\boldsymbol{\kappa}(\mathbf{X}) \neq \boldsymbol{\kappa}(\mathbf{Y})$ for any configuration $\boldsymbol{\kappa}$. Accordingly any configuration $\boldsymbol{\kappa}$ must be a bijection (i.e., one-to-one correspondence) as a map from B onto its range $\boldsymbol{\kappa}(B)$. For any pair of configurations $\boldsymbol{\kappa}$ and $\boldsymbol{\mu}$, it is assumed that the ranges $\boldsymbol{\kappa}(B)$ and $\boldsymbol{\mu}(B)$ are open subsets of $\mathcal{E}$ and that the bijection[6]

$$\mathbf{d} := \boldsymbol{\mu} \circ \boldsymbol{\kappa}^{-1} : \boldsymbol{\kappa}(B) \rightarrow \boldsymbol{\mu}(B) \tag{2.4.2}$$

is of class C^1. Any such map is termed a *deformation* of B.

A *motion* of B is a one-parameter family of configurations, parametrised by time, for some time interval I. If $\boldsymbol{\chi}(.,t)$ denotes the member of this family at time[7] $t \in I$

[4] Any starred section, subsection, or item may be skipped without affecting subsequent unstarred discussions.

[5] Cf., e.g., Gurtin [1], Truesdell & Noll [2], and Chadwick [3].

[6] Property (2.4.2) endows B with the structure of a C^1 differentiable manifold whose charts are configurations.

[7] Time t is usually regarded as present time, and $\boldsymbol{\chi}(.,t)$ is described as the *current configuration*.

and $\mathbf{X} \in B$, then $\chi(\mathbf{X},t)$ is the location (a point in $\mathcal{E}$) of $\mathbf{X}$ at time t in this motion and $\dot{\chi}(\mathbf{X},t)$ (where $\dot{\chi} := \partial\chi/\partial t$) is its *velocity* at this time. Given configurations κ and $\chi(.,t)$, from (2.4.2) with $\mu = \chi(.,t)$, the deformation

$$\chi_{\kappa}(.,t) := \chi(.,t) \circ \kappa^{-1} \tag{2.4.3}$$

which maps $\kappa(B)$ onto $\chi(B,t) \subset \mathcal{E}$ is (spatially) of class C^1 (here t is considered fixed) and is termed the *deformation of B at time t with respect to configuration* κ. Function

$$\chi_{\kappa} : \kappa(B) \times I \longrightarrow \mathcal{E} \tag{2.4.4}$$

is termed the *motion relative to configuration* κ. Region

$$_{\chi}B_t := \chi(B,t) \subset \mathcal{E} \tag{2.4.5}$$

is that *region occupied by the body at time t* in motion χ, and the *trajectory* associated with this motion is

$$_{\chi}\mathcal{T}_B := \{(\mathbf{x},t) : \mathbf{x} \in {_{\chi}B_t} \text{ with } t \in I\}. \tag{2.4.6}$$

[Cf. (2.2.1) and (2.3.1).] Since from (2.4.3)

$$\chi(.,t) = \chi_{\kappa}(\cdot,t) \circ \kappa, \tag{2.4.7}$$

the velocity of $\mathbf{X}$ at time t is

$$\dot{\chi}(\mathbf{X},t) = \frac{\partial}{\partial t}\{\chi_{\kappa}(\kappa(\mathbf{X}),t)\} =: \dot{\chi}_{\kappa}(\kappa(\mathbf{X}),t). \tag{2.4.8}$$

The *velocity field* $\mathbf{v}$ on $_{\chi}\mathcal{T}_B$ is defined by

$$\mathbf{v}(\mathbf{x},t) := \dot{\chi}(\mathbf{X},t), \qquad \text{where } \mathbf{x} = \chi(\mathbf{X},t). \tag{2.4.9}$$

That is, the velocity at the geometrical point $\mathbf{x} \in {_{\chi}B_t}$ at time t is the velocity of that material point which is located at $\mathbf{x}$ at time t. Similarly, the *acceleration field* $\mathbf{a}$ on $_{\chi}\mathcal{T}_B$ is defined by

$$\mathbf{a}(\mathbf{x},t) := \ddot{\chi}(\mathbf{X},t), \qquad \text{where } \mathbf{x} = \chi(\mathbf{X},t). \tag{2.4.10}$$

In view of the bijective nature of κ, to each point $\hat{\mathbf{x}}$ in region $\kappa(B)$ corresponds a unique material point and vice versa. Accordingly, points in $\kappa(B)$ are identifiable with material points, and definitions (2.4.9) and (2.4.10) can be expressed in terms of physically accessible entities [namely, points $\hat{\mathbf{x}}$ in $\kappa(B)$] via (2.4.8) as

$$\mathbf{v}(\mathbf{x},t) = \dot{\chi}_{\kappa}(\hat{\mathbf{x}},t) \qquad \text{and} \qquad \mathbf{a}(\mathbf{x},t) = \ddot{\chi}_{\kappa}(\hat{\mathbf{x}},t), \tag{2.4.11}$$

where

$$\mathbf{x} = \chi_{\kappa}(\hat{\mathbf{x}},t). \tag{2.4.12}$$

In this context κ is termed a *reference configuration*. Choosing $\kappa = \chi(.,t_0)$ and writing, for $\hat{\mathbf{x}} \in {_{\chi}B_{t_0}}$,

$$\chi_{t_0}(\hat{\mathbf{x}},t) := \chi(\mathbf{X},t), \qquad \text{where } \hat{\mathbf{x}} = \chi(\mathbf{X},t_0), \tag{2.4.13}$$

relations (2.4.11) and (2.4.12) become

$$\mathbf{v}(\mathbf{x},t) = \dot{\chi}_{t_0}(\hat{\mathbf{x}},t) \qquad \text{and} \qquad \mathbf{a}(\mathbf{x},t) = \ddot{\chi}_{t_0}(\hat{\mathbf{x}},t), \tag{2.4.14}$$

where

$$\mathbf{x} = \chi_{t_0}(\hat{\mathbf{x}},t). \tag{2.4.15}$$

Remark 2.4.1. Relations (2.4.14) and (2.4.15) are precisely (2.3.3) and (2.3.4). In order for the formal approach to be applied to actual and specific material behaviour, any observer/experimentalist must decide how to monitor changes in position and shape for the system of interest, as discussed in Section 2.3.

2.5 Mass Conservation

Consider the motion of a body B over a time interval I, and let $t_0 \in I$. The motion corresponding to the situation at time t_0 is a map [see (2.3.5) and (2.2.1)]

$$\chi_{t_0}(.,.) : \{(\hat{\mathbf{x}},t) \text{ with } \hat{\mathbf{x}} \in B_{t_0}, t \in I\} \to B_t, \tag{2.5.1}$$

where

$$B_t := \chi_{t_0}(B_{t_0},t). \tag{2.5.2}$$

If R denotes a subregion of B_{t_0}, then we can consider

$$R_t := \chi_{t_0}(R,t) \tag{2.5.3}$$

and compare the mass $m(R,t_0)$ in R at time t_0 [see (2.2.2)] with the mass $m(R_t,t)$ in R_t at time t. In the case of B being a solid body, those points on the boundary ∂R of R at time t_0 can be identified in terms of a set of clusters of neighbouring molecules which will, at time t, delineate the boundary ∂R_t of R_t. [See the discussion preceding (2.3.2).] Those molecules within R at time t_0 will be those within R_t at time t, and hence we postulate that

$$m(R,t_0) = m(R_t,t). \tag{2.5.4}$$

Thus by M.D.2 [see (2.2.2)],

$$\int_R \rho(\hat{\mathbf{x}},t_0)dV_{\hat{\mathbf{x}}} = \int_{R_t} \rho(\mathbf{x},t)dV_{\mathbf{x}}. \tag{2.5.5}$$

Since $\mathbf{x} = \chi_{t_0}(\hat{\mathbf{x}},t)$ and $\chi_{t_0}(.,t)$ is a class C^1 bijection[8],

$$\int_{R_t} \rho(\mathbf{x},t)dV_{\mathbf{x}} = \int_{\chi_{t_0}(R_{t_0})} \rho(\chi_{t_0}(\hat{\mathbf{x}},t),t)dV_{\mathbf{x}} = \int_R \rho(\chi_{t_0}(\hat{\mathbf{x}},t),t)J(\hat{\mathbf{x}},t)dV_{\hat{\mathbf{x}}}. \tag{2.5.6}$$

Here $J(.,t)$ denotes the Jacobian of the map $\chi_{t_0}(.,t)$.

Comparing (2.5.5) and (2.5.6), we have

$$\int_R \{\rho(\hat{\mathbf{x}},t_0) - \rho(\chi_{t_0}(\hat{\mathbf{x}},t),t)J(\hat{\mathbf{x}},t)\}dV_{\hat{\mathbf{x}}} = 0. \tag{2.5.7}$$

[8] See Appendix B, Theorem B.6.2, with $R = R_t, f = \rho(\cdot,t)$ and $\mathbf{d} = \chi_{t_0}(\cdot,t)$.

Continuity of the integrand, together with the arbitrary nature of region R, allows us to deduce[9] that

$$\rho(\mathbf{x},t)J(\hat{\mathbf{x}},t) = \rho(\boldsymbol{\chi}_{t_0}(\hat{\mathbf{x}},t),t)J(\hat{\mathbf{x}},t) = \rho(\hat{\mathbf{x}},t_0). \tag{2.5.8}$$

Differentiating with respect to time, keeping $\hat{\mathbf{x}}$ fixed, and using the result[10]

$$\frac{\partial J}{\partial t} = ((\operatorname{div}\mathbf{v}) \circ \boldsymbol{\chi}_{t_0})J, \tag{2.5.9}$$

we have

$$\frac{\partial}{\partial t}\{\rho(\boldsymbol{\chi}_{t_0}(\hat{\mathbf{x}},t),t)\}J(\hat{\mathbf{x}},t) + \rho(\boldsymbol{\chi}_{t_0}(\hat{\mathbf{x}},t),t)\operatorname{div}_{\mathbf{x}}\{\mathbf{v}(\mathbf{x},t)\}J(\hat{\mathbf{x}},t) = 0. \tag{2.5.10}$$

The bijective nature of $\boldsymbol{\chi}_{t_0}$ ensures that J is never zero. Accordingly, (2.5.10) implies that

$$\dot{\rho} + \rho \operatorname{div}\mathbf{v} = 0, \tag{2.5.11}$$

where all fields are evaluated at $(\mathbf{x},t)$, and the *material time derivative* $\dot{\rho}$ of ρ is defined by

$$\dot{\rho}(\mathbf{x},t) := \frac{\partial}{\partial t}\{\rho(\boldsymbol{\chi}_{t_0}(\hat{\mathbf{x}},t),t)\}. \tag{2.5.12}$$

Here $\mathbf{x} = \boldsymbol{\chi}_{t_0}(\hat{\mathbf{x}},t)$, and $\hat{\mathbf{x}}$ is held fixed in computing the partial derivative. Since[11]

$$\frac{\partial}{\partial t}\{\rho(\boldsymbol{\chi}_{t_0}(\hat{\mathbf{x}},t),t)\} = (\nabla_{\mathbf{x}}\rho)(\mathbf{x},t).\dot{\boldsymbol{\chi}}_{t_0}(\hat{\mathbf{x}},t) + \frac{\partial \rho}{\partial t}(\mathbf{x},t), \tag{2.5.13}$$

from (2.3.3)

$$\dot{\rho} = \nabla\rho.\mathbf{v} + \frac{\partial \rho}{\partial t}. \tag{2.5.14}$$

(Here $\mathbf{x}$ is held fixed in the partial derivative.) Thus (2.5.11) may be written as

$$\frac{\partial \rho}{\partial t} + \nabla\rho.\mathbf{v} + \rho\operatorname{div}\mathbf{v} = 0, \tag{2.5.15}$$

that is [see Appendix B.7, identity (B.7.38)],

$$\frac{\partial \rho}{\partial t} + \operatorname{div}\{\rho\mathbf{v}\} = 0. \tag{2.5.16}$$

Relation (2.5.16) is often termed the *continuity equation.*

Remark 2.5.1 Postulate (2.5.4) of mass conservation was well motivated for solids. For fluids, whose motions are in practice monitored by immersed objects, such a hypothesis does not appear quite so physically obvious. However, once postulate (2.5.4) is made, regularity assumptions result in (2.5.8) together with deductions (2.5.14) and (2.5.16) therefrom. Since J represents the local volume magnification factor [see Appendix B.5, specifically (B.5.6)] in going from B_{t_0} to B_t via motion map $\boldsymbol{\chi}_{t_0}(.,t)$, relation (2.5.8) is intuitively 'correct'. The molecular viewpoint delivers

[9] See Appendix B, Theorem B.6.1.
[10] See Appendix B, Corollary B.5.1, (B.5.18) and (B.7.1).
[11] See Appendix B.5, Result B.5.3, with $\mathbf{x}(t) = \boldsymbol{\chi}_{t_0}(\hat{\mathbf{x}},t)$.

ultimate precision in this respect and results directly in (2.5.16) with very little effort. Wait for Chapter 4!

Remark 2.5.2 (*Mass conservation in the formal approach*). In the approach of Section 2.4, a mass density function is assigned to each configuration $\boldsymbol{\kappa}$ and is assumed to take strictly positive values on the range $\boldsymbol{\kappa}(B)$ of $\boldsymbol{\kappa}$. For any region $R \subset \mathcal{E}$,

$$m_{\kappa}(R) := \int_R \rho_{\kappa}(\hat{\mathbf{x}})dV_{\hat{\mathbf{x}}} \tag{2.5.17}$$

is defined to be the mass in R in configuration $\boldsymbol{\kappa}$. Given any pair $\boldsymbol{\kappa}, \boldsymbol{\mu}$ of configurations, it is assumed that the deformation

$$\mathbf{d} := \boldsymbol{\mu} \circ \boldsymbol{\kappa}^{-1} \tag{2.5.18}$$

conserves mass. That is, for any region $R \subset \boldsymbol{\kappa}(B)$,

$$m_{\kappa}(R) = m_{\mu}(\mathbf{d}(R)). \tag{2.5.19}$$

Thus, from (2.5.17) and (2.5.19),

$$\int_R \rho_{\kappa}(\hat{\mathbf{x}})dV_{\hat{\mathbf{x}}} = \int_{\mathbf{d}(R)} \rho_{\mu}(\bar{\mathbf{x}})dV_{\bar{\mathbf{x}}}, \tag{2.5.20}$$

where

$$\bar{\mathbf{x}} := \mathbf{d}(\hat{\mathbf{x}}). \tag{2.5.21}$$

If $J_{\mathbf{d}}$ denotes the Jacobian associated with $\mathbf{d}$, then (2.5.20) may be written as

$$\int_R \rho_{\kappa}(\hat{\mathbf{x}})dV_{\hat{\mathbf{x}}} = \int_R \rho_{\mu}(\mathbf{d}(\hat{\mathbf{x}}))J_{\mathbf{d}}(\hat{\mathbf{x}})dV_{\hat{\mathbf{x}}}. \tag{2.5.22}$$

The arbitrary nature of region R and continuity of integrands [cf. (2.5.7) and (2.5.8)] yields

$$\rho_{\kappa} = (\rho_{\mu} \circ \mathbf{d})J_{\mathbf{d}}. \tag{2.5.23}$$

Setting $\boldsymbol{\kappa} = \boldsymbol{\chi}_{t_0}(.,t_0)$ (which is the identity map on B_{t_0}) and $\mathbf{d} = \boldsymbol{\chi}_{t_0}(.,t)$ yields (2.5.8) et seq.

An important consequence of mass conservation is the following:

Transport/Material Time Derivative Theorem

If f denotes a field on $\mathcal{T}_B$ [see (2.3.1)] which is of class C^1 in space and time, then

$$\frac{d}{dt}\left\{\int_{R_t} \rho f \, dV\right\} = \int_{R_t} \rho \dot{f} \, dV. \tag{2.5.24}$$

Here $R_t := \boldsymbol{\chi}_{t_0}(R,t)$, where R is any subregion of B_{t_0}, and the *material time derivative* $\dot{f}$ of f [cf. (2.5.12)] is defined, for any $\mathbf{x} \in R_t$, by

$$\dot{f}(\mathbf{x},t) := \frac{\partial}{\partial t}\{f(\boldsymbol{\chi}_{t_0}(\hat{\mathbf{x}},t),t)\}, \tag{2.5.25}$$

where

$$\mathbf{x} = \boldsymbol{\chi}_{t_0}(\hat{\mathbf{x}},t) \tag{2.5.26}$$

and $\hat{\mathbf{x}}$ is held fixed in (2.5.25).

It follows that if f is scalar-valued, then (see Appendix B.5, Result B.5.3)

$$\dot{f} = \frac{\partial f}{\partial t} + \nabla f \, . \, \mathbf{v}. \tag{2.5.27}$$

If f is a tensor field of order 1 or greater, then (see Appendix B.5, Remark B.5.4)

$$\dot{f} = \frac{\partial f}{\partial t} + (\nabla f)\mathbf{v}. \tag{2.5.28}$$

2.6 Dynamics I: Global Relations

2.6.1 Introduction

Forces which act on any given body and derive from external agencies are essentially of two kinds: *contact* forces, which act on its boundary, and *body* forces, which have 'action at a distance' character. The former include the effect of contiguous bodies (e.g., atmospheric pressure and reactions from bodies which constrain the location of the body in some way), while the latter derive from external gravitational and/or electromagnetic influence. Such force systems give rise to motions of the body. In many situations a solid body may undergo no appreciable change in shape but merely translate and rotate. For such situations the body is said to be *rigid*, and any associated motion $\boldsymbol{\chi}_{t_0}(.,t)$ is an isometry[12] for all $t \in I$. The simple nature of isometries allows a complete dynamical description to be obtained[13] which involves knowledge of the body only through its mass density distribution. Such knowledge suffices to determine its mass, centre of mass G, and inertia tensor with respect to G. If, however, shape changes occur (as is natural for fluids in particular), then the body is manifesting an internal character which must be taken into account.

Hereafter in this chapter we discuss global dynamical relations for *all* bodies, consider the implications for rigid bodies, address local dynamical behaviour within non-rigid bodies, and outline both global and local thermomechanical considerations.

2.6.2 Linear Momentum Balance

At time t a body occupies a region B_t with boundary ∂B_t. Let $\mathbf{t}$ denote the *traction* field on ∂B_t and $\mathbf{b}$ the *body force* field in B_t due to agencies external to the body. The resultant forces to which each field gives rise are, respectively,

$$\mathbf{F}_{\text{surface}} := \int_{\partial B_t} \mathbf{t}\, dA \qquad \text{and} \qquad \mathbf{F}_{\text{body}} := \int_{B_t} \mathbf{b}\, dV. \tag{2.6.1}$$

It is postulated (as a generalisation of Newton's second law) that the sum of the preceding resultants is the rate of change of the total momentum of the body computed in an (any) inertial frame. Given the *momentum density*

$$\mathbf{p} := \rho \mathbf{v} \tag{2.6.2}$$

[12] That is, for all $\hat{\mathbf{x}}$ and $\hat{\mathbf{y}}$ in B_{t_0} and all $t \in I$, $\|\boldsymbol{\chi}_{t_0}(\hat{\mathbf{y}},t) - \boldsymbol{\chi}_{t_0}(\hat{\mathbf{x}},t)\| = \|\hat{\mathbf{y}} - \hat{\mathbf{x}}\|$. See Appendix B.3.2.
[13] Cf., e.g., Goldstein [7].

we thus have, in any inertial frame,

$$\mathbf{F}_{\text{surface}} + \mathbf{F}_{\text{body}} = \frac{d}{dt}\left\{\int_{B_t} \rho \mathbf{v}\, dV\right\}. \tag{2.6.3}$$

Accordingly (note all fields are to be regarded a priori as time-dependent),

$$\textbf{glmb} \qquad \int_{\partial B_t} \mathbf{t}\, dS + \int_{B_t} \mathbf{b}\, dV = \frac{d}{dt}\left\{\int_{B_t} \rho \mathbf{v}\, dV\right\}. \tag{2.6.4}$$

Relation (2.6.4) is known as the *global form of linear momentum balance* for the body. If $\mathbf{t}$ and $\mathbf{b}$ are known, then (2.6.4) is an *evolution equation for linear momentum.*

The *mass centre* of a body at any instant is that point $\mathbf{x}_G(t)$ defined (here $\mathbf{x}_0 \in \mathcal{E}$ is arbitrary) by

$$\mathbf{x}_G(t) := \mathbf{x}_0 + M^{-1} \int_{B_t} \rho(\mathbf{x},t)\{\mathbf{x} - \mathbf{x}_0\} dV_{\mathbf{x}}, \tag{2.6.5}$$

where the *total mass of the body* is

$$M := \int_{B_t} \rho(\mathbf{x},t) dV. \tag{2.6.6}$$

Remark 2.6.1. In order to make sense, $\mathbf{x}_G$ should be independent of choice $\mathbf{x}_0$ in (2.6.5). If selection of another point $\mathbf{x}'_0$ had been made with corresponding mass centre $\mathbf{x}'_G$ defined by the analogue of (2.6.5), then

$$\mathbf{x}'_G(t) - \mathbf{x}_G(t) = (\mathbf{x}'_0 - \mathbf{x}_0) + M^{-1} \int_{B_t} \rho(\mathbf{x},t)\{\mathbf{x}_0 - \mathbf{x}'_0\} dV_{\mathbf{x}}. \tag{2.6.7}$$

Exercise 2.6.1. Show that $\mathbf{x}'_G = \mathbf{x}_G$ from (2.6.7) and (2.6.6).

Result 2.6.1. If we take $\mathbf{x}_0 = \mathbf{x}_G(t)$ in (2.6.5), we see that

$$\int_{B_t} \rho(\mathbf{x},t)\{\mathbf{x} - \mathbf{x}_G(t)\} dV_{\mathbf{x}} = \mathbf{0}. \tag{2.6.8}$$

Result 2.6.2.

$$\int_{B_t} \rho \mathbf{v} dV = M \mathbf{v}_G, \tag{2.6.9}$$

where

$$\mathbf{v}_G := \dot{\mathbf{x}}_G. \tag{2.6.10}$$

Proof. From (2.6.5),

$$M(\mathbf{x}_G(t) - \mathbf{x}_0) = \int_{B_t} \rho(\mathbf{x},t)\{\mathbf{x} - \mathbf{x}_0\} dV_{\mathbf{x}}.$$

Differentiating with respect to time and using (2.5.24) with

$$f(\mathbf{x},t) := \mathbf{x}(t) - \mathbf{x}_0, \qquad \text{so } \dot{f}(\mathbf{x},t) = \frac{\partial}{\partial t}\{\boldsymbol{\chi}_{t_0}(\hat{\mathbf{x}},t)\} = \dot{\boldsymbol{\chi}}_{t_0}(\hat{\mathbf{x}},t) = \mathbf{v}(\mathbf{x},t),$$

(here $\mathbf{x}_0$ is assumed to be stationary) we have the result

$$M \mathbf{v}_G(t) = \int_{B_t} \rho(\mathbf{x},t) \mathbf{v}(\mathbf{x},t) dV_{\mathbf{x}}.$$

Result 2.6.3. From (2.6.4) and (2.6.9),

$$\int_{\partial B_t} \mathbf{t}\, dA + \int_{B_t} \mathbf{b}\, dV = M\mathbf{a}_G, \tag{2.6.11}$$

where the acceleration of the mass centre in any inertial frame

$$\mathbf{a}_G := \dot{\mathbf{v}}_G. \tag{2.6.12}$$

2.6.3 Rotational Momentum Balance

Complementing (2.6.4), a *g*lobal *r*otational (angular) *m*omentum *b*alance is postulated which relates the manner in which $\mathbf{t}$ and $\mathbf{b}$ vary over ∂B_t and within B_t, respectively, to rotational motion of the body. In so doing, account has to be taken of the possibility of couples being exerted on ∂B_t and within B_t. The source of such couples could be external electromagnetic fields, which induce dipoles, or microstructure, such as that of nematic liquid crystalline phases. In order to accommodate such phenomena, it is postulated that in any inertial frame

grmb $$\int_{\partial B_t} \{\mathbf{r}\wedge\mathbf{t}+\mathbf{M}\}dA + \int_{B_t} \{\mathbf{r}\wedge\mathbf{b}+\mathbf{J}\}dV = \frac{d}{dt}\int_{B_t} \{\mathbf{r}\wedge\rho\mathbf{v}+\rho\mathbf{S}\}dV. \tag{2.6.13}$$

Here, for $\mathbf{x} \in B_t$ and $\mathbf{x}_0 \in \mathcal{E}$ arbitrary and stationary,

$$\mathbf{r}(\mathbf{x}) := \mathbf{x} - \mathbf{x}_0. \tag{2.6.14}$$

Fields $\mathbf{M}$ on ∂B_t and $\mathbf{J}$ within B_t denote surface couple and body couple densities, respectively, and $\rho\mathbf{S}$ denotes an intrinsic internal contribution to the total angular momentum density. The wedge product has been employed in (2.6.13) [see Appendix A.8, (A.8.20)] and $\mathbf{M}, \mathbf{J}$, and $\mathbf{S}$ take skew-symmetric values. Alternatively, (2.6.13) can be written in terms of a corresponding relation in which fields take axial-vector values (see Appendix A.15).

Remark 2.6.2. Choosing another point $\mathbf{x}_0'$ in (2.6.14), writing down the corresponding form of (2.6.13), and subtracting the new version of (2.6.13) from the old, we have

$$\int_{\partial B_t} (\mathbf{x}_0' - \mathbf{x}_0) \wedge \mathbf{t}\, dA + \int_{B_t} (\mathbf{x}_0' - \mathbf{x}_0) \wedge \mathbf{b}\, dV = \frac{d}{dt}\left\{\int_{B_t} (\mathbf{x}_0' - \mathbf{x}_0) \wedge \rho\mathbf{v}\, dV\right\}. \tag{2.6.15}$$

Thus, noting $\mathbf{x}_0' - \mathbf{x}_0 \in \mathcal{V}$ is arbitrary, glmb follows from grmb as a consequence of assuming that grmb holds for *any* $\mathbf{x}_0 \in \mathcal{E}$. An equivalent formulation of the content of the two axioms would be the postulation of glmb and a version of grmb in which $\mathbf{x}_0$ is a designated (hence distinguished) point. It then would follow that grmb should hold for any other fixed point $\mathbf{x}_0'$ by virtue of glmb. (Show this!) Of course, the only physically distinguished choice of $\mathbf{x}_0$ would be the instantaneous location $\mathbf{x}_G$ of the mass centre G of the body.

The first term on the right-hand side of (2.6.13) may be re-written in a manner which highlights separate contributions from mass centre motion and motion relative to the mass centre. Specifically,

$$\begin{aligned}
&\int_{B_t} (\mathbf{x}-\mathbf{x}_0)\wedge\rho(\mathbf{x},t)\mathbf{v}(\mathbf{x},t)dV_{\mathbf{x}} \\
&= \int_{B_t} \{(\mathbf{x}-\mathbf{x}_G(t))+(\mathbf{x}_G(t)-\mathbf{x}_0)\}\wedge\rho(\mathbf{x},t)\{(\mathbf{v}(\mathbf{x},t)-\mathbf{v}_G(t))+\mathbf{v}_G(t)\}dV_{\mathbf{x}} \\
&= \boldsymbol{\mathcal{A}}(t)+\left(\int_{B_t}(\mathbf{x}-\mathbf{x}_G(t))\rho(\mathbf{x},t)dV_{\mathbf{x}}\right)\wedge\mathbf{v}_G(t) \\
&\quad +(\mathbf{x}_G(t)-\mathbf{x}_0)\wedge\int_{B_t}\rho(\mathbf{x},t)\{\mathbf{v}(\mathbf{x},t)-\mathbf{v}_G(t)\}dV_{\mathbf{x}} \\
&\quad +(\mathbf{x}_G(t)-\mathbf{x}_0)\wedge\left[\int_{B_t}\rho(\mathbf{x},t)dV_{\mathbf{x}}\right]\mathbf{v}_G(t),
\end{aligned} \tag{2.6.16}$$

where $$\boldsymbol{\mathcal{A}}(t) := \int_{B_t}(\mathbf{x}-\mathbf{x}_G(t))\wedge\rho(\mathbf{x},t)\{\mathbf{v}(\mathbf{x},t)-\mathbf{v}_G(t)\}dV_{\mathbf{x}}. \tag{2.6.17}$$

The second and third terms on the right-hand side of (2.6.16) vanish by virtue of (2.6.8), (2.6.9), and (2.6.6). We thus have, via (2.6.6), the following

Result 2.6.4.

$$\int_{B_t}\mathbf{r}\wedge\rho\mathbf{v}\,dV = \boldsymbol{\mathcal{A}}+(\mathbf{x}_G-\mathbf{x}_0)\wedge M\mathbf{v}_G. \tag{2.6.18}$$

Differentiating with respect to time yields

$$\frac{d}{dt}\left\{\int_{B_t}\mathbf{r}\wedge\rho\mathbf{v}\,dV\right\} = \dot{\boldsymbol{\mathcal{A}}}+(\mathbf{x}_G-\mathbf{x}_0)\wedge M\mathbf{a}_G \tag{2.6.19}$$

via (2.6.12) and noting that

$$\frac{d}{dt}\{(\mathbf{x}_G-\mathbf{x}_0)\}\wedge M\mathbf{v}_G = \mathbf{v}_G\wedge M\mathbf{v}_G = \mathbf{O}. \tag{2.6.20}$$

Use of (2.6.19), and (2.5.24) with $f=\mathbf{S}$, enable balance (2.6.13) to be written as

$$\int_{\partial B_t}\{\mathbf{r}\wedge\mathbf{t}+\mathbf{M}\}dA+\int_{B_t}\{\mathbf{r}\wedge\mathbf{b}+\mathbf{J}\}dV = \dot{\boldsymbol{\mathcal{A}}}+(\mathbf{x}_G-\mathbf{x}_0)\wedge M\mathbf{a}_G+\int_{B_t}\rho\dot{\mathbf{S}}dV. \tag{2.6.21}$$

Choosing $\mathbf{x}_0$ to be the instantaneous location $\mathbf{x}_G(t)$ of G yields

$$\int_{\partial B_t}\{(\mathbf{x}-\mathbf{x}_G)\wedge\mathbf{t}+\mathbf{M}\}dA+\int_{B_t}\{(\mathbf{x}-\mathbf{x}_G)\wedge\mathbf{b}+\mathbf{J}\}dV = \dot{\boldsymbol{\mathcal{A}}}+\int_{B_t}\rho\dot{\mathbf{S}}\,dV. \tag{2.6.22}$$

As is well known, $\boldsymbol{\mathcal{A}}$ may be simplified if the body undergoes only rigid body motions. In such case, when $\mathbf{S}=\mathbf{O}$ relation (2.6.22) becomes an evolution equation for the angular velocity of the body. In the next subsection we derive this evolution equation: see (2.6.72).

2.6.4 Rigid Body Dynamics

A *motion* is said to be *rigid* if it preserves distances between pairs of points and is physically possible. The latter requirement might seem to be redundant but is inserted so as to exclude[14] 'reflections', 'inversions', or (loosely speaking) 'turning a body inside out'. At any time $t \in I$ [see (2.3.5)] we thus have

$$\|\chi_{t_0}(\hat{\mathbf{y}},t) - \chi_{t_0}(\hat{\mathbf{x}},t)\| = \|\hat{\mathbf{y}} - \hat{\mathbf{x}}\| \tag{2.6.23}$$

for any pair of points $\hat{\mathbf{x}},\hat{\mathbf{y}} \in B_{t_0}$. Further, the exclusion of non-achievable situations corresponds to preservation of 'orientation' in the following sense. Let $\hat{\mathbf{x}} \in B_{t_0}$ and consider neighbouring points $\hat{\mathbf{x}}_i := \hat{\mathbf{x}} + s\mathbf{e}_i$ $(i = 1,2,3)$ for some $s \neq 0 \in \mathbb{R}$ and choice $\{\mathbf{e}_1,\mathbf{e}_2,\mathbf{e}_3\}$ of an ordered orthonormal basis for $\mathcal{V}$. Then, if $\mathbf{x} := \chi_{t_0}(\hat{\mathbf{x}},t)$ and $\mathbf{x}_i := \chi_{t_0}(\hat{\mathbf{x}}_i,t)$, we require that the triple scalar product

$$(\mathbf{x}_1 - \mathbf{x}) \times (\mathbf{x}_2 - \mathbf{x}) . (\mathbf{x}_3 - \mathbf{x}) = (\hat{\mathbf{x}}_1 - \hat{\mathbf{x}}) \times (\hat{\mathbf{x}}_2 - \hat{\mathbf{x}}_2) . (\hat{\mathbf{x}}_3 - \hat{\mathbf{x}}). \tag{2.6.24}$$

[Convince yourself that (2.6.24) excludes reflections and inversions.] From Appendix B.3, specifically (B.3.23) with $\mathbf{i} := \chi_{t_0}(.,t)$, it follows from (2.6.23) that

$$\chi_{t_0}(\hat{\mathbf{y}},t) - \chi_{t_0}(\hat{\mathbf{x}},t) = \mathbf{Q}(\hat{\mathbf{y}} - \hat{\mathbf{x}}), \tag{2.6.25}$$

where $\mathbf{Q}$ is an orthogonal linear transformation on $\mathcal{V}$ (see Appendix A.16). Further, satisfaction of (2.6.24) requires that $\det \mathbf{Q} = 1$ since the triple scalar product is an alternating trilinear form, and hence (2.6.24) and (2.6.25) define $\det \mathbf{Q}$ [see Appendix A.12.3, specifically (A.12.20) with $\mathbf{L} = \mathbf{Q}$] via

$$(\mathbf{x}_1 - \mathbf{x}) \times (\mathbf{x}_2 - \mathbf{x}) . (\mathbf{x}_3 - \mathbf{x}) = (\det \mathbf{Q})(\hat{\mathbf{x}}_1 - \hat{\mathbf{x}}) \times (\hat{\mathbf{x}}_2 - \hat{\mathbf{x}}) . (\hat{\mathbf{x}}_3 - \hat{\mathbf{x}}). \tag{2.6.26}$$

Of course, $\mathbf{Q}$ in (2.6.25) is time-dependent. Making this explicit, re-labelling $\hat{\mathbf{y}}$ as $\hat{\mathbf{x}}$, $\hat{\mathbf{x}}$ as $\hat{\mathbf{x}}_0$, and defining

$$\mathbf{c}(t) := \chi_{t_0}(\hat{\mathbf{x}}_0,t), \tag{2.6.27}$$

relation (2.6.25) may be re-expressed as

$$\chi_{t_0}(\hat{\mathbf{x}},t) = \mathbf{c}(t) + \mathbf{Q}(t)(\hat{\mathbf{x}} - \hat{\mathbf{x}}_0), \tag{2.6.28}$$

where

$$\det \mathbf{Q}(t) = 1. \tag{2.6.29}$$

Definition. A *body* is said to be *rigid* if the only motions it can undergo are rigid.

Notice that in any rigid motion (2.6.25) the Jacobian is

$$J = \det \mathbf{Q} = 1. \tag{2.6.30}$$

Accordingly [see (2.5.8)]

$$\rho(\mathbf{x},t) = \rho(\hat{\mathbf{x}},t_0), \tag{2.6.31}$$

where

$$\mathbf{x} = \chi_{t_0}(\hat{\mathbf{x}},t). \tag{2.6.32}$$

[14] Suppose at some instant a point $\mathbf{x}$ is mapped into a point $\mathbf{x}'$. If (x_1,x_2,x_3) and (x'_1,x'_2,x'_3) denote their co-ordinates in a Cartesian reference system, then $x'_1 = -x_1, x'_2 = x_2, x'_3 = x_3$ would constitute a reflection (in the plane $x_1 = 0$), while $x'_i = -x_i (i = 1,2,3)$ defines an inversion (with respect to the origin of co-ordinates).

Hence, from (2.6.5), the mass centre location at time t is

$$\begin{aligned}\mathbf{x}_G(t) &= \mathbf{x}_0 + M^{-1}\int_{B_t} \rho(\mathbf{x},t)\{\mathbf{x}-\mathbf{x}_0\}dV_{\mathbf{x}} \\ &= \mathbf{x}_0 + M^{-1}\int_{B_{t_0}} \rho(\hat{\mathbf{x}},t_0)\{\chi_{t_0}(\hat{\mathbf{x}},t)-\mathbf{x}_0\}.1\,dV_{\hat{\mathbf{x}}} \\ &= \mathbf{x}_0 + M^{-1}\int_{B_{t_0}} \rho(\hat{\mathbf{x}},t_0)\{\mathbf{c}(t)-\mathbf{x}_0+\mathbf{Q}(t)(\hat{\mathbf{x}}-\hat{\mathbf{x}}_0)\}dV_{\hat{\mathbf{x}}} \\ &= \mathbf{x}_0 + \mathbf{c}(t) - \mathbf{x}_0 + M^{-1}\mathbf{Q}(t)\int_{B_{t_0}} \rho(\hat{\mathbf{x}},t_0)(\hat{\mathbf{x}}-\hat{\mathbf{x}}_0)dV_{\hat{\mathbf{x}}}.\end{aligned}$$

Thus, recalling (2.6.5) with $t = t_0$ and $\mathbf{x}_0 = \hat{\mathbf{x}}_0$ and (2.6.28),

$$\mathbf{x}_G(t) = \mathbf{c}(t) + \mathbf{Q}(t)(\mathbf{x}_G(t_0) - \hat{\mathbf{x}}_0) = \chi_{t_0}(\mathbf{x}_G(t_0),t), \tag{2.6.33}$$

and we have

Result 2.6.5. A rigid-body motion χ_{t_0} preserves the mass centre of the body in the sense that

$$\mathbf{x}_G(t) = \chi_{t_0}(\mathbf{x}_G(t_0),t). \tag{2.6.34}$$

Differentiating (2.6.33) with respect to time,

$$\mathbf{v}_G(t) := \dot{\mathbf{x}}_G(t) = \dot{\mathbf{c}}(t) + \dot{\mathbf{Q}}(t)(\mathbf{x}_G(t_0) - \hat{\mathbf{x}}_0). \tag{2.6.35}$$

However, noting that $\mathbf{Q}(t)$ is invertible (Why?), (2.6.33) yields

$$\mathbf{x}_G(t_0) - \hat{\mathbf{x}}_0 = \mathbf{Q}^{-1}(t)(\mathbf{x}_G(t) - \mathbf{c}(t)), \tag{2.6.36}$$

whence from (2.6.35), and noting that $\mathbf{Q}^{-1} = \mathbf{Q}^T$ [see Appendix A.16, (A.16.4)],

$$\mathbf{v}_G(t) = \dot{\mathbf{c}}(t) + \dot{\mathbf{Q}}(t)\mathbf{Q}^T(t)(\mathbf{x}_G(t) - \mathbf{c}(t)). \tag{2.6.37}$$

That is (suppressing time dependence),

$$\mathbf{v}_G = \dot{\mathbf{c}} + \mathbf{W}(\mathbf{x}_G - \mathbf{c}), \tag{2.6.38}$$

where

$$\mathbf{W} := \dot{\mathbf{Q}}\mathbf{Q}^T. \tag{2.6.39}$$

In the same way, time differentiation of (2.6.28) yields

$$\mathbf{v}(\mathbf{x},t) := \dot{\chi}_{t_0}(\hat{\mathbf{x}},t) = \dot{\mathbf{c}}(t) + \mathbf{W}(t)(\mathbf{x} - \mathbf{c}(t)), \tag{2.6.40}$$

noting (2.6.32).

Exercise 2.6.2. Prove (2.6.40).

Subtraction of (2.6.38) from (2.6.40) yields

$$\mathbf{v}(\mathbf{x},t) - \mathbf{v}_G(t) = \mathbf{W}(t)(\mathbf{x} - \mathbf{x}_G(t)). \tag{2.6.41}$$

With the aim of simplifying (2.6.22), we note that use of (2.6.41) yields

$$\begin{aligned}\mathcal{A}(t) &= \int_{B_t} (\mathbf{x} - \mathbf{x}_G(t)) \wedge \rho(\mathbf{x},t)\{\mathbf{v}(\mathbf{x},t) - \mathbf{v}_G(t)\} dV_{\mathbf{x}} \\ &= \int_{B_t} (\mathbf{x} - \mathbf{x}_G(t)) \wedge \rho(\mathbf{x},t)\mathbf{W}(t)(\mathbf{x} - \mathbf{x}_G(t)) dV_{\mathbf{x}}. \end{aligned} \tag{2.6.42}$$

Exercise 2.6.3.

(i) Noting that for any linear transformation $\mathbf{A}$ on $\mathcal{V}$, $\mathbf{Ak}.\mathbf{l} = \mathbf{k}.\mathbf{A}^T\mathbf{l}$ for any $\mathbf{k},\mathbf{l} \in \mathcal{V}$ (see Appendix A.8), show that if $\mathbf{A}$ is a differentiable function of time, then

$$\dot{\widehat{\mathbf{A}^T}} = (\dot{\mathbf{A}})^T. \tag{2.6.43}$$

(ii) Noting that for any orthogonal linear transformation $\mathbf{Q}$ on $\mathcal{V}$ [see Appendix A.16, (A.16.3)]

$$\mathbf{QQ}^T = \mathbf{1},$$

show that if $\mathbf{Q}$ is a differentiable function of time, then

$$\dot{\mathbf{Q}}\mathbf{Q}^T + \mathbf{Q}\dot{\widehat{\mathbf{Q}^T}} = \mathbf{O}. \tag{2.6.44}$$

Deduce from (2.6.43) and (2.6.44) that $\mathbf{W}$ defined by (2.6.39) takes skew-symmetric values; that is,

$$\mathbf{W}^T = -\mathbf{W}. \tag{2.6.45}$$

(iii) Noting that, for any $\mathbf{a},\mathbf{b} \in \mathcal{V}$, $\mathbf{a} \otimes \mathbf{b}$ is that linear transformation on $\mathcal{V}$ defined by [see Appendix A.8, (A.8.8)]

$$(\mathbf{a} \otimes \mathbf{b})\mathbf{v} := (\mathbf{b}.\mathbf{v})\mathbf{a} \tag{2.6.46}$$

for any $\mathbf{v} \in \mathcal{V}$, show that if $\mathbf{A}$ is a linear transformation on $\mathcal{V}$, then

$$(\mathbf{Aa}) \otimes \mathbf{b} = \mathbf{A}(\mathbf{a} \otimes \mathbf{b}) \qquad \text{and} \qquad \mathbf{a} \otimes (\mathbf{Ab}) = (\mathbf{a} \otimes \mathbf{b})\mathbf{A}^T. \tag{2.6.47}$$

(iv) Show that if $\mathbf{W}$ is skew, then [see Appendix A.8, (A.8.21)]

$$\begin{aligned}\mathbf{a} \wedge \mathbf{Wa} &= \mathbf{a} \otimes \mathbf{Wa} - \mathbf{Wa} \otimes \mathbf{a} \\ &= (\mathbf{a} \otimes \mathbf{a})\mathbf{W}^T - \mathbf{W}(\mathbf{a} \otimes \mathbf{a}) \\ &= -\{(\mathbf{a} \otimes \mathbf{a})\mathbf{W} + \mathbf{W}(\mathbf{a} \otimes \mathbf{a})\}. \end{aligned} \tag{2.6.48}$$

Deduce that (not surprisingly – why?)

$$\mathbf{a} \wedge \mathbf{Wa} \tag{2.6.49}$$

is skew.

The skew-symmetric linear transformation $\mathbf{W}(t)$ given by (2.6.39) is termed the *spin* of the body at time t.

In view of (2.6.49) with $\mathbf{a} = \mathbf{x} - \mathbf{x}_G(t)$, it follows from (2.6.42) that

$$\mathcal{A} \textit{ is a skew-symmetric function of time.} \tag{2.6.50}$$

Further, from (2.6.48) and (2.6.42),

$$\mathcal{A} = -(\mathcal{I}\mathbf{W} + \mathbf{W}\mathcal{I}), \tag{2.6.51}$$

where
$$\mathcal{I}(t) := \int_{B_t} (\mathbf{x} - \mathbf{x}_G(t)) \otimes \rho(\mathbf{x},t)(\mathbf{x} - \mathbf{x}_G(t))dV_{\mathbf{x}}. \tag{2.6.52}$$

Exercise 2.6.4. Convince yourself of (2.6.51).

Symmetric linear transformation $\mathcal{I}(t)$ is termed the *second moment of mass tensor about G* at time t.

We may write, from (2.6.52), (2.6.32) and (2.6.34), and then (2.6.25),

$$\begin{aligned}\mathcal{I}(t) &= \int_{B_{t_0}} (\chi_{t_0}(\hat{\mathbf{x}},t) - \chi_{t_0}(\mathbf{x}_G(t_0),t) \otimes \rho(\hat{\mathbf{x}},t_0)(\chi_{t_0}(\hat{\mathbf{x}},t) - \chi_{t_0}(\mathbf{x}_G(t_0),t))dV_{\hat{\mathbf{x}}} \\ &= \int_{B_{t_0}} \mathbf{Q}(t)(\hat{\mathbf{x}} - \mathbf{x}_G(t_0)) \otimes \rho(\hat{\mathbf{x}},t_0)\mathbf{Q}(t)(\hat{\mathbf{x}} - \mathbf{x}_G(t_0))dV_{\hat{\mathbf{x}}}.\end{aligned} \tag{2.6.53}$$

That is,
$$\mathcal{I}(t) = \mathbf{Q}(t)\mathcal{I}(t_0)\mathbf{Q}^T(t). \tag{2.6.54}$$

Result (2.6.54) follows from (2.6.47) with $\mathbf{A} = \mathbf{Q}(t), \mathbf{a} = \hat{\mathbf{x}} - \mathbf{x}_G(t_0)$ and $\mathbf{b} = \rho(\hat{\mathbf{x}},t_0)(\hat{\mathbf{x}} - \mathbf{x}_G(t_0))$, on noting that $\mathbf{Q}(t)$ is spatially constant.

Noting that $\mathcal{I}(t_0)$ is independent of t, differentiation of (2.6.54) yields, on suppressing variable t,

$$\begin{aligned}\dot{\mathcal{I}} &= \dot{\mathbf{Q}}\mathcal{I}(t_0)\mathbf{Q}^T + \mathbf{Q}\mathcal{I}(t_0)\dot{\widehat{\mathbf{Q}^T}} \\ &= \dot{\mathbf{Q}}\mathbf{Q}^T(\mathbf{Q}\mathcal{I}(t_0)\mathbf{Q}^T) + \mathbf{Q}\mathcal{I}(t_0)\mathbf{Q}^T(\mathbf{Q}\dot{\widehat{\mathbf{Q}^T}}).\end{aligned}$$

That is, via (2.6.44), (2.6.54) and (2.6.39),

$$\dot{\mathcal{I}} = \mathbf{W}\mathcal{I} - \mathcal{I}\mathbf{W}. \tag{2.6.55}$$

Recalling that our motivation is simplification of (2.6.22) for rigid motions, from (2.6.51) and (2.6.55) we have

$$\begin{aligned}\dot{\mathcal{A}} &= -\dot{\mathcal{I}}\mathbf{W} - \mathcal{I}\dot{\mathbf{W}} - \dot{\mathbf{W}}\mathcal{I} - \mathbf{W}\dot{\mathcal{I}} \\ &= -\{\mathbf{W}\mathcal{I} - \mathcal{I}\mathbf{W}\}\mathbf{W} - \mathcal{I}\dot{\mathbf{W}} - \dot{\mathbf{W}}\mathcal{I} - \mathbf{W}\{\mathbf{W}\mathcal{I} - \mathcal{I}\mathbf{W}\}.\end{aligned}$$

That is,
$$\dot{\mathcal{A}} = -(\mathcal{I}\dot{\mathbf{W}} + \dot{\mathbf{W}}\mathcal{I}) + (\mathcal{I}\mathbf{W}^2 - \mathbf{W}^2\mathcal{I}). \tag{2.6.56}$$

Exercise 2.6.5. Prove that both bracketted terms in (2.6.56) take skew-symmetric values.

Now that we have obtained the form of rotational momentum balance (2.6.22) appropriate to rigid motions [via (2.6.56)], we wish to write this relation (in which each term takes skew-symmetric values) in terms of more familiar axial vectors. Here we need to note the one-to-one correspondence between skew-symmetric linear transformations $\mathbf{A}$ and axial (or pseudo-) vectors $\mathbf{a}$ given by (see Appendix A.15)

$$\mathbf{A}\mathbf{v} = \mathbf{a} \times \mathbf{v} \qquad \text{for any } \mathbf{v} \in \mathcal{V}. \tag{2.6.57}$$

Further [see Appendix A.8, (A.8.22) and Appendix A.15, (A.15.9)],

$$\mathbf{a} \wedge \mathbf{b} \qquad \text{has axial vector} \qquad -\mathbf{a} \times \mathbf{b}. \tag{2.6.58}$$

If $-\mathbf{m}, -\mathbf{j}, -\boldsymbol{a}$ and $-\mathbf{s}$ denote the axial vector counterparts of $\mathbf{M}, \mathbf{J}, \boldsymbol{\mathcal{A}}$ and $\mathbf{S}$, respectively, then (2.6.22) is equivalent to

$$\int_{\partial B_t} \{(\mathbf{x}-\mathbf{x}_G)\times\mathbf{t}+\mathbf{m}\}dA + \int_{B_t} \{(\mathbf{x}-\mathbf{x}_G)\times\mathbf{b}+\mathbf{j}\}dV = \dot{\boldsymbol{a}} + \int_{B_t} \rho\dot{\mathbf{s}}\,dV. \tag{2.6.59}$$

We now need to obtain a representation of $\boldsymbol{a}$ in terms of the *angular velocity* $\boldsymbol{\omega}$ which is the axial vector of $\mathbf{W}$. In fact, we have

Result 2.6.6.

$$\boldsymbol{a} = \mathcal{I}_G\boldsymbol{\omega} \tag{2.6.60}$$

where the *inertia tensor* (see Goldstein *et al.* [7], p. 191)

$$\mathcal{I}_G := (\operatorname{tr}\mathcal{I})\mathbf{1} - \mathcal{I}. \tag{2.6.61}$$

Proof: For any $\mathbf{u}, \mathbf{v} \in \mathcal{V}$, noting that $\boldsymbol{a}$ is the axial vector corresponding to $-\boldsymbol{\mathcal{A}}$, and that $\mathcal{I}$ takes symmetric values,

$$\boldsymbol{a}.(\mathbf{u}\times\mathbf{v}) = \mathbf{v}.(\boldsymbol{a}\times\mathbf{u}) = \mathbf{v}.\{\mathcal{I}\mathbf{W}+\mathbf{W}\mathcal{I}\}\mathbf{u} \tag{2.6.62}$$
$$= \mathbf{v}.\mathcal{I}(\mathbf{W}\mathbf{u}) + \mathbf{v}.\mathbf{W}(\mathcal{I}\mathbf{u})$$
$$= \mathbf{v}.\mathcal{I}(\boldsymbol{\omega}\times\mathbf{u}) + \mathbf{v}.(\boldsymbol{\omega}\times\mathcal{I}\mathbf{u}) \tag{2.6.63}$$
$$= (\mathcal{I}\mathbf{v}.(\boldsymbol{\omega}\times\mathbf{u}) + \mathbf{v}.(\boldsymbol{\omega}\times\mathcal{I}\mathbf{u}) + \mathbf{v}.(\mathcal{I}\boldsymbol{\omega}\times\mathbf{u})) - \mathbf{v}.(\mathcal{I}\boldsymbol{\omega}\times\mathbf{u})$$
$$= (\operatorname{tr}\mathcal{I})\mathbf{v}.(\boldsymbol{\omega}\times\mathbf{u}) - \mathbf{v}.(\mathcal{I}\boldsymbol{\omega}\times\mathbf{u}) \tag{2.6.64}$$
$$= (\operatorname{tr}\mathcal{I})\boldsymbol{\omega}.(\mathbf{u}\times\mathbf{v}) - (\mathcal{I}\boldsymbol{\omega}).(\mathbf{u}\times\mathbf{v}).$$

Here we have used the definition of $\boldsymbol{a}$ and (2.6.51) for (2.6.62), the definition of $\boldsymbol{\omega}$ in (2.6.63), and, to obtain (2.6.64), chosen $\mathbf{u}$ and $\mathbf{v}$ to be linearly independent of $\boldsymbol{\omega}$ and used the definition of the trace operation tr together with the alternating trilinear form provided by the triple scalar product (see Appendix A.12.1). The invariance of this triple product to cyclic permutations has also been employed. Of course, the arbitrary natures of $\mathbf{u}, \mathbf{v}$ in the first and last expressions (and hence the arbitrary nature of $\mathbf{u}\times\mathbf{v}$) establishes result (2.6.60).

Noting from (2.6.54) that

$$\operatorname{tr}\{\mathcal{I}(t)\} = \operatorname{tr}\{\mathbf{Q}(t)\mathcal{I}(t_0)\mathbf{Q}^T(t)\} = \operatorname{tr}\{\mathbf{Q}^T(t)\mathbf{Q}(t)\mathcal{I}(t_0)\},$$

whence

$$\operatorname{tr}\{\mathcal{I}(t)\} = \operatorname{tr}\{\mathcal{I}(t_0)\} \tag{2.6.65}$$

and thus

$$\operatorname{tr}\{\dot{\mathcal{I}}\} = \frac{d}{dt}\{\operatorname{tr}\{\mathcal{I}(t)\}\} = \frac{d}{dt}\{\operatorname{tr}\{\mathcal{I}(t_0)\}\} = 0, \tag{2.6.66}$$

we have, from (2.6.60), (2.6.61), and (2.6.55),

$$\dot{\boldsymbol{a}} = \mathcal{I}_G\dot{\boldsymbol{\omega}} + \dot{\mathcal{I}}_G\boldsymbol{\omega} = \mathcal{I}_G\dot{\boldsymbol{\omega}} - \dot{\mathcal{I}}\boldsymbol{\omega}$$
$$= \mathcal{I}_G\dot{\boldsymbol{\omega}} - (\mathbf{W}\mathcal{I} - \mathcal{I}\mathbf{W})\boldsymbol{\omega} = \mathcal{I}_G\dot{\boldsymbol{\omega}} - \boldsymbol{\omega}\times\mathcal{I}\boldsymbol{\omega} + \mathcal{I}(\boldsymbol{\omega}\times\boldsymbol{\omega}). \tag{2.6.67}$$

However, from (2.6.61),

$$-\boldsymbol{\omega}\times\mathcal{I}\boldsymbol{\omega} = \boldsymbol{\omega}\times\{\mathcal{I}_G\boldsymbol{\omega} - (\operatorname{tr}\mathcal{I})\boldsymbol{\omega}\} = \boldsymbol{\omega}\times\mathcal{I}_G\boldsymbol{\omega}. \tag{2.6.68}$$

Thus we have

Result 2.6.7.
$$\dot{\boldsymbol{a}} = \frac{d}{dt}\{\mathcal{I}_G\boldsymbol{\omega}\} = \mathcal{I}_G\dot{\boldsymbol{\omega}} + \boldsymbol{\omega}\times\mathcal{I}_G\boldsymbol{\omega}. \tag{2.6.69}$$

At this point we have shown that the motion of a rigid body is given by the motion of its mass centre G prescribed by the general result (2.6.11) for *any* body, namely

$$\int_{\partial B_t}\mathbf{t}\,dA + \int_{B_t}\mathbf{b}\,dV = M\dot{\mathbf{v}}_G, \tag{2.6.11}$$

together with prescription of the velocity $\mathbf{v}$ of any other point of the body via (2.6.41), namely

$$\mathbf{v} = \mathbf{v}_G + \boldsymbol{\omega}\times\mathbf{r}, \tag{2.6.70}$$

where
$$\mathbf{r}(\mathbf{x},t) := \mathbf{x} - \mathbf{x}_G(t). \tag{2.6.71}$$

From (2.6.59) and (2.6.69),

$$\int_{\partial B_t}\{\mathbf{r}\times\mathbf{t}+\mathbf{m}\}dA + \int_{B_t}\{\mathbf{r}\times\mathbf{b}+\mathbf{j}\}dV = \mathcal{I}_G\dot{\boldsymbol{\omega}} + \boldsymbol{\omega}\times\mathcal{I}_G\boldsymbol{\omega} + \int_{B_t}\rho\dot{\mathbf{s}}\,dV. \tag{2.6.72}$$

Relation (2.6.11) can be regarded as an evolution equation for $\mathbf{v}_G$, and if $\dot{\mathbf{s}} = \mathbf{0}$, (2.6.72) serves as an evolution equation for $\boldsymbol{\omega}$. Once instantaneous values of $\mathbf{v}_G$ and $\boldsymbol{\omega}$ are known, (2.6.70) delivers the velocity of any point of the body at the instant in question.

The *power* expended by the external forces and couples is

$$P := \int_{\partial B_t}\{\mathbf{t}.\mathbf{v}+\mathbf{m}.\boldsymbol{\omega}\}dA + \int_{B_t}\{\mathbf{b}.\mathbf{v}+\mathbf{j}.\boldsymbol{\omega}\}dV. \tag{2.6.73}$$

Result 2.6.8.
$$P = \dot{K}, \tag{2.6.74}$$

where the *total kinetic energy* of the body is (assuming $\dot{\mathbf{s}} = \mathbf{0}$)

$$K := \frac{1}{2}M\mathbf{v}_G^2 + \frac{1}{2}\mathcal{I}_G\boldsymbol{\omega}.\boldsymbol{\omega}. \tag{2.6.75}$$

Proof: Using (2.6.70), (2.6.11), and (2.6.72) with $\dot{\mathbf{s}} = \mathbf{0}$,

$$\begin{aligned}
P &= \int_{\partial B_t}\{\mathbf{t}.(\mathbf{v}_G+\boldsymbol{\omega}\times\mathbf{r})+\mathbf{m}.\boldsymbol{\omega}\}dA + \int_{B_t}\{\mathbf{b}.(\mathbf{v}_G+\boldsymbol{\omega}\times\mathbf{r})+\mathbf{j}.\boldsymbol{\omega}\}dV \\
&= \left\{\int_{\partial B_t}\mathbf{t}\,dA + \int_{B_t}\mathbf{b}\,dV\right\}.\mathbf{v}_G + \int_{\partial B_t}\{\mathbf{r}\times\mathbf{t}+\mathbf{m}\}dA.\boldsymbol{\omega} + \int_{B_t}\{\mathbf{r}\times\mathbf{b}+\mathbf{j}\}dV.\boldsymbol{\omega} \\
&= M\dot{\mathbf{v}}_G.\mathbf{v}_G + \{\mathcal{I}_G\dot{\boldsymbol{\omega}} + \boldsymbol{\omega}\times\mathcal{I}_G\boldsymbol{\omega}\}.\boldsymbol{\omega} \\
&= \dot{K} - \frac{1}{2}\dot{\mathcal{I}}_G\boldsymbol{\omega}.\boldsymbol{\omega}.
\end{aligned} \tag{2.6.76}$$

However, from (2.6.66), (2.6.61), and (2.6.55),

$$\dot{\mathcal{I}}_G\boldsymbol{\omega}.\boldsymbol{\omega} = -\dot{\mathcal{I}}\boldsymbol{\omega}.\boldsymbol{\omega} = (\mathcal{I}\mathbf{W} - \mathbf{W}\mathcal{I})\boldsymbol{\omega}.\boldsymbol{\omega} = \mathcal{I}(\boldsymbol{\omega}\times\boldsymbol{\omega}).\boldsymbol{\omega} - (\boldsymbol{\omega}\times\mathcal{I}\boldsymbol{\omega}).\boldsymbol{\omega} = 0. \tag{2.6.77}$$

Thus (2.6.74) holds, and it remains only to show that

$$K = \frac{1}{2}\int_{B_t} \rho \mathbf{v}^2\, dV, \tag{2.6.78}$$

so that K is the total kinetic energy of the body. Now, from (2.6.70),

$$\begin{aligned}\int_{B_t} \rho \mathbf{v}^2 dV &= \int_{B_t} \rho(\mathbf{v}_G + \boldsymbol{\omega} \times \mathbf{r})^2 dV \\ &= \int_{B_t} \rho\{\mathbf{v}_G^2 + 2\mathbf{v}_G . (\boldsymbol{\omega} \times \mathbf{r}) + (\boldsymbol{\omega} \times \mathbf{r})^2\} dV.\end{aligned}$$

Further,

$$\int_{B_t} \rho \mathbf{v}_G^2 dV = \left(\int_{B_t} \rho dV\right) \mathbf{v}_G^2 = M\mathbf{v}_G^2, \tag{2.6.79}$$

and

$$\begin{aligned}\int_{B_t} \rho \mathbf{v}_G . (\boldsymbol{\omega} \times \mathbf{r}) dV &= \int_{B_t} \rho \mathbf{r} . (\mathbf{v}_G \times \boldsymbol{\omega}) dV \\ &= \left(\int_{B_t} \rho \mathbf{r}\, dV\right) . (\mathbf{v}_G \times \boldsymbol{\omega}) = 0\end{aligned} \tag{2.6.80}$$

via (2.6.8). It remains to show that

$$\mathcal{I}_G \boldsymbol{\omega} . \boldsymbol{\omega} = \int_{B_t} \rho(\boldsymbol{\omega} \times \mathbf{r})^2 dV. \tag{2.6.81}$$

Now, from (2.6.61) and (2.6.52),

$$\begin{aligned}\mathcal{I}_G \boldsymbol{\omega} . \boldsymbol{\omega} &= (\mathrm{tr}\, \mathcal{I}) \boldsymbol{\omega} . \boldsymbol{\omega} - \mathcal{I} \boldsymbol{\omega} . \boldsymbol{\omega} \\ &= \int_{B_t} \rho \mathbf{r}^2 \boldsymbol{\omega} . \boldsymbol{\omega} - \rho(\mathbf{r} \otimes \mathbf{r}) \boldsymbol{\omega} . \boldsymbol{\omega}\, dV \\ &= \int_{B_t} \rho(\mathbf{r}^2 \boldsymbol{\omega}^2 - (\mathbf{r} . \boldsymbol{\omega})^2) dV = \int_{B_t} \rho(\boldsymbol{\omega} \times \mathbf{r})^2 dV.\end{aligned}$$

QED

Exercise 2.6.6. Use (2.6.78) and (2.6.41) to show that

$$K = \frac{1}{2} M\mathbf{v}_G^2 + \frac{1}{2}\mathcal{I} \cdot \mathbf{W}^T \mathbf{W}. \tag{2.6.82}$$

2.7 Dynamics II: Local Relations

If a body is not rigid, then the *external loading* (represented by $\mathbf{t}$ and $\mathbf{M}$ on ∂B_t and by $\mathbf{b}$ and $\mathbf{J}$ within B_t) results in *distortion*; that is, deformation prescribed by a motion that is not rigid. In order to understand such distortion we consider the motion of matter in subregions of the body, treating matter in each subregion as a body in its own right and taking account of the effect thereon of the rest of the whole body. Let R denote a subregion of B_{t_0}. We can trace the motion of matter in R at time t_0 by the motion map $\boldsymbol{\chi}_{t_0}$ [see (2.3.5)]. Thus with

$$R_t := \boldsymbol{\chi}_{t_0}(R, t) \tag{2.7.1}$$

we postulate as the *interior form of linear momentum balance*

ilmb
$$\int_{\partial R_t} \mathbf{t}\, dA + \int_{R_t} \mathbf{b}\, dV = \frac{d}{dt}\left\{\int_{R_t} \rho\mathbf{v}\, dV\right\}. \tag{2.7.2}$$

The interpretation of $\mathbf{t}$ now differs from that in the global form of balance (2.6.4): $\mathbf{t}$ represents the force per unit area of ∂R_t exerted by that part of the body exterior to R_t (namely, $B_t - R_t$) upon material in R_t if R_t lies completely within B_t (that is, $\partial R_t \cap \partial B_t$ is empty). Further, $\mathbf{b}$ now may include, in addition to volumetric effects from outside B_t, a contribution which derives from the bulk effect of matter in $B_t - R_t$. (This would be the case in dealing with really massive bodies, such as the Earth, or electromagnetic phenomena.)

Similarly, the *interior form of rotational momentum balance* is postulated to be

irmb
$$\int_{\partial R_t} \{\mathbf{r}\wedge\mathbf{t}+\mathbf{M}\}dA + \int_{R_t} \{\mathbf{r}\wedge\mathbf{b}+\mathbf{J}\}dV = \frac{d}{dt}\left\{\int_{R_t} \{\mathbf{r}\wedge\rho\mathbf{v}+\rho\mathbf{S}\}dV\right\}. \tag{2.7.3}$$

Here, if R_t lies in the interior of B_t, $\mathbf{M}$ derives from the effect of matter in $B_t - R_t$ upon that in R_t exerted across ∂R_t, $\mathbf{J}$ is the composite of body couple densities exerted by the world outside B_t and by matter in $B_t - R_t$ upon matter in R_t, and $\rho\mathbf{S}$ accounts for possible 'internal' moment of momentum.

The right-hand sides of (2.7.2) and (2.7.3) can be simplified using the transport/material time derivative theorem (2.5.24). With $f = \mathbf{v}$, we have

$$\frac{d}{dt}\left\{\int_{R_t} \rho\mathbf{v}\, dV\right\} = \int_{R_t} \rho\dot{\mathbf{v}}\, dV, \tag{2.7.4}$$

where [see (2.5.25)]

$$\dot{\mathbf{v}}(\mathbf{x},t) := \frac{\partial}{\partial t}\{\mathbf{v}(\chi_{t_0}(\hat{\mathbf{x}},t),t)\} \tag{2.7.5}$$

with
$$\mathbf{x} := \chi_{t_0}(\hat{\mathbf{x}},t). \tag{2.7.6}$$

However, from (2.3.3) and (2.3.4),

$$\frac{\partial}{\partial t}\{\mathbf{v}(\chi_{t_0}(\hat{\mathbf{x}},t),t)\} = \frac{\partial}{\partial t}\{\dot{\chi}_{t_0}(\hat{\mathbf{x}},t)\} = \ddot{\chi}_{t_0}(\hat{\mathbf{x}},t) = \mathbf{a}(\mathbf{x},t). \tag{2.7.7}$$

Thus
$$\dot{\mathbf{v}} = \mathbf{a}. \tag{2.7.8}$$

Similarly,
$$\frac{d}{dt}\left\{\int_{R_t} \{\mathbf{r}\wedge\rho\mathbf{v}+\rho\mathbf{S}\}dV\right\} = \int_{R_t} \rho\{\dot{\widehat{\mathbf{r}\wedge\mathbf{v}}}+\dot{\mathbf{S}}\}dV. \tag{2.7.9}$$

Now, from (2.5.25) and (2.6.14),

$$\begin{aligned}
\dot{\widehat{\mathbf{r}\wedge\mathbf{v}}}(\mathbf{x},t) &:= \frac{\partial}{\partial t}\{(\chi_{t_0}(\hat{\mathbf{x}},t)-\mathbf{x}_0)\wedge\dot{\chi}_{t_0}(\hat{\mathbf{x}},t)\}\\
&= \dot{\chi}_{t_0}(\hat{\mathbf{x}},t)\wedge\dot{\chi}_{t_0}(\hat{\mathbf{x}},t) + (\chi_{t_0}(\hat{\mathbf{x}},t)-\mathbf{x}_0)\wedge\ddot{\chi}_{t_0}(\hat{\mathbf{x}},t)\\
&= \mathbf{O} + (\mathbf{r}\wedge\mathbf{a})(\mathbf{x},t).
\end{aligned}$$

That is,

$$\widehat{\mathbf{r}\wedge\mathbf{v}} = \mathbf{r}\wedge\mathbf{a}. \tag{2.7.10}$$

Accordingly, from (2.7.2) and (2.7.3), we can write

$$\int_{\partial R_t} \mathbf{t}\,dA = \int_{R_t} \{\rho\mathbf{a} - \mathbf{b}\}dV \tag{2.7.11}$$

and

$$\int_{\partial R_t} \{\mathbf{r}\wedge\mathbf{t} + \mathbf{M}\}dA = \int_{R_t} \{\mathbf{r}\wedge(\rho\mathbf{a} - \mathbf{b}) + \rho\dot{\mathbf{S}} - \mathbf{J}\}dV. \tag{2.7.12}$$

Relations (2.7.11) and (2.7.12), which are assumed to hold for any subregion R_t of B_t which lies strictly inside B_t, give rise to local (otherwise known as *point*) forms provided fields **t** and **M** depend, in a sufficiently regular manner, upon orientation and position. This is a consequence of the following

Theorem 2.7.1. Let g be a tensor field of order n defined throughout a region B, and let R be an arbitrary subregion of B with boundary $\partial R =: \mathcal{S}$. Suppose that there exists an nth order tensor field $f_{\mathcal{S}}$ on $\mathcal{S}$ such that

$$\int_{\mathcal{S}} f_{\mathcal{S}}\,dA = \int_{R} g\,dV. \tag{2.7.13}$$

If dependence of $f_{\mathcal{S}}$ upon $\mathcal{S}$ satisfies, for all $\mathbf{x}\in\mathcal{S}$,

$$f_{\mathcal{S}}(\mathbf{x}) = \hat{f}_{\mathcal{S}}(\mathbf{x};\mathbf{n}(\mathbf{x})), \tag{2.7.14}$$

where **n** denotes the outward unit normal field on $\mathcal{S}$, and dependence of $\hat{f}_{\mathcal{S}}$ upon location **x** is continuous, then there exists a tensor field F of order $(n+1)$ such that

$$f_{\mathcal{S}} = F\mathbf{n}. \tag{2.7.15}$$

In the event that F is continuously differentiable, then use of the divergence theorem (see Appendix B.7) for tensor fields of order $(n+1)$ yields

$$\int_{\mathcal{S}} f_{\mathcal{S}}\,dA = \int_{R} \operatorname{div} F\,dV. \tag{2.7.16}$$

Finally, if g is continuous, then from (2.7.13) and (2.7.16), continuity of $\operatorname{div} F$, and the arbitrary nature of R,

$$\operatorname{div} F = g. \tag{2.7.17}$$

Remark 2.7.1. We do not here prove this theorem but note that it is a simple generalisation of a standard result[15] for vectorial fields together with the divergence theorem for such fields. The generalisation of the divergence theorem to second- and third-order tensor fields is given in Appendix B.7. Proof of (2.7.17) derives from noting that if the integral of a real-valued continuous function of position over arbitrary subregions vanishes, then so too must this function. This follows by noting that its value at a point is $>0, =0$ or <0. If non-zero at point **x**, then it must be either positive or negative, respectively, throughout an open neighbourhood, $N(\mathbf{x})$ say, of

[15] See, for example, Truesdell [8].

$\mathbf{x}$ as a consequence of continuity. Thus the integral over $B \cap N(\mathbf{x})$ is non-zero, contradicting the initial hypothesis. The result follows for a tensor-valued field of any order by considering the real-valued Cartesian components of this field individually.

Corollary 2.7.1. If choices $f_{\mathcal{S}} = \mathbf{t}$ and $g = \rho\mathbf{a} - \mathbf{b}$ satisfy the hypotheses of the theorem, then there exists a second-order tensor field $\mathbf{T}$ defined in the interior of B_t such that

$$\mathbf{t} = \mathbf{T}\mathbf{n}, \tag{2.7.18}$$

$$\int_{\partial R_t} \mathbf{t}\, dA = \int_{\partial R_t} \mathbf{T}\mathbf{n}\, dA = \int_{R_t} \operatorname{div} \mathbf{T}\, dV, \tag{2.7.19}$$

and

$$\textbf{llmb} \qquad \operatorname{div} \mathbf{T} + \mathbf{b} = \rho\mathbf{a}. \tag{2.7.20}$$

Relation (2.7.20) is the l*ocal form of* l*inear* m*omentum* b*alance.*

The theorem also can be employed to derive analogous results in respect of balance (2.7.12). However, it is first necessary to re-write the first term in (2.7.12) as a volume integral over R_t. The result follows from

Lemma 2.7.1.

$$\int_{\partial R_t} \mathbf{r} \otimes \mathbf{t}\, dA = \int_{\partial R_t} \mathbf{r} \otimes \mathbf{T}\mathbf{n}\, dA = \int_{R_t} \{\mathbf{r} \otimes \operatorname{div} \mathbf{T} + \mathbf{T}^T\} dV. \tag{2.7.21}$$

Proof: For any pair of vectors $\mathbf{k}, \mathbf{l} \in \mathcal{V}$ [see Appendix A.8, (A.8.8)],

$$(\mathbf{r} \otimes \mathbf{T}\mathbf{n})\mathbf{k}\,.\,\mathbf{l} = (\mathbf{T}\mathbf{n}\,.\,\mathbf{k})(\mathbf{r}\,.\,\mathbf{l}) = (\mathbf{r}\,.\,\mathbf{l})(\mathbf{T}^T\mathbf{k})\,.\,\mathbf{n}. \tag{2.7.22}$$

Thus, by the divergence theorem for vectorial fields,

$$\left(\int_{\partial R_t} \mathbf{r} \otimes \mathbf{T}\mathbf{n}\, dA\right)\mathbf{k}\,.\,\mathbf{l} = \int_{\partial R_t} (\mathbf{r}\,.\,\mathbf{l})\mathbf{T}^T\mathbf{k}\,.\,\mathbf{n}\, dA = \int_{R_t} \operatorname{div}\{(\mathbf{r}\,.\,\mathbf{l})\mathbf{T}^T\mathbf{k}\} dV. \tag{2.7.23}$$

However [see Appendix B.7, (B.7.28)],

$$\begin{aligned}
\operatorname{div}\{\mathbf{r}\,.\,\mathbf{l})\mathbf{T}^T\mathbf{k}\} &= (\mathbf{r}\,.\,\mathbf{l})\operatorname{div}\{\mathbf{T}^T\mathbf{k}\} + \nabla(\mathbf{r}\,.\,\mathbf{l})\,.\,\mathbf{T}^T\mathbf{k} \\
&= (\mathbf{r}\,.\,\mathbf{l})(\operatorname{div} \mathbf{T})\,.\,\mathbf{k} + \mathbf{l}\,.\,\mathbf{T}^T\mathbf{k} \\
&= \{\mathbf{r} \otimes \operatorname{div} \mathbf{T} + \mathbf{T}^T\}\mathbf{k}\,.\,\mathbf{l}.
\end{aligned} \tag{2.7.24}$$

The result follows from (2.7.23) and (2.7.24) on noting the arbitrary nature of $\mathbf{k}$ and $\mathbf{l}$.
Noting that $\mathbf{r} \wedge \mathbf{t} := \mathbf{r} \otimes \mathbf{t} - (\mathrm{r} \otimes \mathbf{t})^T$ [see Appendix A.8, (A.8.21)], from (2.7.21) we have

$$\int_{\partial R_t} \mathbf{r} \wedge \mathbf{t}\, dA = \int_{R_t} \{\mathbf{r} \wedge \operatorname{div} \mathbf{T} + \mathbf{T}^T - \mathbf{T}\} dV. \tag{2.7.25}$$

Accordingly, (2.7.12) may be written in the form

$$\begin{aligned}
\int_{\partial R_t} \mathbf{M}\, dA &= \int_{R_t} \{\mathbf{r} \wedge (\rho\mathbf{a} - \mathbf{b} - \operatorname{div} \mathbf{T}) + \mathbf{T} - \mathbf{T}^T + \rho\dot{\mathbf{S}} - \mathbf{J}\} dV \\
&= \int_{R_t} \{\mathbf{T} - \mathbf{T}^T + \rho\dot{\mathbf{S}} - \mathbf{J}\} dV
\end{aligned} \tag{2.7.26}$$

by virtue of (2.7.20). Relation (2.7.26) has the form of (2.7.13), so if $\mathbf{M} = f_{\mathcal{S}}$ and $\mathbf{T} - \mathbf{T}^T + \rho\dot{\mathbf{S}} - \mathbf{J} = g$ satisfy the conditions of Theorem 2.7.1, then there exists a third-order tensor field $\mathbf{C}$ on the interior of B_t such that

$$\mathbf{M} = \mathbf{C}\mathbf{n}, \tag{2.7.27}$$

$$\int_{\partial R_t} \mathbf{M}\, dA = \int_{R_t} \operatorname{div} \mathbf{C}\, dV, \tag{2.7.28}$$

and

$$\textbf{lrmb} \qquad \operatorname{div} \mathbf{C} + \mathbf{T}^T - \mathbf{T} + \mathbf{J} = \rho\dot{\mathbf{S}}. \tag{2.7.29}$$

Relation (2.7.29) is the l*ocal form of* r*otational* m*omentum* b*alance.*

In view of (2.7.18) and (2.7.27), balances (2.7.2) and (2.7.3) may be written as

$$\int_{\partial R_t} \mathbf{T}\mathbf{n}\, dA + \int_{R_t} \mathbf{b}\, dV = \frac{d}{dt}\left\{\int_{R_t} \rho\mathbf{v}\, dV\right\} \tag{2.7.30}$$

and

$$\int_{\partial R_t} \{\mathbf{r} \wedge \mathbf{T}\mathbf{n} + \mathbf{C}\mathbf{n}\} dA + \int_{R_t} \{\mathbf{r} \wedge \mathbf{b} + \mathbf{J}\} dV$$
$$= \frac{d}{dt}\left\{\int_{R_t} \{\mathbf{r} \wedge \rho\mathbf{v} + \rho\mathbf{S}\} dV\right\}, \tag{2.7.31}$$

respectively.

Fields $\mathbf{T}$ and $\mathbf{C}$ are termed the *(Cauchy) stress tensor* and *couple stress tensor fields*, respectively.

Each term in balance (2.7.31) takes skew-symmetric values and hence can be expressed alternatively as the corresponding axial vector field. Writing $ax\{\mathbf{A}\}$ to denote the axial vector corresponding to skew-symmetric linear transformation $\mathbf{A}$, so that [see (2.6.57)]

$$\mathbf{A}\mathbf{v} = ax\{\mathbf{A}\} \times \mathbf{v} \qquad \text{for any} \qquad \mathbf{v} \in \mathcal{V}, \tag{2.7.32}$$

and recalling (2.6.58), namely

$$ax\{\mathbf{a} \wedge \mathbf{b}\} = -\mathbf{a} \times \mathbf{b}, \tag{2.7.33}$$

we have

$$ax\{\mathbf{r} \wedge \mathbf{T}\mathbf{n}\} = -\mathbf{r} \times \mathbf{T}\mathbf{n}, \quad ax\{\mathbf{r} \wedge \mathbf{b}\} = -\mathbf{r} \times \mathbf{b} \quad \text{and} \quad ax\{\mathbf{r} \wedge \rho\mathbf{v}\} = -\mathbf{r} \times \rho\mathbf{v}. \tag{2.7.34}$$

Further, defining

$$\mathcal{C}\mathbf{n} := -ax\{\mathbf{C}\mathbf{n}\}, \quad \mathbf{j} := -ax\{\mathbf{J}\}, \quad \mathbf{s} := -ax\{\mathbf{S}\}, \tag{2.7.35}$$

and using (2.7.34), balance (2.7.31) may be written as

$$\int_{\partial R_t} \{\mathbf{r} \times \mathbf{T}\mathbf{n} + \mathcal{C}\mathbf{n}\} dA + \int_{R_t} \{\mathbf{r} \times \mathbf{b} + \mathbf{j}\} dV = \frac{d}{dt}\left\{\int_{R_t} \{\mathbf{r} \times \rho\mathbf{v} + \rho\mathbf{s}\} dV\right\}. \tag{2.7.36}$$

Exercise 2.7.1. Show that $ax\{\mathbf{C}\mathbf{n}\}$ is linear in $\mathbf{n}$ and hence that $\mathcal{C}$ takes linear transformation values.

Remark 2.7.2 In the absence of couple stress, external body couples, and 'internal' rotational momentum, (2.7.29) reduces to

$$\mathbf{T}^T = \mathbf{T}. \tag{2.7.37}$$

That is, in such cases the Cauchy stress tensor takes symmetric values.

Remark 2.7.3. Motivation for expressing rotational momentum balance (2.7.31) in the equivalent axial vector form (2.7.36) is twofold. Firstly, such form is that most usually employed, and secondly, we shall in the following subsection postulate an energy balance which involves axial vectors. If we consider a couple, represented by an axial vector $\mathbf{c}$, applied to a rigid body which has angular velocity $\boldsymbol{\omega}$ (an axial vector), then the power it expends is $\mathbf{c}.\boldsymbol{\omega}$. In generalising this to deformable bodies we shall first postulate an energy balance in which the power terms associated with couple stress and body couples are expressed in terms of axial vectors. The equivalent balance involving skew tensors then follows.

2.8 Thermomechanics

2.8.1 Global Balance of Energy

A body can exchange energy with its environment by both mechanical and thermal agencies. Generally, external tractions and couples, together with body forces and couples, will contribute to the rate at which the total energy of the body is changing in a given motion. Similarly, this rate also will depend upon the rate at which heat is supplied to the body, both by conduction across its boundary and by the penetrative (bulk) effect of radiation.

Global balance of energy is postulated to take the form

$$\int_{\partial B_t} \left\{ q + \mathbf{t}.\mathbf{v} + \mathbf{m}.\mathbf{w} \right\} dV + \int_{B_t} \left\{ r + \mathbf{b}.\mathbf{v} + \mathbf{j}.\mathbf{w} \right\} dV = \frac{d}{dt} \left\{ \int_{B_t} \rho \left(e + \frac{1}{2}\mathbf{v}^2 + k \right) dV \right\}. \tag{2.8.1}$$

Here q denotes the conductive heat supply rate to the body per unit area of its boundary, as before $\mathbf{t}$ is the applied external traction, $\mathbf{m}$ is the applied external couple per unit boundary area, $\mathbf{v}$ is the velocity field, and $\mathbf{w}$ is the spin vector field. (Both $\mathbf{m}$ and $\mathbf{w}$ are axial vector fields.) Term r represents the external radiative heat supply rate density, $\mathbf{b}$ is the external body force density, and $\mathbf{j}$ is the (axial vector-valued) external body couple density. The energy stored within the body derives from densities of internal energy ρe, macroscopic kinetic energy $\frac{1}{2}\rho\mathbf{v}^2$, and microstructural kinetic energy ρk.

2.8.2 Aside on the Spin Vector Field $\mathbf{w}$ and Power Expended by Couples

The powers expended in a rigid body motion by surface couple field $\mathbf{m}$ and body couple field $\mathbf{j}$ are given by appropriate integrals of $\mathbf{m}.\boldsymbol{\omega}$ and $\mathbf{j}.\boldsymbol{\omega}$, where $\boldsymbol{\omega}$ denotes angular velocity [see (2.6.73)]. The appropriate generalisation to deformable body motions requires generalisation of $\boldsymbol{\omega}$. Recall that for a rigid body motion $\boldsymbol{\omega}$ is the axial vector associated with the skew-symmetric linear transformation [see (2.6.28) and (2.6.39)]

$$W := \dot{\mathbf{Q}}\mathbf{Q}^T. \tag{2.8.2}$$

Note that the velocity gradient $\mathbf{L} := \nabla \mathbf{v}$ [see Appendix B.5, (B.5.19)] is, from (2.6.40), given by

$$\mathbf{L} = \mathbf{W}. \tag{2.8.3}$$

Exercise 2.8.1. Prove (2.8.3) from (2.6.40) by noting that upon suppressing time dependence,

$$\mathbf{v}(\mathbf{x}) = \dot{\mathbf{c}} + \mathbf{W}(\mathbf{x} - \mathbf{c}),$$

so for any vector $\mathbf{h}$,

$$\mathbf{v}(\mathbf{x} + \mathbf{h}) - \mathbf{v}(\mathbf{x}) = \mathbf{W}\mathbf{h}.$$

Of course, since $\mathbf{W}$ is skew symmetric, (2.8.3) may be written as

$$\mathbf{W} = \frac{1}{2}(\mathbf{L} - \mathbf{L}^T). \tag{2.8.4}$$

In a *general* motion, $\mathbf{W}$ defined by (2.8.4) [see Appendix B.5, (B.5.26)] is the *spin tensor* field. It is the axial vector $\mathbf{w}$ of $\mathbf{W}$ which is the generalisation we seek, and we term this the *spin vector*. As proved in Appendix B.5 [see (B.5.28) and (B.5.29)],

$$\mathbf{w} = \frac{1}{2} \operatorname{curl} \mathbf{v}. \tag{2.8.5}$$

2.8.3 Local Balance of Energy

As with balances of linear and rotational momentum, any 'part' of the body lying strictly within B_t can be regarded as a body in its own right. We thus postulate, as the analogue of (2.8.1),

$$\int_{\partial R_t} \left\{ q + \mathbf{t}.\mathbf{v} + \mathbf{m}.\mathbf{w} \right\} dV + \int_{R_t} \left\{ r + \mathbf{b}.\mathbf{v} + \mathbf{j}.\mathbf{w} \right\} dV = \frac{d}{dt} \left\{ \int_{R_t} \rho \left(e + \frac{1}{2}\mathbf{v}^2 + k \right) dV \right\}. \tag{2.8.6}$$

Terms q, $\mathbf{t}$ and $\mathbf{m}$ are associated here with the influence on matter in R_t from that in $B_t - R_t$. It should be noted that each of $r, \mathbf{b}$ and $\mathbf{j}$ may have contributions both from the material world external to B_t *and* matter in $B_t - R_t$. Recalling (2.7.18), (2.6.73) and $(2.7.35)_1$, we have

$$\mathbf{t}.\mathbf{v} = \mathbf{T}\mathbf{n}.\mathbf{v} \qquad \text{and} \qquad \mathbf{m}.\mathbf{w} = \mathcal{C}\mathbf{n}.\mathbf{w}. \tag{2.8.7}$$

Further, replacing axial vectors by the corresponding skew-symmetric linear transformations, and invoking (A.15.13) of Appendix A.15, we have, from $(2.7.35)_{1,2}$,

$$\mathcal{C}\mathbf{n}.\mathbf{w} = -\frac{1}{2}\,\mathbf{C}\mathbf{n}\cdot\mathbf{W} \qquad \text{and} \qquad \mathbf{j}.\mathbf{w} = -\frac{1}{2}\,\mathbf{J}\cdot\mathbf{W}. \tag{2.8.8}$$

Thus (2.8.6) becomes, on also using the transport theorem (2.5.24),

$$\int_{\partial R_t} \left\{ q + \mathbf{T}\mathbf{n}.\mathbf{v} - \frac{1}{2}\mathbf{C}\mathbf{n}\cdot\mathbf{W} \right\} dA + \int_{R_t} \left\{ r + \mathbf{b}.\mathbf{v} - \frac{1}{2}\mathbf{J}\cdot\mathbf{W} \right\} dV = \int_{R_t} \rho \left\{ \dot{e} + \mathbf{v}.\mathbf{a} + \dot{k} \right\} dV. \tag{2.8.9}$$

Now

$$\int_{\partial R_t} \mathbf{T}\mathbf{n}.\mathbf{v}\, dA = \int_{\partial R_t} \mathbf{n}.\mathbf{T}^T\mathbf{v}\, dA = \int_{R_t} \operatorname{div}\{\mathbf{T}^T\mathbf{v}\} dV, \tag{2.8.10}$$

and [see Appendix A.19, and specifically (A.19.43)]

$$\int_{\partial R_t} \mathbf{Cn}\cdot\mathbf{W}\,dA = \int_{\partial R_t} \mathbf{n}.\{\mathbf{C}^{\sim}:\mathbf{W}\}\,dA = \int_{R_t} \text{div}\,\{\mathbf{C}^{\sim}:\mathbf{W}\}dV. \tag{2.8.11}$$

Accordingly (2.8.9) may be written as

$$\int_{\partial R_t} q\,dA = \int_{\partial R_t} g\,dV, \tag{2.8.12}$$

where

$$g := \rho\{\dot{e} + \mathbf{v}.\mathbf{a} + \dot{k}\} - \text{div}\{\mathbf{T}^T\mathbf{v} - \mathbf{C}^{\sim}:\mathbf{W}\} - r - \mathbf{b}.\mathbf{v} + \frac{1}{2}\mathbf{J}\cdot\mathbf{W}. \tag{2.8.13}$$

If $q = f_S$ and g satisfy the conditions of Theorem 2.7.1, then there exists a vector field $-\mathbf{q}$ such that

$$q = -\mathbf{q}.\mathbf{n}. \tag{2.8.14}$$

Field $\mathbf{q}$ is termed the *heat flux vector* field.

Remark 2.8.1. Term

$$\int_{\partial R_t} q\,dA = \int_{\partial R_t} -\mathbf{q}.\mathbf{n}\,dA = \int_{\partial R_t} \mathbf{q}.(-\mathbf{n})dA. \tag{2.8.15}$$

The reason for employing the negative sign is that the quantity represented by any of the integrals in (2.8.15) is the rate at which heat is conducted *into* R_t across its boundary: recall that $\mathbf{n}$ denotes the outward unit normal field on ∂R_t. For any unit vector $\hat{\mathbf{u}}$, $\mathbf{q}.\hat{\mathbf{u}}$ represents the local conductive heat supply rate, per unit area of any surface to which $\hat{\mathbf{u}}$ is locally a normal, in the direction of $\hat{\mathbf{u}}$.

From (2.8.14) and (2.8.9) we obtain the *interior form of energy balance*

ieb

$$\int_{R_t} \left\{-\mathbf{q}.\mathbf{n} + \mathbf{Tn}.\mathbf{v} - \frac{1}{2}\mathbf{Cn}\cdot\mathbf{W}\right\} dA + \int_{R_t} \left\{r + \mathbf{b}.\mathbf{v} - \frac{1}{2}\mathbf{J}\cdot\mathbf{W}\right\} dV$$

$$= \frac{d}{dt}\left\{\int_{R_t} \rho\,\{e + \frac{1}{2}\mathbf{v}^2 + k\}dV\right\}. \tag{2.8.16}$$

Noting (2.8.11) and using the divergence theorem, this relation becomes

$$\int_{R_t} \left\{\text{div}(-\mathbf{q} + \mathbf{T}^T\mathbf{v} - \frac{1}{2}\mathbf{C}^{\sim}:\mathbf{W}) + r + \mathbf{b}.\mathbf{v} - \frac{1}{2}\mathbf{J}\cdot\mathbf{W}\right\} dV$$

$$= \int_{R_t} \rho\,\{\dot{e} + \mathbf{v}.\mathbf{a} + \dot{k}\}dV. \tag{2.8.17}$$

However (see Appendix B.7, (B.7.31) and (B.7.36)),

$$\text{div}\{\mathbf{T}^T\mathbf{v}\} = (\text{div}\,\mathbf{T}).\mathbf{v} + \mathbf{T}\cdot\nabla\mathbf{v} \tag{2.8.18}$$

and

$$\text{div}\{\mathbf{C}^{\sim}:\mathbf{W}\} = (\text{div}\,\mathbf{C})\cdot\mathbf{W} + \mathbf{C}\cdot\nabla\mathbf{W}. \tag{2.8.19}$$

Thus, using (2.8.18) and (2.8.19) and re-arranging (2.8.17), we have

$$\int_{R_t}\Big\{-\operatorname{div}\mathbf{q}+(\operatorname{div}\mathbf{T}+\mathbf{b}-\rho\mathbf{a}).\mathbf{v}+\Big(-\frac{1}{2}\operatorname{div}\mathbf{C}-\frac{1}{2}\mathbf{J}\Big)\cdot\mathbf{W}+r \\ +\mathbf{T}\cdot\mathbf{L}-\frac{1}{2}\mathbf{C}\cdot\nabla\mathbf{W}-\rho(\dot{e}+\dot{k})\Big\}\,dV=0. \tag{2.8.20}$$

If the integrand is continuous, then, on recalling the local forms of linear and rotational momentum (2.7.20) and (2.7.29), the arbitrary nature of R_t gives the corresponding local form

$$r-\operatorname{div}\mathbf{q}+0+\frac{1}{2}(\mathbf{T}^T-\mathbf{T}-\rho\dot{\mathbf{S}})\cdot\mathbf{W}+\mathbf{T}\cdot\mathbf{L}-\frac{1}{2}\mathbf{C}\cdot\nabla\mathbf{W}=\rho(\dot{e}+\dot{k}). \tag{2.8.21}$$

Noting

$$\begin{aligned}\mathbf{T}\cdot\mathbf{L}&=\mathbf{T}\cdot\Big(\frac{1}{2}(\mathbf{L}-\mathbf{L}^T)+\frac{1}{2}(\mathbf{L}+\mathbf{L}^T)\Big)\\&=\mathbf{T}\cdot\mathbf{W}+\mathbf{T}\cdot\mathbf{D}\\&=\frac{1}{2}(\mathbf{T}-\mathbf{T}^T)\cdot\mathbf{W}+\mathbf{T}\cdot\mathbf{D},\end{aligned}$$

where the *stretching tensor* field [see Appendix B.5, (B.5.27)]

$$\mathbf{D}:=\frac{1}{2}(\mathbf{L}+\mathbf{L}^T), \tag{2.8.22}$$

relation (2.8.21) simplifies to the *local form of energy balance*

$$\textbf{leb}\qquad r-\operatorname{div}\mathbf{q}+\mathbf{T}\cdot\mathbf{D}-\frac{1}{2}\mathbf{C}\cdot\nabla\mathbf{W}=\rho(\dot{e}+\dot{k}+\frac{1}{2}\dot{\mathbf{S}}\cdot\mathbf{W}). \tag{2.8.23}$$

In the absence of couple stress, internal rotational momentum, and microstructural kinetic energy, (2.8.23) reduces to

$$r-\operatorname{div}\mathbf{q}+\mathbf{T}\cdot\mathbf{D}=\rho\dot{e}. \tag{2.8.24}$$

Remark 2.8.2. The reason for the minus sign associated with $\mathbf{q}$ in balances (2.8.16), (2.8.23) and (2.8.24) was given in Remark 2.8.1. The minus signs attached to the power terms associated with couples in (2.8.16) ultimately derive from (2.6.58) and definitions (2.7.35) which yielded the axial-vector form (2.7.36) of interior rotational momentum balance (2.7.31).

3 Motivation for Seeking a Molecular Scale-Dependent Perspective on Continuum Modelling

3.1 Preamble

The continuum viewpoint is consistent with our physical prejudices, engendered by sensory evidence. However such a perspective gives rise to some fundamental conceptual and physical difficulties which involve questions of scale, interpretation, and reproducibility of phenomena. Here we outline several of these difficulties and indicate how one is forced to take account of the fundamentally discrete nature of matter and the spatial scales at which physical systems are monitored.

3.2 The Natural Continuum Prejudice

We unconsciously adopt a continuum viewpoint when observing and interacting with the world about us. For example, we regard the air we breathe as tangible (when feeling the wind on our faces or filling our lungs) and as permeating the space about us. If we pour some water into a glass, then this water appears to take up a definite shape, determined by the sides and base of the glass and by the free water surface. The water seems to fill this shape, apart from possible visible bubbles of trapped air or immersed foreign particles. The glass itself appears to occupy a definite region, delineated by its bounding surfaces, with the possible exception of visible imperfections. However, while we can see 'inside' water and glass, this is not the case for opaque objects, for which only the external boundary is amenable to direct observation. Nevertheless, we often regard opaque objects to be full of matter in the sense of occupying all space within their perceived external boundaries. Consider, for example, an eraser, dinner plate, or steel ingot. Our intuitive view is to regard such objects as composed of matter which is present throughout the regions these objects appear to occupy. Of course, breaking any such object into several pieces gives some evidence as to whether the intuitive view is correct: the parts may 'fit together' or, if not, draw attention to the prior presence of voids within the object. However, such partial investigation is destructive, and elicits only limited information. From a pragmatic standpoint we assume complete spatial occupation in the absence of knowledge to the contrary.

3.3 The Continuum Viewpoint on Mass Density ρ

The continuum viewpoint outlined in Chapter 2 can be adopted for any observable entity which has, at any time of interest, both mass and the appearance of occupying a definite spatial region. Of course, this region may change with time, such as is seen, for example, in the distortion of an eraser or water rippling in a glass and/or evaporating. As discussed in Section 2.2, both the distribution of matter and the occupied region are described in terms of a mass density field ρ. For convenience, we recapitulate the two fundamental attributes of ρ given in Chapter 2, essentially the two *m*ass *d*ensity properties delineated in (2.2.1) and (2.2.2):

M.D.1′. ρ takes non-negative values; the region B_t occupied by the body at time t consists of those points $\mathbf{x}$ in Euclidean space $\mathcal{E}$ for which $\rho(\mathbf{x},t) > 0$, and

M.D.2′. The mass of material in any region R at any time is delivered by the integral of ρ over R at this time.

Conceptual difficulties arise with both M.D.1′ and M.D.2′.

3.4 Boundaries and the Scale Dependence of ρ

Since the region occupied by a body is intimately related to the mass density function (see M.D.1′), it follows that the *boundary* of the body is delineated by ρ. Specifically, the boundary ∂B_t of the body at time t is the boundary of the region in which ρ takes positive values at this time. [Of course, $\rho(\mathbf{x},t) = 0$ for any point $\mathbf{x}$ *outside* B_t at time t.] However, it is a matter of common experience that the perceived boundary of a body differs according to the manner in which it is observed. Indeed, the closer an observer approaches a body, the more do additional surface features become apparent. Use of a magnifying glass or microscope renders further visible topographic detail.

The foregoing can be appreciated more precisely by considering images within a pinhole camera, the screen of which serves as a simple model of a retina or film. By the *spatial resolution d* of such a camera we mean the smallest separation of point pairs on the screen which can be distinguished. If the screen–pinhole distance is ℓ, then the smallest object directly in front of the camera that can be resolved (i.e., have an image with discernible dimensions and area) at a distance s from the pinhole will have a visible span (see Figure 3.1) $D_{\min}$, where

$$\frac{D_{\min}}{s} = \frac{d}{\ell}. \tag{3.4.1}$$

If λ is the angle subtended at the pinhole by an image of height d directly behind the pinhole, then

$$\tan\left(\frac{\lambda}{2}\right) = \frac{d}{2\ell}. \tag{3.4.2}$$

Accordingly,

$$D_{\min} = \frac{sd}{\ell} = 2s\tan\left(\frac{\lambda}{2}\right). \tag{3.4.3}$$

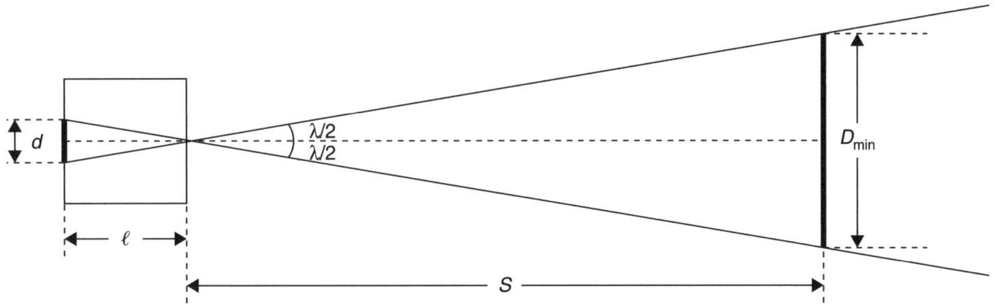

Figure 3.1. If d is the smallest separation of points detectable on the pinhole camera screen, then the smallest detectable dimension $D_{\min}$ of an object at a distance s from pinhole is given by $D_{\min} = (sd/\ell) = 2s\tan(\lambda/2) \simeq \lambda s$ for small λ).

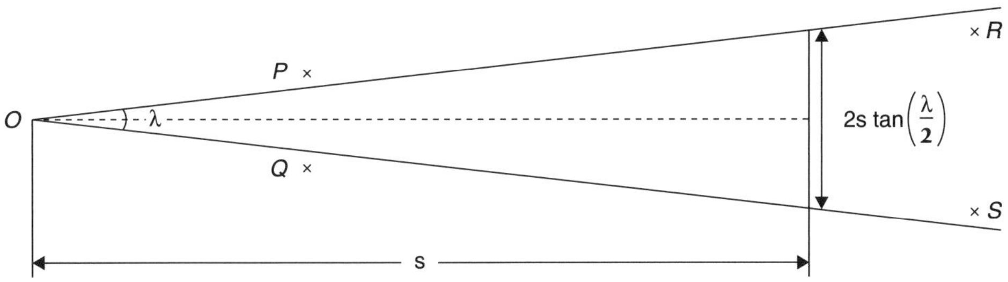

Figure 3.2. An observational instrument with angular resolution λ located at O can detect features of an object which subtend an angle of λ or greater at O. Accordingly, points P and Q can be distinguished, while R and S cannot, although distance $PQ <$ distance RS.

Angle λ might be termed the *angular resolution* of the camera. Any object which lies directly in front of the camera and *within* a cone having apex at, and axis directly away from, the pinhole, and with apex semi-angle $\lambda/2$, will have an image with no discernible dimensions and so appear as a 'dot'. From (3.4.3) it is clear that $D_{\min}$ decreases linearly with decrease in distance s, so more detail of any observed object becomes apparent as the camera gets closer to the object. In a more sophisticated imaging instrument (e.g., an eye or a microscope incorporating a lens or system of lenses), light received from a point of the object under scrutiny is focussed upon a point of the retina or film, no matter where it impinges on the first lens. Accordingly, there is still a one-to-one relation[1] between 'visible' points on the object and points on the retina or film. Such relation will not, of course, yield results as simple as (3.4.1) and (3.4.2), but will depend upon instrument geometry.

A general, formal approach to observation is to regard any observational instrument to be located at a point O and to be able to distinguish features of an object which subtend an angle λ or greater at O. Angle λ is termed the *angular resolution* of the instrument. Accordingly (see Figure 3.2), given a separation s between O and an object, an image of the boundary of this object is obtained in which any features

[1] At least in an ideal imaging instrument: in practice account must be taken of refraction (for non-monochromatic light), aberration (due to imperfect lenses and consequent image-focussing deficiencies), and diffraction (associated with apertures). See, e.g., Born & Wolf [9].

whose characteristic dimension exceed $2s\tan(\lambda/2)$ can become apparent, but features whose dimensions are smaller than $2s\tan(\lambda/2)$ cannot be resolved. Thus the smaller λ or s, the finer is the scale of possible distinguishable boundary topography. The foregoing remarks serve to establish the scale-dependent nature of the boundary of a body: $2s\tan(\lambda/2)$ may be regarded as the *limiting observational scale* associated with angular resolution λ and separation s.

Since the boundary of a body depends upon scale, as evidenced by common observational experience (and formalised in the preceding discussion), and since such boundary is delineated by the mass density function, we are led to conclude that *the mass density function ρ must be scale-dependent.*

3.5 Continuity of ρ and the Discrete Nature of Matter

Recall (2.2.2) which was here recapitulated in M.D.2′: the mass at time t associated with material lying in a region R is

$$m(R,t) = \int_R \rho(\mathbf{x},t)dV_{\mathbf{x}}. \tag{2.2.2}$$

Note also (see Remark 2.2.1) that ρ is assumed to be spatially continuous at any time.

Now suppose that $\mathbf{x}_0$ is any point which lies strictly within B_t. Continuity of ρ at time t requires that for any given number $\varepsilon > 0$, however small, there should exist a number δ (depending upon ε in general) such that

$$|\rho(\mathbf{x},t) - \rho(\mathbf{x}_0,t)| < \varepsilon \qquad \text{wherever} \qquad \|\mathbf{x} - \mathbf{x}_0\| < \delta. \tag{3.5.1}$$

Accordingly,

$$\begin{aligned}\left|\int_{B_\delta(\mathbf{x}_0)} \{\rho(\mathbf{x},t) - \rho(\mathbf{x}_0,t)\}dV_{\mathbf{x}}\right| &\leq \int_{B_\delta(\mathbf{x}_0)} |\rho(\mathbf{x},t) - \rho(\mathbf{x}_0,t)|dV_{\mathbf{x}} \\ &< \varepsilon \times \mathrm{vol}(B_\delta(\mathbf{x}_0)),\end{aligned} \tag{3.5.2}$$

where $B_\delta(\mathbf{x}_0)$ denotes the interior of that spherical region (a *ball*) centred at $\mathbf{x}_0$ with radius δ and volume $\mathrm{vol}(B_\delta(\mathbf{x}_0))$ $(= 4\pi\delta^3/3)$. Accordingly, on dividing by $4\pi\delta^3/3$, (3.5.2) yields

$$\left|\frac{1}{\mathrm{vol}(B_\delta(\mathbf{x}_0))}\left(\int_{B_\delta(\mathbf{x}_0)} \rho(\mathbf{x},t)dV_{\mathbf{x}}\right) - \rho(\mathbf{x}_0,t)\right| < \varepsilon \tag{3.5.3}$$

for sufficiently small δ. Equivalently,

$$\lim_{\delta\to 0}\left\{\frac{1}{\mathrm{vol}(B_\delta(\mathbf{x}_0))}\int_{B_\delta(\mathbf{x}_0)} \rho(\mathbf{x},t)dV_{\mathbf{x}}\right\} = \rho(\mathbf{x}_0,t). \tag{3.5.4}$$

Hence, from (2.2.2),

$$\rho(\mathbf{x}_0,t) = \lim_{\delta\to 0}\left\{\frac{m(B_\delta(\mathbf{x}_0),t)}{\mathrm{vol}(B_\delta(\mathbf{x}_0))}\right\}. \tag{3.5.5}$$

That is, the mass density at any point $\mathbf{x}_0$ within the region occupied by the body at time t is the limit, as δ tends to zero, of the mass in $B_\delta(\mathbf{x}_0)$ at time t divided by the volume of this ball. The problem here is the indubitable fact that matter is distributed discretely in space in the form of molecules, atoms and ions. These, in turn, are composed of nuclei and electrons. To gain some idea of how little space is occupied by matter, it should be noted that if a typical nucleus were scaled up to have a characteristic dimension of order 1 cm, then the nearest electron would be of order 50 m away. Accordingly, most of space is unoccupied by matter, and if δ is much less than the smallest typical nucleus–electron separation, then for nearly all points $\mathbf{x}_0$, $\rho(\mathbf{x}_0, t)$ given by (3.5.5) would be zero. If $\mathbf{x}_0$ were located *within* a nucleus at time t, then the ratio in (3.5.5) would become enormous for δ approaching the characteristic dimension of the nucleus.[2] Since this is not at all what we have in mind for ρ (e.g., we think of the density of water being of order 10^3 kg m^{-3}), it is clear that we must either revise interpretation (2.2.2) or give up the assumption of continuity. Retaining continuity requires a revision along the lines of including a proviso that R not be too small. The key questions is, 'How small?' *Thus we have encountered another conceptual difficulty associated with spatial scale*, the first being the scale dependence of boundaries.

3.6 Velocity

As discussed in Section 2.3., our intuitive notion of macroscopic velocity depends upon whether the body is in a solid or fluid phase. It is helpful here to be aware of the main features of individual molecular mass centre motions in the three main phases in which matter is encountered.

In a moderately rarefied gas, molecules interact with each other only occasionally, and only very rarely are more than two molecules involved in simultaneous mutual interactions. Molecular trajectories, for the most part, are rectilinear: departures therefrom are the consequence of interactions and are termed 'collisions'. As a gas becomes more dense, collisions become more frequent and begin to involve more than just pairs of molecules. In solid phases individual molecules undergo rapid chaotic localised motions centred on points which change position on a much slower, macroscopic time scale. (The erratic motions typically involve speeds of order 10^3 ms^{-1} and are confined to regions of typical span 10^{-10} m, so the time scale to bear in mind for such motions is of order 10^{-13} s.) Further, near-neighbours remain near-neighbours or, said differently, molecular diffusion is (almost!) non-existent in solids. Liquid phase molecules undergo continuously changing, smooth but erratic trajectories governed by simultaneous interactions with near-neighbours. While liquid densities are of a similar order to those of solids (think of water and ice), in liquids the molecules have sufficient kinetic energy to 'escape' from any protracted link with a near-neighbour. Figure 3.3 shows a simulated liquid trajectory in which one can see a molecule enjoying the company of some near neighbours for a short time before diffusing on to the next 'party', and so on. Molecular speeds are typically of order 10^3 ms^{-1}.

Exercise 3.6.1. Consider water emerging from the nozzle of a fire hose at 10 ms^{-1}. If individual water molecules all have speed 10^3 ms^{-1}, estimate the percentage of molecules actually moving in the direction of the emerging flow at any instant.

[2] Nuclear masses and radii are of order 10^{-26} kg and 10^{-4} Å, respectively, giving notional densities of order 10^{16} kg m^{-3}.

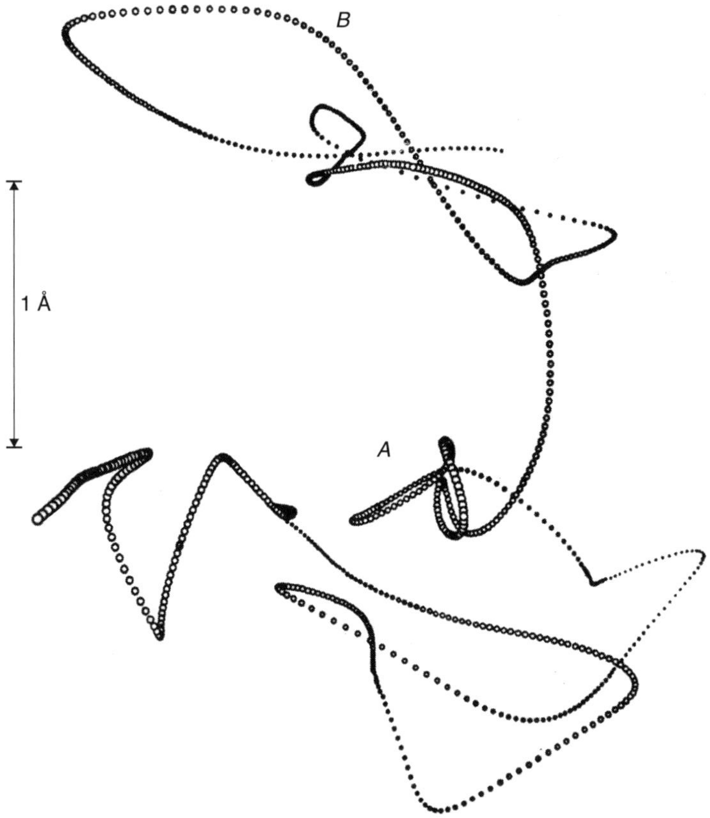

Figure 3.3. Molecular dynamics trajectory of a liquid Argon molecule showing vibratory (A) and diffusive (B) components. (Illustration reproduced from a figure by O. B. Verbeke and W. Brems.)

Hint: Using spherical polar co-ordinates associated with a Cartesian frame in which the z axis is in the emerging flow direction, the velocity of a molecule is of the form

$$\mathbf{v} = 10^3(\sin\theta\cos\phi\mathbf{i} + \sin\theta\sin\phi\mathbf{j} + \cos\theta\mathbf{k}) + 10\mathbf{k}\ \mathrm{ms}^{-1}$$

at any instant. The downstream component is positive if

$$10^3\cos\theta + 10 > 0.$$

Show that this holds for approximately 50.3 per cent of all angles θ in the range $0 \leq \theta \leq \pi$ and deduce, on the basis of assuming a random direction for $\mathbf{v} - 10\mathbf{k}$, that this is the required percentage.

(Accordingly, at any instant, 49.7 per cent of the molecules are actually moving further from the hose nozzle exit plane, and one sees that from a molecular perspective the macroscopic $10\,\mathrm{ms}^{-1}$ represents a relatively slow 'drift'.)

In view of the preceding, it is solid phases which furnish the simplest interpretation of bulk (i.e., macroscopic) velocity $\mathbf{v}$ by associating this with the motion of groups of neighbouring molecules. Indeed, one might think (at least in principle) of doping individual molecules. This would yield a reasonable indication of the value of $\mathbf{v}$ at molecular locations, provided that the erratic motions were somehow discounted (possibly by time-averaging over intervals which are long compared with the erratic vibrational time scales). Values of $\mathbf{v}$ at other than time-averaged molecular mass centre locations then could be obtained by some form of interpolation.

As noted in Section 2.3, fluid motion can be monitored by the insertion of foreign objects such as bubbles or suspended particles in liquids and smoke particles or balloons in gases. In such cases the observed velocity of any immersed 'tracer' object depends upon its size and mass and reflects the co-operative effect of interactions upon molecules of the object by fluid molecules. Since molecular interactions are of short range (typically of order $10\,\text{Å} = 10^{-9}\,\text{m}$), the fluid molecules involved at any instant are just those within this range of the object boundary. Accordingly, for a spherical immersed object of radius R and with molecular interactions of range r, the number of fluid molecules involved at any instant, divided by the number of fluid molecules which would have occupied that region filled by the object in its absence, is roughly of order $4\pi R^2 r \div 4\pi R^3/3 = 3r/R$. Said differently, the number of fluid molecules involved (and whose behaviour is thereby sampled) corresponds roughly to those in a sphere of radius $(3R^2 r)^{1/2}$ rather than R. (Convince yourself of this!)

Exercise 3.6.2. In water macroscopically at rest, immersed spherical objects of radius R and the same (bulk) density as water begin to exhibit erratic (Brownian) motion (Figure 3.4) when $R = 10^{-6}\,\text{m}$. If the range of molecular interactions is $r = 3\times 10^{-10}\,\text{m}$ show that the chaotic behaviour might only reflect the behaviour of liquid occupying spherical regions of radii less than about $10^{-7}\,\text{m}$.

[The motivation here is to draw attention to the possibility that continuum theory (for which $\mathbf{v}$ is here zero) might apply for water down to scales of order $10^{-7}\,\text{m}$, while Brownian motion of immersed particles begins to become evident at one order of magnitude greater.]

Remark 3.6.1. It should be clear at this point that the velocity field for a given fluid depends upon scale: specifically, it depends upon the scale at which the fluid motion is monitored. Further, while experimentation may associate velocity values with those of immersed objects at the scale of interest, it would be more satisfying to have an *intrinsic* definition; that is, a definition relating solely to the fluid.

Remark 3.6.2. The continuity equation (2.5.16) is usually motivated by considering the time rate of change of mass $m(R,t)$ within any fixed closed region R at time t. If mass is conserved, then this rate of change must correspond to the *net* rate at which mass flows into R across its entire boundary ∂R. Such observation then is asserted to be equivalent [see (2.2.2)] to

$$\frac{\partial}{\partial t}\left\{\int_R \rho(\mathbf{x},t)dV_{\mathbf{x}}\right\} = \int_{\partial R} \rho\mathbf{v}\,.\,(-\mathbf{n})dA, \tag{3.6.1}$$

where $\mathbf{n}$ denotes the 'outward' unit normal field on ∂R. This argument identifies $\rho\mathbf{v}\,.\,(-\mathbf{n})\,\Delta A$ with the rate at which mass enters R across a small subsurface of ∂R having area ΔA and unit normal $\mathbf{n}$. (Here $\rho\mathbf{v}$ and $\mathbf{n}$ are assumed to vary insignificantly over the subsurface.) The plausibility and motivation for this identification are based upon considering what happens in a short time interval, of duration Δt, say. If $\mathbf{v}\,.\,(-\mathbf{n}) > 0$ in this time interval, then a volume $\Delta V = \mathbf{v}\Delta t\,.\,(-\mathbf{n})\Delta A$ of fluid is argued to have entered R by crossing the subsurface. This quantity is the volume

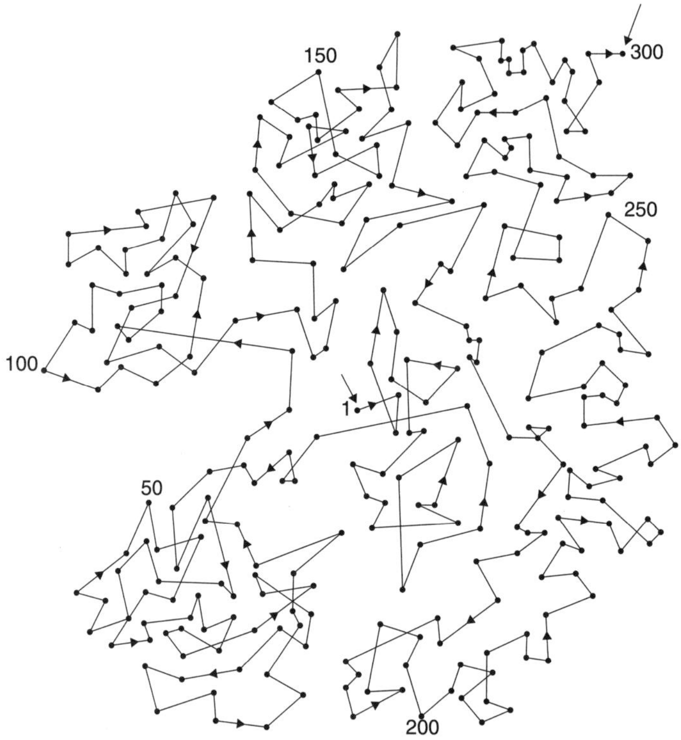

Figure 3.4. Brownian motion of a fine particle. Positions of a tiny dust particle suspended in water at 25°C are noted using a projection microscope of magnification ×100. Locations at 10 second intervals are connected by straight lines. The actual trajectory resembles that of Figure 3.3. (Reproduced from Statistical Mechanics and Properties of Matter, E.S.R. Gopal, Ellis Horwood, Chichester, 1976.)

ΔV of a tilted prism with base area ΔA and generators parallel to $\mathbf{v}$ of length $\|\mathbf{v}\|\Delta t$. Thus, in time lapse Δt, the mass Δm which has entered R by crossing the subsurface is approximated by $\rho\Delta V$; that is, by $\rho(\mathbf{v}\Delta t.\mathbf{n})\Delta A$. Consequently,

$$\frac{\Delta m}{\Delta t} \sim \rho(\mathbf{v}.(-\mathbf{n}))\Delta A, \tag{3.6.2}$$

so 'justifying' the identification.

Remark 3.6.3. The foregoing gives another perspective on velocity. Consider that mass Δm of fluid which has crossed a small planar surface of area ΔA and with unit normal $\hat{\mathbf{u}}$ over a time interval of duration Δt. Writing

$$\alpha(\hat{\mathbf{u}}) := \frac{\Delta m}{\rho\Delta A\Delta t}, \tag{3.6.3}$$

it follows from identification (3.6.2) that $\alpha(\hat{\mathbf{u}})$ should have a maximum value at some $\hat{\mathbf{u}} = \hat{\mathbf{u}}_0$, say. Accordingly, the velocity could be identified as

$$\mathbf{v} := \alpha(\hat{\mathbf{u}}_0)\hat{\mathbf{u}}_0. \tag{3.6.4}$$

In view of the differences in monitoring what might be regarded as velocity in solid and fluid phases, its evident scale dependence, and its link with mass conservation via mass density and the continuity equation, *one is motivated to seek an intrinsic molecular interpretation of velocity which is explicitly scale-dependent and applicable to matter in any state.*

3.7 The Pressure in a Gas

The relationship between molecular behaviour and continuum concepts is often first studied in the context of the kinetic theory of gases. In particular, the pressure on the inner bounding surface of a closed vessel containing gas is explained in terms of gas molecules, modelled as point masses, 'bouncing' off this surface. Each 'impact' of a molecule upon the surface imparts an impulse to this surface equal to the negative of the change in its momentum due to this 'collision'. Summing contributions of those impulses experienced over a portion of the surface and during some period of time and dividing this sum by both the area of surface involved and the duration of the time interval, one obtains the average *stress* (i.e., average force per unit area) exerted by the gas upon this subsurface during the time interval in question. If the subsurface is essentially planar, $\mathbf{n}$ denotes that unit normal to the associated plane directed into the gas, and the average stress is $-p\mathbf{n}(p > 0)$, then this stress is described as a *pressure* p. (Usual arguments consider plane walls and 'elastic' impacts; that is, collisions in which the molecular velocities have their components perpendicular to the wall reversed without change in magnitude and tangential components remain unchanged. In such a case the stress is always a pressure.)

Now consider linear momentum balance (2.7.2) for any closed region R lying strictly *within* a closed vessel containing gas. Here boundary ∂R encloses the geometric region R but as a hypothetical entity has no mechanical attributes. In this case the usual interpretation of

$$\int_{\partial R} \mathbf{t}(\cdot, t) \tag{3.7.1}$$

is of the resultant *force* (at time t) exerted by gas molecules outside R upon those within R. Recall from Section 3.5 that if the gas is moderately rarefied (e.g., air molecules in the atmosphere), then individual molecules experience forces due to the proximity of other molecules only rarely. Further, when they do so, they modify each other's trajectories, and such brief interactions are described as *collisions*. Thus, for a good proportion of the time a molecule is essentially force-free (if the effect of gravity is neglected) and accordingly undergoes rectilinear motion. Indeed, this gives rise to the notion of *mean free path*, which is the average distance a molecule may be expected to travel between collisions. It turns out, as a consequence of the foregoing considerations, that for any subsurface $\mathcal{S}$ of ∂R,

$$\int_{\mathcal{S}} \mathbf{t}(\cdot, t) \tag{3.7.2}$$

is very small. However, it is also assumed in continuum mechanics that if $\mathbf{x} \in \mathcal{S}$, then in 'equilibrium' situations $\mathbf{t}(\mathbf{x}, t)$ is normal to $\mathcal{S}$ and has magnitude p, where p is the pressure on the inner boundary of the containing vessel. The latter viewpoint is borne out by measurements of pressure inside the container. *There thus appears to be a paradox within the continuum viewpoint: the force represented by (3.7.2) is essentially negligible (on the basis of kinetic theory) yet is apparently non-negligible (on the basis of measurement).*

3.8 Reproducibility

The understanding of natural phenomena represented by the current state of science has come about in large measure as the consequence of patterns observed in animate and inanimate (i.e., material) behaviour. Such patterns are recognised when relevant conditions are replicated. For example, bodies thrown up into the air are always observed to fall. More specifically, bodies released from rest in an evacuated enclosure fall in what appears to be exactly the same way. Such behaviour is thus *reproducible*; that is, it can be *replicated*. In posing the question of why such reproducibility occurs we are led to the notions of *cause* and *effect*. In the preceding example, Newtonian dynamics furnishes the cause in terms of force (in particular, the force on the falling body due to gravity) and the effect (or fall) in terms of the motion of the body (in particular, the acceleration of its centre of mass). Any model of reproducible behaviour in which precise knowledge of its cause yields exact information about this behaviour (i.e., the relevant effect) is termed *deterministic*. Most continuum theories are deterministic.

Now consider a sequence of experiments in which a steel sphere of constant density is released from rest in a large container of stationary water at a constant temperature. If the radius of the sphere is 1 cm, the motions appear to be reproducible to the extent of being essentially vertical descents of the spheres. It turns out that an entirely satisfactory continuum description is possible when the water is modelled as a Newtonian incompressible liquid and the sphere as a uniform rigid body. However, if the experiments are repeated with spheres of radius less than 10^{-6} m, the behaviour is quite different: the spheres undergo erratic, irreproducible trajectories and, although tending to sink to the bottom of the container, never come to prolonged 'rest'. Accordingly, the deterministic continuum theory is seen to be inappropriate at a length scale far in excess of molecular size (which in this context is $3\,\text{Å} = 3 \times 10^{-10}$ m). The preceding erratic motions are described as *Brownian* and derive from the inhomogeneous nature of the behaviour of water molecules at macroscopically small, but microscopically large, length scales.

Remark 3.8.1. The question of scale has arisen yet again in the context of the applicability of continuum concepts.

3.9 Summary of Conceptual Problems

While continuum modelling is highly successful, there are a number of conceptual problems which arise therefrom. In particular, we have drawn attention to the following.

C.P.1. Consideration of boundaries leads us to conclude that the mass density function ρ is scale-dependent (Section 3.4).

C.P.2. The actual discrete distribution of matter, together with the assumption that ρ be continuous, means that $\int_R \rho\, dV$ is not the mass in R for very small regions R (Section 3.5).

C.P.3. Velocity is scale-dependent (Section 3.6).
C.P.4. The stress in a rarefied gas at rest (namely, its pressure) cannot be understood in terms of a force per unit area (Section 3.7).
C.P.5. Deterministic continuum modelling may be inappropriate at small length scales (Section 3.8).

3.10 Motivation for Space-Time Averaging of Molecular Quantities

Given the foregoing difficulties and their association with the fundamentally discrete nature of the distribution of matter, it becomes desirable to link continuum concepts and relations to microscopic considerations. The aim of such an endeavour is to gain a deeper insight into continuum modelling: specifically, to appreciate assumptions about molecular behaviour *implicit* in the continuum approach and to ascertain the role played by length scales. Further motivation is provided by the realisation that any measurement made on a material system involves sampling the system at some length scale *and* some time scale.[3] Accordingly, in some sense a measurement value must reflect a space-time average involving molecules of the system. Since measurements are usually identified with continuum field values, one is motivated to attempt to relate such values to space-time averages of molecular quantities.

[3] For example, a photographic snapshot reveals information limited by both exposure time and spatial resolution.

4 Spatial Localisation, Mass Conservation, and Boundaries

4.1 Preamble

Upon modelling molecules as point masses, volumetric densities ρ_w of mass and $\mathbf{p}_w$ of momentum are defined as local spatial averages of molecular masses and momenta using a weighting function w which, while possessing certain essential features, is otherwise unspecified and general. Partial (time) differentiation of ρ_w yields the continuity equation (2.5.16) in which the velocity field $\mathbf{v}_w := \mathbf{p}_w/\rho_w$. The physical interpretations of $\rho_w, \mathbf{p}_w$ and $\mathbf{v}_w$ depend crucially upon the choice of w. Several physically distinguished classes of weighting function are discussed. Emphasis is placed upon a particular class because the corresponding interpretations of the mass density and velocity fields, and of the boundary, associated with any body are particularly simple. The conceptual problems C.P.1, C.P.2, and C.P.3 listed in Section 3.8 are addressed and completely resolved.

4.2 Weighted Averages and the Continuity Equation

The mass density $\rho(\mathbf{x},t)$ at a given location $\mathbf{x}$ (a geometrical point) and time t is a local measure of 'mass per unit volume'. The key questions here are 'What mass?' and 'What volume?'

The mass of any given body of matter derives ultimately from that of its constituent fundamental discrete entities (i.e., electrons and atomic nuclei). While any such fundamental entity could be modelled as a point mass whose location is that of its mass centre, for the purposes of this chapter we adopt a molecular viewpoint. Specifically, we choose here to regard a *material system* (or *body*) $\mathcal{M}$ to be a fixed, identifiable set of (N, say) molecules modelled as point masses. Labelling these as $P_i (i = 1,2,\ldots,N)$, the mass, location and velocity of P_i at time t will be denoted by m_i, $\mathbf{x}_i(t)$ and $\mathbf{v}_i(t)$. Location $\mathbf{x}_i(t)$ is to be identified with that of the mass centre of P_i at time t.

Loosely speaking, $\rho(\mathbf{x},t)$ is the mass at time t in some region containing the geometrical point $\mathbf{x}$ divided by the volume of this region. Such concept can be generalised and made precise by the introduction of a *weighting function* w. Consider

$$\rho_w(\mathbf{x},t) := \sum_{i=1}^{N} m_i w(\mathbf{x}_i(t) - \mathbf{x}), \tag{4.2.1}$$

where w is a scalar-valued function of displacement and whose values have physical dimension L^{-3}. In such case the sum in (4.2.1), taken over *all* molecules of $\mathcal{M}$, has the physical dimension ML^{-3} of mass density. The contribution from an individual molecule is 'weighted' by the value of w at the displacement of the location of this molecule at time t from the geometrical point $\mathbf{x}$. In order that $\rho_w(\mathbf{x},t)$ be local, emphasis should be placed on molecules near $\mathbf{x}$; equivalently, w should ascribe greater weighting to small, as compared to large, displacements.

Remark 4.2.1. Note that any spatial and temporal differentiability properties of ρ_w in (4.2.1) stem from the differentiability of w (as a function from $\mathcal{V}$ into $\mathbb{R}$) together, for temporal differentiation, with differentiability of $\mathbf{x}_i$ (via use of the chain rule).

If ρ_w is to accord with mass density as employed in continuum mechanics, then it is *necessary* that its integral over all space should yield the total mass of the system. That is, it is necessary that

$$\int_{\mathcal{E}} \rho_w(\mathbf{x},t)dV_{\mathbf{x}} = \sum_{i=1}^{N} m_i. \tag{4.2.2}$$

From (4.2.1), the integral in (4.2.2) is the sum of integrals of the form

$$\int_{\mathcal{E}} m_i w(\mathbf{x}_i(t)-\mathbf{x})dV_{\mathbf{x}} = m_i \int_{\mathcal{E}} w(\mathbf{x}_i(t)-\mathbf{x})dV_{\mathbf{x}}. \tag{4.2.3}$$

Clearly, (4.2.2) will hold *if*, for all $i = 1,2,\ldots,N$ and any given t,

$$\int_{\mathcal{E}} w(\mathbf{x}_i(t)-\mathbf{x})dV_{\mathbf{x}} = 1. \tag{4.2.4}$$

However, since $\mathbf{x}$ in (4.2.4) ranges over all locations in $\mathcal{E}$, it follows that, for any given molecule and any time t, displacement $\mathbf{x}_i(t)-\mathbf{x}$ will range over all $\mathbf{u} \in \mathcal{V}$. Accordingly,

$$\int_{\mathcal{E}} w(\mathbf{x}_i(t)-\mathbf{x})dV_{\mathbf{x}} = \int_{\mathcal{V}} w(\mathbf{u})d\mathbf{u}, \tag{4.2.5}$$

and (4.2.4) may be written as

$$\int_{\mathcal{V}} w(\mathbf{u})d\mathbf{u} = 1. \tag{4.2.6}$$

Remark 4.2.2. Note that (4.2.6) is a restriction upon w alone and, if satisfied, delivers (4.2.2) no matter how the molecules are distributed.

Summarising, satisfaction of restriction (4.2.6) is *sufficient* to ensure (4.2.2) for any material system $\mathcal{M}$.

Exercise 4.2.1. Show, for the degenerate case in which $\mathcal{M}$ consists of a single molecule, that satisfaction of (4.2.2) implies that restriction (4.2.6) must *necessarily* hold.

Accordingly we see that restriction (4.2.6) is both a necessary *and* sufficient condition for (4.2.2) to hold. This restriction is termed a *normalisation condition*, and w is said to be *normalised*.

Remark 4.2.3. Since molecular masses are positive and mass density is non-negative, it is tempting to require that w be non-negative. Although often mandated (cf., e.g., Hardy [10]), such restriction rules out a class of weighting functions which corresponds to the requirement that averaging ρ_w in (4.2.1) using w should yield nothing new. That is, requiring that

$$\int_{\mathcal{E}} \rho_w(\mathbf{y},t)w(\mathbf{y}-\mathbf{x})dV_{\mathbf{y}} = \rho_w(\mathbf{x},t) \tag{4.2.7}$$

imposes a restriction upon w which results in negative values of w for some displacements. Explicit forms of w which satisfy (4.2.7), with ρ given by (4.2.1), will be derived in subsection 4.4.2.

At this juncture we list our requirements of *any* physically sensible *w*eighting *f*unction w and therefrom derive the corresponding continuity equation by temporal differentiation of (4.2.1).

W.F.1. $w : \mathcal{V} \to \mathbb{R}$ is a scalar-valued function of displacements with associated physical dimension L^{-3}.

W.F.2. w assigns greater values to small displacements than to large displacements.

W.F.3. w is continuously differentiable on $\mathcal{V}$.

W.F.4. $\int_{\mathcal{V}} w(\mathbf{u})d\mathbf{u} = 1.$

Lemma 4.2.1.

$$\frac{\partial}{\partial t}\{w(\mathbf{x}_i(t)-\mathbf{x})\} = -\operatorname{div}\{w(\mathbf{x}_i(t)-\mathbf{x})\mathbf{v}_i(t)\}. \tag{4.2.8}$$

Proof. Writing $\mathbf{u}(\mathbf{x},t) := \mathbf{x}_i(t)-\mathbf{x}$, and using the chain rule,

$$\frac{\partial}{\partial t}\{w(\mathbf{x}_i(t)-\mathbf{x})\} = \frac{\partial}{\partial t}\{w(\mathbf{u}(\mathbf{x},t))\} = \nabla w \cdot \frac{\partial \mathbf{u}}{\partial t} = \nabla w.\dot{\mathbf{x}}_i = \nabla w.\mathbf{v}_i. \tag{4.2.9}$$

On the other hand, since $\mathbf{v}_i(t)$ is independent of location $\mathbf{x}$ (and hence $\operatorname{div}\mathbf{v}_i = 0$),

$$\begin{aligned}\operatorname{div}\{w(\mathbf{x}_i(t)-\mathbf{x})\mathbf{v}_i(t)\} &= w(\mathbf{x}_i(t)-\mathbf{x})\operatorname{div}\{\mathbf{v}_i(t)\} + \nabla_{\mathbf{x}}\, w(\mathbf{x}_i(t)-\mathbf{x}).\mathbf{v}_i(t) \\ &= \nabla_{\mathbf{x}} w.\mathbf{v}_i. \end{aligned} \tag{4.2.10}$$

Result (4.2.8) follows from (4.2.9) and (4.2.10) upon noting that[1]

$$\nabla_{\mathbf{x}}\, w = -\nabla w. \tag{4.2.11}$$

It follows from (4.2.1) and Lemma 4.2.1 that[2]

$$\begin{aligned}\frac{\partial \rho_w}{\partial t} &= \frac{\partial}{\partial t}\left\{\sum_{i=1}^{N} m_i\, w(\mathbf{x}_i(t)-\mathbf{x})\right\} \\ &= \sum_{i=1}^{N} m_i \frac{\partial}{\partial t}\{w(\mathbf{x}_i(t)-\mathbf{x}\} = -\sum_{i=1}^{N}\operatorname{div}\{m_i\mathbf{v}_i(t)w(\mathbf{x}_i(t)-\mathbf{x})\} \\ &= -\operatorname{div}\left\{\sum_{i=1}^{N} m_i\,\mathbf{v}_i(t)w(\mathbf{x}_i(t)-\mathbf{x})\right\}. \end{aligned} \tag{4.2.12}$$

[1] Here ∇w denotes the derivative with respect to the argument $\mathbf{u}$ of w (see W.F.1), while $\nabla_{\mathbf{x}}\, w$ is the derivative computed with respect to $\mathbf{x}$. Since $\mathbf{u} = \mathbf{x}_i - \mathbf{x}$, (4.2.11) follows by the chain rule.

[2] Here it is to be understood that $\partial\rho_w/\partial t$ is evaluated at point $\mathbf{x}$ and time t.

Writing

$$\mathbf{p}_w(\mathbf{x},t) := \sum_{i=1}^{N} m_i \mathbf{v}_i(t) w(\mathbf{x}_i(t) - \mathbf{x}), \tag{4.2.13}$$

relation (4.2.12) may be expressed as

$$\frac{\partial \rho_w}{\partial t} + \operatorname{div} \mathbf{p}_w = 0. \tag{4.2.14}$$

Field $\mathbf{p}_w$ has values which are local volumetric averages computed in respect of the additive (or 'extensive') molecular quantity of linear momentum precisely as for the additive molecular quantity of mass in (4.2.1). We term ρ_w and $\mathbf{p}_w$ the (w-based) *mass* and *linear momentum density* fields, respectively.

If we define the w-velocity field $\mathbf{v}_w$ by

$$\mathbf{v}_w := \mathbf{p}_w / \rho_w \tag{4.2.15}$$

wherever and whenever $\rho_w \neq 0$, then relation (4.2.14) takes the form of the continuity equation [see (2.5.16)], namely

$$\frac{\partial \rho_w}{\partial t} + \operatorname{div}\{\rho_w \mathbf{v}_w\} = 0. \tag{4.2.16}$$

Integration of (4.2.16) over a fixed bounded regular[3] region R yields

$$\int_R \frac{\partial \rho_w}{\partial t} dV = -\int_R \operatorname{div}\{\rho_w \mathbf{v}_w\} dV,$$

and hence use of the divergence theorem yields

$$\frac{\partial}{\partial t}\left\{\int_R \rho_w \, dV\right\} = -\int_{\partial R} \rho_w \mathbf{v}_w . \mathbf{n} \, dA. \tag{4.2.17}$$

Of course, $\mathbf{n}$ here denotes the outward unit normal field on the boundary ∂R of R.

Remark 4.2.4. Notice that the integral relation (4.2.17) is *derived* from the local (or 'point') relation (4.2.16), in contrast to the usual heuristic postulation of (3.6.1) for arbitrary R and consequent deduction of (2.5.16) via the assumed continuity of integrands. Comparison also can be made with the argumentation of Section 2.5 which is based upon deriving the notion of 'motion' from supposed knowledge of the velocity field.

At this point it is hopefully clear that precise definitions of density fields of mass and linear momentum [namely (4.2.1) and (4.2.13)] have been given on the basis of point mass modelling of molecules (*once selection of a weighting function has been made*) and that these fields have been *proved* to satisfy the form of the continuity equation with velocity field defined by (4.2.15). However, such mathematical precision can only take on physical significance when the form of w is made explicit. Accordingly, we now discuss possible physically sensible choices of w.

[3] This is a technical assumption to ensure that use may be made of the divergence theorem. See, for example, Gurtin [11], p. 13.

4.3 The Simplest Choice w_ϵ of Weighting Function

4.3.1 Definition of w_ϵ

Let

$$S_\epsilon(\mathbf{x}) := \{\mathbf{y} \in \mathcal{E} : \|\mathbf{y} - \mathbf{x}\| < \epsilon\}. \tag{4.3.1}$$

That is, $S_\epsilon(\mathbf{x})$ consists of all points within the spherical region centred at $\mathbf{x}$ with radius ϵ. We can compute the total mass and total momentum of the molecules lying within $S_\epsilon(\mathbf{x})$ at time t and obtain the corresponding densities by dividing by the volume of $S_\epsilon(\mathbf{x})$. This can be achieved by use of the weighting function w_ϵ, where

$$\left.\begin{array}{lll} w_\epsilon(\mathbf{u}) = V_\epsilon^{-1} & \text{if} & \|\mathbf{u}\| < \epsilon \\ w_\epsilon(\mathbf{u}) = 0 & \text{if} & \|\mathbf{u}\| \geq \epsilon \end{array}\right\}, \tag{4.3.2}$$

with

$$V_\epsilon^{-1} = \frac{3}{4\pi\epsilon^3}. \tag{4.3.3}$$

Specifically, we define

$$\rho_\epsilon(\mathbf{x},t) := \sum_{i=1}^{N} m_i w_\epsilon(\mathbf{x}_i(t) - \mathbf{x}) \tag{4.3.4}$$

and

$$\mathbf{p}_\epsilon(\mathbf{x},t) := \sum_{i=1}^{N} m_i \mathbf{v}_i(t) w_\epsilon(\mathbf{x}_i(t) - \mathbf{x}). \tag{4.3.5}$$

Remark 4.3.1. The essential scale-dependent simplicity of the physical interpretations of (4.3.4) and (4.3.5) is enhanced by the corresponding interpretation of

$$\mathbf{v}_\epsilon := \mathbf{p}_\epsilon / \rho_\epsilon, \tag{4.2.15}$$

since $\mathbf{v}_\epsilon(\mathbf{x},t)$ is thereby the velocity of the mass centre of those molecules which lie within $S_\epsilon(\mathbf{x})$ at time t.

4.3.2 The Boundary Corresponding to w_ϵ

Recall from Section 2.2 (specifically mass density assumption M.D.1) that the region B_t occupied by a body at time t is that region in which $\rho(.,t)$ takes positive values at this time. Choice $\rho = \rho_\epsilon$ thus yields

$$B_t = B_t^\epsilon := \bigcup_{i=1}^{N} S_\epsilon(\mathbf{x}_i(t)). \tag{4.3.6}$$

That is, B_t^ϵ consists of all geometrical points that lie within a distance of ϵ from at least one molecule of $\mathcal{M}$ at time t. It follows that if a point $\mathbf{x}$ belongs to the boundary ∂B_t^ϵ of B_t^ϵ at time t, then at this time

(i) $\mathbf{x}$ is distant ϵ from at least one molecule, and
(ii) no molecule is distant less than ϵ from $\mathbf{x}$.

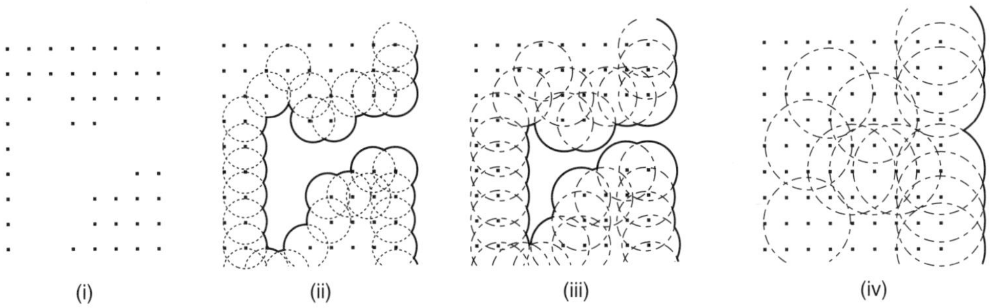

Figure 4.1. Boundary ∂B^ϵ for a fissure in a lattice (i) at three scales: (ii) $\epsilon = \epsilon_0$, (iii) ϵ slightly greater than ϵ_0, and (iv) $\epsilon = 2\epsilon_0$, where ϵ_0 is the lattice spacing.

Remark 4.3.2. Clearly, ∂B_t^ϵ is scale-dependent. Figure 4.1 illustrates[4] several different boundaries for a lattice-like distribution of molecules which contains a 'fissure'.

Remark 4.3.3. If ϵ is smaller than any nearest-neighbour separation, then B_t^ϵ will be the union of disjoint open spherical regions, each one of which is centred at a molecular location. Of course, we do not apply continuum modelling at such scales. More generally, it may happen that B_t^ϵ contains one or more isolated spherical regions. In this case the molecules at which these spheres are centred might be termed *isolated at scale* ϵ and *time t*. Such molecules would be found for systems which have unconfined gaseous phases. At this juncture we do not discuss 'free' liquid boundaries: in such case the molecules of the liquid do not constitute a fixed set because migration between the liquid and its vapour occurs. Time-dependent material systems will be addressed in Chapter 8.

Remark 4.3.4. The notion of the boundary of a material system is of most use for *solid* bodies. Recalling the nature of thermal molecular motion in solids (see Section 3.6), any solid boundary ∂B_t^ϵ must be expected to undergo rapid chaotic localised motions (i.e., to fluctuate) with associated molecular-level scales of length ℓ and time τ ($\ell < 1\,\text{Å}$, $\tau < 10^{-13}\,\text{s}$). Of course, these are macroscopically negligible.

Remark 4.3.5. For porous media (to be discussed in Chapter 10) the notion of 'porosity' is introduced as a measure of 'empty space' within a body. To make this notion precise, it is necessary to recognise that the region 'occupied' by the body is being considered at two different length scales. At the smaller scale, ϵ_1 say, pore structure is evident, while at the larger scale, ϵ_2 say, this is no longer manifest. More precisely, we can define the porosity $\mathcal{P}_{\epsilon_1,\epsilon_2}(\mathbf{x},t)$ by considering

$$\mathcal{V}_{\epsilon_1,\epsilon_2}(\mathbf{x},t) := \frac{\text{vol}\{S_{\epsilon_2}(\mathbf{x}) \cap B_t^{\epsilon_1}\}}{\text{vol}\{S_{\epsilon_2}(x) \cap B_t^{\epsilon_2}}\}. \tag{4.3.7}$$

The numerator and denominator in (4.3.7) are the volumes of regions occupied by the body which lie within $S_{\epsilon_2}(\mathbf{x})$ at time t when the boundary of the body is delineated

[4] Necessarily all illustrations here are two-dimensional. The reader hopefully will be able to visualise three-dimensional counterparts. Also, for ease of illustration, choices of ϵ close to lattice spacings have been selected.

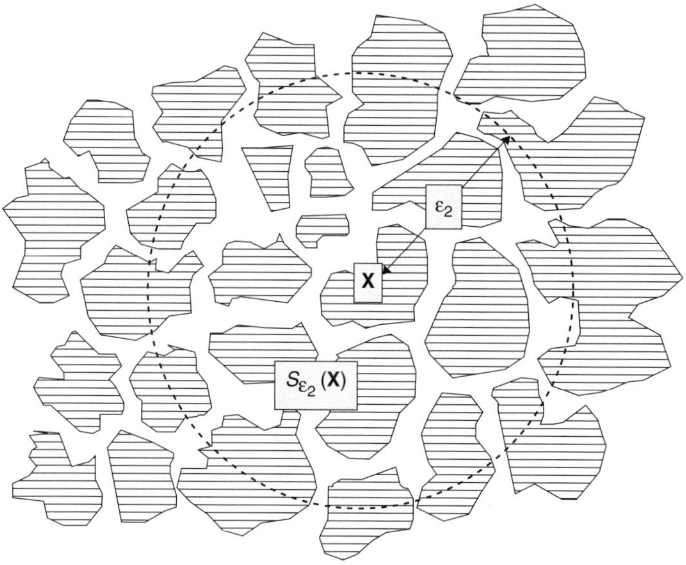

Figure 4.2. Porosity at scales (ϵ_1, ϵ_2). The porous body occupies a region (*shaded*) whose boundary is delineated at scale ϵ_1: the porosity $\mathcal{P}_{\epsilon_1,\epsilon_2}(\mathbf{x})$ is that fraction of the volume within sphere $S_{\epsilon_2}(x)$ which is not occupied by the body at scale ϵ_1.

at scales ϵ_1 and ϵ_2, respectively. Accordingly,

$$\mathcal{P}_{\epsilon_1,\epsilon_2}(\mathbf{x},t) := 1 - \mathcal{V}_{\epsilon_1,\epsilon_2}(\mathbf{x},t). \tag{4.3.8}$$

If $B_t^{\epsilon 2}$ is simply connected, then for points in the ϵ_2-*scale strict interior*

$$^{-}B_t^{\epsilon 2} := \{\mathbf{x} \in B_t^{\epsilon 2} : d(\mathbf{x}, \partial B_t^{\epsilon 2}) \geq \epsilon_2\} \tag{4.3.9}$$

we have

$$S_{\epsilon_2}(\mathbf{x}) \cap B_t^{\epsilon 2} = S_{\epsilon_2}(\mathbf{x}). \tag{4.3.10}$$

(Convince yourself of this!) In such case [note (4.3.3)]

$$\mathcal{P}_{\epsilon_1,\epsilon_2}(\mathbf{x},t) = \frac{(V_{\epsilon_2} - \mathrm{vol}\{S_{\epsilon_2}(\mathbf{x}) \cap B_t^{\epsilon 1}\})}{V_{\epsilon_2}}. \tag{4.3.11}$$

See Figure 4.2.

In order to characterise *boundary molecules* (at a given length scale ϵ) one can proceed as follows. Visualise a moveable spherical balloon of radius ϵ, remote from any molecule of $\mathcal{M}$. Now bring this balloon towards $\mathcal{M}$ molecules until it makes 'contact' with such a molecule on its surface. The totality of 'first contact' molecules obtained in this way constitutes the set of ϵ-scale boundary molecules. See Figure 4.3 for the boundary molecules at several scales for a lattice which contains a fissure.

When boundary molecules lie on a plane (as illustrated in Figure 4.3: for any ϵ value greater than $3\epsilon_0/2$ nearest-neighbour separations the set of boundary molecules remains the same – convince yourself of this!), the distance of ∂B_t^{ϵ} from this plane will increase with any increase in ϵ. This motivates an attempt to define a *geometric boundary*, at any given scale, whose distance from boundary molecules does not increase with any increase in ϵ.

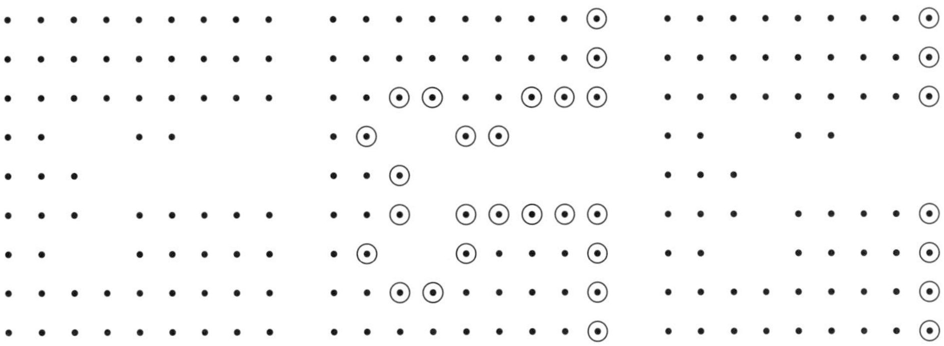

(i) (ii) (iii)

Figure 4.3. Boundary molecules for the fissure in (i) at two scales: (ii) $\epsilon = \epsilon_0$ and (iii) $\epsilon = 4_{\epsilon_0}$ (lattice spacing ϵ_0).

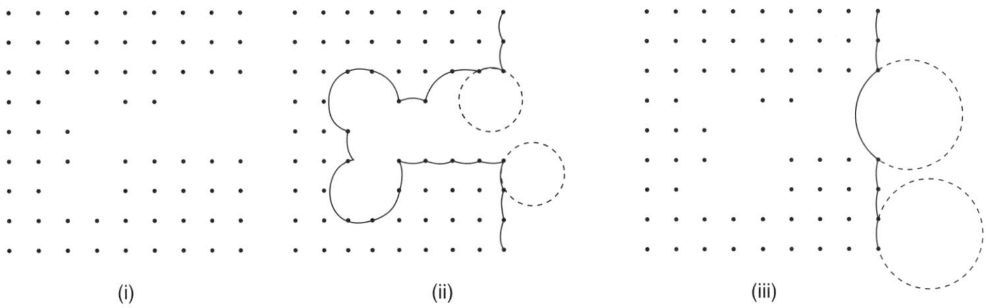

(i) (ii) (iii)

Figure 4.4. Geometric boundary for the fissure in (i) at two scales: (ii) $\epsilon = \epsilon_0$ and (iii) $\epsilon = 4_{\epsilon_0}$.

Let $\mathbf{x}$ denote any point which is distant at least ϵ from any molecular location at time t. Hence no molecules will lie within $S_\epsilon(\mathbf{x})$ at this time. Accordingly

$$R^\epsilon_{\text{ext}}(\mathcal{M},t) := \bigcup_{\mathbf{x}} S_\epsilon(\mathbf{x}), \tag{4.3.12}$$

where the union is taken over all such $\mathbf{x}$ in Euclidean space $\mathcal{E}$, can be regarded to be the geometric region *exterior* to $\mathcal{M}$ at scale ϵ at time t. Correspondingly, the *geometric region occupied by $\mathcal{M}$ at scale ϵ and time t* can be defined meaningfully as[5]

$$R^\epsilon(\mathcal{M},t) := \mathcal{E} - R^\epsilon_{\text{ext}}(\mathcal{M},t). \tag{4.3.13}$$

Of course, the boundary $\partial R^\epsilon(\mathcal{M},t)$ of $R^\epsilon(\mathcal{M},t)$ is to be regarded as the *geometric boundary at scale ϵ*.

Another (related) geometrically based definition of a scale-dependent boundary can be obtained as follows. Consider a moveable spherical balloon as in the discussion following Remark 4.3.5, in particular its 'first encounter' with a molecule as it approaches $\mathcal{M}$ from remote separation. If the balloon is 'locked' onto such a

[5] As defined, R^ϵ_{ext} is the union of open sets [since $S_\epsilon(\mathbf{x})$ is open for any point $\mathbf{x}$], and hence R^ϵ is closed. One may wish for a region of occupation to be open and, if so, define this to be the interior of R^ϵ.

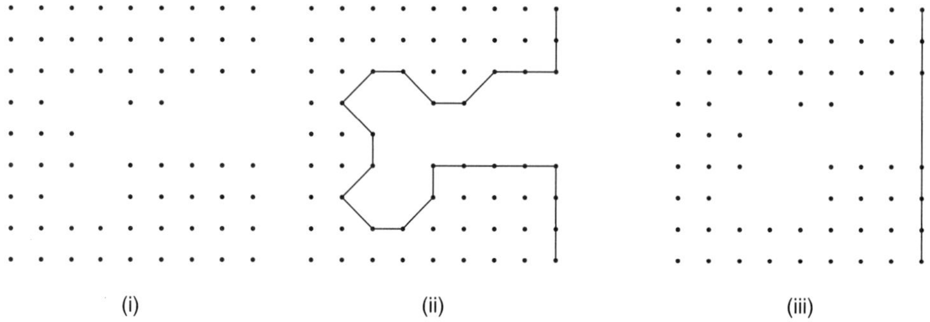

Figure 4.5. Triangulated boundary for the fissure in (i) at two scales: (ii) $\epsilon=\epsilon_0$ and (iii) $\epsilon=2\epsilon_0$.

molecular location, it is free to pivot about this location (a geometrical point). In so doing, the balloon will in general experience a 'second encounter' and also can be 'locked' at this second location. Such additional constraint restricts any further motion of the balloon to one of rotation about the line joining the two encounter locations. Generally, such rotation will result in a third encounter, which, upon also 'locking' its location, allows no further motion of the balloon.

Let $\mathbf{x}\in\mathcal{E}$ denote the location of the centre of any such 'fully locked' balloon. With each such point $\mathbf{x}$ are associated three 'locking' points which uniquely define a triangle, $\Delta_\epsilon(\mathbf{x})$ say. Then the *(triangulated) geometric boundary at scale* ϵ *and* time t is

$$\Lambda^{\epsilon}_{\text{geom}}(\mathcal{M},t) := \bigcup_{\mathbf{x}} \Delta_\epsilon(\mathbf{x}), \tag{4.3.14}$$

where the union is taken over the centres $\mathbf{x}$ of all 'locked' balloons at time t. In general, the collection of triangles of which $\Lambda^{\epsilon}_{\text{geom}}(\mathcal{M},t)$ is composed will form the boundary of one or more many-sided polyhedra, each having triangular faces. The region lying within this polyhedron (or these polyhedra) is what we term the *ϵ-scale triangulated polyhedral region occupied by $\mathcal{M}$ at time t.*

Figure 4.5 illustrates triangulated boundaries at several scales for a lattice which contains a fissure.

4.3.3 Integration of ρ_ϵ and $\mathbf{p}_\epsilon$ over a Region

From (4.2.1), for any regular region R,

$$\int_R \rho_\epsilon(\mathbf{x},t)dV_{\mathbf{x}} = \sum_{i=1}^{N} m_i \int_R w_\epsilon(\mathbf{x}_i(t)-\mathbf{x})dV_{\mathbf{x}}. \tag{4.3.15}$$

Further, from the definition (4.3.2) of w_ϵ,

$$\int_R w_\epsilon(\mathbf{x}_i(t)-\mathbf{x})dV_{\mathbf{x}} = \int_{N_i(t)} w_\epsilon(\mathbf{x}_i(t)-\mathbf{x})dV_{\mathbf{x}} = \frac{V_i(t)}{V_\epsilon}, \tag{4.3.16}$$

where
$$N_i(t) := R\cap S_\epsilon(\mathbf{x}_i(t)) \tag{4.3.17}$$

and
$$V_i(t) := \text{vol}(N_i(t)). \tag{4.3.18}$$

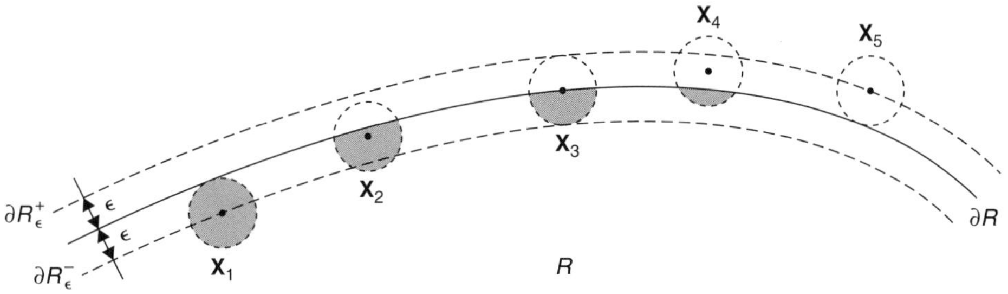

Figure 4.6. Contributions to integrals of ρ_ϵ and $\mathbf{p}_\epsilon$ over region R. Surface ∂R_ϵ^- (∂R_ϵ^+) lies inside (outside) R and is distant ϵ from boundary ∂R of R. A molecule P_i which lies at a point $\mathbf{x_i}$ between ∂R_ϵ^- and ∂R_ϵ^+ has its contribution weighted by $\alpha_i := V_i/V_e$. Here, $\alpha_1 = 1$, $\frac{1}{2} < \alpha_2 < 1$, $\alpha_3 = \frac{1}{2}$, $0 < \alpha_4 < \frac{1}{2}$, $\alpha_5 = 0$.

Note that if P_i lies in R_ϵ^-, at time t, where

$$R_\epsilon^- := \{\mathbf{x} \in R : d(\mathbf{x}, \partial R) \geq \epsilon\}, \tag{4.3.19}$$

then $V_i(t) = V_\epsilon$. Accordingly,

$$\int_R \rho_\epsilon(\mathbf{x},t)dV_{\mathbf{x}} = \left(\sum_{\substack{P_i \in R_\epsilon^- \\ \text{at time } t}} m_i \right) + \left(\sum_{\substack{P_i \notin R_\epsilon^- \\ \text{at time } t}} V_i(t) m_i \right) V_\epsilon^{-1}. \tag{4.3.20}$$

Remark 4.3.6. From (4.3.20) we see that, for choice w_ϵ, the integral of the corresponding mass density function ρ_ϵ over region R is not exactly the total mass of molecules within R (at any given time). In general its value will differ from this total mass because any molecule P_i for which $\mathbf{x}_i(t)$ lies within a distance ϵ of its boundary ∂R contributes only that fraction of the volume of the sphere, of radius ϵ and centre $\mathbf{x}_i(t)$, which lies within R. See Figure 4.6.

An identical argument to the foregoing yields

$$\int_R \mathbf{p}_\epsilon(\mathbf{x},t)dV_{\mathbf{x}} = \left(\sum_{\substack{P_i \in R_\epsilon^- \\ \text{at time } t}} m_i \mathbf{v}_i(t) \right) + \left(\sum_{\substack{P_i \notin R_\epsilon^- \\ \text{at time } t}} m_i \mathbf{v}_i(t) V_i(t) \right) V_\epsilon^{-1}. \tag{4.3.21}$$

That is, this integral represents the total momentum of molecules within R_ϵ^- at time t, together with further contributions from molecules which lie within a distance ϵ from ∂R, each weighted by the volume fraction $V_i(t)/V_\epsilon$.

4.3.4 A Wrinkle to Be Resolved: Use of a Mollifier

The foregoing discussions, concerning the definitions and integration of fields ρ_ϵ and $\mathbf{p}_\epsilon$, and delineation of scale-ϵ boundaries, are mathematically precise. However, a problem arises in respect of the continuity equation due to jump discontinuities suffered by w: these occur [see (4.3.2)] at displacements $\mathbf{u}$ for which $\|\mathbf{u}\| = \epsilon$. Accordingly, w_ϵ fails to be differentiable for such values of $\mathbf{u}$, and the proof of Lemma 4.2.1

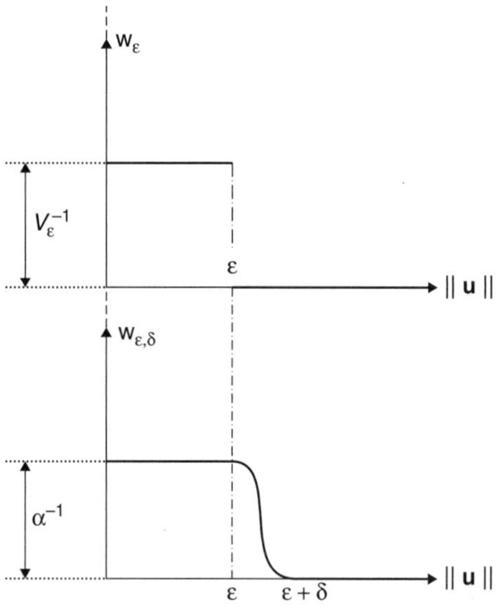

Figure 4.7. Mollification of weighting function w_ϵ.

fails at time t for points $\mathbf{x}$ distant ϵ from point $\mathbf{x}_i(t)$. It follows that *ρ_ϵ and $\mathbf{p}_\epsilon$ suffer jump discontinuities at time t at all points distant ϵ from any single molecular location at this time, and at such points the continuity equation fails to hold.* Ouch!

Resolution of this 'wrinkle' can be effected by 'mollifying' w_ϵ in the neighbourhood of the critical ($\|\mathbf{u}\| = \epsilon$) displacements as follows. Let φ denote any continuously differentiable, real-valued, monotonic decreasing function defined on $[0,1]$ for which $\varphi(0) = 1, \varphi(1) = 0$ and $\varphi'(0) = 0 = \varphi'(1)$. [Here $\varphi'(0)$ and $\varphi'(1)$ denote one-sided derivatives.] Now define

$$\left.\begin{aligned} w_{\epsilon,\delta}(\mathbf{u}) &:= \frac{1}{\alpha} && \text{if} \quad \|\mathbf{u}\| < \epsilon, \\ w_{\epsilon,\delta}(\mathbf{u}) &:= \frac{\varphi(\lambda)}{\alpha} && \text{if} \quad \|\mathbf{u}\| = \epsilon + \lambda\delta \ (\lambda \in [0,1]), \\ w_{\epsilon,\delta}(\mathbf{u}) &:= 0 && \text{if} \quad \|\mathbf{u}\| > \epsilon + \delta. \end{aligned}\right\} \tag{4.3.22}$$

Here $\delta > 0$. Constant α is determined by the normalisation requirement W.F.4. [see (4.2.6)], namely

$$4\pi \int_0^\infty \hat{w}_{\epsilon,\delta}(u) u^2 du = 1, \tag{4.3.23}$$

where

$$u := \|\mathbf{u}\| \tag{4.3.24}$$

and

$$\hat{w}_{\epsilon,\delta}(u) := w_{\epsilon,\delta}(\mathbf{u}). \tag{4.3.25}$$

See Figure 4.7.

Exercise 4.3.1. Prove (4.3.23) by identifying elements $\mathbf{u}$ of $\mathcal{V}$ with points $\mathbf{x} \in \mathcal{E}$ via relation $\mathbf{x} - \mathbf{x}_0 = \mathbf{u}$, in which $\mathbf{x}_0 \in \mathcal{E}$ is an arbitrary point, and employing a spherical polar co-ordinate system with origin at $\mathbf{x}_0$.

Mollification (4.3.22) yields the following:

Result 4.3.1 Mollified weighting function $w_{\epsilon,\delta}$ is continuously differentiable.

**Proof.* Since $w_{\epsilon,\delta}(\mathbf{u})$ is constant for $\|\mathbf{u}\| < \epsilon$ and $\|\mathbf{u}\| > \epsilon + \delta$, *it has zero derivatives of all orders for such values of* $\mathbf{u}$. From (4.3.25), (4.3.24), and (4.3.22$_2$), for $\epsilon \le \|\mathbf{u}\| \le \epsilon + \delta$

$$w_{\epsilon,\delta}(\mathbf{u}) = \hat{w}_{\epsilon,\delta}(u) = \frac{1}{\alpha}\,\varphi(\lambda), \tag{4.3.26}$$

where

$$\lambda := \frac{1}{\delta}\,(u - \epsilon). \tag{4.3.27}$$

Hence, by the chain rule,

$$\nabla w_{\epsilon,\delta}(\mathbf{u}) = \frac{1}{\alpha}\,\varphi'(\lambda)\nabla_{\mathbf{u}}\lambda. \tag{4.3.28}$$

Now from (4.3.27)

$$\nabla_{\mathbf{u}}(\lambda) = \frac{1}{\delta}\,\nabla_{\mathbf{u}}\,u. \tag{4.3.29}$$

However, since

$$u^2 = \mathbf{u}\,.\,\mathbf{u}$$

we have[6]

$$2u\nabla_{\mathbf{u}}\,u = \nabla_{\mathbf{u}}(\mathbf{u}\,.\,\mathbf{u}) = 2\mathbf{u}. \tag{4.3.30}$$

Accordingly

$$\nabla_{\mathbf{u}}\,u = \frac{1}{u}\,\mathbf{u}, \tag{4.3.31}$$

and from (4.3.28), (4.3.29) and (4.3.31),

$$\nabla w_{\epsilon,\delta}(\mathbf{u}) = \frac{1}{\alpha u\delta}\,\varphi'(\lambda)\mathbf{u}. \tag{4.3.32}$$

Since, by hypothesis, φ' is a continuous function of λ on $[\epsilon, \epsilon + \delta]$ (one-sidedly at endpoints), λ is (trivially!) an analytic function of u for all u, and u^{-1} is of class C^∞ for $u \neq 0$. It follows from (4.3.32) that $w_{\epsilon,\delta}(\mathbf{u})$ is differentiable for $\epsilon \le \|\mathbf{u}\| \le \epsilon + \delta$ with zero derivatives whenever $\|\mathbf{u}\| = \epsilon$ or $\epsilon + \delta$ [because $\varphi'(0) = 0 = \varphi'(1)$ by hypothesis]. This latter property, together with the vanishing of $\nabla w_{\epsilon,\delta}$ for $\|\mathbf{u}\| < \epsilon$ or $\|\mathbf{u}\| > \epsilon + \delta$, establishes the result.

Corollary 4.3.1. If φ has derivatives of all orders up to and including $n(\ge 2)$, and these derivatives all vanish one-sidedly at $\lambda = 0$ and $\lambda = 1$, then $\nabla w_{\epsilon,\delta}$ has derivatives of all orders up to and including n everywhere.

Proof. This result follows from repeated differentiation of (4.3.32).

Remark 4.3.7. Replacing w_ϵ in (4.2.1) and (4.2.13) by $w_{\epsilon,\delta}$ we obtain mass and momentum density fields $\rho_{\epsilon,\delta}$ and $\mathbf{p}_{\epsilon,\delta}$ which are continuously differentiable as a consequence of Result 4.3.1. Accordingly result (4.2.14) follows with ρ_w and $\mathbf{p}_w$ replaced by $\rho_{\epsilon,\delta}$ and $\mathbf{p}_{\epsilon,\delta}$. Defining

$$\mathbf{v}_{\epsilon,\delta} := \mathbf{p}_{\epsilon,\delta}/\rho_{\epsilon,\delta} \tag{4.3.33}$$

yields the corresponding continuity equation

$$\frac{\partial \rho_{\epsilon,\delta}}{\partial t} + \operatorname{div}\{\rho_{\epsilon,\delta}\,\mathbf{v}_{\epsilon,\delta}\} = 0. \tag{4.3.34}$$

[6] $\nabla_{\mathbf{u}}(\mathbf{u}\,.\,\mathbf{u})\,.\,\mathbf{h} + o(\mathbf{h}) = (\mathbf{u}+\mathbf{h}).(\mathbf{u}+\mathbf{h}) - \mathbf{u}\,.\,\mathbf{u} = 2\mathbf{u}\,.\,\mathbf{h} + o(\mathbf{h})$ as $\mathbf{h} \to \mathbf{0}$. Thus $\nabla_{\mathbf{u}}\,(\mathbf{u}\,.\,\mathbf{u}) = 2\mathbf{u}$.

4.3.5 Further Mollification Considerations

Mollification (4.3.22) results in the desired continuity equation, previously unobtainable due to discontinuities in w_ϵ, and accordingly removes the 'wrinkle' flagged up in the preceding subsection. However, in addition to removing the *technical* 'wrinkle', the purpose of mollification was to obtain smooth mass and momentum fields *with essentially the same physical interpretations as* ρ_ϵ *and* $\mathbf{p}_\epsilon$. This can be accomplished by choosing δ to be very small. To see this, we first prove the following:

Result 4.3.2.

$$\frac{1}{\alpha} = \frac{1}{V_\epsilon} + O\left(\frac{\delta}{\epsilon}\right) \quad \text{as } \delta \to 0. \tag{4.3.35}$$

**Proof.* (*Exercise*) Deduce from (4.3.23) and (4.3.22) that

$$\frac{4\pi}{\alpha}\left\{\frac{\epsilon^3}{3} + A\right\} = 1, \tag{4.3.36}$$

where

$$A := \delta \int_0^1 \varphi(\lambda)(\epsilon + \lambda\delta)^2 d\lambda. \tag{4.3.37}$$

Noting that φ is monotone (by hypothesis) and thus takes values in [0, 1], show that

$$0 < A < \delta(\epsilon^2 + \epsilon\delta + \delta^2/3). \tag{4.3.38}$$

Writing

$$k := \frac{3A}{\epsilon^3}, \tag{4.3.39}$$

show further that from (4.3.36) and (4.3.38)

$$\alpha = V_\epsilon(1+k), \tag{4.3.40}$$

where

$$0 < k < \frac{3\delta}{\epsilon}\left\{1 + \frac{\delta}{\epsilon} + \frac{\delta^2}{3\epsilon^2}\right\}. \tag{4.3.41}$$

Choosing

$$\delta << \epsilon, \tag{4.3.42}$$

note that (4.3.41) implies $0 < k < 1$. Deduce from (4.3.40) that

$$\frac{1}{\alpha} = \frac{1}{V_\epsilon}(1 - k + o(k)) \quad \text{as } k \to 0, \tag{4.3.43}$$

where

$$k = O\left(\frac{\delta}{\epsilon}\right) \quad \text{as } \delta \to 0, \tag{4.3.44}$$

and hence that result (4.3.35) holds.

Now consider [see (4.3.4) and Remark 4.3.7]

$$\rho_{\epsilon,\delta}(\mathbf{x},t) - \rho_\epsilon(\mathbf{x},t) = \sum_{i=1}^{N} m_i[w_{\epsilon,\delta}(\mathbf{x}_i(t) - \mathbf{x}) - w_\epsilon(\mathbf{x}_i(t) - \mathbf{x})]. \tag{4.3.45}$$

The sum can be decomposed into one over those molecules within $S_\epsilon(\mathbf{x})$ at time t and those in the shell-like region $S_{\epsilon,\delta}(\mathbf{x})$ between $S_\epsilon(\mathbf{x})$ and $S_{\epsilon+\delta}(\mathbf{x})$. Terming these $\sum_i^{(1)}$ and $\sum_i^{(2)}$ yields, from (4.3.22) and (4.3.2),

$$\rho_{\epsilon,\delta}(\mathbf{x},t) - \rho_\epsilon(\mathbf{x},t) = D_1 + D_2, \tag{4.3.46}$$

where
$$D_1 := \left(\frac{1}{\alpha} - \frac{1}{V_\epsilon}\right)\sum_i^{(1)} m_i \tag{4.3.47}$$

and
$$D_2 := \sum_i^{(2)} m_i w_{\epsilon,\delta}(\mathbf{x}_i(t) - \mathbf{x}). \tag{4.3.48}$$

Now, from (4.3.4),
$$\sum_i^{(1)} m_i = \rho_\epsilon(\mathbf{x},t)V_\epsilon, \tag{4.3.49}$$

so (4.3.47) and (4.3.35) yield
$$D_1 = \left(\frac{V_\epsilon}{\alpha} - 1\right)\rho_\epsilon(\mathbf{x},t) = O\left(\frac{\delta}{\epsilon}\right) \quad \text{as } \delta \to 0. \tag{4.3.50}$$

Further, from $(4.3.22)_2$,
$$D_2 < \frac{\text{mass in } S_{\epsilon,\delta}(\mathbf{x})}{\alpha} = \frac{\text{mass in } S_{\epsilon,\delta}(\mathbf{x})}{V_\epsilon}\left\{1 + O\left(\frac{\delta}{\epsilon}\right)\right\}. \tag{4.3.51}$$

Since D_1 is of order $O(\delta/\epsilon)$ as $\delta \to 0$ and the mass in $S_{\epsilon,\delta}(\mathbf{x}) := S_{\epsilon+\delta}(\mathbf{x}) - S_\epsilon(\mathbf{x})$ also will tend to zero as δ tends to zero, it follows from (4.3.46) that we have the following[7]:

Result 4.3.3.
$$\rho_{\epsilon,\delta}(\mathbf{x},t) - \rho_\epsilon(\mathbf{x},t) \to 0 \quad \text{as} \quad \delta \to 0. \tag{4.3.52}$$

Remark 4.3.8. An essentially identical argument yields
$$\mathbf{p}_{\epsilon,\delta}(\mathbf{x},t) - \mathbf{p}_\epsilon(\mathbf{x},t) \to \mathbf{0} \quad \text{as} \quad \delta \to 0. \tag{4.3.53}$$

It follows that $\mathbf{v}_{\epsilon,\delta}$ defined in (4.3.33) has essentially the same local mass centre velocity interpretation as $\mathbf{v}_\epsilon$ provided that $\delta << \epsilon$. Further, the region occupied by the body at scale ϵ, given by (4.3.6), is now modified to that in which $\rho_{\epsilon,\delta} > 0$. Accordingly the boundary of the body for $\delta << \epsilon$ is essentially that of the region in which $\rho_\epsilon > 0$, and hence our previous discussion of boundaries needs no essential revision.

[7] The mathematical purist would like the result to hold uniformly in $\mathbf{x}$. In this respect we observe that (4.3.35) is independent of $\mathbf{x}$ and uniformity is guaranteed if $\rho_\epsilon(\mathbf{x},t)$ is bounded and the mass in $S_{\epsilon,\delta}(\mathbf{x})$ tends uniformly to zero as δ tends to zero.

*4.3.6. Regularity of Mollified Fields: Polynomial Mollifiers

Consider any additive molecular quantity $G_i(t)$, such as the mass, momentum, and kinetic energy of, or resultant force on, P_i at time t. The corresponding spatial density field at scale ϵ is given by

$$G_\epsilon(\mathbf{x},t) := \sum_{i=1}^{N} G_i(t) w_\epsilon(\mathbf{x}_i(t) - \mathbf{x}). \tag{4.3.54}$$

The associated mollified field is

$$G_{\epsilon,\delta}(\mathbf{x},t) := \sum_{i=1}^{N} G_i(t) w_{\epsilon,\delta}(\mathbf{x}_i(t) - \mathbf{x}). \tag{4.3.55}$$

As a consequence of Corollary 4.3.1., $G_{\epsilon,\delta}$ inherits the spatial differentiability properties of φ. Specifically, $G_{\epsilon,\delta}$ is spatially of class C^n everywhere if φ is of class C^n on $[0,1]$ and derivatives of all orders up to and including n vanish at endpoints 0 and 1. Further, it follows from use of the chain rule that $G_{\epsilon,\delta}$ has continuous temporal derivatives of all orders up to and including $\min(n,p)$ whenever G_i and $\mathbf{x}_i$ are of class C^p for all P_i.

Clearly, there are infinitely many mollifiers φ on [0, 1] on noting that the only requirements are those of monotonicity, existence and continuity of φ', together with values

$$\varphi(0) = 1, \qquad \varphi(1) = 0 \qquad \text{and} \qquad \varphi'(0) = 0 = \varphi'(1). \tag{4.3.56}$$

The simplest choice is the cubic polynomial defined uniquely by restrictions (4.3.56).

Exercise 4.3.2. Show that if a cubic polynomial satisfies (4.3.56), then it must take the form

$$\varphi_3(\lambda) := 2\lambda^3 - 3\lambda^2 + 1 \tag{4.3.57}$$

Verify that

$$\varphi_3'(\lambda) = 6\lambda(\lambda - 1) \tag{4.3.58}$$

is monotonic decreasing for $0 \le \lambda \le 1$.

In the same way there is a unique polynomial φ_{2n+1} of degree $2n+1$ which is a class C^n mollifier. The explicit form of φ_{2n+1} will now be derived.

Exercise 4.3.3. Note that for a class C^2 mollifier[8] we require, in addition to monotonicity and (4.3.56), that $\varphi''(0) = 0 = \varphi''(1)$. Show that the unique polynomial $\varphi_5(\lambda)$ of degree 5 which fits the bill is

$$\varphi_5(\lambda) := -6\lambda^5 + 15\lambda^4 - 10\lambda^3 + 1$$

and that

$$\varphi_5'(\lambda) = -30\lambda^2(\lambda - 1)^2. \tag{4.3.59}$$

Show further that the unique degree 7 polynomial $\varphi_7(\lambda)$ which is a class C^3 mollifier is

$$\varphi_7(\lambda) := 20\lambda^7 - 70\lambda^6 + 84\lambda^5 - 35\lambda^4 + 1$$

[8] That is, a mollifier which leads to class C^2 mollified fields everywhere in $\mathcal{E}$.

so that
$$\varphi_7'(\lambda) = 140\lambda^3(\lambda-1)^3. \tag{4.3.60}$$

Verify that φ_5 and φ_7 are monotonic decreasing on [0, 1].

At this point a pattern has emerged in expressions for derivatives: compare (4.3.58), (4.3.59), and (4.3.60). We thus are led to assert that the polynomial mollifier of least degree to guarantee class C^n mollified fields everywhere is φ_{2n+1}, where

$$\varphi_{2n+1}'(\lambda) = c_n\lambda^n(\lambda-1)^n. \tag{4.3.61}$$

Repeated use of the product and chain rules yields the required vanishing of all derivatives up to and including n at $\lambda = 0$ and $\lambda = 1$. It remains to ensure that $\varphi_{2n+1}(0) = 1$ and $\varphi_{2n+1}(1) = 0$. The former is satisfied by

$$\varphi_{2n+1}(\lambda) := c_n \int_0^\lambda s^n(s-1)^n ds + 1. \tag{4.3.62}$$

Satisfaction of $\varphi_{2n+1}(1) = 0$ requires that

$$c_n I_{n,n} + 1 = 0, \tag{4.3.63}$$

where, for any non-negative integers m and n,

$$I_{m,n} := \int_0^1 s^m(s-1)^n ds. \tag{4.3.64}$$

Exercise 4.3.4. Show, by integrating by parts, that

$$I_{m,n} = -\frac{m}{(n+1)} I_{m-1,n+1} \tag{4.3.65}$$

for $m \geq 1$. Deduce that

$$I_{n,n} = \frac{(-1)^n(n!)^2}{(2n)!} I_{0,2n}. \tag{4.3.66}$$

Prove that

$$I_{0,2n} = \frac{1}{(2n+1)} \tag{4.3.67}$$

and hence that

$$I_{n,n} = \frac{(-1)^n(n!)^2}{(2n+1)!}. \tag{4.3.68}$$

Show further from (4.3.62), (4.3.63), and (4.3.68) that

$$\varphi_{2n+1}(\lambda) = 1 + \frac{(-1)^{n+1}(2n+1)!}{(n!)^2} \int_0^\lambda s^n(s-1)^n ds \tag{4.3.69}$$

and

$$\varphi_{2n+1}'(\lambda) = \frac{(-1)^{n+1}(2n+1)!}{(n!)^2} \lambda^n(\lambda-1)^n. \tag{4.3.70}$$

Deduce that φ_{2n+1} is monotonic decreasing for $0 \leq \lambda \leq 1$.

4.3.7 Mollification as a Natural Consequence of Spatial Imprecision

While the distance between any pair of geometrical points is (axiomatically) a unique non-negative real number, *measurement* of such distance involves unavoidable lack of such precision. If distance measurements are only precise to within $\delta/2$ ($\delta > 0$), then any such measurement u_m which satisfies

$$\epsilon - \delta/2 < u_m < \epsilon + \delta/2$$

cannot be regarded to be unequivocally less or greater than distance ϵ ($\epsilon > \delta/2$). Accordingly, in definition (4.3.2) of the original and natural choice w_ϵ of ϵ-scale weighting function, we have

$$\left.\begin{array}{lllr} w_\epsilon(u_m) & = & V_\epsilon^{-1} & \text{if} \quad 0 \le u_m < \epsilon - \delta/2 \\ w_\epsilon(u_m) & = & 0 & \text{if} \quad u_m > \epsilon + \delta/2 \end{array}\right\}. \tag{4.3.71}$$

For values in the range $\epsilon - \delta/2 \le u_m \le \epsilon + \delta/2$ we can interpolate values of w_ϵ between V_ϵ^{-1} and 0 using a smooth monotonic decreasing function on $[\epsilon - \delta/2, \epsilon + \delta/2]$. Once this is accomplished, w_ϵ is defined for all non-negative real numbers and can be multiplied by a scaling factor determined by the normalisation requirement (4.2.6). The resulting smooth weighting function is essentially a mollified version of w_ϵ given by (4.3.2).

Remark 4.3.9. The imprecision of distance measurements means that, from a practical standpoint, membership of $S_\epsilon(\mathbf{x})$ [see (4.3.1)] is also imprecise. The exact characteristic (or 'membership') function for $S_\epsilon(\mathbf{x})$ is

$$\chi : \mathcal{E} \to \{0,1\}, \tag{4.3.72}$$

where

$$\left.\begin{array}{ll} \chi(\mathbf{y}) = 1 & \text{if} \quad \|\mathbf{y} - \mathbf{x}\| < \epsilon \\ \chi(\mathbf{y}) = 0 & \text{if} \quad \|\mathbf{y} - \mathbf{x}\| \ge \epsilon \end{array}\right\}. \tag{4.3.73}$$

The corresponding practical version χ_m of χ satisfies

$$\left.\begin{array}{ll} \chi_m(\mathbf{y}) = 1 & \text{if} \quad \|\mathbf{y} - \mathbf{x}\|_m < \epsilon - \delta/2 \\ \chi_m(\mathbf{y}) = 0 & \text{if} \quad \|\mathbf{y} - \mathbf{x}\|_m > \epsilon + \delta/2 \end{array}\right\} \tag{4.3.74}$$

and

$$\chi_m(\mathbf{y}) = \phi(\lambda) \quad \text{if} \quad \epsilon - \delta/2 \le \lambda = \|\mathbf{y} - \mathbf{x}\|_m \le \epsilon + \delta/2. \tag{4.3.75}$$

Here $\|\mathbf{y} - \mathbf{x}\|_m$ denotes the *measured* distance between $\mathbf{x}$ and $\mathbf{y}$, and φ is monotonic decreasing on $[\epsilon - \delta/2, \epsilon + \delta/2]$ with $\varphi(\epsilon - \delta/2) = 1$ and $\varphi(\epsilon + \delta/2) = 0$.

Remark 4.3.10. Relation (4.3.74) coupled with (4.3.75) defines a *fuzzy set* (cf., e.g., Zadeh [12]). Mollification of (4.3.2) can be regarded as the natural consequence of our ultimately imprecise (or 'fuzzy') measurement of distances. Such fuzziness will be encountered again in Chapter 8 when dealing with systems whose material content changes with time.

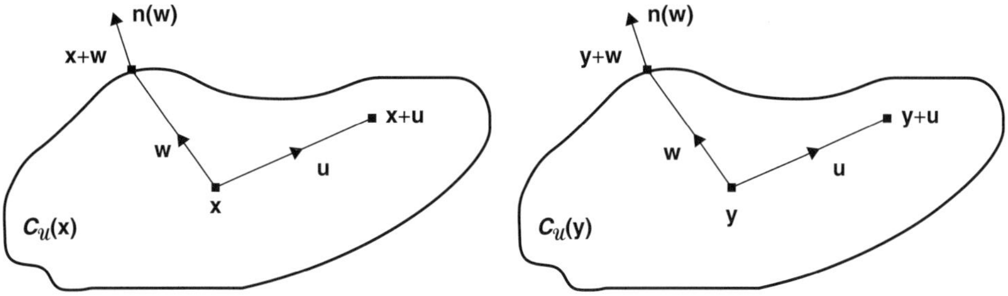

Figure 4.8. Two $\mathcal{U}$-cells, centred at $\mathbf{x}$ and $\mathbf{y}$, with outward unit normal field $\mathbf{n}$ on cell boundaries.

4.4 Other Choices of Weighting Function

4.4.1 Cellular Averaging

Any weighting function of form

$$w(\mathbf{u}) = \hat{w}(u) \tag{4.4.1}$$

is *isotropic* in the sense that there is no dependence upon the *direction* of $\mathbf{u}$. In particular, w_ϵ (and any mollification $w_{\epsilon,\delta}$ thereof) is isotropic. It is natural to choose isotropic weighting functions when computing spatial density fields for material systems in which there is no known geometrical bias in the distribution of molecular locations. However, such bias is encountered in systems with non-isotropic symmetries (such as crystals) or directionally biased inhomogeneity (to be expected at or near boundaries[9]). In such case w can be chosen to yield averages over spatial cells whose geometry reflects such symmetry or bias. Such *a cell is a moveable, simply connected region of characteristic shape, orientation, and size*, and may be defined in terms of its centroid and the set $\mathcal{U}$ of displacements therefrom to each of its points. Specifically, for any point $\mathbf{x}$, the $\mathcal{U}$-cell centred at $\mathbf{x}$ is

$$C_{\mathcal{U}}(\mathbf{x}) := \{\mathbf{x} + \mathbf{u} : \mathbf{u} \in \mathcal{U}\}. \tag{4.4.2}$$

Of course, any two such cells have exactly the same shape, size, orientation, and volume

$$V_{\mathcal{U}} := \mathrm{vol}\{C_{\mathcal{U}}(\mathbf{x})\}, \tag{4.4.3}$$

where $\mathbf{x}$ is arbitrary. See Figure 4.8.

The corresponding choice of weighting function is given by

$$\left.\begin{array}{ll} w(\mathbf{u}) = V_{\mathcal{U}}^{-1} & \text{if} \quad \mathbf{u} \in \mathcal{U} \\ w(\mathbf{u}) = 0 & \text{if} \quad \mathbf{u} \notin \mathcal{U} \end{array}\right\}. \tag{4.4.4}$$

It follows that $\rho_w(\mathbf{x},t)$ and $\mathbf{p}_w(\mathbf{x},t)$ represent the total mass and momentum of those molecules within $C_{\mathcal{U}}(\mathbf{x})$ at time t divided by the cell volume, and [see (4.2.15)] $\mathbf{v}_w(\mathbf{x},t)$ denotes the velocity of the mass centre of these molecules at time t.

[9] For example, if $\mathbf{x}$ is a point on a bounding surface $\mathcal{S}$, then $\mathcal{S}$ will divide $S_\epsilon(\mathbf{x})$ into two regions: one of these will contain few, if any, molecules. Comments on boundary and interfacial regions are made in Subsection 15.3.1.

Remark 4.4.1 The shape, orientation and size of a $\mathcal{U}$-cell is determined by $\mathcal{U}$: the size may be defined as the *span* of the cell (i.e., the maximum distance between any pair of points on its boundary) or, alternatively, the supremum $\sup\{\|\mathbf{u}\| : \mathbf{u} \in \mathcal{U}\}$. For example, if $\mathbf{e}_i \in \mathcal{V}(i = 1,2,3)$ are unit, mutually orthogonal vectors and $\epsilon_i \in \mathbb{R}^+$ $(i = 1,2,3)$, then

$$\mathcal{U} := \{\mathbf{u} = u_1\mathbf{e}_1 + u_2\mathbf{e}_2 + u_3\mathbf{e}_3 : -\epsilon_i < u_i < \epsilon_i\}$$

defines a cell which is an open rectangular box with edges parallel to $\mathbf{e}_i$ and of lengths $2\epsilon_i$, respectively. The span of any cell is $s = 2(\epsilon_1^2 + \epsilon_2^2 + \epsilon_3^2)^{1/2}$, while $\sup\{\|\mathbf{u}\|\} = s/2$.

Remark 4.4.2. Spatial averaging for porous media is implemented over so-called *representative elementary volumes* (or REVs) (cf., e.g., Bear [13]). These 'volumes' are cells whose size is much larger than the length scale at which pore structure is evident. Porous media modelling is addressed in Chapter 10.

Remark 4.4.3. Cellular averaging is employed in Chapter 11 to obtain global forms of balance relations by partitioning any macroscopic region or subregion 'occupied' by a body into many so-called ϵ-cells, defining field values at cell centroids in terms of molecular-based cellular averages, and identifying partition sums with corresponding Riemann integrals.

4.4.2 Choices Associated with Repeated Averaging

Let $G_i(t)$ denote any additive molecular quantity (e.g., its mass, momentum, or kinetic energy). The associated spatial density field corresponding to a weighting function w is given by

$$G_w(\mathbf{x},t) := \sum_{i=1}^{N} G_i(t)w(\mathbf{x}_i(t) - \mathbf{x}). \tag{4.4.5}$$

Such averaging of discretely defined quantities using w can be generalised to the averaging of continuous functions F of location by introducing

$$F_w(\mathbf{x}) := \int_{\mathcal{E}} F(\mathbf{y})w(\mathbf{y} - \mathbf{x})dV_{\mathbf{y}}. \tag{4.4.6}$$

Function F_w is termed the (spatial) *w-average of F*.

Remark 4.4.4. F_w has the same physical dimensions as F. (Why?)

Remark 4.4.5. Note that the discrete (*mic*roscopic) distribution of the quantity represented by G_i may be written formally as

$$G^{\text{mic}}(\mathbf{x},t) := \sum_{i=1}^{N} G_i(t)\delta(\mathbf{x}_i(t) - \mathbf{x}), \tag{4.4.7}$$

where δ denotes the Dirac delta 'function' which has the property that, for any function f of location,

$$\int_{\mathcal{E}} f(\mathbf{y})\delta(\mathbf{y} - \mathbf{x})dV_{\mathbf{y}} = f(\mathbf{x}). \tag{4.4.8}$$

Exercise 4.4.1. Show that (formally)

$$(G^{\text{mic}})_w(\mathbf{x},t) = G_w(\mathbf{x},t). \tag{4.4.9}$$

From (4.4.6) it follows that any molecular-based weighted average of form (4.4.5) may be further averaged using w. It is natural to enquire how such an average $(G_w)_w$ is related to the average G_w. In particular, we may investigate whether or not

$$(G_w)_w = G_w. \tag{4.4.10}$$

That is, we may ask whether repeated averaging yields anything new. It turns out that if we *require* that (4.4.10) holds, then there is an associated restriction upon w which defines the *form* of this function. Said differently, if (4.4.10) is a result we wish to hold, then we must choose a very specific form of weighting function: the form of such function depends upon whether or not the molecular system is confined to a bounded region.

In an unbounded region the convolution form of (4.4.6) suggests use of Fourier transforms. Indeed, with the Fourier transform of any square-integrable real-valued function F on $\mathcal{V}$ defined by

$$\mathcal{F}\{F\}(\mathbf{k}) := \int_{\mathcal{V}} F(\mathbf{u})e^{-i\mathbf{u}.\mathbf{k}}d\mathbf{u}, \tag{4.4.11}$$

and re-writing (4.4.6) in convolution notation as

$$F_w := F * w, \tag{4.4.12}$$

we have

$$\mathcal{F}\{F_w\} = \mathcal{F}\{F * w\} = \mathcal{F}\{F\}\mathcal{F}\{w\}. \tag{4.4.13}$$

Further,

$$\mathcal{F}\{(F_w)_w\} = \mathcal{F}\{F_w * w\} = \mathcal{F}\{F_w\}\mathcal{F}\{w\}. \tag{4.4.14}$$

From (4.4.13) and (4.4.14)

$$\mathcal{F}\{(F_w)_w\} = \mathcal{F}(F)(\mathcal{F}(w))^2. \tag{4.4.15}$$

If we require that

$$(F_w)_w = F_w, \tag{4.4.16}$$

then taking transforms and using (4.4.13) and (4.4.15) yield

$$\mathcal{F}\{F\}(\mathcal{F}\{w\})^2 = \mathcal{F}\{F\}\mathcal{F}\{w\}. \tag{4.4.17}$$

This can only be ensured for general choice of F if

$$(\mathcal{F}\{w\})^2 = \mathcal{F}\{w\}. \tag{4.4.18}$$

Accordingly $\mathcal{F}\{w\}$ can only take the values 0 and 1. Spatially isotropic averaging (i.e., w depends on $\mathbf{u}$ only via $\|\mathbf{u}\|$) with notional associated scale ϵ can be accomplished by choosing

$$\bar{w}(\mathbf{k}) := \mathcal{F}\{w\}(\mathbf{k}) := H(1 - \epsilon\|\mathbf{k}\|), \tag{4.4.19}$$

where H denotes the Heaviside unit step function. That is,

$$\bar{w}(\mathbf{k}) = 1 \;\text{ if }\; \|\mathbf{k}\| < \epsilon^{-1} \tag{4.4.20}$$

and zero otherwise. Noting that (4.4.11) in respect of w may be written as

$$\bar{w}(\mathbf{k}) = \int_{\mathcal{V}} w(\mathbf{u})e^{-i\mathbf{u}.\mathbf{k}}d\mathbf{u}, \tag{4.4.21}$$

the inverse transform yields

$$w(\mathbf{u}) = \frac{1}{(2\pi)^3}\int_{\mathcal{V}} \bar{w}(\mathbf{k})e^{i\mathbf{u}.\mathbf{k}}d\mathbf{k} \tag{4.4.22}$$

$$= \frac{1}{(2\pi)^3}\int_{\mathcal{V}} H(1-\epsilon\|\mathbf{k}\|)e^{i\mathbf{k}.\mathbf{u}}d\mathbf{k}. \tag{4.4.23}$$

For a given $\mathbf{u} \in \mathcal{V}$ we may choose spherical polar co-ordinates in $\mathbf{k}$ space with the third axis in the direction of $\mathbf{u}$. Writing

$$u = \|\mathbf{u}\| \tag{4.4.24}$$

we have

$$\begin{aligned}
w(\mathbf{u}) &= \frac{1}{(2\pi)^3}\int_0^\infty\int_0^\pi\int_0^{2\pi} H(1-\epsilon k)e^{iku\cos\theta}k^2\sin\theta\, dk\, d\theta\, d\phi \\
&= \frac{1}{(2\pi)^2}\int_0^\infty H(1-\epsilon k)\left[-\frac{e^{iku\cos\theta}}{iu}\cdot k\right]_0^\pi dk \\
&= \frac{i}{(2\pi)^2 u}\int_0^{\epsilon^{-1}} k(e^{-iku}-e^{iku})dk \\
&= \frac{1}{2\pi^2 u}\int_0^{\epsilon^{-1}} k\sin(ku)dk \\
&= \frac{1}{2\pi^2 u}\left\{\left[-\frac{k\cos(ku)}{u}\right]_0^{\epsilon^{-1}} + \int_0^{\epsilon^{-1}}\frac{\cos(ku)}{u}dk\right\} \\
&= \frac{1}{2\pi^2 u}\left[-\frac{k\cos(ku)}{u}+\frac{\sin(ku)}{u^2}\right]_0^{\epsilon^{-1}}.
\end{aligned}$$

That is, after simplifying,

$$w(u) = \frac{1}{2\pi^2 u^3}\left\{\sin\left(\frac{u}{\epsilon}\right) - \frac{u}{\epsilon}\cos\left(\frac{u}{\epsilon}\right)\right\}. \tag{4.4.25}$$

The counterpart of (4.4.19) for non-isotropic averaging with notional scales ϵ_1, ϵ_2 and ϵ_3 associated with unit mutually orthogonal vectors $\mathbf{e}_1$, $\mathbf{e}_2$, and $\mathbf{e}_3$ is

$$\bar{w}(\mathbf{k}) = H(1-\epsilon_1|k_1|)H(1-\epsilon_2|k_2|)H(1-\epsilon_3|k_3|), \tag{4.4.26}$$

where

$$\mathbf{k} = k_1\mathbf{e}_1 + k_2\mathbf{e}_2 + k_3\mathbf{e}_3. \tag{4.4.27}$$

Thus, from (4.4.22) and (4.4.26),

$$\begin{aligned}
w(\mathbf{u}) &= \frac{1}{(2\pi)^3}\int_{-\infty}^{+\infty}\int_{-\infty}^{+\infty}\int_{-\infty}^{+\infty}\bar{w}(\mathbf{k})e^{i(u_1k_1+u_2k_2+u_3k_3)}dk_1dk_2dk_3 \\
&= I_1 I_2 I_3,
\end{aligned} \tag{4.4.28}$$

where $(p = 1,2,3)$

$$I_p := \frac{1}{2\pi}\int_{-\epsilon_p^{-1}}^{\epsilon_p^{-1}} 1.e^{iu_p k_p}\,dk_p = \frac{1}{\pi u_p}\sin\left(\frac{u_p}{\epsilon_p}\right). \tag{4.4.29}$$

That is,

$$w(\mathbf{u}) = \frac{1}{\pi^3 u_1 u_2 u_3}\sin\left(\frac{u_1}{\epsilon_1}\right)\sin\left(\frac{u_2}{\epsilon_2}\right)\sin\left(\frac{u_3}{\epsilon_3}\right). \tag{4.4.30}$$

Now suppose that the material system is known to lie in a rectangular box of dimensions $2L_1 \times 2L_2 \times 2L_3$. Here use of Fourier series is appropriate. We consider first the one-dimensional analogue and then the actual case in question.

If f is a continuous real-valued function on $[-L,L]$, then

$$f(x) = \sum_{k=-\infty}^{+\infty} c_k\, e^{ik\pi x/L}, \tag{4.4.31}$$

where

$$c_k := \frac{1}{2L}\int_{-L}^{L} f(y)e^{-ik\pi y/L}dy. \tag{4.4.32}$$

Accordingly, if we wish to obtain the approximation to $f(x)$ using terms in (4.4.31) which involve only wavelengths in excess of a given spatial scale ϵ, then the values of k in (4.4.31) must be restricted by

$$k < [2L\epsilon^{-1}] =: N, \tag{4.4.33}$$

where $[2L\epsilon^{-1}]$ denotes the smallest positive integer less than $2L\epsilon^{-1}$. The approximation required is thus, from (4.4.31), (4.4.32), and (4.4.33),

$$f_\epsilon(x) := \int_{-L}^{L} f(y)w(y-x)dy, \tag{4.4.34}$$

where

$$w(y-x) := \frac{1}{2L}\sum_{k=-N}^{N} e^{ik\pi(x-y)/L}. \tag{4.4.35}$$

Exercise 4.4.2. Notice that (4.4.35) may be expressed as

$$2Lw(y-x) = 1 + \sum_{k=1}^{N} e^{ik\theta} + \sum_{k=1}^{N} e^{-ik\theta}, \tag{4.4.36}$$

where

$$\theta := \pi(y-x)/L. \tag{4.4.37}$$

The two sums in (4.4.36) are geometric progressions. Use this observation to show that

$$2Lw(y-x) = 1 + \frac{e^{i\theta}(1-e^{iN\theta})}{(1-e^{i\theta})} + \frac{e^{-i\theta}(1-e^{-iN\theta})}{(1-e^{-i\theta})},$$

and simplify this to obtain

$$w(y-x) = \frac{1}{2L}\cdot\frac{\sin(N+\frac{1}{2})\theta}{\sin(\frac{1}{2}\theta)}. \tag{4.4.38}$$

The foregoing shows that, for a function f defined on $[-L, L]$, it is possible to obtain a truncated Fourier series approximation to f which involves only wavelengths in excess of a chosen length scale ϵ using a weighting function. Specifically, (4.4.34) holds with [see (4.4.38); we set $u := y - x$]

$$w(u) := \frac{1}{2L} \cdot \frac{\sin\left(\frac{(N+\frac{1}{2})\pi u}{L}\right)}{\sin\left(\frac{\pi u}{2L}\right)} \quad \text{with } N := [2L\epsilon^{-1}]. \tag{4.4.39}$$

Returning to the three-dimensional situation of system confinement within a rectangular box of dimensions $2L_1 \times 2L_2 \times 2L_3$, we can employ multiple Fourier series based upon Cartesian axes with origin at the centre of the box and axes parallel to the edges. In particular, denoting points by co-ordinates (x_1, x_2, x_3), and choosing only wavelengths greater than ϵ_i in the x_i direction $(i = 1,2,3)$, the weighting function

$$w(\mathbf{u}) := \frac{1}{8L_1L_2L_3} \cdot \frac{\sin(a_1\theta_1)\sin(a_2\theta_2)\sin(a_3\theta_3)}{\sin\theta_1 \sin\theta_2 \sin\theta_3}, \tag{4.4.40}$$

where $$a_i := N_i + \frac{1}{2}, \qquad \theta_i := \frac{\pi u_i}{L_i}, \qquad \text{and} \qquad N_i := \left[\frac{2L_i}{\epsilon_i}\right]. \tag{4.4.41}$$

Remark 4.4.6. Satisfaction of (4.4.16) is immediate upon noting that the Fourier series of any truncated series is precisely the truncated series.

Averages computed with weighting functions which satisfy (4.4.16) are particularly important if a system is to be investigated at several scales. Consider, in particular, consecutive averaging via choices w_1 and w_2 of form (4.4.25) corresponding to different scales ϵ_1 and ϵ_2. Then [cf. (4.4.14)]

$$\mathcal{F}\{(F_{w_1})_{w_2}\} = \mathcal{F}\{F_{w_1} * w_2\} = \mathcal{F}\{F_{w_1}\}\mathcal{F}\{w_2\} = \mathcal{F}\{F\}\mathcal{F}\{w_1\}\mathcal{F}\{w_2\}. \tag{4.4.42}$$

Thus $$\mathcal{F}\{(F_{w_1})_{w_2}\} = \mathcal{F}\{F_w\} = \mathcal{F}(F)\mathcal{F}\{w\}, \tag{4.4.43}$$

where $$\mathcal{F}\{w\}(\mathbf{k}) := H(1 - \epsilon_1\|\mathbf{k}\|)H(1 - \epsilon_2\|\mathbf{k}\|). \tag{4.4.44}$$

That is, $$\mathcal{F}\{w\}(\mathbf{k}) = H(1 - \epsilon\|\mathbf{k}\|), \tag{4.4.45}$$

where $$\epsilon := \max\{\epsilon_1, \epsilon_2\}. \tag{4.4.46}$$

(Convince yourself of this!) It follows that, for unbounded systems, spatial averaging associated with weighting functions of form (4.4.25) at a scale ϵ_1, followed by a further such averaging at scale ϵ_2, is equivalent to a single averaging at scale max $\{\epsilon_1, \epsilon_2\}$. Similar conclusions can be drawn for weighting functions of forms (4.4.30) and (4.4.40).

Exercise 4.4.3. Consider the precise natures of the aforementioned 'similar conclusions'.

Remark 4.4.7. The foregoing choices of weighting functions are clearly of relevance in comparing continuum descriptions at different scales, such as fluid flow through porous media. At sufficiently small scale (ϵ_1 say) pore structure is delineated (cf. Remark 4.3.5), and fluid flow within pore space can be modelled at this scale. Such modelling can be related to that at a larger (ϵ_2 say) scale at which pore space is no longer manifest. Fluid flow at ϵ_2 scale (usually described in terms of Darcy's 'law') can be related to flow within pores via w_{ϵ_2} averaging of ϵ_1-scale relations and definitions, and also identified with direct w_{ϵ_2}-averaging of the molecular description. Such an approach was adopted by Murdoch & Kubik [14] and Murdoch & Hassanizadeh [15] and will be discussed in Chapter 10.

Remark 4.4.8. While taking predominantly positive values, the weighting functions defined in (4.4.25), (4.4.30), and (4.4.40) also take negative values. This belies the common assertion (cf., e.g., Hardy [10]) that a weighting function should only take non-negative values.

Remark 4.4.9. Choices (4.4.25), (4.4.30), and (4.4.39) of weighting functions *guarantee* that repeated spatial averages at *any* specified length scale should yield nothing new, *no matter what material behaviour is being considered.* However, the behaviour of interest may well give rise to spatial averages which vary negligibly over a range of length scales. (This is often an *assumption* made in connection with continuum modelling; cf., e.g., Paterson [5], III, §1.) Consistent with such behaviour, field values may vary little over displacements whose magnitudes are commensurate with the averaging length scale. In such case repeated spatial averaging at this scale will yield no appreciably different values, no matter which weighting function is chosen. Said differently, property (4.4.10) *may* hold at a given scale, irrespective of which weighting function is chosen, as a consequence of the material behaviour in question.

4.4.3 Other Choices

The simplest choice of weighting function given in Section 4.3, and the choices associated with cellular averaging and with repeated averaging, were all motivated by *physical* considerations. If, however, only mathematical aspects are considered, then choices can be governed by the simplicity of the form of w and/or the desired regularity of the fields defined in terms of w (see Subsection 4.3.6.). Hardy [10] gave, as an example, the *G*aussian

$$w(\mathbf{u}) = \hat{w}_G(\mathbf{u}.\mathbf{u}) := \pi^{-3/2}\epsilon^{-3}\exp\left\{-\frac{u^2}{\epsilon^2}\right\}, \tag{4.4.47}$$

which is analytic (and hence, in particular, of class C^∞). Although $\hat{w}_G$ decays rapidly for $u > \epsilon$, there is no definite 'cut-off' value of u. Polynomial-based weighting functions with a specific cut-off at $u = \epsilon$ can be obtained simply. For example, we may construct a polynomial-based weighting function w_n defined in terms of a degree $(2n+1)$ polynomial P_{2n+1} as follows. Define

$$\left.\begin{array}{lll} w(\mathbf{u}) := w_n(u) = P_{2n+1}(u) & \text{if} & 0 \le u \le \epsilon \\ w(\mathbf{u}) = 0 & \text{if} & u > \epsilon \end{array}\right\}, \tag{4.4.48}$$

where the first n derivatives of w_n all vanish at both $u=0$ and $u=\epsilon$,

$$w_n(\epsilon)=0, \tag{4.4.49}$$

and w_n satisfies the normalisation condition (4.3.23). That is,

$$P_{2n+1}(\epsilon)=0 \tag{4.4.50}$$

and
$$4\pi\int_0^{\epsilon} P_{2n+1}(u)u^2du=1. \tag{4.4.51}$$

Notice that the vanishing of derivatives, together with (4.4.50) and (4.4.51), constitute $(2n+2)$ conditions which suffice to determine P_{2n+1} uniquely.

Exercise 4.4.4. Show that for $n=1$,

$$w_1(u)=\frac{15}{4\pi\epsilon^3}\left(2\left(\frac{u}{\epsilon}\right)^3-3\left(\frac{u}{\epsilon}\right)^2+1\right) \tag{4.4.52}$$

and that this function is monotonic decreasing on $[0,\epsilon]$ (and hence accords greater weighting to small u than large u).

**Exercise 4.4.5.* Recalling the exercises of Subsection 4.3.6, note that vanishing of the first n derivatives of P_{2n+1} at $u=0$ and $u=\epsilon$ suggests that

$$w_n'(u)=P_{2n+1}'(u)=c_nu^n(u-\epsilon)^n. \tag{4.4.53}$$

Show, by integrating (4.4.51) by parts and using (4.4.50), that

$$-\frac{4\pi}{3}\int_0^{\epsilon} P_{2n+1}'(u)u^3du=1, \tag{4.4.54}$$

and hence from (4.4.53) that c_n is given by

$$-\frac{4\pi c_n}{3}\int_0^{\epsilon} u^{n+3}(u-\epsilon)^n du=1. \tag{4.4.55}$$

Recalling (4.3.64) and (4.3.65), show that

$$-\frac{4\pi\epsilon^{2n+4}c_n}{3}I_{n+3,n}=1, \tag{4.4.56}$$

and
$$I_{n+3,n}=\frac{(-1)^{n+3}(n+3)!n!}{(2n+3)!}I_{0,2n+3}. \tag{4.4.57}$$

Noting
$$I_{0,2n+3}=-\frac{1}{(2n+4)}, \tag{4.4.58}$$

deduce that
$$c_n=-\frac{3(-1)^n(2n+4)!}{4\pi\epsilon^{2n+4}n!(n+3)!}. \tag{4.4.59}$$

Hence show from (4.4.53), (4.4.59), and (4.4.50) that

$$w_n(u)=P_{2n+1}(u)=\frac{3(-1)^n(2n+4)!}{4\pi\epsilon^{2n+4}n!(n+3)!}\int_u^{\epsilon} s^n(s-\epsilon)^n ds. \tag{4.4.60}$$

[Notice we chose the limits here so that $w_n(\epsilon)=0$ automatically and $w_n'(u)$ satisfies (4.4.53).] Verify that w_n is monotonic decreasing for $0\le u\le\epsilon$.

*4.5 Temporal Fluctuations

Consider definitions (4.2.1) and (4.2.13) of $\rho_w(\mathbf{x},t)$ and $\mathbf{p}_w(\mathbf{x},t)$ for a given location $\mathbf{x}$. The molecules which contribute to the sums in these definitions change with time. The simplest form of weighting function $w = w_{\epsilon,\delta}$ [see (4.3.22)] involves contributions $m_i w_{\epsilon,\delta}(\mathbf{x}_i(t) - \mathbf{x})$ and $m_i \mathbf{v}_i(t) w_{\epsilon,\delta}(\mathbf{x}_i(t) - \mathbf{x})$ from molecule P_i. If $\mathbf{x}_i(t) \in S_\epsilon(\mathbf{x})$, then these contributions are $m_i \alpha^{-1}$ and $m_i \mathbf{v}_i(t)\alpha^{-1}$, respectively, while such contributions vanish if $\mathbf{x}_i(t) \notin S_{\epsilon+\delta}(\mathbf{x})$. Accordingly, if P_i crosses $S_{\epsilon,\delta}(\mathbf{x}) := S_{\epsilon+\delta}(\mathbf{x}) - S_\epsilon(\mathbf{x})$, then there are 'blips' (fluctuations) in the values at $\mathbf{x}$ of ρ_w and $\mathbf{p}_w$ [and consequently in the values of $\mathbf{v}_w$, given by (4.3.33)] due to this crossing. Notice, however, that the smooth nature of $w_{\epsilon,\delta}$ ensures that such blips are smooth. Consideration of molecular thermal motions indicates that $\rho_w, \mathbf{p}_w$ and $\mathbf{v}_w$ (with $w = w_{\epsilon,\delta}$) are all subject to fine-scale fluctuations on a time scale much shorter than that over which these fields undergo significant macroscopic change, as is also the case for fields defined in terms of cellular-based weighting functions.

Exercise 4.5.1. Molecules in the atmosphere at sea level have a density of order $1.2\,\mathrm{kg\,m^{-3}}$ at $0°$ C, have a root mean square (rms) speed of order $5 \times 10^2\,\mathrm{ms^{-1}}$, and the number to be found in $1\,\mathrm{m}^3$ is of order 2.7×10^{25}. Show that the number of molecules between concentric spheres of radii ϵ and $\epsilon + \delta$ ($\delta << \epsilon$) is of order $3.4\epsilon^2\delta \times 10^{26}$. If a molecule crosses the region between these spheres, moving radially outward at rms speed, show that

(i) the corresponding fractional change in the density is of order $9\epsilon^{-3} \times 10^{-27}$,
(ii) such crossing occurs in a time $2\delta \times 10^{-3}\,\mathrm{s}$, and
(iii) the associated density change divided by the time taken is of order $5\epsilon^{-3}\delta^{-1} \times 10^{-24}\,\mathrm{kg\,m^{-3}\,s^{-1}}$.

[Thus, if $\epsilon = 10^{-6}\,\mathrm{m}$ and $\delta = 10^{-11}\,\mathrm{m}$, then (i), (ii), and (iii) involve orders $9 \times 10^{-9}, 2 \times 10^{-14}\,\mathrm{s}$, and $5 \times 10^5\,\mathrm{kg\,m^{-3}\,s^{-1}}$, respectively.]

Remark 4.5.1. The crude estimates in the foregoing exercise give an indication of how a continuum field value such as mass density may suffer macroscopically negligible changes over very short time intervals (temporal blips) if weighting functions of form $w_{\epsilon,\delta}$ (or analogous, mollified, cell-based forms of w) are employed. Such fluctuations derive from the ϵ-scale (or cell size) sharp cut-off nature of the weighting functions and are far less pronounced if the weighting functions of Subsections 4.4.2 and 4.4.3 are used. What is at first surprising is the fluctuation in what is essentially $\partial\rho/\partial t$ due to a single crossing [see (iii)]. Of course, consideration of the changing molecular population of any interspherical region $S_{\epsilon+\delta}(\mathbf{x}) - S_\epsilon(\mathbf{x})$ over a time interval of duration $2\delta \times 10^{-3}\,\mathrm{s}$ indicates that the *net* fluctuation in $\partial\rho/\partial t$ over such time lapse is much smaller. (Convince yourself of this!) A further time averaging (implemented in Chapter 8) smoothes these fluctuations. Indeed, such additional averaging is in part motivated by the desirability of eliminating fluctuations. (Complementary motivation is the need to identify field values in terms of measurement and recognition that measurements have associated scales of both length *and* time.)

4.6 Summary

Here use of weighting functions has been shown to provide precise and specific links between the discrete (molecular) nature of matter and continuum fields. Selection of a particular weighting function w, made on the basis of physical (or strictly mathematical) criteria, yields scale-dependent density fields of mass and linear momentum ρ_w and $\mathbf{p}_w$ and corresponding velocity field $\mathbf{v}_w$ [via (4.2.1), (4.2.13), and (4.2.15)] which satisfy the continuity equation (2.5.16). Accordingly, conceptual problems C.P.1 and C.P.3 of Section 3.8 disappear. Further, in respect of C.P.2, the integral of ρ_w over a region R does not yield exactly the mass of the molecules which lie within R: the difference between these quantities depends upon both the form and length scale associated with w and involves weighted contributions from molecules near the boundary ∂R of R. Particular attention has been paid to choice w_ϵ of weighting function in view of the very simple interpretations of field values and clear delineation of both physical and geometric boundaries at any choice of length scale ϵ. Another category of weighting function, relevant to multi-scale modelling, has been explored. This allows two continuum descriptions, corresponding to ϵ_1- and ϵ_2-scale averaging (with $\epsilon_2 > \epsilon_1$, say), of extensive molecular attributes to be compared. The ϵ_2-scale continuum description has, for this category, been *proved* to be the ϵ_2-scale average of the ϵ_1-scale continuum description.

5 Motions, Material Points, and Linear Momentum Balance

5.1 Preamble

The kinematic behaviour of any material system is here established, at any given spatial scale, in terms of a motion map whose time evolution is described by a local balance of linear momentum. The velocity field $\mathbf{v}_w$ introduced in Chapter 4 is used to define a corresponding motion which describes the gross dynamic distortion of the system/body at the scale embodied in the weighting function w. Visualisation of such motion is effected in terms of fictitious 'material points'. The concepts of velocity, motion, and material point are considered in the context of a non-reacting binary mixture to emphasise the simplicity and clarity of the methodology.

After a brief discussion of the subatomic origin of molecular interactions, a local form of linear momentum balance is established by computation of local weighted spatial averages of equations which govern the motions of individual molecules in any inertial frame. This balance relation involves an interaction force density $\mathbf{f}_w$. The usual form of balance is obtained by determination of an interaction stress tensor field $\mathbf{T}_w^-$ for which $\operatorname{div}\mathbf{T}_w^- = \mathbf{f}_w$. The Cauchy stress tensor $\mathbf{T}_w$ is thereupon defined to be $\mathbf{T}_w^- - \mathcal{D}_w$, where $\mathcal{D}_w$ is a symmetric tensor of kinematic character, $-\mathcal{D}_w$ is pressure-like, and the trace of $\mathcal{D}_w$ is a measure of heat energy density. The relative contributions of $\mathbf{T}_w^-$ and $-\mathcal{D}_w$ to $\mathbf{T}_w$ in gases, liquids, and solids are compared.

Since, by definition, $\mathbf{T}_w^-$ can only be unique to within a divergence-free tensor field, this non-uniqueness is studied. Three classes of candidate interaction stress tensors are considered. It is at this stage that the symmetry, or otherwise, of the Cauchy stress tensor emerges. At the level of generality here adopted, such symmetry is not to be expected for all systems (cf., e.g., asymmetric stresses in liquid crystalline phases) and is not found here except for one class of candidate in the case of 'central' interactions. Calculations of candidate interaction stresses for the simplest form of weighting function (emphasised in Chapter 4) are made and compared, as also are their integrals over planar surfaces. What at first is surprising is the equality of the integrals considered (given the non-trivial and distinct calculations involved). However, it is then simply shown that (modulo very weak assumptions concerning interactions) for *any* choice of weighting function w, the integral over an infinite plane is independent of the candidate $\mathbf{T}_w^-$ chosen, and, for the simplest form of w, the value of the integral has a simple geometrical interpretation.

5.2 Motions and Material Points

In Chapter 4 the link between the microscopic (molecular) and macroscopic (continuum) description of matter and its motion was effected via scale-dependent weighting functions. In particular, to any point $\mathbf{x}$ in the region 'occupied' by a material system (at a given scale) at a given time t, was ascribed a velocity value $\mathbf{v}_w(\mathbf{x},t)$. Modulo the choice of weighting function w, the values of $\mathbf{v}_w$ are *intrinsic* to the system since they depend only upon molecular attributes [specifically, molecular masses and velocities: see (4.2.15), (4.2.1) and (4.2.13)]. The form of w determines the length scale and physical interpretation to be ascribed to $\mathbf{v}_w$, irrespective of whether the system is in a solid or fluid phase. Indeed, the system might consist of a particular molecular constituent in a fluid mixture of several molecular species: in such a mixture each constituent has its own fields of mass and momentum densities and hence also of velocity. Such precision clearly improves upon the discussion of Chapter 2, Section 2.3, wherein considerations were heuristic (and certainly inapplicable to mixtures). Having chosen a weighting function w, and thus obtained the corresponding velocity field $\mathbf{v}_w$, the *motion map* $\boldsymbol{\chi}_{w,t_0}$ *corresponding to the situation at any given time* t_0 can be constructed in the manner of Section 2.3. To this end we write

$$B_{w,t} := \{\mathbf{x} \in \mathcal{E} : \rho_w(\mathbf{x},t) > 0\} \tag{5.2.1}$$

to denote the region occupied by the material system/body at time t when using weighting function w. Then

$$\boldsymbol{\chi}_{w,t_0}(\cdot,t) : B_{w,t_0} \to \mathcal{E} \tag{5.2.2}$$

is to satisfy, for any $\hat{\mathbf{x}} \in B_{w,t_0}$,

$$\dot{\boldsymbol{\chi}}_{w,t_0}(\hat{\mathbf{x}},t) = \mathbf{v}_w(\boldsymbol{\chi}_{w,t_0}(\hat{\mathbf{x}},t),t), \tag{5.2.3}$$

where (see Figure 5.1.)

$$\boldsymbol{\chi}_{w,t_0}(\hat{\mathbf{x}},t_0) = \hat{\mathbf{x}}. \tag{5.2.4}$$

Remark 5.2.1. Choice $w = w_\epsilon$ [see (4.3.2) and the last paragraph of subsection 4.3.5] yields a motion prescribed by local ϵ-scale mass centre trajectories. More precisely, given any point $\hat{\mathbf{x}}$ for which $\rho_\epsilon(\hat{\mathbf{x}},t_0) > 0$ (so that there must be at least one molecule within a distance ϵ from $\hat{\mathbf{x}}$ at time t_0), to determine $\boldsymbol{\chi}_{w_\epsilon,t_0}(\hat{\mathbf{x}},t)$ we must start from point $\hat{\mathbf{x}}$ at time t_0 and thereafter move with the velocity of the mass centre of those molecules

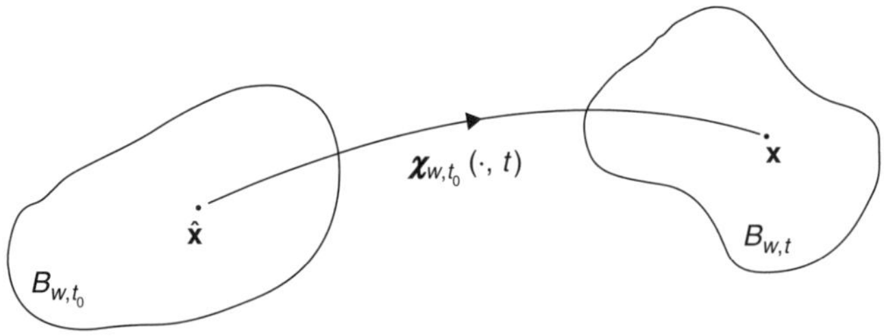

Figure 5.1. Motion map $\boldsymbol{\chi}_{w,t_0}$ corresponding to the situation at time t_0 and choice w of weighting function.

instantaneously located within a distance ϵ of our current location. More generally, for *any* choice w, and any point $\hat{\mathbf{x}}$ at which $\rho_w(\hat{\mathbf{x}}, t_0) > 0$, $\boldsymbol{\chi}_w(\hat{\mathbf{x}}, t)$ is determined by starting from $\hat{\mathbf{x}}$ at time t_0 and moving at any time $t \geq t_0$ with the molecular w-average velocity $\mathbf{v}_w$ computed at the current location at this time.

Remark 5.2.2. It follows from Remark 5.2.1 that

(i) if, at some time $\tau \geq t_0$, $\boldsymbol{\chi}_{w,t_0}(\hat{\mathbf{x}}, \tau) = \boldsymbol{\chi}_{w,t_0}(\hat{\mathbf{y}}, \tau)$, then for all subsequent times $t > \tau$ we must have $\boldsymbol{\chi}_{w,t_0}(\hat{\mathbf{x}}, t) = \boldsymbol{\chi}_{w,t_0}(\hat{\mathbf{y}}, t)$, and further that
(ii) if $\hat{\mathbf{x}} \neq \hat{\mathbf{y}}$, then the closer $\boldsymbol{\chi}_{w,t_0}(\hat{\mathbf{x}}, t)$ is to $\boldsymbol{\chi}_{w,t_0}(\hat{\mathbf{y}}, t)$, the closer will be the relevant values of $\mathbf{v}_w$.

Properties (i) and (ii) indicate that we should expect $\boldsymbol{\chi}_{w,t_0}(\cdot, t)$ to be 1:1; that is,

$$\text{if} \qquad \boldsymbol{\chi}_{w,t_0}(\hat{\mathbf{x}}, t) = \boldsymbol{\chi}_{w,t_0}(\hat{\mathbf{y}}, t), \qquad \text{then} \qquad \hat{\mathbf{x}} = \hat{\mathbf{y}}. \tag{5.2.5}$$

To each $\hat{\mathbf{x}} \in B_{w,t_0}$ can be ascribed a *w-based material point* whose location at time t is the point

$$\mathbf{x} = \boldsymbol{\chi}_{w,t_0}(\hat{\mathbf{x}}, t). \tag{5.2.6}$$

In view of (5.2.5), it follows that, for a given material system, no two different material points can ever occupy the same location at the same time.

Remark 5.2.3. The dependence of the notion of material point upon choice w of weighting function is of crucial importance. Specifically, this notion is explicitly linked both to a choice of scale *and* to molecular behaviour. The utility of the notion is in tracking local gross molecular behaviour at any selected length scale in an explicit manner via the associated motion map. For example, in modelling the flow of air over an aerofoil or round a skyscraper, one might work with averages at a scale in the range 10^{-3} to 1 m, while meteorological models of wind motion could involve scales in excess of 10^2 m. Each specific choice of scale yields the corresponding velocity field together with the associated form of motion map which identifies the relevant set of material points. Summarising, *the notion of material point is a very useful mathematical artefact which derives from the physically defined (and scale-dependent) velocity field and which, together with the associated motion map, delineates the distortion of the body at the relevant scale.* In no way does a material point have the physical reality of a molecule: it cannot be observed and has no mass.[1]

Relations (5.2.3) and (5.2.4) link the *physically defined* velocity field with the *induced* notion of motion (corresponding to the situation at some chosen time t_0). If $\mathbf{x} \in B_{w,t}$, then these relations may be rephrased as

$$\mathbf{v}_w(\mathbf{x}, t) := \dot{\boldsymbol{\chi}}_{w,t_0}(\hat{\mathbf{x}}, t), \tag{5.2.7}$$

$$\text{where} \qquad \mathbf{x} = \boldsymbol{\chi}_{w,t_0}(\hat{\mathbf{x}}, t) \qquad \text{and} \qquad \boldsymbol{\chi}_{w,t_0}(\hat{\mathbf{x}}, t_0) = \hat{\mathbf{x}}. \tag{5.2.8}$$

[1] Material points are more commonly termed fluid *elements* or *particles* in fluid dynamics texts. The latter usage is somewhat misleading because 'particle' is normally reserved for the notion of point mass in mechanics or elementary discrete entity in physics, and in both cases the particle has mass.

The corresponding *acceleration* field on $B_{w,t}$ is $\mathbf{a}_w$, where [cf. (2.3.4) and (2.4.10)]

$$\mathbf{a}_w(\mathbf{x},t) := \ddot{\boldsymbol{\chi}}_{w,t_0}(\hat{\mathbf{x}},t). \tag{5.2.9}$$

Thus

$$\begin{aligned}\mathbf{a}_w(\mathbf{x},t) &= \frac{\partial}{\partial t}\{\dot{\boldsymbol{\chi}}_{w,t_0}(\hat{\mathbf{x}},t)\}\\ &= \frac{\partial}{\partial t}\{\mathbf{v}_w(\boldsymbol{\chi}_{w,t_0}(\hat{\mathbf{x}},t),t\}\\ &= \nabla_{\mathbf{x}}\mathbf{v}_w(\mathbf{x},t)\dot{\boldsymbol{\chi}}_{w,t_0}(\hat{\mathbf{x}},t) + \frac{\partial \mathbf{v}_w}{\partial t},\end{aligned} \tag{5.2.10}$$

on using the chain rule. That is,

$$\mathbf{a}_w = (\nabla \mathbf{v}_w)\mathbf{v}_w + \frac{\partial \mathbf{v}_w}{\partial t}. \tag{5.2.11}$$

5.3 Motions and Material Points for Non-Reacting Binary Mixtures

In the material systems $\mathcal{M}$ considered so far all molecules have been given individual status. Thus in principle all masses m_i could be different. Of course, in practice molecules may be grouped into 'species'. If only one such species is present in $\mathcal{M}$, then the system is regarded to be a 'pure substance'. More generally, a material system composed of a number of molecular species may be expected to behave differently according to the relative numbers and spatial distributions of these species. Such considerations motivate looking at each species in the 'mixture' (of two or more distinct species) as a material system in its own right. Consider the simplest example of a binary, non-reacting mixture; that is, one in which only two distinct molecular species are involved, and each molecule preserves its integrity. Let (cf. Section 4.2)

$$\mathcal{M} := \mathcal{M}_\alpha \cup \mathcal{M}_\beta, \tag{5.3.1}$$

where $\mathcal{M}_\alpha = \{P_{\alpha_i} : i = 1,2,\ldots,N_\alpha\}$ and $\mathcal{M}_\beta = \{P_{\beta_j} : j = 1,2,\ldots,N_\beta\}$. (5.3.2)

If $m_{\alpha_i}(=m_\alpha), \mathbf{x}_{\alpha_i}, m_{\beta_j}(=m_\beta)$ and $\mathbf{x}_{\beta_j}$ denote the masses and locations of molecules P_{α_i} and P_{β_j}, then the mass density fields ρ_w^α and ρ_w^β associated with the choice w of weighting function [cf. (4.2.1)] are given by

$$\rho_w^\alpha(\mathbf{x},t) := \sum_{i=1}^{N_\alpha} m_\alpha w(\mathbf{x}_{\alpha_i}(t) - \mathbf{x}) \tag{5.3.3}$$

and

$$\rho_w^\beta(\mathbf{x},t) := \sum_{j=1}^{N_\beta} m_\beta w(\mathbf{x}_{\beta_j}(t) - \mathbf{x}). \tag{5.3.4}$$

The corresponding momentum density fields [cf. (4.2.13)] are

$$\mathbf{p}_w^\alpha(\mathbf{x},t) := \sum_{i=1}^{N_\alpha} m_\alpha \mathbf{v}_{\alpha_i}(t) w(\mathbf{x}_{\alpha_i}(t) - \mathbf{x}) \tag{5.3.5}$$

and
$$\mathbf{p}_w^\beta(\mathbf{x},t) := \sum_{j=1}^{N_\beta} m_\beta \mathbf{v}_{\beta_j}(t) w(\mathbf{x}_{\beta_j}(t) - \mathbf{x}). \tag{5.3.6}$$

Accordingly, the mass and momentum fields for $\mathcal{M}$ [cf. (5.3.1)] are

$$\rho(\mathbf{x},t) = \sum_{\text{all particles}} m_i w(\mathbf{x}_i(t) - \mathbf{x}) = \rho_w^\alpha(\mathbf{x},t) + \rho_w^\beta(\mathbf{x},t) \tag{5.3.7}$$

and
$$\mathbf{p}(\mathbf{x},t) = \sum_{\text{all particles}} m_i \mathbf{v}_i(t) w(\mathbf{x}_i(t) - \mathbf{x}) = \mathbf{p}_w^\alpha(\mathbf{x},t) + \mathbf{p}_w^\beta(\mathbf{x},t). \tag{5.3.8}$$

The velocity fields $\mathbf{v}_w^\alpha, \mathbf{v}_w^\beta$ and $\mathbf{v}_w$ for $\mathcal{M}_\alpha, \mathcal{M}_\beta$ and $\mathcal{M}$ are defined by

$$\mathbf{v}_w^\alpha := \mathbf{p}_w^\alpha/\rho_w^\alpha, \qquad \mathbf{v}_w^\beta := \mathbf{p}_w^\beta/\rho_w^\beta, \qquad \text{and} \qquad \mathbf{v}_w := \mathbf{p}_w/\rho_w. \tag{5.3.9}$$

Thus
$$\rho_w = \rho_w^\alpha + \rho_w^\beta \tag{5.3.10}$$

and
$$\rho_w \mathbf{v}_w = \rho_w^\alpha \mathbf{v}_w^\alpha + \rho_w^\beta \mathbf{v}_w^\beta. \tag{5.3.11}$$

A binary mixture is termed *non-diffusive* if $\mathbf{v}_w^\alpha = \mathbf{v}_w^\beta$. This would be expected if $\mathcal{M}$ were a 'solid' body. If $\mathcal{M}$ is a fluid then diffusion is possible: this is evident when a dye disperses in a liquid – the dye 'moves' through the liquid, indicating that relative mass transport is taking place, and hence $\mathbf{v}_{\text{dye}} \neq \mathbf{v}_{\text{liquid}}$. Mass conservation relations for $\mathcal{M}_\gamma (\gamma = \alpha, \beta)$ may be derived precisely as in Section 4.2 to yield

$$\frac{\partial \rho_w^\gamma}{\partial t} + \operatorname{div}\{\rho_w^\gamma \mathbf{v}_w^\gamma\} = 0. \tag{5.3.12}$$

Exactly as in Section 5.2, motion maps for $\mathcal{M}_\gamma$ can be defined and visualised in terms of trajectories of γ-material points. Specifically, the motion $\boldsymbol{\chi}_{w,t_0}^\gamma$ of $\mathcal{M}_\gamma$ relative to the situation at time t_0 is the solution to the initial-value problem

$$\dot{\boldsymbol{\chi}}_{w,t_0}^\gamma(\hat{\mathbf{x}},t) = \mathbf{v}_w^\gamma(\boldsymbol{\chi}_{w,t_0}^\gamma(\hat{\mathbf{x}},t),t), \tag{5.3.13}$$

where
$$\boldsymbol{\chi}_{w,t_0}^\gamma(\hat{\mathbf{x}},t_0) = \hat{\mathbf{x}}, \tag{5.3.14}$$

with
$$\hat{\mathbf{x}} \in B_{w,t_0}^\gamma \tag{5.3.15}$$

and
$$B_{w,t}^\gamma := \{\mathbf{x} \in \mathcal{E} : \rho_w^\gamma(\mathbf{x},t) > 0\}. \tag{5.3.16}$$

Remark 5.3.1. If $\mathbf{x} \in B_{w,t}^\alpha \cap B_{w,t}^\beta$, then $\mathbf{x}$ can be considered to be the simultaneous location of three material points, namely those from each of $\mathcal{M}^\alpha, \mathcal{M}^\beta$, and $\mathcal{M}$ which at time t_0 were located at points $\hat{\mathbf{x}}_\alpha, \hat{\mathbf{x}}_\beta$ and $\hat{\mathbf{x}}$ determined by

$$\mathbf{x} = \boldsymbol{\chi}_{w,t_0}^\alpha(\hat{\mathbf{x}}_\alpha,t) = \boldsymbol{\chi}_{w,t_0}^\beta(\hat{\mathbf{x}}_\beta,t) = \boldsymbol{\chi}_{w,t_0}(\hat{\mathbf{x}},t). \tag{5.3.17}$$

In particular, there is no conceptual problem associated with such simultaneous 'occupancy' of location $\mathbf{x}$ by three material points, since these are mathematical constructs and in no way are to be confused with point masses.

5.4 Linear Momentum Balance Preliminaries: Intermolecular Forces

Having established the macroscopic kinematic behaviour of a material system in terms of the velocity field and associated concept of motion, the next step is to explore the *cause* of such behaviour. Accordingly in this section we review the Newtonian dynamics of a point mass and then discuss molecular dynamics.

In any inertial frame, a point mass P of mass m and located at point $\mathbf{x}(t)$ at time t is said to be subject to a force

$$\mathbf{f} := m\mathbf{a}, \tag{5.4.1}$$

where

$$\mathbf{a} := \ddot{\mathbf{x}} \tag{5.4.2}$$

denotes the acceleration in this, and any other, inertial frame. This force is attributed to an agency, or agencies, which give rise to a change in velocity $\mathbf{v} := \dot{\mathbf{x}}$. That is, such agency (agencies) is (are) considered to be the *cause* of this change (its *effect*). Two such agencies, $\mathcal{A}'$ and $\mathcal{A}''$ say, are *independent* if, when both agencies are present, (5.4.1) holds with

$$\mathbf{f} = \mathbf{f}' + \mathbf{f}'' \qquad \text{and} \qquad \mathbf{a} = \mathbf{a}' + \mathbf{a}''. \tag{5.4.3}$$

Here $\mathbf{f}'$ and $\mathbf{a}'$ denote the force and acceleration in the presence of $\mathcal{A}'$ alone, and similarly for $\mathbf{f}''$ and $\mathbf{a}''$ in the presence of $\mathcal{A}''$ alone. Of course, the foregoing generalises to any number of agencies and results in linear superposition of forces attributable to independent agencies.

Now consider the motion of a molecule modelled as a point mass P_i in a material system $\mathcal{M}$ (see Section 4.2). In any inertial frame, we assume that

$$\sum_{j \neq i} \mathbf{f}_{ij} + \mathbf{b}_i = m_i \mathbf{a}_i, \tag{5.4.4}$$

where

$$\mathbf{a}_i := \ddot{\mathbf{x}}_i = \dot{\mathbf{v}}_i. \tag{5.4.5}$$

Here $\mathbf{f}_{ij}$ represents the force exerted on P_i by P_j (the sum is taken over all other molecules P_j of $\mathcal{M}$) and $\mathbf{b}_i$ denotes the resultant force on P_i due to the material universe outwith $\mathcal{M}$. (In most contexts this is due solely to gravitation, and $\mathbf{b}_i = m_i \mathbf{g}$, where $\mathbf{g}$ denotes the local value of the acceleration due to gravity.) Relation (5.4.4) has the simplest interpretation

Interpretation 1. Each of $\mathbf{f}_{ij} (j \neq i)$ and $\mathbf{b}_i$ are associated with independent agencies, and (5.4.4) is a generalised version of (5.4.1) and (5.4.3).

According to *Interpretation* 1, the force exerted on molecule P_i by P_j is independent of all other molecules of $\mathcal{M}$. In such case Noll [16] has indicated that, on the basis of objective considerations (i.e., observer agreement: see Chapter 12), such a force must depend upon the distance between P_i and P_j, and be directed along the line which joins their locations at any instant. Consistent with this interpretation are models of interactions governed by separation-dependent pair potentials such as those of Lennard–Jones form (cf. e.g., Atkins [17]).

There is, however, a fundamental objection to the foregoing. Each molecule is a composite system of fundamental subatomic discrete entities (atomic nuclei and electrons), and the behaviour of such a system must be expected to be somewhat complex. In particular, the resultant force between any two molecules must be expected to depend upon the behaviour of their constituent nuclei and electrons. *We now show how (5.4.4) can be re-interpreted to take account of such substructure.* The outcome is an expected dependence of $\mathbf{f}_{ij}$ both upon molecules close to P_i and upon molecules close to P_j.

We model fundamental subatomic discrete entities as point masses. Let P_{i_p} and P_{j_q} denote such entities associated with molecules P_i and P_j, having masses m_{i_p}, m_{j_q} and locations $\mathbf{x}_{i_p}(t), \mathbf{x}_{j_q}(t)$ at time t. We assume that each force $\mathbf{f}_{i_p j_q}$ exerted by P_{j_q} upon $P_{i_p}(i \neq j$ or, if $i = j, p \neq q)$, and resultant external force $\mathbf{b}_{i_p}$, are all independent. Then the motion of P_{i_p} in an inertial frame is governed by (here $\mathbf{v}_{i_p} := \dot{\mathbf{x}}_{i_p}$)

$$\sum_{\substack{i_{p'} \\ p' \neq p}} \mathbf{f}_{i_p i_{p'}} + \sum_{j \neq i} \sum_{j_q} \mathbf{f}_{i_p j_q} + \mathbf{b}_{i_p} = m_{i_p} \dot{\mathbf{v}}_{i_p}. \tag{5.4.6}$$

Summing over all subentities P_{i_p} of P_i yields

$$\mathbf{f}_i + \sum_{j \neq i} \mathbf{f}_{ij} + \mathbf{b}_i = m_i \dot{\mathbf{v}}_i, \tag{5.4.7}$$

where

$$\mathbf{f}_i := \sum_{p' \neq p} \sum \mathbf{f}_{i_p i_{p'}} = \frac{1}{2} \sum_{p' \neq p} \sum (\mathbf{f}_{i_p i_{p'}} + \mathbf{f}_{i_{p'} i_p}),$$

$$\mathbf{f}_{ij} := \sum_{i_p} \sum_{j_q} \mathbf{f}_{i_p j_q}, \qquad \mathbf{b}_i := \sum_{i_p} \mathbf{b}_{i_p}, \tag{5.4.8}$$

and

$$\mathbf{v}_i := \sum_{i_p} m_{i_p} \mathbf{v}_{i_p} / m_i \qquad \text{with} \quad m_i := \sum_{i_p} m_{i_p}. \tag{5.4.9}$$

Thus we obtain (5.4.4) and its interpretations if $\mathbf{f}_i = \mathbf{0}$ and [as follows from relations (5.4.9)] $\mathbf{v}_i$ is the velocity of the mass centre of P_i.

At this stage $\mathbf{f}_{ij}$, given by (5.4.8), is sensitive to rapid and complex motions of electrons in the vicinity of the constituent atomic nuclei of each of the molecules P_i and P_j, and also to the motions of these nuclei. *Any discussion in which a molecule–molecule interaction is labelled* $\mathbf{f}_{ij}$ *without the detail explicit in (5.4.8) must in some way represent an average value associated with this subatomic detail.* Here we implement a formal time-averaging procedure which is intended to smooth out such fine detail.

The Δ-*time average* of any continuous function f of time is[2]

$$f_\Delta(t) := \frac{1}{\Delta} \int_{t-\Delta}^{t} f(\tau) d\tau. \tag{5.4.10}$$

[2] That is, $f_\Delta(t)$ is the mean value of f computed over a time interval of duration Δ which ends at time t.

If f is continuously differentiable, then

$$\begin{aligned}(\dot{f})_\Delta &= \frac{1}{\Delta}\int_{t-\Delta}^{t} \dot{f}(\tau)d\tau \\ &= \frac{1}{\Delta}\{f(t) - f(t-\Delta)\} = (\dot{\widehat{f_\Delta}})(t).\end{aligned}$$

That is,

$$(\dot{f})_\Delta = \dot{\widehat{f_\Delta}}. \tag{5.4.11}$$

It follows, on taking the Δ-time average of (5.4.7), that

$$(\mathbf{f}_i)_\Delta + \sum_{j\neq i}(\mathbf{f}_{ij})_\Delta + (\mathbf{b}_i)_\Delta = m_i(\dot{\mathbf{v}}_i)_\Delta = m_i\dot{\widehat{(\mathbf{v}_i)}}_\Delta. \tag{5.4.12}$$

Exercise 5.4.1. Suppose that interactions between fundamental subatomic discrete entities are pairwise balanced; that is, at any instant

$$\mathbf{f}_{i_{p'}\,i_p} = -\mathbf{f}_{i_p\,i_{p'}} \qquad \text{and} \qquad \mathbf{f}_{j_q\,i_p} = -\mathbf{f}_{i_p\,j_q}. \tag{5.4.13}$$

Deduce from $(5.4.8)_{1,2}$ that

$$(\mathbf{f}_i)_\Delta = \mathbf{0} \qquad \text{and} \qquad (\mathbf{f}_{ji})_\Delta = -(\mathbf{f}_{ij})_\Delta. \tag{5.4.14}$$

Remark 5.4.1. Interactions $\mathbf{f}_{i_p i_{p'}}$ and $\mathbf{f}_{i_p j_q}$ are of electromagnetic nature and are accordingly transmitted at the speed of light. Thus relations (5.4.13) cannot be expected to be exact. (Why?) The time-averaged counterparts (5.4.14), while consequences of (5.4.13), may hold to within a higher degree of accuracy *in their own right* as a consequence of an averaging out both in time and over the assembly of molecular subentities. Notice that the vanishing of $(\mathbf{f}_i)_\Delta$ merely corresponds to an assumption that the Δ-time-average velocity of the mass centre of P_i is insensitive to interactions between its subatomic particles. This average velocity is, of course, sensitive to interactions with constituent subentities of *other* molecules. If $(5.4.14)_1$ holds, then (5.4.12) takes the form of (5.4.4) wherein all terms are interpreted as Δ-time averages. If $(5.4.14)_2$ holds then interactions in (5.4.4) satisfy $\mathbf{f}_{ji} = -\mathbf{f}_{ij}$ and are described as *pairwise balanced.*

At this stage we have arrived at a second interpretation of (5.4.4), namely (5.4.12) with the first term absent. Specifically, we have

Interpretation 2. Interaction $\mathbf{f}_{ij}$ is a time average of the resultant force the fundamental constituent electrons and nuclei of molecule P_j exert upon those of P_i (with scale Δ macroscopically small, but large compared with that of subatomic motions), $\mathbf{b}_i$ is a Δ-time average of the resultant force on the constituent particles of P_i from sources other than molecules of $\mathcal{M}$, and $\mathbf{a}_i$ is the time derivative of the Δ-time average of the mass centre velocity of the P_i particles.

Remark 5.4.2. Implicit in the foregoing is the assumption that there exists a time scale Δ which renders a meaningful interpretation of $(\mathbf{f}_{ij})_\Delta$ as an intermolecular force without further need to consider finer-scale detail. Here it is relevant to note that the time scales associated with subatomic motions are very small. For example, an electron moving at a speed of 10^3 ms^{-1} would travel a typical atomic radius of order 1 Å ($= 10^{-10}$ m) in 10^{-13} s. Accordingly, a *notional* value of $\Delta = 10^{-6}$ s would not seem to be unreasonable.

Remark 5.4.3. In computing the force on P_i due to P_j as a time average of resultant interactions between the constituent electrons and nuclei of these molecules, we must note that the trajectories of these fundamental particles must be expected to be influenced by those associated with molecules near to P_i and near to P_j. Specifically, the trajectories of 'outer' electrons of P_i will be influenced by the behaviour of outer electrons of those molecules in the immediate vicinity of P_i, and similarly for P_j. Accordingly, $\mathbf{f}_{ij}$ in (5.4.4), interpreted as $(\mathbf{f}_{ij})_\Delta$ in (5.4.12), must be expected to be sensitive to molecules other than P_i and P_j. *Consequently, it cannot be argued on objective grounds that $\mathbf{f}_{ij}$ given by (5.4.8) be directed along the line joining the instantaneous locations of the mass centres of P_i and P_j, and we cannot assert that $(\mathbf{f}_{ij})_\Delta$ must be directed along the line joining the time-averaged locations of the mass centres of P_i and P_j. Nevertheless, Remark 5.4.1. provides support for the assumption of pairwise balance $\mathbf{f}_{ji} = -\mathbf{f}_{ij}$.*

Remark 5.4.4. Interactions between molecules in general will depend upon their relative orientation, unless in some sense they are 'spherical' and hence isotropic. Further, the submolecular situations described here did not include the case of systems with 'free' electrons. Such considerations are essential for understanding the behaviour of 'structured' media (such as liquid crystalline phases), and of electrical conductors, respectively. These will be discussed later.

Remark 5.4.5. Discussion of subatomic behaviour is usually treated in terms of quantum mechanics. In this context it is necessary to write down the Hamiltonian associated with the nuclei and electrons of P_i and P_j, together with an additional potential term to account for the effect of neighbouring molecules. Analysis of the resulting highly complex Schrödinger equation requires simplification by additional assumptions. In principle, the state function for the system (namely, the solution of the Schrödinger equation) delivers the interaction $\mathbf{f}_{ij}$ in terms of expectations. Such a probabilistic nature, explicit within the quantum mechanical approach, can be interpreted in terms of time averaging via a 'frequentist' perspective. For example, if an electron is to be found in an atomically small region of volume dV for a time dt within a time interval of duration Δ, then the candidate probability density value at the region location is $(dt/\Delta) \cdot (1/dV)$. The resulting probability density function for such an electron describes its 'orbital'; that is, the region within which it is to be found, and the associated probability density. This is often depicted in terms of an electron 'cloud' in which darker shading indicates higher probability.

5.5 Linear Momentum Balance

5.5.1 Derivation of the Balance Relation

Multiplication of (5.4.4) by $w(\mathbf{x}_i(t) - \mathbf{x})$ and then summing over *all* molecules of $\mathcal{M}$ yields

$$\mathbf{f}_w + \mathbf{b}_w = \sum_{i=1}^{N} m_i \dot{\mathbf{v}}_i \, w(\mathbf{x}_i(t) - \mathbf{x}), \tag{5.5.1}$$

where the *interaction force density* field $\mathbf{f}_w$ is given by

$$\mathbf{f}_w(\mathbf{x},t) := \sum\sum_{i \neq j} \mathbf{f}_{ij}(t) w(\mathbf{x}_i(t) - \mathbf{x}), \tag{5.5.2}$$

and the *external body force* field $\mathbf{b}_w$ is given by

$$\mathbf{b}_w(\mathbf{x},t) := \sum_i \mathbf{b}_i(t) w(\mathbf{x}_i(t) - \mathbf{x}). \tag{5.5.3}$$

In (5.5.2) the sums involve, for each and every molecule P_i, all other molecules P_j of $\mathcal{M}$, and in (5.5.3) the sum is taken over all molecules of $\mathcal{M}$. Such notation will be used in what follows.

Exercise 5.5.1. If $w = w_\epsilon$ [see (4.3.2)] note that $\mathbf{f}_w(\mathbf{x},t)$ represents the resultant force exerted on those molecules which lie inside $S_\epsilon(\mathbf{x})$ at time t by other molecules of $\mathcal{M}$, whether inside $S_\epsilon(\mathbf{x})$ or not. If the resultant of interactions between all pairs of molecules which lie within $S_\epsilon(\mathbf{x})$ at time t vanishes [i.e., what might be called the 'self-force' associated with molecules in $S_\epsilon(\mathbf{x})$ vanishes], deduce that $\mathbf{f}_w(\mathbf{x},t)$ *is the resultant force exerted on molecules inside* $S_\epsilon(\mathbf{x})$ *by those outside or on the boundary of* $S_\epsilon(\mathbf{x})$ *at time t, divided by the volume of* $S_\epsilon(\mathbf{x})$. Prove further that if interactions are pairwise balanced (i.e., $\mathbf{f}_{ji} = -\mathbf{f}_{ij}$), then the $S_\epsilon(\mathbf{x})$ net self-force necessarily vanishes.

Exercise 5.5.2. If $\mathbf{b}_i(t) = m_i \mathbf{g}$, where $\mathbf{g}$ denotes gravitational acceleration, show that $\mathbf{b}_w = \rho_w \mathbf{g}$.

Here (to be consistent with *Interpretation 2* of Section 5.4) $\mathbf{x}_i(t)$ denotes the (Δ-) time-averaged location of the mass centre of molecule P_i. To simplify the right-hand side of (5.5.1), we note (here we suppress time dependence for brevity) that from (4.2.8) and (4.2.11)

$$m_i \dot{\mathbf{v}}_i \, w(\mathbf{x}_i - \mathbf{x}) = \frac{\partial}{\partial t}\{m_i \mathbf{v}_i \, w(\mathbf{x}_i - \mathbf{x})\} - m_i \mathbf{v}_i (\nabla w . \mathbf{v}_i). \tag{5.5.4}$$

Further, from (4.2.11), and noting[3] that $\mathbf{v}_i \otimes \mathbf{v}_i$ is independent of $\mathbf{x}$,

$$\begin{aligned} m_i \mathbf{v}_i . (\nabla w)\mathbf{v}_i &= m_i(\mathbf{v}_i \otimes \mathbf{v}_i)\nabla w \\ &= -m_i(\mathbf{v}_i \otimes \mathbf{v}_i)\nabla_{\mathbf{x}} w = -\text{div}\{m_i \mathbf{v}_i \otimes \mathbf{v}_i \, w\}. \end{aligned} \tag{5.5.5}$$

[3] If $\mathbf{a},\mathbf{b}$, and $\mathbf{v}$ are vectors, then $\mathbf{a} \otimes \mathbf{b}$ is that linear transformation which maps $\mathbf{v}$ into $(\mathbf{b}.\mathbf{v})\mathbf{a}$. See Appendix A.8.

Thus, from (5.5.1), (5.5.4), (5.5.5), (4.2.13), and (4.2.15),

$$\mathbf{f}_w + \mathbf{b}_w = \frac{\partial}{\partial t}\{\rho_w \mathbf{v}_w\} + \operatorname{div}\mathcal{D}_w^+, \tag{5.5.6}$$

where

$$\mathcal{D}_w^+(\mathbf{x},t) := \sum_i m_i \mathbf{v}_i(t) \otimes \mathbf{v}_i(t) w(\mathbf{x}_i(t) - \mathbf{x}). \tag{5.5.7}$$

Writing

$$\hat{\mathbf{v}}_i(t;\mathbf{x}) := \mathbf{v}_i(t) - \mathbf{v}_w(\mathbf{x},t) \tag{5.5.8}$$

we have, from (4.2.13), (4.2.1), and (4.2.15),

$$\sum_{i=1}^N m_i \hat{\mathbf{v}}_i(t;\mathbf{x}) w(\mathbf{x}_i(t) - \mathbf{x}) = \mathbf{p}_w(\mathbf{x},t) - \rho_w(\mathbf{x},t)\mathbf{v}_w(\mathbf{x},t) = \mathbf{0}. \tag{5.5.9}$$

Hence, from (5.5.7) and (5.5.8),

$$\begin{aligned}\mathcal{D}_w^+ &= \sum_i m_i(\mathbf{v}_w + \hat{\mathbf{v}}_i) \otimes (\mathbf{v}_w + \hat{\mathbf{v}}_i) w \\ &= \left(\sum_i m_i w)\mathbf{v}_w \otimes \mathbf{v}_w + \sum_i m_i \hat{\mathbf{v}}_i w \otimes \mathbf{v}_w + \mathbf{v}_w \otimes \sum_i m_i \hat{\mathbf{v}}_i w\right) + \mathcal{D}_w, \end{aligned} \tag{5.5.10}$$

where

$$\mathcal{D}_w(\mathbf{x},t) := \sum_i m_i \hat{\mathbf{v}}_i(t;\mathbf{x}) \otimes \hat{\mathbf{v}}_i(t;\mathbf{x}) w(\mathbf{x}_i(t) - \mathbf{x}). \tag{5.5.11}$$

Use of (5.5.9) reduces (5.5.10) to

$$\mathcal{D}_w^+ = \mathcal{D}_w + \rho_w \mathbf{v}_w \otimes \mathbf{v}_w. \tag{5.5.12}$$

At this stage (5.5.6) may be re-written as

$$\mathbf{f}_w + \mathbf{b}_w = \frac{\partial}{\partial t}\{\rho_w \mathbf{v}_w\} + \operatorname{div}\{\rho_w \mathbf{v}_w \otimes \mathbf{v}_w\} + \operatorname{div}\mathcal{D}_w. \tag{5.5.13}$$

Exercise 5.5.3. Show that, as a consequence of the continuity equation,

$$\frac{\partial}{\partial t}\{\rho_w \mathbf{v}_w\} + \operatorname{div}\{\rho_w \mathbf{v}_w \otimes \mathbf{v}_w\} = \rho_w \left\{\frac{\partial \mathbf{v}_w}{\partial t} + (\nabla \mathbf{v}_w)\mathbf{v}_w\right\}. \tag{5.5.14}$$

[Note that (see Appendix B.7.29)

$$\operatorname{div}\{\rho_w \mathbf{v}_w \otimes \mathbf{v}_w\} = \operatorname{div}\{\mathbf{v}_w \otimes \rho_w \mathbf{v}_w\} = (\nabla \mathbf{v}_w)\rho_w \mathbf{v}_w + (\operatorname{div}\{\rho_w \mathbf{v}_w\})\mathbf{v}_w.\ ]$$

From (5.5.13) and (5.5.14) we have

$$-\operatorname{div}\mathcal{D}_w + \mathbf{f}_w + \mathbf{b}_w = \rho_w \mathbf{a}_w, \tag{5.5.15}$$

where [see (5.2.11)]

$$\mathbf{a}_w := \frac{\partial \mathbf{v}_w}{\partial t} + (\nabla \mathbf{v}_w)\mathbf{v}_w \tag{5.5.16}$$

denotes the *acceleration* field appropriate to velocity $\mathbf{v}_w$.

Relation (5.5.15) is to be compared with the usual local form [see (2.7.20)]

$$\operatorname{div}\mathbf{T}+\mathbf{b}=\rho\mathbf{a} \tag{2.7.30}$$

of linear momentum balance. Identifications of $\mathbf{b}$ with $\mathbf{b}_w, \rho$ with ρ_w, and $\mathbf{a}$ with $\mathbf{a}_w$ are entirely natural. (Why?) Accordingly we are forced to

$$\textit{identify} \quad \operatorname{div}\mathbf{T} \quad \textit{with} \quad -\operatorname{div}\boldsymbol{\mathcal{D}}_w+\mathbf{f}_w. \tag{5.5.17}$$

In order to study such an identification it is necessary to express $\mathbf{f}_w$ as the divergence of a tensor-valued field. If $\mathbf{T}_w^-$ is such a field, so that

$$\mathbf{f}_w=\operatorname{div}\mathbf{T}_w^-, \tag{5.5.18}$$

then identification (5.5.17) leads us to

$$\textit{identify} \quad \mathbf{T} \quad \textit{with} \quad \mathbf{T}_w, \tag{5.5.19}$$

where

$$\mathbf{T}_w:=\mathbf{T}_w^- -\boldsymbol{\mathcal{D}}_w. \tag{5.5.20}$$

We term $\mathbf{T}_w^-$ an *interaction stress tensor* field.

Remark 5.5.1. From (5.5.18) $\mathbf{T}_w^-$, and hence $\mathbf{T}_w$, can only be unique to within a divergence-less second-order tensor field.

Remark 5.5.2. The physical interpretations of $\mathbf{T}_w^-$ and $\mathbf{T}_w$ depend both upon the choice of w and which solution to (5.5.18) is selected.

5.5.2 The Thermal Nature of $\boldsymbol{\mathcal{D}}_w$

Before addressing solutions to (5.5.18) we consider $\boldsymbol{\mathcal{D}}_w$ given by (5.5.11). Choice $w=w_\epsilon$ (or, more precisely, $w_{\epsilon,\delta}$: recall Subsections 4.3.1 and 4.3.4) yields $\boldsymbol{\mathcal{D}}_w(\mathbf{x},t)$ as a local volumetric average which involves the velocities of individual molecules in $S_\epsilon(\mathbf{x})$ relative to the mass centre velocity $\mathbf{v}(\mathbf{x},t)$ of these molecules. Such velocities are known to be chaotic (no matter whether the material system is in gaseous, liquid, or solid phase) and are at the heart of the kinetic theory of heat (cf. Brush [6]). In particular, if interactions are negligible (so that $\mathbf{f}_w=\mathbf{0}$) and $\boldsymbol{\mathcal{D}}_w$ is a pressure P_w (i.e., $\boldsymbol{\mathcal{D}}_w=P_w\mathbf{1}$), then taking the trace of (5.5.11) yields

$$3P_w(\mathbf{x},t)=\operatorname{tr}\{\boldsymbol{\mathcal{D}}_w(\mathbf{x},t)\}=\sum_i m_i\hat{\mathbf{v}}_i^2(t;\mathbf{x})w(\mathbf{x}_i(t)-\mathbf{x}) \tag{5.5.21}$$

$$=\left(\sum_{\substack{P_i\in S_\epsilon(\mathbf{x})\\ \text{at time } t}} m_i\hat{\mathbf{v}}_i^2(t;\mathbf{x})\right)V_\epsilon^{-1}. \tag{5.5.22}$$

Thus

$$P_w(\mathbf{x},t)V_\epsilon=\frac{1}{3}\sum_{\substack{P_i\in S_\epsilon(\mathbf{x})\\ \text{at time } t}} m_i\hat{\mathbf{v}}_i^2(t;\mathbf{x}). \tag{5.5.23}$$

Relation (5.5.23) corresponds to a rarefied gas (in which molecular interactions occur only through rare binary 'collisions' and $\mathbf{f}_w$ is negligible) in a state of pressure P_w. This relation may be compared with the ideal gas 'law'

$$PV = NkT \tag{5.5.24}$$

which pertains to a system of N identical, non-interacting molecules (each of mass m, say) in a state of macroscopic equilibrium ($\mathbf{v} = \mathbf{0}$) which occupies a region of volume V and gives rise to a pressure P at absolute temperature T. Under these conditions, $m_i = m$ and $\hat{\mathbf{v}}_i = \mathbf{v}_i$ in (5.5.23), so this relation is seen to be a local version of (5.5.24) upon identification of kT with two-thirds of the kinetic energy of molecules in $S_\epsilon(\mathbf{x})$ divided by the number of molecules in $S_\epsilon(\mathbf{x})$. Accordingly, $\mathcal{D}_w$ given by (5.5.11) can be regarded as a generalisation of (5.5.24) to local, non-static situations for moderately rarefied gaseous mixtures (since masses need not be identical). More generally, the Cauchy stress has been identified in (5.5.19) and (5.5.20) as having distinct contributions from molecular interactions, via $\mathbf{T}_w^-$ [see (5.5.18)], and from $\mathcal{D}_w$. The special case of a moderately rarefied gas has illustrated the essentially thermal nature of $\mathcal{D}_w$. Indeed, the kinetic theory of heat is based upon the chaotic nature of molecular trajectories, whether the material in question is gaseous, liquid, or solid. Such chaotic behaviour is obtained by considering individual molecular motion relative to the local macroscopic motion. Specifically, the *thermal velocity of a molecule* (associated with choice w of weighting function) is

$${}_w\tilde{\mathbf{v}}_i(t) := \mathbf{v}_i(t) - \mathbf{v}_w(\mathbf{x}_i(t), t). \tag{5.5.25}$$

For molecules in the vicinity of point $\mathbf{x}$, $\mathbf{v}_w(\mathbf{x}_i(t), t)$ is approximated by $\mathbf{v}_w(\mathbf{x}, t)$. Accordingly, $\hat{\mathbf{v}}_i(t; \mathbf{x})$ approximates ${}_w\tilde{\mathbf{v}}_i(t)$ for such molecules. We term $\hat{\mathbf{v}}_i(t; \mathbf{x})$ the *notional thermal velocity of* P_i *(with respect to location* $\mathbf{x}$*)*. Correspondingly, $\mathcal{D}_w$ is termed the *thermokinetic stress tensor* (associated with choice w).

Exercise 5.5.4. Note that $\mathcal{D}_w$ takes symmetric tensor values. (Why?) Show further that for $w = w_\epsilon$ [see (4.3.2)] $-\mathcal{D}_w$ is pressure-like in that

$$-\mathcal{D}_w \mathbf{n} \,.\, \mathbf{n} < 0 \tag{5.5.26}$$

unless $\hat{\mathbf{v}}_i = \mathbf{0}$ for all molecules. (Such a case corresponds to zero absolute temperature at which $\mathcal{D}_w = \mathbf{O}$.)

Remark 5.5.3. Definition (5.5.25) may be compared with its statistical mechanical analogue in which the random (or chaotic, or thermal) velocity of a molecule P_i is defined by (5.5.25) with ensemble average $\langle \mathbf{v} \rangle$ in place of spatial average $\mathbf{v}_w$.

5.5.3 Comparison of Contributions $\mathbf{T}_w^-$ and $\mathcal{D}_w$ to $\mathbf{T}_w$

We have seen that in a rarefied gas $\mathbf{T}_w^-$ is negligible, and hence the Cauchy stress $\mathbf{T}_w = -\mathcal{D}_w$. For dense gases $\mathbf{T}_w^-$ contributes to $\mathbf{T}_w$, but the dominant contribution is that of $-\mathcal{D}_w$. For liquids and solids both $\mathbf{T}_w^-$ and $-\mathcal{D}_w$ play significant roles. Consider, for example, a liquid or solid body which occupies a given region at some temperature. An increase in this temperature gives rise, in general,[4] to a volume increase in

[4] An anomaly here is ice near its freezing point. (Such anomaly is manifest in the floating of ice in water.)

the region occupied if no constraint is imposed. If the body is rigidly enclosed, then a temperature rise results in a stress upon the walls of the confining vessel which tends to increase its volume. [Such stress may not, strictly speaking, be a pressure in the case of anisotropic materials, but merely pressure-like in the sense of (5.5.26).] The source of such 'thermal' stress is precisely the contribution $-\mathcal{D}_w$ to $\mathbf{T}_w$. Such stresses can be very large, as exemplified by the buckling of railway track in hot weather if insufficient gaps are left between individual rails to allow for thermal expansion.

Remark 5.5.4. The pressure in a rarefied gas has been shown to be essentially of thermokinetic nature since in such case molecular interactions are negligible. Accordingly, such pressure cannot be regarded as a force per unit area. This resolves the conceptual problem C.P.4. of Section 3.9.

A solid may, if unconstrained, be 'stress-free' (i.e., $\mathbf{T}_w = \mathbf{O}$) over a range of temperatures. In such case $\mathcal{D}_w$ will be non-negligible and balanced by $\mathbf{T}_w^-$ in the sense that, at each such temperature,

$$\mathbf{T}_w^- = \mathcal{D}_w. \tag{5.5.27}$$

If a solid is in a state of 'uniaxial tension', so that

$$\alpha := \mathbf{Tn}.\mathbf{n} > 0 \tag{5.5.28}$$

for some unit vector $\mathbf{n}$, then from (5.5.20) and (5.5.26)

$$\mathbf{T}_w^-\mathbf{n}.\mathbf{n} - \mathcal{D}_w\mathbf{n}.\mathbf{n} = \alpha, \tag{5.5.29}$$

and hence

$$\mathbf{T}_w^-\mathbf{n}.\mathbf{n} > \alpha. \tag{5.5.30}$$

For solids in states of large uniaxial tension the contribution of $\mathbf{T}_w^-$ to α in (5.5.29) can far exceed that of $\mathcal{D}_w^-$.

5.6 Determination of Candidate Interaction Stress Tensors

5.6.1 Preamble

Interaction force density $\mathbf{f}_w$ is defined by (5.5.2). Here we examine solutions $\mathbf{T}_w^-$ to (5.5.18). Each solution is *a* candidate interaction stress tensor. As noted in Remark 5.5.1, non-uniqueness is to be expected.

5.6.2 Simple Form

On suppressing time dependence, (5.5.18) and (5.5.2) yield

$$(\mathrm{div}\,\mathbf{T}_w^-)(\mathbf{x}) = \sum\sum_{i \neq j} \mathbf{f}_{ij}\, w(\mathbf{x}_i - \mathbf{x}) = \mathbf{f}_w(\mathbf{x}). \tag{5.6.1}$$

Now suppose that $\mathbf{a}_i$ is a vector field which satisfies

$$(\mathrm{div}\,\mathbf{a}_i)(\mathbf{x}) = w(\mathbf{x}_i - \mathbf{x}). \tag{5.6.2}$$

Since $\mathbf{f}_{ij}$ does not depend upon $\mathbf{x}$, it follows that

$$\operatorname{div}\{\mathbf{f}_{ij} \otimes \mathbf{a}_i\} = \mathbf{f}_{ij} \operatorname{div} \mathbf{a}_i = \mathbf{f}_{ij}\, w. \tag{5.6.3}$$

Accordingly
$$\operatorname{div}\left\{ \sum\sum_{i \neq j} \mathbf{f}_{ij} \otimes \mathbf{a}_i \right\} = \sum\sum_{i \neq j} \mathbf{f}_{ij}\, w = \mathbf{f}_w, \tag{5.6.4}$$

and a *s*imple solution to (5.5.18) is

$${}_s\mathbf{T}_w^-(\mathbf{x},t) := \sum\sum_{i \neq j} \mathbf{f}_{ij}(t) \otimes \mathbf{a}_i(\mathbf{x},t). \tag{5.6.5}$$

5.6.3 Form for Pairwise-Balanced Interactions

If interactions are *pairwise balanced* (cf. Remark 5.4.1), that is

$$\mathbf{f}_{ji} = -\mathbf{f}_{ij}, \tag{5.6.6}$$

then (5.5.2) may be written, on suppressing time dependence, as

$$\begin{aligned}\mathbf{f}_w(\mathbf{x}) = \sum\sum_{i \neq j} \mathbf{f}_{ij}\, w(\mathbf{x}_i - \mathbf{x}) &= \frac{1}{2} \sum\sum_{i \neq j} \{\mathbf{f}_{ij}\, w(\mathbf{x}_i - \mathbf{x}) + \mathbf{f}_{ji}\, w(\mathbf{x}_j - \mathbf{x})\} \\ &= \frac{1}{2} \sum\sum_{i \neq j} \mathbf{f}_{ij} \{w(\mathbf{x}_i - \mathbf{x}) - w(\mathbf{x}_j - \mathbf{x})\}.\end{aligned} \tag{5.6.7}$$

Accordingly any vector field $\mathbf{b}_{ij}$ for which

$$(\operatorname{div} \mathbf{b}_{ij})(\mathbf{x}) = \frac{1}{2}\{w(\mathbf{x}_i - \mathbf{x}) - w(\mathbf{x}_j - \mathbf{x})\} \tag{5.6.8}$$

yields
$$\operatorname{div}\left\{ \sum\sum_{i \neq j} \mathbf{f}_{ij} \otimes \mathbf{b}_{ij} \right\} = \sum\sum_{i \neq j} \mathbf{f}_{ij} \operatorname{div} \mathbf{b}_{ij} = \mathbf{f}_w. \tag{5.6.9}$$

Thus
$${}_b\mathbf{T}_w^-(\mathbf{x},t) := \sum\sum_{i \neq j} \mathbf{f}_{ij}(t) \otimes \mathbf{b}_{ij}(\mathbf{x},t) \tag{5.6.10}$$

satisfies (5.5.18). Here prefix b indicates the interaction *b*alancing requirement.

For the remainder of this section explicit dependence of fields upon location $\mathbf{x}$ and time t will be omitted wherever and whenever possible, for brevity.

5.6.4 Simple Choice of $\mathbf{b}_{ij}$ for Pairwise-Balanced Interactions

Given any solution $\mathbf{a}_i$ to (5.6.2), it immediately follows that

$${}_s\mathbf{b}_{ij} := \frac{1}{2}(\mathbf{a}_i - \mathbf{a}_j) \tag{5.6.11}$$

is a *s*imple solution to (5.6.8). Hence a *s*imple candidate stress tensor consistent with *b*alancing property (5.6.6) is

$$ {}_{sb}\mathbf{T}_w^- := \sum_{i \neq j}\sum \mathbf{f}_{ij} \otimes \frac{1}{2}(\mathbf{a}_i - \mathbf{a}_j). \tag{5.6.12}$$

5.6.5 Hardy-Type Choice of $\mathbf{b}_{ij}$ for Pairwise-Balanced Interactions

Motivated by virial considerations (see Murdoch [18]), we seek solutions to (5.6.8) of the form

$$\mathbf{b}_{ij}(\mathbf{x}) = \frac{1}{2}\hat{b}_{ij}(\mathbf{x})(\mathbf{x}_j - \mathbf{x}_i), \tag{5.6.13}$$

where $\hat{b}_{ij}$ is scalar-valued. In such case (5.6.8) requires that

$$(\mathbf{x}_j - \mathbf{x}_i).\nabla \hat{b}_{ij}(\mathbf{x}) = w(\mathbf{x}_i - \mathbf{x}) - w(\mathbf{x}_j - \mathbf{x}). \tag{5.6.14}$$

A solution to (5.6.14) is *Hardy's bond function* (see Hardy [10])

$$\hat{b}_{ij}^H(\mathbf{x}) := \int_0^1 w(\lambda(\mathbf{x}_j - \mathbf{x}_i) + (\mathbf{x}_i - \mathbf{x}))d\lambda. \tag{5.6.15}$$

Exercise 5.6.1. Prove (5.6.15) by noting that

$$\nabla \hat{b}_{ij}^H(\mathbf{x}) = \int_0^1 -\nabla w(\lambda(\mathbf{x}_j - \mathbf{x}_i) + (\mathbf{x}_i - \mathbf{x}))d\lambda$$

and
$$\frac{\partial}{\partial \lambda}\{w(\lambda(\mathbf{x}_j - \mathbf{x}_i) + (\mathbf{x}_i - \mathbf{x}))\} = \nabla w(\lambda(\mathbf{x}_j - \mathbf{x}_i) + (\mathbf{x}_i - \mathbf{x})) \cdot (\mathbf{x}_j - \mathbf{x}_i).$$

Remark 5.6.1. Since

$$\lambda(\mathbf{x}_j - \mathbf{x}_i) + (\mathbf{x}_i - \mathbf{x}) \equiv \lambda(\mathbf{x}_j - \mathbf{x}) + (1 - \lambda)(\mathbf{x}_i - \mathbf{x}) \qquad \text{and} \qquad 0 < \lambda < 1,$$

contributions at time t to the integral in (5.6.15) derive solely from values of w computed for displacements between $\mathbf{x}$ and points on the line segment joining locations $\mathbf{x}_i(t)$ and $\mathbf{x}_j(t)$.

From (5.6.10) and (5.6.13) we obtain the *Hardy-type interaction stress tensor*

$${}_H\mathbf{T}_w^- := \frac{1}{2}\sum_{i \neq j}\sum \mathbf{f}_{ij} \otimes \hat{b}_{ij}^H(\mathbf{x}_j - \mathbf{x}_i). \tag{5.6.16}$$

Remark 5.6.2. If $\mathbf{f}_{ij}$ is parallel to $(\mathbf{x}_j - \mathbf{x}_i)$ at all times (note, however, Remark 5.4.3), then each term in (5.6.16) is symmetric and so ${}_H\mathbf{T}_w^-$ is a symmetric tensor field. Since $\mathcal{D}_w$ takes symmetric values [cf. Exercise (5.5.4)], the associated Hardy-type stress tensor field

$${}_H\mathbf{T}_w := {}_H\mathbf{T}_w^- - \mathcal{D}_w \tag{5.6.17}$$

is also symmetric.

5.6.6 Noll-Type Choice of $\mathbf{b}_{ij}$ for Pairwise-Balanced Interactions

Noll's Lemma 1 of [16] leads to another solution to (5.6.8), namely

$$\mathbf{b}_{ij}^N(\mathbf{x}) := -\frac{1}{2}\int_{\mathcal{V}}\int_0^1 \mathbf{u}w(\mathbf{x}_i - \mathbf{x} - \alpha\mathbf{u})w(\mathbf{x}_j - \mathbf{x} + (1-\alpha)\mathbf{u})d\alpha d\mathbf{u}. \tag{5.6.18}$$

This is proved to be a solution in Section 5.3 of Murdoch [18]. The conditions of Noll's lemma are satisfied provided that the weighting function has compact support or tends to zero faster than $\|\mathbf{u}\|^{-3}$ as $\|\mathbf{u}\| \to \infty$. [It is also necessary to note that $\mathbf{b}_{ji}^N = -\mathbf{b}_{ij}^N$. This follows from changing variables of integration in (5.6.18).] Relations (5.6.10) and (5.6.18) yield the *Noll-type interaction stress tensor*

$$_N\mathbf{T}_w^-(\mathbf{x}) := -\frac{1}{2}\sum_{i\neq j}\sum \mathbf{f}_{ij} \otimes \int_{\mathcal{V}}\int_0^1 \mathbf{u}w(\mathbf{x}_i - \mathbf{x} - \alpha\mathbf{u})w(\mathbf{x}_j - \mathbf{x} + (1-\alpha)\mathbf{u})d\alpha d\mathbf{u}. \tag{5.6.19}$$

5.6.7 Conclusions

To each *c*andidate interaction stress tensor $_c\mathbf{T}_w^-$ corresponds a candidate *C*auchy stress tensor $_c\mathbf{T}_w$ via [see (5.5.20)]

$$_c\mathbf{T}_w := {_c\mathbf{T}_w^-} - \mathcal{D}_w. \tag{5.6.20}$$

Clearly, each candidate $_c\mathbf{T}_w$ depends firstly upon the choice w of weighting function, and then upon solution $\mathbf{a}_i$ to (5.6.2) in the case of general interactions, and upon solution $\mathbf{b}_{ij}$ to (5.6.8) for pairwise-balanced interactions.

Since $\mathcal{D}_w$ takes symmetric values (cf. Exercise 5.5.4), the symmetry or otherwise of $_c\mathbf{T}_w$ depends upon whether or not $_c\mathbf{T}_w^-$ is symmetric. Here explicit forms (5.6.5), (5.6.12), (5.6.16), and (5.6.19) provide the verdict for the choices they represent. Said differently, considerations of linear momentum balance, together with choices of weighting function w and related fields $\mathbf{a}_i$ and $\mathbf{b}_{ij}$, yield explicit expressions for candidate interaction stress tensors whose symmetry or not can be checked (recall Remark 5.6.2). Standard texts on continuum mechanics often 'prove' symmetry of the Cauchy stress tensor by *postulating* a moment of momentum balance [cf. (2.7.3)] in which couple-stress $\mathbf{C}$, body couple density $\mathbf{J}$ and internal angular momentum density $\mathbf{S}$ are absent. Accordingly (2.7.29) yields the symmetry of stress $\mathbf{T}$. Unless the absence of $\mathbf{C},\mathbf{J}$ or $\mathbf{S}$ is noted (as in Truesdell & Noll [2] and Chadwick [3]), one can be misled into thinking that $\mathbf{T}$ is symmetric for all materials. This is manifestly incorrect because adequate modelling of liquid-crystalline phases requires the inclusion of couple stress and body couple density (cf., e.g., Carlsson & Leslie [19]). In Chapter 6 a balance of rotational momentum will be derived (from molecular considerations) in which, a priori, terms $\mathbf{C},\mathbf{J}$, and $\mathbf{S}$ appear naturally. Thus, from a general perspective, asymmetry of stress is to be expected, and the degree of such asymmetry depends upon the size of the effects represented by $\mathbf{C},\mathbf{J}$ and $\mathbf{S}$.

5.7 Calculation of Interaction Stresses for the Simplest Form of Weighting Function w_ϵ

5.7.1 Determination of $\mathbf{a}_i$ and Calculation of ${}_s\mathbf{T}^-_{w_\epsilon}$ and ${}_{sb}\mathbf{T}^-_{w_\epsilon}$

Here we seek a solution $\mathbf{a}_i$ to

$$\operatorname{div}\mathbf{a}_i = w(\mathbf{x}_i - \mathbf{x}) \tag{5.6.2}$$

for any (isotropic) weighting function w of the form

$$w(\mathbf{x}_i - \mathbf{x}) = \hat{w}(\|\mathbf{x}_i - \mathbf{x}\|) \tag{5.7.1}$$

and then consider the special case $w = w_\epsilon$. In such case it is natural to seek a solution $\mathbf{a}_i$ which is also isotropic in the sense that

$$\mathbf{a}_i(\mathbf{x}) = \hat{a}(u)\mathbf{u}, \tag{5.7.2}$$

where

$$\mathbf{u} := \mathbf{x}_i - \mathbf{x}. \tag{5.7.3}$$

Accordingly

$$\operatorname{div}\mathbf{a}_i = \nabla_{\mathbf{x}}\hat{a}(u)\,.\,\mathbf{u} + \hat{a}(u)\operatorname{div}_{\mathbf{x}}\mathbf{u}$$

$$= \hat{a}'(u)\nabla_{\mathbf{x}}u\,.\,\mathbf{u} + \hat{a}(u)\operatorname{div}_{\mathbf{x}}\mathbf{u}. \tag{5.7.4}$$

Now

$$\mathbf{u}\,.\,\mathbf{u} = u^2, \tag{5.7.5}$$

so

$$2(\nabla_{\mathbf{x}}\mathbf{u})^T\mathbf{u} = 2u\nabla_{\mathbf{x}}u. \tag{5.7.6}$$

However, from (5.7.3)

$$\nabla_{\mathbf{x}}\mathbf{u} = -\mathbf{1}, \tag{5.7.7}$$

so

$$\operatorname{div}_{\mathbf{x}}\mathbf{u} = -3, \tag{5.7.8}$$

and, from (5.7.6) and (5.7.7),

$$\nabla_{\mathbf{x}}u = -\frac{1}{u}\mathbf{u}. \tag{5.7.9}$$

Thus, from (5.7.4), (5.7.8), and (5.7.9),

$$\operatorname{div}\mathbf{a}_i = -u\hat{a}'(u) - 3\hat{a}(u) = -\frac{1}{u^2}\frac{d}{du}\{u^3\hat{a}(u)\}. \tag{5.7.10}$$

Accordingly, from (5.6.2) and (5.7.10),

$$-\frac{1}{u^2}\frac{d}{du}\{u^3\hat{a}(u)\} = \hat{w}(u), \tag{5.7.11}$$

and hence
$$u^3\hat{a}(u) = -\int_0^u s^2\hat{w}(s)ds. \tag{5.7.12}$$

In the case of the mollified weighting function $w_{\epsilon,\delta}$ given by relations (4.3.22) we have

$$u^3\hat{a}(u) = -\int_0^u \frac{s^2}{\alpha}ds = -\frac{u^3}{3\alpha} \quad \text{if} \quad 0 \le u \le \epsilon, \tag{5.7.13}$$

$$u^3\hat{a}(u) = -\int_0^\epsilon \frac{s^2}{\alpha}ds - \int_\epsilon^u \frac{s^2}{\alpha}\varphi\left(\frac{s-\epsilon}{\delta}\right)ds \quad \text{if} \quad \epsilon < u \le \epsilon+\delta, \tag{5.7.14}$$

and
$$u^3\hat{a}(u) = -\int_0^\epsilon \frac{s^2}{\alpha}ds - \int_\epsilon^{\epsilon+\delta} \frac{s^2}{\alpha}\varphi\left(\frac{s-\epsilon}{\delta}\right)ds \quad \text{if} \quad u > \epsilon+\delta. \tag{5.7.15}$$

Recall that in relations (4.3.22) φ is monotonic decreasing on $[0,1], \varphi(0) = 1, \varphi(1) = 0$, and $\varphi(\lambda) = 0$ for $\lambda > 1$. It follows that for any $u > \epsilon$ we have

$$\left|u^{-3}\int_\epsilon^u s^2\varphi\left(\frac{s-\epsilon}{\delta}\right)ds\right| < \epsilon^{-3}\left|\int_\epsilon^{\epsilon+\delta} s^2.1\,ds\right| = \frac{\delta}{\epsilon}\left\{1+\frac{\delta}{\epsilon}+\frac{1}{3}\left(\frac{\delta}{\epsilon}\right)^2\right\}. \tag{5.7.16}$$

Exercise 5.7.1. Check the details which led to (5.7.16).

In Subsection 4.3.5 the values of $\rho_{\epsilon,\delta}$ and $\mathbf{p}_{\epsilon,\delta}$ were identified with those of ρ_ϵ and $\mathbf{p}_\epsilon$ upon choosing δ to be very small and, in particular, much smaller than ϵ (see Result 4.3.3 and Remark 4.3.8). Such choice also resulted in the value of α being essentially V_ϵ ([$= (4\pi\epsilon^3/3)$; see Result 4.3.2]. Further, such a choice of δ renders [from (5.7.16)] negligible the integrals involving φ in (5.7.14) and (5.7.15) in comparison with α^{-1}. Accordingly from (5.7.13), (5.7.14), and (5.7.15) we have, consistent with the foregoing identifications, that essentially

$$\hat{a}(u) = -\frac{1}{4\pi\epsilon^3} \quad \text{if} \quad 0 \le u \le \epsilon \tag{5.7.17}$$

and
$$\hat{a}(u) = -\frac{1}{4\pi u^3} \quad \text{if} \quad u > \epsilon. \tag{5.7.18}$$

It follows from (5.6.5) and (5.7.2) that the corresponding stress tensor ${}_s\mathbf{T}_w$ is given (on suppressing time dependence) by

$$\begin{aligned} {}_s\mathbf{T}^-_{w_\epsilon}(\mathbf{x}) &= \sum_{i\neq j}\sum \mathbf{f}_{ij} \otimes \mathbf{a}_i(\mathbf{x}) \\ &= \sum_{i\neq j}\sum \mathbf{f}_{ij} \otimes \hat{a}_i(\|\mathbf{x}_i - \mathbf{x}\|)(\mathbf{x}_i - \mathbf{x}) && (5.7.19) \\ &= -\frac{1}{4\pi\epsilon^3}\sum_{P_i\in S_\epsilon(\mathbf{x})}\sum_{j\neq i} \mathbf{f}_{ij} \otimes (\mathbf{x}_i - \mathbf{x}) - \frac{1}{4\pi}\sum_{P_i\notin S_\epsilon(\mathbf{x})}\sum_{j\neq i} \mathbf{f}_{ij} \otimes \frac{(\mathbf{x}_i - \mathbf{x})}{\|\mathbf{x}_i - \mathbf{x}\|^3}. && (5.7.20) \end{aligned}$$

Similarly, from (5.6.12), (5.7.2), (5.7.17), and (5.7.18),

$$ {}_{sb}\mathbf{T}_w^-(\mathbf{x}) = \sum\sum_{i\neq j} \mathbf{f}_{ij} \otimes \frac{1}{2}\{\hat{a}_i(\mathbf{x})(\mathbf{x}_i-\mathbf{x}) - \hat{a}_j(\mathbf{x})(\mathbf{x}_j-\mathbf{x})\} \tag{5.7.21} $$

$$ \begin{aligned} &= -\frac{1}{8\pi\epsilon^3} \sum\sum_{\substack{P_i,P_j\in S_\epsilon(\mathbf{x})\\ j\neq i}} \mathbf{f}_{ij}\otimes(\mathbf{x}_i-\mathbf{x}_j) \\ &\quad -\frac{1}{8\pi} \sum_{P_i\in S_\epsilon(\mathbf{x})}\sum_{P_j\notin S_\epsilon(\mathbf{x})} \mathbf{f}_{ij}\otimes\left\{\frac{(\mathbf{x}_i-\mathbf{x})}{\epsilon^3} - \frac{(\mathbf{x}_j-\mathbf{x})}{\|\mathbf{x}_j-\mathbf{x}\|^3}\right\} \\ &\quad -\frac{1}{8\pi} \sum_{P_i\notin S_\epsilon(\mathbf{x})}\sum_{P_j\in S_\epsilon(\mathbf{x})} \mathbf{f}_{ij}\otimes\left\{\frac{(\mathbf{x}_i-\mathbf{x})}{\|\mathbf{x}_i-\mathbf{x}\|^3} - \frac{(\mathbf{x}_j-\mathbf{x})}{\epsilon^3}\right\} \\ &\quad -\frac{1}{8\pi} \sum\sum_{\substack{P_i,P_j\notin S_\epsilon(\mathbf{x})\\ j\neq i}} \mathbf{f}_{ij}\otimes\left\{\frac{(\mathbf{x}_i-\mathbf{x})}{\|\mathbf{x}_i-\mathbf{x}\|^3} - \frac{(\mathbf{x}_j-\mathbf{x}}{\|\mathbf{x}_j-\mathbf{x}\|^3}\right\}. \end{aligned} \tag{5.7.22} $$

Remark 5.7.1. If $\mathbf{f}_{ij}$ is parallel to $(\mathbf{x}_j-\mathbf{x}_i)$, as would be the case if such an interaction were to depend only upon the locations of P_i and P_j, then the first double sum in (5.7.22) is symmetric. However, the presence of the remaining terms indicates that ${}_{sb}\mathbf{T}_w^-$ is not to be expected a priori to be symmetric.

5.7.2 Determination of $\hat{b}_{ij}^H$ and Calculation of ${}_H\mathbf{T}_{w_\epsilon}^-$

If $w = w_\epsilon$, then (5.6.16) becomes

$$ {}_H\mathbf{T}_{w_\epsilon}^-(\mathbf{x}) = \frac{1}{2}\sum\sum_{i\neq j} \mathbf{f}_{ij}\otimes \hat{b}_{ij}^H(\mathbf{x})(\mathbf{x}_j-\mathbf{x}_i), \tag{5.7.23} $$

where, from (5.6.15),

$$ \hat{b}_{ij}^H(\mathbf{x}) := \int_0^1 w_\epsilon(\lambda(\mathbf{x}_j-\mathbf{x}_i)+(\mathbf{x}_i-\mathbf{x}))d\lambda. \tag{5.7.24} $$

From Remark 5.6.1 and the definition of w_ϵ, the only contributions to this integral come from displacements with magnitudes less than ϵ which are directed from $\mathbf{x}$ to points on the line segment which joins $\mathbf{x}_i$ to $\mathbf{x}_j$. Thus such contributions derive from points on the line joining $\mathbf{x}_i$ to $\mathbf{x}_j$ which lie both between these points *and* within $S_\epsilon(\mathbf{x})$. If the length of that segment of the line joining $\mathbf{x}_i$ to $\mathbf{x}_j$ which lies within $S_\epsilon(\mathbf{x})$ is denoted by $\ell_{ij}(\mathbf{x})$, then

$$ \hat{b}_{ij}^H(\mathbf{x}) = \frac{\ell_{ij}(\mathbf{x})}{\|\mathbf{x}_i-\mathbf{x}_j\|}\cdot\frac{1}{V_\epsilon}. \tag{5.7.25} $$

Accordingly (5.7.23) simplifies to

$$ {}_H\mathbf{T}_{w_\epsilon}^- = \frac{1}{2V_\epsilon}\sum\sum_{i\neq j} \mathbf{f}_{ij}\otimes\alpha_{ij}(\mathbf{x})(\mathbf{x}_j-\mathbf{x}_i), \tag{5.7.26} $$

where
$$\alpha_{ij}(\mathbf{x}) := \frac{\ell_{ij}(\mathbf{x})}{\|\mathbf{x}_i - \mathbf{x}_j\|}. \tag{5.7.27}$$

Of course, $\alpha_{ij}(\mathbf{x})(\mathbf{x}_j - \mathbf{x}_i)$ denotes that portion of the displacement from $\mathbf{x}_i$ to $\mathbf{x}_j$ which lies within $S_\epsilon(\mathbf{x})$.

Exercise 5.7.2. Sketch situations in which (i) $\mathbf{x}_i$ and $\mathbf{x}_j$ both lie outside $S_\epsilon(\mathbf{x})$ and the line joining these intersects $S_\epsilon(\mathbf{x})$, (ii) $\mathbf{x}_i \in S_\epsilon(\mathbf{x})$ and $\mathbf{x}_j \notin S_\epsilon(\mathbf{x})$, and (iii) $\mathbf{x}_i$ and $\mathbf{x}_j$ lie inside $S_\epsilon(\mathbf{x})$. Indicate the line segment of length $\ell_{ij}(\mathbf{x})$ in each case.

Remark 5.7.2. The *net* contribution to ${}_H\mathbf{T}^-_{w_\epsilon}(x)$ from P_i and P_j is

$$\frac{1}{V_\epsilon}\mathbf{f}_{ij} \otimes \alpha_{ij}(\mathbf{x})(\mathbf{x}_j - \mathbf{x}_i). \tag{5.7.28}$$

5.7.3 The Geometrical Complexity of $\mathbf{b}^N_{ij}$

If $w = w_\epsilon$, then, from (5.6.18) and (5.6.19),

$${}_N\mathbf{T}^-_{w_\epsilon} = -\frac{1}{2}\sum_{i\neq j}\sum \mathbf{f}_{ij} \otimes \mathbf{b}^N_{ij}, \tag{5.7.29}$$

where
$$\mathbf{b}^N_{ij}(\mathbf{x}) = -\frac{1}{2}\int_{\mathcal{V}}\int_0^1 \mathbf{u} w_\epsilon(\mathbf{x}_i - [\mathbf{x} + \alpha\mathbf{u}]) w_\epsilon(\mathbf{x}_j - [\mathbf{x} - (1-\alpha)\mathbf{u}])\,d\alpha\, d\mathbf{u}. \tag{5.7.30}$$

Non-zero contributions derive from any point $\mathbf{x} + \alpha\mathbf{u}$ within a distance ϵ of $\mathbf{x}_i$ for which point $\mathbf{x} - (1-\alpha)\mathbf{u}$ lies within a distance ϵ of $\mathbf{x}_j$. Since points $\mathbf{x} + \alpha\mathbf{u}$ and $\mathbf{x} - (1-\alpha)\mathbf{u}$ lie on a line segment parallel to $\mathbf{u}$ of length $\|\mathbf{u}\|$ which contains $\mathbf{x}$, it follows that this line segment should intersect both $S_\epsilon(\mathbf{x}_i)$ *and* $S_\epsilon(\mathbf{x}_j)$. In particular, we can deduce that $\mathbf{b}^N_{ij}(\mathbf{x})$ will vanish if no such line segment exists. This would be the case if no line through $\mathbf{x}$ can be found which is common to the two solid cones $\mathcal{C}_i(\mathbf{x})$ and $\mathcal{C}_j(\mathbf{x})$, where $\mathcal{C}_i(\mathbf{x})$ has vertex at $\mathbf{x}$ and generators tangential to the surface of sphere $S_\epsilon(\mathbf{x}_i)$, and similarly for $\mathcal{C}_j(\mathbf{x})$. The arbitrary nature of $\mathbf{u}$ and the independent values of $\alpha \in [0,1]$ imply that there *will* be contributions to $\mathbf{b}^N_{ij}(\mathbf{x})$ for *each* line through $\mathbf{x}$ which passes through *both* cones. These contributions derive from the intersection of any such line with each of $S_\epsilon(\mathbf{x}_i)$ and $S_\epsilon(\mathbf{x}_j)$. The geometrical complexity involved here makes any general calculation of $\mathbf{b}^N_{ij}(\mathbf{x})$ difficult.

Remark 5.7.3. In the limit as $\epsilon \to 0$ the only contributions to the corresponding interaction stress at $\mathbf{x}$ derive from molecules located at points $\mathbf{x}_i$ and $\mathbf{x}_j$, where $\mathbf{x}_i, \mathbf{x}_j$, and $\mathbf{x}$ are collinear, and each such contribution is parallel to $\mathbf{x}_i - \mathbf{x}_j$.

5.8 Comparison of Interaction Stress Tensors for the Simplest Form of Weighting Function w_ϵ

5.8.1 Values for Two Simple Geometries

If $\mathbf{x}_i = \mathbf{x} - \epsilon\mathbf{e}$ and $\mathbf{x}_j = \mathbf{x} + \epsilon\mathbf{e}$, where $\|\mathbf{e}\| = 1$, then, from (5.7.17), $\hat{a}(\|\mathbf{x}_i - \mathbf{x}\|) = -(4\pi\epsilon^3)^{-1} = \hat{a}(\|\mathbf{x}_j - \mathbf{x}\|)$, and the contribution of P_i to ${}_s\mathbf{T}^-_{w_\epsilon}(\mathbf{x})$ is, from

(5.6.5) and (5.7.2), $(4\pi\epsilon^2)^{-1}\sum_{k\neq i}\mathbf{f}_{ik}\otimes\mathbf{e}$. In respect of ${}_{sb}\mathbf{T}^-_{w_\epsilon}(\mathbf{x})$ the *net* contribution from P_i *and* P_j is, from (5.6.12) and (5.7.2), $(2\pi\epsilon^2)^{-1}\mathbf{f}_{ij}\otimes\mathbf{e}$. The net contribution of P_i and P_j to ${}_H\mathbf{T}^-_{w_\epsilon}(\mathbf{x})$ is, from (5.7.28) and (5.7.27), $3(2\pi\epsilon^2)^{-1}\mathbf{f}_{ij}\otimes\mathbf{e}$. A non-trivial calculation yields the corresponding net contribution to ${}_N\mathbf{T}^-_{w_\epsilon}(\mathbf{x})$ as $7(2\pi\epsilon^2)^{-1}\mathbf{f}_{ij}\otimes\mathbf{e}$.

If $\mathbf{e}_1$ and $\mathbf{e}_2$ denote two unit orthogonal vectors, and $\mathbf{x}_i = \mathbf{x}+\epsilon\mathbf{e}_1$, $\mathbf{x}_j = \mathbf{x}+\epsilon\mathbf{e}_2$, then the P_i contribution to ${}_s\mathbf{T}^-_{w_\epsilon}(\mathbf{x})$ is $-(4\pi\epsilon^2)^{-1}\sum_{k\neq i}\mathbf{f}_{ik}\otimes\mathbf{e}_1$. The net contributions of P_i and P_j to ${}_{sb}\mathbf{T}^-_{w_\epsilon}(\mathbf{x})$, ${}_H\mathbf{T}^-_{w_\epsilon}(\mathbf{x})$ and ${}_N\mathbf{T}^-_{w_\epsilon}(\mathbf{x})$ turn out to be $(4\pi\epsilon^2)^{-1}\mathbf{f}_{ij}\otimes(\mathbf{e}_2-\mathbf{e}_1)$, $3(4\pi\epsilon^2)^{-1}\mathbf{f}_{ij}\otimes(\mathbf{e}_2-\mathbf{e}_1)$ and $(2\pi\epsilon^2)^{-1}\mathbf{f}_{ij}\otimes(\mathbf{e}_2-\mathbf{e}_1)$, respectively.

Exercise 5.8.1. Prove the preceding assertions for ${}_s\mathbf{T}_{w_\epsilon}$, ${}_{sb}\mathbf{T}_{w_\epsilon}$, and ${}_H\mathbf{T}^-_{w_\epsilon}$.

Remark 5.8.1. It is clear from the preceding results that all three interaction stress tensors for pairwise-balanced interactions are distinct. If $\mathbf{f}_{ij}$ is parallel to $(\mathbf{x}_j-\mathbf{x}_i)$, then, in the preceding simple geometries, the net contributions from P_i and P_j to these stresses are symmetric. However, with the exception of ${}_H\mathbf{T}^-_w$ (which is symmetric for *any* choice of w), such 'central' forces do not yield symmetric values of ${}_{sb}\mathbf{T}^-_{w_\epsilon}$ and ${}_N\mathbf{T}_{w_\epsilon}$ in general. For example, the net contribution to ${}_{sb}\mathbf{T}^-_{w_\epsilon}(\mathbf{x})$ from P_i and P_j if $P_i\in S_\epsilon(\mathbf{x})$ and $P_j\notin S_\epsilon(\mathbf{x})$ is

$$-\frac{1}{4\pi}\mathbf{f}_{ij}\otimes\left(\frac{(\mathbf{x}_i-\mathbf{x})}{\epsilon^3}-\frac{(\mathbf{x}_j-\mathbf{x})}{\|\mathbf{x}_j-\mathbf{x}\|^3}\right).$$

If $\mathbf{x}_i-\mathbf{x}=-\epsilon\,\mathbf{e}_2$, $\mathbf{x}_j-\mathbf{x}=2\epsilon(\mathbf{e}_1+\mathbf{e}_2)$ and $\mathbf{f}_{ij}$ is parallel to $\mathbf{x}_j-\mathbf{x}_i$, then this net contribution is not symmetric. (Show this!)

5.8.2 Integration over Planar Surfaces

Although pointwise values of $\mathbf{T}^-_w$ and $\mathbf{T}_w$ are of interest because of constitutive considerations (usually couched in terms of the dependence of the Cauchy stress $\mathbf{T}_w$ upon measures of distortion and temperature), it is surface integrals of form

$$\mathrm{Surf}(\mathbf{T}_w,S,\mathbf{n}) := \int_S \mathbf{T}_w\mathbf{n}\,dA \tag{5.8.1}$$

which have the most physical interpretation (cf. Section 2.7). Here S is a smooth surface with a unit normal field $\mathbf{n}$. In this connection we examine the integrals of ${}_{sb}\mathbf{T}^-_{w_\epsilon}$, ${}_H\mathbf{T}^-_{w_\epsilon}$, and ${}_N\mathbf{T}^-_{w_\epsilon}$ over

$$S = \Pi_{\mathbf{n}}(\mathbf{x}_0), \tag{5.8.2}$$

where $\mathbf{x}_0\in\mathcal{E}$, $\mathbf{n}$ is a unit vector, and

$$\Pi_{\mathbf{n}}(\mathbf{x}_0) := \{\mathbf{x}\in\mathcal{E} : (\mathbf{x}-\mathbf{x}_0).\mathbf{n}=0\}. \tag{5.8.3}$$

That is, $\Pi_{\mathbf{n}}(\mathbf{x}_0)$ is that plane through point $\mathbf{x}_0$ to which $\mathbf{n}$ is a unit normal. Specifically, we consider *net* contributions from particle pairs P_i and P_j to $\mathrm{Surf}(\mathbf{T}^-_{w_\epsilon},S,\mathbf{n})$ with each of the three choices of $\mathbf{T}^-_{w_\epsilon}$ associated with pairwise-balanced interactions. These net contributions will be denoted by ${}_{sb}\mathbf{c}_{ij}$, ${}_H\mathbf{c}_{ij}$, and ${}_N\mathbf{c}_{ij}$.

Consider ${}_H\mathbf{c}_{ij}$. From (5.7.23) and (5.7.24)

$${}_H\mathbf{T}^-_{w_\epsilon}(\mathbf{x}) = \sum\sum_{i\neq j}\mathbf{f}_{ij}\otimes\mathbf{b}_{ij}(\mathbf{x}), \tag{5.8.4}$$

where, from (5.6.13) and (5.6.15),

$$\mathbf{b}_{ij}(\mathbf{x}) = \frac{1}{2}(\mathbf{x}_j - \mathbf{x}_i)\hat{b}_{ij}^H(\mathbf{x}) \tag{5.8.5}$$

with

$$\hat{b}_{ij}^H(\mathbf{x}) := \int_0^1 w_\epsilon(\lambda(\mathbf{x}_j - \mathbf{x}_i) + (\mathbf{x}_i - \mathbf{x}))d\lambda. \tag{5.8.6}$$

Notice that from (5.6.8)

$$\mathbf{b}_{ji} = -\mathbf{b}_{ij}, \tag{5.8.7}$$

so from (5.8.5)

$$\hat{b}_{ji}^H = \hat{b}_{ij}^H. \tag{5.8.8}$$

Exercise 5.8.2. Show that (5.8.8) also can be proved directly from (5.8.6) by a change of variable to $\mu := 1 - \lambda$.

It follows from (5.8.4), (5.8.5), (5.8.6), and (5.8.8) that the *net* contribution of P_i and P_j to ${}_H\mathbf{T}^-_{w_\epsilon}(\mathbf{x})\mathbf{n}$ is

$$\mathbf{f}_{ij}((\mathbf{x}_j - \mathbf{x}_i).\mathbf{n}) \int_0^1 w_\epsilon([\mathbf{x}_i + \lambda(\mathbf{x}_j - \mathbf{x}_i)] - \mathbf{x})d\lambda, \tag{5.8.9}$$

where, without loss of generality, we may assume that

$$(\mathbf{x}_j - \mathbf{x}_i).\mathbf{n} \geq 0. \tag{5.8.10}$$

[Convince yourself of the generality of assumption (5.8.10).] Note that for $0 \leq \lambda \leq 1$ the point

$$\mathbf{y}(\lambda) := \mathbf{x}_i + \lambda(\mathbf{x}_j - \mathbf{x}_i) \tag{5.8.11}$$

lies on the line segment which joins $\mathbf{x}_i$ and $\mathbf{x}_j$. Thus the contribution to $\mathrm{Surf}({}_H\mathbf{T}^-_{w_\epsilon}, S, \mathbf{n})$ of P_i and P_j is, from (5.8.9) and (5.8.11),

$$\begin{aligned} {}_H\mathbf{c}_{ij} &= \mathbf{f}_{ij}((\mathbf{x}_j - \mathbf{x}_i).\mathbf{n}) \int_S \int_0^1 w_\epsilon(\mathbf{y}(\lambda) - \mathbf{x})d\lambda\, dA_\mathbf{x} \\ &= \mathbf{f}_{ij}((\mathbf{x}_j - \mathbf{x}_i).\mathbf{n}) \int_0^1 \int_S w_\epsilon(\mathbf{x} - \mathbf{y}(\lambda))dA_\mathbf{x}\, d\lambda, \end{aligned} \tag{5.8.12}$$

on changing the order of integration and noting $w_\epsilon(\mathbf{y}(\lambda) - \mathbf{x}) = w_\epsilon(\mathbf{x} - \mathbf{y}(\lambda))$. Now $w_\epsilon(\mathbf{x} - \mathbf{y}(\lambda))$ vanishes unless $\mathbf{x}$ lies within a distance ϵ of $\mathbf{y}(\lambda)$, in which case it has value V_ϵ^{-1}. Accordingly the only non-zero contributions to the integral over S come from points $\mathbf{y}(\lambda)$ within a distance ϵ of S. Let

$$z(\lambda) := (\mathbf{y}(\lambda) - \mathbf{x}_0).\mathbf{n} \tag{5.8.13}$$

and

$$z_i := (\mathbf{x}_i - \mathbf{x}_0).\mathbf{n} \quad \text{and} \quad z_j := (\mathbf{x}_j - \mathbf{x}_0).\mathbf{n}. \tag{5.8.14}$$

Thus

$$(\mathbf{x}_j - \mathbf{x}_i).\mathbf{n} = z_j - z_i, \tag{5.8.15}$$

and from (5.8.11), (5.8.13), and (5.8.14),

$$z(\lambda) = z_i + \lambda(z_j - z_i). \tag{5.8.16}$$

Hence (5.8.12) may be written, on changing variables from λ to z, as

$$_H\mathbf{c}_{ij} = \mathbf{f}_{ij} \int_{z_i}^{z_j} \int_S w_\epsilon(\mathbf{x} - \hat{\mathbf{y}}(z)) dA_\mathbf{x}\, dz, \tag{5.8.17}$$

with $\hat{\mathbf{y}}(z)$ that point $\mathbf{y}$ on the line segment joining $\mathbf{x}_i$ and $\mathbf{x}_j$ for which

$$(\mathbf{y} - \mathbf{x}_0).\mathbf{n} = z. \tag{5.8.18}$$

Points $\mathbf{x}$ contribute to the S integral if and only if $\hat{\mathbf{y}}(z)$ lies within a distance ϵ of S; that is, if and only if

$$-\epsilon < z < \epsilon. \tag{5.8.19}$$

Further, if z satisfies (5.8.19) and $\|\mathbf{x} - \hat{\mathbf{y}}(\mathbf{z})\| < \epsilon$, then the integrand is V_ϵ^{-1} (and is otherwise zero) and in this case the area of S involved is that of the intersection of $S_\epsilon(\mathbf{y})$ with S. This intersection is a circle of radius $(\epsilon^2 - z^2)^{1/2}$. Hence

$$A(z) := \int_S w_\epsilon(\mathbf{x} - \hat{\mathbf{y}}(z)) dA_\mathbf{x} = \frac{\pi(\epsilon^2 - z^2)}{V_\epsilon} \quad \text{if} \quad -\epsilon < z < \epsilon, \tag{5.8.20}$$

and

$$A(z) = 0 \quad \text{if} \quad |z| \geq \epsilon. \tag{5.8.21}$$

It follows from (5.8.17) that

$$_H\mathbf{c}_{ij} = \mathbf{f}_{ij} \int_{z_i}^{z_j} A(z) dz = \mathbf{f}_{ij} \left(\int_{\max(z_i,-\epsilon)}^{\min(z_j,\epsilon)} \pi(\epsilon^2 - z^2) dz \right) V_\epsilon^{-1}. \tag{5.8.22}$$

At this point it might seem natural to explore the different possibilities concerning the locations of P_i and P_j.

Exercise 5.8.3. Show that if $z_j > \epsilon$ and $z_i < -\epsilon$, then $_H\mathbf{c}_{ij} = \mathbf{f}_{ij}$. Show also that if $z_i = z_j$, then $_H\mathbf{c}_{ij} = \mathbf{0}$ for *all* possibilities [cf. (5.8.17)].

There is, however, an immediate geometrical interpretation of the integral in (5.8.22): it is the volume of that portion of a sphere of radius ϵ with centre on $z = 0$ which lies between the planes $z = \max(z_i, -\epsilon)$ and $z = \min(z_j, \epsilon)$. Further, if (here $k = i$ or j) V_k^- denotes the volume of that part of $S_\epsilon(\mathbf{x}_k)$ which lies 'below' S and $V_k^+ = V_\epsilon - V_k^-$ is the volume of that part of $S_\epsilon(\mathbf{x}_k)$ which lies 'above' S, then for all possible locations of $\mathbf{x}_i$ and $\mathbf{x}_j$ we have

$$_H\mathbf{c}_{ij} = \left(\frac{V_j^+ - V_i^+}{V_\epsilon} \right) \mathbf{f}_{ij} = \left(\frac{V_i^- - V_j^-}{V_\epsilon} \right) \mathbf{f}_{ij}. \tag{5.8.23}$$

Figure 5.2. depicts a situation in which $-\epsilon < z_i < z_j < \epsilon$, and consequently the multiple of $\mathbf{f}_{ij}$ involved in $_H\mathbf{c}_{ij}$ is

$$(V_\epsilon - V_i^+ - V_j^-)V_\epsilon^{-1} = ((V_\epsilon - V_j^-) - V_i^+)V_\epsilon^{-1} = (V_j^+ - V_i^+)V_\epsilon^{-1}.$$

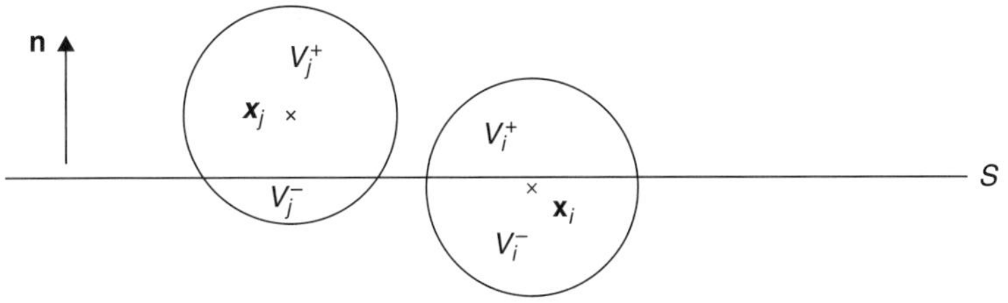

Figure 5.2. The contribution of molecules P_i and P_j to the value of the integral of ${}_H\mathbf{T}^-_{w_\epsilon}$ over S is $c_{ij}\mathbf{f}_{ij}$, where $c_{ij} = (V_j^+ - V_i^+)/V_\epsilon = (V_i^- - V_j^-)/V_\epsilon$.

Exercise 5.8.4. Consider all other possibilities (namely, $z_i < -\epsilon$ and $z_j < -\epsilon$; $z_i < -\epsilon$ and $-\epsilon < z_j < \epsilon$; $z_i < -\epsilon$ and $z_j > \epsilon$; $-\epsilon < z_i < \epsilon$ and $z_j > \epsilon$), and convince yourself that (5.8.23) holds.

Calculation of ${}_{sb}\mathbf{c}_{ij}$ and ${}_H\mathbf{c}_{ij}$, the comparable net contributions of P_i and P_j to $\mathrm{Surf}({}_{sb}\mathbf{T}^-_{w_\epsilon}, S, \mathbf{n})$ and $\mathrm{Surf}({}_N\mathbf{T}^-_{w\epsilon}, S, \mathbf{n})$, is rather tedious. Full details can be found in Murdoch [18] in respect of ${}_{sb}\mathbf{c}_{ij}$ and in Murdoch [20] for ${}_N\mathbf{c}_{ij}$. *What is very much a surprise, a priori, is that*

$$ {}_{sb}\mathbf{c}_{ij} = {}_H\mathbf{c}_{ij} = {}_N\mathbf{c}_{ij}, \tag{5.8.24}$$

and hence we have the following

Results 5.8.1.

$$ \int_S {}_{sb}\mathbf{T}^-_{w_\epsilon}\,\mathbf{n}\,dS = \int_S {}_H\mathbf{T}^-_{w_\epsilon}\,\mathbf{n}\,dS = \int_S {}_N\mathbf{T}^-_{w_\epsilon}\,\mathbf{n}\,dS \tag{5.8.25}$$

$$ = \sum_{\substack{i \neq j \\ (\mathbf{x}_j - \mathbf{x}_i).\mathbf{n} \geq 0}} \sum \left(\frac{V_j^+ - V_i^+}{V_\epsilon} \right) \mathbf{f}_{ij}. \tag{5.8.26}$$

Completely unexpected, equalities (5.8.25) appear to constitute an amazing coincidence. However, the geometrical interpretation (5.8.26) gives a hint that something more general or fundamental might lie behind these results. The appearance of volume ratios brings to mind the considerations of Subsection 4.3.3 and motivates, in particular, investigation of the integral of interaction force density $\mathbf{f}_w$ over a region R.

5.9 Integrals of General Interaction Stress Tensors over the Boundaries of Regular Regions

5.9.1 Results for a General Choice of Weighting Function

In the manner of Subsection 4.3.3, we consider the integral of the interaction force density $\mathbf{f}_w$ over a regular region[5] R with boundary ∂R and outward unit normal $\mathbf{n}$. For any candidate interaction stress tensor $\mathbf{T}^-_w$ which corresponds to choice w of

[5] That is, a bounded closed region for which the divergence theorem holds: see Kellogg [21], p. 113.

weighting function [see (5.5.18)],

$$\int_{\partial R} \mathbf{T}_w^- \mathbf{n}\, dA = \int_R \operatorname{div} \mathbf{T}_w^-\, dA = \int_R \mathbf{f}_w\, dV. \tag{5.9.1}$$

Accordingly the value of the surface integral is independent of which candidate interaction stress tensor is involved. (Why?) Further, from (5.5.2) this value is, on suppressing time dependence,

$$\int_R \mathbf{f}_w\, dV = \int_R \sum\sum_{i \neq j} \mathbf{f}_{ij}\, w(\mathbf{x}_i - \mathbf{x}) dV_{\mathbf{x}} \tag{5.9.2}$$

$$= \sum\sum_{i \neq j} v_i^w(R) \mathbf{f}_{ij}, \tag{5.9.3}$$

where

$$v_i^w(R) := \int_R w(\mathbf{x}_i - \mathbf{x}) dV_{\mathbf{x}}. \tag{5.9.4}$$

Thus we have, from (5.9.1) and (5.9.3), the following

Result 5.9.1.

$$\int_{\partial R} \mathbf{T}_w^- \mathbf{n}\, dA = \sum\sum_{i \neq j} v_i^w(R) \mathbf{f}_{ij}. \tag{5.9.5}$$

In order to link (5.9.4) with (5.8.25) and (5.8.26) it is necessary to choose a region R which has a boundary with a planar component. Consider the two hemispherical regions formed by the intersection of plane $\Pi_{\mathbf{n}_0}(\mathbf{x}_0)$ [see (5.8.3)] with the spherical region $S_r(\mathbf{x}_0)$ [see (4.3.1)]. (Here $\mathbf{n}_0$ is a unit vector and $\mathbf{x}_0$ an arbitrary point.) In particular, let

$$H_r^-(\mathbf{x}_0; \mathbf{n}_0) := \{\mathbf{x} \in S_r(\mathbf{x}_0) : (\mathbf{x} - \mathbf{x}_0).\mathbf{n}_0 \leq 0\}. \tag{5.9.6}$$

The boundary of H_r^- is

$$\partial H_r^- = \Pi_r \cup S_r^-, \tag{5.9.7}$$

where

$$\Pi_r(\mathbf{x}_0; \mathbf{n}_0) := \{\mathbf{x} \in \mathcal{E} : (\mathbf{x} - \mathbf{x}_0).\mathbf{n}_0 = 0 \quad \text{and} \quad \|\mathbf{x} - \mathbf{x}_0\| < r\}, \tag{5.9.8}$$

and

$$S_r^-(\mathbf{x}_0; \mathbf{n}_0) := \{\mathbf{x} \in \mathcal{E} : (\mathbf{x} - \mathbf{x}_0).\mathbf{n}_0 \leq 0 \quad \text{and} \quad \|\mathbf{x} - \mathbf{x}_0\| = r\}. \tag{5.9.9}$$

(See Figure 5.3.) It follows from (5.9.7) that

$$\int_{\partial H_r^-(\mathbf{x}_0; \mathbf{n}_0)} \mathbf{T}_w^- \mathbf{n}\, dA = \int_{\Pi_r(\mathbf{x}_0; \mathbf{n}_0)} \mathbf{T}_w^- \mathbf{n}_0\, dA + \int_{S_r^-(\mathbf{x}_0; \mathbf{n}_0)} \mathbf{T}_w^- \mathbf{n}\, dA. \tag{5.9.10}$$

Exercise 5.9.1. Note that if $\mathbf{x} \in S_r^-(\mathbf{x}_0; \mathbf{n}_0)$, then $\mathbf{n}(\mathbf{x}) = (\mathbf{x} - \mathbf{x}_0)/r$, and if $\mathbf{x} \in \Pi_r(\mathbf{x}_0; \mathbf{n}_0)$, then $\mathbf{n}(\mathbf{x}) = \mathbf{n}_0$.

Since [see (5.8.3)]

$$\Pi_{\mathbf{n}_0}(\mathbf{x}_0) = \lim_{r \to \infty} \{\Pi_r(\mathbf{x}_0; \mathbf{n}_0)\}, \tag{5.9.11}$$

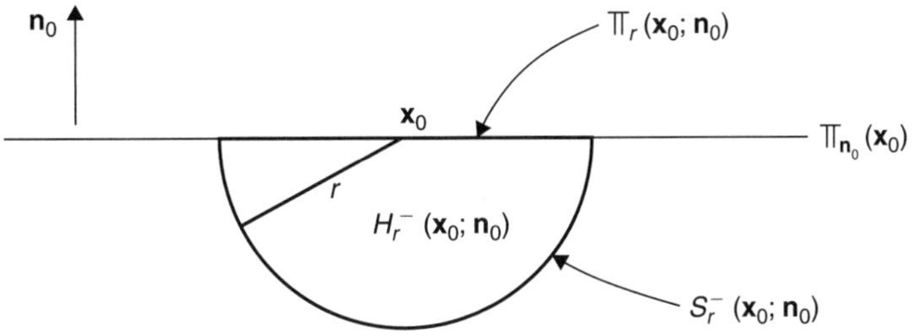

Figure 5.3. Hemispherical region $H_r^-(\mathbf{x}_0;\mathbf{n}_0)$ centred at $\mathbf{x}_0$ with radius r and lying 'below' oriented plane $\Pi_{\mathbf{n}_0}(\mathbf{x}_0)$ through $\mathbf{x}_0$ with unit normal $\mathbf{n}_0$. Boundary $\partial H_r^- = \Pi_r \cup S_r^-$ with $\Pi_r \subset \Pi_{\mathbf{n}_0}$.

from (5.9.10) we have the following:

Result 5.9.2. If

$$\lim_{r\to\infty}\left\{\int_{S_r^-(\mathbf{x}_0;\mathbf{n}_0)} \mathbf{T}_w^- \mathbf{n}\, dA\right\} = \mathbf{0}, \tag{5.9.12}$$

then

$$\lim_{r\to\infty}\left\{\int_{\partial H_r^-(\mathbf{x}_0;\mathbf{n}_0)} \mathbf{T}_w^- \mathbf{n}\, dA\right\} = \int_{\Pi_{\mathbf{n}_0}(\mathbf{x}_0)} \mathbf{T}_w^- \mathbf{n}_0\, dA. \tag{5.9.13}$$

Corollary. If (5.9.12) holds, then

$$\int_{\Pi_{\mathbf{n}_0}(\mathbf{x}_0)} \mathbf{T}_w^- \mathbf{n}_0\, dA = \sum\sum_{i\neq j} \mathbf{f}_{ij}\, \nu_i^w(\mathcal{E}^-(\mathbf{x}_0;\mathbf{n}_0)), \tag{5.9.14}$$

where

$$\mathcal{E}^-(\mathbf{x}_0;\mathbf{n}_0) := \{\mathbf{x} : (\mathbf{x}-\mathbf{x}_0).\mathbf{n}_0 \leq 0\}. \tag{5.9.15}$$

This result follows from (5.9.11) upon choosing $R = H_r^-(\mathbf{x}_0;\mathbf{n}_0)$ in (5.9.5).

Remark 5.9.1. In the event that (5.9.12) holds, (5.9.14) equates the integral of $\mathbf{T}_w^-\mathbf{n}_0$, taken over the whole plane through $\mathbf{x}_0$ with unit normal $\mathbf{n}_0$, with a double sum taken over weighted contributions of interactions. The weighting factor for $\mathbf{f}_{ij}$ is the integral of $w(\mathbf{x}-\mathbf{x}_i)$ taken over the half-space $\mathcal{E}^-(\mathbf{x}_0;\mathbf{n}_0)$. *In particular, the value of the integral is independent of the choice* $\mathbf{T}_w^-$ *of w-based interaction stress tensor.*

Remark 5.9.2. Given a choice w of weighting function, for verification or otherwise of (5.9.12) it is necessary to consider candidates ${}_s\mathbf{T}_w^-$ and ${}_b\mathbf{T}_w^-$ [see (5.6.5) and (5.6.10)] separately. In particular, verification of (5.9.12) for simple solution ${}_s\mathbf{T}_w^-$ requires consideration of

$$\int_{S_r^-(\mathbf{x}_0;\mathbf{n}_0)} \sum\sum_{i\neq j} \mathbf{f}_{ij}(\mathbf{a}_i.\mathbf{n})dA = \sum\sum_{i\neq j} \mathbf{f}_{ij} A_i(r), \tag{5.9.16}$$

where (see Exercise 5.9.1)

$$A_i(r) := \frac{1}{r}\int_{S_r^-(\mathbf{x}_0;\mathbf{n}_0)} \mathbf{a}_i(\mathbf{x}).(\mathbf{x}-\mathbf{x}_0)dA_{\mathbf{x}}. \tag{5.9.17}$$

Similarly, the validity of (5.9.12) for balanced interactions involves analysis of

$$\int_{S_r^-(\mathbf{x}_0;\mathbf{n}_0)} \sum_{i\neq j}\sum \mathbf{f}_{ij}(\mathbf{b}_{ij}\,.\,\mathbf{n})dA = \sum_{i\neq j}\sum \mathbf{f}_{ij}B_{ij}(r), \tag{5.9.18}$$

where
$$B_{ij}(r) := \frac{1}{r}\int_{S_r^-(\mathbf{x}_0;\mathbf{n}_0)} \mathbf{b}_{ij}(\mathbf{x})\,.\,(\mathbf{x}-\mathbf{x}_0)dA_{\mathbf{x}}. \tag{5.9.19}$$

Accordingly, if interactions $\mathbf{f}_{ij}$ are bounded, then (5.9.12) holds for

(i) ${}_s\mathbf{T}_w^-$, provided $\lim_{r\to\infty}\{A_i(r)\} = 0$ for all i, and
(ii) ${}_b\mathbf{T}_w^-$, provided $\lim_{r\to\infty}\{B_{ij}(r)\} = 0$ for all i and j.

These are appropriate *sufficient* conditions which ensure (5.9.14) in respect of ${}_s\mathbf{T}_w^-$ and ${}_b\mathbf{T}_w^-$. However, such restrictions upon $A_i(r)$ and $B_{ij}(r)$ may not be *necessary* to ensure (5.9.12) [and hence (5.9.14)]. For example, if $\lim_{r\to\infty}\{A_i(r)\} = a$ or $\lim\{B_{ij}(r)\} = b$, and if the net self-force $\sum_{i\neq j}\sum \mathbf{f}_{ij}$ associated with all molecules of $\mathcal{M}$ is zero, then (5.9.12) holds. (Convince yourself, noting that a and b are independent of i.)

5.9.2 Results for Choice $w = w_\epsilon$

If $w = w_\epsilon$, then result (5.9.5) implies that, for any interaction stress tensor $\mathbf{T}_{w_\epsilon}^-$ and any regular region R,

$$\int_{\partial R} \mathbf{T}_{w_\epsilon}^- \mathbf{n}\, dA = \sum_{i\neq j}\sum \nu_i^{w_\epsilon}(R)\mathbf{f}_{ij}, \tag{5.9.20}$$

where [see (5.9.4); notice we are here repeating steps (4.3.16)]

$$\nu_i^{w_\epsilon}(R) = \int_R w_\epsilon(\mathbf{x}_i - \mathbf{x})dV_{\mathbf{x}} = \int_{R\cap S_\epsilon(\mathbf{x}_i)} V_\epsilon^{-1}dV. \tag{5.9.21}$$

Thus
$$\nu_i^{w_\epsilon}(R) = \frac{\mathrm{vol}(R\cap S_\epsilon(\mathbf{x}_i))}{V_\epsilon}. \tag{5.9.22}$$

This is that fraction of the volume of the sphere, with centre at $\mathbf{x}_i$ and radius ϵ, which lies within R.

Notice that if $\mathbf{T}_{w_\epsilon}^-$ satisfies (5.9.12) (see Remark 5.9.2 for a number of possible conditions which ensure this), then

$$\int_{\Pi_{\mathbf{n}_0}(\mathbf{x}_0)} \mathbf{T}_{w_\epsilon}^- \mathbf{n}_0\, dA = \sum_{i\neq j}\sum \left(\frac{V_i^-}{V_\epsilon}\right)\mathbf{f}_{ij}. \tag{5.9.23}$$

Here [see (5.8.23)] V_i^- denotes the volume of that part of $S_\epsilon(\mathbf{x}_i)$ which lies in the half-space $\mathcal{E}^-(\mathbf{x}_0;\mathbf{n}_0)$ [see (5.9.15)]. If interactions are pairwise balanced, then the net contribution from P_i and P_j to the right-hand side of (5.9.23) is

$$\frac{1}{V_\epsilon}\{V_i^-\mathbf{f}_{ij} + V_j^-\mathbf{f}_{ji}\} = \frac{(V_i^- - V_j^-)}{V_\epsilon}\mathbf{f}_{ij}. \tag{5.9.24}$$

Taken together (5.9.23) and (5.9.24) yield

$$\int_{\Pi_{\mathbf{n}_0}(\mathbf{x}_0)} {}_b\mathbf{T}_{w_\epsilon}\mathbf{n}_0\,dA = \frac{1}{2}\sum_{i\neq j}\sum \frac{(V_i^- - V_j^-)}{V_\epsilon}\mathbf{f}_{ij} \tag{5.9.25}$$

for any interaction stress tensor of form (5.6.10) for which $\lim_{r\to\infty}\{B_{ij}(r)\} = 0$ holds with B_{ij} given by (5.9.19) [see (ii) of Remark 5.9.2]. *Accordingly, Results 5.8.1 are no longer surprising. What the individual calculations of* ${}_{sb}\mathbf{c}_{ij}$, ${}_H\mathbf{c}_{ij}$ *and* ${}_N\mathbf{c}_{ij}$ *[see (5.8.24)] accomplished were indirectly to establish the vanishing of limit (5.9.12).*

Exercise 5.9.2. Convince yourself of the preceding statement. [Masochists also could verify $\lim_{r\to\infty}\{B_{ij}(r)\} = 0$ directly, case by case!]

It remains to examine whether result (5.9.23) holds for the choice ${}_s\mathbf{T}^-_{w_\epsilon}$ [see (5.6.5)] with $\mathbf{a}_i$ given by (5.7.2) together with (5.7.3), (5.7.17), and (5.7.18). In such case [see (5.9.17)]

$$A_i(r) = \frac{1}{r}\int_{S_r^-(\mathbf{x}_0;\mathbf{n}_0)} \hat{a}(\|\mathbf{x}_i - \mathbf{x}\|)(\mathbf{x}_i - \mathbf{x}).(\mathbf{x}-\mathbf{x}_0)dA_{\mathbf{x}}. \tag{5.9.26}$$

If all molecules P_i lie in a bounded region, then, for sufficiently large r, $\|\mathbf{x}_i - \mathbf{x}\| > \epsilon$ for all $\mathbf{x} \in S_r^-(\mathbf{x}_0;\mathbf{n}_0)$. Accordingly, for such r we have, from (5.7.18),

$$A_i(r) = -\frac{1}{4\pi r}\int_{S_r^-(\mathbf{x}_0;\mathbf{n}_0)} \frac{(\mathbf{x}_i - \mathbf{x})}{\|\mathbf{x}_i - \mathbf{x}\|^3}\cdot(\mathbf{x}-\mathbf{x}_0)dA_{\mathbf{x}}. \tag{5.9.27}$$

As $r \to \infty$, so $\mathbf{x} - \mathbf{x}_i \sim \mathbf{x} - \mathbf{x}_0$ as a consequence of the bounded locations of P_i (taken together, $\mathbf{x}_0$ and all locations $\mathbf{x}_i$ will lie in a bounded region). Since $\|\mathbf{x} - \mathbf{x}_0\| = r$ in (5.9.26) (Why?), the integrand is asymptotically of value r^{-1}, and the value of the integral is of order $r^{-1} \times 4\pi r^2 = 4\pi r$. Accordingly, for large r, $A_i \sim -1$. Thus, since $\lim_{r\to\infty}\{A_i(r)\} = -1$ for each and every molecule P_i, from (5.6.5) and (5.9.17)

$$\int_{S_r^-(\mathbf{x}_0;\mathbf{n}_0)} {}_s\mathbf{T}^-_{w_\epsilon}\mathbf{n}\,dA = \int_{S_r^-(\mathbf{x}_0;\mathbf{n}_0)} \sum_{i\neq j}\sum \mathbf{f}_{ij}(\mathbf{a}_i.\mathbf{n})dA = \sum_{i\neq j}\sum \mathbf{f}_{ij}A_i(r), \tag{5.9.28}$$

and hence

$$\lim_{r\to\infty}\left\{\int_{S_r^-(\mathbf{x}_0;\mathbf{n}_0)} {}_s\mathbf{T}^-_{w_\epsilon}\mathbf{n}\,dA\right\} = -\sum_{i\neq j}\sum \mathbf{f}_{ij}. \tag{5.9.29}$$

The right-hand side of (5.9.29) represents the net self-force associated with all $\mathcal{M}$ molecules. Accordingly we have, from (5.9.12), (5.9.13), and (5.9.23), the following

Result 5.9.3. If the net self-force associated with molecules of $\mathcal{M}$ vanishes, then

$$\int_{\Pi_{\mathbf{n}_0}(\mathbf{x}_0)} {}_s\mathbf{T}^-_{w_\epsilon}\mathbf{n}_0\,dA = \sum_{i\neq j}\sum\left(\frac{V_i^-}{V_\epsilon}\right)\mathbf{f}_{ij}. \tag{5.9.30}$$

Remark 5.9.3. Vanishing of the net self-force for $\mathcal{M}$ is equivalent to the inability of $\mathcal{M}$ molecules to influence the motion of their mass centre. It is difficult to conceive of a weaker restriction upon interactions.

5.9.3 Further Remarks for Choice $w = w_\epsilon$

Remark 5.9.4. Relation (5.9.20) may be written as

$$\int_{\partial R} \mathbf{T}^-_{w_\epsilon}\mathbf{n}\,dA = \sum_i \nu_i^{w_\epsilon}(R)\mathbf{F}_i, \tag{5.9.31}$$

where the resultant force on P_i due to all other molecules of $\mathcal{M}$ is

$$\mathbf{F}_i := \sum_{j\neq i} \mathbf{f}_{ij}. \tag{5.9.32}$$

If the location $\mathbf{x}_i$ of P_i lies further than ϵ from the nearest point of R, then its contribution to sum (5.9.31) is zero. (Why?) Further, if $\mathbf{x}_i$ lies inside R and is further than ϵ from the nearest point on its boundary ∂R, then its contribution is $\mathbf{F}_i$. (Why?)

Remark 5.9.5. If interactions are pairwise-balanced, then (5.9.20) may be written as

$$\int_{\partial R} \mathbf{T}^-_{w_\epsilon}\mathbf{n}\,dA = \frac{1}{2}\sum\sum_{i\neq j}(\nu_i^{w_\epsilon}(R) - \nu_j^{w_\epsilon}(R))\mathbf{f}_{ij}, \tag{5.9.33}$$

and the *net* contribution of P_i and P_j to the sum is

$$\mathbf{c}_{ij}^{w_\epsilon}(R) := (\nu_i^{w_\epsilon}(R) - \nu_j^{w_\epsilon}(R))\mathbf{f}_{ij}. \tag{5.9.34}$$

Exercise 5.9.3. Show that if $\mathbf{x}_i$ and $\mathbf{x}_j$ are distant greater than ϵ from ∂R, then

(i) $\mathbf{c}_{ij}^{w_\epsilon}(R) = \mathbf{0}$ if both $\mathbf{x}_i$ and $\mathbf{x}_j$ lie within R or both lie outside R, and
(ii) $\mathbf{c}_{ij}^{w_\epsilon}(R) = \mathbf{f}_{ij}$ if one of $\mathbf{x}_i, \mathbf{x}_j$ lies within R and the other outside R.

Show further that if $\mathbf{x}_i$ and $\mathbf{x}_j$ lie within a distance ϵ of ∂R, then $\mathbf{c}_{ij}^{w_\epsilon}(R) = \alpha\mathbf{f}_{ij}$ with $|\alpha| < 1$. [It is helpful to draw sketches for (i) and (ii) and note for the third result that $0 < \nu_k^{w_\epsilon}(R) < 1$, where $k = i$ or j.]

Remark 5.9.6. Results (5.9.25) and (5.9.30) show that any smooth displacement of plane $\Pi_{\mathbf{n}_0}(\mathbf{x}_0)$ parallel to itself gives rise to a smooth change in the value of the integral of $\mathbf{T}^-_{w_\epsilon}\mathbf{n}_0$ over the displaced plane since the values of V_i^- and V_j^- change smoothly.

Remark 5.9.7. Consider any interaction stress ${}_b\mathbf{T}^-_{w_\epsilon}$ which satisfies (5.9.12) and hence yields (5.9.25). (Of course, this includes ${}_{sb}\mathbf{T}^-_{w_\epsilon}$, ${}_H\mathbf{T}^-_{w_\epsilon}$, and ${}_N\mathbf{T}^-_{w_\epsilon}$ in particular.) In such case the integral of ${}_b\mathbf{T}^-_{w_\epsilon}$ over $\Pi_{\mathbf{n}_0}(\mathbf{x}_0)$ has non-zero net contributions from pairs of molecules which lie *on the same side* of this plane surface. For example, if $z_i = \epsilon/2, z_j = \epsilon$ [see (5.8.14)] then $V_j^- = 0$, $V_i^- = (5/32)\,V_\epsilon$, and hence the net contribution $\mathbf{c}_{ij}$ of P_i and P_j to the integral is $(5/32)\,\mathbf{f}_{ij}$ from (5.9.25). [Here we have noted that if $z = z_k$ for P_k, then

$$V_k^- := \pi\int_{-\epsilon}^{-z_k}(\epsilon^2 - s^2)ds = \frac{\pi}{3}(z_k^3 - 3\epsilon^2 z_k + 2\epsilon^3). \tag{5.9.35}$$

Thus $V_i^- = 5\pi\epsilon^3/24$.] Further, net contributions $\mathbf{c}_{ij}$ from molecules P_i and P_j which lie close to $\Pi_{\mathbf{n}_0}(\mathbf{x}_0)$ are 'de-emphasised' in the following sense. Suppose that $z_j = \alpha\epsilon$ and $z_i = -\alpha\epsilon$ $(0 \leq \alpha \leq 1)$. Then $V_i^- - V_j^- = (1/2)\,\alpha(3-\alpha^2)V_\epsilon$ [show this via (5.9.35)], and $\mathbf{c}_{ij} = (1/2)\,\alpha(3-\alpha^2)\mathbf{f}_{ij} \to \mathbf{0}$ as $\alpha \to 0$.

Remark 5.9.8. Molecular interactions decay rapidly with separation as a consequence of each molecule being an assembly of charged particles with zero net charge. Consequently both individual and co-operative intermolecular effects are negligible at separations in excess of a 'range', σ say. (Interactions are regarded as of extremely long range if $\sigma \sim 1000$ Å $= 10^{-7}$ m: see Israelachvili [22]). It follows that the integral of $\mathbf{T}^-_{w_\epsilon}\mathbf{n}$ over S derives effectively only from molecules within a distance $\sigma + \epsilon$ of S. (Convince yourself of this!)

Remark 5.9.9. The value of the surface integral Surf($\mathbf{T}_w, S, \mathbf{n}$) [see (5.8.1)] associated with the Cauchy stress $\mathbf{T}_w$ is usually interpreted to be the resultant force exerted by molecules P_j for which $z_j > 0$ upon molecules P_i for which $z_i < 0$ [see (5.8.14)]. Our considerations modify this interpretation in the following respects:

(i) The contribution of $-\mathcal{D}_w$ to $\mathbf{T}_w$ [see (5.5.20)] does not involve forces, and hence this is true also of its integral over S. Thus, unless $\mathcal{D}_w = \mathbf{O}$ (which corresponds to a temperature of absolute zero), the integral incorporates a thermokinetic effect.
(ii) The interaction contribution to $\mathbf{T}_w$, namely $\mathbf{T}^-_w$, does not derive solely from interactions between pairs of molecules which lie on opposite sides of S (see Remark 5.9.7).
(iii) Interactions are weighted.

Notice that these comments are quite general, and are relevant to any choice of weighting function.

Remark 5.9.10. Computation of integrals of $\mathbf{T}^-_{w_\epsilon}$ over a *bounded* subsurface of a plane can be effected. In such a case result (5.8.26) does not hold since the individual choices ${}_{sb}\mathbf{T}^-_{w_\epsilon}$, ${}_H\mathbf{T}^-_{w_\epsilon}$, and ${}_N\mathbf{T}^-_{w_\epsilon}$ give rise to different contributions near the 'edges' of the subsurface. Examples of these are given in Murdoch [20] in respect of ${}_N\mathbf{T}^-_{w_\epsilon}$ for a general bounded planar surface $\mathcal{S}$.

Remark 5.9.11. Although point masses employed here have so far modelled only molecules, it is possible to consider a material system $\mathcal{M}$ which consists of ions of a single polarity (i.e., having the same net, non-zero electric charge). With notable exceptions, such as magnetically confined plasmas, such ions are found in electrically neutral mixtures of differing ionic species. In such cases our derivations are unchanged, but $\mathbf{b}_w$ will incorporate a force density which derives from the effect of those ions which do not belong to $\mathcal{M}$. Notice that interaction stress tensors are defined without formal change of argumentation. There will, however, be constitutive implications. In particular, $\mathbf{T}^-_w(\mathbf{x})$ must be expected to depend upon ions remote from $\mathbf{x}$, and, accordingly constitutive relations for $\mathbf{T}_w$ and $\mathbf{b}_w$ must be non-local.

6 Balance of Energy

6.1 Preamble

Here local forms of energy balance, comparable with its most usually accepted form [cf., e.g., Truesdell & Noll [2], (7.9.3) and Chadwick [3], (39)]

$$r - \operatorname{div}\mathbf{q} + \mathbf{T}\cdot\mathbf{D} = \rho\dot{e}, \tag{2.8.24}$$

are derived. Such forms are appropriate to material systems for which couple stress, internal rotational momentum, and microstructural kinetic energy are negligible[1]. The forms of balance are obtained from the equation which governs the motion of an individual molecule in an inertial frame, namely

$$\sum_{j\neq i}\mathbf{f}_{ij} + \mathbf{b}_i = m_i\dot{\mathbf{v}}_i, \tag{6.1.1}$$

[see (5.4.4) and (5.4.5)] by multiplying each term scalarly by $\mathbf{v}_i w(\mathbf{x}_i - \mathbf{x})$ and then summing over all particles. Upon writing $\mathbf{v}_i = \hat{\mathbf{v}}_i + \mathbf{v}_w$ [see (5.5.8)], a decomposition of resulting fields into mechanical (and kinematic) and thermal (and thermokinetic) contributions is effected. The latter are characterised by the presence of $\hat{\mathbf{v}}_i$ in their definitions. The analysis then is re-examined from a subatomic perspective consistent with the discussion of Section 5.4.

6.2 Derivation of Energy Balances

Scalar multiplication of (6.1.1) by $\mathbf{v}_i\, w(\mathbf{x}_i - \mathbf{x})$ followed by summation over all particles P_i yields

$$\sum\sum_{i\neq j}\mathbf{f}_{ij}\,.\,\mathbf{v}_i\, w(\mathbf{x}_i-\mathbf{x}) + \sum_i \mathbf{b}_i\,.\,\mathbf{v}_i\, w(\mathbf{x}_i-\mathbf{x}) = \sum_i m_i\dot{\mathbf{v}}_i\,.\,\mathbf{v}_i\, w(\mathbf{x}_i-\mathbf{x})$$
$$= \sum_i \frac{d}{dt}\left\{\frac{1}{2}m_i\mathbf{v}_i^2\right\} w(\mathbf{x}_i-\mathbf{x}). \tag{6.2.1}$$

[1] The more general situation in which these effects are manifest is discussed in Chapter 7. There such finer detail is explored by first deriving local balances of moments both of mass and momentum, and then obtaining the corresponding local form of energy balance (see section 2.8.3).

Writing [see (5.5.8)]

$$\mathbf{v}_i = \hat{\mathbf{v}}_i + \mathbf{v}_w, \tag{6.2.2}$$

it follows that

$$\sum\sum_{i\neq j} \mathbf{f}_{ij} . \mathbf{v}_i \, w(\mathbf{x}_i - \mathbf{x}) = Q_w + \mathbf{f}_w . \mathbf{v}_w, \tag{6.2.3}$$

where [see also (5.5.2)]

$$Q_w(\mathbf{x},t) := \sum\sum_{i\neq j} \mathbf{f}_{ij}(t) . \hat{\mathbf{v}}_i(t;\mathbf{x}) w(\mathbf{x}_i(t) - \mathbf{x}). \tag{6.2.4}$$

Similarly [see (5.5.3)],

$$\sum_i \mathbf{b}_i . \mathbf{v}_i \, w(\mathbf{x}_i - \mathbf{x}) = r_w + \mathbf{b}_w . \mathbf{v}_w, \tag{6.2.5}$$

where the *external heat supply rate*

$$r_w(\mathbf{x},t) := \sum_i \mathbf{b}_i(t) . \hat{\mathbf{v}}_i(t;\mathbf{x}) w(\mathbf{x}_i(t) - \mathbf{x}). \tag{6.2.6}$$

Now $$\sum_i \frac{d}{dt}\left\{\frac{1}{2} m_i \mathbf{v}_i^2\right\} w = \frac{\partial}{\partial t}\left\{\sum_i \frac{1}{2} m_i \mathbf{v}_i^2 w\right\} - \sum_i \frac{1}{2} m_i \mathbf{v}_i^2 \nabla w . \mathbf{v}_i, \tag{6.2.7}$$

and $$-\sum_i \frac{1}{2} m_i \mathbf{v}_i^2 \nabla w . \mathbf{v}_i = +\sum_i \frac{1}{2} m_i \mathbf{v}_i^2 \nabla_{\mathbf{x}} w . \mathbf{v}_i = \operatorname{div}\left\{\sum_i \frac{1}{2} m_i \mathbf{v}_i^2 \mathbf{v}_i \, w\right\}. \tag{6.2.8}$$

Further, with the aim of distinguishing between macroscopic kinematic and microscopic thermokinetic contributions, we have, from (5.5.8),

$$\sum_i \frac{1}{2} m_i \mathbf{v}_i^2 \mathbf{v}_i \, w = \sum_i \frac{1}{2} m_i \mathbf{v}_i^2 (\hat{\mathbf{v}}_i + \mathbf{v}_w). \tag{6.2.9}$$

Thus, from (6.2.7), (6.2.8) and (6.2.9), and writing

$$\rho_w K_w := \sum_i \frac{1}{2} m_i \mathbf{v}_i^2 w, \tag{6.2.10}$$

$$\sum_i \frac{d}{dt}\left\{\frac{1}{2} m_i \mathbf{v}_i^2\right\} w = \frac{\partial}{\partial t}\{\rho_w K_w\} + \operatorname{div}\{\rho_w K_w \mathbf{v}_w\} + \operatorname{div}\left\{\sum_i \frac{1}{2} m_i \mathbf{v}_i^2 \hat{\mathbf{v}}_i \, w\right\}. \tag{6.2.11}$$

Continuing the procedure of separating kinematic and thermal contributions, we observe from (6.2.10) and (5.5.8) that

$$\rho_w K_w = \sum_i \frac{1}{2} m_i (\hat{\mathbf{v}}_i^2 + 2\hat{\mathbf{v}}_i . \mathbf{v}_w + \mathbf{v}_w^2) w = \rho_w (h_w + \frac{1}{2} \mathbf{v}_w^2) \tag{6.2.12}$$

on noting (5.5.9) and with the *heat energy density*

$$\rho_w(\mathbf{x},t) h_w(\mathbf{x},t) := \sum_i \frac{1}{2} m_i \hat{\mathbf{v}}_i^2 (t;\mathbf{x}) w(\mathbf{x}_i(t) - \mathbf{x}). \tag{6.2.13}$$

Accordingly, from (5.5.8), (5.5.11), and (5.5.9),

$$\sum_i \frac{1}{2} m_i \mathbf{v}_i^2 \hat{\mathbf{v}}_i w = \sum_i \frac{1}{2} m_i (\hat{\mathbf{v}}_i^2 + 2\hat{\mathbf{v}}_i \,.\, \mathbf{v}_w + \mathbf{v}_w^2)\hat{\mathbf{v}}_i w$$

$$= \sum_i \frac{1}{2} m_i \hat{\mathbf{v}}_i^2 \hat{\mathbf{v}}_i w + \left(\sum_i m_i (\hat{\mathbf{v}}_i \otimes \hat{\mathbf{v}}_i) w\right) \mathbf{v}_w + \frac{1}{2}\left(\sum_i m_i \hat{\mathbf{v}}_i w\right) \mathbf{v}_w^2$$

$$= \boldsymbol{\kappa}_w + \boldsymbol{\mathcal{D}}_w \mathbf{v}_w + \mathbf{0}, \tag{6.2.14}$$

where

$$\boldsymbol{\kappa}_w(\mathbf{x},t) := \sum_i \frac{1}{2} m_i \hat{\mathbf{v}}_i^2(t;\mathbf{x}) \hat{\mathbf{v}}_i(t;\mathbf{x}) w(\mathbf{x}_i(t) - \mathbf{x}). \tag{6.2.15}$$

Thus (6.2.11) and (6.2.14) yield

$$\sum_i \frac{d}{dt}\left\{\frac{1}{2} m_i \mathbf{v}_i^2\right\} w = \frac{\partial}{\partial t}\{\rho_w K_w\} + \operatorname{div}\{\rho_w K_w \mathbf{v}_w\} + \operatorname{div}\{\boldsymbol{\kappa}_w + \boldsymbol{\mathcal{D}}_w \mathbf{v}_w\}. \tag{6.2.16}$$

Exercise 6.2.1. Show that as a consequence of the continuity equation (4.2.16)

$$\frac{\partial}{\partial t}\{\rho_w K_w\} + \operatorname{div}\{\rho_w K_w \mathbf{v}_w\} = \rho_w \dot{K}_w, \tag{6.2.17}$$

where [see (2.5.25) and (2.5.27)]

$$\dot{K}_w := \frac{\partial K_w}{\partial t} + (\nabla K_w) \,.\, \mathbf{v}_w \tag{6.2.18}$$

denotes the material time derivative of K_w.

Finally, the continuum form of (6.2.1) may be written, via (6.2.3), (6.2.5), (6.2.11), (6.2.17), and (6.2.14), as

$$Q_w + r_w + (\mathbf{f}_w + \mathbf{b}_w) \,.\, \mathbf{v}_w = \rho_w \dot{K}_w + \operatorname{div}\{\boldsymbol{\kappa}_w + \boldsymbol{\mathcal{D}}_w \mathbf{v}_w\}. \tag{6.2.19}$$

Noting that from (6.2.12)

$$\dot{K}_w = \dot{h}_w + \mathbf{v}_w \,.\, \dot{\mathbf{v}}_w = \dot{h}_w + \mathbf{v}_w \,.\, \mathbf{a}_w, \tag{6.2.20}$$

and [see Appendix B.7, (B.7.31) and recall from (5.5.11) that $\boldsymbol{\mathcal{D}}_w = \boldsymbol{\mathcal{D}}_w^T$]

$$\operatorname{div}\{\boldsymbol{\mathcal{D}}_w \mathbf{v}_w\} = \operatorname{div}\{\boldsymbol{\mathcal{D}}_w^T \mathbf{v}_w\} = (\operatorname{div} \boldsymbol{\mathcal{D}}_w) \cdot \mathbf{v}_w + \boldsymbol{\mathcal{D}}_w \,.\, \nabla \mathbf{v}_w, \tag{6.2.21}$$

(6.2.19) may be written as

$$Q_w + r_w + (\mathbf{f}_w + \mathbf{b}_w) \,.\, \mathbf{v}_w = \rho_w(\dot{h}_w + \mathbf{a}_w \,.\, \mathbf{v}_w) + (\operatorname{div} \boldsymbol{\mathcal{D}}_w) \,.\, \mathbf{v}_w + \boldsymbol{\mathcal{D}}_w \cdot \nabla \mathbf{v}_w + \operatorname{div} \boldsymbol{\kappa}_w. \tag{6.2.22}$$

Writing

$$\mathbf{L}_w := \nabla \mathbf{v}_w \tag{6.2.23}$$

for the *velocity gradient*, and recalling linear momentum balance (5.5.15), relation (6.2.22) reduces to

$$-\operatorname{div} \boldsymbol{\kappa}_w + Q_w + r_w - \boldsymbol{\mathcal{D}}_w \cdot \mathbf{L}_w = \rho_w \dot{h}_w. \tag{6.2.24}$$

Remark 6.2.1. All fields in (6.2.24) are thermal (in the sense of involving the notional thermal velocity $\hat{\mathbf{v}}_i$ in their definitions) with the exception of the velocity gradient $\mathbf{L}_w$. This relation is at the same level of generality as form (5.5.15) of linear momentum balance. We now proceed to derive equivalent forms which introduce candidate stress tensors with the aim of obtaining a balance more directly comparable with the usual form (2.8.24).

From (6.2.3) and (5.6.2),

$$Q_w + \mathbf{f}_w . \mathbf{v}_w = \sum_{i \neq j}\sum \mathbf{f}_{ij} . \mathbf{v}_i w = \sum_{i \neq j}\sum \mathbf{f}_{ij} . \mathbf{v}_i \operatorname{div} \mathbf{a}_i = \operatorname{div}\left\{ \sum_{i \neq j}\sum (\mathbf{a}_i \otimes \mathbf{f}_{ij})\mathbf{v}_i \right\}. \tag{6.2.25}$$

However, from (5.5.8) and (5.6.5),

$$\sum_{i \neq j}\sum (\mathbf{a}_i \otimes \mathbf{f}_{ij})\mathbf{v}_i = \sum_{i \neq j}\sum (\mathbf{a}_i \otimes \mathbf{f}_{ij})(\hat{\mathbf{v}}_i + \mathbf{v}_w) = -{}_s\mathbf{q}_w^- + ({}_s\mathbf{T}_w^-)^T \mathbf{v}, \tag{6.2.26}$$

where
$${}_s\mathbf{q}_w^-(\mathbf{x},t) := -\sum_{i \neq j}\sum (\mathbf{f}_{ij}(t) . \hat{\mathbf{v}}_i(t;\mathbf{x}))\mathbf{a}_i(\mathbf{x},t). \tag{6.2.27}$$

From (6.2.25) and (6.2.26), balance (6.2.22) may be written as

$$\operatorname{div}\{-{}_s\mathbf{q}_w^- + ({}_s\mathbf{T}_w^-)^T \mathbf{v}_w\} + r_w + \mathbf{b}_w . \mathbf{v}_w = \rho_w(\dot{h}_w + \mathbf{a}_w . \mathbf{v}_w) + \operatorname{div} \boldsymbol{\kappa}_w + \boldsymbol{\mathcal{D}}_w \cdot \mathbf{L}_w + (\operatorname{div} \boldsymbol{\mathcal{D}}_w) . \mathbf{v}_w. \tag{6.2.28}$$

Equivalently, recalling (5.5.21) in respect of the simplest solution [(5.6.5) to (5.5.18)],

$$r_w - \operatorname{div} {}_s\mathbf{q}_w + (\operatorname{div} {}_s\mathbf{T}_w + \mathbf{b}_w - \rho_w \mathbf{a}_w) . \mathbf{v}_w + {}_s\mathbf{T}_w \cdot \mathbf{L}_w = \rho_w \dot{h}_w, \tag{6.2.29}$$

where
$${}_s\mathbf{q}_w := {}_s\mathbf{q}_w^- + \boldsymbol{\kappa}_w. \tag{6.2.30}$$

Since by its construction [see (5.5.15), (5.6.1) and (5.5.20)] ${}_s\mathbf{T}_w$ satisfies

$$\operatorname{div} {}_s\mathbf{T}_w + \mathbf{b}_w = \rho_w \mathbf{a}_w, \tag{6.2.31}$$

(6.2.29) reduces to
$$r_w - \operatorname{div} {}_s\mathbf{q}_w + {}_s\mathbf{T}_w \cdot \mathbf{L}_w = \rho_w \dot{h}_w. \tag{6.2.32}$$

Remark 6.2.2. The interaction contributions to ${}_s\mathbf{q}_w$ and ${}_s\mathbf{T}_w$ (namely ${}_s\mathbf{q}_w^-$ and ${}_s\mathbf{T}_w^-$) involve only resultant forces

$$\mathbf{F}_i := \sum_{j \neq i} \mathbf{f}_{ij} \tag{6.2.33}$$

on individual molecules P_i of the system/body by all other system molecules [see (6.2.27) and (5.6.5)]. Accordingly, for situations in which one wishes, or is only able, to work with such force resultants, forms (6.2.31) and (6.2.32) are the appropriate balances of linear momentum and energy.

A further balance can be obtained if interactions $\mathbf{f}_{ij}$ are pairwise balanced [see (5.6.6)]. In such case [see (6.2.3)]

$$Q_w + \mathbf{f}_w . \mathbf{v}_w = \sum_{i \neq j}\sum \mathbf{f}_{ij} . \mathbf{v}_i w = \sum_{i \neq j}\sum \mathbf{f}_{ij} . \mathbf{v}_i w(\mathbf{x}_i - \mathbf{x}). \tag{6.2.34}$$

Now

$$\sum_{i \neq j}\sum \mathbf{f}_{ij} . \mathbf{v}_i w(\mathbf{x}_i - \mathbf{x}) = \frac{1}{2} \sum_{i \neq j}\sum \{\mathbf{f}_{ij} . \mathbf{v}_i w(\mathbf{x}_i - \mathbf{x}) + \mathbf{f}_{ji} . \mathbf{v}_j w(\mathbf{x}_j - \mathbf{x})\}$$

$$= A_w + B_w, \tag{6.2.35}$$

where

$$A_w(\mathbf{x},t) := \frac{1}{2} \sum_{i \neq j}\sum \mathbf{f}_{ij}(t) . (\mathbf{v}_i(t) - \mathbf{v}_j(t)) w(\mathbf{x}_i(t) - \mathbf{x}) \tag{6.2.36}$$

and

$$B_w(\mathbf{x},t) := \frac{1}{2} \sum_{i \neq j}\sum \mathbf{f}_{ij}(t) . \mathbf{v}_j(t) \{w(\mathbf{x}_i(t) - \mathbf{x}) - w(\mathbf{x}_j(t) - \mathbf{x})\} \tag{6.2.37}$$

$$= \frac{1}{2} \sum_{i \neq j}\sum \mathbf{f}_{ij}(t) . \mathbf{v}_i(t) \{w(\mathbf{x}_i(t) - \mathbf{x}) - w(\mathbf{x}_j(t) - \mathbf{x})\}. \tag{6.2.38}$$

Exercise 6.2.2. Derive (6.2.35) with A_w and B_w given by (6.2.36) and (6.2.37), using (5.6.6). Obtain (6.2.38) from (6.2.37) by relabelling *(ij)* as *(ji)* and using (5.6.6).

Given any solution $\mathbf{b}_{ij}$ to (5.6.8),

$$\begin{aligned} B_w &= \frac{1}{2} \sum_{i \neq j}\sum \mathbf{f}_{ij} . \mathbf{v}_i \{w(\mathbf{x}_i - \mathbf{x}) - w(\mathbf{x}_j - \mathbf{x})\} \\ &= \sum_{i \neq j}\sum (\mathbf{f}_{ij} . \mathbf{v}_i) \operatorname{div} \mathbf{b}_{ij} = \sum_{i \neq j}\sum \operatorname{div}\{(\mathbf{f}_{ij} . \mathbf{v}_i) \mathbf{b}_{ij}\} \\ &= \sum_{i \neq j}\sum \operatorname{div}\{(\mathbf{b}_{ij} \otimes \mathbf{f}_{ij}) \mathbf{v}_i\}. \end{aligned} \tag{6.2.39}$$

However, from (6.2.2) and (5.6.10),

$$\begin{aligned} \sum_{i \neq j}\sum (\mathbf{b}_{ij} \otimes \mathbf{f}_{ij}) \mathbf{v}_i &= \sum_{i \neq j}\sum (\mathbf{b}_{ij} \otimes \mathbf{f}_{ij})(\hat{\mathbf{v}}_i + \mathbf{v}_w) \\ &= -{}_b\mathbf{q}_w^- + ({}_b\mathbf{T}_w^-)^T \mathbf{v}_w, \end{aligned} \tag{6.2.40}$$

where

$${}_b\mathbf{q}_w^-(\mathbf{x},t) := -\sum_{i \neq j}\sum (\mathbf{f}_{ij}(t) . \hat{\mathbf{v}}_i(t;\mathbf{x})) \mathbf{b}_{ij}(\mathbf{x},t). \tag{6.2.41}$$

Hence, from (6.2.39) and (6.2.40),

$$B_w = \operatorname{div}\{-{}_b\mathbf{q}_w^- + ({}_b\mathbf{T}_w^-)^T \mathbf{v}_w\}. \tag{6.2.42}$$

From (6.2.34), (6.2.35), and (6.2.42),

$$Q_w + \mathbf{f}_w . \mathbf{v}_w = A_w + \operatorname{div}\{-{}_b\mathbf{q}_w^- + ({}_b\mathbf{T}_w^-)^T \mathbf{v}_w\}. \tag{6.2.43}$$

Substituting from (6.2.43) into balance (6.2.22) yields

$$\begin{aligned} r_w + A_w + \operatorname{div}\{-{}_b\mathbf{q}_w^- + ({}_b\mathbf{T}_w^-)^T \mathbf{v}_w\} + \mathbf{b}_w . \mathbf{v}_w \\ = \rho_w \dot{h}_w + (\rho_w \mathbf{a}_w + \operatorname{div} \mathcal{D}_w) . \mathbf{v}_w + \mathcal{D}_w \cdot \mathbf{L}_w + \operatorname{div} \boldsymbol{\kappa}_w. \end{aligned} \tag{6.2.44}$$

Simplifying,

$$r_w + A_w - \operatorname{div}{}_b\mathbf{q}_w + \operatorname{div}\{{}_b\mathbf{T}_w^T\mathbf{v}_w\} + \mathbf{b}_w\,.\,\mathbf{v}_w = \rho_w(\dot{h}_w + \mathbf{a}_w\,.\,\mathbf{v}_w), \tag{6.2.45}$$

where

$${}_b\mathbf{T}_w := {}_b\mathbf{T}_w^- - \mathcal{D}_w \tag{6.2.46}$$

and

$${}_b\mathbf{q}_w := {}_b\mathbf{q}_w^- + \kappa_w. \tag{6.2.47}$$

Since by its construction [see (5.5.15), (5.5.18), (5.6.10), and (5.5.20)] ${}_b\mathbf{T}_w$ satisfies

$$\operatorname{div}{}_b\mathbf{T}_w + \mathbf{b}_w = \rho_w\,\mathbf{a}_w, \tag{6.2.48}$$

(6.2.45) reduces to the form

$$A_w + r_w - \operatorname{div}{}_b\mathbf{q}_w + {}_b\mathbf{T}_w \cdot \mathbf{L}_w = \rho_w\dot{h}_w \tag{6.2.49}$$

on using identity (B.7.31) of Appendix B.7.

Notice that from its definition (6.2.36), term A_w represents a time rate of change of energy associated with work done by interactions in relative motions of molecules. This term can be simplified if interactions are delivered by separation-dependent pair potentials. In such a case there exist scalar-valued functions $\hat{\varphi}_{ij}(r_{ij})$, where

$$r_{ij} := \|\mathbf{x}_i - \mathbf{x}_j\|, \tag{6.2.50}$$

for which

$$\mathbf{f}_{ij} = \nabla_{\mathbf{x}_j}\hat{\varphi}_{ij}(r_{ij}) = \hat{\varphi}'_{ij}(r_{ij})\,\frac{(\mathbf{x}_j - \mathbf{x}_i)}{r_{ij}}. \tag{6.2.51}$$

Scalar $\hat{\varphi}_{ij}(r_{ij})$ denotes the work done (by the force $\mathbf{f}_{ji}$ on P_j due to P_i) in bringing P_j up to its current location from an infinitely remote location at which, without loss of generality, $\hat{\varphi}_{ij}$ can be assumed to be zero (see Figure 6.1).

The total work done by interactions in assembling molecules in their current locations is thus

$$W = \sum_{j>1}\hat{\varphi}_{1j}(r_{1j}) + \sum_{j>2}\hat{\varphi}_{2j}(r_{2j}) + \sum_{j>3}\hat{\varphi}_{3j}(r_{3j}) + \cdots. \tag{6.2.52}$$

The overall sum in (6.2.52) corresponds to locating first P_1 in its current position, starting from a situation in which all molecules are infinitely remote from their current situations (this involves zero work; why?), and then P_2, then P_3, etc.

Exercise 6.2.3. Noting that in this discussion interactions are pairwise balanced [see (5.6.6)], deduce that

$$\hat{\varphi}'_{ij}(r_{ij}) = \hat{\varphi}'_{ji}(r_{ji}).$$

Show that if potentials vanish for infinite separations, then

$$\hat{\varphi}_{ij} = \hat{\varphi}_{ji}. \tag{6.2.53}$$

Show further that as a consequence

$$W = \sum_{j>i}\sum\hat{\varphi}_{ij}(r_{ij}) = \frac{1}{2}\sum_{i\neq j}\sum\hat{\varphi}_{ij}(r_{ij}). \tag{6.2.54}$$

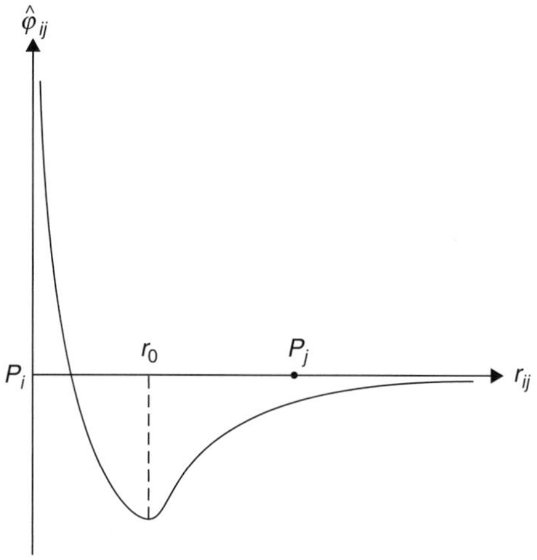

Figure 6.1. A typical separation-dependent pair potential function $\hat{\varphi}_{ij}$. With P_j distant r_{ij} from P_i, the molecular interation is attractive if $r_{ij} > r_0$ (where $\hat{\varphi}'_{ij} > 0$) and repulsive if $r_{ij} < r_0$ (where $\hat{\varphi}'_{i_j} < 0$).

Remark 6.2.3. Result (6.2.54) demonstrates that W is independent of the order of assembly of molecules in their current locations.

The *energy of assembly density* is

$$(\rho_w \beta_w)(\mathbf{x},t) := \frac{1}{2} \sum\sum_{i \neq j} \hat{\varphi}_{ij}(r_{ij}(t)) w(\mathbf{x}_i(t) - \mathbf{x}). \tag{6.2.55}$$

Thus
$$\frac{\partial}{\partial t}\{\rho_w \beta_w\} = \frac{1}{2} \sum\sum_{i \neq j} \hat{\varphi}'_{ij} \dot{r}_{ij}\, w(\mathbf{x}_i - \mathbf{x}) + \frac{1}{2} \sum\sum_{i \neq j} \hat{\varphi}_{ij} \nabla w \,.\, \mathbf{v}_i. \tag{6.2.56}$$

Since [see (6.2.50)]

$$r_{ij}^2 = (\mathbf{x}_j - \mathbf{x}_i)\,.\,(\mathbf{x}_j - \mathbf{x}_i), \tag{6.2.57}$$

$$2 r_{ij} \dot{r}_{ij} = 2(\mathbf{x}_j - \mathbf{x}_i)\,.\,(\mathbf{v}_j - \mathbf{v}_i), \tag{6.2.58}$$

and hence, from (6.2.56), (6.2.51), (6.2.36), and (5.5.8),

$$\begin{aligned}
\frac{\partial}{\partial t}\{\rho_w \beta_w\} &= \frac{1}{2} \sum\sum_{i \neq j} \hat{\varphi}'_{ij} \frac{(\mathbf{x}_j - \mathbf{x}_i)}{r_{ij}} \cdot (\mathbf{v}_j - \mathbf{v}_i) w - \frac{1}{2} \sum\sum_{i \neq j} \hat{\varphi}_{ij} \nabla_{\mathbf{x}} w \,.\, \mathbf{v}_i \\
&= \frac{1}{2} \sum\sum_{i \neq j} \mathbf{f}_{ij}\,.\,(\mathbf{v}_j - \mathbf{v}_i) w - \operatorname{div}\left\{ \frac{1}{2} \sum\sum_{i \neq j} \hat{\varphi}_{ij} \mathbf{v}_i w \right\} \qquad (6.2.59) \\
&= -A_w - \operatorname{div}\left\{ \frac{1}{2} \sum\sum_{i \neq j} \hat{\varphi}_{ij} (\hat{\mathbf{v}}_i + \mathbf{v}_w) w \right\}. \qquad (6.2.60)
\end{aligned}$$

Thus
$$\frac{\partial}{\partial t}\{\rho_w \beta_w\} = -A_w - \operatorname{div}{}_2\mathbf{q}_w - \operatorname{div}\{\rho_w \beta_w \mathbf{v}_w\}, \tag{6.2.61}$$

where $$ {}_2\mathbf{q}_w(\mathbf{x},t) := \frac{1}{2} \sum\sum_{i \neq j} \hat{\phi}_{ij}(r_{ij}(t))\hat{\mathbf{v}}_i(t;\mathbf{x})w(\mathbf{x}_i(t)-\mathbf{x}). \tag{6.2.62}$$

From the continuity equation and (6.2.61)

$$\rho_w \dot{\beta}_w = -A_w - \operatorname{div} {}_2\mathbf{q}_w, \tag{6.2.63}$$

and energy balance (6.2.49) may be written as

$$r_w - \operatorname{div} {}_b\mathbf{q}_w^+ + {}_b\mathbf{T}_w \cdot \mathbf{L}_w = \rho_w \dot{e}_w, \tag{6.2.64}$$

where the *heat flux vector* [see (6.2.47)]

$${}_b\mathbf{q}_w^+ := {}_b\mathbf{q}_w + {}_2\mathbf{q}_w = {}_b\mathbf{q}_w^- + {}_2\mathbf{q}_w + \kappa_w, \tag{6.2.65}$$

and the *specific internal energy density*

$$e_w = \beta_w + h_w. \tag{6.2.66}$$

If ${}_b\mathbf{T}_w$ is symmetric (equivalently, if ${}_b\mathbf{T}_w^-$ is symmetric; why?), then

$${}_b\mathbf{T}_w \cdot \mathbf{L}_w = {}_b\mathbf{T}_w \cdot \mathbf{D}_w, \tag{6.2.67}$$

where $$\mathbf{D}_w := \frac{1}{2}(\mathbf{L}_w + \mathbf{L}_w^T), \tag{6.2.68}$$

and (6.2.64) becomes

$$r_w - \operatorname{div} {}_b\mathbf{q}_w^+ + {}_b\mathbf{T}_w \cdot \mathbf{D}_w = \rho_w \dot{e}_w. \tag{6.2.69}$$

Remark 6.2.4. Here a hierarchy of energy balance relations has been established which relates to the possible forms of linear momentum balance. If only resultant interaction forces $\mathbf{F}_i$ [see (6.2.33)] are considered, then only form (5.5.15) of linear momentum balance is available, and the corresponding energy balance is (6.2.24). Given any solution $\mathbf{a}_i$ to (5.6.2), the appropriate corresponding balances are (6.2.31) and (6.2.32). Further, any solution $\mathbf{b}_{ij}$ to (5.6.8), together with the assumption of pairwise-balanced interactions, leads to balances (6.2.48) and (6.2.49). Finally, the additional assumption (6.2.51) of the existence of separation-dependent pair potentials yields balances (6.2.48) and (6.2.64).

Remark 6.2.5. Since the usual interpretation of specific internal energy is that of energy 'stored' within the body, in the forms of both 'strain energy' (purely mechanical) and 'heat' (purely thermal), it is balance (6.2.64) that merits most attention. In particular, (6.2.66) delineates the separation of specific internal energy density into contribution β_w from assembly of the system (which may be termed 'binding' energy) and contribution h_w from heat.

Caveat: At this point, note that thermal quantities have so far been identified as those involving *notional* thermal molecular velocities $\hat{\mathbf{v}}_i$ [see (5.5.8)]. More precisely, thermal quantities are those which involve thermal velocities $\tilde{\mathbf{v}}_i$ [see (5.5.25) et seq.]. This distinction will be discussed further in Chapter 7, in which a finer-scale viewpoint is adopted, and a higher-order approximation to $\tilde{\mathbf{v}}_i$ is made.

Remark 6.2.6. Also noteworthy in balances (6.2.64) and (6.2.69) are the contributions ${}_b\mathbf{q}_w^-, \kappa_w$, and ${}_2\mathbf{q}_w$ to the heat flux vector ${}_b\mathbf{q}_w^+$ [see (6.2.65)]. In a rarefied gas interactions are negligible, so ${}_b\mathbf{q}_w^-$ and ${}_2\mathbf{q}_w$ are both zero. Accordingly the heat flux is associated with the diffusion of thermal kinetic energy [see (6.2.15)]. More generally there are additional contributions which involve molecular interactions ${}_b\mathbf{q}_w^-$ [see (6.2.41)] and diffusion of binding energy ${}_2\mathbf{q}_w$ [see (6.2.62)].

Remark 6.2.7. Local balance (6.2.69) takes the precise *form* of that usually considered, namely (2.8.24). Such a form also can be obtained without the assumption of interaction potentials. To achieve this, we note that the definition of A_w can be framed in 'material' (or 'referential') format as

$$A_w(\chi_{w,t_0}(\hat{\mathbf{x}},t),t) = \frac{1}{2}\sum\sum_{i\neq j}\mathbf{f}_{ij}(t)\,.\,(\mathbf{v}_i(t)-\mathbf{v}_j(t))w(\mathbf{x}_i(t)-\chi_{w,t_0}(\hat{\mathbf{x}},t)), \tag{6.2.70}$$

where (see Section 5.2)

$$\chi_{w,t_0}(\hat{\mathbf{x}},t) = \mathbf{x} \quad \text{and} \quad \chi_{w,t_0}(\hat{\mathbf{x}},t_0) = \hat{\mathbf{x}}. \tag{5.2.8}$$

Writing

$$A_w = \rho_w \mathcal{A}_w, \tag{6.2.71}$$

consider

$$\beta_{w,t_0}(\mathbf{x},t) := \int_{t_0}^{t} \mathcal{A}_w(\chi_{w,t_0}(\hat{\mathbf{x}},\tau),\tau)d\tau, \tag{6.2.72}$$

where $\mathbf{x}$ and $\hat{\mathbf{x}}$ are related at time t by (5.2.8). Differentiation with respect to time with $\hat{\mathbf{x}}$ held fixed yields

$$\dot{\beta}_{w,t_0}(\mathbf{x},t) = \mathcal{A}_w(\chi_{w,t_0}(\hat{\mathbf{x}},t),t) = \mathcal{A}_w(\mathbf{x},t). \tag{6.2.73}$$

Accordingly, from (6.2.71)

$$A_w = \rho_w \dot{\beta}_{w,t_0}, \tag{6.2.74}$$

and (6.2.49) may be written as

$$r_w - \operatorname{div}{}_b\mathbf{q}_w + {}_b\mathbf{T}_w \cdot \mathbf{L}_w = \rho_w \dot{e}_w. \tag{6.2.75}$$

Here

$$e_w := \beta_{w,t_0} + h_w, \tag{6.2.76}$$

and β_{w,t_0} represents a w-scale measure of stored energy per unit mass. Notice that choice of the arbitrary reference time t_0 means that β_{w,t_0} is unique to within a constant. (Why?)

6.3 A Subatomic Perspective

Recall the discussion in Section 5.4 concerning the motions of electrons and nuclei associated with molecules of a material system of interest. Specifically, the motion of any such fundamental subatomic discrete entity P_{i_p} associated with molecule P_i, in any inertial frame, is governed [see (5.4.6)] by

$$\mathbf{f}_{i_p i} + \sum_{j\neq i}\mathbf{f}_{i_p j} + \mathbf{b}_{i_p} = m_{i_p}\dot{\mathbf{v}}_{i_p}, \tag{6.3.1}$$

where $$\mathbf{f}_{i_p i} := \sum_{p' \neq p} \mathbf{f}_{i_p i_{p'}} \quad \text{and} \quad \mathbf{f}_{i_p j} := \sum_{\substack{q \\ j \neq i}} \mathbf{f}_{i_p j_q}. \tag{6.3.2}$$

Thus $\mathbf{f}_{i_p i}$ denotes the resultant force exerted on P_{i_p} by all other nuclei and electrons of P_i, and $\mathbf{f}_{i_p j}$ represents the resultant force exerted on P_{i_p} by constituent nuclei and electrons of molecule P_j. Scalar multiplication of (6.3.1) by $\mathbf{v}_{i_p}$ yields

$$\mathbf{f}_{i_p i} . \mathbf{v}_{i_p} + \sum_{j \neq i} \mathbf{f}_{i_p j} . \mathbf{v}_{i_p} + \mathbf{b}_{i_p} . \mathbf{v}_{i_p} = \frac{d}{dt} \left\{ \frac{1}{2} m_{i_p} \mathbf{v}_{i_p}^2 \right\}. \tag{6.3.3}$$

Writing [see (5.4.9)] $$\check{\mathbf{v}}_{i_p} := \mathbf{v}_{i_p} - \mathbf{v}_i, \tag{6.3.4}$$

and summing relations (6.3.3) over all nuclei and electrons P_{i_p} of P_i, yield

$$q_i + \mathbf{f}_i . \mathbf{v}_i + \sum_{j \neq i} (q_{ij} + \mathbf{f}_{ij} . \mathbf{v}_i) + r_i + \mathbf{b}_i . \mathbf{v}_i = \frac{d}{dt} \left\{ \frac{1}{2} m_i \mathbf{v}_i^2 \right\} + \dot{h}_i. \tag{6.3.5}$$

Here $$q_i := \sum_{i_p} \mathbf{f}_{i_p} . \check{\mathbf{v}}_{i_p}, \quad q_{ij} := \sum_{i_p} \mathbf{f}_{i_p j} . \check{\mathbf{v}}_{i_p}, \tag{6.3.6}$$

$$r_i := \sum_{i_p} \mathbf{b}_{i_p} . \check{\mathbf{v}}_{i_p}, \quad h_i := \sum_{i_p} \frac{1}{2} m_{i_p} \check{\mathbf{v}}_{i_p}^2, \tag{6.3.7}$$

and $\mathbf{f}_i, \mathbf{f}_{ij}$ and $\mathbf{b}_i$ are defined in (5.4.8). Terms $\mathbf{f}_i$ and q_i represent what might be called the *net self-force* and *net self-heating* associated with molecule P_i.

In order to obtain the energetic counterpart of (5.4.12) it is necessary to time average (6.3.5). Here, for simplicity of representation, we change notation and write [see (5.4.10)]

$$\bar{f} := f_\Delta. \tag{6.3.8}$$

Further, the associated (Δ-) *time fluctuation* in f is

$$f' := f - \bar{f}. \tag{6.3.9}$$

At this point we make two sets of assumptions.

Assumption A.1. $$\bar{q}_i = 0 \quad \text{and} \quad \mathbf{f}_i = \mathbf{0}, \tag{6.3.10}$$

and

Assumption A.2. Time averaging satisfies, for any functions of time here considered,

$$\bar{\bar{f}} = \bar{f} \quad \text{and} \quad \overline{\bar{f} g'} = 0. \tag{6.3.11}$$

Exercise 6.3.1. Show that

$$\overline{f'} = 0 \tag{6.3.12}$$

and
$$\overline{fg} = \bar{f}\bar{g} + \overline{f'g'}. \tag{6.3.13}$$

Noting that (5.4.11) may be written as

$$\bar{\dot{f}} = \dot{\bar{f}}, \tag{6.3.14}$$

from Assumptions A.1 and A.2, and (6.3.13), the time average of (6.3.5) takes the form

$$\sum_{j\neq i}(\bar{q}_{ij} + \bar{\mathbf{f}}_{ij}.\bar{\mathbf{v}}_i + \overline{\mathbf{f}'_{ij}.\mathbf{v}'_i}) + \bar{r}_i + \bar{\mathbf{b}}_i.\bar{\mathbf{v}}_i + \overline{\mathbf{b}'_i.\mathbf{v}'_i}$$
$$= \frac{d}{dt}\left\{\overline{\frac{1}{2}m_i\mathbf{v}_i^2}\right\} + \dot{\bar{h}}_i = \frac{d}{dt}\left\{\frac{1}{2}m_i\bar{\mathbf{v}}_i^2 + \frac{1}{2}m_i\overline{(\mathbf{v}'_i)^2}\right\} + \dot{\bar{h}}_i. \tag{6.3.15}$$

Equivalently,
$$\sum_{j\neq i}\bar{\mathbf{f}}_{ij}.\bar{\mathbf{v}}_i + \bar{\mathbf{b}}_i.\bar{\mathbf{v}}_i + \sum_{j\neq i}Q_{ij} + R_i = \frac{d}{dt}\left\{\frac{1}{2}m_i\bar{\mathbf{v}}_i^2\right\} + \dot{H}_i, \tag{6.3.16}$$

where
$$Q_{ij} := \bar{q}_{ij} + \sum_{j\neq i}\overline{\mathbf{f}'_{ij}.\mathbf{v}'_i}, \tag{6.3.17}$$

$$R_i := \bar{r}_i + \overline{\mathbf{b}'_i.\mathbf{v}'_i}, \tag{6.3.18}$$

and
$$H_i := \bar{h}_i + \frac{1}{2}m_i\overline{(\mathbf{v}'_i)^2}. \tag{6.3.19}$$

Remark 6.3.1. Relation (6.3.16) constitutes an energy balance for molecule P_i and is to be compared with the equivalent relation in which subatomic structure is neglected, namely (6.1.1) multiplied scalarly by $\bar{\mathbf{v}}_i$:

$$\bar{\mathbf{f}}_{ij}.\bar{\mathbf{v}}_i + \bar{\mathbf{b}}_i.\bar{\mathbf{v}}_i = \frac{d}{dt}\left\{\frac{1}{2}m_i\bar{\mathbf{v}}_i^2\right\}. \tag{6.3.20}$$

[Here one needs to recall that (6.1.1), the consequence of (5.4.4) and (5.4.5), was identified with subatomic considerations in (5.4.12).] While (6.1.1) is entirely consistent with subatomic structure, via the interpretation of terms in (5.4.12), relation (6.3.20) 'misses' the heat rates of supply to P_i from P_j (term Q_{ij}) and from the work done by external agency in the thermal motions prescribed by $\check{\mathbf{v}}_{i_p}$ and $\mathbf{v}'_i$ (term R_i) and also does not account for the molecular heat content of P_i associated with these thermal motions (term H_i).

Omitting the superposed bars in (6.3.16) (but retaining the interpretations of $\mathbf{f}_{ij}, \mathbf{b}_i, \mathbf{v}_i$ as time averages) yields

$$\sum_{j\neq i}\mathbf{f}_{ij}.\mathbf{v}_i + \mathbf{b}_i.\mathbf{v}_i + \sum_{j\neq i}Q_{ij} + R_i = \frac{d}{dt}\left(\frac{1}{2}m_i\mathbf{v}_i^2\right) + \dot{H}_i. \tag{6.3.21}$$

We now proceed as in Section 6.2; that is, we multiply (6.3.21) by $w(\mathbf{x}_i - \mathbf{x})$ and sum over all molecules P_i. (Here, to be consistent, $\mathbf{x}_i$ is to be regarded as $\bar{\mathbf{x}}_i$.) It follows that

$$Q_w^+ + \mathbf{f}_w.\mathbf{v}_w + r_w^+ + \mathbf{b}_w.\mathbf{v}_w = \rho_w\dot{K}_w^+ + \mathrm{div}\{\boldsymbol{\kappa}_w + \mathbf{k}_w + \boldsymbol{\mathcal{D}}_w\mathbf{v}_w\}, \tag{6.3.22}$$

where
$$Q_w^+(\mathbf{x},t) := \sum_{i \neq j}\sum \{Q_{ij}(t) + \mathbf{f}_{ij}(t)\,.\,\hat{\mathbf{v}}_i(t;\mathbf{x})\}w(\mathbf{x}_i(t) - \mathbf{x}), \tag{6.3.23}$$

$$r_w^+(\mathbf{x},t) := \sum_i \{R_i(t) + \mathbf{b}_i(t)\,.\,\hat{\mathbf{v}}_i(t;\mathbf{x})\}w(\mathbf{x}_i(t) - \mathbf{x}), \tag{6.3.24}$$

$$K_w^+ := H_w + h_w + \frac{1}{2}\mathbf{v}_w^2 \tag{6.3.25}$$

with
$$(\rho_w H_w)(\mathbf{x},t) := \sum_i H_i(t)w(\mathbf{x}_i(t) - \mathbf{x}), \tag{6.3.26}$$

and
$$\mathbf{k}_w(\mathbf{x},t) := \sum_i H_i(t)\hat{\mathbf{v}}_i(t;\mathbf{x})w(\mathbf{x}_i(t) - \mathbf{x}). \tag{6.3.27}$$

The definitions of $\boldsymbol{\kappa}_w$ and $\mathcal{D}_w$ remain unchanged. It follows from linear momentum balance that
$$-\operatorname{div}\boldsymbol{\kappa}_w^+ + Q_w^+ + r_w^+ - \mathcal{D}_w \cdot \mathbf{L}_w = \rho_w \dot{h}_w^+, \tag{6.3.28}$$

where
$$\boldsymbol{\kappa}_w^+ := \boldsymbol{\kappa}_w + \mathbf{k}_w \tag{6.3.29}$$

and
$$h_w^+ := H_w + h_w. \tag{6.3.30}$$

Remark 6.3.2. Relation (6.3.28) is formally the same as (6.2.24) but now includes the subatomic contributions $\mathbf{k}_w$ to $\boldsymbol{\kappa}_w^+, \sum_{i\neq j}\sum Q_{ij}\, w$ to $Q_w^+, \sum_i R_i\, w$ to r_w^+, and H_w to h_w^+: this is, of course, why the superscript plus sign has been employed.

The analysis of Section 6.2 can be repeated to derive from (6.3.28) other, more familiar, forms of energy balance. This will not be attempted here: the intention has been merely to indicate how, from the viewpoint of classical mechanics, subatomic behaviour contributes to macroscopic field values. In so doing, the role of time averaging has been crucial: in particular, several time scales have entered the reckoning. At this point it is helpful to consider the very notion of 'time'.

Remark 6.3.3. (Time and length scales) The passage of time is detectable only through change: if no clock were to register a change (in orientation of its hands or in its digital display), were our hearts not to beat, nor our environment to alter in any way, then both 'time' and 'motion' would be meaningless concepts. From a kinematic viewpoint, the detection of change depends upon the smallest length scale (λ, say) associated with the description of a physical system. The corresponding time scale is the smallest time interval (Δ_λ, say) necessary for change to be detected at scale λ. For example, configuration changes in stellar constellations depend upon the accuracy of observational equipment, but unaided human observation would seem to indicate a time scale in excess of several thousand years, while distances are measured in lightyears. For short-distance athletic track events, times may be electronically measured to an accuracy of 10^{-3} s, which corresponds (at speeds of $10\,\mathrm{ms}^{-1}$) to a spatial resolution of $\lambda = 1$ cm. Atomic dimensions are of order $1\,\text{Å}\,(= 10^{-10}$ m), while those of nuclei and electrons[2] are of orders 10^{-14} m and 10^{-15} m, respectively.

[2] The figure of 10^{-15} m for electrons is the accepted value associated with the classical viewpoint here adopted.

In envisaging motions of such entities it is thus reasonable to adopt length scales $\lambda_{\mathrm{at}} = 10^{-10}$ m, $\lambda_{\mathrm{nuc}} = 10^{-14}$ m, and $\lambda_{\mathrm{el}} = 10^{-15}$ m. If $\Delta_{\mathrm{at}}, \Delta_{\mathrm{nuc}}$ and Δ_{el} denote the corresponding time scales for motions, then it seems reasonable to assume that

$$\Delta_{\mathrm{el}} < \Delta_{\mathrm{nuc}} << \Delta_{\mathrm{at}} < \Delta_{\mathrm{mol}}, \tag{6.3.31}$$

where Δ_{mol} is the time scale associated with molecular mass centre motion. Consideration of atomic vibrations in crystals indicates that $\Delta_{\mathrm{at}} \sim 10^{-13}$ s, while recent studies suggest $\Delta_{\mathrm{el}} \sim 10^{-18}$ s.

In view of Remark 6.3.3 [and specifically inequalities (6.3.31)] the time scales associated with electron and nucleus velocities $\mathbf{v}_{i_p}$ are Δ_{el} and Δ_{nuc}. Accordingly the subatomic thermal velocities $\check{\mathbf{v}}_{i_p}$, computed with respect to molecular mass centre velocities [see 5.4.9)], are detectable over time intervals of duration Δ_{nuc}. The time scale utilised in (5.4.12) to obtain the term $(\mathbf{f}_{ij})_{\Delta}$ (here re-labelled $\bar{\mathbf{f}}_{ij}$) is associated both with averaging electronic motions in such a way as to establish a probabilistic density distribution (often described in terms of an electron 'cloud': see Remark 5.4.5) *and* also with averaging atomic vibrations. Thus Δ must be selected to exceed Δ_{at}: it is such a scale that appears in inequalities (6.3.31) as Δ_{mol}. It follows that in (5.4.4), and (5.4.5) $\mathbf{v}_i$ is (via Interpretation 2) actually $(\mathbf{v}_i)_{\Delta_{\mathrm{mol}}}$; that is, in these relations $\mathbf{v}_i$ is associated with time scale Δ_{mol}. Of course, $\mathbf{v}_i'$ [see (6.3.9)] is associated with time scale Δ_{at}. From (5.5.8), $\hat{\mathbf{v}}_i$ and $\mathbf{v}_i$ share the same time scale, namely Δ_{at}, on noting that in this relation $\mathbf{v}_i$ represents $\bar{\mathbf{v}}_i\,[= (\mathbf{v}_i)_{\Delta_{\mathrm{mol}}}]$. The relative motions associated with $\mathbf{v}_{i_p} - \mathbf{v}_i$, $\mathbf{v}_i - \bar{\mathbf{v}}_i$, and $\bar{\mathbf{v}}_i - \mathbf{v}_w$ are all regarded to be chaotic, and hence, since they represent submacroscopic activity, are of thermal character. The associated time scales are of orders Δ_{el} (or Δ_{nuc}: just which depends upon whether P_{i_p} is an electron or nucleus), Δ_{at}, and Δ_{mol}, respectively. The corresponding contributions to the heat energy density $\rho_w h_w^+$ are $\rho_w H_w$ [thermal motions of electrons, nuclei, and atoms: see (6.3.26), (6.3.19), and $(6.3.7)_2$] and $\rho_w h_w$ (thermal motions of molecular mass centres).

Remark 6.3.4. Assumption A.2 [see (6.3.11)] is formal. It embodies assumptions concerning the separation of time scales associated with subatomic kinematics and atomic vibrations from the time scale appropriate to describing molecular mass centre motions.

7 Fine-Scale Considerations: Moments, Couple Stress, Inhomogeneity, and Energetics

7.1 Preamble

The main aim in this chapter is to explore the molecular basis for the local form of rotational momentum balance

$$\operatorname{div}\mathbf{C} + \mathbf{T}^T - \mathbf{T} + \mathbf{J} = \rho\dot{\mathbf{S}}. \tag{2.7.29}$$

In Section 2.7 this relation was obtained from the postulated interior form of balance (2.7.3) which was couched in terms of skew tensor-valued fields. Accordingly $\operatorname{div}\mathbf{C}, \mathbf{J}$ and $\mathbf{S}$ in (2.7.29) take skew tensor values. Of course, such considerations can be phrased in terms of axial vectors [see (2.7.32) through (2.7.36)]. However, a more general perspective is now adopted which results in a full (not merely skew) tensor-valued relation, termed *balance of generalised moment of momentum*. The utility of such balance emerges when the corresponding *moment of mass balance* is established. The latter balance details the evolution of a measure of inhomogeneity and involves the divergence of a full tensor-valued moment of momentum density whose time evolution is prescribed by the generalised moment of momentum balance. These two moment balances (of mass and momentum) are 'fine-scale' counterparts of the continuity equation and linear momentum balance. The balance of energy is re-examined in the light of fine-scale considerations by adopting an appropriate finer-scale approximation to molecular thermal velocities.

7.2 Generalised Moment of Momentum Balance

Consider the equation governing the motion of molecule P_i in an inertial frame:

$$\sum_{j\neq i}\mathbf{f}_{ij} + \mathbf{b}_i = \frac{d}{dt}\{m_i\mathbf{v}_i\}. \tag{6.1.1}$$

Tensorial pre-multiplication by $(\mathbf{x}_i - \mathbf{x})w(\mathbf{x}_i - \mathbf{x})$ followed by summation over all molecules P_i yield

$$\mathbf{c}_w + \mathbf{J}_w = \sum_i (\mathbf{x}_i - \mathbf{x}) \otimes \frac{d}{dt}\{m_i\mathbf{v}_i\}w(\mathbf{x}_i - \mathbf{x}), \tag{7.2.1}$$

where $$\mathbf{c}_w(\mathbf{x},t) := \sum_{i\neq j}\sum (\mathbf{x}_i(t)-\mathbf{x})\otimes \mathbf{f}_{ij}(t)w(\mathbf{x}_i(t)-\mathbf{x}) \tag{7.2.2}$$

and $$\mathbf{J}_w(\mathbf{x},t) := \sum_i (\mathbf{x}_i(t)-\mathbf{x})\otimes \mathbf{b}_i(t)w(\mathbf{x}_i(t)-\mathbf{x}). \tag{7.2.3}$$

Now

$$(\mathbf{x}_i-\mathbf{x})\otimes \frac{d}{dt}\{m_i\mathbf{v}_i\}w(\mathbf{x}_i-\mathbf{x}) = \frac{\partial}{\partial t}\{(\mathbf{x}_i-\mathbf{x})\otimes m_i\mathbf{v}_i\, w(\mathbf{x}_i-\mathbf{x})\} - \mathbf{R}_i, \tag{7.2.4}$$

where $$\mathbf{R}_i := \mathbf{v}_i\otimes m_i\,\mathbf{v}_i\, w(\mathbf{x}_i-\mathbf{x}) + ((\mathbf{x}_i-\mathbf{x})\otimes m_i\,\mathbf{v}_i)\nabla w\,.\,\mathbf{v}_i. \tag{7.2.5}$$

Further[1],

$$\begin{aligned}((\mathbf{x}_i-\mathbf{x})\otimes m_i\,\mathbf{v}_i)\nabla w\,.\,\mathbf{v}_i &= -((\mathbf{x}_i-\mathbf{x})\otimes m_i\,\mathbf{v}_i)\nabla_{\mathbf{x}}\, w\,.\,\mathbf{v}_i \\ &= -((\mathbf{x}_i-\mathbf{x})\otimes m_i\,\mathbf{v}_i)\mathrm{div}\{\mathbf{v}_i\, w\} \\ &= -\mathbf{v}_i\otimes m_i\,\mathbf{v}_i\, w - \mathrm{div}\{(\mathbf{x}_i-\mathbf{x})\otimes m_i\,\mathbf{v}_i\otimes \mathbf{v}_i\, w\}.\end{aligned} \tag{7.2.6}$$

Hence, from (7.2.5),

$$\mathbf{R}_i = -\mathrm{div}\{(\mathbf{x}_i-\mathbf{x})\otimes m_i\,\mathbf{v}_i\otimes \mathbf{v}_i\, w\}. \tag{7.2.7}$$

It follows from (7.2.4) and (7.2.7) that [see (7.2.1)]

$$\sum_i (\mathbf{x}_i-\mathbf{x})\otimes \frac{d}{dt}\{m_i\,\mathbf{v}_i\}w = \frac{\partial}{\partial t}\{\rho_w\mathbf{B}_w\} + \mathrm{div}\,\mathbf{M}_w, \tag{7.2.8}$$

where the *tensor-valued moment of momentum density*

$$(\rho_w\mathbf{B}_w)(\mathbf{x},t) := \sum_i (\mathbf{x}_i(t)-\mathbf{x})\otimes m_i\,\mathbf{v}_i(t)w(\mathbf{x}_i(t)-\mathbf{x}) \tag{7.2.9}$$

and $$\mathbf{M}_w(\mathbf{x},t) := \sum_i (\mathbf{x}_i(t)-\mathbf{x})\otimes m_i\,\mathbf{v}_i(t)\otimes \mathbf{v}_i(t)w(\mathbf{x}_i(t)-\mathbf{x}). \tag{7.2.10}$$

Thus (7.2.1) and (7.2.8) yield the balance relation

$$\mathbf{c}_w + \mathbf{J}_w = \frac{\partial}{\partial t}\{\rho_w\mathbf{B}_w\} + \mathrm{div}\,\mathbf{M}_w. \tag{7.2.11}$$

From (7.2.10) and (5.5.8)

$$\begin{aligned}\mathbf{M}_w &= \sum_i (\mathbf{x}_i-\mathbf{x})\otimes m_i\,\mathbf{v}_i\otimes(\hat{\mathbf{v}}_i+\mathbf{v}_w)w \\ &= \hat{\mathbf{M}}_w + \mathbf{B}_w\otimes\rho_w\,\mathbf{v}_w,\end{aligned} \tag{7.2.12}$$

[1] In Cartesian tensor notation, $[\mathrm{div}\{\mathbf{a}\otimes\mathbf{b}\otimes\mathbf{c}\}]_{ij} = (a_ib_jc_k)_{,k}$. Here we have used [see Appendix B.7, (B.7.32)]

$$\mathrm{div}\{\mathbf{a}\otimes\mathbf{b}\otimes\mathbf{c}\} = (\nabla\mathbf{a})\mathbf{c}\otimes\mathbf{b} + \mathbf{a}\otimes(\nabla\mathbf{b})\mathbf{c} + (\mathrm{div}\,\mathbf{c})(\mathbf{a}\otimes\mathbf{b})$$

with $\mathbf{a} := (\mathbf{x}_i-\mathbf{x})$, $\mathbf{b} = m_i\mathbf{v}_i$, $\mathbf{c} = \mathbf{v}_i w$ and noted that $\nabla\mathbf{a} = -\mathbf{1}$ and $\nabla\mathbf{b} = \mathbf{O}$.

where $$\hat{\mathbf{M}}_w(\mathbf{x},t) := \sum_i (\mathbf{x}_i(t) - \mathbf{x}) \otimes m_i \mathbf{v}_i(t) \otimes \hat{\mathbf{v}}_i(t;\mathbf{x}) w(\mathbf{x}_i(t) - \mathbf{x}). \tag{7.2.13}$$

Since [see Appendix B.7, (B.7.33) with $\mathbf{A} = \mathbf{B}_w$ and $\mathbf{v} = \rho_w \mathbf{v}_w$]

$$\operatorname{div}\{\mathbf{B}_w \otimes \rho_w \mathbf{v}_w\} = (\nabla \mathbf{B}_w)\rho_w \mathbf{v}_w + (\operatorname{div}\{\rho_w \mathbf{v}_w\})\mathbf{B}_w, \tag{7.2.14}$$

$$\begin{aligned}\frac{\partial}{\partial t}\{\rho_w \mathbf{B}_w\} + \operatorname{div}\{\mathbf{B}_w \otimes \rho_w \mathbf{v}_w\} &= \left(\frac{\partial \rho_w}{\partial t} + \operatorname{div}\{\rho_w \mathbf{v}_w\}\right)\mathbf{B}_w + \rho_w\left(\frac{\partial \mathbf{B}_w}{\partial t} + (\nabla \mathbf{B}_w)\mathbf{v}_w\right) \\ &= \rho_w \dot{\mathbf{B}}_w.\end{aligned} \tag{7.2.15}$$

Here the continuity equation (4.2.16) has been invoked, and the material time derivative [cf. (2.5.25)] of $\mathbf{B}_w$ is noted to be [see (2.5.28)]

$$\begin{aligned}\dot{\mathbf{B}}_w(\mathbf{x},t) &:= \frac{\partial}{\partial t}\{\mathbf{B}_w(\chi_{t_0}(\hat{\mathbf{x}},t),t\} \\ &= \nabla_{\mathbf{x}} \mathbf{B}_w(\mathbf{x},t)\frac{\partial}{\partial t}\{\chi_{t_0}(\hat{\mathbf{x}},t)\} + \frac{\partial \mathbf{B}_w}{\partial t}(\mathbf{x},t) \\ &= \nabla \mathbf{B}_w(\mathbf{x},t)\mathbf{v}_w(\mathbf{x},t) + \frac{\partial \mathbf{B}_w}{\partial t}(\mathbf{x},t).\end{aligned} \tag{7.2.16}$$

Accordingly, from (7.2.11), (7.2.12), and (7.2.15),

$$\mathbf{c}_w + \mathbf{J}_w = \rho_w \dot{\mathbf{B}}_w + \operatorname{div}\hat{\mathbf{M}}_w. \tag{7.2.17}$$

If $\mathbf{a}_i$ denotes any solution to (5.6.2), then, on suppressing time dependence, from (7.2.2) and (5.6.5)

$$\begin{aligned}\mathbf{c}_w(\mathbf{x}) &= \sum_{i \neq j}\sum (\mathbf{x}_i - \mathbf{x}) \otimes \mathbf{f}_{ij} \operatorname{div} \mathbf{a}_i \\ &= \sum_{i \neq j}\sum \{\operatorname{div}\{(\mathbf{x}_i - \mathbf{x}) \otimes \mathbf{f}_{ij} \otimes \mathbf{a}_i + \mathbf{a}_i \otimes \mathbf{f}_{ij} - \mathbf{O}\} \\ &= (\operatorname{div}{}_s\mathbf{C}_w^-)(\mathbf{x}) + ({}_s\mathbf{T}_w^-)^T,\end{aligned} \tag{7.2.18}$$

where the *simple interaction couple stress tensor*

$${}_s\mathbf{C}_w^-(\mathbf{x},t) := \sum_{i \neq j}\sum (\mathbf{x}_i(t) - \mathbf{x}) \otimes \mathbf{f}_{ij}(t) \otimes \mathbf{a}_i(\mathbf{x},t). \tag{7.2.19}$$

In deriving (7.2.18) use has been made of the identity in footnote 1 [with $\mathbf{u} = (\mathbf{x}_i - \mathbf{x})$, $\mathbf{v} = \mathbf{f}_{ij}$, $\mathbf{w} = \mathbf{a}_i$: notice $\nabla_{\mathbf{x}}\{\mathbf{x}_i - \mathbf{x}\} = -\mathbf{1}$ and $\nabla_{\mathbf{x}}\mathbf{f}_{ij} = \mathbf{O}$] and identity [see Appendix A.8, (A.8.10)] $\mathbf{a} \otimes \mathbf{b} = (\mathbf{b} \otimes \mathbf{a})^T$ (with $\mathbf{a} = \mathbf{f}_{ij}$ and $\mathbf{b} = \mathbf{a}_i$).

Accordingly (7.2.17) may be re-written in the form

$$\operatorname{div}{}_s\mathbf{C}_w + ({}_s\mathbf{T}_w^-)^T + \mathbf{J}_w = \rho_w \dot{\mathbf{B}}_w, \tag{7.2.20}$$

where $${}_s\mathbf{C}_w := {}_s\mathbf{C}_w^- - \hat{\mathbf{M}}_w. \tag{7.2.21}$$

Remark 7.2.1. Notice that the interaction contribution ${}_s\mathbf{C}_w^-$ to ${}_s\mathbf{C}_w$ involves only resultant forces $\mathbf{F}_i$ [see (6.2.33), (7.2.19), and Remark 6.2.2].

In the event that interactions are pairwise balanced [see (5.6.6)], and on suppressing time dependence, (7.2.2) may be re-written as

$$
\begin{aligned}
\mathbf{c}_w(\mathbf{x}) :=& \sum_{i \neq j}\sum (\mathbf{x}_i - \mathbf{x}) \otimes \mathbf{f}_{ij}\, w(\mathbf{x}_i - \mathbf{x}) \\
=& \frac{1}{2} \sum_{i \neq j}\sum \{(\mathbf{x}_i - \mathbf{x}) \otimes \mathbf{f}_{ij}\, w(\mathbf{x}_i - \mathbf{x}) + (\mathbf{x}_j - \mathbf{x}) \otimes \mathbf{f}_{ji}\, w(\mathbf{x}_j - \mathbf{x})\} \\
=& \frac{1}{2} \sum_{i \neq j}\sum (\mathbf{x}_i - \mathbf{x}) \otimes \mathbf{f}_{ij} \{w(\mathbf{x}_i - \mathbf{x}) - w(\mathbf{x}_j - \mathbf{x})\} \\
&+ \sum_{i \neq j}\sum (\mathbf{x}_i - \mathbf{x}_j) \otimes \mathbf{f}_{ij}\, w(\mathbf{x}_j - \mathbf{x}) \\
=& \sum_{i \neq j}\sum (\mathbf{x}_i - \mathbf{x}) \otimes \mathbf{f}_{ij} (\operatorname{div} \mathbf{b}_{ij}) + {}_2\mathbf{C}_w,
\end{aligned}
\tag{7.2.22}
$$

where $\mathbf{b}_{ij}$ denotes any solution to (5.6.8), and

$$
{}_2\mathbf{C}_w(\mathbf{x}, t) := \frac{1}{2} \sum_{i \neq j}\sum (\mathbf{x}_i(t) - \mathbf{x}_j(t)) \otimes \mathbf{f}_{ij}(t) w(\mathbf{x}_j(t) - \mathbf{x}). \tag{7.2.23}
$$

However, using the identity in footnote 1 as in the derivation of (7.2.18) and recalling (5.6.10),

$$
\begin{aligned}
&\sum_{i \neq j}\sum (\mathbf{x}_i - \mathbf{x}) \otimes \mathbf{f}_{ij} (\operatorname{div} \mathbf{b}_{ij}) \\
&= \sum_{i \neq j}\sum \operatorname{div}\{(\mathbf{x}_i - \mathbf{x}) \otimes \mathbf{f}_{ij} \otimes \mathbf{b}_{ij}\} + \sum_{i \neq j}\sum \mathbf{b}_{ij} \otimes \mathbf{f}_{ij} - \mathbf{O} \\
&= \operatorname{div} {}_b\mathbf{C}_w^- + ({}_b\mathbf{T}_w^-)^T,
\end{aligned}
\tag{7.2.24}
$$

where the *pairwise-balanced form of interaction couple stress tensor*

$$
{}_b\mathbf{C}_w^-(\mathbf{x}, t) := \sum_{i \neq j}\sum (\mathbf{x}_i(t) - \mathbf{x}) \otimes \mathbf{f}_{ij}(t) \otimes \mathbf{b}_{ij}(\mathbf{x}, t). \tag{7.2.25}
$$

Thus, from (7.2.22) and (7.2.24),

$$
\mathbf{c}_w = \operatorname{div} {}_b\mathbf{C}_w^- + ({}_b\mathbf{T}_w^-)^T + {}_2\mathbf{C}_w, \tag{7.2.26}
$$

and balance (7.2.17) can be re-expressed in the form

$$
\operatorname{div} {}_b\mathbf{C}_w + ({}_b\mathbf{T}_w^-)^T + {}_2\mathbf{C}_w + \mathbf{J}_w = \rho_w \dot{\mathbf{B}}_w, \tag{7.2.27}
$$

where the *couple stress tensor for balanced interactions*

$$
{}_b\mathbf{C}_w := {}_b\mathbf{C}_w^- - \hat{\mathbf{M}}_w. \tag{7.2.28}
$$

Noting that, on suppressing time dependence, (7.2.23) may be written as

$$
{}_2\mathbf{C}_w(\mathbf{x}) = \frac{1}{2} \sum_{i \neq j}\sum (\mathbf{x}_i - \mathbf{x}_j) \otimes \mathbf{f}_{ij}\, w(\mathbf{x}_i - \mathbf{x}), \tag{7.2.29}
$$

and denoting by $\mathbf{a}_i$ any solution to (5.6.2), we have

$$_2\mathbf{C}_w = \text{div}\ _2\boldsymbol{\mathcal{C}}_w, \qquad (7.2.30)$$

where

$$_2\boldsymbol{\mathcal{C}}_w(\mathbf{x},t) := \frac{1}{2} \sum_{i \neq j}\sum (\mathbf{x}_i(t) - \mathbf{x}_j(t)) \otimes \mathbf{f}_{ij}(t) \otimes \mathbf{a}_i(\mathbf{x},t). \qquad (7.2.31)$$

Exercise 7.2.1. Prove relations (7.2.29) and (7.2.30).

From (7.2.31) it follows that balance (7.2.27) may be written in the form

$$\text{div}_b\mathbf{C}_w^+ + (_b\mathbf{T}_w^-)^T + \mathbf{J}_w = \rho_w \dot{\mathbf{B}}_w, \qquad (7.2.32)$$

where

$$_b\mathbf{C}_w^+ := {}_b\mathbf{C}_w + {}_2\boldsymbol{\mathcal{C}}_w = {}_b\mathbf{C}_w^- + {}_2\boldsymbol{\mathcal{C}}_w - \hat{\mathbf{M}}_w. \qquad (7.2.33)$$

Remark 7.2.2. Here the derived forms of moment of momentum balance (7.2.17), (7.2.20), (7.2.27), and (7.2.32) are equations which involve second-order tensor fields. Specifically, $\mathbf{c}_w, \mathbf{J}_w, \mathbf{B}_w, {}_s\mathbf{T}_w^-$, and ${}_b\mathbf{T}_w^-$ are all second-order tensor fields, as also are $\text{div}\hat{\mathbf{M}}_w$, $\text{div}_s\mathbf{C}_w$, $\text{div}_b\mathbf{C}_w$ and $\text{div}_b\mathbf{C}_w^+$ (of course, $\hat{\mathbf{M}}_w, {}_s\mathbf{C}_w, {}_b\mathbf{C}_w$, and ${}_b\mathbf{C}_w^+$ are third-order tensor fields). Since the usual statement of moment of momentum balance involves only axial vector fields (or, equivalently, skew tensor fields: see Appendix A.15), we observe that

(i) the derived balances are generalisations of the usual postulated balance, and
(ii) it is the skew part of the derived balances which should be compared with the usual postulated balance.

In order to make comparisons it is helpful to define, for any second-order tensor $\mathbf{A}$,

$$\mathbf{A}^{\text{sk}} := \mathbf{A} - \mathbf{A}^T. \qquad (7.2.34)$$

That is, $\mathbf{A}^{\text{sk}}$ denotes twice the skew part of $\mathbf{A}$. In particular,

$$(\mathbf{a} \otimes \mathbf{b})^{\text{sk}} = \mathbf{a} \otimes \mathbf{b} - \mathbf{b} \otimes \mathbf{a} = \mathbf{a} \wedge \mathbf{b}. \qquad (7.2.35)$$

Accordingly, from (7.2.17), (7.2.20), (7.2.27), and (7.2.32), it follows that

$$\mathbf{c}_w^{\text{sk}} + \mathbf{J}_w^{\text{sk}} = \rho_w \widehat{\mathbf{B}_w^{\text{sk}}} + (\text{div}\,\hat{\mathbf{M}}_w)^{\text{sk}}, \qquad (7.2.36)$$

$$(\text{div}\ {}_s\mathbf{C}_w)^{\text{sk}} + ({}_s\mathbf{T}_w)^T - {}_s\mathbf{T}_w + \mathbf{J}_w^{\text{sk}} = \rho_w \widehat{\mathbf{B}_w^{\text{sk}}}, \qquad (7.2.37)$$

$$(\text{div}\ {}_b\mathbf{C}_w)^{\text{sk}} + ({}_b\mathbf{T}_w)^T - {}_b\mathbf{T}_w + {}_2\mathbf{C}_w^{\text{sk}} + \mathbf{J}_w^{\text{sk}} = \rho_w \widehat{\mathbf{B}_w^{\text{sk}}}, \qquad (7.2.38)$$

and

$$(\text{div}\ {}_b\mathbf{C}_w^+)^{\text{sk}} + ({}_b\mathbf{T}_w)^T - {}_b\mathbf{T}_w + \mathbf{J}_w^{\text{sk}} = \rho_w \widehat{\mathbf{B}_w^{\text{sk}}}, \qquad (7.2.39)$$

respectively.

Exercise 7.2.2. Show that

$$({}_s\mathbf{T}_w^-)^{\text{sk}} = {}_s\mathbf{T}_w - {}_s\mathbf{T}_w^T \quad \text{and} \quad ({}_b\mathbf{T}_w^-)^{\text{sk}} = {}_b\mathbf{T}_w - {}_b\mathbf{T}_w^T. \qquad (7.2.40)$$

Recalling the usual postulated form of balance (2.7.29), it is relation (7.2.39) which is directly comparable. Clearly, the identifications to be made are (noting that $\rho_w \leftrightarrow \rho$)

$$ {}_b\mathbf{T}_w \leftrightarrow \mathbf{T},\ \mathbf{J}_w^{\text{sk}} \leftrightarrow \mathbf{J},\ \mathbf{B}_w^{\text{sk}} \leftrightarrow \mathbf{S}, \tag{7.2.41} $$

and

$$ (\text{div}\, {}_b\mathbf{C}_w^+)^{\text{sk}} \longleftrightarrow \text{div}\,\mathbf{C}. \tag{7.2.42} $$

Notice in this respect (on suppressing time dependence) that [see (7.2.3) and (7.2.9)]

$$ \mathbf{J}_w^{\text{sk}}(\mathbf{x}) = \sum_i (\mathbf{x}_i - \mathbf{x}) \wedge \mathbf{b}_i\, w(\mathbf{x}_i - \mathbf{x}), \tag{7.2.43} $$

and

$$ (\rho_w \mathbf{B}_w^{\text{sk}})(\mathbf{x}) = \sum_i (\mathbf{x}_i - \mathbf{x}) \wedge m_i \mathbf{v}_i\, w(\mathbf{x}_i - \mathbf{x}). \tag{7.2.44} $$

Further, employing Cartesian tensor notation, note that for any third-order tensor field $\mathbf{C}$ [see Appendix B, (B.7.22)]

$$ \begin{aligned} ((\text{div}\,\mathbf{C})^{\text{sk}})_{ij} &= (\text{div}\,\mathbf{C})_{ij} - (\text{div}\,\mathbf{C})_{ji} \\ &= C_{ijk,k} - C_{jik,k}. \end{aligned} $$

Accordingly

$$ ({}_b\mathbf{C}_w^+)^{\text{sk}} \longleftrightarrow \mathbf{C}, \tag{7.2.45} $$

where

$$ ({}_b\mathbf{C}_w^+)^{\text{sk}} := ({}_b\mathbf{C}_w^-)^{\text{sk}} - \hat{\mathbf{M}}_w^{\text{sk}} + {}_2\mathcal{C}_w^{\text{sk}} \tag{7.2.46} $$

with

$$ ({}_b\mathbf{C}_w^-)^{\text{sk}}(\mathbf{x}) := \sum\sum_{i \neq j} (\mathbf{x}_i - \mathbf{x}) \wedge \mathbf{f}_{ij} \otimes \mathbf{b}_{ij}(\mathbf{x}), \tag{7.2.47} $$

$$ (\hat{\mathbf{M}}_w)^{\text{sk}}(\mathbf{x}) := \sum_i (\mathbf{x}_i - \mathbf{x}) \wedge m_i \mathbf{v}_i \otimes \hat{\mathbf{v}}_i(\mathbf{x})\, w(\mathbf{x}_i - \mathbf{x}), \tag{7.2.48} $$

and

$$ {}_2\mathcal{C}_w^{\text{sk}}(\mathbf{x}) := \frac{1}{2} \sum\sum_{i \neq j} (\mathbf{x}_i - \mathbf{x}_j) \wedge \mathbf{f}_{ij} \otimes \mathbf{a}_i(\mathbf{x}). \tag{7.2.49} $$

Here we have written, for any vectors $\mathbf{u}, \mathbf{v}$, and $\mathbf{w}$,

$$ (\mathbf{u} \wedge \mathbf{v} \otimes \mathbf{w})_{ijk} := u_i v_j w_k - u_j v_i w_k. \tag{7.2.50} $$

Remark 7.2.3. Recall (see Remark 5.6.2) that if $\mathbf{f}_{ij}$ is parallel to $(\mathbf{x}_j - \mathbf{x}_i)$ at all times, then the Hardy choice ${}_H\mathbf{T}_w^-$ of interaction stress tensor is symmetric-valued. It follows that ${}_H\mathbf{T}_w$ is also symmetric-valued (Why?). Further, in such case ${}_2\mathbf{C}_w$ [see (7.2.29)] takes symmetric values and hence [see also (7.2.30)] ${}_2\mathbf{C}_w^{\text{sk}} = \mathcal{O} = {}_2\mathcal{C}_w$. Thus, when the Hardy choice [see (5.6.13) and (5.6.15)] to the solution of (5.6.8) is made, (7.2.38) reduces to

$$ (\text{div}\, {}_b\mathbf{C}_w)^{\text{sk}} + \mathbf{J}_w^{\text{sk}} = \rho_w \widehat{\dot{\mathbf{B}}_w^{\text{sk}}}. \tag{7.2.51} $$

Here

$$ (\text{div}\, {}_b\mathbf{C}_w)^{\text{sk}} = \text{div}\, {}_b\mathbf{C}_w^{\text{sk}}, \tag{7.2.52} $$

where [see (7.2.46), (7.2.47), and (7.2.48)]

$$ {}_b\mathbf{C}_w^{\text{sk}} := ({}_b\mathbf{C}_w^{-})^{\text{sk}} - \hat{\mathbf{M}}_w^{\text{sk}}. \tag{7.2.53}$$

That is, in such case the balances of linear and moment of momentum [skew-valued version (7.2.51)] uncouple in the sense that the only field common to both balances is ρ_w.

7.3 Inhomogeneity and Moment of Mass Conservation

The considerations of Section 7.2 indicate that, a priori, couple stress is to be expected. However, in most common contexts it is negligible, as evidenced by the success of theories (termed 'non-polar') in which it does not appear (cf. e.g., Truesdell & Noll [2] and Gurtin, Fried & Anand [23]). Nevertheless, adequate descriptions of materials with significant microstructure (e.g., liquid-crystalline phases: cf. De Gennes [24]) or inhomogeneity (cf., e.g., Toupin [25], Mindlin & Tiersten [26]) require balances of form (7.2.20)/(7.2.32), supplemented by appropriate constitutive equations and boundary conditions. To see how generalised moment of momentum balance is linked to inhomogeneity, consider the *moment of mass density*

$$(\rho_w \mathbf{d}_w)(\mathbf{x},t) := \sum_i (\mathbf{x}_i(t) - \mathbf{x}) m_i w(\mathbf{x}_i(t) - \mathbf{x}). \tag{7.3.1}$$

For the simplest choice w_ϵ of weighting function (see Section 4.3) $\mathbf{d}_w(\mathbf{x},t)$ denotes the displacement from $\mathbf{x}$ of the mass centre of those molecules in $S_\epsilon(\mathbf{x})$ at time t. (Convince yourself of this!) Accordingly, keeping $\mathbf{x}$ fixed in (7.3.1) and differentiating with respect to time,

$$\frac{\partial}{\partial t}\{\rho_w \mathbf{d}_w\} = \sum_i m_i \mathbf{v}_i w + \sum_i m_i (\mathbf{x}_i - \mathbf{x}) \nabla w \,.\, \mathbf{v}_i. \tag{7.3.2}$$

Now, using identity (B.7.28) of Appendix B.7 with $\phi = w, \mathbf{v} = \mathbf{v}_i$ and then identity (B.7.29) of Appendix B.7 with $\mathbf{u} = (\mathbf{x}_i - \mathbf{x}), \mathbf{v} = m_i w \mathbf{v}_i$,

$$\begin{aligned}
\sum_i m_i (\mathbf{x}_i - \mathbf{x}) \nabla w \,.\, \mathbf{v}_i &= -\sum_i m_i (\mathbf{x}_i - \mathbf{x}) \nabla_{\mathbf{x}} w \,.\, \mathbf{v}_i \\
&= -\sum_i m_i (\mathbf{x}_i - \mathbf{x}) \operatorname{div}\{\mathbf{v}_i w\} \\
&= -\sum_i \Big(\operatorname{div}\{(\mathbf{x}_i - \mathbf{x}) \otimes m_i \mathbf{v}_i w\} - \nabla\{m_i (\mathbf{x}_i - \mathbf{x})\} \mathbf{v}_i w \Big) \\
&= -\operatorname{div}\{\rho_w \mathbf{B}_w\} - \sum_i m_i \mathbf{v}_i w.
\end{aligned} \tag{7.3.3}$$

[See (7.2.9).] Thus (7.3.2) and (7.3.3) yield

$$\frac{\partial}{\partial t}\{\rho_w \mathbf{d}_w\} + \operatorname{div}\{\rho_w \mathbf{B}_w\} = \mathbf{0}. \tag{7.3.4}$$

This may be re-written, upon noting that from the continuity equation (4.2.16)

$$\begin{aligned}\frac{\partial}{\partial t}\{\rho_w \mathbf{d}_w\} &= \frac{\partial \rho_w}{\partial t}\mathbf{d}_w + \rho_w \frac{\partial \mathbf{d}_w}{\partial t} = -(\mathrm{div}\{\rho_w \mathbf{v}_w\})\mathbf{d}_w + \rho_w \frac{\partial \mathbf{d}_w}{\partial t} \\ &= (\nabla \mathbf{d}_w)\rho_w \mathbf{v}_w - \mathrm{div}\{\mathbf{d}_w \otimes \rho_w \mathbf{v}_w\} + \rho_w \frac{\partial \mathbf{d}_w}{\partial t} \\ &= \rho_w \dot{\mathbf{d}}_w - \mathrm{div}\{\mathbf{d}_w \otimes \rho_w \mathbf{v}_w\},\end{aligned} \tag{7.3.5}$$

where

$$\dot{\mathbf{d}}_w := \frac{\partial \mathbf{d}_w}{\partial t} + (\nabla \mathbf{d}_w)\mathbf{v}_w \tag{7.3.6}$$

denotes the material time derivative of $\mathbf{d}_w$ [see Appendix (B.5.40)] and use has been made of identity (B.7.29) of Appendix B.7. From (7.3.4) and (7.3.5)

$$\rho_w \dot{\mathbf{d}}_w + \mathrm{div}\{\rho_w(\mathbf{B}_w - \mathbf{d}_w \otimes \mathbf{v}_w)\} = \mathbf{0}. \tag{7.3.7}$$

Note that from (7.2.9) and (7.3.1), at point $\mathbf{x}$ (time is suppressed)

$$\begin{aligned}\rho_w(\mathbf{B}_w - \mathbf{d}_w \otimes \mathbf{v}_w) &= \sum_i \Big((\mathbf{x}_i - \mathbf{x}) \otimes m_i \mathbf{v}_i - \mathbf{d}_w(\mathbf{x}) \otimes m_i \mathbf{v}_i\Big) w(\mathbf{x}_i - \mathbf{x}) \\ &= \sum_i (\mathbf{x}_i - \mathbf{x} - \mathbf{d}_w(\mathbf{x})) \otimes m_i \mathbf{v}_i\, w(\mathbf{x}_i - \mathbf{x}).\end{aligned} \tag{7.3.8}$$

That is, at any instant the value of $\rho_w(\mathbf{B}_w - \mathbf{d}_w \otimes \mathbf{v}_w)$ at location $\mathbf{x}$ is the resultant (generalised) moment of momentum density of molecules in $S_\epsilon(\mathbf{x})$ about their mass centre.

Exercise 7.3.1. Show also that at point $\mathbf{x}$

$$\begin{aligned}\rho_w(\mathbf{B}_w - \mathbf{d}_w \otimes \mathbf{v}_w) &= \sum_i (\mathbf{x}_i - \mathbf{x}) \otimes m_i \hat{\mathbf{v}}_i(\mathbf{x}) w(\mathbf{x}_i - \mathbf{x}) \\ &\left[= \sum_i (\mathbf{x}_i - \mathbf{x} - \mathbf{d}_w(\mathbf{x})) \otimes m_i \hat{\mathbf{v}}_i(\mathbf{x}) w(\mathbf{x}_i - \mathbf{x})\right].\end{aligned} \tag{7.3.9}$$

Remark 7.3.1. Balances (7.2.11), (7.2.20), (7.2.27), and (7.2.32) of generalised moment of momentum constitute evolution equations for $\mathbf{B}_w$ and thus are intimately linked to considerations of inhomogeneity, in view of the connection between measure $\mathbf{d}_w$ of such inhomogeneity and $\mathbf{B}_w$ in moment of mass relations (7.3.4) and (7.3.7).

Remark 7.3.2. The definitions of $\mathbf{c}_w, \mathbf{J}_w, \mathbf{B}_w, \hat{\mathbf{M}}_w,$ and $\mathbf{d}_w$ all involve factors $\boldsymbol{\alpha}_i := (\mathbf{x}_i - \mathbf{x})w(\mathbf{x}_i - \mathbf{x})$. Given the localising nature[2] of w (see W.F.2 in Section 4.2), these fields elicit finer-scale information about a given material system than is delivered by fields which appear in the balances of linear momentum and energy so far considered. Accordingly these balances will now be re-examined for what has so far been an *implicit* contribution of fine-scale behaviour. The key here is to introduce a finer-scale approximation $\check{\mathbf{v}}_i(t;\mathbf{x})$ to the thermal velocity of molecules P_i in the

[2] Notice that if $w = w_\epsilon$ [see (4.3.2)], then $\|\boldsymbol{\alpha}_i\| < \epsilon V_\epsilon^{-1}$.

vicinity of $\mathbf{x}$. Specifically [see (5.5.8) and the discussion associated with (5.5.25)],

$$\check{\mathbf{v}}_i(t;\mathbf{x}) := \mathbf{v}_i(t) - \mathbf{v}_w(\mathbf{x},t) - \mathbf{L}_w(\mathbf{x},t)(\mathbf{x}_i(t) - \mathbf{x}) \tag{7.3.10}$$

$$= \hat{\mathbf{v}}_i(t;\mathbf{x}) - \mathbf{L}_w(\mathbf{x},t)(\mathbf{x}_i(t) - \mathbf{x}). \tag{7.3.11}$$

Here $\mathbf{L}_w$ denotes the spatial gradient of $\mathbf{v}_w$ [see (6.2.23)], and $\check{\mathbf{v}}_i(t;\mathbf{x})$ is seen to be a higher-order approximation [for $\mathbf{x}_i(t)$ near to $\mathbf{x}$] than $\hat{\mathbf{v}}_i(t;\mathbf{x})$ to the actual thermal velocity of P_i, namely

$$\tilde{\mathbf{v}}_i(t) := \mathbf{v}_i(t) - \mathbf{v}_w(\mathbf{x}_i(t),t). \tag{5.5.25}$$

7.4 Fine-Scale Energetics

The balances of linear momentum and energy which have been derived are *exact*. However, the distinction between fields of thermal and mechanical natures was made on the basis of the appearance of notional thermal velocities $\hat{\mathbf{v}}_i$ in the definitions of thermal fields. A finer-scale viewpoint requires revision of this distinction. The first appearance of a thermal term was that of $\mathcal{D}_w$ in (5.5.11) (see Subsection 5.5.2). Introduction of $\mathcal{D}_w$ enabled balance (5.5.6) to be expressed in the form (5.5.15): this latter form served as the basis for the subsequent standard form of balance (2.7.30) upon selection of an interaction stress tensor $\mathbf{T}_w^-$ [see (5.5.18)]. From (5.5.11) and (7.3.11), upon suppressing time dependence,

$$\begin{aligned}\mathcal{D}_w(\mathbf{x}) &= \sum_i m_i(\check{\mathbf{v}}_i + \mathbf{L}_w(\mathbf{x}_i - \mathbf{x})) \otimes (\check{\mathbf{v}}_i + \mathbf{L}_w(\mathbf{x}_i - \mathbf{x}))w(\mathbf{x}_i - \mathbf{x}) \\ &= \check{\mathcal{D}}_w(\mathbf{x}) + \mathbf{L}_w\left(\sum_i (\mathbf{x}_i - \mathbf{x}) \otimes m_i\,\check{\mathbf{v}}_i\,w(\mathbf{x}_i - \mathbf{x})\right) \\ &\quad + \left(\sum_i m_i\check{\mathbf{v}}_i \otimes (\mathbf{x}_i - \mathbf{x})w(\mathbf{x}_i - \mathbf{x})\right)\mathbf{L}_w^T + \rho_w\,\mathbf{L}_w\,\mathbf{I}_w\,\mathbf{L}_w^T,\end{aligned} \tag{7.4.1}$$

where
$$\check{\mathcal{D}}_w(\mathbf{x},t) := \sum_i m_i\,\check{\mathbf{v}}_i(t;\mathbf{x}) \otimes \check{\mathbf{v}}_i(t;\mathbf{x})w(\mathbf{x}_i(t) - \mathbf{x}), \tag{7.4.2}$$

and the *second moment of mass density*

$$(\rho_w\mathbf{I}_w)(\mathbf{x},t) := \sum_i (\mathbf{x}_i(t) - \mathbf{x}) \otimes m_i(\mathbf{x}_i(t) - \mathbf{x})w(\mathbf{x}_i(t) - \mathbf{x}). \tag{7.4.3}$$

Thus
$$\mathcal{D}_w = \check{\mathcal{D}}_w + \rho_w\,\mathbf{L}_w\,\mathcal{A}_w + \rho_w\,\mathcal{A}_w^T\,\mathbf{L}_w^T + \rho_w\,\mathbf{L}_w\,\mathbf{I}_w\,\mathbf{L}_w^T, \tag{7.4.4}$$

where
$$(\rho_w\mathcal{A}_w)(\mathbf{x},t) := \sum_i (\mathbf{x}_i(t) - \mathbf{x}) \otimes m_i\,\check{\mathbf{v}}_i(t;\mathbf{x})w(\mathbf{x}_i(t) - \mathbf{x}). \tag{7.4.5}$$

Exercise 7.4.1. Show that, from (7.3.10), (7.2.9), (7.3.1), and (7.4.3),

$$\mathcal{A}_w = \mathbf{B}_w - \mathbf{d}_w \otimes \mathbf{v}_w - \mathbf{I}_w\,\mathbf{L}_w^T. \tag{7.4.6}$$

Notice that (7.3.7) may be written via (7.4.6) as

$$\rho_w\,\dot{\mathbf{d}}_w + \operatorname{div}\{\rho_w(\mathcal{A}_w + \mathbf{I}_w\,\mathbf{L}_w^T)\} = \mathbf{0}. \tag{7.4.7}$$

Remark 7.4.1. On adopting the finer-scale viewpoint, it can be seen from (7.4.4) that the contribution $\mathcal{D}_w$ to any candidate Cauchy stress [see (5.5.20)] is not entirely thermal. Here both $\breve{\mathcal{D}}_w$ and $\mathcal{A}_w$ are notionally thermal, while $\rho_w \mathbf{L}_w \mathbf{I}_w \mathbf{L}_w^T$ is a fine-scale kinetic quantity.

Now consider energy balance (6.2.64),

$$r_w - \text{div}\, {}_b\mathbf{q}_w^+ + {}_b\mathbf{T}_w \cdot \mathbf{L}_w = \rho_w \dot{e}_w. \tag{7.4.8}$$

The fine-scale contribution to ${}_b\mathbf{T}_w$ is given, from (6.2.46) and (7.4.4), by

$${}_b\mathbf{T}_w = {}_b\mathbf{T}_w^- - \mathcal{D}_w = {}_b\mathbf{T}_w^- - \breve{\mathcal{D}}_w - \rho_w \mathbf{L}_w \mathcal{A}_w - \rho_w \mathcal{A}_w^T \mathbf{L}_w^T - \rho_w \mathbf{L}_w \mathbf{I}_w \mathbf{L}_w^T. \tag{7.4.9}$$

In the same way, the remaining terms in (7.4.8) will be re-written to exhibit the *implicit* fine-scale contributions. Thus, from (6.2.6) and (7.2.3),

$$r_w = \sum_i \mathbf{b}_i \,.\, \hat{\mathbf{v}}_i\, w = \sum_i \mathbf{b}_i \,.\, (\breve{\mathbf{v}}_i + \mathbf{L}_w(\mathbf{x}_i - \mathbf{x}))_w = \breve{r}_w + \mathbf{J}_w \cdot \mathbf{L}_w^T, \tag{7.4.10}$$

where

$$\breve{r}_w(\mathbf{x},t) := \sum_i \mathbf{b}_i(t) \,.\, \breve{\mathbf{v}}_i(t;\mathbf{x}) w(\mathbf{x}_i(t) - \mathbf{x}). \tag{7.4.11}$$

Contribution ${}_b\mathbf{q}_w^-$ to ${}_b\mathbf{q}_w^+$ [see (6.2.65)] is, from (6.2.41) and (7.3.11),

$$\begin{aligned} {}_b\mathbf{q}_w^- &= -\sum\sum_{i \neq j} \mathbf{f}_{ij} \,.\, (\breve{\mathbf{v}}_i + \mathbf{L}_w(\mathbf{x}_i - \mathbf{x}))\mathbf{b}_{ij} = {}_b\breve{\mathbf{q}}_w^- - \sum\sum_{i \neq j} (\mathbf{b}_{ij} \otimes \mathbf{f}_{ij}) \mathbf{L}_w(\mathbf{x}_i - \mathbf{x}) \\ &= {}_b\breve{\mathbf{q}}_w^- - ({}_b\mathbf{C}_w^-)^\sim : \mathbf{L}_w^T. \end{aligned} \tag{7.4.12}$$

Here

$${}_b\breve{\mathbf{q}}_w^-(\mathbf{x},t) := -\sum\sum_{i \neq j} \mathbf{f}_{ij}(t) \,.\, \breve{\mathbf{v}}_i(t;\mathbf{x}) \mathbf{b}_{ij}(\mathbf{x},t). \tag{7.4.13}$$

The final step in obtaining (7.4.12) follows from (7.2.25), together with the identity, for vectors $\mathbf{a}, \mathbf{b}$, and $\mathbf{c}$ and linear transformation $\mathbf{L}$,

$$(\mathbf{c} \otimes \mathbf{b})\mathbf{L}\mathbf{a} = (\mathbf{c} \otimes \mathbf{b} \otimes \mathbf{a}) : \mathbf{L}^T = (\mathbf{a} \otimes \mathbf{b} \otimes \mathbf{c})^\sim : \mathbf{L}^T, \tag{7.4.14}$$

with $\mathbf{a} = (\mathbf{x}_i - \mathbf{x})$, $\mathbf{b} = \mathbf{f}_{ij}$, $\mathbf{c} = \mathbf{b}_{ij}$ and $\mathbf{L} = \mathbf{L}_w$ (see Appendix B.8). Further, from (7.4.12) and (B.7.36) of Appendix B.7 with $\mathbf{C} = {}_b\mathbf{C}_w^-, \mathbf{A} = \mathbf{L}^T$,

$$\text{div}\, {}_b\mathbf{q}_w^- = \text{div}\, {}_b\breve{\mathbf{q}}_w^- - (\text{div}\, {}_b\mathbf{C}_w^-) \cdot \mathbf{L}_w^T - {}_b\mathbf{C}_w^- \cdot \nabla(\mathbf{L}_w^T). \tag{7.4.15}$$

Contribution ${}_2\mathbf{q}_w$ to ${}_b\mathbf{q}_w^+$ [see (6.2.65)] is, from (6.2.62) and (7.3.11),

$${}_2\mathbf{q}_w = \frac{1}{2} \sum\sum_{i \neq j} \hat{\phi}_{ij}(\breve{\mathbf{v}}_i + \mathbf{L}_w(\mathbf{x}_i - \mathbf{x}))w = {}_2\breve{\mathbf{q}}_w + \mathbf{L}_w \mathbf{i}_w, \tag{7.4.16}$$

where

$${}_2\breve{\mathbf{q}}_w(\mathbf{x},t) := \frac{1}{2} \sum\sum_{i \neq j} \hat{\phi}_{ij}(t) \breve{\mathbf{v}}_i(t;\mathbf{x}) w(\mathbf{x}_i(t) - \mathbf{x}) \tag{7.4.17}$$

and

$$\mathbf{i}_w(\mathbf{x},t) := \frac{1}{2} \sum\sum_{i \neq j} \hat{\phi}_{ij}(t)(\mathbf{x}_i(t) - \mathbf{x}) w(\mathbf{x}_i(t) - \mathbf{x}). \tag{7.4.18}$$

The final contribution $\boldsymbol{\kappa}_w$ to ${}_b\mathbf{q}_w^+$ [see (6.2.65)] is, from (6.2.15) and (7.3.11), given by

$$2\boldsymbol{\kappa}_w = \sum_i m[(\check{\mathbf{v}}_i + \mathbf{L}_w(\mathbf{x}_i - \mathbf{x})) . (\check{\mathbf{v}}_i + \mathbf{L}_w(\mathbf{x}_i - \mathbf{x}))](\check{\mathbf{v}}_i + \mathbf{L}_w(\mathbf{x}_i - \mathbf{x}))w \tag{7.4.19}$$

$$\begin{aligned}
&= \sum_i m_i(\check{\mathbf{v}}_i^2 + 2\check{\mathbf{v}}_i . \mathbf{L}_w(\mathbf{x}_i - \mathbf{x}) + \mathbf{L}_w(\mathbf{x}_i - \mathbf{x}) . \mathbf{L}_w(\mathbf{x}_i - \mathbf{x}))(\check{\mathbf{v}}_i + \mathbf{L}_w(\mathbf{x}_i - \mathbf{x}))w \\
&= \sum_i m_i \check{\mathbf{v}}_i^2 \check{\mathbf{v}}_i w + \mathbf{L}_w \sum_i m_i \check{\mathbf{v}}_i^2 (\mathbf{x}_i - \mathbf{x})w + 2\sum_i m_i(\check{\mathbf{v}}_i \otimes \check{\mathbf{v}}_i)\mathbf{L}_w(\mathbf{x}_i - \mathbf{x})w \\
&\quad + 2\mathbf{L}_w\left(\sum_i m_i(\check{\mathbf{v}}_i . \mathbf{L}_w(\mathbf{x}_i - \mathbf{x}))(\mathbf{x}_i - \mathbf{x})w\right) \\
&\quad + \sum_i m_i(\mathbf{L}_w(\mathbf{x}_i - \mathbf{x}) . \mathbf{L}_w(\mathbf{x}_i - \mathbf{x}))\check{\mathbf{v}}_i w \\
&\quad + \mathbf{L}_w \sum_i m_i(\mathbf{L}_w(\mathbf{x}_i - \mathbf{x}) . \mathbf{L}_w(\mathbf{x}_i - \mathbf{x}))(\mathbf{x}_i - \mathbf{x})w.
\end{aligned} \tag{7.4.20}$$

That is[3],

$$\begin{aligned}
\boldsymbol{\kappa}_w = \check{\boldsymbol{\kappa}}_w &+ \frac{1}{2}\mathbf{L}_w(\check{\mathbf{M}}_w : \mathbf{1}) + \check{\mathbf{M}}_w^{\sim} : \mathbf{L}_w^T + \mathbf{L}_w(\mathbf{R}_w : \mathbf{L}_w) \\
&+ \frac{1}{2}\mathbf{R}_w^{\sim} : \mathbf{L}_w^T\mathbf{L}_w + \frac{1}{2}\mathbf{L}_w(\mathbf{S}_w : \mathbf{L}_w^T\mathbf{L}_w).
\end{aligned} \tag{7.4.21}$$

Here

$$\check{\boldsymbol{\kappa}}_w(\mathbf{x},t) := \sum_i \frac{1}{2} m_i(\check{\mathbf{v}}_i(t;\mathbf{x}))^2 \check{\mathbf{v}}_i(t;\mathbf{x})w(\mathbf{x}_i(t) - \mathbf{x}), \tag{7.4.22}$$

$$\check{\mathbf{M}}_w(\mathbf{x},t) := \sum_i m_i(\mathbf{x}_i(t) - \mathbf{x}) \otimes \check{\mathbf{v}}_i(t;\mathbf{x}) \otimes \check{\mathbf{v}}_i(t;\mathbf{x})w(\mathbf{x}_i(t) - \mathbf{x}), \tag{7.4.23}$$

$$\mathbf{R}_w(\mathbf{x},t) := \sum_i m_i(\mathbf{x}_i(t) - \mathbf{x}) \otimes (\mathbf{x}_i(t) - \mathbf{x}) \otimes \check{\mathbf{v}}_i(t;\mathbf{x})w(\mathbf{x}_i(t) - \mathbf{x}), \tag{7.4.24}$$

and the *third moment of mass density*

$$\mathbf{S}_w(\mathbf{x},t) := \sum_i m_i(\mathbf{x}_i(t) - \mathbf{x}) \otimes (\mathbf{x}_i(t) - \mathbf{x}) \otimes (\mathbf{x}_i(t) - \mathbf{x})w(\mathbf{x}_i(t) - \mathbf{x}). \tag{7.4.25}$$

Finally, the fine-scale nature of $\rho_w e_w$ derives [see (6.2.66) and (6.2.55)] from that of $\rho_w h_w$, which, from (6.2.13), (7.4.4), and (5.5.11), is expressible as

$$\rho_w h_w = \frac{1}{2}\,\mathrm{tr}\boldsymbol{\mathcal{D}}_w = \frac{1}{2}\,\mathrm{tr}\{\check{\boldsymbol{\mathcal{D}}}_w + \rho_w\mathbf{L}_w\boldsymbol{\mathcal{A}}_w + \rho_w\boldsymbol{\mathcal{A}}_w^T\mathbf{L}_w^T + \rho_w\mathbf{L}_w\mathbf{I}_w\mathbf{L}_w^T\}. \tag{7.4.26}$$

Thus

$$h_w = \check{h}_w + \frac{1}{2}(\boldsymbol{\mathcal{A}}_w + \boldsymbol{\mathcal{A}}_w^T) \cdot \mathbf{L}_w^T + \frac{1}{2}\mathbf{I}_w\mathbf{L}_w^T \cdot \mathbf{L}_w^T, \tag{7.4.27}$$

where

$$(\rho_w\check{h}_w)(\mathbf{x},t) := \sum_i \frac{1}{2} m_i(\check{\mathbf{v}}_i(t;\mathbf{x}))^2 w(\mathbf{x}_i(t) - \mathbf{x}). \tag{7.4.28}$$

[3] Details of all calculations are given in Appendix B.8.

At this point local balance (7.4.8) can be re-written in such a way that finer-scale features become evident. Specifically, from (7.4.10), (6.2.65), (7.4.12), (7.4.16), (7.4.21), (7.4.9), and (7.4.27), balance (7.4.8) becomes

$$\begin{aligned}
&\breve{r}_w + \mathbf{J}_w \cdot \mathbf{L}_w^T + ({}_b\mathbf{T}_w^- - \breve{\mathcal{D}}_w - \rho_w \mathbf{L}_w \mathcal{A}_w - \rho_w \mathcal{A}_w^T \mathbf{L}_w^T - \rho_w \mathbf{L}_w \mathbf{I}_w \mathbf{L}_w^T) \cdot \mathbf{L}_w \\
&- \operatorname{div}\Big\{ {}_b\breve{\mathbf{q}}_w^- - ({}_b\mathbf{C}_w^-)^\sim : \mathbf{L}_w^T + {}_2\breve{\mathbf{q}}_w + \mathbf{L}_w \mathbf{i}_w + \breve{\kappa}_w + \frac{1}{2}\, \mathbf{L}_w(\breve{\mathbf{M}}_w : \mathbf{1}) + \breve{\mathbf{M}}_w^\sim : \mathbf{L}_w^T \\
&+ \mathbf{L}_w(\mathbf{R}_w : \mathbf{L}_w) + \frac{1}{2}\mathbf{R}_w^\sim : \mathbf{L}_w^T \mathbf{L}_w + \frac{1}{2}\mathbf{L}_w(\mathbf{S}_w : \mathbf{L}_w^T \mathbf{L}_w) \Big\} \\
&= \rho_w \, \overline{(\beta_w + \breve{h}_w +) \frac{1}{2}(A_w + A_w^T) \cdot \mathbf{L_w^T} + \frac{\mathbf{1}}{\mathbf{2}} \mathbf{I_w L_w^T} \cdot \mathbf{L_w^T}}^{\,\bullet}).
\end{aligned} \tag{7.4.29}$$

It is instructive to look at the corresponding moment of momentum balance (7.2.27). To this end, (7.2.13) yields, on suppressing time-dependence,

$$\begin{aligned}
\hat{\mathbf{M}}_w(\mathbf{x}) &= \sum_i (\mathbf{x}_i - \mathbf{x}) \otimes m_i \mathbf{v}_i \otimes \hat{\mathbf{v}}_i(\mathbf{x}) w(\mathbf{x}_i - \mathbf{x}) \\
&= \sum_i (\mathbf{x}_i - \mathbf{x}) \otimes m_i (\breve{\mathbf{v}}_i + \mathbf{v}_w + \mathbf{L}_w(\mathbf{x}_i - \mathbf{x})) \otimes (\breve{\mathbf{v}}_i + \mathbf{L}_w(\mathbf{x}_i - \mathbf{x})) w(\mathbf{x}_i - \mathbf{x}) \\
&= \breve{\mathbf{M}}_w(\mathbf{x}) + \sum_i (\mathbf{x}_i - \mathbf{x}) \otimes m_i \breve{\mathbf{v}}_i \otimes \mathbf{L}_w(\mathbf{x}_i - \mathbf{x}) w(\mathbf{x}_i - \mathbf{x}) \\
&\quad + \sum_i (\mathbf{x}_i - \mathbf{x}) \otimes \mathbf{v}_w \otimes m_i \breve{\mathbf{v}}_i \, w(\mathbf{x}_i - \mathbf{x}) \\
&\quad + \sum_i (\mathbf{x}_i - \mathbf{x}) \otimes \mathbf{v}_w \otimes m_i \, \mathbf{L}_w(\mathbf{x}_i - \mathbf{x}) w(\mathbf{x}_i - \mathbf{x}) \\
&\quad + \sum_i (\mathbf{x}_i - \mathbf{x}) \otimes m_i \, \mathbf{L}_w(\mathbf{x}_i - \mathbf{x}) \otimes \breve{\mathbf{v}}_i \, w(\mathbf{x}_i - \mathbf{x}) \\
&\quad + \sum_i (\mathbf{x}_i - \mathbf{x}) \otimes m_i \, \mathbf{L}_w(\mathbf{x}_i - \mathbf{x}) \otimes \mathbf{L}_w(\mathbf{x}_i - \mathbf{x}) w(\mathbf{x}_i - \mathbf{x}).
\end{aligned} \tag{7.4.30}$$

Thus (see Appendix B.8 for details)

$$\begin{aligned}
\hat{\mathbf{M}}_w = \breve{\mathbf{M}}_w + \mathbf{R}_w^T \mathbf{L}_w^T + (\rho_w \mathcal{A}_w \otimes \mathbf{v}_w)^T + (\rho_w \mathbf{I}_w \otimes \mathbf{v}_w)^T \mathbf{L}_w^T \\
+ (\mathbf{R}_w^T \mathbf{L}_w^T)^T + (\mathbf{S}_w \mathbf{L}_w^T)^T \mathbf{L}_w^T.
\end{aligned} \tag{7.4.31}$$

Here $$\breve{\mathbf{M}}_w(\mathbf{x}, t) := \sum_i (\mathbf{x}_i(t) - \mathbf{x}) \otimes m_i \breve{\mathbf{v}}_i(t; \mathbf{x}) \otimes \breve{\mathbf{v}}_i(t; \mathbf{x}) w(\mathbf{x}_i(t) - \mathbf{x}). \tag{7.4.32}$$

Recall (see Exercise 7.4.1) that

$$\mathcal{A}_w = \mathbf{B}_w - \mathbf{d}_w \otimes \mathbf{v}_w - \mathbf{I}_w \mathbf{L}_w^T. \tag{7.4.33}$$

It follows from (7.2.27), (7.4.31), and (7.4.33) that moment of momentum balance (for pairwise-balanced interactions) may be written in the form

$$\mathrm{div}\{{}_b\mathbf{C}_w^- - \check{\mathbf{M}}_w - \mathbf{R}_w^T\mathbf{L}_w^T - \rho_w(\mathcal{A}_w \otimes \mathbf{v}_w)^T - \rho_w(\mathbf{I}_w \otimes \mathbf{v}_w)^T\mathbf{L}_w^T - (\mathbf{R}_w^T\mathbf{L}_w^T)^T - (\mathbf{S}_w\mathbf{L}_w^T)^T\mathbf{L}_w^T\}$$

$$+({}_b\mathbf{T}_w^-)^T + {}_2\mathbf{C}_w + \mathbf{J}_w = \rho_w\,\dot{\overline{(\mathcal{A}_w + \mathbf{d}_w \otimes \mathbf{v}_w + \mathbf{I}_w\mathbf{L}_w^T)}}\,. \tag{7.4.34}$$

Further, in view of (7.4.33), moment of mass balance (7.3.7) may be written as

$$\rho_w\dot{\mathbf{d}}_w + \mathrm{div}\{\rho_w(\mathcal{A}_w + \mathbf{I}_w\mathbf{L}_w^T\} = \mathbf{0}. \tag{7.4.35}$$

In view of the appearance in (7.4.29) of the material time derivative of $\mathbf{I}_w$, it is interesting to note the following:

Result 7.4.1.

$$\frac{\partial}{\partial t}\{\rho_w\mathbf{I}_w\} + \mathrm{div}\{\mathbf{I}_w \otimes \rho_w\mathbf{v}_w + \mathbf{R}_w + \mathbf{S}_w\mathbf{L}_w^T\} = \mathbf{O}. \tag{7.4.36}$$

Equivalently,

$$\rho_w\dot{\mathbf{I}}_w + \mathrm{div}\{\mathbf{R}_w + \mathbf{S}_w\,\mathbf{L}_w^T\} = \mathbf{O}. \tag{7.4.37}$$

Proof. From (7.4.3)

$$\frac{\partial}{\partial t}\{\rho_w\mathbf{I}_w\} = \sum_i \mathbf{v}_i \otimes m_i(\mathbf{x}_i - \mathbf{x})w + \sum_i (\mathbf{x}_i - \mathbf{x}) \otimes m_i\mathbf{v}_i w$$
$$+\sum_i (\mathbf{x}_i - \mathbf{x}) \otimes m_i(\mathbf{x}_i - \mathbf{x})(\nabla w\,.\,\mathbf{v}_i). \tag{7.4.38}$$

However,

$$(\mathbf{x}_i - \mathbf{x}) \otimes m_i(\mathbf{x}_i - \mathbf{x})(\nabla w\,.\,\mathbf{v}_i) = -(\mathbf{x}_i - \mathbf{x}) \otimes m_i(\mathbf{x}_i - \mathbf{x})(\nabla_{\mathbf{x}} w\,.\,\mathbf{v}_i)$$
$$= -(\mathbf{x}_i - \mathbf{x}) \otimes m_i(\mathbf{x}_i - \mathbf{x})\,\mathrm{div}\{\mathbf{v}_i\,w\}$$
$$= -\,\mathrm{div}\{(\mathbf{x}_i - \mathbf{x}) \otimes m_i(\mathbf{x}_i - \mathbf{x}) \otimes \mathbf{v}_i\,w\} + \nabla\{(\mathbf{x}_i - \mathbf{x}) \otimes m_i(\mathbf{x}_i - \mathbf{x})\}\mathbf{v}_i\,w$$
$$= -\,\mathrm{div}\{(\mathbf{x}_i - \mathbf{x}) \otimes m_i(\mathbf{x}_i - \mathbf{x}) \otimes \mathbf{v}_i\,w\} - \mathbf{v}_i \otimes m_i(\mathbf{x}_i - \mathbf{x})w$$
$$-(\mathbf{x}_i - \mathbf{x}) \otimes m_i\mathbf{v}_i w, \tag{7.4.39}$$

where the last step follows from identity (B.7.32) of Appendix B.7.

Accordingly, summing relation (7.4.39) over all particles P_i, it follows from (7.4.38) that

$$\frac{\partial}{\partial t}\{\rho_w\mathbf{I}_w\} = -\,\mathrm{div}\left\{\sum_i (\mathbf{x}_i - \mathbf{x}) \otimes m_i(\mathbf{x}_i - \mathbf{x}) \otimes \mathbf{v}_i\,w\right\}. \tag{7.4.40}$$

Now

$$\sum_i (\mathbf{x}_i - \mathbf{x}) \otimes m_i(\mathbf{x}_i - \mathbf{x}) \otimes \mathbf{v}_i\,w = \sum_i (\mathbf{x}_i - \mathbf{x}) \otimes m_i(\mathbf{x}_i - \mathbf{x}) \otimes (\check{\mathbf{v}}_i + \mathbf{v}_w + \mathbf{L}_w(\mathbf{x}_i - \mathbf{x}))w$$
$$= \mathbf{R}_w + \rho_w\mathbf{I}_w \otimes \mathbf{v}_w + \mathbf{S}_w\mathbf{L}_w^T, \tag{7.4.41}$$

on recalling (7.4.24), (7.4.3), and (7.4.25). Thus (7.4.36) follows from (7.4.40) and (7.4.41). To obtain (7.4.37), notice that

$$\mathrm{div}\{\mathbf{I}_w \otimes \rho_w\mathbf{v}_w\} = (\nabla\mathbf{I}_w)\rho_w\mathbf{v}_w + \mathrm{div}\{\rho_w\,\mathbf{v}_w\}\mathbf{I}_w \tag{7.4.42}$$

via identity (B.7.33) of Appendix B.7 (with $\mathbf{A} = \mathbf{I}_w$ and $\mathbf{v} = \rho_w \mathbf{v}_w$). Thus (7.4.36) and (7.4.42) yield

$$\frac{\partial \rho_w}{\partial t}\mathbf{I}_w + \rho_w \frac{\partial \mathbf{I}_w}{\partial t} + (\nabla \mathbf{I}_w)\rho_w \mathbf{v}_w + \operatorname{div}\{\rho_w \mathbf{v}_w\}\mathbf{I}_w + \operatorname{div}\{\mathbf{R}_w + \mathbf{S}_w \mathbf{L}_w^T\} = \mathbf{O}. \quad (7.4.43)$$

Result (7.4.37) follows from the continuity equation (4.2.16).

Remark 7.4.2. The foregoing fine-scale relations are complex. Previous analysis, in which a different averaging procedure was used,[4] suggests that simplification can be expected if molecules have identical structure and align with, and deform similarly to, near-neighbours. Indeed, if such deformation is affine, then generalised moment of momentum balance delivers an evolution equation for the common local rank-two tensor-valued measure of such deformation.

Balance of energy (7.4.29) can be re-written upon noting the appearance of terms which also appear in moment of momentum balance. However, the result does not appear to be edifying. As an illustration, consider making the following approximative assumptions:

values of $\mathcal{A}_w, \mathbf{R}_w$, and $\mathbf{S}_w$ are negligible compared with remaining terms. (7.4.44)

In such case energy balance (7.4.29) reduces[5] [see (7.4.15)] to

$$\begin{aligned} \breve{r}_w - \operatorname{div}\, {}_b\breve{\mathbf{q}}_w^+ + [({}_b\mathbf{T}_w^-)^T - \breve{\mathcal{D}}_w - \rho_w \mathbf{L}_w \mathbf{I}_w \mathbf{L}_w^T + \operatorname{div}\{{}_b\breve{\mathbf{C}}_w^- - \breve{\mathbf{M}}_w\} + \mathbf{J}_w] \cdot \mathbf{L}_w^T \\ - \operatorname{div}\{\mathbf{L}_w(\mathbf{i}_w + \frac{1}{2}\breve{\mathbf{M}}_w : \mathbf{1})\} + ({}_b\mathbf{C}_w^- - \breve{\mathbf{M}}_w) \cdot \nabla\{\mathbf{L}_w^T\} = \rho_w \dot{\breve{e}}_w, \end{aligned} \quad (7.4.45)$$

where

$${}_b\breve{\mathbf{q}}_w^+ := {}_b\breve{\mathbf{q}}_w^- + {}_2\breve{\mathbf{q}}_w + \breve{\kappa}_w \quad (7.4.46)$$

and

$$\breve{e}_w := \beta_w + \breve{h}_w + \frac{1}{2}\mathbf{I}_w \mathbf{L}_w^T \cdot \mathbf{L}_w^T. \quad (7.4.47)$$

The corresponding reduced forms of (7.4.34), (7.4.35) and (7.4.37) are

$$\operatorname{div}\{{}_b\mathbf{C}_w^- - \breve{\mathbf{M}}_w - \rho_w(\mathbf{I}_w \otimes \mathbf{v}_w)^T \mathbf{L}_w^T\} + ({}_b\mathbf{T}^-)^T + {}_2\mathbf{C}_w + \mathbf{J}_w = \rho_w \overline{(\mathbf{d}_w \otimes \mathbf{v}_w + \mathbf{I}_w \mathbf{L}_w^T)}^{\boldsymbol{\cdot}}, \quad (7.4.48)$$

$$\rho_w \dot{\mathbf{d}}_w + \operatorname{div}\{\mathbf{I}_w \mathbf{L}_w^T\} = \mathbf{0}, \quad (7.4.49)$$

and

$$\dot{\mathbf{I}}_w = \mathbf{O}. \quad (7.4.50)$$

7.5 Summary and Discussion

The Cauchy stress tensor field $\mathbf{T}$ is often 'proved' to take only symmetric values by postulating a moment of momentum balance in which couple stress, body couple, and

[4] See Murdoch [27]. The cellular averaging procedure adopted in this paper is considered in Chapter 11.

[5] Here it has been noted that for any two linear transformations $\mathbf{A}$ and $\mathbf{B}$, $\mathbf{A} \cdot \mathbf{B} = \mathbf{A}^T \cdot \mathbf{B}^T$, and both $\breve{\mathcal{D}}_w$ and $\mathbf{L}_w \mathbf{I}_w \mathbf{L}_w^T$ are symmetric.

internal angular momentum density are absent. Such absence may not be acknowledged (see Gurtin, Fried & Anand [23], p. 135, and Sedov [28], p. 129), partially acknowledged (see Reddy [29], p. 162), or explicitly acknowledged (see Truesdell & Noll [2], p. 41 and Section 98, and Chadwick [3], p. 93). However, from a molecular perspective, the symmetry (or otherwise) of $\mathbf{T}$ emerges from the nature of molecular interactions together with selection of a candidate interaction stress tensor $\mathbf{T}^{-}$, as discussed in Chapter 5. Accordingly it is natural to investigate the nature of the role played by moment of momentum balance. Here a link with inhomogeneity has been established by introducing a moment of mass conservation relation. This relation [see (7.3.4) or, equivalently, (7.3.7)] delineates the time evolution of measure $\mathbf{d}_w$ of inhomogeneity in terms of a generalised moment of momentum density $\rho_w \mathbf{B}_w$. In turn, the time evolution of $\mathbf{B}_w$ has been shown to be prescribed by a generalised moment of momentum balance which can be expressed in a number of ways [see (7.2.17), (7.2.20) and (7.2.27)], each of which corresponds to a specific form of linear momentum balance [namely (5.5.15), (6.2.31), and (6.2.48), respectively]. Attention has been drawn to the nature of these relations as fine-scale moment counterparts of the continuity equation and linear momentum balance.

The importance of moment of momentum balances derives from situations in which the effects of couple stress and microstructure are manifest at macroscopic scales (cf., e.g., Kröner [30]). The natures of the microscopic sources of these effects are diverse. If the microscopic situation is known, then relevant atomic/molecular detail can be incorporated into the molecular modelling and spatial averaging procedure (see Remark 7.4.2). At the level of generality adopted here it has only been possible to study the effect of a simple measure of inhomogeneity.

8 Time Averaging and Systems with Changing Material Content

8.1 Preamble

Time averaging is shown here to pervade the microscopic foundations of macroscopic modelling, whether in appreciating the nature of intermolecular forces in terms of subatomic dynamics (recall Section 5.4), establishing links between continuum field values and actual measurements (recall Section 3.10), deriving balance relations for systems with changing material content, or exploring the conceptual foundations of statistical mechanics.

After recalling the notion of a Δ-time average introduced in Section 5.4, such averaging is applied directly to the continuity equation and balances of linear momentum and energy established in Chapters 4, 5, and 6. Field values in the resulting relations are thereby identified as local averages of molecular quantities computed jointly in both space and time. Such relations apply to material systems which do not change with time. Any system $\mathcal{M}^+$ whose material content *does* change with time is regarded to be a subset of an unchanging material system $\mathcal{M}$. Relations for $\mathcal{M}^+$ are established by introducing a membership function e_i for each element P_i of $\mathcal{M}$: $e_i(t) = 1$ (or 0) according to whether $P_i \in \mathcal{M}^+$ (or $P_i \notin \mathcal{M}^+$) at time t. An equation which prescribes the time evolution of a single global representative velocity for $\mathcal{M}^+$ is derived utilising time averaging: this serves to establish the methodology to be used later in deriving local balances of mass, linear momentum, and energy, and also to discuss the fundamentals of rocket and jet propulsion. The chapter concludes with detailed derivations of evolution equations for densities of mass, linear momentum, and energy which take full account of changing material content. In the absence of any such change, these relations are the time-averaged versions of relations established in previous chapters.

8.2 Motivation

There are three fundamental *r*easons for *t*ime *a*veraging.

RTA 1. As indicated in Section 5.4, understanding the nature of intermolecular forces requires an appreciation that these derive from interactions between assemblies of electrons and atomic nuclei. Such fine detail is absent when molecules are modelled as point masses, as in Chapters 4 through 7. Accordingly the notion of the force $\mathbf{f}_{ij}(t)$ exerted on molecule P_i by molecule P_j at instant t cannot be modelled as a function of molecular locations $\mathbf{x}_i(t)$ and $\mathbf{x}_j(t)$ [and possibly molecular velocities $\mathbf{v}_i(t)$ and $\mathbf{v}_j(t)$] without further elaboration, since the instantaneous submolecular situation [upon which $\mathbf{f}_{ij}(t)$ depends] is unspecified. To make sense, $\mathbf{f}_{ij}(t)$ must represent an average associated with the fine detail. Given the small time scales associated with subatomic motions, it is possible to identify $\mathbf{f}_{ij}$ values as time averages (see Remark 5.4.2 and the preceding Interpretation 2). In other words, *there is an implicit averaging in time associated with models of molecular interactions.*

RTA 2. In Chapters 4 through 7 continuum balance equations were derived in which field values were related to local scale-dependent spatial averages of molecular quantities. However, as remarked in Section 3.10, actual observations or measurements of material behaviour involve averages jointly in space *and* time. Specifically, no observation or measurement can be made at a single geometrical point, nor at a given instant in time. Said differently, any observation or measurement is limited by both the spatial *and* temporal sensitivity of the monitoring procedure and apparatus: recall the 'snapshot' example of Section 3.10. Accordingly, if field values are to be identified with the results of observation or measurement, then the molecular averages which define such values should be computed jointly in space and time.

RTA 3. Physical systems of interest may have a changing material content; for example, molecules on one side of a permeable membrane, water molecules in a drinking glass, molecules constituting a rocket and its fuel, or a specific constituent in a reacting mixture. The derivation of equations of balance for such time-dependent systems requires time averaging.

Over and above the fundamental and practical considerations outlined in RTA 1 through 3, time averaging plays a central role in conceiving the link between discrete behaviour and continuum modelling of its consequences. An extreme and striking example is the case of a single molecule in a rectangular box. Consider a small part of one face of the box, of area ΔS say, and monitor the collisions of the molecule with this part over a time interval of duration T. Each collision delivers an impulse to the wall. Summing all impulses over the time interval and dividing by $T \times \Delta S$ gives a quantity that can be considered to be the average pressure associated with this part of the box over the time interval. Of course, such pressure *fluctuates* with time (i.e., suffers rapid random changes, each due to a single collision), but if T is sufficiently large, such fluctuations may contribute negligibly small changes about some definite value. In such case this 'definite value' (accurate to within what is to be regarded as negligible) is the *continuum pressure value.* The ultimate example of time averaging would seem to be cosmological modelling of intergalactic matter as 'gaseous', with suns, planets and their satellites, and asteroids as constituent 'particles'.

Further appreciation of the importance of time averaging can be gained by considering statistical mechanical approaches to continuum modelling in which a fundamental *assumption* is the existence of a 'probability density' to be associated with the behaviour of a system of particles. To motivate and understand such assumption, consider a small region of volume ΔV so small that at most one molecule (more precisely, one molecular mass centre) can lie inside this region at any given time. Observe over a time interval of duration T the proportion of this time the region actually contains a molecular mass centre. If such a proportion tends to a limit as T becomes larger and larger, then this limit is to be regarded to be the probability of finding a molecule in this 'macroscopically infinitesimal' region. The involvement of unbounded time intervals here might seem to limit physical relevance to particle systems in 'equilibrium' at the macroscopic level. However, the foregoing time-occupation fraction may stabilise for time intervals of duration $T = T_0$, in the sense of merely undergoing negligible fluctuation about some fixed value. In such case this value is interpreted to be the probability of finding a particle in the region consistent with macroscopic behaviour that is manifest in temporal evolution on time scales large compared with T_0.

8.3 Time Averaging

Recall from Section 5.4 that the Δ-time average f_Δ of a continuous function f of time was the mean value of f computed over that time interval of duration Δ ending at current time. Specifically,

$$f_\Delta(t) := \frac{1}{\Delta}\int_{t-\Delta}^{t} f(\tau)d\tau. \tag{5.4.10}$$

In the event that f is continuously differentiable it was shown that

$$(\dot{f})_\Delta = \widehat{\dot{f}_\Delta}. \tag{5.4.11}$$

In (5.4.10) equal weighting is accorded to all values of f in the interval $t - \Delta \leq \tau \leq t$. A more general approach, strictly analogous to the use of spatial weighting functions introduced in Section 4.2, is to define the *α-time average of f at time t* by

$$f_\alpha(t) := \int_{-\infty}^{t} f(\tau)\alpha(t-\tau)d\tau. \tag{8.3.1}$$

Here $\alpha : [0,\infty) \to \mathbb{R}$ is a *temporal* weighting function. In order to ensure that the α-time average of a constant function of time yields the same constant function it is necessary that

$$\int_{-\infty}^{t} \alpha(t-\tau)d\tau = 1. \tag{8.3.2}$$

Exercise 8.3.1. Prove (8.3.2), and show that this is equivalent to the normalisation condition

$$\int_{0}^{\infty} \alpha(s)ds = 1. \tag{8.3.3}$$

Show also that (8.3.1) may be written in the form

$$f_\alpha(t) = \int_{0}^{\infty} f(t-s)\alpha(s)ds. \tag{8.3.4}$$

Exercise 8.3.2. Show that (5.4.10) is a special case of (8.3.1) with

$$\alpha(s) := \Delta^{-1}(H(s) - H(s-\Delta)),$$

where H denotes the Heaviside 'step' function. [That is, $H(s) = 0$ if $s < 0$, $H(s) = 1$ if $s \geq 0$.] Notice that this choice satisfies (8.3.3).

If α is continuously differentiable and, for some $s_0(> 0), \alpha(s) = 0$ whenever $s \geq s_0$, then we have a direct analogue of (5.4.11). In such case, if f is continuous, then

$$f_\alpha(t) = \int_{t-s_0}^{t} f(\tau)\alpha(t-\tau)d\tau, \tag{8.3.5}$$

so (see Apostol [91], Theorem 9-38)

$$\begin{aligned} \overset{\bullet}{\widehat{f_\alpha}} &= \frac{d}{dt}\left\{\int_{t-s_0}^{t} f(\tau)\alpha(t-\tau)d\tau\right\} \\ &= [f(\tau)\alpha(t-\tau)]_{t-s_0}^{t} + \int_{t-s_0}^{t} f(\tau)\dot{\alpha}(t-\tau)d\tau. \end{aligned} \tag{8.3.6}$$

If f is continuously differentiable, then, on integrating by parts,

$$\begin{aligned} (\dot{f})_\alpha(t) &= \int_{t-s_0}^{t} \dot{f}(\tau)\alpha(t-\tau)d\tau \\ &= [f(\tau)\alpha(t-\tau)]_{t-s_0}^{t} - \int_{t-s_0}^{t} f(\tau)\frac{\partial}{\partial\tau}\{\alpha(t-\tau)\}d\tau. \end{aligned} \tag{8.3.7}$$

Noting that $(\partial/\partial\tau)\{\alpha(t-\tau)\} = -\dot{\alpha}(t-\tau)$, we have, from (8.3.6) and (8.3.7),

$$(\dot{f})_\alpha = \overset{\bullet}{\widehat{f_\alpha}} \tag{8.3.8}$$

if f is continuously differentiable.

Remark 8.3.1. Noting that measurements are taken over finite time intervals, and that identification of field values was a main motivation for time averaging, the assumption that α vanish outside $[0, s_0]$ for some $s_0 > 0$ is entirely realistic.

Now suppose that f is a class C^1 field. Then, in addition to (8.3.8), it follows that

$$\begin{aligned} (\nabla f_\alpha)(\mathbf{x},t) &= \nabla\left\{\int_{t-s_0}^{t} f(\mathbf{x},\tau)\alpha(t-\tau)d\tau\right\} \\ &= \int_{t-s_0}^{t} \nabla f(\mathbf{x},\tau)\alpha(t-\tau)d\tau = (\nabla f)_\alpha(\mathbf{x},t). \end{aligned} \tag{8.3.9}$$

Thus if $f = f_w$, a weighted spatial average of class C^1, then

$$\nabla\{(f_w)_\alpha\}(\mathbf{x},t) = (\nabla f_w)_\alpha(\mathbf{x},t). \tag{8.3.10}$$

For vector-valued weighted averages $\mathbf{f}_w$ of class C^1 it follows, by taking the trace of (8.3.10), that

$$\operatorname{div}\{(\mathbf{f}_w)_\alpha\} = (\operatorname{div}\mathbf{f}_w)_\alpha. \tag{8.3.11}$$

Similarly, for tensor-valued weighted averages $\mathbf{A}_w$ of class C^1,

$$\operatorname{div}\{(\mathbf{A}_w)_\alpha\} = (\operatorname{div}\mathbf{A}_w)_\alpha. \tag{8.3.12}$$

Exercise 8.3.3. Show that (8.3.12) holds for tensor fields of orders 2 or 3. [*Hint*: Consider definitions (B.7.8) and (B.7.16) in Appendix B.7.]

8.4 The Time-Averaged Continuity Equation

On taking the α-time average of continuity equation (4.2.14), and invoking (8.3.8) in respect of $\partial\rho_w/\partial t$ and (8.3.11) for $\mathbf{p}_w$, it follows that

$$\frac{\partial}{\partial t}\{(\rho_w)_\alpha\} + \operatorname{div}\{(\mathbf{p}_w)_\alpha\} = 0. \tag{8.4.1}$$

Writing

$$\mathbf{v}_{w;\alpha} := (\mathbf{p}_w)_\alpha / (\rho_w)_\alpha \tag{8.4.2}$$

for the corresponding velocity field yields

$$\frac{\partial}{\partial t}\{(\rho_w)_\alpha\} + \operatorname{div}\{(\rho_w)_\alpha \mathbf{v}_{w;\alpha}\} = 0. \tag{8.4.3}$$

Remark 8.4.1. Notice that

$$\mathbf{v}_{w;\alpha} \neq (\mathbf{v}_w)_\alpha. \tag{8.4.4}$$

(Why?)

Remark 8.4.2. Consider how the first term in (8.4.3) is evaluated. At location $\mathbf{x}$ and time t,

$$\frac{\partial}{\partial t}\{(\rho_w)_\alpha\}(\mathbf{x},t) = \frac{\partial}{\partial t}\left\{\int_{-\infty}^{t} \rho_w(\mathbf{x},\tau)\alpha(t-\tau)d\tau\right\}. \tag{8.4.5}$$

That is, *location* $\mathbf{x}$ *is fixed as time averaging is implemented.* This observation has several profound consequences:

(i) In principle the values of $(\rho_w)_\alpha$ and $(\mathbf{p}_w)_\alpha$ at $(\mathbf{x},t)$ can be identified with measurements made at the 'fixed' location $\mathbf{x}$. Here 'fixed' must refer to an observer's frame of reference. However, a fixed location for one observer is not so for another observer in relative motion to the first, even if both utilise inertial frames of reference. This is not quite as unsettling as it may first appear, since any measurement is undertaken by a single observer. Any other observer, aware of such measuring process, must appreciate just how the measurement is made, and how values of $(\rho_w)_\alpha$ and $(\mathbf{p}_w)_\alpha$ are identified at $(\mathbf{x},t)$ with field point $\mathbf{x}$ fixed in the frame of the measuring observer.

(ii) While (i) establishes a firm link between measurements and the local space-time averages $(\rho_w)_\alpha$ and $(\mathbf{p}_w)_\alpha$, the observer-dependent nature of such a link can only be appreciated if observers are aware of measurements being undertaken (and by whom!). This observation motivates a search for an *intrinsic* time-averaging procedure; that is, an averaging procedure which directly involves only material behaviour, and is distinct from considerations of measurement. Such intrinsic averaging is obtained by, roughly speaking, following the motion prescribed by the spatially defined velocity. More precisely, for any spatially averaged field f_w, define

$$(f_w^{\text{int}})_\alpha(\mathbf{x},t) := \int_{t-s_0}^{t} f_w(\chi_0(\hat{\mathbf{x}},\tau),\tau)\alpha(t-\tau)d\tau, \tag{8.4.6}$$

where [see (5.2.7) and (5.2.8)]

$$\dot{\chi}_0(\hat{\mathbf{x}},t) := \mathbf{v}_w(\chi_0(\hat{\mathbf{x}},t),t), \tag{8.4.7}$$

with

$$\chi_0(\hat{\mathbf{x}},t_0) = \hat{\mathbf{x}}. \tag{8.4.8}$$

Such averaging ensures that, instant by instant, the molecules involved in the computation of a space-time-averaged field value $(f_w^{\text{int}})_\alpha(\mathbf{x},t)$ are the same for all observers.

Remark 8.4.3. Definition (8.4.6), which ensures intrinsic time averaging, has been effected by adopting a referential[1] approach. As proved in Appendix B.5, Result B.5.2, the deformation gradient associated with a motion χ_0 with respect to the situation at time t_0, namely

$$\mathbf{F}_w(\mathbf{x},t) := \nabla\chi_0(\hat{\mathbf{x}},t), \tag{8.4.9}$$

satisfies [see (5.2.7) and (5.2.8)]

$$\dot{J}_w = (\operatorname{div}\mathbf{v}_w)J_w, \tag{8.4.10}$$

where

$$J_w := \det\mathbf{F}_w. \tag{8.4.11}$$

Thus

$$\dot{\overline{\rho_w J_w}} = \dot{\rho}_w J_w + \rho_w \dot{J}_w = (\dot{\rho}_w + \rho_w \operatorname{div}\mathbf{v}_w)J_w. \tag{8.4.12}$$

However, noting that the continuity equation (4.2.16) may be written as

$$\dot{\rho}_w + \rho_w \operatorname{div}\mathbf{v}_w = 0, \tag{8.4.13}$$

it follows from (8.4.12) that an equivalent formulation is

$$\dot{\overline{\rho_w J_w}} = 0. \tag{8.4.14}$$

That is,

$$\rho_w(\chi_0(\hat{\mathbf{x}},t),t)(\det\mathbf{F}(\hat{\mathbf{x}},t)) \quad \textit{is constant in time.} \tag{8.4.15}$$

Accordingly, although for any point $\hat{\mathbf{x}}$ in B_{t_0} (said differently, for any w-defined material point) the values of ρ_w and $\det\mathbf{F}_w$ may fluctuate in time, their product remains constant.

8.5 Time-Averaged Forms of Linear Momentum Balance

In this and the next section, any time average of a field f will be denoted by $\bar{f}$. It is assumed that such time averaging is employed either in the manner of (5.4.10) or (8.3.1), and that specific choices of Δ or α, respectively, have been made. Then properties (5.4.11) [or (8.3.8), (8.3.10), (8.3.11), and (8.3.12)] take the forms

$$\dot{\bar{f}} = \bar{\dot{f}}, \tag{8.5.1}$$

$$\nabla\bar{f} = \overline{\nabla f}, \tag{8.5.2}$$

$$\operatorname{div}\bar{\mathbf{f}} = \overline{\operatorname{div}\mathbf{f}}, \tag{8.5.3}$$

[1] Often termed 'material' or 'Lagrangian'.

and $$\operatorname{div}\bar{\mathbf{A}} = \overline{\operatorname{div}\mathbf{A}}. \tag{8.5.4}$$

Now recall the first and most general form of linear momentum balance:

$$\mathbf{f}_w + \mathbf{b}_w = \frac{\partial}{\partial t}\{\rho_w \mathbf{v}_w\} + \operatorname{div}\{\rho_w \mathbf{v}_w \otimes \mathbf{v}_w\} + \operatorname{div}\mathcal{D}_w. \tag{5.5.13}$$

On time averaging, and invoking (8.5.1) and (8.5.4),

$$\bar{\mathbf{f}}_w + \bar{\mathbf{b}}_w = \frac{\partial}{\partial t}\{\bar{\mathbf{p}}_w\} + \operatorname{div}\{\overline{\rho_w \mathbf{v}_w \otimes \mathbf{v}_w}) + \operatorname{div}\bar{\mathcal{D}}_w. \tag{8.5.5}$$

Writing $$\bar{\mathbf{v}}_w := \bar{\mathbf{p}}_w / \bar{\rho}_w \tag{8.5.6}$$

[but noting that $\bar{\mathbf{v}}_w$ is not the time average of $\mathbf{v}_w$: see (8.4.4)], relation (8.5.5) may be written as

$$\bar{\mathbf{f}}_w + \bar{\mathbf{b}}_w = \frac{\partial}{\partial t}\{\bar{\rho}_w \bar{\mathbf{v}}_w\} + \operatorname{div}\{\bar{\rho}_w \bar{\mathbf{v}}_w \otimes \bar{\mathbf{v}}_w\} + \operatorname{div}\bar{\mathcal{D}}_w^{+}, \tag{8.5.7}$$

where $$\bar{\mathcal{D}}_w^{+} := \bar{\mathcal{D}}_w + \overline{\rho_w \mathbf{v}_w \otimes \mathbf{v}_w} - \bar{\rho}_w \bar{\mathbf{v}}_w \otimes \bar{\mathbf{v}}_w. \tag{8.5.8}$$

That is, recalling that (8.4.3) may be written as

$$\frac{\partial \bar{\rho}_w}{\partial t} + \operatorname{div}\{\bar{\rho}_w \bar{\mathbf{v}}_w\} = 0, \tag{8.5.9}$$

$$-\operatorname{div}\bar{\mathcal{D}}_w^{+} + \bar{\mathbf{f}}_w + \bar{\mathbf{b}}_w = \bar{\rho}_w \bar{\mathbf{a}}_w, \tag{8.5.10}$$

where $$\bar{\mathbf{a}}_w := \dot{\bar{\mathbf{v}}}_w := \frac{\partial \bar{\mathbf{v}}_w}{\partial t} + (\nabla \bar{\mathbf{v}}_w)\bar{\mathbf{v}}_w. \tag{8.5.11}$$

Exercise 8.5.1. Prove (8.5.10) using (8.5.9).

Remark 8.5.1. Note that $\bar{\mathbf{a}}_w$ is the acceleration field associated with the motion defined by $\bar{\mathbf{v}}_w$: this is a direct analogue of (5.2.11).

Given any solution $\mathbf{T}_w^{-}$ to (5.5.18), relation (8.5.10) may be written as

$$\operatorname{div}\overline{\mathbf{T}}_w + \bar{\mathbf{b}}_w = \bar{\rho}_w \bar{\mathbf{a}}_w, \tag{8.5.12}$$

where $$\bar{\mathbf{T}}_w := \bar{\mathbf{T}}_w^{-} - \bar{\mathcal{D}}_w^{+}. \tag{8.5.13}$$

This form of balance relates to time averaging at fixed points in an inertial frame, and hence field values can be associated with measurements made at such points (see Remark 8.4.2). Intrinsic time averaging [see (8.4.6)] can be implemented if linear momentum balance is first couched in *r*eferential form. Specifically (cf., e.g., Gurtin [1], p. 179, and Truesdell & Noll [2], p. 124),

$$\operatorname{div}\mathbf{S}_w + {}_r\mathbf{b}_w = {}_r\rho_w \ddot{\mathbf{u}}_w. \tag{8.5.14}$$

Here $$\mathbf{S}_w := (\det \mathbf{F}_w)\mathbf{T}_w \mathbf{F}_w^{-T} \tag{8.5.15}$$

denotes the Piola-Kirchhoff stress tensor, ${}_r\rho_w$ is the (time-independent) mass density in the *r*eference configuration, $\mathbf{u}_w$ is the displacement field, given by

$$\mathbf{u}_w(\hat{\mathbf{x}},t) := \boldsymbol{\chi}_0(\hat{\mathbf{x}},t) - \hat{\mathbf{x}}, \tag{8.5.16}$$

and [see (8.4.11)]

$${}_r\mathbf{b}_w(\hat{\mathbf{x}},t) := \mathbf{b}_w(\boldsymbol{\chi}_0(\hat{\mathbf{x}},t),t)J_w(\hat{\mathbf{x}},t). \tag{8.5.17}$$

Using (8.5.4), and (8.5.1) twice, the time average of (8.5.14) is

$$\operatorname{div}\bar{\mathbf{S}}_w + {}_r\bar{\mathbf{b}}_w = {}_r\rho_w\,\ddot{\bar{\mathbf{u}}}_w. \tag{8.5.18}$$

Remark 8.5.2. A more fundamental approach to obtaining a form of linear momentum balance in which all fields are related to local averages in both space and time is given in Subsection 8.9.2. Although Section 8.9 is concerned with systems whose material content changes with time, the terms which express such change may be omitted to yield balances for systems with constant material content. Specifically, the relevant forms of linear momentum balance are (8.9.40) and (8.9.47) with ${}^{ext}\mathbf{f}, \mathbf{P}^{in}$, and $\mathbf{P}^{out}$ all zero, or (8.9.54) with ${}^{ext}\mathbf{f}, \mathbf{I}^{in}$, and $\mathbf{I}^{out}$ all zero. The difference between the forms obtained here and corresponding forms which appear in Subsection 8.9.2 derives from the different thermal velocities employed (which enter via the thermal contribution to stress $-\mathcal{D}_w$): here $\mathcal{D}_w\ (\mathbf{x},\tau)$ is defined in terms of $\hat{\mathbf{v}}_i(\tau;\mathbf{x}) := \mathbf{v}_i(\tau) - \mathbf{v}_w(\mathbf{x},\tau)$, while in Section 8.9 the notional thermal velocities employed are

$$\hat{\mathbf{v}}_i(\tau;\mathbf{x},t) := \mathbf{v}_i(\tau) - \mathbf{v}_{w,\Delta}(\mathbf{x},t), \qquad \text{where } \mathbf{v}_{w,\Delta} := \mathbf{p}_{w,\Delta}/\rho_{w,\Delta}.$$

Here $\mathbf{p}_{w,\Delta}$ and $\mathbf{v}_{w,\Delta}$ denote the Δ-time averages of $\mathbf{p}_w$ and ρ_w.

8.6 Time-Averaged Forms of Energy Balance

Consider the most general form of energy balance

$$-\operatorname{div}\boldsymbol{\kappa}_w + Q_w + r_w - \mathcal{D}_w\cdot\mathbf{L}_w = \rho_w\dot{h}_w. \tag{6.2.24}$$

Invoking (8.5.3) and noting that (Why?)

$$\rho_w\dot{h}_w = \frac{\partial}{\partial t}\{\rho_w h_w\} + \operatorname{div}\{\rho_w h_w \mathbf{v}_w\}, \tag{8.6.1}$$

the time average of (6.2.24) is

$$-\operatorname{div}\bar{\boldsymbol{\kappa}}_w + \bar{Q}_w + \bar{r}_w - \overline{\mathcal{D}_w\cdot\mathbf{L}_w} = \frac{\partial}{\partial t}\{\overline{\rho_w h_w}\} + \operatorname{div}\{\overline{\rho_w h_w \mathbf{v}_w}\}. \tag{8.6.2}$$

At this point it does not appear that simplification can be achieved without further hypothesis. Recalling assumptions (6.3.11), suppose that the material behaviour is such that, at the space-time scales associated with choices w and α (or Δ), fields f_w considered here yield time-averaged fields which vary negligibly over time intervals of duration associated with the time averaging procedure. In such case, then effectively, for any pair of fields f_w and g_w,

$$\bar{\bar{f}}_w = \bar{f}_w, \qquad \bar{\bar{g}}_w = \bar{g}_w, \tag{8.6.3}$$

and
$$\overline{f_w g_w} = \bar{f}_w \bar{g}_w + \overline{f'_w g'_w}, \tag{8.6.4}$$

where the *fluctuations* in f_w and g_w are

$$f'_w := f_w - \bar{f}_w \qquad \text{and} \qquad g'_w := g_w - \bar{g}_w. \tag{8.6.5}$$

Accordingly,

$$\overline{\rho_w h_w} = \bar{\rho}_w \bar{h}_w + \overline{\rho'_w h'_w}, \tag{8.6.6}$$

and [recall (8.5.6)]

$$\begin{aligned}\overline{\rho_w h_w \mathbf{v}_w} &= \overline{\rho_w \mathbf{v}_w}\, \bar{\mathbf{h}}_w + \overline{(\rho_w \mathbf{v}_w)' h'_w} \\ &= \bar{\rho}_w \bar{\mathbf{v}}_w \bar{h}_w + \overline{(\rho_w \mathbf{v}_w)' h'_w}.\end{aligned} \tag{8.6.7}$$

Thus
$$\begin{aligned}\frac{\partial}{\partial t}\{\overline{\rho_w h_w}\} + \operatorname{div}\{\overline{\rho_w h_w \mathbf{v}_w}\} &= \frac{\partial}{\partial t}\{\bar{\rho}_w \bar{h}_w\} + \operatorname{div}\{\bar{\rho}_w \bar{\mathbf{v}}_w \bar{h}_w\} \\ &\quad + \frac{\partial}{\partial t}\{\overline{\rho'_w h'_w}\} + \operatorname{div}\{\overline{(\rho_w \mathbf{v}_w)' h'_w}\} \qquad (8.6.8) \\ &= \bar{\rho}_w \dot{\bar{h}}_w + \frac{\partial}{\partial t}\{\overline{\rho'_w h'_w}\} + \operatorname{div}\{\overline{(\rho_w \mathbf{v}_w)' h'_w}\}. \qquad (8.6.9)\end{aligned}$$

Here (8.4.3) has been invoked and

$$\dot{\bar{h}}_w := \frac{\partial \bar{h}_w}{\partial t} + (\nabla \bar{h}_w)\bar{\mathbf{v}}_w \tag{8.6.10}$$

denotes the time derivative with respect to the motion prescribed by field $\bar{\mathbf{v}}_w$ [see (8.5.6) and (2.5.28)]. It follows from (8.6.2), (8.6.4) and (8.6.9) that

$$\begin{aligned}&-\operatorname{div} \bar{\boldsymbol{\kappa}}_w + \bar{Q}_w + \bar{r}_w - \bar{\mathcal{D}}_w \cdot \bar{\mathbf{L}}_w - \overline{\mathcal{D}'_w \cdot \mathbf{L}'_w} \\ &\quad = \bar{\rho}_w \dot{\bar{h}}_w + \left[\frac{\partial}{\partial t}\{\overline{\rho'_w h'_w}\} + \operatorname{div}\{\overline{(\rho_w \mathbf{v}_w)' h'_w}\}\right].\end{aligned} \tag{8.6.11}$$

Identical reasoning yields the time-averaged counterpart of form of balance (6.2.32) as

$$\begin{aligned}&\bar{r}_w - \operatorname{div}{}_s \bar{\mathbf{q}}_w + {}_s\bar{\mathbf{T}}_w \cdot \bar{\mathbf{L}}_w + \overline{{}_s\mathbf{T}'_w \cdot \mathbf{L}'_w} \\ &\quad = \bar{\rho}_w \dot{\bar{h}}_w + \left[\frac{\partial}{\partial t}\{\overline{\rho'_w h'_w}\} + \operatorname{div}\{\overline{(\rho_w \mathbf{v}_w)' h'_w}\}\right],\end{aligned} \tag{8.6.12}$$

and the corresponding result for balance (6.2.64) as

$$\begin{aligned}&\bar{r}_w - \operatorname{div}{}_b \overline{\mathbf{q}_w^+} + \overline{{}_b\mathbf{T}_w} \cdot \bar{\mathbf{L}}_w + \overline{{}_b\mathbf{T}'_w \cdot \mathbf{L}'_w} \\ &\quad = \bar{\rho}_w \dot{\bar{e}}_w + \left[\frac{\partial}{\partial t}\{\overline{\rho'_w e'_w} + \operatorname{div}\{\overline{(\rho_w \mathbf{v}_w)' e'_w}\}\right].\end{aligned} \tag{8.6.13}$$

Here [cf. (8.6.10)]

$$\dot{\bar{e}}_w := \frac{\partial \bar{e}_w}{\partial t} + (\nabla \bar{e}_w).\bar{\mathbf{v}}_w \tag{8.6.14}$$

denotes the time derivative associated with the motion prescribed by $\bar{\mathbf{v}}_w$.

The starting point for intrinsic averaging is the referential version of any energy balance. Here only the referential version of the most familiar form (6.2.64) is considered, namely [see Chadwick [3], p. 112, and Carlson [31], eq. (3.6)],

$$_r r_w - \operatorname{div}{}_{rb} \mathbf{q}_w^+ + \mathbf{S}_w \cdot \dot{\mathbf{F}} = {}_r\rho_w\, {}_r\dot{e}_w. \tag{8.6.15}$$

Here
$$ {}_r r_w(\hat{\mathbf{x}},t) := J(\hat{\mathbf{x}},t) r_w(\mathbf{x},t), \tag{8.6.16} $$
$$ {}_{rb}\mathbf{q}_w^+(\hat{\mathbf{x}},t) := J(\hat{\mathbf{x}},t)(\mathbf{F}(\hat{\mathbf{x}},t))^{-1}\, {}_b\mathbf{q}_w^+(\mathbf{x},t), \tag{8.6.17} $$
and
$$ {}_r e_w(\hat{\mathbf{x}},t) := J(\hat{\mathbf{x}},t) e_w(\mathbf{x},t), \tag{8.6.18} $$
where
$$ \mathbf{x} := \chi_0(\hat{\mathbf{x}},t). \tag{8.6.19} $$

Noting that ${}_r\rho_w$ is time-independent, averaging (8.6.15) in time yields
$$ {}_r\bar{r}_w - \mathrm{div}\, \overline{{}_{rb}\mathbf{q}_w^+} + \bar{\mathbf{S}}_w \cdot \dot{\bar{\mathbf{F}}} + \overline{\mathbf{S}_w' \cdot \mathbf{F}'} = {}_r\rho_w\, {}_r\dot{\bar{e}}_w. \tag{8.6.20} $$

Remark 8.6.1. Different forms of energy balance are derived in Subsection 8.9.3 which can be compared with (8.6.11), (8.6.12), and (8.6.13) when terms associated with the temporal change of material content of the system are omitted. These are (8.9.89), (8.9.98), and (8.9.115) with ${}^{\mathrm{ext}}Q, H$, and $\mathcal{G}$ omitted. The differences between the balances here and those derived later stem from the different notional thermal velocities employed [see Remark 8.5.2 and (8.9.31)].

8.7 Systems with Changing Material Content I: General Global Considerations

Consider a material system $\mathcal{M}$ as in Section 4.2; that is, a fixed, identifiable, set of fundamental discrete entities (atoms, ions or molecules). Suppose that at any instant τ system $\mathcal{M}$ may be regarded as the union of two mutually disjoint time-*dependent* systems, $\mathcal{M}^+(\tau)$ and $\mathcal{M}^-(\tau)$, say. For example, $\mathcal{M}$ could denote the molecules of a crystal/melt system in which $\mathcal{M}^+(\tau)$ denotes those molecules in the solid phase at instant τ [so that $\mathcal{M}^-(\tau)$ would denote those molecules in the liquid phase at this instant], or $\mathcal{M}$ could consist of a rocket together with its initial fuel, and $\mathcal{M}^+(\tau)$ could represent the rocket together with unexpended fuel at instant τ [What is $\mathcal{M}^-(\tau)$ here?]. Clearly, by definition, at any time τ
$$ \mathcal{M} = \mathcal{M}^+(\tau) \cup \mathcal{M}^-(\tau) \qquad \text{and} \qquad \mathcal{M}^+(\tau) \cap \mathcal{M}^-(\tau) = \phi. \tag{8.7.1} $$

The fixed set of fundamental entities is denoted by $P_i (i = 1,2,\ldots,N)$ say.

Relations (8.7.1) require that, at any given instant τ, P_i must belong to just one of $\mathcal{M}^+(\tau)$ and $\mathcal{M}^-(\tau)$. Focussing attention on $\mathcal{M}^+$, define the *membership* (for $\mathcal{M}^+$) *function* e_i for P_i by
$$ \left.\begin{array}{lll} e_i(\tau) = 1 & \text{if} & P_i \in \mathcal{M}^+(\tau) \text{ at instant } \tau \\ e_i(\tau) = 0 & \text{if} & P_i \notin \mathcal{M}^+(\tau) \text{ at instant } \tau \end{array}\right\}. \tag{8.7.2} $$

Remark 8.7.1. Note that the membership function for P_i in respect of $\mathcal{M}^-$ is $1 - e_i$.

The total mass and total momentum associated with $\mathcal{M}^+$ at instant τ are (recall the mass and velocity of P_i are denoted by m_i and $\mathbf{v}_i$)
$$ m(\tau) := \sum_{i=1}^{N} m_i e_i(\tau) \tag{8.7.3} $$
and
$$ \mathbf{p}(\tau) := \sum_{i=1}^{N} m_i \mathbf{v}_i(\tau) e_i(\tau). \tag{8.7.4} $$

Thus the instantaneous mass centre velocity is

$$\mathbf{v}(\tau) := \mathbf{p}(\tau)/m(\tau). \tag{8.7.5}$$

The aim is to obtain a relation between the time derivative of any time average $\bar{\mathbf{p}}$ of $\mathbf{p}$ and the agencies (including external forces) which determine $\dot{\bar{\mathbf{p}}}$. To this end, consider the motion of P_i in an inertial frame. This motion is governed by [see (5.4.4) and (5.4.5)]

$$\sum_{j \neq i} \mathbf{f}_{ij} + \mathbf{b}_i = m_i \dot{\mathbf{v}}_i. \tag{8.7.6}$$

Here $\mathbf{f}_{ij}$ denotes the force exerted on P_i by P_j, the sum is taken over *all* entities P_j in $\mathcal{M}$, and $\mathbf{b}_i$ is the resultant force on P_i due to all agencies outwith $\mathcal{M}$. Multiplication of (8.7.6) by e_i and summing over all P_i in $\mathcal{M}$ yield (on suppressing time dependence)

$$\sum_{i \neq j} \sum \mathbf{f}_{ij} e_i + \sum_i \mathbf{b}_i e_i = \sum_i m_i \dot{\mathbf{v}}_i e_i. \tag{8.7.7}$$

Now

$$\sum_{j \neq i} \sum \mathbf{f}_{ij} e_i = \sum_{j \neq i} \sum \mathbf{f}_{ij} e_i e_j + \sum_{j \neq i} \sum \mathbf{f}_{ij} e_i (1 - e_j) \tag{8.7.8}$$

is an instantaneous decomposition of the resultant force upon $\mathcal{M}^+$ particles due to *all* $\mathcal{M}$ particles into the resultant self-force exerted by $\mathcal{M}^+$ particles on themselves [the first term on the right-hand side of (8.7.8)] together with the resultant force on $\mathcal{M}^+$ particles exerted by $\mathcal{M}^-$ particles [the final term in (8.7.8)]. Noting that

$$\sum_{j \neq i} \sum \mathbf{f}_{ij} e_i e_j = \frac{1}{2} \sum_{i \neq j} \sum (\mathbf{f}_{ij} + \mathbf{f}_{ji}) e_i e_j, \tag{8.7.9}$$

it follows that this term vanishes if interactions are pairwise balanced [see (5.6.6)]. Assuming such pairwise balance [or, more generally, if the middle term in (8.7.8) vanishes; i.e., the *resultant* $\mathcal{M}^+ - \mathcal{M}^+$ self-force vanishes at any instant], then (8.7.7) reduces [via (8.7.8)] to

$$\sum_{i \neq j} \sum \mathbf{f}_{ij} e_i (1 - e_j) + \sum_i \mathbf{b}_i e_i = \sum_i m_i \dot{\mathbf{v}}_i e_i. \tag{8.7.10}$$

For simplicity the Δ-time average [see (5.4.10)] of (8.7.10) is considered. (Formally identical results can be obtained for α-time averages in similar fashion.) Specifically, such averaging yields

$$\mathbf{f}_\Delta(t) + \mathbf{b}_\Delta(t) = \frac{1}{\Delta} \sum_i m_i \int_{t-\Delta}^{t} \dot{\mathbf{v}}_i(\tau) e_i(\tau) d\tau, \tag{8.7.11}$$

where

$$\mathbf{f}_\Delta(t) := \frac{1}{\Delta} \sum_{i \neq j} \sum \int_{t-\Delta}^{t} \mathbf{f}_{ij}(\tau) e_i(\tau)(1 - e_j(\tau)) d\tau \tag{8.7.12}$$

and

$$\mathbf{b}_\Delta(t) := \frac{1}{\Delta} \sum_i \int_{t-\Delta}^{t} \mathbf{b}_i(\tau) e_i(\tau) d\tau. \tag{8.7.13}$$

In order to understand the last term in (8.7.11) we consider a number of possibilities from which a general pattern emerges:

(i) If $P_i \in \mathcal{M}^+$ at time $t-\Delta$, stays a member of $\mathcal{M}^+$ until time $t_1 < t$, and then remains in $\mathcal{M}^-$ until time t, then

$$e_i(\tau) = 1 \quad \text{if } t-\Delta \le \tau \le t, \qquad \text{and} \qquad e_i(\tau) = 0 \quad \text{if } t_1 < \tau \le t.$$

Accordingly, the contribution of P_i to the right-hand side of (8.7.11) is

$$\frac{m_i}{\Delta}\left\{\int_{t-\Delta}^{t_1} \dot{\mathbf{v}}_i(\tau).1\,d\tau + \int_{t_1}^{t} \dot{\mathbf{v}}(\tau).0\,d\tau\right\} = \frac{m_i}{\Delta}\{\mathbf{v}_i(t_1) - \mathbf{v}_i(t-\Delta)\}. \tag{8.7.14}$$

(ii) If $P_i \in \mathcal{M}^-$ for $t-\Delta \le \tau < t_2 < t$, becomes a member of $\mathcal{M}^+$ at time t_2, and remains in $\mathcal{M}^+$ for $t_2 < t \le t$, then the contribution of P_i is

$$\frac{m_i}{\Delta}\int_{t_2}^{t} \dot{\mathbf{v}}_i(\tau)d\tau = \frac{m_i}{\Delta}\{\mathbf{v}_i(t) - \mathbf{v}_i(t_2)\}. \tag{8.7.15}$$

(iii) If $P_i \in \mathcal{M}^+$ for times

$$\tau \in (t_{i_1}, t_{i_2}) \cup (t_{i_3}, t_{i_4}) \cup \cdots \cup (t_{i_{2N_i-1}}, t_{i_{2N_i}}), \tag{8.7.16}$$

where $$t-\Delta \le t_{i_1} < t_{i_2} < t_{i_3} < t_{i_4} < \cdots < t_{i_{2N_i-1}} < t_{i_{2N_i}} \le t-\Delta, \tag{8.7.17}$$

then the contribution from P_i is

$$\frac{m_i}{\Delta}\left\{\int_{t_{i_1}}^{t_{i_2}} \dot{\mathbf{v}}_i\,d\tau + \ldots + \int_{t_{i_{2N_i-1}}}^{t_{i_{2N_i}}} \dot{\mathbf{v}}_i\,d\tau\right\} = \frac{m_i}{\Delta}\Big\{(\mathbf{v}_i(t_{i_2}) + \mathbf{v}_i(t_{i_4}) + \ldots + \mathbf{v}_i(t_{i_{2N_i}}))$$
$$-(\mathbf{v}_i(t_{i_1}) + \mathbf{v}_i(t_{i_3}) + \ldots + \mathbf{v}_i(t_{i_{2N_i-1}}))\Big\}. \tag{8.7.18}$$

In all cases that can be considered there is a contribution $\pm(m_i/\Delta)\,\mathbf{v}_i(t_{i_k})$ whenever P_i 'enters' or 'leaves' $\mathcal{M}^+$: the plus sign corresponds to 'leaving' contributions and the minus sign to 'entering' contributions. Further, if P_i is in $\mathcal{M}^+$ at time $t-\Delta$, then there is a contribution $-m_i\mathbf{v}_i(t-\Delta)$, while if P_i is in $\mathcal{M}^+$ at time t there is a contribution $+m_i\mathbf{v}_i(t)$. Accordingly, consideration of all possible cases yields

$$\frac{1}{\Delta}\int_{t-\Delta}^{t}\sum_i m_i\dot{\mathbf{v}}_i(\tau)e_i(\tau)d\tau = \mathcal{F}^{\Delta}_{\text{out}}(t) - \mathcal{F}^{\Delta}_{\text{in}}(t)$$
$$+\frac{1}{\Delta}\left\{\sum_i m_i\mathbf{v}_i(t)e_i(t) - \sum_i \mathbf{v}_i(t-\Delta)e_i(t-\Delta)\right\}. \tag{8.7.19}$$

Here $$\mathcal{F}^{\text{out}}_{\Delta}(t) := \frac{1}{\Delta}\sum_i\sum_{i_\ell} m_i\mathbf{v}_i(t_{i_\ell}) \tag{8.7.20}$$

and $$\mathcal{F}^{\text{in}}_{\Delta}(t) := \frac{1}{\Delta}\sum_i\sum_{i_e} m_i\mathbf{v}_i(t_{i_e}). \tag{8.7.21}$$

The sums in (8.7.20) and (8.7.21) are taken over all $P_i \in \mathcal{M}$, all times t_{i_ℓ} at which P_i *l*eaves $\mathcal{M}^+$, and all times t_{i_e} at which P_i *e*nters $\mathcal{M}^+$, during the time interval $(t-\Delta, t)$. Since the last term in (8.7.19) is[2] *formally* [see (8.7.4) and (5.4.11)]

$$\frac{1}{\Delta}\{\mathbf{p}(t)-\mathbf{p}(t-\Delta)\} = \frac{1}{\Delta}\int_{t-\Delta}^{t} \dot{\mathbf{p}}(\tau)d\tau = (\dot{\mathbf{p}})_\Delta(t) = \widehat{\mathbf{p}_\Delta}(t), \tag{8.7.22}$$

relation (8.7.11) may be written as

$$\mathbf{f}_\Delta + \mathbf{b}_\Delta = \widehat{\mathbf{p}_\Delta} + \mathcal{F}^{\text{out}}_\Delta - \mathcal{F}^{\text{in}}_\Delta. \tag{8.7.23}$$

Here $\mathbf{f}_\Delta$ denotes the Δ-time averaged resultant force exerted on system $\mathcal{M}^+$ due to interactions with system $\mathcal{M}^-$, $\mathbf{b}_\Delta$ denotes the Δ-time averaged resultant external force on $\mathcal{M}^+$, $\widehat{\mathbf{p}_\Delta}$ is the time rate of change of Δ-averaged total $\mathcal{M}^+$ momentum, $\mathcal{F}^\Delta_{\text{out}}$ is the momentum loss due to particles leaving $\mathcal{M}^+$ at times during the time interval (of duration Δ) in question, divided by Δ, and $\mathcal{F}^\Delta_{\text{in}}$ is the momentum gain due to particles joining $\mathcal{M}^+$ over this time interval, divided by Δ.

Relation (8.7.23) may be re-written upon noting that (*formally*: see footnote 2) from (8.7.3) and (5.4.11)

$$\frac{1}{\Delta}\{m(t)-m(t-\Delta)\} = \frac{1}{\Delta}\int_{t-\Delta}^{t} \dot{m}(\tau)d\tau = (\dot{m})_\Delta(t) = \widehat{m_\Delta}(t). \tag{8.7.24}$$

Thus, upon writing

$$\mathbf{v}_\Delta := \frac{\mathbf{p}_\Delta}{m_\Delta}, \tag{8.7.25}$$

(8.7.23) may be written as

$$\mathbf{f}_\Delta(t) + \mathbf{b}_\Delta(t) = \frac{1}{\Delta}\{m(t)-m(t-\Delta)\}\mathbf{v}_\Delta(t) + m_\Delta(t)\widehat{\mathbf{v}_\Delta}(t) + \mathcal{F}^{\text{out}}_\Delta(t) - \mathcal{F}^{\text{in}}_\Delta(t). \tag{8.7.26}$$

This simplifies as a consequence of the following observation.

Result 8.7.1.

$$\frac{1}{\Delta}\{m(t)-m(t-\Delta)\}\mathbf{v}_\Delta(t) + \mathcal{F}^{\text{out}}_\Delta(t) - \mathcal{F}^{\text{in}}_\Delta(t) = \tilde{\mathcal{F}}^{\text{out}}_\Delta(t) - \tilde{\mathcal{F}}^{\text{in}}_\Delta(t), \tag{8.7.27}$$

where [cf. (8.7.20) and (8.7.21)]

$$\tilde{\mathcal{F}}^{\text{out}}_\Delta(t) := \frac{1}{\Delta}\sum_i \sum_{t_{i_\ell}} m_i(\mathbf{v}_i(t_{i_\ell}) - \mathbf{v}_\Delta(t)) \tag{8.7.28}$$

and

$$\tilde{\mathcal{F}}^{\text{in}}(t) := \frac{1}{\Delta}\sum_i \sum_{t_{i_e}} m_i(\mathbf{v}_i(t_{i_e}) - \mathbf{v}_\Delta(t)). \tag{8.7.29}$$

Proof. Since there are contributions to $\tilde{\mathcal{F}}^{\text{out}}_\Delta(t)$ for any time $t_{i_\ell} \in [t-\Delta, t]$ at which P_i leaves $\mathcal{M}^+$, the contribution of P_i to $\tilde{\mathcal{F}}^{\text{out}}_\Delta(t)$ differs from its contribution to $\mathcal{F}^{\text{out}}_\Delta(t)$ by the number of such leavings in this time interval *multiplied* by $-\Delta^{-1}m_i\mathbf{v}_\Delta(t)$. A

[2] See Remark 8.7.3. for explanation and modification to the argument which renders the result mathematically precise.

similar remark applies to P_i contributions to $\tilde{\mathcal{F}}^{\text{in}}_{\Delta}(t)$ and $\mathcal{F}^{\text{in}}_{\Delta}(t)$, but here the difference is the *number* of *entry* times in the time interval $[t-\Delta, t]$. Now consider the following distinct and exhaustive possibilities for P_i:

(i) P_i is in $\mathcal{M}^+$ at both times $t-\Delta$ and t,
(ii) P_i is not in $\mathcal{M}^+$ at either times $t-\Delta$ or t,
(iii) P_i is in $\mathcal{M}^+$ at time $t-\Delta$ but not at time t, and
(iv) P_i is not in $\mathcal{M}^+$ at time $t-\Delta$ but is in $\mathcal{M}^+$ at time t.

In cases (i) and (ii) the numbers of entries and leavings must be equal. (Convince yourself of this!). In both of these cases P_i does not yield a net contribution to $m(t)-m(t-\Delta)$, and the contributions of P_i to $\mathcal{F}^{\text{out}}_{\Delta}(t)$ and $\tilde{\mathcal{F}}^{\text{out}}_{\Delta}(t)$ are the same, as are the contributions of P_i to $\mathcal{F}^{\text{in}}_{\Delta}(t)$ and $\tilde{\mathcal{F}}^{\text{in}}(t)$. Thus the P_i contributions to both sides of (8.7.27) are equal in cases (i) and (ii).

In case (iii) there must be one more leaving time than entry time. It follows that the P_i contribution to $\tilde{\mathcal{F}}^{\text{out}}_{\Delta}(t) - \tilde{\mathcal{F}}^{\text{in}}_{\Delta}(t)$ differs from its contribution to $\mathcal{F}^{\text{out}}_{\Delta}(t) - \mathcal{F}^{\text{in}}_{\Delta}(t)$ by the term $-\Delta^{-1} m_i \mathbf{v}_{\Delta}(t)$. However, this is also the contribution of P_i to $\Delta^{-1}\{m(t)-m(t-\Delta)\}$ in this case.

Similarly, in case (iv) the number of entry times must exceed the number of leaving times by one. It follows that the P_i contributions to both sides of (8.7.27) are the same. (Prove this!)

From (8.7.26) and (8.7.27) relation (8.7.23) takes the equivalent form

$$\mathbf{f}_{\Delta} + \mathbf{b}_{\Delta} + \tilde{\mathcal{F}}^{\text{net}}_{\Delta} = m_{\Delta}\dot{\mathbf{v}}_{\Delta}, \tag{8.7.30}$$

where

$$\tilde{\mathcal{F}}^{\text{net}}_{\Delta} := \tilde{\mathcal{F}}^{\text{in}}_{\Delta} - \tilde{\mathcal{F}}^{\text{out}}_{\Delta}. \tag{8.7.31}$$

Thus $\tilde{\mathcal{F}}^{\text{net}}_{\Delta}(t)$ represents the net momentum transfer into $\mathcal{M}^+$ from $\mathcal{M}^-$ in time interval $[t-\Delta, t]$ associated with particle *relative* velocities $\mathbf{v}_i - \mathbf{v}_{\Delta}(t)$, divided by Δ.

Remark 8.7.2. Velocity $\mathbf{v}_{\Delta}$ is not the Δ-time average of the mass centre velocity but is the natural choice appropriate to the Δ-time global averages m_{Δ} and $\mathbf{p}_{\Delta}$ of mass and momentum.

Remark 8.7.3. Equations (8.7.22) and (8.7.24) are formal and suggestive. However, they are mathematically incorrect since neither $\mathbf{p}$ nor m is differentiable at times associated with instantaneous transitions between $\mathcal{M}^+$ and $\mathcal{M}^-$: the membership functions suffer jump discontinuities at each and every transition, and hence so too do $\mathbf{p}$ and m. This lack of rigour can be rectified by mollifying membership functions at transition times (cf. Subsection 4.3.4). Specifically, suppose that $\gamma > 0$ is chosen and

$$\varphi : [-\gamma, \gamma] \longrightarrow [0,1] \tag{8.7.32}$$

is continuously differentiable (one-sided at endpoints) with

$$\varphi(-\gamma) = 0,\ \varphi(\gamma) = 1 \quad \text{and} \quad \varphi'(-\gamma) = 0 = \varphi'(\gamma). \tag{8.7.33}$$

Now consider a typical 'entry–leaving' time interval $[t_{i_e}, t_{i_\ell}]$ for P_i, where $t_{i\ell}$ denotes the time P_i next leaves $\mathcal{M}^+$ after entering at time t_{i_e}. Membership function e_i is

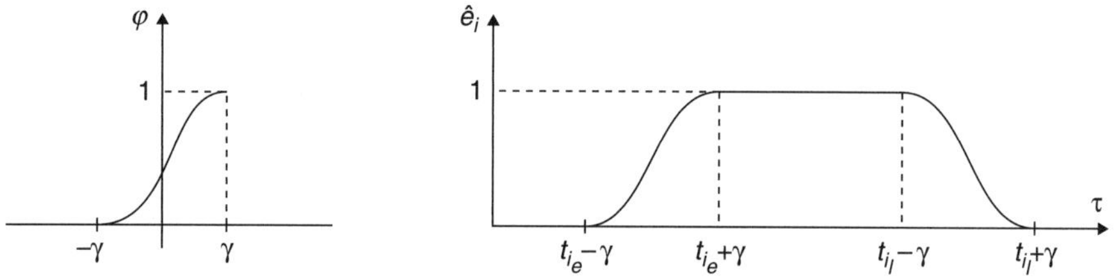

Figure 8.1. Mollifier φ and mollification of $\hat{e}_i$ for typical entry–leaving time pair t_{i_e} and t_{i_ℓ}.

mollified at entry times using φ and at leaving times using $1-\varphi$. Thus the mollification $\hat{e}_i$ of e_i satisfies (see Figure 8.1)

$$\left.\begin{array}{lll} \hat{e}_i(\tau) = \varphi(\tau - t_{i_e}) & \text{for} & t_{i_e} - \gamma \leq \tau \leq t_{i_e} + \gamma \\ \hat{e}_i(\tau) = 1(= e_i(\tau)) & \text{for} & t_{i_e} + \gamma \leq \tau \leq t_{i_\ell} - \gamma \\ \hat{e}_i(\tau) = 1 - \varphi(\tau - t_{i_\ell}) & \text{for} & t_{i_\ell} - \gamma \leq \tau \leq t_{i_\ell} + \gamma \end{array}\right\}. \tag{8.7.34}$$

It may happen that one or more mollifying subintervals overlap for a specific choice of γ; that is, a leaving time may be followed by an entry time less than 2γ time units later. In such cases mollification on the overlap is preserved simply by adding the two mollifiers involved.

Physical quantities $m, \mathbf{p}, \mathbf{f}_\Delta$ and $\mathbf{b}_\Delta$ [see (8.7.3), (8.7.4), (8.7.12), and (8.7.13)] now can be re-defined by replacing e_i by $\hat{e}_i$. In particular,

$$m(t) := \sum_i m_i \hat{e}_i(t), \tag{8.7.35}$$

so we may define [cf. (5.4.10)]

$$m_\Delta(t) := \frac{1}{\Delta} \int_{t-\Delta}^{t} m(\tau) d\tau. \tag{8.7.36}$$

Since m is now continuously differentiable as a consequence of mollification, from (5.4.11) with $f = m$

$$(\dot{m})_\Delta = \dot{\widehat{m_\Delta}}, \tag{8.7.37}$$

where
$$(\dot{m})_\Delta(t) := \frac{1}{\Delta} \int_{t-\Delta}^{t} \dot{m}(\tau) d\tau = \frac{1}{\Delta}\{m(t) - m(t-\Delta)\}. \tag{8.7.38}$$

Thus the mollified form of m satisfies

$$\dot{\widehat{m_\Delta}}(t) = \frac{1}{\Delta}\{m(t) - m(t-\Delta)\}. \tag{8.7.39}$$

That is, all steps in (8.7.24) are now justified upon mollification of m.

Now consider the argument leading to (8.7.11) when all membership functions have been mollified. The contribution to the right-hand side from P_i is

$$\boldsymbol{\alpha}_i := \frac{1}{\Delta} \int_{t-\Delta}^{t} m_i \dot{\mathbf{v}}_i(\tau) \hat{e}_i(\tau) d\tau. \tag{8.7.40}$$

Let $\mathbf{c}_i(t_{i_e}, t_{i_\ell})$ denote the contribution to $\boldsymbol{\alpha}_i$ associated with a typical entry–leaving time pair t_{i_e} and t_{i_ℓ} with $t - \Delta \leq t_{i_e} - \gamma$ and $t_{i_\ell} + \gamma \leq t$. Then

$$\begin{aligned}\mathbf{c}_i(t_{i_e}, t_{i_\ell}) = & \frac{1}{\Delta}\left\{\int_{t_{ie}-\gamma}^{t_{ie}+\gamma} m_i\dot{\mathbf{v}}_i(\tau)\varphi(\tau - t_{i_e})d\tau + \int_{t_{ie}+\gamma}^{t_{i_\ell}-\gamma} m_i\dot{\mathbf{v}}_i(\tau).1 d\tau\right. \\ & \left. + \int_{t_{i_\ell}-\gamma}^{t_{i_\ell}+\gamma} m_i\dot{\mathbf{v}}_i(\tau)[1 - \varphi(\tau - t_{i_\ell})]d\tau\right\} \\ = & \frac{1}{\Delta}\left\{\int_{-\gamma}^{\gamma} m_i\dot{\mathbf{v}}_i(s + t_{i_e})\varphi(s)ds + [m_i\mathbf{v}_i(\tau)]_{t_{ie}+\gamma}^{t_{i_\ell}-\gamma}\right. \\ & \left. + \int_{-\gamma}^{\gamma} m_i\dot{\mathbf{v}}_i(s + t_{i_\ell})[1 - \varphi(s)]ds\right\}. \end{aligned} \tag{8.7.41}$$

Exercise 8.7.1. Show that (8.7.41) simplifies to

$$\begin{aligned}\mathbf{c}_i(t_{i_e}, t_{i_\ell}) = \frac{m_i}{\Delta}&\left\{\mathbf{v}_i(t_{i_\ell} + \gamma) - \mathbf{v}_i(t_{i_e} + \gamma)\right. \\ & \left. + \int_{-\gamma}^{\gamma} [\dot{\mathbf{v}}_i(s + t_{i_e}) - \dot{\mathbf{v}}_i(s + t_{i_\ell})]\varphi(s)ds\right\}. \end{aligned} \tag{8.7.42}$$

Show further that if $\dot{\mathbf{v}}_i$ is continuous (and hence bounded on any closed time interval of interest, say $\|\dot{\mathbf{v}}_i(\tau)\| \leq \beta_i$), then

$$\left\|\int_{-\gamma}^{\gamma} [\dot{\mathbf{v}}_i(s + t_{i_e}) - \dot{\mathbf{v}}_i(s + t_{i_\ell})]\varphi(s)ds\right\| \leq 4\gamma\beta_i. \tag{8.7.43}$$

Notice that if we interpret the mollification procedure to be associated with imprecision in deciding upon membership of $\mathcal{M}^+$ (see Subsection 4.3.7 and the 'fuzzy set' interpretation), then in (8.7.42) times $t_{i_e} + \gamma$ and $t_{i_\ell} + \gamma$ are those at which we are sure that P_i has entered and left $\mathcal{M}^+$, respectively.

Since γ may be chosen to be arbitrarily small, then from (8.7.42), (8.7.43), and the continuity of $\dot{\mathbf{v}}_i$,

$$\mathbf{c}_i(t_{i_e}, t_{i_\ell}) \qquad \textit{differs negligibly from} \qquad \frac{m_i}{\Delta}\{\mathbf{v}_i(t_{i_\ell}) - \mathbf{v}_i(t_{i_e})\} \tag{8.7.44}$$

if γ is chosen to be suitably small [cf. (8.7.16) through (8.7.18)].

There remain contributions to $\boldsymbol{\alpha}_i$ associated with other possibilities. For example, if $P_i \in \mathcal{M}^+$ for $t - \Delta - \gamma \leq \tau \leq t_{i_\ell} \leq t - \gamma$, then the contribution is

$$\begin{aligned}&\frac{m_i}{\Delta}\int_{t-\Delta}^{t_{i_\ell}+\gamma} \dot{\mathbf{v}}_i(\tau)\hat{e}_i(\tau)d\tau = \frac{m_i}{\Delta}\left\{\int_{t-\Delta}^{t_{i_\ell}-\gamma} \dot{\mathbf{v}}_i(\tau).1 d\tau + \int_{t_{i_\ell}-\gamma}^{t_{i_\ell}+\gamma} \dot{\mathbf{v}}_i(\tau)\hat{e}_i(\tau)d\tau\right\} \\ &= \frac{m_i}{\Delta}\left\{\mathbf{v}_i(t_{i_\ell} - \gamma) - \mathbf{v}_i(t - \Delta) + \textit{ term of order } O(2\gamma\beta_i)\right\}. \end{aligned} \tag{8.7.45}$$

If terms of order $2\gamma\beta_i$ are neglected, then this term differs negligibly from $m_i\Delta^{-1}(\mathbf{v}_i(t_{i_\ell}) - \mathbf{v}_i(t - \Delta))$. Considering all possibilities, and noting the boundedness of accelerations $\dot{\mathbf{v}}_i$ on any closed time interval of interest, the right-hand side of (8.7.11) (when membership functions are mollified) is negligibly different [cf. (8.7.19) and (8.7.4)] from

$$\mathcal{F}_{\Delta}^{\text{out}}(t) - \mathcal{F}_{\Delta}^{\text{in}}(t) + \frac{1}{\Delta}\{\mathbf{p}(t) - \mathbf{p}(t - \Delta)\}, \tag{8.7.46}$$

where
$$\mathbf{p}(\tau) := \sum_i m_i \mathbf{v}_i(\tau)\hat{e}_i(\tau). \tag{8.7.47}$$

Since $\mathbf{p}$ is continuously differentiable, all steps in (8.7.22) are now justified: specifically,

$$(\dot{\mathbf{p}})_\Delta(t) := \frac{1}{\Delta}\int_{t-\Delta}^{t} \dot{\mathbf{p}}(\tau)d\tau = \frac{1}{\Delta}\{\mathbf{p}(t) - \mathbf{p}(t-\Delta)\} = \dot{\widehat{\mathbf{p}_\Delta}}(t), \tag{8.7.48}$$

where
$$\mathbf{p}_\Delta(t) := \frac{1}{\Delta}\int_{t-\Delta}^{t} \mathbf{p}(\tau)d\tau. \tag{8.7.49}$$

8.8 Systems with Changing Material Content II: Specific Global Examples

8.8.1 Rocketry

Here $\mathcal{M}$ consists of the molecules which constitute a rocket together with its initial fuel. Since we are to examine the motion of the rocket, the system $\mathcal{M}^+(t)$ of interest at any given time t consists of those molecules which constitute the rocket at this time (note these may change if it is of multi-stage design) together with those molecules of on-board fuel. Accordingly $\mathbf{f}_\Delta$ derives from intermolecular forces between molecules of ejected fuel and molecules of the rocket plus unexpended fuel.

If no ejected/burnt fuel molecules ever find their way back into $\mathcal{M}^+$, then both $\mathcal{F}_\Delta^{\text{in}}$, and $\tilde{\mathcal{F}}_\Delta^{\text{in}}$ vanish. Further, if $\mathbf{f}_\Delta$ is negligible, then (8.7.30) reduces to

$$-\tilde{\mathcal{F}}_\Delta^{\text{out}} + \mathbf{b}_\Delta = m_\Delta \dot{\mathbf{v}}_\Delta. \tag{8.8.1}$$

Of course, $\mathbf{b}_\Delta$ is the Δ-time average of the resultant external force on the rocket and will incorporate the effects of gravitational force(s) and, if the rocket is in the terrestrial atmosphere, air resistance. The term $-\tilde{\mathcal{F}}_w^{\text{out}}$ represents the thrust 'force'.

If all ejected fuel molecules leave the rocket with the same relative velocity $\mathbf{V}$, so that in (8.7.28)

$$\mathbf{v}_i(t_{i_\ell}) - \mathbf{v}_\Delta(t) = \mathbf{V}, \tag{8.8.2}$$

then
$$\tilde{\mathcal{F}}_{\text{out}}^{\Delta} = -\dot{m}_\Delta \mathbf{V}. \tag{8.8.3}$$

This is a consequence of (8.7.27), (8.7.24), and cases (i), (ii) and (iii) considered in the proof of Result 8.7.1.

Exercise 8.8.1. Convince yourself of result (8.8.3).

It follows from (8.8.1) and (8.8.3) that

$$\dot{m}_\Delta \mathbf{V} + \mathbf{b}_\Delta = m_\Delta \dot{\mathbf{v}}_\Delta. \tag{8.8.4}$$

This is the so-called rocket equation if $\mathbf{b}_\Delta$ is negligible (cf., e.g., Ohanian [32], p. 214). *In fact (8.8.2) is not a reasonable assumption*, since expelled fuel is gaseous and at a high temperature. Accordingly, molecular motions will involve high speeds and significant randomness.

Of course, from (8.7.30) and (8.7.31) it follows that if $\mathbf{f}_\Delta = \mathbf{0}$, then an equation having the *form* of (8.8.4) holds, where $\mathbf{V}$ is re-interpreted via

$$\mathbf{V} := (\dot{m}_\Delta)^{-1}\tilde{\mathcal{F}}_\Delta^{\text{net}}. \tag{8.8.5}$$

Remark 8.8.1. If $\hat{\mathbf{u}}$ denotes a unit vector in the direction of $\dot{\mathbf{v}}_\Delta(\dot{\mathbf{v}}_\Delta \neq \mathbf{0})$, then from (8.8.4)

$$\dot{m}_\Delta \mathbf{V}.\hat{\mathbf{u}} + \mathbf{b}_\Delta.\hat{\mathbf{u}} > 0.$$

If, as we should hope, $|\dot{m}_\Delta \mathbf{V}.\hat{\mathbf{u}}| > |\mathbf{b}_\Delta.\hat{\mathbf{u}}|$ (otherwise gravitation, air resistance, or a combination of these would overcome propulsion of the rocket in its intended direction), then

$$\dot{m}_\Delta \mathbf{V}.\hat{\mathbf{u}} > 0.$$

Since $\dot{m}_\Delta < 0$ (Why?) this means that $\mathbf{V}.\hat{\mathbf{u}} < 0$. This is not surprising: we expect the rocket to have to expel gas in the direction opposite to that in which it is intended to move. While this may seem 'obvious', the following example illustrates the essence of rocket propulsion.

Example 8.8.1. An Eskimo is stranded at rest on a horizontal sheet of frictionless sea ice. How can the Eskimo reach a shoreline in the direction of a unit vector $\hat{\mathbf{u}}$? If the Eskimo has two boots, each of mass m, and without boots has mass M, then he/she can throw the boots in the direction of $-\hat{\mathbf{u}}$. Suppose that the first boot is thrown with speed V relative to the Eskimo. Since the system consisting of Eskimo plus boots experiences no resultant horizontal force (we suppose that there is no wind nor air resistance), the horizontal component of system momentum remains unchanged (and is hence zero). If $\mathbf{v}$ is the velocity of the Eskimo and remaining boot, then it follows that

$$(m+M)\mathbf{v} + m(\mathbf{v} - V\hat{\mathbf{u}}) = \mathbf{0}.$$

Accordingly

$$\mathbf{v} = \frac{mV}{(M+2m)}\hat{\mathbf{u}}.$$

Exercise 8.8.2. If the Eskimo throws the remaining boot in the same direction as the first, again with relative speed V, show that the velocity of the Eskimo changes to $\mathbf{v}'$, where

$$\mathbf{v}' = \frac{m(2M+3m)V}{(M+m)(M+2m)}\hat{\mathbf{u}}.$$

If both boots were thrown at the same time with relative speed V, show that the velocity of the Eskimo would be $\mathbf{v}''$, where

$$\mathbf{v}'' = \frac{2mV}{(M+2m)}\hat{\mathbf{u}}.$$

Show also that $|\mathbf{v}'| - |\mathbf{v}''| = m^2V/(M+m)(M+2m) > 0.$

Comment: The final result leads one to suspect that higher speeds could be achieved by cutting each boot into smaller pieces and then throwing these sequentially at relative speed V. The ultimate limit thus would correspond to ejecting individual boot molecules sequentially at relative speed V. Our enterprising Eskimo (thoughtfully carrying a sharp knife) would have discovered the fundamentals of rocket propulsion (and survived – frostbite of foot is preferable to death by hypothermia!).

8.8.2 Jet Propulsion

Let $\mathcal{M}$ denote the set of molecules which constitute a jet aircraft, its initial fuel, and the atmosphere, and let $\mathcal{M}^+(\tau)$ denote the molecules of the aircraft, fuel on board, and air on board and within its engine(s) at time τ. Neglecting interactions between $\mathcal{M}^+$ and the expelled fuel molecules and air outside the aircraft and engines (namely $\mathcal{M}^-$), in (8.7.30) the term $\mathbf{f}_\Delta$ reduces to the Δ-time averaged resultant force on $\mathcal{M}^+$ by atmospheric molecules (usually resolved into 'drag' and 'lift' components). If $\mathbf{b}_i = m_i\mathbf{g}$ in (8.7.6), where $\mathbf{g}$ denotes gravitational acceleration, then from (8.7.13) and (8.7.3)

$$\mathbf{b}_\Delta = m_\Delta \mathbf{g}. \tag{8.8.6}$$

Thus (8.7.30) becomes

$$\text{drag} + \text{lift} + m_\Delta \mathbf{g} + \tilde{\mathcal{F}}_\Delta^{\text{net}} = m_\Delta \dot{\mathbf{v}}_\Delta, \tag{8.8.7}$$

and $\tilde{\mathcal{F}}_\Delta^{\text{net}}$ is seen to be the propulsive agency. The difference with rocketry stems from a significant intake of air, which suggests separate book-keeping for 'intakes' and 'exhausts'. A further consideration is whether the intake is enhanced by rotation of turbine-driven blades (a *turbo-jet*) or derives only from motion of the aircraft through the air (a *ram jet*). Writing ${}^{\text{in}}\tilde{\mathcal{F}}_\Delta^{\text{in}}$ and ${}^{\text{ex}}\tilde{\mathcal{F}}_\Delta^{\text{in}}$ for the separate contributions to [see (8.7.31)] $\tilde{\mathcal{F}}_\Delta^{\text{in}}$ associated, respectively, with air entering via *in*take and *ex*haust,

$$\tilde{\mathcal{F}}_\Delta^{\text{in}} = {}^{\text{in}}\tilde{\mathcal{F}}_\Delta^{\text{in}} + {}^{\text{ex}}\tilde{\mathcal{F}}_\Delta^{\text{in}}. \tag{8.8.8}$$

(Implicit here is the assumption that no ejected fuel molecules find their way into intakes.) For expelled air and fuel we may write

$$\tilde{\mathcal{F}}_\Delta^{\text{out}} = {}^{\text{ex}}\tilde{\mathcal{F}}_\Delta^{\text{out}} + {}^{\text{in}}\tilde{\mathcal{F}}_\Delta^{\text{out}}, \tag{8.8.9}$$

where ${}^{\text{ex}}\tilde{\mathcal{F}}_\Delta^{\text{out}}$ is associated with air and fuel molecules leaving engine exhausts and ${}^{\text{in}}\tilde{\mathcal{F}}_\Delta^{\text{out}}$ with air molecules leaving engine intakes. A simplistic picture is obtained by assuming that no air molecules leave by an intake and no fuel or air molecules enter by an exhaust (in particular, an expelled fuel molecule does not re-enter an exhaust); that is,

$${}^{\text{in}}\tilde{\mathcal{F}}_\Delta^{\text{out}} = \mathbf{0} = {}^{\text{ex}}\tilde{\mathcal{F}}_\Delta^{\text{in}}. \tag{8.8.10}$$

In such case (8.8.7) becomes [via (8.7.31)]

$$\text{drag} + \text{lift} + \rho_\Delta \mathbf{g} + {}^{\text{in}}\tilde{\mathcal{F}}_\Delta^{\text{in}} - {}^{\text{ex}}\tilde{\mathcal{F}}_\Delta^{\text{out}} = m_\Delta \dot{\mathbf{v}}_\Delta. \tag{8.8.11}$$

If all ejected molecules leave at the same velocity $\mathbf{V}$ relative to the aircraft, and all air molecules enter at the same velocity $\mathbf{U}$ relative to the aircraft, then

$${}^{\text{ex}}\tilde{\mathcal{F}}_\Delta^{\text{out}} = \frac{M_\Delta^{\text{ex}}\mathbf{V}}{\Delta}, \tag{8.8.12}$$

where M_Δ^{ex} is the mass of air and fuel ejected over a time interval of duration Δ and Similarly,

$${}^{\text{in}}\tilde{\mathcal{F}}_\Delta^{\text{in}} = \frac{M_\Delta^{\text{in}}\mathbf{U}}{\Delta}, \tag{8.8.13}$$

where M_Δ^{in} is the mass of air taken in over this time interval. It follows from (8.7.24) that

$$\dot{m}_\Delta = \frac{M_\Delta^{\text{in}} - M_\Delta^{\text{ex}}}{\Delta}. \tag{8.8.14}$$

Hence

$$\begin{aligned} {}^{\text{in}}\tilde{\mathcal{F}}_\Delta^{\text{in}} - {}^{\text{ex}}\tilde{\mathcal{F}}_\Delta^{\text{out}} &= \frac{M_\Delta^{\text{in}}\mathbf{U} - M_\Delta^{\text{ex}}\mathbf{V}}{\Delta} \\ &= \dot{m}_\Delta \mathbf{V} + \frac{M_\Delta^{\text{in}}(\mathbf{U}-\mathbf{V})}{\Delta}, \end{aligned} \tag{8.8.15}$$

and (8.8.11) becomes

$$\text{drag} + \text{lift} + \rho_\Delta \mathbf{g} + \dot{m}_\Delta \mathbf{V} + \frac{M_\Delta^{\text{in}}(\mathbf{U}-\mathbf{V})}{\Delta} = m_\Delta \dot{\mathbf{v}}_\Delta. \tag{8.8.16}$$

As to be expected, this relation simplifies to the rocket equation when there is no intake of air (so $M_\Delta^{\text{in}} = 0$). Notice that there are now *two* propulsive agencies, represented by terms $\dot{m}_\Delta \mathbf{V}$ and $M_\Delta^{\text{in}}(\mathbf{U}-\mathbf{V})/\Delta$. Indeed, if $\mathbf{U}$ and $\mathbf{V}$ are parallel (say, $\mathbf{U} = U\hat{\mathbf{u}}$ and $\mathbf{V} = V\hat{\mathbf{u}}$, where $\hat{\mathbf{u}}$ is a unit vector and $U > 0, V > 0$), then

$$M_\Delta^{\text{in}}(\mathbf{U}-\mathbf{V}) = M_\Delta^{\text{in}}(U-V)\hat{\mathbf{u}}. \tag{8.8.17}$$

Now $V > U$ (Why?), and $\hat{\mathbf{u}} \,.\, \mathbf{v}_\Delta < 0$ since the direction of travel of the aircraft and the direction in which exhaust gas molecules are ejected are essentially opposite. Thus (8.8.17) may be written as

$$M_\Delta^{\text{in}}(\mathbf{U}-\mathbf{V}) = M_\Delta^{\text{in}}(V-U)(-\hat{\mathbf{u}}), \tag{8.8.18}$$

where $M_\Delta^{\text{in}} > 0$ (Why?) and $(V-U) > 0$. This quantity clearly represents a propulsive agency in (8.8.16), and this is also true of term $\dot{m}_\Delta \mathbf{V} = (-\dot{m}_\Delta)V(-\hat{\mathbf{u}})$ since $V > 0$ and $-\dot{m}_\Delta > 0$. (Why?)

The same *form* (8.8.16) of evolution equation, more realistic than the foregoing, is obtained by *defining*

$$\mathbf{V} := \frac{\left({}^{\text{ex}}\tilde{\mathcal{F}}_\Delta^{\text{out}} - {}^{\text{ex}}\tilde{\mathcal{F}}_\Delta^{\text{in}}\right)\Delta}{M_\Delta^{\text{ex}}} \tag{8.8.19}$$

and

$$\mathbf{U} := \frac{\left({}^{\text{in}}\tilde{\mathcal{F}}_\Delta^{\text{in}} - {}^{\text{in}}\tilde{\mathcal{F}}_\Delta^{\text{out}}\right)\Delta}{M_\Delta^{\text{in}}}. \tag{8.8.20}$$

Remark 8.8.2. In the case of a ram jet in still air

$$\mathbf{U} = -\mathbf{v}_\Delta. \tag{8.8.21}$$

Remark 8.8.3. Equations (8.8.4) and (8.8.16) may be obtained by simplistic arguments based upon assumption (8.8.2) [for (8.8.4)] and the constant relative velocity assumptions of entering air molecules and ejected fuel and air molecules which lead to (8.8.12) and (8.8.13) [for (8.8.16)]. However, the molecular viewpoint enables the more realistic interpretations of $\mathbf{V}$ in (8.8.5), and of $\mathbf{U}$ and $\mathbf{V}$ in (8.8.19) and (8.8.20), to be made. Of course, our model would not appear to allow for the vital combustion process, particularly in the case of jet propulsion. This objection can be overcome by re-interpreting the 'particles' of $\mathcal{M}, \mathcal{M}^+$ and $\mathcal{M}^-$ as fundamental subatomic discrete entities, namely electrons and atomic nuclei: these retain their identity in the chemical combustion reactions that change the molecular species involved.

Remark 8.8.4. Time averaging played a central role in the foregoing discussions. This is of particular help when considering *pulse jets*: here the intake of air and expulsion of exhaust gases are not continuous but occur cyclically over mutually exclusive time intervals. For most purposes it is sufficient to choose Δ well in excess of the cycle period. This is also the situation in the case of jet propulsion as employed by living creatures such as squid or certain micro-organisms. Here liquid may not be taken in and ejected at the same time. In such case (8.8.16) may be interpreted appropriately for *any* given choice of Δ. At sufficiently large time scales $\dot{m}_\Delta$ will be negligible, since the mass of such creatures remains essentially constant, and the liquid they can contain has variable, but bounded, mass. (Why?) Thus the long-time propulsive agency is seen to be $M_\Delta^{\text{in}}(\mathbf{U} - \mathbf{V})\Delta^{-1}$.

8.8.3 Falling Raindrop

Let $\mathcal{M}$ denote the molecules which constitute a particular raindrop together with all other raindrops and moist air in some closed region, and let $\mathcal{M}^+(\tau)$ denote the raindrop molecules at time τ. Here $\mathbf{f}_\Delta$ represents the Δ-time-averaged resultant force exerted on the raindrop by contiguous damp air ('viscous drag') and $\mathbf{b}_\Delta$ the time-averaged weight, namely $m_\Delta\, \mathbf{g}$. Terms $\mathcal{F}_\Delta^{\text{in}}$ and $\mathcal{F}_\Delta^{\text{out}}$ [see (8.7.21) and (8.7.20)] represent Δ-time averages of net momentum gain and loss, respectively, associated with mass gain ('accretion') and mass loss ('evaporation').

8.9 Systems with Changing Material Content III: Local Evolution Equations at Specific Scales of Length and Time

8.9.1 Mass Balance

For any given spatial localisation function w the mass and momentum density functions for $\mathcal{M}^+$ are

$$\rho_w(\mathbf{x},t) := \sum_i m_i e_i(t) w(\mathbf{x}_i(t) - \mathbf{x}) \tag{8.9.1}$$

and

$$\mathbf{p}_w(\mathbf{x},t) := \sum_i m_i \mathbf{v}_i(t) e_i(t) w(\mathbf{x}_i(t) - \mathbf{x}). \tag{8.9.2}$$

Here the sums are over all point masses P_i in $\mathcal{M}$. The presence of factor $e_i(t)$ limits the contributions to these sums to only those P_i which belong to $\mathcal{M}^+$ at time t. The presence of factor $w(\mathbf{x}_i(t) - \mathbf{x})$ further limits contributions to those P_i for which $w(\mathbf{x}_i(t) - \mathbf{x}) \neq 0$, ensuring that the values of $\rho_w(\mathbf{x},t)$ and $\mathbf{p}_w(\mathbf{x},t)$ are spatial averages localised at the geometrical point $\mathbf{x}$ at time t.

Remark 8.9.1. In order to make rigorous the arguments used throughout Section 8.9, membership functions e_i [cf. (8.9.1) and (8.9.2)] should be replaced by their mollifications, as discussed in Remark 8.7.3. However, since the results can be shown to differ negligibly from those of the formal arguments which follow (by suitably small choice of mollifying subintervals, as in Remark 8.7.3), we omit details in order to focus on the most important features of the analysis.

The Δ-time averaged counterparts of (8.9.1) and (8.9.2) are

$$\rho_{w,\Delta}(\mathbf{x},t) := \frac{1}{\Delta}\int_{t-\Delta}^{t} \rho_w(\mathbf{x},\tau)d\tau \tag{8.9.3}$$

and

$$\mathbf{p}_{w,\Delta}(\mathbf{x},t) := \frac{1}{\Delta}\int_{t-\Delta}^{t} \mathbf{p}_w(\mathbf{x},\tau)d\tau. \tag{8.9.4}$$

(As in Sections 8.7 and 8.8, Δ-time averaging is employed here for simplicity: α-time-average computations are strictly analogous.)

From (8.9.3)

$$\begin{aligned}\frac{\partial\rho_{w,\Delta}}{\partial t}(\mathbf{x},t) &= \frac{1}{\Delta}\{\rho_w(\mathbf{x},t)-\rho_w(\mathbf{x},t-\Delta)\}\\ &= \frac{1}{\Delta}\sum_i m_i\{e_i(t)w(\mathbf{x}_i(t)-\mathbf{x})-e_i(t-\Delta)w(\mathbf{x}_i(t-\Delta)-\mathbf{x})\}. \end{aligned}\tag{8.9.5}$$

Further, from (8.9.4), (8.9.2), and (8.3.11),

$$\begin{aligned}(\operatorname{div}\mathbf{p}_{w,\Delta})(\mathbf{x},t) &= \frac{1}{\Delta}\int_{t-\Delta}^{t}(\operatorname{div}\mathbf{p}_w)(\mathbf{x},\tau)d\tau\\ &= \frac{1}{\Delta}\sum_i m_i\int_{t-\Delta}^{t} e_i(\tau)\operatorname{div}_{\mathbf{x}}\{\mathbf{v}_i(\tau)w(\mathbf{x}_i(\tau)-\mathbf{x})\}d\tau\\ &= -\frac{1}{\Delta}\sum_i\int_{t-\Delta}^{t} e_i(\tau)\mathbf{v}_i(\tau).\nabla w(\mathbf{x}_i(\tau)-\mathbf{x})d\tau\\ &= -\frac{1}{\Delta}\sum_i\int_{t-\Delta}^{t} e_i(\tau)\frac{\partial}{\partial\tau}\{w(\mathbf{x}_i(\tau)-\mathbf{x})\}d\tau. \end{aligned}\tag{8.9.6}$$

Recall the analysis in Section 8.7 concerning evaluation of the integrals in (8.7.11). These may be compared with the last integrals in (8.9.6). Specifically, the comparable integrals are

$$\int_{t-\Delta}^{t} e_i(\tau)\dot{\mathbf{v}}_i(\tau)d\tau \longleftrightarrow \int_{t-\Delta}^{t} e_i(\tau)\frac{\partial}{\partial\tau}\{w(\mathbf{x}_i(\tau)-\mathbf{x})\}d\tau. \tag{8.9.7}$$

Proceeding exactly as in Section 8.7, it follows that [cf. (8.7.19)]

$$\begin{aligned}&\frac{1}{\Delta}\sum_i\int_{t-\Delta}^{t} m_i e_i(\tau)\frac{\partial}{\partial\tau}\{w(\mathbf{x}_i(\tau)-\mathbf{x})\}d\tau\\ &= \frac{1}{\Delta}\left\{\sum_i m_i w(\mathbf{x}_i(t)-\mathbf{x})e_i(t)-\sum_i m_i w(\mathbf{x}_i(t-\Delta)-\mathbf{x})e_i(t-\Delta)\right\}\\ &\qquad + \mathcal{G}^{\text{out}}_{w,\Delta}(\mathbf{x},t)-\mathcal{G}^{\text{in}}_{w,\Delta}(\mathbf{x},t), \end{aligned}\tag{8.9.8}$$

where $$\mathcal{G}^{\text{out}}_{w,\Delta}(\mathbf{x},t) := \frac{1}{\Delta}\sum_i \sum_{i_\ell} m_i w(\mathbf{x}_i(t_{i_\ell}) - \mathbf{x}) \tag{8.9.9}$$

and $$\mathcal{G}^{\text{in}}_{w,\Delta}(\mathbf{x},t) := \frac{1}{\Delta}\sum_i \sum_{i_e} m_i w(\mathbf{x}_i(t_{i_e}) - \mathbf{x}). \tag{8.9.10}$$

In (8.9.9) the inner sum is taken over all times t_{i_ℓ} during the time interval $(t-\Delta,t)$ that P_i leaves $\mathcal{M}^+$, while the inner sum in (8.9.10) is taken over all times in this interval that P_i enters $\mathcal{M}^+$. There are no contributions from point masses P_i which do not undergo $\mathcal{M}^+ \to \mathcal{M}^-$ or $\mathcal{M}^- \to \mathcal{M}^+$ transitions in time interval $(t-\Delta,t)$. Thus $\mathcal{G}^{\text{out}}_{w,\Delta}$ is a local measure of the rate of mass density loss over a time lapse of duration Δ due to point masses leaving $\mathcal{M}^+$, and $\mathcal{G}^{\text{in}}_{w,\Delta}$ is the corresponding rate of mass density gain due to point masses joining $\mathcal{M}^+$.

From (8.9.5), (8.9.6), and (8.9.8) it follows that

$$\frac{\partial \rho_{w,\Delta}}{\partial t} = \mathcal{G}^{\text{in}}_{w,\Delta} - \mathcal{G}^{\text{out}}_{w,\Delta} - \operatorname{div}\mathbf{p}_{w,\Delta}. \tag{8.9.11}$$

Equivalently [cf. (8.4.2) and (8.4.3)],

$$\frac{\partial \rho_{w,\Delta}}{\partial t} + \operatorname{div}\{\rho_{w,\Delta}\mathbf{v}_{w,\Delta}\} = \mathcal{G}^{\text{net}}_{w,\Delta}, \tag{8.9.12}$$

where $$\mathbf{v}_{w,\Delta} := \mathbf{p}_{w,\Delta}/\rho_{w,\Delta} \tag{8.9.13}$$

and $$\mathcal{G}^{\text{net}}_{w,\Delta} := \mathcal{G}^{\text{in}}_{w,\Delta} - \mathcal{G}^{\text{out}}_{w,\Delta}. \tag{8.9.14}$$

Remark 8.9.2. Integration of (8.9.12) over a region R whose boundary ∂R has unit outward normal $\mathbf{n}$ yields, on suppressing suffices,

$$\frac{\partial}{\partial t}\left\{\int_R \rho\, dV\right\} = -\int_{\partial R} \rho\mathbf{v}.\mathbf{n}\, dA + \int_R \mathcal{G}\, dV. \tag{8.9.15}$$

That is, the rate of change of mass in R has two contributions, the first from migration of $\mathcal{M}^+$ mass into R across its boundary and the second from net 'production' of $\mathcal{M}^+$ mass due to transitions between $\mathcal{M}^-$ and $\mathcal{M}^+$ within R. Of course, all field values depend upon both the spatial scale embodied in choice w and the time-averaging scale Δ.

Remark 8.9.3. Field $\mathbf{v}_{w,\Delta}$ has here emerged as the natural fundamental kinematic variable from which the concepts of motion, deformation, and material point may be derived, precisely as in Section 5.2 (see also Section 2.3). Specifically, consider

$$B_{w,\Delta,t} := \{\mathbf{x} \in \mathcal{E} : \rho_{w,\Delta}(\mathbf{x},t) > 0\}. \tag{8.9.16}$$

Then the *motion map corresponding to spatial and temporal scales embodied in choices w and Δ, and relative to the situation at time t_0, is χ_{w,Δ,t_0}*, where

$$\chi_{w,\Delta,t_0}(\cdot,t) : B_{w,\Delta,t_0} \longrightarrow \mathcal{E} \tag{8.9.17}$$

and, for each $\hat{\mathbf{x}} \in B_{w,\Delta,t_0}$, this map is the solution to the initial-value problem

$$\dot{\chi}_{w,\Delta,t_0}(\hat{\mathbf{x}},t) = \mathbf{v}_{w,\Delta}(\chi_{w,\Delta,t_0}(\hat{\mathbf{x}},t),t), \tag{8.9.18}$$

where $$\chi_{w,\Delta,t_0}(\hat{\mathbf{x}},t_0) = \hat{\mathbf{x}}. \tag{8.9.19}$$

Each map $\chi_{w,\Delta,t_0}(\cdot,t)$ is termed the *deformation* at time t associated with the motion, and with each geometrical point $\hat{\mathbf{x}} \in B_{w,\Delta,t_0}$ can be identified a (fictional!) material point $\mathbf{X}$ of $\mathcal{M}^+$ whose location at time t is $\chi_{w,\Delta,t_0}(\hat{\mathbf{x}},t)$. Thus, starting from $\hat{\mathbf{x}}$ at time t_0, the trajectory of $\mathbf{X}$ is obtained by moving in such a way that, at any subsequent time t, the velocity is the value of $\mathbf{v}_{w,\Delta}(\cdot,t)$ at the location of $\mathbf{X}$ at this time.

Remark 8.9.4. In tracking the motion of $\mathbf{X}$ (identified by $\hat{\mathbf{x}} \in B_{w,\Delta,t_0}$) it may happen that $\rho_{w,\Delta}(\chi_{w,\Delta,t_0}(\hat{\mathbf{x}},t),t)$ vanishes at some time after $t = t_0$. This would be the consequence of $\mathcal{M}^+ \to \mathcal{M}^-$ transitions resulting in a complete local absence of point masses in $\mathcal{M}^+$. Accordingly at such locations and times $\mathbf{v}_{w,\Delta}$ would be undefined, as would χ_{w,Δ,t_0}, and material point $\mathbf{X}$ would have disappeared! More precisely, the trajectory of $\mathbf{X}$ would terminate. Of course, the reverse situation could occur: locally, point masses in $\mathcal{M}^+$ could appear where none were to be found previously (as a consequence of $\mathcal{M}^- \to \mathcal{M}^+$ transitions). If this were manifest at $\hat{\mathbf{x}}'$ at time t_0', then another motion map could be constructed as in (8.9.17) et seq, based upon the situation at time t_0'. Notice that in such case, and for times $t > t_0'$, the trajectories associated with χ_{w,Δ,t_0} and $\chi_{w,\Delta,t_0'}$ would coincide at points common to $\chi_{w,\Delta,t_0}(B_{w,\Delta,t_0},t)$ and $\chi_{w,\Delta,t_0'}(B_{w,\Delta,t_0'},t)$. (Why?) The disappearance and appearance of material points thus is seen to be a consequence of local, 'complete', transitions from $\mathcal{M}^+$ into $\mathcal{M}^-$ and $\mathcal{M}^-$ into $\mathcal{M}^+$, respectively.

Remark 8.9.5. The foregoing concentrated on $\mathcal{M}^+$. Of course, an identical analysis applies to $\mathcal{M}^-$: one only needs to change e_i to $(1-e_i)$ throughout.

8.9.2 Linear Momentum Balance

Recall the governing equation for the motion of $P_i \in \mathcal{M}$ in an inertial frame: from (5.4.4) and (5.4.5) this is

$$\sum_{j\neq i} \mathbf{f}_{ij} + \mathbf{b}_i = m_i\dot{\mathbf{v}}_i. \tag{8.9.20}$$

Multiplying by $e_i(\tau)w(\mathbf{x}_i(\tau)-\mathbf{x})$ and summing over all $P_i \in \mathcal{M}$ yield

$$\sum_i \sum_{j\neq i} \mathbf{f}_{ij}(\tau)e_i(\tau)w(\mathbf{x}_i(\tau)-\mathbf{x}) + \sum_i \mathbf{b}_i(\tau)e_i(\tau)w(\mathbf{x}_i(\tau)-\mathbf{x})$$
$$= \sum_i m_i e_i(\tau)w(\mathbf{x}_i(\tau)-\mathbf{x})\dot{\mathbf{v}}_i(\tau). \tag{8.9.21}$$

Effecting a decomposition as in (8.7.8), but with the inclusion of a spatially localising weighting factor,

$$\sum_i \sum_{j\neq i} \mathbf{f}_{ij}(\tau)e_i(\tau)w(\mathbf{x}_i(\tau)-\mathbf{x}) = {}^{\text{int}}\mathbf{f}_w(\mathbf{x},\tau) + {}^{\text{ext}}\mathbf{f}_w(\mathbf{x},\tau), \tag{8.9.22}$$

where

$${}^{\text{int}}\mathbf{f}_w(\mathbf{x},\tau) := \sum_i \sum_{j\neq i} \mathbf{f}_{ij}(\tau)e_i(\tau)e_j(\tau)w(\mathbf{x}_i(\tau)-\mathbf{x}) \tag{8.9.23}$$

and

$${}^{\text{ext}}\mathbf{f}_w(\mathbf{x},\tau) := \sum_i \sum_{j\neq i} \mathbf{f}_{ij}(\tau)e_i(\tau)(1-e_j(\tau))w(\mathbf{x}_i(\tau)-\mathbf{x}). \tag{8.9.24}$$

The Δ-time average of the left-hand side of (8.9.21) yields

$$^{\text{int}}\mathbf{f}_{w,\Delta} + {}^{\text{ext}}\mathbf{f}_{w,\Delta} + \mathbf{b}_{w,\Delta}, \tag{8.9.25}$$

where

$$^{\text{int}}\mathbf{f}_{w,\Delta}(\mathbf{x},t) := \frac{1}{\Delta}\int_{t-\Delta}^{t} {}^{\text{int}}\mathbf{f}_{w}(\mathbf{x},\tau)d\tau, \tag{8.9.26}$$

$$^{\text{ext}}\mathbf{f}_{w,\Delta}(\mathbf{x},t) := \frac{1}{\Delta}\int_{t-\Delta}^{t} {}^{\text{ext}}\mathbf{f}_{w}(\mathbf{x},\tau)d\tau, \tag{8.9.27}$$

and

$$\mathbf{b}_{w,\Delta}(\mathbf{x},t) := \frac{1}{\Delta}\int_{t-\Delta}^{t} \mathbf{b}_{w}(\mathbf{x},\tau)d\tau. \tag{8.9.28}$$

Remark 8.9.6. Interpretation of ${}^{\text{int}}\mathbf{f}_{w}(\mathbf{x},\tau)$ and ${}^{\text{ext}}\mathbf{f}_{w}(\mathbf{x},\tau)$ is particularly simple in the case of pairwise-balanced interactions and choice $w = w_\epsilon$ [see (4.3.2) and (4.3.3)]. In such case, suppressing time dependence,

$$\begin{aligned} {}^{\text{int}}\mathbf{f}_{w}(\mathbf{x}) &= \frac{1}{2}\sum_{i\neq j}\sum \{\mathbf{f}_{ij}e_i e_j w(\mathbf{x}_i - \mathbf{x}) + \mathbf{f}_{ji}e_j e_i w(\mathbf{x}_j - \mathbf{x})\} \\ &= \frac{1}{2}\sum_{i\neq j}\sum \mathbf{f}_{ij}e_i e_j \{w(\mathbf{x}_i - \mathbf{x}) - w(\mathbf{x}_j - \mathbf{x})\}. \end{aligned} \tag{8.9.29}$$

Pairs P_i and P_j which both lie within $S_\epsilon(\mathbf{x})$ yield zero net contribution, as do pairs which both lie outside $S_\epsilon(\mathbf{x})$. It follows that the net contribution of P_i and P_j, with one in $S_\epsilon(\mathbf{x})$ and the other outside $S_\epsilon(\mathbf{x})$, is [without loss of generality suppose that $P_i \in S_\epsilon(\mathbf{x})$] precisely $\mathbf{f}_{ij}\, e_i e_j\, w(\mathbf{x}_i - \mathbf{x})$. Thus ${}^{\text{int}}\mathbf{f}_{w}(\mathbf{x},\tau)$ denotes the resultant force exerted by $\mathcal{M}^+$ point masses (at time τ) *outside* $S_\epsilon(\mathbf{x})$ upon $\mathcal{M}^+$ point masses *inside* $S_\epsilon(\mathbf{x})$ at this time, divided by the volume V_ϵ of $S_\epsilon(\mathbf{x})$. By way of contrast, ${}^{\text{ext}}\mathbf{f}_{w}(\mathbf{x},\tau)$ represents the resultant force exerted on $\mathcal{M}^+$ point masses inside $S_\epsilon(\mathbf{x})$ by $\mathcal{M}^-$ point masses both inside *and* outside $S_\epsilon(\mathbf{x})$, divided by the volume V_ϵ of $S_\epsilon(\mathbf{x})$. Of course, ${}^{\text{int}}\mathbf{f}_{w,\Delta}(\mathbf{x},t)$ and ${}^{\text{ext}}\mathbf{f}_{w,\Delta}(\mathbf{x},t)$ denote the Δ-time averages of these resultant force densities at time t.

From (8.9.21) expression (8.9.25) is to be equated with

$$\begin{aligned} \frac{1}{\Delta}\int_{t-\Delta}^{t}\sum_i m_i e_i(\tau) w(\mathbf{x}_i(\tau)-\mathbf{x})\dot{\mathbf{v}}_i(\tau)d\tau &= \frac{1}{\Delta}\int_{t-\Delta}^{t}\sum_i m_i e_i(\tau) \\ \left\{\frac{\partial}{\partial\tau}\{\mathbf{v}_i(\tau)w(\mathbf{x}_i(\tau)-\mathbf{x})\} + \text{div}_{\mathbf{x}}\{\mathbf{v}_i(\tau)\otimes\mathbf{v}_i(\tau)w(\mathbf{x}_i(\tau)-\mathbf{x})\}\right\}d\tau. \end{aligned} \tag{8.9.30}$$

Use has here been made of (B.7.30) of Appendix B.7, with $\phi = w$ and $\mathbf{A} = \mathbf{v}_i \otimes \mathbf{v}_i$.

Defining the *notional thermal velocity* [cf. (5.5.8)] by

$$\hat{\mathbf{v}}_i(\tau;\mathbf{x},t) := \mathbf{v}_i(\tau) - \mathbf{v}_{w,\Delta}(\mathbf{x},t) \tag{8.9.31}$$

yields [in connection with the divergence term in (8.9.30)]

$$\frac{1}{\Delta}\sum_i \int_{t-\Delta}^{t} m_i e_i(\tau)\mathbf{v}_i(\tau)\otimes\mathbf{v}_i(\tau)w(\mathbf{x}_i(\tau)-\mathbf{x})d\tau$$

$$= \frac{1}{\Delta}\int_{t-\Delta}^{t}\sum_i m_i e_i(\tau)[\hat{\mathbf{v}}_i(\tau;\mathbf{x},t)+\mathbf{v}_{w,\Delta}(\mathbf{x},t)]\otimes[\hat{\mathbf{v}}_i(\tau;\mathbf{x},t)+\mathbf{v}_{w,\Delta}(\mathbf{x},t)]w(\mathbf{x}_i(\tau)-\mathbf{x})d\tau$$

$$= \mathcal{D}_{w,\Delta}(\mathbf{x},t)+\rho_{w,\Delta}(\mathbf{x},t)\mathbf{v}_{w,\Delta}(\mathbf{x},t)\otimes\mathbf{v}_w(\mathbf{x},t). \tag{8.9.32}$$

Here $$\mathcal{D}_{w,\Delta}(\mathbf{x},t) := \frac{1}{\Delta}\sum_i\int_{t-\Delta}^{t} m_i e_i(\tau)\hat{\mathbf{v}}_i(\tau;\mathbf{x},t)\otimes\hat{\mathbf{v}}_i(\tau;\mathbf{x},t)w(\mathbf{x}_i(\tau)-\mathbf{x}), \tag{8.9.33}$$

and in obtaining (8.9.32) we have noted that

$$\frac{1}{\Delta}\int_{t-\Delta}^{t} m_i e_i(\tau)\hat{\mathbf{v}}_i(\tau;\mathbf{x},t)w(\mathbf{x}_i(\tau)-\mathbf{x})d\tau = \mathbf{p}_{w,\Delta}(\mathbf{x},t)-\rho_{w,\Delta}(\mathbf{x},t)\mathbf{v}_{w,\Delta}(\mathbf{x},t)$$

$$= \mathbf{0}. \tag{8.9.34}$$

Exercise 8.9.1. Prove that (8.9.34) follows from (8.9.31) together with the definitions of $\rho_{w,\Delta}, \mathbf{p}_{w,\Delta}$, and $\mathbf{v}_{w,\Delta}$.

There remains computation of

$$\frac{1}{\Delta}\int_{t-\Delta}^{t}\sum_i m_i e_i(\tau)\frac{\partial}{\partial\tau}\{\mathbf{v}_i(\tau)w(\mathbf{x}_i(\tau)-\mathbf{x})\}d\tau. \tag{8.9.35}$$

Consideration of all possibilities, exactly as in Section 8.7 in respect of the right-hand side of (8.7.11), together with the simplification (8.9.8) of the right-hand side of (8.9.6), enable expression (8.9.35) to be written as

$$\frac{1}{\Delta}\sum_i m_i\{e_i(t)\mathbf{v}_i(t)w(\mathbf{x}_i(t)-\mathbf{x})-e_i(t-\Delta)\mathbf{v}_i(t-\Delta)w(\mathbf{x}_i(t-\Delta)-\mathbf{x})\}$$

$$+\mathbf{P}^{\text{out}}_{w,\Delta}(\mathbf{x},t)-\mathbf{P}^{\text{in}}_{w,\Delta}(\mathbf{x},t), \tag{8.9.36}$$

where $$\mathbf{P}^{\text{out}}_{w,\Delta}(\mathbf{x},t) := \frac{1}{\Delta}\sum_i\sum_{i_\ell} m_i\mathbf{v}_i(t_{i_\ell})w(\mathbf{x}_i(t_{i_\ell})-\mathbf{x}) \tag{8.9.37}$$

and $$\mathbf{P}^{\text{in}}_{w,\Delta}(\mathbf{x},t) := \frac{1}{\Delta}\sum_i\sum_{i_e} m_i\mathbf{v}_i(t_{i_e})w(\mathbf{x}_i(t_{i_e})-\mathbf{x}). \tag{8.9.38}$$

The sums in (8.9.37) and (8.9.38) involve precisely the same times t_{i_ℓ} and t_{i_e} which appear in (8.9.9) and (8.9.10). Thus $\mathbf{P}^{\text{out}}_{w,\Delta}(\mathbf{x},t)$ represents a local rate of change of momentum density associated with point masses leaving $\mathcal{M}^+$ (and entering $\mathcal{M}^-$), while $\mathbf{P}^{\text{in}}_{w,\Delta}(\mathbf{x},t)$ is the corresponding rate of change for point masses entering $\mathcal{M}^+$ (and leaving $\mathcal{M}^-$).

Since expression (8.9.36) may be written as

$$\frac{1}{\Delta}\{\mathbf{p}_w(\mathbf{x},t)-\mathbf{p}_w(\mathbf{x},t-\Delta)\}+\mathbf{P}^{\text{out}}_{w,\Delta}(\mathbf{x},t)-\mathbf{P}^{\text{in}}_{w,\Delta}(\mathbf{x},t)$$

$$\left(=\left(\frac{\partial}{\partial t}\{\mathbf{p}_{w,\Delta}\}+\mathbf{P}^{\text{out}}_{w,\Delta}-\mathbf{P}^{\text{in}}_{w,\Delta}\right)(\mathbf{x},t)\right), \tag{8.9.39}$$

the Δ-time average of (8.9.21) is, from (8.9.25), (8.9.30), and (8.9.32), on suppressing arguments $\mathbf{x}$ and t,

$$^{\text{int}}\mathbf{f}_{w,\Delta} + {}^{\text{ext}}\mathbf{f}_{w,\Delta} + \mathbf{b}_{w,\Delta} = \frac{\partial}{\partial t}\{\mathbf{p}_{w,\Delta}\} + \mathbf{P}^{\text{out}}_{w,\Delta} - \mathbf{P}^{\text{in}}_{w,\Delta} + \text{div}\{\mathcal{D}_{w,\Delta} + \rho_{w,\Delta}\mathbf{v}_{w,\Delta} \otimes \mathbf{v}_{w,\Delta}\}. \tag{8.9.40}$$

Given solution $\mathbf{a}_i$ to (5.6.2), and proceeding as in Subsection 5.6.2, from (8.9.23)

$$^{\text{int}}\mathbf{f}_w = \text{div}\{{}_s\mathbf{T}^-_w\}, \tag{8.9.41}$$

where the *s*imple $\mathcal{M}^+ - \mathcal{M}^+$ *interaction stress tensor*

$${}_s\mathbf{T}^-_w(\mathbf{x},\tau) := \sum_{i \neq j}\sum \mathbf{f}_{ij}(\tau) \otimes \mathbf{a}_i(\mathbf{x},\tau)e_i(\tau)e_j(\tau). \tag{8.9.42}$$

If interactions are pairwise balanced, then, noting

$$^{\text{int}}\mathbf{f}_w(\mathbf{x},\tau) = \frac{1}{2}\sum_{i \neq j}\sum \mathbf{f}_{ij}(\tau)e_i(\tau)e_j(\tau)\{w(\mathbf{x}_i(\tau) - \mathbf{x}) - w(\mathbf{x}_j(\tau) - \mathbf{x}\} \tag{8.9.43}$$

and proceeding as in Subsection 5.6.3,

$$^{\text{int}}\mathbf{f}_w = \text{div}\{{}_b\mathbf{T}^-_w\}, \tag{8.9.44}$$

where the *b*alanced $\mathcal{M}^+ - \mathcal{M}^+$ *interaction stress tensor*

$${}_b\mathbf{T}^-_w(\mathbf{x},\tau) := \sum_{i \neq j}\sum \mathbf{f}_{ij}(\tau) \otimes \mathbf{b}_{ij}(\mathbf{x},\tau)e_i(\tau)e_j(\tau). \tag{8.9.45}$$

Of course, Δ-time averages of (8.9.41) and (8.9.44) enable $\mathbf{f}^{\text{int}}_{w,\Delta}$ to be expressed as the divergence of an interaction stress tensor which involves only $\mathcal{M}^+ - \mathcal{M}^+$ interactions. Specifically [and recalling (8.3.12)],

$$^{\text{int}}\mathbf{f}_{w,\Delta} = \text{div}\{{}_s\mathbf{T}^-_{w,\Delta}\} = \text{div}\{{}_b\mathbf{T}^-_{w,\Delta}\}, \tag{8.9.46}$$

where ${}_s\mathbf{T}^-_{w,\Delta}$ and ${}_b\mathbf{T}^-_{w,\Delta}$ denote the Δ-time averages of ${}_s\mathbf{T}^-_w$ and ${}_b\mathbf{T}^-_w$, respectively.

It follows from (8.9.40) and (8.9.41) or (8.9.44) that linear momentum balance takes the form, upon suppressing explicit appearance of choices w and Δ,

$$\text{div}\,\mathbf{T} + {}^{\text{ext}}\mathbf{f} + \mathbf{P}^{\text{in}} - \mathbf{P}^{\text{out}} + \mathbf{b} = \frac{\partial}{\partial t}\{\rho\mathbf{v}\} + \text{div}\{\rho\mathbf{v} \otimes \mathbf{v}\}, \tag{8.9.47}$$

where

$$\mathbf{T} := \mathbf{T}^-_{w,\Delta} - \mathcal{D}_{w,\Delta} \tag{8.9.48}$$

and

$$\mathbf{T}^-_{w,\Delta} = {}_s\mathbf{T}_{w,\Delta} \quad \text{or} \quad {}_b\mathbf{T}_{w,\Delta}. \tag{8.9.49}$$

In the same subscript-free format, mass balance (8.9.12) is

$$\frac{\partial \rho}{\partial t} + \text{div}\{\rho\mathbf{v}\} = \mathcal{G}. \tag{8.9.50}$$

Thus [cf. (5.5.14) and (5.5.16)] (8.9.47) may be written in the form

$$\text{div}\,\mathbf{T} + {}^{\text{ext}}\mathbf{f} + \mathbf{b} + \mathbf{P} = \rho\dot{\mathbf{v}} + \mathcal{G}\mathbf{v}, \tag{8.9.51}$$

where
$$\mathbf{P} = \mathbf{P}^{\text{in}} - \mathbf{P}^{\text{out}} \tag{8.9.52}$$

and
$$\dot{\mathbf{v}} := \frac{\partial \mathbf{v}}{\partial t} + (\nabla \mathbf{v})\mathbf{v} \tag{8.9.53}$$

denotes the *acceleration* field.

Exercise 8.9.2. Derive (8.9.51) from (8.9.47) and (8.9.50).

An alternative and equivalent form of (8.9.51) is

$$\operatorname{div}\mathbf{T} + {}^{\text{ext}}\mathbf{f} + \mathbf{b} + \mathbf{I}^{\text{in}} - \mathbf{I}^{\text{out}} = \rho\dot{\mathbf{v}}, \tag{8.9.54}$$

where [cf. (8.9.31)]

$$\mathbf{I}^{\text{in}}(\mathbf{x},t) := \frac{1}{\Delta}\sum_{i}\sum_{i_e} m_i\, \hat{\mathbf{v}}_i(t_{i_e};\mathbf{x},t)w(\mathbf{x}_i(t_{i_e}) - \mathbf{x}) \tag{8.9.55}$$

and
$$\mathbf{I}^{\text{out}}(\mathbf{x},t) := \frac{1}{\Delta}\sum_{i}\sum_{i_\ell} m_i\, \hat{\mathbf{v}}_i(t_{i_\ell};\mathbf{x},t)w(\mathbf{x}_i(t_{i_\ell}) - \mathbf{x}). \tag{8.9.56}$$

Exercise 8.9.3. Derive (8.9.54) by showing that

$$\mathbf{I}^{\text{in}} = \mathbf{P}^{\text{in}} - \mathcal{G}^{\text{in}}\mathbf{v} \qquad \text{and} \qquad \mathbf{I}^{\text{out}} = \mathbf{P}^{\text{out}} - \mathcal{G}^{\text{out}}\mathbf{v}. \tag{8.9.57}$$

[See the reduction of (8.7.26) to (8.7.30) via Result 8.7.1.]

Remark 8.9.7. Integration of (8.9.54) over a region R (cf. Remark 8.9.2) yields (on writing $\mathbf{I} := \mathbf{I}^{\text{in}} - \mathbf{I}^{\text{out}}$)

$$\int_{\partial R} \mathbf{T}\mathbf{n}\, dA + \int_R \{{}^{\text{ext}}\mathbf{f} + \mathbf{b} + \mathbf{I}\}dV = \frac{d}{d\tau}\left\{\int_{R_\tau} \rho\mathbf{v}\, dV\right\}\bigg|_{\tau=t}. \tag{8.9.58}$$

Here R_τ deforms with the motion prescribed by $\mathbf{v}$ and $R_t = R$. Relation (8.9.58) indicates the distinct contributions to the time rate of change of $\mathcal{M}^+$ momentum within the deforming region R_τ at time t. The surface integral has contributions deriving from corpuscular interaction forces exerted on $\mathcal{M}^+$ material in R by $\mathcal{M}^+$ material outside R [$\mathbf{T}^-$: see (8.9.48)] together with a supply rate of $\mathcal{M}^+$ momentum into R associated with diffusion across R [$-\mathcal{D}$: see (8.9.48)]. The volume contribution has three sources: the resultant force density ${}^{\text{ext}}\mathbf{f}$ exerted by $\mathcal{M}^-$ material *everywhere* (both inside and outside R) upon $\mathcal{M}^+$ material in R, the resultant external force density $\mathbf{b}$ upon $\mathcal{M}^+$ material in R, and the resultant supply rate $\mathbf{I}$ of $\mathcal{M}^+$ momentum to R as a consequence of net transitions from $\mathcal{M}^-$ into $\mathcal{M}^+$ within R.

8.9.3 Energy Balance

Consider equation (8.9.20) which governs the motion of any $P_i \in \mathcal{M}$. Multiplication of each term scalarly by $\Delta^{-1}e_i(\tau)w(\mathbf{x}_i(\tau) - \mathbf{x})\mathbf{v}_i(\tau)$ and integrating with respect to τ over the interval $[t - \Delta, t]$ yield from the first term

$$\frac{1}{\Delta}\int_{t-\Delta}^{t} \sum\sum_{i\neq j}\{\mathbf{f}_{ij}(\tau)e_i(\tau)e_j(\tau) + \mathbf{f}_{ij}(\tau)e_i(\tau)(1 - e_j(\tau))\}\cdot\mathbf{v}_i(\tau)w(\mathbf{x}_i(\tau) - \mathbf{x}). \tag{8.9.59}$$

Writing [see (8.9.31)]

$$\mathbf{v}_i(\tau) = \hat{\mathbf{v}}_i(\tau;\mathbf{x},t) + \mathbf{v}_{w,\Delta}(\mathbf{x},t) \tag{8.9.60}$$

reduces expression (8.9.59) to

$$({}^{\text{int}}Q_{w,\Delta} + {}^{\text{int}}\mathbf{f}_{w,\Delta}\,.\,\mathbf{v}_{w,\Delta} + {}^{\text{ext}}Q_{w,\Delta} + {}^{\text{ext}}\mathbf{f}_{w,\Delta}\,.\,\mathbf{v}_{w,\Delta})(\mathbf{x},t). \tag{8.9.61}$$

Here

$${}^{\text{int}}Q_{w,\Delta}(\mathbf{x},t) := \frac{1}{\Delta}\int_{t-\Delta}^{t} \sum\sum_{i\neq j} e_i(\tau)e_j(\tau)\mathbf{f}_{ij}(\tau)\,.\,\hat{\mathbf{v}}_i(\tau;\mathbf{x},t)w(\mathbf{x}_i(\tau)-\mathbf{x})d\tau \tag{8.9.62}$$

and

$${}^{\text{ext}}Q_{w,\Delta}(\mathbf{x},t) := \frac{1}{\Delta}\int_{t-\Delta}^{t} \sum\sum_{i\neq j} e_i(\tau)(1-e_j(\tau))\mathbf{f}_{ij}(\tau)\,.\,\hat{\mathbf{v}}_i(\tau;\mathbf{x},t)w(\mathbf{x}_i(\tau)-\mathbf{x})d\tau. \tag{8.9.63}$$

The second term in (8.9.20) yields the expression

$$\begin{aligned} &\frac{1}{\Delta}\int_{t-\Delta}^{t} \sum_i e_i(\tau)\mathbf{b}_i(\tau)\,.\,(\hat{\mathbf{v}}_i(\tau;\mathbf{x},t) + \mathbf{v}_{w,\Delta}(\mathbf{x},t))w(\mathbf{x}_i(\tau)-\mathbf{x})d\tau \\ &\qquad = \mathbf{r}_{w,\Delta}(\mathbf{x},t) + \mathbf{b}_{w,\Delta}(\mathbf{x},t)\,.\,\mathbf{v}_{w,\Delta}(\mathbf{x},t), \end{aligned} \tag{8.9.64}$$

where $$r_{w,\Delta}(\mathbf{x},t) := \frac{1}{\Delta}\int_{t-\Delta}^{t} \sum_i e_i(\tau)\mathbf{b}_i(\tau)\,.\,\hat{\mathbf{v}}_i(\tau;\mathbf{x},t)w(\mathbf{x}_i(\tau)-\mathbf{x})d\tau. \tag{8.9.65}$$

The right-hand side of (8.9.20) produces

$$\begin{aligned} &\frac{1}{\Delta}\int_{t-\Delta}^{t} \sum_i m_i\dot{\mathbf{v}}_i(\tau)\,.\,\mathbf{v}_i(\tau)e_i(\tau)w(\mathbf{x}_i(\tau)-\mathbf{x})d\tau \\ &= \frac{1}{\Delta}\int_{t-\Delta}^{t} \sum_i m_i \frac{\partial}{\partial\tau}\left\{\frac{1}{2}\mathbf{v}_i^2(\tau)\right\} e_i(\tau)w(\mathbf{x}_i(\tau)-\mathbf{x})d\tau \\ &= R_{w,\Delta}(\mathbf{x},t) + S_{w,\Delta}(\mathbf{x},t), \end{aligned} \tag{8.9.66}$$

where $$R_{w,\Delta}(\mathbf{x},t) := \frac{1}{\Delta}\int_{t-\Delta}^{t} \sum_i m_i e_i(\tau)\frac{\partial}{\partial\tau}\left\{\frac{1}{2}\mathbf{v}_i^2(\tau)w(\mathbf{x}_i(\tau)-\mathbf{x})\right\}d\tau \tag{8.9.67}$$

and $$S_{w,\Delta}(\mathbf{x},t) := \text{div}_{\mathbf{x}}\left\{\frac{1}{\Delta}\int_{t-\Delta}^{t} \sum_i \frac{1}{2} m_i e_i(\tau)\mathbf{v}_i^2(\tau)\mathbf{v}_i(\tau)w(\mathbf{x}_i(\tau)-\mathbf{x})d\tau\right\}. \tag{8.9.68}$$

The integrand in (8.9.67) may be compared with those in (8.7.11) and (8.9.35). It follows as in Section 8.7 in respect of (8.7.11) that

$$R_{w,\Delta}(\mathbf{x},t) = \frac{1}{\Delta}\sum_i \frac{m_i}{2}\{e_i(t)\mathbf{v}_i^2(t)w(\mathbf{x}_i(t)-\mathbf{x}) - e_i(t-\Delta)\mathbf{v}_i^2(t-\Delta)w(\mathbf{x}_i(t-\Delta)-\mathbf{x})\}$$

$$+\frac{1}{\Delta}\sum_i\sum_{i_\ell}\frac{1}{2}m_i\mathbf{v}_i^2(t_{i_\ell})w(\mathbf{x}_i(t_{i_\ell})-\mathbf{x})$$

$$-\frac{1}{\Delta}\sum_i\sum_{i_e}\frac{1}{2}m_i\mathbf{v}_i^2(t_{i_e})w(\mathbf{x}_i(t_{i_e})-\mathbf{x}). \tag{8.9.69}$$

Using (8.9.60) and noting (8.9.34), the first term on the right-hand side of (8.9.69) reduces to

$$\frac{\partial}{\partial t}\{\rho_{w,\Delta}h_{w,\Delta}+\frac{1}{2}\rho_{w,\Delta}\mathbf{v}_{w,\Delta}^2\}(\mathbf{x},t), \tag{8.9.70}$$

where $$(\rho_{w,\Delta}h_{w,\Delta})(\mathbf{x},t):=\frac{1}{\Delta}\int_{t-\Delta}^{t}\sum_i e_i(\tau)\frac{m_i}{2}\hat{\mathbf{v}}_i^2(\tau;\mathbf{x},t)w(\mathbf{x}_i(\tau)-\mathbf{x})d\tau \tag{8.9.71}$$

denotes the local $\mathcal{M}^+$ *heat energy density* [cf. (6.2.13)]. A similar decomposition of $\mathbf{v}_i$ in the remaining terms on the right-hand side of (8.9.69) yields a remaining contribution to $R_{w,\Delta}(\mathbf{x},t)$ of

$$(H_{w,\Delta}^{\text{out}}-H_{w,\Delta}^{\text{in}}+(\mathbf{I}_{w,\Delta}^{\text{out}}-\mathbf{I}_{w,\Delta}^{\text{in}}).\mathbf{v}_{w,\Delta}+\frac{1}{2}(\mathcal{G}_{w,\Delta}^{\text{out}}-\mathcal{G}_{w,\Delta}^{\text{in}})\mathbf{v}_{w,\Delta}^2)(\mathbf{x},t). \tag{8.9.72}$$

Here $$H_{w,\Delta}^{\text{out}}(\mathbf{x},t):=\frac{1}{\Delta}\sum_i\sum_{i_\ell}\frac{1}{2}m_i\hat{\mathbf{v}}_i^2(t_{i_\ell};\mathbf{x},t)w(\mathbf{x}_i(t_{i_\ell})-\mathbf{x}) \tag{8.9.73}$$

and $$H_{w,\Delta}^{\text{in}}(\mathbf{x},t):=\frac{1}{\Delta}\sum_i\sum_{i_e}\frac{1}{2}m_i\hat{\mathbf{v}}_i^2(t_{i_e};\mathbf{x},t)w(\mathbf{x}_i(t_{i_e})-\mathbf{x}). \tag{8.9.74}$$

Thus, from (8.9.70) and (8.9.72), and suppressing arguments and subscripts,

$$R=\frac{\partial}{\partial t}\left\{\rho(h+\frac{1}{2}\mathbf{v}^2)\right\}+H^{\text{out}}-H^{\text{in}}+(\mathbf{I}^{\text{out}}-\mathbf{I}^{\text{in}}).\mathbf{v}+\frac{1}{2}(\mathcal{G}^{\text{out}}-\mathcal{G}^{\text{in}})\mathbf{v}^2. \tag{8.9.75}$$

The first contribution to R is the only term remaining if $\mathcal{M}^+$ has unchanging material content and is the sum of the macroscopic kinetic energy density and the heat energy density [see (6.2.12)]. Contribution $H^{\text{out}}-H^{\text{in}}$ represents the density of the rate at which $\mathcal{M}^+$ heat energy is lost as a consequence of $\mathcal{M}^-\leftrightarrow\mathcal{M}^+$ transitions, while $\frac{1}{2}(\mathcal{G}^{\text{out}}-\mathcal{G}^{\text{in}})\mathbf{v}^2$ is the density associated with the loss of macroscopic kinetic energy due to $\mathcal{M}^-\leftrightarrow\mathcal{M}^+$ transitions. Term $(\mathbf{I}^{\text{out}}-\mathbf{I}^{\text{in}}).\mathbf{v}$ is a 'hybrid' rate loss density associated with what might be termed 'notional relative (or thermal) momentum': recall the definition of $\hat{\mathbf{v}}_i(t;\mathbf{x})$ as an approximation to thermal velocity $\tilde{\mathbf{v}}_i(t)$ given in (5.5.25) et seq.

The integrand involved in definition (8.9.68) of $S_{w,\Delta}$ is merely the $\mathcal{M}^+$ version of the quantity in (6.2.9). There is, however, a 'wrinkle' to be resolved. To see this we note that

$$\begin{aligned}\frac{1}{2}\sum_i m_ie_i\mathbf{v}_i^2\mathbf{v}_iw(\mathbf{x}_i-\mathbf{x})&=\frac{1}{2}\sum_i m_ie_i(\hat{\mathbf{v}}_i^2+2\hat{\mathbf{v}}_i.\mathbf{v}_{w,\Delta}+\mathbf{v}_{w,\Delta}^2)(\hat{\mathbf{v}}_i+\mathbf{v}_{w,\Delta})w(\mathbf{x}_i-\mathbf{x})\\&=\boldsymbol{\kappa}_w+\rho_wh_w\mathbf{v}_{w,\Delta}+\boldsymbol{\mathcal{D}}_w\mathbf{v}_{w,\Delta}+(\mathbf{v}_{w,\Delta}\otimes\mathbf{v}_{w,\Delta})(\mathbf{p}_w-\rho_w\mathbf{v}_{w,\Delta})\\&\quad+\mathbf{v}_{w,\Delta}^2(\mathbf{p}_w-\rho_w\mathbf{v}_{w,\Delta})+\frac{1}{2}\rho_w\mathbf{v}_{w,\Delta}^2\mathbf{v}_{w,\Delta}.\end{aligned} \tag{8.9.76}$$

Here $\boldsymbol{\kappa}_w, \rho_w, \boldsymbol{\mathcal{D}}_w$, and $\mathbf{p}_w$ are the $\mathcal{M}^+$ counterparts of the corresponding definitions of Chapter 6. Specifically, these are given by (6.2.15), (4.2.1), (5.5.11), and (4.2.13). The only difference here is the presence in all these definitions of membership factor e_i and that, in the appropriate version of (6.2.15) and (5.5.11), $\hat{\mathbf{v}}_i$ is given by (8.9.31). Notice that this involves the macroscopic velocity $\mathbf{v}_{w,\Delta}$ [see (8.9.13)] rather than $\mathbf{v}_w$. Simplification of the term $S_{w,\Delta}$ requires that the right-hand side of (8.9.76) be Δ-time-averaged. *If we make the assumption that the behaviour of the system is such that $\mathbf{v}_{w,\Delta}$ is essentially constant in such time averaging*, then the Δ-time average we seek is

$$\begin{aligned}\boldsymbol{\kappa}_{w,\Delta} + \rho_{w,\Delta} h_{w,\Delta}\mathbf{v}_{w,\Delta} + \boldsymbol{\mathcal{D}}_{w,\Delta}\mathbf{v}_{w,\Delta} + (\mathbf{v}_{w,\Delta}\otimes\mathbf{v}_{w,\Delta})(\mathbf{p}_{w,\Delta} - \rho_{w,\Delta}\mathbf{v}_{w,\Delta})\\ + \frac{1}{2}\mathbf{v}^2_{w,\Delta}(\mathbf{p}_{w,\Delta} - \rho_{w,\Delta}\mathbf{v}_{w,\Delta}) + \frac{1}{2}\rho_{w,\Delta}\mathbf{v}^2_{w,\Delta}\mathbf{v}_{w,\Delta},\end{aligned} \tag{8.9.77}$$

and hence, from (8.9.68) and (8.9.13),

$$S_{w,\Delta} = \operatorname{div}\{\boldsymbol{\kappa}_{w,\Delta} + \rho_{w,\Delta} h_{w,\Delta}\mathbf{v}_{w,\Delta} + \boldsymbol{\mathcal{D}}_{w,\Delta}\mathbf{v}_{w,\Delta} + \frac{1}{2}\rho_{w,\Delta}\mathbf{v}^2_{w,\Delta}\mathbf{v}_{w,\Delta}\}. \tag{8.9.78}$$

Thus, from (8.9.61), (8.9.64), (8.9.75), and (8.9.78), energy balance takes the form (on suppressing subscripts)

$$\begin{aligned}&{}^{\text{int}}Q + {}^{\text{int}}\mathbf{f}\,.\,\mathbf{v} + {}^{\text{ext}}Q + {}^{\text{ext}}\mathbf{f}\,.\,\mathbf{v} + r + \mathbf{b}\,.\,\mathbf{v}\\ &= \frac{\partial}{\partial t}\left\{\rho(h + \frac{1}{2}\mathbf{v}^2)\right\} + H^{\text{out}} - H^{\text{in}} + (\mathbf{I}^{\text{out}} - \mathbf{I}^{\text{in}})\,.\,\mathbf{v} + \frac{1}{2}(\mathcal{G}^{\text{out}} - \mathcal{G}^{\text{in}})\mathbf{v}^2\\ &\qquad + \operatorname{div}\left\{\boldsymbol{\kappa} + \rho h\mathbf{v} + \boldsymbol{\mathcal{D}}\mathbf{v} + \frac{1}{2}\rho\mathbf{v}^2\mathbf{v}\right\}.\end{aligned} \tag{8.9.79}$$

This relation simplifies upon invoking mass balance (8.9.12) and linear momentum balance (8.9.40) in the manner of the derivation of (6.2.24) from (6.2.19) via (6.2.17). To this end, the subscript-free versions of (8.9.12) and (8.9.40) are

$$\frac{\partial\rho}{\partial t} + \operatorname{div}\{\rho\mathbf{v}\} = \mathcal{G} \tag{8.9.80}$$

and [see also (8.9.52) and (8.9.14)]

$${}^{\text{int}}\mathbf{f} + {}^{\text{ext}}\mathbf{f} + \mathbf{b} + \mathbf{P} - \operatorname{div}\boldsymbol{\mathcal{D}} = \frac{\partial}{\partial t}\{\rho\mathbf{v}\} + \operatorname{div}\{\rho\mathbf{v}\otimes\mathbf{v}\} = \rho\dot{\mathbf{v}} + \mathcal{G}\mathbf{v}, \tag{8.9.81}$$

respectively.

Exercise 8.9.4. Show that the last step in (8.9.81), namely

$$\frac{\partial}{\partial t}\{\rho\mathbf{v}\} + \operatorname{div}\{\rho\mathbf{v}\otimes\mathbf{v}\} = \mathcal{G}\mathbf{v} + \rho\dot{\mathbf{v}}, \tag{8.9.82}$$

where $\dot{\mathbf{v}}$ is given by (8.9.53), follows from (8.9.80) [cf. (5.5.14)]. Show also that, for any class C^1 scalar field α,

$$\frac{\partial}{\partial t}\{\rho\alpha\} + \operatorname{div}\{\rho\alpha\mathbf{v}\} := \rho\dot{\alpha} + \mathcal{G}\alpha, \tag{8.9.83}$$

where

$$\dot{\alpha} := \frac{\partial\alpha}{\partial t} + \nabla\alpha\,.\,\mathbf{v}. \tag{8.9.84}$$

Energy balance (8.9.79) may be written [using (8.9.83) with $\alpha := h + \frac{1}{2}\mathbf{v}^2$] as

$$r - \operatorname{div}\boldsymbol{\kappa} + {}^{\text{int}}Q + {}^{\text{ext}}Q + H - \mathcal{D}^T \cdot \nabla\mathbf{v}$$
$$+({}^{\text{int}}\mathbf{f} + {}^{\text{ext}}\mathbf{f} + \mathbf{b} - \operatorname{div}\mathcal{D}^T + \mathbf{I}).\mathbf{v} = \rho\{\dot{h} + \mathbf{v}.\dot{\mathbf{v}}\} + \mathcal{G}h, \tag{8.9.85}$$

where
$$H := H^{\text{in}} - H^{\text{out}}. \tag{8.9.86}$$

Use of (8.9.81), upon noting $\mathcal{D}^T = \mathcal{D}$ and writing [cf. (6.2.23)] $\mathbf{L} := \nabla\mathbf{v}$, reduces (8.9.85) to

$$r - \operatorname{div}\boldsymbol{\kappa} + {}^{\text{int}}Q + {}^{\text{ext}}Q + H - \mathcal{D}\cdot\mathbf{L} + (\mathcal{G}\mathbf{v} - \mathbf{P} + \mathbf{I}).\mathbf{v} = \rho\dot{h} + \mathcal{G}h. \tag{8.9.87}$$

However, from (8.9.57)
$$\mathbf{I} - \mathbf{P} + \mathcal{G}\mathbf{v} = \mathbf{0} \tag{8.9.88}$$

so that [cf. (6.2.24)]
$$r - \operatorname{div}\boldsymbol{\kappa} + {}^{\text{int}}Q + {}^{\text{ext}}Q + H - \mathcal{D}\cdot\mathbf{L} = \rho\dot{h} + \mathcal{G}h. \tag{8.9.89}$$

Remark 8.9.8. Here r represents the external heat supply rate density for $\mathcal{M}^+$ material, $-\boldsymbol{\kappa}$ is the $\mathcal{M}^+$ diffusive heat flux vector, ${}^{\text{int}}Q$ is the $\mathcal{M}^+$ self-heating energy density, ${}^{\text{ext}}Q$ is the heating rate density due to interactions with $\mathcal{M}^-$, H is the net heating rate density due to $\mathcal{M}^- \leftrightarrow \mathcal{M}^+$ transitions, and h is the heat energy density per unit mass.

Balance (8.9.89) is the changing material content version of (6.2.24). The special case of systems with constant material content corresponds to the vanishing of ${}^{\text{ext}}Q$, H, and $\mathcal{G}$.

We now proceed to obtain the generalisations of energy balances (6.2.32) and (6.2.64) to systems with time-dependent material content. To this end, note that from (8.9.62), (8.9.23), (8.9.26), and (8.9.60),

$$({}^{\text{int}}Q + {}^{\text{int}}\mathbf{f}.\mathbf{v})(\mathbf{x},t) = \frac{1}{\Delta}\int_{t-\Delta}^{t} \sum\sum_{i\neq j} e_i(\tau)e_j(\tau)\mathbf{f}_{ij}(\tau).\mathbf{v}_i(\tau)w(\mathbf{x}_i(\tau)-\mathbf{x})d\tau. \tag{8.9.90}$$

Given any solution $\mathbf{a}_i$ to (5.6.2) gives [cf. (6.2.25)]

$$({}^{\text{int}}Q + {}^{\text{int}}\mathbf{f}.\mathbf{v})(\mathbf{x},t) = \frac{1}{\Delta}\int_{t-\Delta}^{t} \sum\sum_{i\neq j} e_i(\tau)e_j(\tau)\operatorname{div}_{\mathbf{x}}\{\mathbf{a}_i(\mathbf{x},\tau)\otimes\mathbf{f}_{ij}(\tau))\mathbf{v}_i(\tau)\}d\tau$$

$$= \frac{1}{\Delta}\int_{t-\Delta}^{t} \operatorname{div}_{\mathbf{x}}\left\{\sum\sum_{i\neq j} e_i(\tau)e_j(\tau)(\mathbf{a}_i(\mathbf{x},\tau)\otimes\mathbf{f}_{ij}(\tau))\right\}[\hat{\mathbf{v}}_i(\tau;\mathbf{x},t) + \mathbf{v}_{w,\Delta}(\mathbf{x},t)]d\tau \tag{8.9.91}$$

$$= -\operatorname{div}\{{}_s\mathbf{q}^-_{w,\Delta}\}(\mathbf{x},t) + \operatorname{div}\{({}_s\mathbf{T}^-_{w,\Delta})^T\mathbf{v}_{w,\Delta}\}. \tag{8.9.92}$$

Here [cf. (6.2.27)]

$${}_s\mathbf{q}^-_{w,\Delta}(\mathbf{x},t) := -\frac{1}{\Delta}\int_{t-\Delta}^{t} \sum\sum_{i\neq j} e_i(\tau)e_j(\tau)(\mathbf{a}_i(\mathbf{x},\tau)\otimes\mathbf{f}_{ij}(\tau))\hat{\mathbf{v}}_i(\tau;\mathbf{x},t)d\tau, \tag{8.9.93}$$

and [cf. (5.6.5)]

$$ {}_s\mathbf{T}^-_{w,\Delta}(\mathbf{x},t) := \frac{1}{\Delta}\int_{t-\Delta}^{t} \sum\sum_{i\neq j} e_i(\tau)e_j(\mathbf{f}_{ij}(\tau)\otimes\mathbf{a}_i(\mathbf{x},\tau))d\tau. \tag{8.9.94} $$

It follows from (8.9.85) and (8.9.92) (noting $\mathcal{D}^T = \mathcal{D}$), and omitting subscripts w and Δ, that

$$ \begin{aligned} r - \operatorname{div}{}_s\mathbf{q} + {}^{\text{ext}}Q + H - \mathcal{D}\cdot\mathbf{L} + \operatorname{div}\{({}_s\mathbf{T}^-)^T\mathbf{v}\} \\ + ({}^{\text{ext}}\mathbf{f} + \mathbf{b} - \operatorname{div}\mathcal{D} + \mathbf{I} - \rho\dot{\mathbf{v}}).\mathbf{v} = \rho\dot{h} + \mathcal{G}h, \end{aligned} \tag{8.9.95} $$

where [cf. (6.2.30)]

$$ {}_s\mathbf{q} := {}_s\mathbf{q}^- + \boldsymbol{\kappa}. \tag{8.9.96} $$

Noting that linear momentum balance (8.9.51) [see also (8.9.49)] takes the form

$$ \operatorname{div}{}_s\mathbf{T} + {}^{\text{ext}}\mathbf{f} + \mathbf{P} + \mathbf{b} = \rho\dot{\mathbf{v}} + \mathcal{G}\mathbf{v}, \tag{8.9.97} $$

with ${}_s\mathbf{T}$ defined by (8.9.48), balance (8.9.95) reduces to

$$ r - \operatorname{div}{}_s\mathbf{q} + {}^{\text{ext}}Q + H + {}_s\mathbf{T}\cdot\mathbf{L} + (\mathbf{I} - \mathbf{P} + \mathcal{G}\mathbf{v}).\mathbf{v} = \rho\dot{h} + \mathcal{G}h. $$

In view of (8.9.88) this becomes [cf. (6.2.32)]

$$ r - \operatorname{div}{}_s\mathbf{q} + {}^{\text{ext}}Q + H + {}_s\mathbf{T}\cdot\mathbf{L} = \rho\dot{h} + \mathcal{G}h. \tag{8.9.98} $$

To obtain the form of energy balance appropriate to pair-wise balanced interactions delivered by pair potential functions we proceed along the lines of (6.2.34) et seq. In particular, from (8.9.90) and following (6.2.34) through (6.2.38), we obtain

$$ ({}^{\text{int}}Q + {}^{\text{int}}\mathbf{f}.\mathbf{v})(\mathbf{x},t) = \frac{1}{\Delta}\int_{t-\Delta}^{t} A_w(\mathbf{x},\tau) + B_w(\mathbf{x},\tau)d\tau, \tag{8.9.99} $$

where

$$ A_w(\mathbf{x},\tau) := \frac{1}{2}\sum\sum_{i\neq j} e_i(\tau)e_j(\tau)\mathbf{f}_{ij}(\tau).(\mathbf{v}_i(\tau) - \mathbf{v}_j(\tau))w(\mathbf{x}_i(\tau) - \mathbf{x}) \tag{8.9.100} $$

and

$$ B_w(\mathbf{x},\tau) := \frac{1}{2}\sum\sum_{i\neq j} e_i(\tau)e_j(\tau)\mathbf{f}_{ij}(\tau).\mathbf{v}_i(\tau)\{w(\mathbf{x}_i(\tau) - \mathbf{x}) - w(\mathbf{x}_j(\tau) - \mathbf{x})\}. \tag{8.9.101} $$

Given any solution $\mathbf{b}_{ij}$ to (5.6.8),

$$ B_w(\mathbf{x},\tau) = \sum\sum_{i\neq j} e_i(\tau)e_j(\tau)\operatorname{div}_{\mathbf{x}}\{(\mathbf{b}_{ij}(\mathbf{x},\tau)\otimes\mathbf{f}_{ij}(\tau))(\hat{\mathbf{v}}_i(\tau;\mathbf{x},t) + \mathbf{v}_{w,\Delta}(\mathbf{x},t))\} $$

[cf. (6.2.39) and (6.2.40)]

$$ = (\operatorname{div}\{-{}_b\mathbf{q}^-_w + ({}_b\mathbf{T}^-_w)^T\mathbf{v}_{w,\Delta}(\mathbf{x},t)\})(\mathbf{x},\tau), \tag{8.9.102} $$

where we have recalled (8.9.45) and

$$ {}_b\mathbf{q}_w^-(\mathbf{x},\tau) := -\sum\sum_{i\neq j} e_i(\tau)e_j(\tau)\mathbf{b}_{ij}(\mathbf{x},\tau)\otimes\mathbf{f}_{ij}(\tau)\hat{\mathbf{v}}_i(\tau;\mathbf{x},t). \tag{8.9.103} $$

The corresponding contribution to (8.9.99) is thus

$$ \operatorname{div}\{-{}_b\mathbf{q}_{w,\Delta}^- + ({}_b\mathbf{T}_{w,\Delta}^-)^T\mathbf{v}_{w,\Delta}(\mathbf{x},t)\}(\mathbf{x},t), \tag{8.9.104} $$

where

$$ {}_b\mathbf{q}_{w,\Delta}^-(\mathbf{x},t) := \frac{1}{\Delta}\int_{t-\Delta}^{t} {}_b\mathbf{q}_w^-(\mathbf{x},\tau)d\tau. \tag{8.9.105} $$

[In obtaining the second term in (8.9.104) we have noted that $\mathbf{v}_{w,\Delta}(\mathbf{x},t)$ does not change as the integral is taken with respect to τ.]

Turning to A_w, the analysis of interactions delivered by pair potentials in Chapter 6, specifically (6.2.51) et seq., yields

$$ A_w(\mathbf{x},\tau) = \left(-\frac{\partial}{\partial\tau}\{\rho_w\beta_w\} - \operatorname{div}\{\rho_w\beta_w\mathbf{v}_{w,\Delta}(\cdot,t)\} - \operatorname{div}{}_2\mathbf{q}_w\right)(\mathbf{x},\tau), \tag{8.9.106} $$

where

$$ (\rho_w\beta_w)(\mathbf{x},\tau) := \frac{1}{2}\sum\sum_{i\neq j} e_i(\tau)e_j(\tau)\hat{\phi}_{ij}(r_{ij}(\tau))w(\mathbf{x}_i(\tau)-\mathbf{x}) \tag{8.9.107} $$

and

$$ {}_2\mathbf{q}_w(\mathbf{x},\tau) := \frac{1}{2}\sum\sum_{i\neq j} e_i(\tau)e_j(\tau)\hat{\phi}_{ij}(r_{ij}(\tau))\hat{\mathbf{v}}_i(\tau;\mathbf{x},t)w(\mathbf{x}_i(\tau)-\mathbf{x}). \tag{8.9.108} $$

Taking the Δ-time average of (8.9.106) and defining

$$ \beta_{w,\Delta} := (\rho_w\beta_w)_{,\Delta}/\rho_{w,\Delta} \tag{8.9.109} $$

we have

$$ \begin{aligned} A_{w,\Delta} &= -\frac{\partial}{\partial t}\{(\rho_w\beta_w)_{,\Delta}\} - \operatorname{div}\{(\rho_w\beta_w)_{,\Delta}\mathbf{v}_{w,\Delta}\} - \operatorname{div}{}_2\mathbf{q}_{w,\Delta} \\ &= -\frac{\partial}{\partial t}\{\rho_{w,\Delta}\beta_{w,\Delta} - \operatorname{div}\{\rho_{w,\Delta}\beta_{w,\Delta}\mathbf{v}_{w,\Delta}\} - \operatorname{div}{}_2\mathbf{q}_{w,\Delta} \\ &= -\rho_{w,\Delta}\dot{\beta}_{w,\Delta} - \mathcal{G}_{w,\Delta}\beta_{w,\Delta} - \operatorname{div}_2\mathbf{q}_{w,\Delta} \end{aligned} \tag{8.9.110} $$

upon invoking mass balance (8.9.12), and where

$$ \dot{\beta}_{w,\Delta} := \frac{\partial}{\partial t}\{\beta_{w,\Delta}\} + \nabla\beta_{w,\Delta}\,.\,\mathbf{v}_{w,\Delta} \tag{8.9.111} $$

denotes the appropriate material time derivative of $\beta_{w,\Delta}$. Taken together, (8.9.110) and (8.9.104) yield from (8.9.99)

$$ {}^{\text{int}}Q + {}^{\text{int}}\mathbf{f}\,.\,\mathbf{v} = -\rho\dot{\beta} - \mathcal{G}\beta - \operatorname{div}{}_2\mathbf{q} + \operatorname{div}\{-{}_b\mathbf{q}^- + ({}_b\mathbf{T}^-)^T\mathbf{v}\}, \tag{8.9.112} $$

upon omitting subscripts w and Δ. Accordingly (8.9.85) may be written as (recall $\mathcal{D}^T = \mathcal{D}$ and $\mathbf{L} := \nabla\mathbf{v}$)

$$ \begin{aligned} &r - \operatorname{div}\boldsymbol{\kappa} + {}^{\text{ext}}Q + H - \rho\dot{\beta} - \mathcal{G}\beta - \operatorname{div}{}_2\mathbf{q} - \operatorname{div}{}_b\mathbf{q}^- + {}_b\mathbf{T}\cdot\mathbf{L} \\ &+(\operatorname{div}{}_b\mathbf{T} + {}^{\text{ext}}\mathbf{f} + \mathbf{b} + \mathbf{I} - \rho\dot{\mathbf{v}})\,.\,\mathbf{v} = \rho\dot{h} + \mathcal{G}h. \end{aligned} \tag{8.9.113} $$

Equivalently, invoking (8.9.51) and (8.9.88),

$$r - \mathrm{div}\,{}_b\mathbf{q}^+ + {}^{\mathrm{ext}}Q + H + \mathbf{T}_b \cdot \mathbf{L} = \rho\dot{e} + \mathcal{G}e, \tag{8.9.114}$$

where

$${}_b\mathbf{q}^+ := {}_b\mathbf{q}^- + {}_2\mathbf{q} + \boldsymbol{\kappa} \tag{8.9.115}$$

and

$$e := \beta + h. \tag{8.9.116}$$

Balance (8.9.114) is the generalisation of (6.2.64) to materials with changing content.

8.9.4 Concluding Remarks

Remark 8.9.9. Omission of terms associated with change of material content (equivalently, setting $e_i \equiv 1$ for all i so that $\mathcal{M}^+ = \mathcal{M}$) yields relations strictly comparable with those derived in Chapters 4, 5 and 6. Field values in these relations are now local averages in both space *and* time. Specifically, (8.9.12) with $\mathcal{G}^{\mathrm{net}}_{w,\Delta}$ omitted corresponds to (4.2.16); (8.9.18) and (8.9.19) correspond to (5.2.3) and (5.2.4); (8.9.40) with ${}^{\mathrm{ext}}\mathbf{f}_{w,\Delta}, \mathbf{P}^{\mathrm{out}}_{w,\Delta}$, and $\mathbf{P}^{\mathrm{in}}_{w,\Delta}$ absent corresponds to (5.5.13) while (8.9.51), with omission of these terms and also of $\mathcal{G}$, corresponds to (6.2.31) or (6.2.48); and (8.9.89) with ${}^{\mathrm{ext}}Q, H$, and $\mathcal{G}$ absent corresponds to (6.2.24), while the same omissions yield (8.9.98) and (8.9.114) as the counterparts of (6.2.32) and (6.2.69).

Remark 8.9.10. Throughout Section 8.9, field values at a typical point $\mathbf{x}$ and time t were defined in terms of local space-time averages in which $\mathbf{x}$ was held fixed during time averaging. Since the balance relations were derived on the basis of equations governing the motion of individual molecules in inertial frames, it follows that the final forms of balance relate to such frames and that field values which appear therein involve temporal averaging at fixed points in an inertial frame of choice. Accordingly, if $f(\mathbf{x},t)$ and $f^*(\mathbf{x}^*,t)$ denote a particular field value as computed by two inertial observers O and O^* at $\mathbf{x} = \mathbf{x}^*$ at time t, then the molecules involved instant by instant in the temporal averaging will differ if O and O^* are in relative motion. This is not a problem if $f(\mathbf{x},t)$ is to be associated with a measurement at $\mathbf{x}$ monitored by O at time t. If O^* is in communication with O or observes the measurement, then he/she can appreciate the identification with $f(\mathbf{x},t)$. [In such case $f^*(\mathbf{x}^*,t)$ corresponds to a different measuring process for the same quantity.] If an intrinsic (i.e., material-based) formulation is desired, then we can proceed as outlined in the referential considerations of Sections 8.4, 8.5 and 8.6. That is, the balance relations obtained by purely spatial averaging should be drafted in referential format before implementing temporal averaging. The resulting referential descriptions then can be reformulated to yield balance relations in which fields correspond to the current (rather than referential) situation. The balances differ in the definition of thermal velocity (see Remark 8.5.2) and the appearance of an extra term in energy balance [the counterpart of $\overline{\mathbf{S}'_w \cdot \mathbf{F}'}$ in (8.6.20)]. However, it *is* possible to obtain fields values defined via temporal averaging in a manner which is observer-independent. This is discussed later, after Remark 12.2.2.

8.10 Summary

Time averaging has been shown to play a vital role, not only in the context of systems with changing material content (for which such averaging is manifestly essential), but also in considerations both practical (such averaging complements spatial averaging to yield fields which correspond to specific scales of length and time and thus may be related to measurements at such scales) and conceptual (as in understanding the source of intermolecular forces, and the link with probabilistic approaches). We now list the specific issues that were raised and addressed in this chapter.

1. The notion of a Δ-time average, introduced in Section 5.4, was generalised, and the operations of differentiation and temporal averaging were shown to commute. (Section 8.3.)
2. Direct time averaging of the continuity equation (4.2.14) established a formally identical counterpart (8.4.1) in which both the mass and momentum densities are molecular averages jointly in space and time. However, such time averaging is implemented at points fixed relative to an observer. Accordingly, observers in relative motion compute temporal averages associated with sets of molecules which may differ, observer by observer, at each instant in the averaging period except current time. While this may accord with experimental practice and observers in mutual communication, an alternative *intrinsic* procedure was considered in which, roughly speaking, time averaging is implemented following the motion prescribed by the local (spatially defined) velocity field. Such averaging ensures that all observers who adopt this procedure compute field values which contain exactly the same molecular information. (Section 8.4.)
3. Direct time averaging of each of the different local forms of linear momentum balance was implemented, and intrinsic averaging (see point 2 above) effected for the standard referential form of this balance. Such an approach also was applied to local forms of energy balance. Upon assuming that time-averaged fields vary negligibly over time intervals of duration associated with the time averaging procedure, balances were obtained which incorporated extra terms involving time averages of products of fluctuations [see (8.6.11), (8.6.12), (8.6.13), and (8.6.20)]. (Sections 8.5 and 8.6.)
4. Any system $\mathcal{M}^+$ with changing corpuscular content was regarded to be a subset of a fixed (i.e., unchanging) set $\mathcal{M}$ of point masses, and each element $P_i \in \mathcal{M}$ was assigned a membership function $e_i : e_i(t) = 1$ or 0 according to whether or not $P_i \in \mathcal{M}^+$ at time t. The equation governing the motion of each P_i was multiplied by e_i, such equations summed over *all* $P_i \in \mathcal{M}$, and the resulting equation averaged in time. The equation so obtained prescribes the time evolution of a single velocity associated with the global motion of $\mathcal{M}^+$. This velocity is the time-averaged total momentum of $\mathcal{M}^+$ divided by the time-averaged total mass of $\mathcal{M}^+$ (see Remark 8.7.1). The agencies driving velocity evolution are threefold: time-averaged interactions on $\mathcal{M}^+$ point masses due to those in $\mathcal{M} - \mathcal{M}^+$, time averages of forces on $\mathcal{M}^+$ due to external sources, and a contribution associated with each and every transition between $\mathcal{M}^+$ and $\mathcal{M} - \mathcal{M}^+$ in the time-averaging period. The last contribution involves net momentum transfer linked to all transitions, computed in terms of individual corpuscular velocities relative to the velocity associated

with the global motion of $\mathcal{M}^+$. Attention was drawn to a shortcoming in the derivation of the preceding results: specifically, for each $P_i \in \mathcal{M}$, the membership function e_i should have been continuously differentiable. This shortcoming was removed in Remark 8.7.2 by mollifying membership functions appropriately without significantly changing results or physical interpretations. (Section 8.7.)

5. The analysis outlined in point 4 above was applied in turn to rocket and jet propulsion, and the resulting equations were compared with those obtained on the basis of simplistic assumptions. (Section 8.8.)
6. Local evolutions of mass, and the various forms of balance of linear momentum and energy, were derived for systems with changing material content, taking full account of molecular migration. All fields involved were related to local averages in both space and time. Concluding remarks were made on simplification of the foregoing for systems with unchanging material content, the link with measurements made in inertial frames, and how intrinsic formulations may be obtained. (Sections 8.9 and 8.10.)

9 Elements of Mixture Theory

9.1 Preamble

Any material system $\mathcal{M}$ which consists of more than one type of molecule may be regarded to be a composite (or *mixture*) of individual systems, each of which consists of a single molecular type (or *species*). Accordingly each family of identical molecules can be regarded as a subsystem (or *constituent*) of $\mathcal{M}$. Here we derive concepts and relations for individual constituent systems in a mixture (taking due account of the effects of other constituents) and investigate how in combination these are related to the corresponding concepts and relations for $\mathcal{M}$ as a whole.

9.2 Mass Conservation and Material Points for a Non-Reacting Mixture Constituent

Recall the discussion of Section 5.3. Here this is trivially generalised to a system $\mathcal{M}$ which consists of any number of molecular species. Specifically, we consider

$$\mathcal{M} = \bigcup_{\alpha \in \mathcal{A}} \mathcal{M}_\alpha, \tag{9.2.1}$$

where $\mathcal{A}$ is an indexing set, and subsystem

$$\mathcal{M}_\alpha := \{P_{\alpha_i} : i = 1,2,\ldots,N_\alpha\}. \tag{9.2.2}$$

Denoting the mass and location of molecule P_{α_i} by $m_{\alpha_i}(= m_\alpha)$ and $\mathbf{x}_{\alpha_i}$, the mass and momentum density fields for subsystem $\mathcal{M}_\alpha$ associated with choice w of weighting function are given by

$$\rho_w^\alpha(\mathbf{x},t) := \sum_{\alpha_i} m_\alpha \, w(\mathbf{x}_{\alpha_i}(t) - \mathbf{x}) \tag{9.2.3}$$

and

$$\mathbf{p}_w^\alpha(\mathbf{x},t) := \sum_{\alpha_i} m_\alpha \mathbf{v}_{\alpha_i}(t) w(\mathbf{x}_{\alpha_i}(t) - \mathbf{x}). \tag{9.2.4}$$

The corresponding fields for the whole system $\mathcal{M}$ are

$$\rho_w(\mathbf{x},t) := \sum_{\alpha} \sum_{\alpha_i} m_\alpha \, w(\mathbf{x}_{\alpha_i}(t) - \mathbf{x}) \tag{9.2.5}$$

and $$\mathbf{p}_w(\mathbf{x},t) := \sum_{\alpha}\sum_{\alpha_i} m_\alpha\, \mathbf{v}_{\alpha_i}(t) w(\mathbf{x}_{\alpha_i}(t) - \mathbf{x}). \tag{9.2.6}$$

Accordingly,

$$\rho_w = \sum_{\alpha} \rho_w^\alpha \quad and \quad \mathbf{p}_w = \sum_{\alpha} \mathbf{p}_w^\alpha. \tag{9.2.7}$$

The velocity field for $\mathcal{M}_\alpha$ is [cf. $(5.3.9)_{1,2}$]

$$\mathbf{v}_w^\alpha := \mathbf{p}_w^\alpha / \rho_w^\alpha, \tag{9.2.8}$$

while that for $\mathcal{M}$ is given by (4.2.15). Thus from (4.2.15), $(9.2.7)_2$, and (9.2.8) [cf. (5.3.11)],

$$\rho_w \mathbf{v}_w = \sum_{\alpha} \rho_w^\alpha \mathbf{v}_w^\alpha. \tag{9.2.9}$$

Arguing in what should now be a familiar manner (cf. Lemma 4.2.1 et seq.),

$$\begin{aligned} \partial\rho_w^\alpha/\partial t &= \sum_{\alpha_i} m_\alpha \nabla w . \mathbf{v}_{\alpha_i} = \sum_{\alpha_i} m_\alpha(-\nabla_{\mathbf{x}} w). \mathbf{v}_{\alpha_i} \\ &= -\sum_{\alpha_i} \operatorname{div}\{m_\alpha \mathbf{v}_{\alpha_i} w\} = -\operatorname{div}\mathbf{p}_w^\alpha. \end{aligned} \tag{9.2.10}$$

Hence, from (9.2.8),

$$\frac{\partial \rho_w^\alpha}{\partial t} + \operatorname{div}\{\rho_w^\alpha \mathbf{v}_w^\alpha\} = 0. \tag{9.2.11}$$

As detailed in Sections 5.2 and 5.3, any velocity field $\mathbf{v}$ gives rise to a corresponding motion which we describe as the *motion prescribed by* $\mathbf{v}$. In the context of mixtures, the *motion* $\boldsymbol{\chi}_{w,t_0}^\alpha$ *of* $\mathcal{M}_\alpha$ *relative to the situation at time* t_0 is the solution to the initial-value problem

$$\dot{\boldsymbol{\chi}}_{w,t_0}^\alpha(\mathbf{x},t) = \mathbf{v}_w^\alpha(\boldsymbol{\chi}_{w,t_0}^\alpha(\hat{\mathbf{x}},t),t), \tag{9.2.12}$$

where $$\boldsymbol{\chi}_{w,t_0}^\alpha(\hat{\mathbf{x}},t) = \mathbf{x} \quad \text{and} \quad \boldsymbol{\chi}_{w,t_0}^\alpha(\hat{\mathbf{x}},t_0) = \hat{\mathbf{x}}, \tag{9.2.13}$$

with $$\hat{\mathbf{x}} \in B_{w,t_0}^\alpha \tag{9.2.14}$$

and $$B_{w,t}^\alpha := \{\mathbf{x} \in \mathcal{E} : \rho_w^\alpha(\mathbf{x},t) > 0\}. \tag{9.2.15}$$

Exercise 9.2.1. Consider the corresponding derived notions of material point. Specifically, note that with each geometrical point $\hat{\mathbf{x}}$ in the region $\bigcap_\alpha B_{w,t_0}^\alpha$ can be associated material points for each system $\mathcal{M}_\alpha$ and also $\mathcal{M}$, $\mathbf{X}^\alpha$ and $\mathbf{X}$, say. The locations of $\mathbf{X}^\alpha$ and $\mathbf{X}$ at subsequent time t are $\boldsymbol{\chi}_{w,t_0}^\alpha(\hat{\mathbf{x}},t)$ and $\boldsymbol{\chi}_{w,t_0}(\hat{\mathbf{x}},t)$, respectively.

A mixture is described as *non-diffusive* if velocity fields are the same for all constituents; that is, if for any $\alpha_1 \neq \alpha_2$ and $\mathbf{x} \in B_{w,t}^{\alpha_1} \cap B_{w,t}^{\alpha_2}$, then $\mathbf{v}_w^{\alpha_1}(\mathbf{x},t) = \mathbf{v}_w^{\alpha_2}(\mathbf{x},t)$.

The $\mathcal{M}_\alpha$ *intrinsic material time derivative* of a scalar field f_α is [cf. (2.5.27) and (2.5.28)]

$$\grave{f}_\alpha := \frac{\partial f_\alpha}{\partial t} + (\nabla f_\alpha).\mathbf{v}_\alpha \tag{9.2.16}$$

and of a vector field $\mathbf{u}_\alpha$ is

$$\grave{\mathbf{u}}_\alpha := \frac{\partial \mathbf{u}_\alpha}{\partial t} + (\nabla \mathbf{u}_\alpha)\mathbf{v}_\alpha. \tag{9.2.17}$$

Exercise 9.2.2. Show that (9.2.11) may be written as

$$\grave{\rho}_w^\alpha + \rho_w^\alpha \operatorname{div} \mathbf{v}_w^\alpha = 0. \tag{9.2.18}$$

The *concentration* (or *mass fraction*) for constituent system $\mathcal{M}_\alpha$ in the whole system $\mathcal{M}$ (corresponding to choice w of weighting function) is

$$c_w^\alpha := \frac{\rho_w^\alpha}{\rho_w}. \tag{9.2.19}$$

Exercise 9.2.3. Show from $(9.2.7)_1$ that

$$\sum_\alpha c_w^\alpha = 1 \tag{9.2.20}$$

and from $(9.2.7)_2$ that

$$\sum_\alpha c_w^\alpha \mathbf{v}_w^\alpha = \mathbf{v}. \tag{9.2.21}$$

Using (9.2.11) and (4.2.16), show further that

$$\rho_w \grave{c}_w^\alpha + c_w^\alpha \operatorname{div}\{\rho_w(\mathbf{v}_w^\alpha - \mathbf{v}_w)\} = 0. \tag{9.2.22}$$

9.3 Linear Momentum Balance for a Non-Reacting Mixture Constituent

The motion of P_{α_i} in an inertial frame takes the form

$$\sum_{j \neq i} \mathbf{f}_{\alpha_i \alpha_j} + \sum_{\beta \neq \alpha} \sum_{\beta_k} \mathbf{f}_{\alpha_i \beta_k} + \mathbf{b}_{\alpha_i} = \frac{d}{dt}\{m_\alpha \mathbf{v}_{\alpha_i}\}. \tag{9.3.1}$$

Here $\mathbf{f}_{\alpha_i \alpha_j}$ denotes the force exerted on P_{α_i} by P_{α_j} and $\mathbf{f}_{\alpha_i \beta_k}$ the force on P_{α_i} due to P_{β_k}. The first term in (9.3.1) represents the resultant force on P_{α_i} due to all other $\mathcal{M}_\alpha$ molecules, while the second term denotes the resultant force on P_{α_i} due to molecules of all the *other* constituents of $\mathcal{M}$. Multiplication of (9.3.1) by $w(\mathbf{x}_{\alpha_i}(t) - \mathbf{x})$, followed by summation over all $P_{\alpha_i} \in \mathcal{M}_\alpha$, yields

$$\mathbf{f}_\alpha + \sum_{\beta \neq \alpha} \mathbf{f}_{\alpha\beta} + \mathbf{b}_\alpha = \sum_{\alpha_i} \frac{d}{dt}\{m_\alpha \mathbf{v}_{\alpha_i}\} w(\mathbf{x}_{\alpha_i} - \mathbf{x}). \tag{9.3.2}$$

Here
$$\mathbf{f}_\alpha(\mathbf{x},t) := \sum_{i\neq j}\sum \mathbf{f}_{\alpha_i\alpha_j}(t)w(\mathbf{x}_{\alpha_i}(t)-\mathbf{x}), \tag{9.3.3}$$

$$\mathbf{f}_{\alpha\beta}(\mathbf{x},t) := \sum_{\alpha_i}\sum_{\beta_k} \mathbf{f}_{\alpha_i\beta_k}(t)w(\mathbf{x}_{\alpha_i}(t)-\mathbf{x}), \tag{9.3.4}$$

and
$$\mathbf{b}_\alpha(\mathbf{x},t) := \sum_{\alpha_i} \mathbf{b}_{\alpha_i}(t)w(\mathbf{x}_{\alpha_i}(t)-\mathbf{x}). \tag{9.3.5}$$

Remark 9.3.1. Explicit dependence upon the choice w of weighting function has been omitted for ease of notation.

Simplification of the right-hand side of (9.3.2) is effected in exactly the same way as detailed in (5.5.4) and (5.5.5), and enables (9.3.2) to be written as

$$\mathbf{f}_\alpha + \sum_{\beta\neq\alpha}\mathbf{f}_{\alpha\beta} + \mathbf{b}_\alpha = \frac{\partial}{\partial t}\{\rho_\alpha\mathbf{v}_\alpha\} + \operatorname{div}\mathcal{D}_\alpha^+. \tag{9.3.6}$$

Here
$$\mathcal{D}_\alpha^+(\mathbf{x},t) := \sum_{\alpha_i} m_\alpha \mathbf{v}_{\alpha_i}(t)\otimes\mathbf{v}_{\alpha_i}(t)w(\mathbf{x}_{\alpha_i}(t)-\mathbf{x}), \tag{9.3.7}$$

and
$$\rho_\alpha := \rho_w^\alpha, \quad \mathbf{v}_\alpha := \mathbf{v}_w^\alpha. \tag{9.3.8}$$

Replicating (5.5.10) and (5.5.11) in respect of $\mathcal{M}_\alpha$ and writing

$$\hat{\mathbf{v}}_{\alpha_i}(t;\mathbf{x}) := \mathbf{v}_{\alpha_i}(t) - \mathbf{v}_\alpha(\mathbf{x},t) \tag{9.3.9}$$

yield
$$\mathcal{D}_\alpha^+ = \mathcal{D}_\alpha + \rho_\alpha\mathbf{v}_\alpha\otimes\mathbf{v}_\alpha, \tag{9.3.10}$$

where
$$\mathcal{D}_\alpha(\mathbf{x},t) := \sum_{\alpha_i} m_\alpha\hat{\mathbf{v}}_{\alpha_i}(t;\mathbf{x})\otimes\hat{\mathbf{v}}_{\alpha_i}(t;\mathbf{x})w(\mathbf{x}_{\alpha_i}(t)-\mathbf{x}), \tag{9.3.11}$$

and we have noted that

$$\sum_{\alpha_i} m_\alpha\hat{\mathbf{v}}_{\alpha_i}(t;\mathbf{x})w(\mathbf{x}_{\alpha_i}(t)-\mathbf{x}) = \mathbf{0}. \tag{9.3.12}$$

Exercise 9.3.1. Prove (9.3.12).

From (9.3.10) and (9.3.6)

$$-\operatorname{div}\mathcal{D}_\alpha + \mathbf{f}_\alpha + \sum_{\beta\neq\alpha}\mathbf{f}_{\alpha\beta} + \mathbf{b}_\alpha = \frac{\partial}{\partial t}\{\rho_\alpha\mathbf{v}_\alpha\} + \operatorname{div}\{\rho_\alpha\mathbf{v}_\alpha\otimes\mathbf{v}_\alpha\}. \tag{9.3.13}$$

It follows from the $\mathcal{M}_\alpha$ continuity equation (9.2.11), and the analogue of (5.5.14) for this constituent, that [cf. (5.5.15)]

$$-\operatorname{div}\mathcal{D}_\alpha + \mathbf{f}_\alpha + \sum_{\beta\neq\alpha}\mathbf{f}_{\alpha\beta} + \mathbf{b}_\alpha = \rho_\alpha\mathbf{a}_\alpha, \tag{9.3.14}$$

where the $\mathcal{M}_\alpha$ *acceleration field* [cf. (5.5.16)] is

$$\mathbf{a}_\alpha := \grave{\mathbf{v}}_\alpha := \frac{\partial \mathbf{v}_\alpha}{\partial t} + (\nabla \mathbf{v}_\alpha)\mathbf{v}_\alpha. \tag{9.3.15}$$

Remark 9.3.2. If $w = w_\epsilon$, then neglecting mollification considerations (and suppressing time dependence for brevity), the argument of Exercise 5.5.1 can be employed for system $\mathcal{M}_\alpha$ if $\alpha - \alpha$ interactions are pairwise balanced [see (5.6.6)] or, more generally, if the net $\alpha - \alpha$ self-force associated with α molecules in $S_\epsilon(\mathbf{x})$ vanishes. *In such case* $\mathbf{f}_\alpha(\mathbf{x},t)$ *represents the resultant force exerted on* α *molecules in sphere* $S_\epsilon(\mathbf{x})$ *at time* t *by* α *molecules outside or on the boundary of* $S_\epsilon(\mathbf{x})$ *at this time, divided by the volume of* $S_\epsilon(\mathbf{x})$.

Exercise 9.3.2. Show that, consistent with the considerations of Remark 9.3.2, $\mathbf{f}_{\alpha\beta}(\mathbf{x},t)$ *represents the resultant force exerted on* α *molecules in* $S_\epsilon(\mathbf{x})$ *at time* t *by* β *molecules everywhere* [both inside and outside $S_\epsilon(\mathbf{x})$ or on its boundary], *divided by the volume of* $S_\epsilon(\mathbf{x})$. Show also that if $\mathbf{b}_{\alpha_i}(t) = m_\alpha \mathbf{g}$, then $\mathbf{b}_\alpha(\mathbf{x},t) = \rho_\alpha(\mathbf{x},t)\mathbf{g}$.

Using the analysis of Section 5.6, the $\mathcal{M}_\alpha$ force density field $\mathbf{f}_\alpha$ may be written as the divergence of a linear transformation field. If $\mathbf{a}_{\alpha_i}$ is a vector field which satisfies (on suppressing time dependence)

$$(\operatorname{div} \mathbf{a}_{\alpha_i})(\mathbf{x}) = w(\mathbf{x}_{\alpha_i} - \mathbf{x}), \tag{9.3.16}$$

then [cf. (5.6.2) and (5.6.3)]

$$\operatorname{div}\{\mathbf{f}_{\alpha_i\alpha_j} \otimes \mathbf{a}_{\alpha_i}\} = \mathbf{f}_{\alpha_i\alpha_j} w. \tag{9.3.17}$$

Accordingly

$$\operatorname{div}\left\{\sum_{i \neq j}\sum \mathbf{f}_{\alpha_i\alpha_j} \otimes \mathbf{a}_{\alpha_i}\right\} = \sum_{i \neq j}\sum \mathbf{f}_{\alpha_i\alpha_j} w = \mathbf{f}_\alpha, \tag{9.3.18}$$

and a simple choice of $\alpha - \alpha$ interaction stress tensor is

$${}_s\mathbf{T}_\alpha^-(\mathbf{x},t) := \sum_{i \neq j}\sum \mathbf{f}_{\alpha_i\alpha_j}(t) \otimes \mathbf{a}_{\alpha_i}(\mathbf{x},t). \tag{9.3.19}$$

On noting that from (9.3.19) and (9.3.18)

$$\operatorname{div}{}_s\mathbf{T}_\alpha^- = \mathbf{f}_\alpha, \tag{9.3.20}$$

the corresponding $\alpha - \alpha$ *stress tensor*

$${}_s\mathbf{T}_\alpha := {}_s\mathbf{T}_\alpha^- - \mathcal{D}_\alpha \tag{9.3.21}$$

satisfies [see (9.3.14)] the local form of α-linear momentum balance

$$\operatorname{div}{}_s\mathbf{T}_\alpha + \sum_{\beta \neq \alpha} \mathbf{f}_{\alpha\beta} + \mathbf{b}_\alpha = \rho_\alpha \grave{\mathbf{v}}_\alpha = \rho_\alpha \mathbf{a}_\alpha. \tag{9.3.22}$$

Exercise 9.3.3. Verify that (9.3.22) follows from (9.3.14), (9.3.20), and (9.3.21).

Continuing the methodology of Section 5.6, specifically that of Subsection 5.6.3, if $\mathbf{b}_{\alpha_i\alpha_j}$ is any vector field for which (suppressing time dependence)

$$(\operatorname{div}\mathbf{b}_{\alpha_i\alpha_j})(\mathbf{x}) = \frac{1}{2}\{w(\mathbf{x}_{\alpha_i}-\mathbf{x}) - w(\mathbf{x}_{\alpha_j}-\mathbf{x})\}, \tag{9.3.23}$$

and $\alpha-\alpha$ interactions are pairwise balanced, then

$$_b\mathbf{T}_\alpha^-(\mathbf{x},t) := \sum_{i\neq j}\sum \mathbf{f}_{\alpha_i\alpha_j}(t)\otimes\mathbf{b}_{\alpha_i\alpha_j}(\mathbf{x},t) \tag{9.3.24}$$

satisfies

$$\operatorname{div}{}_b\mathbf{T}_\alpha^- = \mathbf{f}_\alpha. \tag{9.3.25}$$

Exercise 9.3.4. Repeat the analysis of Subsection 5.6.3 to obtain (9.3.25) from (9.3.3) and (9.3.23), modulo pairwise balance for all $\alpha-\alpha$ interactions, namely

$$\mathbf{f}_{\alpha_j\alpha_i} = -\mathbf{f}_{\alpha_i\alpha_j}. \tag{9.3.26}$$

The appropriate $\alpha-\alpha$ *stress tensor* here is

$$_b\mathbf{T}_\alpha := {}_b\mathbf{T}_\alpha^- - \mathcal{D}_\alpha, \tag{9.3.27}$$

and the corresponding local form of α-linear momentum balance is

$$\operatorname{div}{}_b\mathbf{T}_\alpha + \sum_{\beta\neq\alpha}\mathbf{f}_{\alpha\beta} + \mathbf{b}_\alpha = \rho_\alpha\dot{\mathbf{v}}_\alpha = \rho_\alpha\mathbf{a}_\alpha. \tag{9.3.28}$$

Exercise 9.3.5. Verify (9.3.28) using (9.3.27), (9.3.25) and (9.3.14).

Remark 9.3.3. Notice that there are three distinct forms of $_b\mathbf{T}_\alpha^-$ which may be denoted, in the manner of Subsections 5.6.4, 5.6.5, and 5.6.6, by $_{sb}\mathbf{T}_\alpha^-$, $_H\mathbf{T}_\alpha^-$ and $_N\mathbf{T}_\alpha^-$. These correspond to the three distinct solutions to (9.3.23) given in these subsections.

A question that now arises naturally is whether it is possible to express a force density $\mathbf{f}_{\alpha\beta}$ as the divergence of a second-order tensor. In fact, from (9.3.16) and (9.3.4), and suppressing time dependence,

$$\begin{aligned}\mathbf{f}_{\alpha\beta}(\mathbf{x}) &= \sum_{\alpha_i}\sum_{\beta_k}\mathbf{f}_{\alpha_i\beta_k}w(\mathbf{x}_{\alpha_i}-\mathbf{x})\\ &= \sum_{\alpha_i}\sum_{\beta_k}\mathbf{f}_{\alpha_i\beta_k}(\operatorname{div}\mathbf{a}_{\alpha_i})(\mathbf{x}) = \operatorname{div}\left\{\sum_{\alpha_i}\sum_{\beta_k}\mathbf{f}_{\alpha_i\beta_k}\otimes\mathbf{a}_{\alpha_i}\right\}(\mathbf{x}).\end{aligned} \tag{9.3.29}$$

Thus

$$\mathbf{f}_{\alpha\beta} = \operatorname{div}{}_s\mathbf{T}_{\alpha\beta}^-, \tag{9.3.30}$$

where

$$_s\mathbf{T}_{\alpha\beta}^- := \sum_{\alpha_i}\sum_{\beta_k}\mathbf{f}_{\alpha_i\beta_k}\otimes\mathbf{a}_{\alpha_i}. \tag{9.3.31}$$

Similarly,

$$ {}_s\mathbf{T}^{-}_{\beta\alpha} := \sum_{\beta_k} \sum_{\alpha_i} \mathbf{f}_{\beta_k \alpha_i} \otimes \mathbf{a}_{\beta_k} \tag{9.3.32} $$

satisfies

$$ \operatorname{div} {}_s\mathbf{T}^{-}_{\beta\alpha} = \mathbf{f}_{\beta\alpha}. \tag{9.3.33} $$

Accordingly, linear momentum balance for constituent $\mathcal{M}_\alpha$ has local form [via (9.3.30) and (9.3.14)]

$$ \operatorname{div} \mathbf{T}_\alpha + \sum_{\beta \neq \alpha} \operatorname{div} {}_s\mathbf{T}^{-}_{\alpha\beta} + \mathbf{b}_\alpha = \rho_\alpha \dot{\mathbf{v}}_\alpha = \rho_\alpha \mathbf{a}_\alpha, \tag{9.3.34} $$

where $\mathbf{T}_\alpha$ is ${}_s\mathbf{T}_\alpha$ or ${}_b\mathbf{T}_\alpha$ [see (9.3.20) and (9.3.25)].

Remark 9.3.4. Suppose that $\alpha - \beta$ interactions are pairwise balanced; that is, for all α_i and all β_k,

$$ \mathbf{f}_{\beta_k \alpha_i} = -\mathbf{f}_{\alpha_i \beta_k}. \tag{9.3.35} $$

Then, from (9.3.31), (9.3.32), and (9.3.35),

$$ \begin{aligned} {}_s\mathbf{T}^{-}_{\beta\alpha} + {}_s\mathbf{T}^{-}_{\alpha\beta} &= \sum_{\beta_k} \sum_{\alpha_i} \mathbf{f}_{\beta_k \alpha_i} \otimes \mathbf{a}_{\beta_k} + \sum_{\alpha_i} \sum_{\beta_k} \mathbf{f}_{\alpha_i \beta_k} \otimes \mathbf{a}_{\alpha_i} \\ &= \sum_{\alpha_i} \sum_{\beta_k} \mathbf{f}_{\alpha_i \beta_k} \otimes (\mathbf{a}_{\alpha_i} - \mathbf{a}_{\beta_k}). \end{aligned} \tag{9.3.36} $$

Now [see (9.3.16)]

$$ \begin{aligned} (\operatorname{div}\{\mathbf{a}_{\alpha_i} - \mathbf{a}_{\beta_k}\})(\mathbf{x}) &= w(\mathbf{x}_{\alpha_i} - \mathbf{x}) - w(\mathbf{x}_{\beta_k} - \mathbf{x}) \\ &\neq 0 \quad \text{in general.} \end{aligned} \tag{9.3.37} $$

[Why? Draw a sketch of $S_\epsilon(\mathbf{x}_{\alpha_i})$ and $S_\epsilon(\mathbf{x}_{\beta_k})$ to convince yourself.] Thus in general

$$ \mathbf{a}_{\alpha_i} \neq \mathbf{a}_{\beta_k}, \tag{9.3.38} $$

and consequently we must expect that [see (9.3.36)]

$$ {}_s\mathbf{T}^{-}_{\beta\alpha} + {}_s\mathbf{T}^{-}_{\alpha\beta} \neq \mathbf{0}, \tag{9.3.39} $$

despite pairwise balancing (9.3.35). Further insight may be gained by noting that

$$ \operatorname{div}\{{}_s\mathbf{T}^{-}_{\beta\alpha} + {}_s\mathbf{T}^{-}_{\alpha\beta}\} = \mathbf{f}_{\beta\alpha} + \mathbf{f}_{\alpha\beta}. \tag{9.3.40} $$

Here, from Remark 9.3.2 with $w = w_\epsilon, \mathbf{f}_{\beta\alpha}(\mathbf{x}) V_\epsilon$ denotes the resultant force on $\mathcal{M}_\beta$ molecules in $S_\epsilon(\mathbf{x})$ exerted by $\mathcal{M}_\alpha$ molecules *everywhere*, while $\mathbf{f}_{\alpha\beta}(\mathbf{x}) V_\epsilon$ represents the resultant force on $\mathcal{M}_\alpha$ molecules in $S_\epsilon(\mathbf{x})$ exerted by $\mathcal{M}_\beta$ molecules *everywhere*. Since pairwise balancing implies that for molecules *within* $S_\epsilon(\mathbf{x})$ the $\alpha - \beta$ interactions balance $\beta - \alpha$ interactions, it follows that $(\mathbf{f}_{\alpha\beta}(\mathbf{x}) + \mathbf{f}_{\beta\alpha}(\mathbf{x})) V_\epsilon$ *denotes the resultant of forces exerted by* $\mathcal{M}_\beta$ *molecules outside* $S_\epsilon(\mathbf{x})$ *upon* $\mathcal{M}_\alpha$ *molecules inside* $S_\epsilon(\mathbf{x})$ *together with those of* $\mathcal{M}_\alpha$ *molecules outside* $S_\epsilon(\mathbf{x})$ *upon* $\mathcal{M}_\beta$ *molecules inside* $S_\epsilon(\mathbf{x})$.

Such resultant force cannot be expected to vanish. [Convince yourself, by considering the trivial case of one molecule of each constituent within $S_\epsilon(\mathbf{x})$ and one molecule of each constituent outside $S_\epsilon(\mathbf{x})$.] It follows from (9.3.40) that since $\mathbf{f}_{\beta\alpha} + \mathbf{f}_{\alpha\beta} \neq \mathbf{0}$, relation (9.3.39) must hold. Notice further, from Remark 9.3.2 and the preceding, that $\mathbf{f}_\alpha(\mathbf{x}) + \sum_{\beta\neq\alpha} \mathbf{f}_{\alpha\beta}(\mathbf{x})$ represents the resultant force exerted on 'α' molecules in $S_\epsilon(\mathbf{x})$ by *all* molecules [whether inside, outside, or on the boundary of $S_\epsilon(\mathbf{x})$], divided by V_ϵ.

9.4 On Relating Total Mixture Fields to Those of Constituents

Equations (9.2.7) delivered the total mixture densities of mass and momentum as merely the sums of these densities taken over all constituents. Accordingly the continuity equation for $\mathcal{M}$ is obtained simply by summing the individual $\mathcal{M}_\alpha$ continuity equations (9.2.11) over all $\alpha \in \mathcal{A}$. (Convince yourself!) Matters are not so simple for linear momentum balance.

Consider a non-diffusive mixture; that is, a mixture for which $\mathbf{v}_{\alpha_1} = \mathbf{v}_{\alpha_2}$ for all pairs $\alpha_1, \alpha_2 \in \mathcal{A}$. It follows that for all α

$$\mathbf{v}_\alpha = \mathbf{v}. \tag{9.4.1}$$

Exercise 9.4.1. Prove (9.4.1) using (9.2.9) and $(9.2.7)_1$.

Further, non-diffusivity also implies that

$$\mathcal{D} = \sum_\alpha \mathcal{D}_\alpha, \tag{9.4.2}$$

since [cf. (5.5.11) and (5.5.8)]

$$\mathcal{D}(\mathbf{x},t) := \sum_\alpha \sum_{\alpha_i} m_\alpha (\mathbf{v}_{\alpha_i}(t) - \mathbf{v}(\mathbf{x},t)) \otimes (\mathbf{v}_{\alpha_i}(t) - \mathbf{v}(\mathbf{x},t)) w(\mathbf{x}_{\alpha_i}(t) - \mathbf{x}). \tag{9.4.3}$$

Remark 9.4.1. In a mixture of moderately rarefied gases molecular interactions are negligible, and for a mixture of *ideal* gases such interactions are absent (recall Subsection 5.5.2). In equilibrium ($\mathbf{v}_\alpha = \mathbf{0}$ for all α) and in the absence of body force ($\mathbf{b}_\alpha = \mathbf{0}$ for all α),

$$\mathcal{D}_\alpha = P_\alpha \mathbf{1} \quad \text{and} \quad \mathcal{D} = P\mathbf{1}. \tag{9.4.4}$$

That is, the only stresses are thermal in nature and are all pressures. It follows from (9.4.2) and relations (9.4.4) that

$$P = \sum_\alpha P_\alpha. \tag{9.4.5}$$

Result (9.4.5) is known as *Dalton's law of partial pressures.*

Remark 9.4.2. Now consider a *diffusive* mixture of ideal gases. In particular, note that [suppressing time dependence: see also (9.3.11) and (9.3.9)]

$$\mathcal{D}_\alpha(\mathbf{x}) := \sum_{\alpha_i} m_\alpha (\mathbf{v}_{\alpha_i} - \mathbf{v}_\alpha(\mathbf{x})) \otimes (\mathbf{v}_{\alpha_i} - \mathbf{v}_\alpha(\mathbf{x})) w(\mathbf{x}_{\alpha_i} - \mathbf{x})$$

$$= \sum_{\alpha_i} m_\alpha \big((\mathbf{v}_{\alpha_i} - \mathbf{v}(\mathbf{x})) + (\mathbf{v}(\mathbf{x}) - \mathbf{v}_\alpha(\mathbf{x})) \big) \otimes \big((\mathbf{v}_{\alpha_i} - \mathbf{v}(\mathbf{x}))$$

$$+ (\mathbf{v}(\mathbf{x}) - \mathbf{v}_\alpha(\mathbf{x})) \big) w(\mathbf{x}_{\alpha_i} - \mathbf{x})$$

$$= {}_\alpha\mathcal{D}(\mathbf{x}) + \sum_{\alpha_i} m_\alpha (\mathbf{v}_{\alpha_i} - \mathbf{v}(\mathbf{x})) w(\mathbf{x}_{\alpha_i} - \mathbf{x}) \otimes (\mathbf{v}(\mathbf{x}) - \mathbf{v}_\alpha(\mathbf{x}))$$

$$+ (\mathbf{v}(\mathbf{x}) - \mathbf{v}_\alpha(\mathbf{x})) \otimes \sum_{\alpha_i} m_\alpha (\mathbf{v}_{\alpha_i} - \mathbf{v}(\mathbf{x})) w(\mathbf{x}_{\alpha_i} - \mathbf{x})$$

$$+ \rho_\alpha(\mathbf{x})(\mathbf{v}(\mathbf{x}) - \mathbf{v}_\alpha(\mathbf{x})) \otimes (\mathbf{v}(\mathbf{x}) - \mathbf{v}_\alpha(\mathbf{x})), \tag{9.4.6}$$

where
$$ {}_\alpha\mathcal{D}(\mathbf{x}) := \sum_{\alpha_i} m_\alpha (\mathbf{v}_{\alpha_i} - \mathbf{v}(\mathbf{x})) \otimes (\mathbf{v}_{\alpha_i} - \mathbf{v}(\mathbf{x})) w(\mathbf{x}_{\alpha_i} - \mathbf{x}). \tag{9.4.7}$$

However, from (9.2.4), (9.2.3), and (9.2.8),

$$\begin{aligned} \sum_{\alpha_i} m_{\alpha_i} (\mathbf{v}_{\alpha_i} - \mathbf{v}(\mathbf{x})) w(\mathbf{x}_{\alpha_i} - \mathbf{x}) &= \rho_\alpha(\mathbf{x})\mathbf{v}_\alpha(\mathbf{x}) - \rho_\alpha(\mathbf{x})\mathbf{v}(\mathbf{x}) \\ &= \rho_\alpha(\mathbf{x})(\mathbf{v}_\alpha(\mathbf{x}) - \mathbf{v}(\mathbf{x})). \end{aligned} \tag{9.4.8}$$

Together (9.4.6) and (9.4.8) yield

$$\mathcal{D}_\alpha = {}_\alpha\mathcal{D} + \rho_\alpha(\mathbf{v}_\alpha - \mathbf{v}) \otimes (\mathbf{v} - \mathbf{v}_\alpha) + (\mathbf{v} - \mathbf{v}_\alpha) \otimes \rho_\alpha(\mathbf{v}_\alpha - \mathbf{v}) + \rho_\alpha(\mathbf{v} - \mathbf{v}_\alpha) \otimes (\mathbf{v} - \mathbf{v}_\alpha). \tag{9.4.9}$$

That is,
$$\mathcal{D}_\alpha = {}_\alpha\mathcal{D} - \rho_\alpha(\mathbf{v}_\alpha - \mathbf{v}) \otimes (\mathbf{v}_\alpha - \mathbf{v}). \tag{9.4.10}$$

It follows from (9.4.3), (9.4.7), and (9.4.10) that

$$\mathcal{D} = \sum_\alpha {}_\alpha\mathcal{D} = \sum_\alpha \{\mathcal{D}_\alpha + \rho_\alpha(\mathbf{v}_\alpha - \mathbf{v}) \otimes (\mathbf{v}_\alpha - \mathbf{v})\}. \tag{9.4.11}$$

Accordingly, if relations (9.4.4) hold (that is, if $\mathcal{D}$ and $\mathcal{D}_\alpha$ are all pressures), then

$$P\mathbf{1} = \left(\sum_\alpha P_\alpha\right)\mathbf{1} + \sum_\alpha \rho_\alpha(\mathbf{v}_\alpha - \mathbf{v}) \otimes (\mathbf{v}_\alpha - \mathbf{v}). \tag{9.4.12}$$

Remark 9.4.3. Relation (9.4.11) holds quite generally, having been derived directly from basic definitions for *any* mixture. This relation shows that *partial thermokinetic stress tensors $\mathcal{D}_\alpha$ do not sum to the thermokinetic stress $\mathcal{D}$ for the mixture as a whole if constituent diffusion occurs.* Consequently, for stresses in general we should not expect simple additivity. In such case

$$\mathbf{T} = \mathbf{T}^- - \mathcal{D} \qquad \text{and} \qquad \mathbf{T}_\alpha = \mathbf{T}_\alpha^- - \mathcal{D}_\alpha \tag{9.4.13}$$

[see (5.5.20), (9.3.21), and (9.3.27)], and we should expect, at least in the case of constituent diffusion, that

$$\mathbf{T} \neq \sum_\alpha \mathbf{T}_\alpha. \tag{9.4.14}$$

Having analysed thermokinetic contributions to stress and obtained (9.4.11), it is natural to turn to the remaining contributions associated with interactions. Since the basis of any interaction stress $\mathbf{T}^-$ is an interaction force density $\mathbf{f}$ [see (5.5.18)], we consider the interaction density for the mixture as a whole. This is [on suppressing time dependence and with the i and j sums taken over *all* molecules of $\mathcal{M}$: see (5.5.2)]

$$\begin{aligned}\mathbf{f}(\mathbf{x}) &:= \sum_{i \neq j}\sum \mathbf{f}_{ij}\, w(\mathbf{x}_i - \mathbf{x}) \\ &= \sum_{\alpha}\sum_{\alpha_i}\left\{\sum_{\substack{\alpha_j \\ j\neq i}} \mathbf{f}_{\alpha_i\alpha_j} + \sum_{\beta\neq\alpha}\sum_{\beta_k} \mathbf{f}_{\alpha_i\beta_k}\right\} w(\mathbf{x}_{\alpha_i} - \mathbf{x}).\end{aligned}$$

That is [see (9.3.3) and (9.3.4)],

$$\mathbf{f} = \sum_{\alpha} \mathbf{f}_\alpha + \sum_{\alpha\neq\beta}\sum \mathbf{f}_{\alpha\beta}. \tag{9.4.15}$$

Thus, if

$$\mathbf{f} = \operatorname{div}\mathbf{T}^- \quad \text{and} \quad \mathbf{f}_\alpha = \operatorname{div}\mathbf{T}^-_\alpha, \tag{9.4.16}$$

then

$$\operatorname{div}\mathbf{T}^- = \sum_{\alpha} \operatorname{div}\mathbf{T}^-_\alpha + \sum_{\alpha\neq\beta}\sum \mathbf{f}_{\alpha\beta}. \tag{9.4.17}$$

Correspondingly, from (9.4.13), (9.4.11), and (9.4.17),

$$\begin{aligned}\operatorname{div}\mathbf{T} &:= \operatorname{div}\{\mathbf{T}^- - \mathcal{D}\} \\ &= \operatorname{div}\left\{\sum_{\alpha}(\mathbf{T}^-_\alpha - \mathcal{D}_\alpha - \rho_\alpha(\mathbf{v}_\alpha - \mathbf{v})\otimes(\mathbf{v}_\alpha - \mathbf{v}))\right\} + \sum_{\alpha\neq\beta}\sum \mathbf{f}_{\alpha\beta}.\end{aligned} \tag{9.4.18}$$

That is,

$$\operatorname{div}\mathbf{T} = \operatorname{div}\left\{\sum_{\alpha}(\mathbf{T}_\alpha - \rho_\alpha(\mathbf{v}_\alpha - \mathbf{v})\otimes(\mathbf{v}_\alpha - \mathbf{v}))\right\} + \sum_{\alpha\neq\beta}\sum \mathbf{f}_{\alpha\beta}. \tag{9.4.19}$$

Equivalently, with

$${}_\alpha\mathbf{T} := \mathbf{T}^-_\alpha - {}_\alpha\mathcal{D} \tag{9.4.20}$$

it follows from (9.4.10) that

$$\operatorname{div}\mathbf{T} = \operatorname{div}\left\{\sum_{\alpha} {}_\alpha\mathbf{T}\right\} + \sum_{\alpha\neq\beta}\sum \mathbf{f}_{\alpha\beta}. \tag{9.4.21}$$

Remark 9.4.4. Recalling the interpretation of $\mathbf{f}_{\alpha\beta}(\mathbf{x}) + \mathbf{f}_{\beta\alpha}(\mathbf{x})$ for choice $w = w_\epsilon$ given in Remark 9.3.4, it should be clear that in general $\sum_{\alpha\neq\beta}\sum \mathbf{f}_{\alpha\beta} = \frac{1}{2}\sum_{\alpha\neq\beta}\sum(\mathbf{f}_{\alpha\beta} + \mathbf{f}_{\beta\alpha}) \neq \mathbf{0}$.

Equation (9.4.15) delineates the relation between the $\mathcal{M}$ (or *total*) interaction force density $\mathbf{f}$ and *partial* interaction force densities $\mathbf{f}_\alpha$ and $\mathbf{f}_{\alpha\beta}$, while (9.4.19) [or (9.4.21)] details the relationship between the $\mathcal{M}$ (or *total*) interaction stress $\mathbf{T}$ and the *partial* stresses $\mathbf{T}_\alpha$ (or $_\alpha\mathbf{T}$) together with interaction force densities $\mathbf{f}_{\alpha\beta}$ associated with pairs of distinct constituents. The relation for body forces is simply

$$\mathbf{b} = \sum_\alpha \mathbf{b}_\alpha. \tag{9.4.22}$$

(Convince yourself of this!) It requires some manipulation to establish the relationship between the $\rho\mathbf{a}$ term in linear momentum balance for $\mathcal{M}$ [see (5.5.16)] and corresponding constituent terms $\rho_\alpha\mathbf{a}_\alpha$ etc. Recall (5.5.14), and note that together with (5.5.16) we have

$$\rho\mathbf{a} = \frac{\partial}{\partial t}\{\rho\mathbf{v}\} + \operatorname{div}\{\rho\mathbf{v}\otimes\mathbf{v}\}. \tag{9.4.23}$$

Similarly, for each constituent

$$\rho_\alpha\mathbf{a}_\alpha = \frac{\partial}{\partial t}\{\rho_\alpha\mathbf{v}_\alpha\} + \operatorname{div}\{\rho_\alpha\mathbf{v}_\alpha\otimes\mathbf{v}_\alpha\}. \tag{9.4.24}$$

Exercise 9.4.2. Show that from (9.2.9)

$$\sum_\alpha \rho_\alpha(\mathbf{v}_\alpha - \mathbf{v})\otimes(\mathbf{v}_\alpha - \mathbf{v}) = \sum_\alpha \rho_\alpha\mathbf{v}_\alpha\otimes\mathbf{v}_\alpha - \rho\mathbf{v}\otimes\mathbf{v}. \tag{9.4.25}$$

Accordingly, from (9.4.24) and (9.2.9),

$$\begin{aligned}\sum_\alpha \rho_\alpha\mathbf{a}_\alpha &= \frac{\partial}{\partial t}\left\{\sum_\alpha \rho_\alpha\mathbf{v}_\alpha\right\} + \operatorname{div}\left\{\sum_\alpha \rho_\alpha\mathbf{v}_\alpha\otimes\mathbf{v}_\alpha\right\}\\ &= \frac{\partial}{\partial t}\{\rho\mathbf{v}\} + \operatorname{div}\{\rho\mathbf{v}\otimes\mathbf{v}\} + \operatorname{div}\left\{\sum_\alpha \rho_\alpha(\mathbf{v}_\alpha - \mathbf{v})\otimes(\mathbf{v}_\alpha - \mathbf{v})\right\}.\end{aligned} \tag{9.4.26}$$

That is, from (9.4.23)

$$\rho\mathbf{a} = \sum_\alpha \rho_\alpha\mathbf{a}_\alpha - \operatorname{div}\left\{\sum_\alpha \rho_\alpha\mathbf{u}_\alpha\otimes\mathbf{u}_\alpha\right\}, \tag{9.4.27}$$

where

$$\mathbf{u}_\alpha := \mathbf{v}_\alpha - \mathbf{v}. \tag{9.4.28}$$

9.5 A Paradox in Early Continuum Theories of Mixtures

In the approach to mixture theory advocated fifty years ago by Truesdell the balance of linear momentum for a non-reacting mixture constituent α was postulated (see Truesdell & Toupin [33], Truesdell [34], Atkin & Craine [35], and Bowen [36]) to take the form (at time t)

$$\int_{\partial R} \boldsymbol{T}_\alpha\mathbf{n}\,dA + \int_R (\mathbf{F}_\alpha + \mathbf{b}_\alpha)\,dV = \frac{d}{d\tau}\left\{\int_{R_\tau} \rho_\alpha\mathbf{v}_\alpha\,dV\right\}\Bigg|_{\tau=t}. \tag{9.5.1}$$

Here R denotes any regular region strictly within that occupied by the constituent α at any given instant t, $\mathbf{n}$ the outward unit normal to the boundary ∂R of R, and R_τ that region which deforms with the motion of constituent α (as mandated by the velocity field $\mathbf{v}_\alpha$) and coincides with R at instant t. Mass conservation (i.e., satisfaction of the α continuity equation), use of the divergence theorem for second-order tensor fields, and continuity assumptions lead to the local form of (9.5.1) as

$$\operatorname{div} \mathcal{T}_\alpha + \mathbf{F}_\alpha + \mathbf{b}_\alpha = \rho_\alpha \mathbf{a}_\alpha. \tag{9.5.2}$$

Here $\mathbf{b}_\alpha, \rho_\alpha, \mathbf{v}_\alpha$ and $\mathbf{a}_\alpha$ have precisely the interpretations of these symbols as used in Section 9.4. However, in the works cited, the integral of $\mathcal{T}_\alpha \mathbf{n}$ over ∂R is interpreted to be the force exerted by the *whole mixture* outside R upon constituent α inside R. Further, $\mathbf{F}_\alpha$ is taken to represent the force density associated with the effect of all *other* constituents upon constituent α. Gurtin, Oliver & Williams [37] pointed out that the preceding interpretation of $\mathcal{T}_\alpha$ leads to unacceptable consequences. In particular, it leads to the conclusion that if A and C are disjoint but contiguous regions occupied at some instant by a binary mixture, then

$$\mathbf{F}_{\alpha\beta}(A,C) = -\mathbf{F}_{\alpha\beta}(C,A). \tag{9.5.3}$$

Here $\mathbf{F}_{\alpha\beta}(A,C)$ denotes the resultant force exerted upon constituent α in region A by constituent β in region C, and $\mathbf{F}_{\alpha\beta}(C,A)$ represents the resultant force on constituent α in C exerted by constituent β in A. The physical absurdity of (9.5.3) becomes apparent if we consider a one-dimensional binary lattice

$$\begin{array}{ccc} \ldots.\alpha\,\beta\,\alpha\,\beta & | & \beta\,\alpha\,\beta\,\alpha\,\beta\ldots. \\ A & & C \end{array}$$

Here each Greek letter represents a molecule. By considering the case in which nearest-neighbour and next-nearest neighbour interactions predominate, it is clear that (9.5.3) does not hold. Accordingly this relation cannot be expected to hold in general.

Exercise 9.5.1. Convince yourself of (9.5.3) as follows. Let $\mathcal{S} := \partial A \cap \partial C$ denote the surface common to the boundaries $\partial A, \partial C$ of A and C. If $\mathbf{n}$ denotes that unit normal field on $\mathcal{S}$ directed from A into C, then note that

$$\int_{\mathcal{S}} \mathcal{T}_\alpha \mathbf{n}\, dA = \mathbf{F}_{\alpha\alpha}(A,C) + \mathbf{F}_{\alpha\beta}(A,C) \tag{9.5.4}$$

and

$$\int_{\mathcal{S}} \mathcal{T}_\alpha(-\mathbf{n})\, dA = \mathbf{F}_{\alpha\alpha}(C,A) + \mathbf{F}_{\alpha\beta}(C,A). \tag{9.5.5}$$

If

$$\mathbf{F}_{\alpha\alpha}(A,C) + \mathbf{F}_{\alpha\alpha}(C,A) = \mathbf{0} \tag{9.5.6}$$

(as would certainly be the case if $\alpha - \alpha$ interactions were pairwise balanced), deduce that addition of (9.5.4) and (9.5.5) yields (9.5.3).

In later works by Williams [38] and Oliver & Williams [39] it was shown how the paradox (9.5.3) could be obviated by postulating integral representations for $\mathbf{F}_{\alpha\beta}(A,C)$ which lead to a balance of form (9.5.2) but with different interpretations of $\mathcal{T}_\alpha$ and $\mathbf{F}_\alpha$. Murdoch & Morro [40, 41] derived balance relations for non-reacting mixtures from the molecular viewpoint adopted throughout this book using cellular averaging (see Chapter 11). Linear momentum balance in these works takes the form of (9.3.28) with the α constituent stress tensor $\mathbf{T}_\alpha^-$ composed of contributions from $\alpha - \alpha$ molecular interactions together with a diffusive (kinematic) contribution identifiable with $-\mathcal{D}_\alpha$. The effect of other constituents was shown to be given by a bulk force density $\sum_{\beta\neq\alpha}\sum \mathbf{f}_{\alpha\beta}$ exactly as in (9.3.28). That is, modelling molecules as interacting point masses led directly to a form of balance (9.5.2) with

$$\mathcal{T}_\alpha = \mathbf{T}_\alpha^- - \mathcal{D}_\alpha \qquad \text{and} \qquad \mathbf{F}_\alpha = \sum_{\beta\neq\alpha} \mathbf{f}_{\alpha\beta}. \tag{9.5.7}$$

With these interpretations no paradox arises.

Remark 9.5.1. The analysis of Section 9.4 led unequivocally to form (9.5.2) of linear momentum balance with $\mathbf{F}_\alpha = \sum_{\beta\neq\alpha} \mathbf{f}_{\alpha\beta}$. Stress tensor $\mathcal{T}_\alpha$ was shown to be expressible in the form ${}_s\mathbf{T}_\alpha$ [see (9.3.20)] and, if $\alpha - \alpha$ interactions are pairwise balanced, in the equivalent form ${}_b\mathbf{T}_\alpha$ [see (9.3.27)]. In each case there is an $\alpha - \alpha$ interaction stress tensor (namely ${}_s\mathbf{T}_\alpha^-$ or ${}_b\mathbf{T}_\alpha^-$) together with the α thermokinetic stress tensor $-\mathcal{D}_\alpha$. Further, these interpretations are linked with scale dependence via the choice of weighting function. Accordingly, much greater insight is gained into the nature (and consequent physical interpretation) of fields which appear in linear momentum (and other) balances by adopting a molecular perspective than is possible by adopting 'intuitive' macroscopic concepts and prejudices. Indeed, the very framing of balance relations based upon molecular considerations provides a more secure foundation than forms postulated from a purely continuum viewpoint.

9.6 Energy Balances

Multiplying the equation governing the motion of P_{α_i} in an inertial frame [namely (9.3.1)] scalarly by $\mathbf{v}_{\alpha_i}(t)w(\mathbf{x}_{\alpha_i}(t) - \mathbf{x})$ and summing over all other molecules of $\mathcal{M}$ yield (on suppressing dependence upon $\mathbf{x}$ and t)

$$\begin{aligned}\sum_{\alpha_i}\sum_{\substack{\alpha_j\\ i\neq j}} \mathbf{f}_{\alpha_i\alpha_j}\,.\,\mathbf{v}_{\alpha_i}w + \sum_{\alpha_i}\sum_{\beta\neq\alpha}\sum_{\beta_k} \mathbf{f}_{\alpha_i\beta_k}\,.\,\mathbf{v}_{\alpha_i}w + \sum_{\alpha_i} \mathbf{b}_{\alpha_i}\,.\,\mathbf{v}_{\alpha_i}w \\ = \sum_{\alpha_i} \frac{d}{dt}\left\{\frac{1}{2}m_i\mathbf{v}_{\alpha_i}^2\right\} w.\end{aligned} \tag{9.6.1}$$

Writing $\mathbf{v}_{\alpha_i} = \hat{\mathbf{v}}_{\alpha_i} + \mathbf{v}_\alpha$ [see (9.3.9)], this relation may be written in the form

$$\begin{aligned}&Q_\alpha + \mathbf{f}_\alpha\,.\,\mathbf{v}_\alpha + \sum_{\beta\neq\alpha}(Q_{\alpha\beta} + \mathbf{f}_{\alpha\beta}\,.\,\mathbf{v}_\alpha) + r_\alpha + \mathbf{b}_\alpha\,.\,\mathbf{v}_\alpha \\ &= \frac{\partial}{\partial t}\{\rho_\alpha K_\alpha\} + \operatorname{div}\{\rho_\alpha K_\alpha \mathbf{v}_\alpha\} + \operatorname{div}\left\{\sum_{\alpha_i}\frac{1}{2}m_\alpha \mathbf{v}_{\alpha_i}^2\hat{\mathbf{v}}_{\alpha_i}w\right\}.\end{aligned} \tag{9.6.2}$$

Here

$$Q_\alpha(\mathbf{x},t) := \sum_{\alpha_i}\sum_{\substack{\alpha_j\\ i\neq j}} \mathbf{f}_{\alpha_i\alpha_j}(t)\,.\,\hat{\mathbf{v}}_{\alpha_i}(t;\mathbf{x})w(\mathbf{x}_{\alpha_i}(t) - \mathbf{x}), \tag{9.6.3}$$

$$Q_{\alpha\beta}(\mathbf{x},t) := \sum_{\alpha_i}\sum_{\beta_k}\mathbf{f}_{\alpha_i\beta_k}(t)\,.\,\hat{\mathbf{v}}_{\alpha_i}(t;\mathbf{x})w(\mathbf{x}_{\alpha_i}(t)-\mathbf{x}), \tag{9.6.4}$$

$$r_\alpha(\mathbf{x},t) := \sum_{\alpha_i}\mathbf{b}_{\alpha_i}(t)\,.\,\hat{\mathbf{v}}_{\alpha_i}(t;\mathbf{x})w(\mathbf{x}_{\alpha_i}(t)-\mathbf{x}), \tag{9.6.5}$$

and [cf. (6.2.7) through (6.2.11)]

$$\rho_\alpha(\mathbf{x},t)K_\alpha(\mathbf{x},t) := \sum_{\alpha_i}\frac{1}{2}m_\alpha\mathbf{v}^2_{\alpha_i}(t)w(\mathbf{x}_{\alpha_i}(t)-\mathbf{x}). \tag{9.6.6}$$

Repeating manipulation of (6.2.10) into form (6.2.12) for constituent α yields

$$\rho_\alpha K_\alpha = \rho_\alpha(h_\alpha + \frac{1}{2}\mathbf{v}^2_\alpha), \tag{9.6.7}$$

where the α heat energy density is

$$\rho_\alpha(\mathbf{x},t)h_\alpha(\mathbf{x},t) := \sum_{\alpha_i}\frac{1}{2}m_\alpha\hat{\mathbf{v}}^2_{\alpha_i}(t;\mathbf{x})w(\mathbf{x}_{\alpha_i}(t)-\mathbf{x}). \tag{9.6.8}$$

The α analogues of (6.2.17) and (6.2.18) are

$$\frac{\partial}{\partial t}\{\rho_\alpha K_\alpha\} + \operatorname{div}\{\rho_\alpha K_\alpha\mathbf{v}_\alpha\} = \rho_\alpha \grave{K}_\alpha, \tag{9.6.9}$$

where [see (9.2.16)]

$$\grave{K}_\alpha := \frac{\partial K_\alpha}{\partial t} + (\nabla K_\alpha)\,.\,\mathbf{v}_\alpha. \tag{9.6.10}$$

Similarly, the α counterpart of (6.2.14) is

$$\sum_{\alpha_i}\frac{1}{2}m_\alpha\mathbf{v}^2_{\alpha_i}\hat{\mathbf{v}}_{\alpha_i}w = \boldsymbol{\kappa}_\alpha + \mathcal{D}_\alpha\mathbf{v}_\alpha, \tag{9.6.11}$$

where

$$\boldsymbol{\kappa}_\alpha(\mathbf{x},t) := \sum_{\alpha_i}\frac{1}{2}m_\alpha\hat{\mathbf{v}}^2_{\alpha_i}(t;\mathbf{x})\hat{\mathbf{v}}_{\alpha_i}(t;\mathbf{x})w(\mathbf{x}_{\alpha_i}(t)-\mathbf{x}). \tag{9.6.12}$$

From (9.6.2), (9.6.9), (9.6.7), and (9.6.11),

$$Q_\alpha + \sum_{\beta\neq\alpha}Q_{\alpha\beta} + r_\alpha - \operatorname{div}(\boldsymbol{\kappa}_\alpha + \mathcal{D}_\alpha\mathbf{v}_\alpha) + \{\mathbf{f}_\alpha + \sum_{\beta\neq\alpha}\mathbf{f}_{\alpha\beta} + \mathbf{b}_\alpha\}\,.\,\mathbf{v}_\alpha$$

$$= \rho_\alpha(\grave{h}_\alpha + \mathbf{v}_\alpha\,.\,\mathbf{a}_\alpha). \tag{9.6.13}$$

Using linear momentum balance this simplifies [cf. (6.2.24)] to

$$Q_\alpha + \sum_{\beta\neq\alpha}Q_{\alpha\beta} + r_\alpha - \operatorname{div}\boldsymbol{\kappa}_\alpha - \mathcal{D}_\alpha\cdot\mathbf{L}_\alpha = \rho_\alpha\grave{h}_\alpha, \tag{9.6.14}$$

where

$$\mathbf{L}_\alpha := \nabla\mathbf{v}_\alpha. \tag{9.6.15}$$

Exercise 9.6.1. Verify (9.6.13) and (9.6.14).

Repeating steps (6.2.25) and (6.2.26) for constituent α,

$$Q_\alpha + \mathbf{f}_\alpha . \mathbf{v}_\alpha = \sum_{\substack{\alpha_i \\ i \neq j}} \sum_{\alpha_j} \mathbf{f}_{\alpha_i\alpha_j} . \mathbf{v}_{\alpha_i} w = \operatorname{div}\{-{}_s\mathbf{q}_\alpha^- + ({}_s\mathbf{T}_\alpha^-)^T \mathbf{v}_\alpha\}. \tag{9.6.16}$$

Here

$${}_s\mathbf{q}_\alpha^-(\mathbf{x},t) := -\sum_{\substack{\alpha_i \\ i \neq j}} \sum_{\alpha_j} \mathbf{f}_{\alpha_i\alpha_j}(t) . \hat{\mathbf{v}}_{\alpha_i}(t;\mathbf{x}) \mathbf{a}_{\alpha_i}(\mathbf{x},t), \tag{9.6.17}$$

and $\mathbf{a}_{\alpha_i}(\mathbf{x},t)$ is any solution to

$$(\operatorname{div} \mathbf{a}_{\alpha_i})(\mathbf{x},t) = w(\mathbf{x}_{\alpha_i}(t) - \mathbf{x}). \tag{9.6.18}$$

From (9.6.13), (9.6.16), and the linear momentum balance (9.3.22), balance of energy takes the form [cf. (6.2.32)]

$$r_\alpha - \operatorname{div} {}_s\mathbf{q}_\alpha + \sum_{\alpha \neq \beta} Q_{\alpha\beta} + {}_s\mathbf{T}_\alpha \cdot \mathbf{L}_\alpha = \rho_\alpha \grave{h}_\alpha, \tag{9.6.19}$$

where

$${}_s\mathbf{q}_\alpha := {}_s\mathbf{q}_\alpha^- + \boldsymbol{\kappa}_\alpha. \tag{9.6.20}$$

Exercise 9.6.2. Verify (9.6.19).

Following the manipulations of (6.2.34) through (6.2.43) for constituent α yields, when $\alpha - \alpha$ interactions are pairwise balanced,

$$Q_\alpha + \mathbf{f}_\alpha . \mathbf{v}_\alpha = A_\alpha + \operatorname{div}\{-{}_b\mathbf{q}_\alpha^- + ({}_b\mathbf{T}_\alpha^-)^T \mathbf{v}_\alpha\}, \tag{9.6.21}$$

where

$$A_\alpha(\mathbf{x},t) := \frac{1}{2} \sum_{\substack{\alpha_i \\ i \neq j}} \sum_{\alpha_j} \mathbf{f}_{\alpha_i\alpha_j}(t) . (\mathbf{v}_{\alpha_i}(t) - \mathbf{v}_{\alpha_j}(t)) w(\mathbf{x}_{\alpha_i}(t) - \mathbf{x}) \tag{9.6.22}$$

and

$${}_b\mathbf{q}_\alpha^-(\mathbf{x},t) := -\sum_{\substack{\alpha_i \\ i \neq j}} \sum_{\alpha_j} \mathbf{f}_{\alpha_i\alpha_j}(t) . \hat{\mathbf{v}}_{\alpha_i}(t;\mathbf{x}) \mathbf{b}_{\alpha_i\alpha_j}(\mathbf{x},t). \tag{9.6.23}$$

Here $\mathbf{b}_{\alpha_i\alpha_j}$ is any solution to [see (9.3.23)]

$$(\operatorname{div} \mathbf{b}_{\alpha_i\alpha_j})(\mathbf{x},t) = \frac{1}{2}\{w(\mathbf{x}_{\alpha_i}(t) - \mathbf{x}) - w(\mathbf{x}_{\alpha_j}(t) - \mathbf{x})\}. \tag{9.6.24}$$

Exercise 9.6.3. Verify (9.6.21).

From (9.6.13), (9.6.21), and form (9.3.28) of linear momentum balance, the corresponding form of energy balance is [cf. (6.2.49)]

$$A_\alpha + r_\alpha + \sum_{\alpha \neq \beta} Q_{\alpha\beta} - \operatorname{div} {}_b\mathbf{q}_\alpha + {}_b\mathbf{T}_\alpha \cdot \mathbf{L}_\alpha = \rho_\alpha \dot{h}_\alpha, \tag{9.6.25}$$

where

$${}_b\mathbf{q}_\alpha := {}_b\mathbf{q}_\alpha^- + \boldsymbol{\kappa}_\alpha. \tag{9.6.26}$$

Exercise 9.6.4. Verify (9.6.25).

In the event that $\alpha - \alpha$ interactions are governed by separation-dependent pair potentials, then repeating the analysis following (6.2.50) yields the α constituent analogue of (6.2.63), namely

$$A_\alpha = -\text{div}\, {}_2\mathbf{q}_\alpha - \rho_\alpha \grave{\beta}_\alpha. \tag{9.6.27}$$

Here the energy of assembly density for constituent α is

$$(\rho_\alpha \beta_\alpha)(\mathbf{x},t) := \frac{1}{2}\sum_{\alpha_i}\sum_{\substack{\alpha_j \\ i\neq j}} \hat{\phi}_{\alpha_i\alpha_j}(r_{\alpha_i\alpha_j}(t))w(\mathbf{x}_{\alpha_i}(t) - \mathbf{x}), \tag{9.6.28}$$

where $\hat{\phi}_{\alpha_i\alpha_j}$ is the pair potential for P_{α_i} and P_{α_j}; that is,

$$\mathbf{f}_{\alpha_i\alpha_j} = \nabla_{\mathbf{x}_{\alpha_j}} \hat{\phi}_{\alpha_i\alpha_j}(r_{\alpha_i\alpha_j}) = \hat{\phi}'_{\alpha_i\alpha_j}(r_{\alpha_i\alpha_j})\frac{(\mathbf{x}_{\alpha_j} - \mathbf{x}_{\alpha_i})}{r_{\alpha_i\alpha_j}} \tag{9.6.29}$$

with

$$r_{\alpha_i\alpha_j} := \|\mathbf{x}_{\alpha_i} - \mathbf{x}_{\alpha_j}\|. \tag{9.6.30}$$

Accordingly, for constituents with interactions governed by separation-dependent pair potentials, the appropriate form of energy balance is [from (9.6.25) and (9.6.27): cf. (6.2.69)]

$$r_\alpha + \sum_{\beta\neq\alpha} Q_{\alpha\beta} - \text{div}\, {}_b\mathbf{q}_\alpha^+ + {}_b\mathbf{T}_\alpha \cdot \mathbf{L}_\alpha = \rho_\alpha \grave{e}_\alpha. \tag{9.6.31}$$

Here [cf. (6.2.65) and (6.2.66)]

$${}_b\mathbf{q}_\alpha^+ := {}_b\mathbf{q}_\alpha + {}_2\mathbf{q}_\alpha = {}_b\mathbf{q}_\alpha^- + \boldsymbol{\kappa}_\alpha + {}_2\mathbf{q}_\alpha \tag{9.6.32}$$

and

$$e_\alpha := \beta_\alpha + h_\alpha. \tag{9.6.33}$$

Remark 9.6.1. As indicated in Remark 6.2.7, it is not necessary to assume the existence of interaction potentials in order to obtain a local balance of form (9.6.31). Specifically, we write

$$A_\alpha = \rho_\alpha \mathcal{A}_\alpha \tag{9.6.34}$$

and define [see (9.2.12) et seq.: note subscript w is suppressed here]

$$\beta_{\alpha,t_0}(\mathbf{x},t) := \int_{t_0}^{t} \mathcal{A}_\alpha(\chi_{t_0}^\alpha(\hat{\mathbf{x}},\tau),\tau)d\tau, \tag{9.6.35}$$

with [see (9.2.13)]

$$\mathbf{x} = \chi_{t_0}^\alpha(\hat{\mathbf{x}},t) \qquad \text{and} \qquad \hat{\mathbf{x}} = \chi_{t_0}^\alpha(\hat{\mathbf{x}},t_0). \tag{9.6.36}$$

It follows, upon differentiating (9.6.35) with respect to time, keeping $\hat{\mathbf{x}}$ fixed, that

$$\grave{\beta}_{\alpha,t_0}(\mathbf{x},t) = \mathcal{A}_\alpha(\mathbf{x},t) \tag{9.6.37}$$

and hence, from (9.6.34), that

$$A_\alpha = \rho_\alpha \grave{\beta}_{\alpha,t_0}. \tag{9.6.38}$$

Accordingly (9.6.25) may be written as

$$r_\alpha + \sum_{\beta\neq\alpha} Q_{\alpha\beta} - \operatorname{div}{}_b\mathbf{q}_\alpha + {}_b\mathbf{T}_\alpha \cdot \mathbf{L}_\alpha = \rho_\alpha \grave{e}_\alpha, \tag{9.6.39}$$

where

$$e_\alpha := \beta_{\alpha,t_0} + h_\alpha. \tag{9.6.40}$$

Relationships between total mixture thermal fields and those of constituents can be obtained in the manner of Section 9.4. In particular, from (6.2.6) and (9.3.9),

$$\begin{aligned} r &= \sum_\alpha \sum_{\alpha_i} \mathbf{b}_{\alpha_i} \, . \, (\mathbf{v}_{\alpha_i} - \mathbf{v})w \\ &= \sum_\alpha \sum_{\alpha_i} \mathbf{b}_{\alpha_i} \, . \, ((\mathbf{v}_{\alpha_i} - \mathbf{v}_\alpha) + (\mathbf{v}_\alpha - \mathbf{v}))w. \end{aligned} \tag{9.6.41}$$

Hence, from (9.6.5), (9.3.5), and using notation (9.4.28),

$$r = \sum_\alpha (r_\alpha + \mathbf{b}_\alpha \, . \, \mathbf{u}_\alpha). \tag{9.6.42}$$

Similarly, from (6.2.4)

$$Q = \sum_\alpha \sum_{\alpha_i \neq \alpha_j} \sum \mathbf{f}_{\alpha_i\alpha_j} \, . \, (\mathbf{v}_{\alpha_i} - \mathbf{v})w + \sum_{\alpha\neq\beta} \sum \sum_{\alpha_i} \sum_{\beta_k} \mathbf{f}_{\alpha_i\beta_k} \, . \, (\mathbf{v}_{\alpha_i} - \mathbf{v})w. \tag{9.6.43}$$

Writing

$$\mathbf{v}_{\alpha_i} - \mathbf{v} = (\mathbf{v}_{\alpha_i} - \mathbf{v}_\alpha) + \mathbf{u}_\alpha = \hat{\mathbf{v}}_{\alpha_i} + \mathbf{u}_\alpha \tag{9.6.44}$$

and invoking (9.6.3), (9.6.4), (9.3.3), and (9.3.4), relation (9.6.43) may be written as

$$Q = \sum_\alpha (Q_\alpha + \mathbf{f}_\alpha \, . \, \mathbf{u}_\alpha) + \sum_{\alpha\neq\beta} \sum (Q_{\alpha\beta} + \mathbf{f}_{\alpha\beta} \, . \, \mathbf{u}_\alpha). \tag{9.6.45}$$

Exercise 9.6.5. Verify (9.6.45) [recalling (9.4.15)], and deduce that

$$Q + \mathbf{f} \, . \, \mathbf{v} = \sum_\alpha (Q_\alpha + \mathbf{f}_\alpha \, . \, \mathbf{v}_\alpha) + \sum_{\alpha\neq\beta} \sum (Q_{\alpha\beta} + \mathbf{f}_{\alpha\beta} \, . \, \mathbf{v}_\alpha). \tag{9.6.46}$$

Deduce further from (6.2.25), (6.2.26), and (9.6.16) that

$$\operatorname{div}\{-{}_s\mathbf{q}^- + ({}_s\mathbf{T}^-)^T\mathbf{v}\} = \sum_\alpha \operatorname{div}\{-{}_s\mathbf{q}_\alpha^- + ({}_s\mathbf{T}_\alpha^-)^T\mathbf{v}_\alpha\} + \sum_{\alpha\neq\beta} \sum (Q_{\alpha\beta} + \mathbf{f}_{\alpha\beta} \, . \, \mathbf{v}_\alpha). \tag{9.6.47}$$

Exercise 9.6.6. Noting from (6.2.13), (5.5.11), (9.3.11), and (9.6.8) that

$$\rho h = \frac{1}{2} \operatorname{tr} \mathcal{D} \qquad \text{and} \qquad \rho_\alpha h_\alpha = \frac{1}{2} \operatorname{tr} \mathcal{D}_\alpha, \tag{9.6.48}$$

deduce from (9.4.11) that

$$\rho h = \sum_\alpha \rho_\alpha (h_\alpha + \frac{1}{2} \mathbf{u}_\alpha^2). \tag{9.6.49}$$

The remaining terms in the most basic form of energy balances are $\boldsymbol{\kappa}$ in (6.2.24) and $\boldsymbol{\kappa}_\alpha$ in (9.6.14). From (6.2.15)

$$\boldsymbol{\kappa}_W = \sum_\alpha \sum_{\alpha_i} \frac{1}{2} m_\alpha (\mathbf{v}_{\alpha_i} - \mathbf{v})^2 (\mathbf{v}_{\alpha_i} - \mathbf{v}) w. \tag{9.6.50}$$

Use of (9.6.44) yields

$$\boldsymbol{\kappa}_W = \sum_\alpha \sum_{\alpha_i} \frac{1}{2} m_\alpha (\hat{\mathbf{v}}_{\alpha_i}^2 + 2\hat{\mathbf{v}}_{\alpha_i} . \mathbf{u}_\alpha + \mathbf{u}_\alpha^2)(\hat{\mathbf{v}}_{\alpha_i} + \mathbf{u}_\alpha) w. \tag{9.6.51}$$

Exercise 9.6.7. Show that

$$\boldsymbol{\kappa}_W = \sum_\alpha (\boldsymbol{\kappa}_\alpha + \rho_\alpha h_\alpha \, \mathbf{u}_\alpha + \mathcal{D}_\alpha \, \mathbf{u}_\alpha + \frac{1}{2} \rho_\alpha \, \mathbf{u}_\alpha^2 \, \mathbf{u}_\alpha). \tag{9.6.52}$$

9.7 On Reacting Mixtures

9.7.1 General Considerations

The distinguishing feature of a *reacting* mixture is mass exchange between constituents.[1] Consequently, any constituent which experiences such mass transfer must be regarded to be a system with changing material content in the manner of Chapter 8. Reactions are in general complex, and may involve decomposition of constituent molecules into ionic subunits which subsequently combine to form new molecular species. Such decomposition and recombination may be accompanied by heat production or require external supply of heat for initiation. Accordingly the dynamics of submolecular entities must be taken into account, as discussed in Sections 5.4 and 6.3. Here we do not address such general complexity but consider a simple model reaction which involves three constituents. This serves as the first step towards a general approach.

9.7.2 A Simple Model of a Reacting Ternary Mixture

Consider a ternary mixture [see (9.2.1)]

$$\mathcal{M} = \mathcal{M}_\alpha \cup \mathcal{M}_\beta \cup \mathcal{M}_\gamma \tag{9.7.1}$$

in which each $\mathcal{M}_\gamma$ molecule consists of a pair of molecules, one each from $\mathcal{M}_\alpha$ and $\mathcal{M}_\beta$. Thus, in particular (see Section 9.2),

$$m_\gamma = m_\alpha + m_\beta. \tag{9.7.2}$$

Reactions here are those in which $\mathcal{M}_\gamma$ molecules break up into their constituent $\mathcal{M}_\alpha$ and $\mathcal{M}_\beta$ molecules, or $\mathcal{M}_\alpha$ and $\mathcal{M}_\beta$ molecules combine pairwise to form $\mathcal{M}_\gamma$

[1] In our discussion of non-reacting mixtures, both momentum and energy exchange were natural features.

molecules. The local forms of mass balance for constituents, following the analysis of Subsection 8.9.1, are

$$\frac{\partial \rho_\alpha}{\partial t} + \operatorname{div}\{\rho_\alpha \mathbf{v}_\alpha\} = \mathcal{G}_{\alpha\gamma}, \tag{9.7.3}$$

$$\frac{\partial \rho_\beta}{\partial t} + \operatorname{div}\{\rho_\beta \mathbf{v}_\beta\} = \mathcal{G}_{\beta\gamma}, \tag{9.7.4}$$

and

$$\frac{\partial \rho_\gamma}{\partial t} + \operatorname{div}\{\rho_\gamma \mathbf{v}_\gamma\} = \mathcal{G}_{\gamma(\alpha+\beta)}. \tag{9.7.5}$$

(Here subscripts w and Δ, associated with the spatial and temporal averaging necessary to establish these relations, have been omitted for simplicity of expression.) Net supply rate density $\mathcal{G}_{\alpha\gamma}$ of $\mathcal{M}^\alpha$ mass derives from $\gamma \leftrightarrow \alpha(+\beta)$ transitions, $\mathcal{G}_{\beta\gamma}$ of $\mathcal{M}^\beta$ from $\gamma \leftrightarrow \beta(+\alpha)$ transitions, and $\mathcal{G}_{\gamma(\alpha+\beta)}$ of $\mathcal{M}^\gamma$ from $(\alpha+\beta) \leftrightarrow \gamma$ transitions. It follows that

$$\mathcal{G}_{\alpha\gamma} + \mathcal{G}_{\beta\gamma} + \mathcal{G}_{\gamma(\alpha+\beta)} = 0 \tag{9.7.6}$$

and hence that

$$\frac{\partial \rho}{\partial t} + \operatorname{div}\{\rho \mathbf{v}\} = 0 \tag{9.7.7}$$

for the mixture as a whole.

Exercise 9.7.1. Prove (9.7.7) by noting (9.2.7), summing relations (9.7.3) through (9.7.5), and invoking (9.7.6).

The local forms of linear momentum balance are, from the analysis of Section 8.9 [specifically relation (8.9.51)],

$$\operatorname{div}\mathbf{T}_\alpha + \mathbf{f}_{\alpha\beta} + \mathbf{f}_{\alpha\gamma} + \mathbf{b}_\alpha + \mathbf{P}_{\alpha\gamma} = \rho_\alpha \grave{\mathbf{v}}_\alpha + \mathcal{G}_{\alpha\gamma}\mathbf{v}_\alpha, \tag{9.7.8}$$

$$\operatorname{div}\mathbf{T}_\beta + \mathbf{f}_{\beta\alpha} + \mathbf{f}_{\beta\gamma} + \mathbf{b}_\beta + \mathbf{P}_{\beta\gamma} = \rho_\beta \grave{\mathbf{v}}_\beta + \mathcal{G}_{\beta\gamma}\mathbf{v}_\beta, \tag{9.7.9}$$

and

$$\operatorname{div}\mathbf{T}_\gamma + \mathbf{f}_{\gamma\alpha} + \mathbf{f}_{\gamma\beta} + \mathbf{b}_\gamma + \mathbf{P}_{\gamma(\alpha+\beta)} = \rho_\gamma \grave{\mathbf{v}}_\gamma + \mathcal{G}_{\gamma(\alpha+\beta)}\mathbf{v}_\gamma. \tag{9.7.10}$$

In balance (9.7.8) the external force density ${}^{\text{ext}}\mathbf{f}$ of (8.9.51), which represents the effect of $\mathcal{M}^-$ (here $\mathcal{M}_\beta \cup \mathcal{M}_\gamma$) upon $\mathcal{M}^+$ (here $\mathcal{M}_\alpha$), has been decomposed into the separate effects $\mathbf{f}_{\alpha\beta}$ and $\mathbf{f}_{\alpha\gamma}$ of, respectively, $\mathcal{M}_\beta$ and $\mathcal{M}_\gamma$ upon $\mathcal{M}_\alpha$. Thus, if $w = w_\epsilon$ and $\Delta = \Delta_0$, then $\mathbf{f}_{\alpha\beta}(\mathbf{x},t)$ denotes the Δ_0-time-averaged resultant force exerted by β molecules everywhere upon α molecules within $S_\epsilon(\mathbf{x})$, divided by V_ϵ, with time averaging ending at time t. Similar interpretations apply to $\mathbf{f}_{\beta\alpha}, \mathbf{f}_{\beta\gamma}, \mathbf{f}_{\gamma\alpha}$, and $\mathbf{f}_{\gamma\beta}$. Term $\mathbf{P}_{\alpha\gamma}(\mathbf{x},t)$ in (9.7.8) represents the net gain in $\mathcal{M}_\alpha$ momentum within $S_\epsilon(\mathbf{x})$ over time interval $(t-\Delta,t)$, divided both by Δ and V_ϵ, where this gain is associated with the creation of stand-alone α molecules due to the breakup of $\mathcal{M}_\gamma$ molecules and the 'disappearance' of α molecules due to their combination with β molecules to form γ molecules. Similar interpretations apply to $\mathbf{P}_{\beta\gamma}$ in (9.7.9) and $\mathbf{P}_{\gamma(\alpha+\beta)}$ in (9.7.10). (The notation $\mathbf{P}_{\gamma(\alpha+\beta)}$ has been used to indicate that γ transitions involve 'creation' via pairs of molecules, one each from $\mathcal{M}_\alpha$ and $\mathcal{M}_\beta$, and 'depletion' via breakup into such pairs). The term $\mathcal{G}_{\alpha\gamma}(\mathbf{x},t)$ in (9.7.8) denotes the net change of mass of α molecules in $S_\epsilon(\mathbf{x},t)$ over the time interval $(t-\Delta,t)$, divided by Δ and V_ϵ, due to creation and depletion of α molecules associated with creation and breakup of γ molecules. The terms $\mathcal{G}_{\beta\gamma}$ and $\mathcal{G}_{\gamma(\alpha+\beta)}$ have analogous interpretations.

Consider the creation of a γ molecule, P_{γ_ℓ} say, from molecules $P_{\alpha_i} \in \mathcal{M}^\alpha$ and $P_{\beta_k} \in \mathcal{M}^\beta$. If momentum is conserved, then

$$m_\alpha \mathbf{v}_{\alpha_i} + m_\beta \mathbf{v}_{\beta_k} = (m_\alpha + m_\beta)\mathbf{v}_{\gamma_\ell}. \tag{9.7.11}$$

In calculating the contributions of this reaction to momentum density rates $\mathbf{P}_{\alpha\gamma}, \mathbf{P}_{\alpha\beta}$ and $\mathbf{P}_{\gamma(\alpha+\beta)}$, we have a loss of $m_\alpha \mathbf{v}_{\alpha_i}/V_\epsilon \Delta$ in $\mathbf{P}_{\alpha\gamma}$, a loss of $m_\beta \mathbf{v}_{\beta_k}/V_\epsilon \Delta$ in $\mathbf{P}_{\beta\gamma}$, and a gain of $(m_\alpha + m_\beta)\mathbf{v}_{\gamma_\ell}/V_\epsilon \Delta$ in $P_{\gamma(\alpha+\beta)}$. If (9.7.11) also holds for γ molecular breakups, then clearly

$$\mathbf{P}_{\alpha\gamma} + \mathbf{P}_{\beta\gamma} + \mathbf{P}_{\gamma(\alpha+\beta)} = \mathbf{0}. \tag{9.7.12}$$

Exercise 9.7.2. Show that if the mixture is not diffusive (i.e., $\mathbf{v}_\alpha = \mathbf{v}_\beta = \mathbf{v}_\gamma =: \mathbf{v}$), then summation of relations (9.7.7), (9.7.8), and (9.7.9) yields

$$\operatorname{div}\{\mathbf{T}_\alpha + \mathbf{T}_\beta + \mathbf{T}_\gamma\} + (\mathbf{f}_{\alpha\beta} + \mathbf{f}_{\beta\alpha} + \mathbf{f}_{\beta\gamma} + \mathbf{f}_{\gamma\beta} + \mathbf{f}_{\gamma\alpha} + \mathbf{f}_{\alpha\gamma}) + \mathbf{b} = \rho\dot{\mathbf{v}}, \tag{9.7.13}$$

where $\mathbf{b} := \mathbf{b}_\alpha + \mathbf{b}_\beta + \mathbf{b}_\gamma$ and $\rho := \rho_\alpha + \rho_\beta + \rho_\gamma$. Note that (9.4.19) in this non-diffusive context yields

$$\operatorname{div}\mathbf{T} = \operatorname{div}\{\mathbf{T}_\alpha + \mathbf{T}_\beta + \mathbf{T}_\gamma\} + (\mathbf{f}_{\alpha\beta} + \mathbf{f}_{\beta\alpha} + \mathbf{f}_{\beta\gamma} + \mathbf{f}_{\gamma\beta} + \mathbf{f}_{\gamma\alpha} + \mathbf{f}_{\alpha\gamma}).$$

Local energy balances take the form of (8.9.114) for pairwise-balanced interactions, namely

$$r_\alpha - \operatorname{div}\mathbf{q}_\alpha^+ + Q_{\alpha\beta} + Q_{\alpha\gamma} + H_{\alpha\gamma} + \mathbf{T}_\alpha \cdot \mathbf{L}_\alpha = \rho_\alpha \grave{e}_\alpha + \mathcal{G}_{\alpha\gamma} e_\alpha, \tag{9.7.14}$$

$$r_\beta - \operatorname{div}\mathbf{q}_\beta^+ + Q_{\beta\alpha} + Q_{\beta\gamma} + H_{\beta\gamma} + \mathbf{T}_\beta \cdot \mathbf{L}_\beta = \rho_\beta \grave{e}_\beta + \mathcal{G}_{\beta\gamma} e_\beta, \tag{9.7.15}$$

and
$$r_\gamma - \operatorname{div}\mathbf{q}_\gamma^+ + Q_{\gamma\alpha} + Q_{\gamma\beta} + H_{\gamma(\alpha+\beta)} + \mathbf{T}_\gamma \cdot \mathbf{L}_\gamma = \rho_\gamma \grave{e}_\gamma + \mathcal{G}_{\gamma(\alpha+\beta)} \mathbf{e}_\gamma. \tag{9.7.16}$$

[Why is there no term $H_{\alpha\beta}$ in (9.7.14)? Note corresponding 'absences' in (9.7.15) and (9.7.16).]

In (9.7.14) the heat flux vector $\mathbf{q}_\alpha^+ (= {}_b\mathbf{q}_\alpha^- + {}_2\mathbf{q}_\alpha + \boldsymbol{\kappa}_\alpha)$ [see (8.9.115)] has contributions from the work expended by $\alpha - \alpha$ interactions in α thermal motions (${}_b\mathbf{q}_\alpha^-$), diffusive supply of α binding energy (${}_2\mathbf{q}_\alpha$), and diffusive supply of α thermokinetic energy ($\boldsymbol{\kappa}_\alpha$). Analogous interpretations hold for $\mathbf{q}_\beta^+$ and $\mathbf{q}_\gamma^+$ in (9.7.15) and (9.7.16).

In (9.7.14) the term $Q_{\alpha\beta}(\mathbf{x},t)$ denotes a thermal energy supply rate density associated with work expended by interactions (between α molecules in $S_\epsilon(\mathbf{x})$ and β molecules everywhere) in α thermal motions. Taken together with the corresponding supply rate $Q_{\alpha\gamma}$ we obtain ${}^{\mathrm{ext}}Q$ as given by (8.9.63) for constituent α: here $\mathcal{M}^+ = \mathcal{M}_\alpha$ and $\mathcal{M}^- = \mathcal{M}_\beta \cup \mathcal{M}_\gamma$. Similar interpretations apply to $Q_{\beta\alpha} + Q_{\beta\gamma}$ in (9.7.15) and to $Q_{\gamma\alpha} + Q_{\gamma\beta}$ in (9.7.16).

Terms $H_{\alpha\gamma}, H_{\beta\gamma}$, and $H_{\gamma(\alpha+\beta)}$ represent net heating supply rate densities associated with thermokinetic energy transfer associated with $\alpha \leftrightarrow \gamma, \beta \leftrightarrow \gamma$, and $\gamma \leftrightarrow (\alpha+\beta)$ transitions. Finally, e_α, e_β, and e_γ denote internal energy densities of constituents α, β, and γ (per unit mass of these species).

Remark 9.7.1. Constituent γ has been regarded here as a single entity and modelled as a point mass with $m_\gamma = m_\alpha + m_\beta$. In particular, if (9.7.11) holds, then $\mathbf{v}_\gamma$ denotes the velocity of the mass centre of the composite γ molecule. A more realistic model would allow for both 'internal' kinetic and binding energy associated with the degree of freedom given by the separation of α and β molecules when combined to form a γ molecule.

9.8 Concluding Remarks

Remark 9.8.1. While the kinetic theory of gases, together with physical intuition, have provided the basis for postulated balance relations for mixtures, interpretation of individual terms in such relations is not simple and, as indicated in Section 9.5, sometimes has been problematical. Here definitions are precise, physically clear, and explicitly take account of scales of both length and time. Although a general development of reactions has not been presented, the methodology can be extended to take account of specific molecular/ionic/atomic details as appropriate.

Remark 9.8.2. Mixtures of fluids involve molecules of these fluids encountering each other as near neighbours as a consequence of molecular diffusion. Said differently, fluids, if and when they mix, often do so at the molecular level. However, this not always the case in systems of biological or engineering interest. For example, the behaviour of emulsions, suspensions, and fluids containing cellular structures (such as blood), at scales large compared with the relevant substructure, may be approached from the mixture viewpoint. However, such structures usually will maintain their molecular constitution. Studies of composite bodies, granular materials, and fluid flow through porous media, are generally concerned with properties associated with scales greater than those at which the detailed structure of these systems is manifest. Such structure is often evident at length scales well in excess of nearest-neighbour molecular separations, and individual molecular constituents occupy distinct regions at these longer scales. For this reason, averages such as appear in (9.2.3) and (9.2.4), computed at scales large compared with fine structure, are said to be those appropriate to an *immiscible* mixture (cf., e.g., Bedford [42]). In Chapter 10 the methodology developed here is applied to such a mixture, namely flow of fluid through a porous solid.

10 Fluid Flow through Porous Media

10.1 Preamble

Here we are concerned with fluid flow through a body which is accordingly 'porous' in some sense. In order that such flow be possible it is necessary that

(i) there is vacant space available 'within' the body to accommodate fluid, and
(ii) the space in which fluid can reside must be connected in order that the fluid can move *through* the body.

Vacant space within a body is termed *pore space*, a measure of which is *porosity*. Of course, not all pore space may be accessible to fluid: there may be isolated space inaccessible to fluid penetration. Such penetration, associated with connectivity, gives rise to the notion of *permeability*. Consider an insect attempting to crawl through a rectangular block of porous material from the centre of one face to exit through another particular face. This may or may not be possible. It could be that no connected route between the point of entry and the destination face exists, or that the insect is unable to squeeze through available 'gaps' en route. The former snag indicates that permeability is in some sense direction-dependent, while the latter draws attention to the scale dependence of both porosity *and* permeability.

Before addressing technicalities, it is worthwhile to note that the effects of porosity are crucial to our very existence: semipermeable membranes help to govern vital processes throughout our bodies, and the porous nature of bone provides structural strength without undue mass. Further, plant life and soil properties depend in part on relevant porosities, while the presence of subterranean water sources (aquifers) and oil-bearing shale derives from porosity within the surface of the Earth. More mundanely, we utilise sponges for cleaning and filtration systems in our water supplies but can be inconvenienced by dampness in the fabric of buildings and swelling of kitchen worktops due to water ingress. The foregoing serves to illustrate the diversity of porous system effects and the range of associated length scales.

Fluid flow in a porous body involves mass transfer through a region which is usually highly convoluted, many-fold connected[1], and whose detailed geometry is unknown. Accordingly details of the associated meandering flow are usually inaccessible. However, overall mass transfer (such as that out of the end of a porous 'plug') may be monitored. Roughly speaking, two length scales are involved here: that associated with flow within pores (ϵ_1, say), and that at which monitoring is possible (ϵ_2, say). Of course, ϵ_2 will be large compared with characteristic pore size and porous body structural dimension. We choose ϵ_1 to be the scale at which pores are delineated in the manner of Subsection 4.3.2. Accordingly, the boundary of the pores within $S_{\epsilon_2}(\mathbf{x})$ at time t is $\partial B_t^{\epsilon_1} \cap S_{\epsilon_2}(\mathbf{x})$ (see Remark 4.3.5 and Figure 10.2). Recall also that the porosity field corresponding to this choice of a *pair* of length scales is given by

$$\mathcal{P}_{\epsilon_1,\epsilon_2}(\mathbf{x},t) := \frac{V_{\epsilon_2} - \text{vol}\{S_{\epsilon_2}(\mathbf{x}) \cap B_t^{\epsilon_1}\}}{V_{\epsilon_2}}. \tag{4.3.11}$$

Remark 10.1.1. Region $B_t^{\epsilon_1}$ [see (4.3.6)] may be replaced in (4.3.11) by the associated ϵ_1-scale triangular polyhedral region [see (4.3.14) *et seq*].

Exercise 10.1.1. Convince yourself that $0 < \mathcal{P}_{\epsilon_1,\epsilon_2} < 1$ and that the choice of triangular polyhedral region yields a larger value for porosities.

Remark 10.1.2. Scale ϵ_1 is that at which pore boundaries are delineated. *This should not be confused with typical pore size.* Indeed, if s denotes such size, then in what follows we shall assume

$$\epsilon_1 << s. \tag{10.1.1}$$

In such case it is possible to apply local ϵ_1-scale continuum modelling to fluid within pore space.

In what follows, continuum descriptions of fluid flow within a porous body will be obtained at both scale ϵ_1 and scale ϵ_2 using weighting function methodology. Assuming that the ϵ_1-scale flow within pores is incompressible, linearly viscous, and saturates (i.e., 'fills') pore space at this scale, this flow is governed by the Navier–Stokes equations. The results of averaging these equations at scale ϵ_2 are compared with those obtained by direct averaging of molecular behaviour at scale ϵ_2. The widely used Brinkman equation and Darcy's 'law' emerge upon making clear approximative assumptions.

10.2 The General Forms of Mass Conservation and Linear Momentum Balance

Let $\mathcal{M}^f$ and $\mathcal{M}^p$ denote the material systems which consist of *f*luid molecules and of *p*orous body molecules, respectively. The total material system [cf. (9.2.1)]

$$\mathcal{M} := \mathcal{M}^f \cup \mathcal{M}^p \tag{10.2.1}$$

[1] That is, given any pair of points P and Q in this region, there are many classes of smooth curves which link P and Q. Within each class, such a curve can be continuously deformed into any other curve. However, it is impossible to deform a curve in one class continuously into a curve in another class. Said more simply, if Γ_1 and Γ_2 are curves which link P to Q but belong to different classes, then a string in the form of the closed curve $\Gamma_1 \cup \Gamma_2$ cannot be shrunk (by shortening) to a point in such a way that it always remains in the region.

is regarded to be a non-reacting immiscible (see Remark 9.8.2) mixture. Choice of a weighting function w leads, via the analysis of Section 9.2, to the local form [cf. (9.2.11)]

$$\frac{\partial \rho_w^f}{\partial t} + \operatorname{div}\{\rho_w^f \mathbf{v}_w^f\} = 0 \tag{10.2.2}$$

of mass conservation. Of course [cf. (9.2.3), (9.2.8), and (9.2.4)],

$$\rho_w^f(\mathbf{x},t) := \sum_{P_i^f} m_i^f \, w(\mathbf{x}_i^f(t) - \mathbf{x}) \tag{10.2.3}$$

and

$$\mathbf{v}_w^f := \mathbf{p}_w^f / \rho_w^f, \tag{10.2.4}$$

where

$$\mathbf{p}_w^f(\mathbf{x},t) := \sum_{P_i^f} m_i^f \, \mathbf{v}_i^f(t) w(\mathbf{x}_i^f(t) - \mathbf{x}). \tag{10.2.5}$$

Here $m_i^f, \mathbf{x}_i^f(t)$ and $\mathbf{v}_i^f(t)$ denote the mass, location and velocity of fluid molecule P_i^f at time t, respectively, and sums are taken over *all* $P_i^f \in \mathcal{M}^f$.

As discussed in Section 9.3, the corresponding balance of linear momentum can be expressed in several ways. The form corresponding to (9.3.13) is

$$-\operatorname{div} \mathcal{D}_w^f + \mathbf{f}_w^f + \mathbf{f}_w^{fp} + \mathbf{b}_w^f = \frac{\partial}{\partial t}\{\rho_w^f \mathbf{v}_w^f\} + \operatorname{div}\{\rho_w^f \mathbf{v}_w^f \otimes \mathbf{v}_w^f\} \tag{10.2.6}$$

$$= \rho_w^f \mathbf{a}_w^f. \tag{10.2.7}$$

Here [cf. (9.3.11), (9.3.9), (9.3.3), (9.3.4), and (9.3.5)]

$$\mathcal{D}_w^f(\mathbf{x},t) := \sum_{P_i^f} m_i^f \, \hat{\mathbf{v}}_i^f(\mathbf{x};t) \otimes \hat{\mathbf{v}}_i^f(\mathbf{x};t) w(\mathbf{x}_i^f(t) - \mathbf{x}), \tag{10.2.8}$$

with *notional thermal velocity* given by

$$\hat{\mathbf{v}}_i^f(\mathbf{x};t) := \mathbf{v}_i^f(t) - \mathbf{v}_w^f(\mathbf{x},t), \tag{10.2.9}$$

$$\mathbf{f}_w^f(\mathbf{x},t) := \sum_{P_i^f \neq P_j^f} \sum \mathbf{f}_{ij}^{ff}(t) w(\mathbf{x}_i^f(t) - \mathbf{x}), \tag{10.2.10}$$

$$\mathbf{f}_w^{fp}(\mathbf{x},t) := \sum_{P_i^f} \sum_{P_j^p} \mathbf{f}_{ij}^{fp}(t) w(\mathbf{x}_i^f(t) - \mathbf{x}), \tag{10.2.11}$$

and

$$\mathbf{b}_w^f(\mathbf{x},t) := \sum_{P_i^f} \mathbf{b}_i^f(t) w(\mathbf{x}_i^f(t) - \mathbf{x}). \tag{10.2.12}$$

Here $\mathbf{f}_{ij}^{ff}, \mathbf{f}_{ij}^{fp}$, and $\mathbf{b}_i^f$ denote the forces exerted on fluid molecule P_i^f by, respectively, fluid molecule P_j^f, porous body molecule P_j^p, and the material universe outwith $\mathcal{M}$ [see (10.2.1)].

Given any solution $(\mathbf{T}_w^f)^-$ to

$$\operatorname{div}(\mathbf{T}_w^f)^- = \mathbf{f}_w^f, \tag{10.2.13}$$

then (10.2.7) may be re-written in the form [cf. (9.3.34)]

$$\operatorname{div} \mathbf{T}_w^f + \mathbf{f}_w^{fp} + \mathbf{b}_w^f = \frac{\partial}{\partial t}\{\rho_w^f \mathbf{v}_w^f\} + \operatorname{div}\{\rho_w^f \mathbf{v}_w^f \otimes \mathbf{v}_w^f\}, \tag{10.2.14}$$

where the *fluid–fluid stress tensor*

$$\mathbf{T}_w^f := (\mathbf{T}_w^f)^- - \mathcal{D}_w^f. \tag{10.2.15}$$

The *fluid–fluid interaction stress tensor* $(\mathbf{T}_w^f)^-$ is given by either [cf. (9.3.19)]

$$(\mathbf{T}_w^f)^-(\mathbf{x},t) = ({}_s\mathbf{T}_w^f)^-(\mathbf{x},t) := \sum_{P_i^f \neq P_j^f} \sum \mathbf{f}_{ij}^{ff}(t) \otimes \mathbf{a}_i^f(\mathbf{x},t), \tag{10.2.16}$$

where $\mathbf{a}_i^f$ is a vector field which satisfies [cf. (9.3.16)]

$$(\operatorname{div} \mathbf{a}_i^f)(\mathbf{x},t) = w(\mathbf{x}_i^f(t) - \mathbf{x}), \tag{10.2.17}$$

or, in the case of balanced interactions [cf. (9.3.24)],

$$(\mathbf{T}_w^f)^-(\mathbf{x},t) = ({}_b\mathbf{T}_w^f)^-(\mathbf{x},t) := \sum_{P_i^f \neq P_j^f} \sum \mathbf{f}_{ij}^{ff}(t) \otimes \mathbf{b}_{ij}^f(\mathbf{x},t), \tag{10.2.18}$$

where $\mathbf{b}_{ij}^f$ is a vector field which satisfies [cf. (9.3.23)]

$$(\operatorname{div} \mathbf{b}_{ij}^f)(\mathbf{x},t) = \frac{1}{2}\{w(\mathbf{x}_i^f(t) - \mathbf{x}) - w(\mathbf{x}_j^f(t) - \mathbf{x})\}. \tag{10.2.19}$$

Remark 10.2.1. The physical interpretations of all fields defined here depend both upon the *form* of the choice of weighting function and, more crucially, upon the spatial *scale* ϵ associated with w. This is because these fields are only defined, at time t, within the region $B_{w,t}^f$ defined by (9.2.15) with $\alpha = f$. As discussed in the preceding section, choices $\epsilon = \epsilon_1$ and $\epsilon = \epsilon_2$ delineate very different regions: the former yields that portion of the pore space in which fluid is to be found, while the latter yields a region which includes both pore space *and* the porous body, and whose boundary does not reveal the existence of pores. Here it may prove helpful to refer to the figures in Subsection 4.3.2 for visualisation of how pore space or fissure features disappear with an increase in the averaging scale ϵ when choice $w = w_\epsilon$ is made.

Remark 10.2.2. Entirely analogous relations to those which follow (10.2.1) hold for the porous body, and are obtained simply by replacing superscript f by p and vice versa.

Given the choice $w = w_\epsilon$ [see (4.3.2)], the region occupied by a material system $\mathcal{M}$ at time t is [cf. (4.3.6)]

$$B_t^\epsilon(\mathcal{M}) := \bigcup_{P_i \in \mathcal{M}} S_\epsilon(\mathbf{x}_i(t)), \tag{10.2.20}$$

where $\mathbf{x}_i(t)$ denotes the location of molecule P_i at time t. Said differently,

$$B_t^\epsilon(\mathcal{M}) = \{\mathbf{x} \in \mathcal{E} : \rho_{w_\epsilon}(\mathbf{x},t) > 0\}. \tag{10.2.21}$$

Exercise 10.2.1. Show that

$$B_t^\epsilon(\mathcal{M}) \subset B_t^{\epsilon'}(\mathcal{M}) \quad \text{if} \quad \epsilon' > \epsilon. \tag{10.2.22}$$

We write

$${}^fB_t^\epsilon := B_t^\epsilon(\mathcal{M}^f) \ \text{ and } \ {}^pB_t^\epsilon := B_t^\epsilon(\mathcal{M}^p) \tag{10.2.23}$$

and can make the following *o*bservations.

O.1. If the body is immersed in fluid which permeates all pores, then

$$ {}^{p}B_t^{\epsilon_2} \subset {}^{f}B_t^{\epsilon_2} \tag{10.2.24}$$

since if $\mathbf{x} \in {}^{p}B_t^{\epsilon_2}$, then sphere $S_{\epsilon_2}(\mathbf{x},t)$ will contain pore space (recall that ϵ_2 is large compared with the structural dimension of the porous body) and hence fluid: thus $\rho^f_{w_{\epsilon_2}}(\mathbf{x},t) > 0$, and hence $\mathbf{x} \in {}^{f}B_t^{\epsilon_2}$. In particular, fields $\rho^f_{w_{\epsilon_2}}$ and $\mathbf{v}^f_{w_{\epsilon_2}}$ are defined *everywhere* in ${}^{f}B_t^{\epsilon_2}$, which includes all points within the porous body region ${}^{p}B_t^{\epsilon_1}$ delineated at scale ϵ_1 [via (10.2.21) and (10.2.19) with $\mathcal{M} = \mathcal{M}^f$].

O.2. The *interfacial region at scale* ϵ_1 *and time* t is

$$ \mathcal{I}^{pf}_{\epsilon_1}(t) := {}^{p}B_t^{\epsilon_1} \cap {}^{f}B_t^{\epsilon_1}. \tag{10.2.25}$$

Thus, if $\mathbf{x} \in \mathcal{I}^{pf}_{\epsilon_1}(t)$, then sphere $S_{\epsilon_1}(\mathbf{x})$ will contain both porous body *and* fluid molecules. If interactions $\mathbf{f}^{fp}_{ij}$ are short-range,[2] in the sense that

$$ \mathbf{f}^{fp}_{ij}(t) = \mathbf{0} \qquad \text{if} \quad \|\mathbf{x}^f_i(t) - \mathbf{x}^p_i(t)\| > \delta, \tag{10.2.26}$$

and if

$$ \delta < \epsilon_1, \tag{10.2.27}$$

then the effect of porous body molecules upon those of the fluid derives only from interactions between such molecules which lie within $\mathcal{I}^{pf}_{\epsilon_1}(t)$. Indeed, noting that porous body molecules at scale ϵ_1 lie on the triangulated geometric boundary $\Lambda^{\epsilon_1}_{\text{geom}}(\mathcal{M}^p)$ [see (4.3.14)], this effect is confined to fluid molecules within a distance δ of this boundary.

Remark 10.2.3. If fluid exists in both liquid and gaseous phases, then capillary effects are manifest and derive from behaviour within $\mathcal{I}^{pf}_{\epsilon_1}$. Specifically, the effects of surface tension at liquid–vapour interfaces within pores derive chiefly from behaviour at highly convoluted curve-like regions containing fluid molecules in both liquid and gaseous states *and* porous body molecules. These so-called contact lines are essentially porous body–liquid–gas interfacial regions.

In the following sections the interpretations of field values associated with choices $w = w_{\epsilon_1}$ and w_{ϵ_2} will be discussed individually, and then linked in the context of the flow of an incompressible fluid through a porous body that it saturates.

10.3 Linear Momentum Balance at Scale $\epsilon = \epsilon_1$ with $w = w_{\epsilon_1}$

Modulo short-range interactions as described in observation O.2 of the preceding section, the porous body molecules do not give rise to forces on fluid in the region

$$ ({}^{f}B_t^{\epsilon_1})^- := {}^{f}B_t^{\epsilon_1} - \mathcal{I}^{pf}_{\epsilon_1}(t) \tag{10.3.1}$$

at time t. Accordingly, *in this region* $\mathbf{f}^{fp}_{w_{\epsilon_1}}(t) = \mathbf{0}$ [see (10.2.11)], and balance (10.2.14) involves only fluid-related fields. The form of $\mathbf{T}^f_w$ (which now will be denoted by

[2] Molecular interactions are regarded to be long range if they extend to $100\,\text{Å} = 10^{-8}$ m.

$\mathbf{T}^f_{\epsilon_1}$ to emphasise the associated length scale) depends upon whether the fluid is in a gaseous or liquid phase. For a moderately rarefied gaseous phase $\mathbf{T}^f_{\epsilon_1}$ will be a density-dependent pressure (see Subsection 5.5.2). That is,

$$\mathbf{T}^f_{\epsilon_1} = -P(\rho^f_{\epsilon_1})\mathbf{1}. \tag{10.3.2}$$

For incompressible, linearly viscous fluids (see Gurtin, Fried, and Anand [23], p. 261), $\rho^f_{\epsilon_1}$ is constant and

$$\mathbf{T}^f_{\epsilon_1} = -P\mathbf{1} + 2\mu\mathbf{D}^f_{\epsilon_1}, \tag{10.3.3}$$

where viscosity μ is a constant, and

$$\mathbf{D}^f_{\epsilon_1} := \frac{1}{2}(\nabla\mathbf{v}^f_{\epsilon_1} + (\nabla\mathbf{v}^f_{\epsilon_1})^T). \tag{10.3.4}$$

For compressible, linearly viscous fluids (see Gurtin, Fried, and Anand [23], p. 255)

$$\mathbf{T}^f_{\epsilon_1} = -P_{eq}\mathbf{1} + 2\mu\mathbf{D}^f_{\epsilon_1} + \lambda(\mathrm{tr}\,\mathbf{D}^f_{\epsilon_1})\mathbf{1}, \tag{10.3.5}$$

where $-P_{eq}\mathbf{1}$ denotes the pressure in the absence of flow, and P_{eq}, μ and λ depend upon $\rho^f_{\epsilon_1}$.

Remark 10.3.1. The constitutive relations (10.3.2), (10.3.3) and (10.3.5) are those most commonly employed to describe fluids, but by no means do they exhaust possible fluid behaviour. However, since flow in pores is slow, non-Newtonian fluids may not exhibit their full nature.

Remark 10.3.2. The behaviour of fluid in $\mathcal{I}^{pf}_{\epsilon_1}$ is not discussed here. In the next section the *effect* of the porous body upon fluid flow [which we have seen in observation O.2 derives entirely from $\mathcal{I}^{pf}_{\epsilon_1}$ interactions, modulo (10.2.26) and (10.2.27)] will be modelled at scale $\epsilon = \epsilon_1$ via the standard non-slip and non-penetration conditions of fluid dynamics. Behaviour within $\mathcal{I}^{pf}_{\epsilon_1}$ will be otherwise neglected.

10.4 Linear Momentum balance at Scale $\epsilon = \epsilon_2$ with $w = w_{\epsilon_2}$

To emphasise the relevant scale dependence here, we re-write (10.2.14) as

$$\mathrm{div}\,\mathbf{T}^f_{\epsilon_2} + \mathbf{f}^{fp}_{\epsilon_2} + \mathbf{b}^f_{\epsilon_2} = \frac{\partial}{\partial t}\{\rho^f_{\epsilon_2}\mathbf{v}^f_{\epsilon_2}\} + \mathrm{div}\{\rho^f_{\epsilon_2}\mathbf{v}^f_{\epsilon_2}\otimes\mathbf{v}^f_{\epsilon_2}\}. \tag{10.4.1}$$

(This is a little less cumbersome than writing $\mathbf{T}^f_{w_{\epsilon_2}}$ etc.) The stress tensor [cf. (10.2.15)]

$$\mathbf{T}^f_{\epsilon_2} = (\mathbf{T}^f_{\epsilon_2})^- - \mathcal{D}^f_{\epsilon_2}, \tag{10.4.2}$$

where [cf. (10.2.13)]

$$\mathrm{div}(\mathbf{T}^f_{\epsilon_2})^- = \mathbf{f}^f_{\epsilon_2}, \tag{10.4.3}$$

with [cf. (10.2.10)]

$$\mathbf{f}^f_{\epsilon_2}(\mathbf{x},t) := \sum_{\substack{P_i,P_j\in\mathcal{M}^f \\ i\neq j}}\sum \mathbf{f}^{ff}_{ij}(t)w_{\epsilon_2}(\mathbf{x}^f_i(t)-\mathbf{x}). \tag{10.4.4}$$

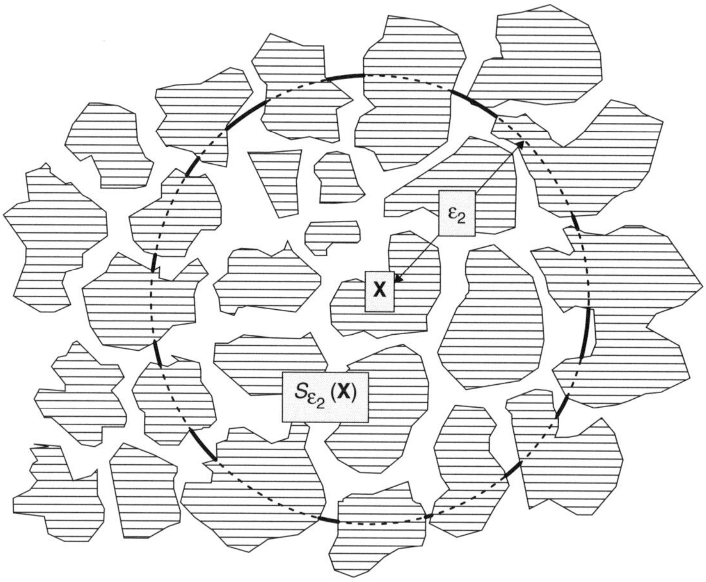

Figure 10.1. Force density $\mathbf{f}^f{}_{\epsilon_2}(\mathbf{x},t)$ represents the effect of fluid outside $S_{\epsilon_2}(\mathbf{x})$ upon fluid inside this sphere and hence stems from fluid–fluid interactions across the boundary of this sphere at locations indicated with an unbroked curve —.

Recalling Remark 9.3.2, if the net self-force associated with fluid molecules within $S_{\epsilon_2}(\mathbf{x})$ vanishes at any time, that is, if

$$\sum_{\substack{\mathbf{x}_i^f,\mathbf{x}_j^f \in S_{\epsilon_2}(\mathbf{x}) \\ i \neq j}} \sum \mathbf{f}_{ij}^{ff} = \mathbf{0} \tag{10.4.5}$$

at any time,[3] then $\mathbf{f}_{\epsilon_2}^f(\mathbf{x},t)$ denotes the force exerted at time t on fluid molecules inside $S_{\epsilon_2}(\mathbf{x})$ by fluid molecules outside or on the boundary of $S_{\epsilon_2}(\mathbf{x})$, divided by the volume V_{ϵ_2} of $S_{\epsilon_2}(\mathbf{x})$ (see Figure 10.1).

Remark 10.4.1. Recall the considerations of Subsection 5.9.2. In the present context we have

$$\int_{\Pi_{\mathbf{n}_0}(\mathbf{x}_0)} (\mathbf{T}_{\epsilon_2}^f)^- \mathbf{n}_0 \, dA = \sum_{i \neq j} \sum \frac{(V_i^f)^-}{V_{\epsilon_2}} \mathbf{f}_{ij}^{ff} \tag{10.4.6}$$

for choice $({}_s\mathbf{T}_{\epsilon_2}^f)^-$ [cf. (5.9.30)]. This also holds for choice $({}_b\mathbf{T}_{\epsilon_2}^f)^-$ in respect of the specific choices of simple balanced, Hardy, or Noll interaction stresses (Convince yourself of this: see Results 5.8.1.) Recall that $\Pi_{\mathbf{n}_0}(\mathbf{x}_0)$ denotes the (infinite) plane through $\mathbf{x}_0$ with unit normal $\mathbf{n}_0$, and $(V_i^f)^-$ is that volume within $S_{\epsilon_2}(\mathbf{x}_i^f)$ which lies 'below' this plane.

The definition of $\mathcal{D}_{\epsilon_2}^f(\mathbf{x},t)$ [given by (10.2.8) with $w = w_{\epsilon_2}$] differs from its ϵ_1-scale counterpart $\mathcal{D}_{\epsilon_1}^f(\mathbf{x},t)$ in two respects. Firstly, the weighting of contributions

[3] Satisfaction of (10.4.5) is guaranteed by individual pairwise balancing of interactions but is a much less restrictive interaction assumption which yields the same interpretation of $\mathbf{f}_{\epsilon_2}^f(\mathbf{x},t)$ (see Exercise 5.5.1).

in its definition is $V_{\epsilon_2}^{-1}$, but fluid molecules are to be found only within pore space inside $S_{\epsilon_2}(\mathbf{x})$. Secondly, the notional thermal velocity [see (10.2.9) with $w = w_{\epsilon_2}$] is computed with respect to $\mathbf{v}_w^f = \mathbf{v}_{w_{\epsilon_2}}^f$ rather than $\mathbf{v}_w^f = \mathbf{v}_{w_{\epsilon_1}}^f$. For creeping flow such a distinction is not significant, since molecular speeds are of order $10^3\,\text{ms}^{-1}$ at room temperature and hence the overwhelmingly dominant contribution to $\mathcal{D}_w^f$ stems from individual molecular velocities. [Said differently, neglect of $\mathbf{v}_{\epsilon_2}^f$ and $\mathbf{v}_{\epsilon_1}^f$ in the appropriate forms of (10.2.9) will affect the values of $\mathcal{D}_w^f$ negligibly, in general.] If the ϵ_1-scale distribution of fluid molecular velocities is essentially uniform within pores over scale ϵ_2, then we should expect [cf. (4.3.8) and (4.3.7)]

$$\mathcal{D}_{\epsilon_2}^f \simeq \mathcal{P}_{\epsilon_1,\epsilon_1} \mathcal{D}_{\epsilon_1}^f. \tag{10.4.7}$$

Term $\mathbf{f}_{\epsilon_2}^{fp}(\mathbf{x},t)$ represents the effect of porous body molecules upon fluid molecules within $S_{\epsilon_2}(\mathbf{x})$. In view of the short-range nature of molecular interactions this effect is localised at the boundaries of pore space occupied by fluid within $S_{\epsilon_2}(\mathbf{x})$. As can be visualised from Figure 10.1, such effect will represent the resistance offered to fluid flow by the porous body. This resistance stems from two sources:

(i) The porous body boundary (at scale ϵ_1) forms an obstacle which constrains flow as a consequence of fluid being unable to penetrate this boundary.
(ii) The porous body boundary exerts a 'drag' on fluid moving over it.

As will be seen in the next section, source (i) takes the form of a 'back-pressure' exerted by the pore space boundary within $S_{\epsilon_2}(\mathbf{x})$ (this is the 'reaction' to the pressure exerted by the fluid on this boundary), while the effect of (ii) is proportional to fluid viscosity.

10.5 Flow of an Incompressible Linearly Viscous Fluid through a Porous Body It Saturates

At scale ϵ_1 pore details are manifest, yet continuum modelling within pore space is sensible, since ϵ_1 is much greater than the scale at which individual molecular behaviour is apparent or significant. Accordingly, as discussed in Section 10.3, in the region occupied by fluid at this scale, but not within the range of molecular interaction between fluid and porous body [namely $({}^fB^{\epsilon_1})^-$: see (10.3.1)], flow of a fluid is governed by the ϵ_1-scale continuity equation (10.2.2) and linear momentum balance (10.2.14), both with $w = w_{\epsilon_1}$, and

$$\mathbf{f}_w^{fp} = \mathbf{0}. \tag{10.5.1}$$

If the fluid is incompressible, then the ϵ_1-scale mass density is

$$\rho_w^f = \rho_0, \tag{10.5.2}$$

where ρ_0 is a constant. Further, if the external body force density derives purely from the effect of gravity, then

$$\mathbf{b}_w^f = \rho_0\, \mathbf{g}. \tag{10.5.3}$$

Accordingly linear momentum balance (10.2.14) in $({}^fB_t^{\epsilon_1})^-$ takes the form, upon suppressing superscripts and subscripts for simplicity,

$$\operatorname{div}\mathbf{T} + \rho_0\,\mathbf{g} = \frac{\partial}{\partial t}\{\rho_0\,\mathbf{v}\} + \operatorname{div}\{\rho_0\,\mathbf{v}\otimes\mathbf{v}\}, \tag{10.5.4}$$

and the corresponding continuity equation (10.2.2) reduces to

$$\operatorname{div}\mathbf{v} = 0. \tag{10.5.5}$$

If the fluid is linearly viscous, then [see (10.3.3) and (10.3.4)]

$$\mathbf{T} = -P\mathbf{1} + \mu(\nabla\mathbf{v} + (\nabla\mathbf{v})^T). \tag{10.5.6}$$

Together (10.5.4), (10.5.5) and (10.5.6) yield the Navier–Stokes equation

$$-\nabla P + \mu\Delta\mathbf{v} + \rho_0\,\mathbf{g} = \frac{\partial}{\partial t}\{\rho_0\,\mathbf{v}\} + \operatorname{div}\{\rho_0\,\mathbf{v}\otimes\mathbf{v}\}. \tag{10.5.7}$$

Exercise 10.5.1. Derive (10.5.7).

In describing flow within pores, account must be taken of the effect of the porous body: this is not explicit in (10.5.7), which holds in $({}^fB^{\epsilon_1})^-$. Indeed, the influence of the pore boundaries on fluid is communicated to fluid in $({}^fB^{\epsilon_1})^-$ by fluid in ${}^{pf}\mathcal{I}_{\epsilon_1}$ [see (10.2.25)]. If ϵ_1 is much smaller than pore size (see Remark 10.1.2), then this influence is a boundary effect for fluid flow modelled at scale ϵ_1. (See also observation O.2 of Subsection 10.2.) Thus at this scale we are considering flow, governed by the Navier-Stokes equation, over a solid boundary and, accordingly adopt the classical non-slip, non-penetration condition $\mathbf{v} = \mathbf{v}_b$ on the 'boundary', where $\mathbf{v}_b$ denotes the velocity of the boundary. The question here arises as to what 'boundary' is to be chosen. The natural candidates here are $\partial^p B^{\epsilon_1}, \partial^f B^{\epsilon_1}, \partial({}^fB^{\epsilon_1})^-$ and [see (4.3.14)] $\Lambda^{\epsilon_1}_{\mathrm{geom}}(\mathcal{M}^p)$. At this point it is helpful to note the following

Remark 10.5.1. Continuum modelling at scale ϵ is only sensible when applied to the behaviour of bodies whose dimensions greatly exceed ϵ. Further, in evaluating ϵ-scale averages, no information concerning behaviour at smaller scales is elicited. Accordingly it is not meaningful to distinguish aspects of behaviour at scales smaller than the modelling scale ϵ.

In view of Remark 10.5.1, $\partial^p B^{\epsilon_1}, \partial^f B^{\epsilon_1}, \partial({}^fB^{\epsilon_1})^-$, and $\Lambda^{\epsilon_1}_{\mathrm{geom}}(\mathcal{M}^p)$ are to be regarded as indistinguishable in describing flows with continuum fields computed via ϵ_1-scale averages. *In what follows, the region occupied at scale ϵ_1 by the fluid within the porous body at time t will be denoted by R_t* without further qualification. Fields associated both with $\mathcal{M}^f$ *and* $\mathcal{M}^p$ will be assumed to be meaningful on ∂R_t, which forms the totality of pore boundaries. In particular, the non-slip, non-penetration boundary condition to be applied to solutions of (10.5.7) is thus

$$\mathbf{v}(\mathbf{x},t) = \mathbf{v}^p(\mathbf{x},t) \quad \text{if} \quad \mathbf{x}\in\partial R_t, \tag{10.5.8}$$

where $\mathbf{v}^p$ denotes the porous body velocity field (at scale ϵ_1).

Question: How can we now average relation (10.5.7) at a scale ϵ_2 large compared with pore size and porous body structural dimension to obtain an ϵ_2-scale continuum description?

Answer: Use weighting function methodology for fields, as in generalisation (4.4.6).

Changing notation, we define the *w-average of a field f* by

$$\langle f\rangle_w(\mathbf{x},t) := \int_{\mathcal{E}} f(\mathbf{y},t)w(\mathbf{y}-\mathbf{x})dV_{\mathbf{y}}. \tag{10.5.9}$$

Thus, in order to obtain the required ϵ_2-scale average of (10.5.7), it is necessary to choose a weighting function w which has an associated scale ϵ_2, evaluate each term of (10.5.7) at $(\mathbf{y},t)$, multiply by $w(\mathbf{y}-\mathbf{x})$, and integrate over all $\mathbf{y}\in\mathcal{E}$. However, in such a procedure it should be noted that (10.5.7) relates only to points in R_t at time t. In the evaluation of integrals over $\mathcal{E}$, we note that the fields which appear in (10.5.7) vanish outside[4] R_t and that the effect of the porous body is encapsulated in (10.5.8).

While averaging (10.5.7) requires computation of $\langle\nabla P\rangle_w, \langle\Delta\mathbf{v}\rangle_w, \langle\rho_0\mathbf{g}\rangle_w$, $\langle\frac{\partial}{\partial t}\{\rho_0\mathbf{v}\}\rangle_w$, and $\langle\operatorname{div}\{\rho_0\mathbf{v}\otimes\mathbf{v}\}\rangle_w$, it is instructive to examine first the w-average of the gradient of any scalar field φ defined on R_t. Suppressing time dependence,

$$\begin{aligned}
\langle\nabla\varphi\rangle_w(\mathbf{x}) &:= \int_{\mathcal{E}} \nabla\varphi(\mathbf{y})w(\mathbf{y}-\mathbf{x})dV_{\mathbf{y}} \\
&= \int_R \nabla\varphi(\mathbf{y})w(\mathbf{y}-\mathbf{x})dV_{\mathbf{y}} \\
&= \int_R \nabla_{\mathbf{y}}\{\varphi w\} - \varphi\nabla_{\mathbf{y}} w\, dV_{\mathbf{y}} \\
&= \int_R \operatorname{div}\{\varphi w\mathbf{1}\} + \varphi\nabla_{\mathbf{x}} w\, dV_{\mathbf{y}} \\
&= \int_{\partial R} \varphi w\mathbf{n}\, dA_{\mathbf{y}} + \nabla_{\mathbf{x}}\left\{\int_R \varphi(\mathbf{y})w(\mathbf{y}-\mathbf{x})dV_{\mathbf{y}}\right\}.
\end{aligned}$$

That is,

$$\langle\nabla\varphi\rangle_w(\mathbf{x}) = \int_{\partial R} \varphi(\mathbf{y})w(\mathbf{y}-\mathbf{x})\mathbf{n}(\mathbf{y})dA_{\mathbf{y}} + \nabla\{\langle\varphi\rangle_w\}(\mathbf{x}). \tag{10.5.10}$$

Thus the difference between the average of a gradient, $\langle\nabla\varphi\rangle_w$, and the gradient of an average, $\nabla\{\langle\varphi\rangle_w\}$, is the integral of $\varphi w\mathbf{n}$ over the pore boundary ∂R with $\mathbf{n}$ the outward unit normal to R. (Notice that $\mathbf{n}$ is directed *into* the porous body.) Clearly this difference derives only from points $\mathbf{y}$ on ∂R for which $w(\mathbf{y}-\mathbf{x}) \neq 0$. Accordingly, if $w = w_{\epsilon_2}$, then only points on the pore boundary which lie within a distance of ϵ_2 from $\mathbf{x}$ contribute to the integral term in (10.5.10). (See Figure 10.2).

Averages in the porous media literature are computed over so-called *r*epresentative *e*lementary *v*olumes (abbreviated to 'REVs'). These are regions whose characteristic dimensions are large compared to pore and structure size. More precisely, an REV is a spatial 'cell', as described in Subsection 4.4.1, of appropriate size. Recall that

$$C_{\mathcal{U}}(\mathbf{x}) := \{\mathbf{x}+\mathbf{u} : \mathbf{u}\in\mathcal{U}\}, \tag{4.4.2}$$

where $\mathcal{U}$ is a fixed set of displacements which form a neighbourhood of $\mathbf{0}\in\mathcal{V}$ (the space of all displacements between pairs of points in space $\mathcal{E}$), and that $\mathbf{x}$ is the centroid of $C_{\mathcal{U}}(\mathbf{x})$ (see Figure 4.8). In the present context we require that if

[4] This follows since the fields in (10.5.7) relate to fluid which is only to be found in pore space R_t.

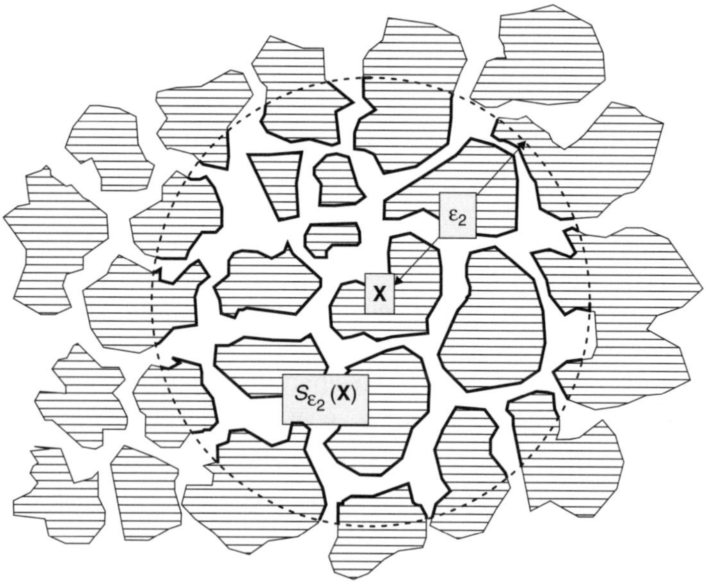

Figure 10.2. Contributions to the surface integral in (10.5.10) derive only from pore boundaries within $S_{\epsilon_2}(\mathbf{x})$ if $w = w_{\epsilon_2}$, indicated here with unbroken curves —.

$\mathbf{y} \in \partial C_{\mathcal{U}}(\mathbf{x})$, then $\|\mathbf{y}-\mathbf{x}\|$ should be of order ϵ_2, and cell boundaries should be 'simple' in the sense of being composed of not too many piece-wise smooth subsurfaces upon each of which principal curvatures should be less than say $2/\epsilon_2$: this excludes 'pathological' possibilities. If $V_{\mathcal{U}}$ is the volume of $C_{\mathcal{U}}(\mathbf{x})$ [see (4.4.3) and note that $V_{\mathcal{U}}$ is independent of $\mathbf{x}$. Why?], then the corresponding choice of weighting function w is a mollified version of $\hat{w}$, where [see (4.4.4)]

$$\left.\begin{array}{lll}\hat{w}(\mathbf{u}) & = (V_{\mathcal{U}})^{-1} & \text{if } \mathbf{u} \in \mathcal{U} \\ \hat{w}(\mathbf{u}) & = 0 & \text{if } \mathbf{u} \notin \mathcal{U}\end{array}\right\}. \tag{10.5.11}$$

Remark 10.5.2. As in subsection 4.3.4, mollification can be implemented over a physically insignificant region so that the values of

$$\int_{\mathcal{E}} f(\mathbf{y})w(\mathbf{y}-\mathbf{x})dV_{\mathbf{y}} \qquad \textit{and} \qquad \int_{\mathcal{E}} f(\mathbf{y})\hat{w}(\mathbf{y}-\mathbf{x})dV_{\mathbf{y}}$$

are physically indistinguishable for the fields considered. Of course, mollification is necessary to ensure the existence of ∇w: notice that $\nabla \hat{w}$ fails to exist on the boundary of set $\mathcal{U}$.

Returning to (10.5.10), we now note that

$$\int_{\partial R} \varphi(\mathbf{y})w(\mathbf{y}-\mathbf{x})\mathbf{n}(\mathbf{y})dA_{\mathbf{y}} = \int_{S(\mathbf{x})} \varphi(\mathbf{y})w(\mathbf{y}-\mathbf{x})\mathbf{n}(\mathbf{y})dA_{\mathbf{y}} \tag{10.5.12}$$

$$= (V_{\mathcal{U}})^{-1} \int_{S(\mathbf{x})} \varphi(\mathbf{y})\mathbf{n}(\mathbf{y})dA_{\mathbf{y}}, \tag{10.5.13}$$

where

$$S(\mathbf{x}) := \partial R \cap \{\mathbf{y} : \mathbf{y}-\mathbf{x} \in \mathcal{U}\}. \tag{10.5.14}$$

Thus $S(\mathbf{x})$ consists of all points on the pore boundary which lie within the that REV cell which has its centroid at $\mathbf{x}$.

Exercise 10.5.2. Check (10.5.12) and (10.5.13).

Accordingly (10.5.10) becomes

$$\langle\nabla\varphi\rangle_w(\mathbf{x}) = \nabla\{\langle\varphi\rangle_w\}(\mathbf{x}) + (V_{\mathcal{U}})^{-1}\int_{S(\mathbf{x})}\varphi(\mathbf{y})\mathbf{n}(\mathbf{y})dA_{\mathbf{y}}. \tag{10.5.15}$$

The averaging of the first term on the right-hand side of (10.5.7) thus yields, with $\varphi = -P$,

$$\langle\nabla(-P)\rangle_w(\mathbf{x}) = -\nabla\{\langle P\rangle_w\}(\mathbf{x}) - (V_{\mathcal{U}})^{-1}\int_{S(\mathbf{x})}P(\mathbf{y})\mathbf{n}(\mathbf{y})dA_{\mathbf{y}}. \tag{10.5.16}$$

In abbreviated form, omitting subscripts and arguments, this may be written as

$$\langle-\nabla P\rangle = -\nabla\{\langle P\rangle\} - \frac{1}{V}\int_S P\mathbf{n}\,dA. \tag{10.5.17}$$

Recalling that S is a subsurface of ∂R, and hence normal $\mathbf{n}$ is directed *into* the porous body, the last term in (10.5.17) represents an REV-volume averaged 'back-pressure' exerted *on* the fluid *by* the porous body. This back pressure force density is the 'reaction' (in the sense of Newton's third law) to the force density associated with the pressure exerted *on* the pore boundary subsurface S.

We now cite results necessary to effect averages of the remaining terms in (10.5.7).

Results 10.5.1. If $\mathbf{u}$ and $\mathbf{B}$ denote, respectively, vector- and tensor-valued fields defined in the region occupied by fluid, and are taken to be zero elsewhere, then (omitting subscripts)

$$\langle\nabla\mathbf{u}\rangle = \nabla\{\langle\mathbf{u}\rangle\} + \frac{1}{V}\int_S \mathbf{u}\otimes\mathbf{n}\,dA, \tag{10.5.18}$$

$$\langle\operatorname{div}\mathbf{u}\rangle = \operatorname{div}\{\langle\mathbf{u}\rangle\} + \frac{1}{V}\int_S \mathbf{u}.\mathbf{n}\,dA, \tag{10.5.19}$$

$$\langle\operatorname{div}\mathbf{B}\rangle = \operatorname{div}\{\langle\mathbf{B}\rangle\} + \frac{1}{V}\int_S \mathbf{B}\mathbf{n}\,dA, \tag{10.5.20}$$

and

$$\langle\frac{\partial\mathbf{u}}{\partial t}\rangle = \frac{\partial}{\partial t}\{\langle\mathbf{u}\rangle\} - \frac{1}{V}\int_S(\mathbf{u}\otimes\mathbf{v})\mathbf{n}\,dA. \tag{10.5.21}$$

Proofs of these results are given in Appendix B.9.

The Navier–Stokes equation (10.5.7) that we wish to average may be written (noting that ρ_0 is constant) as

$$-\nabla P + \left(\frac{\mu}{\rho_0}\right)\Delta\mathbf{p} + \rho_0\,\mathbf{g} = \frac{\partial}{\partial t}\{\mathbf{p}\} + \frac{1}{\rho_0}\operatorname{div}\{\mathbf{p}\otimes\mathbf{p}\}, \tag{10.5.22}$$

where

$$\mathbf{p} := \rho_0\,\mathbf{v}. \tag{10.5.23}$$

Using Results 10.5.1 we have the following

Corollary 10.5.1.

$$\langle \Delta \mathbf{p} \rangle = \Delta\{\langle \mathbf{p} \rangle\} + \frac{1}{V}\operatorname{div}\left\{\int_S \mathbf{p}\otimes\mathbf{n}\,dA\right\} + \frac{1}{V}\int_S (\nabla\mathbf{p})\mathbf{n}\,dA, \tag{10.5.24}$$

$$\langle \frac{\partial \mathbf{p}}{\partial t} \rangle = \frac{\partial}{\partial t}\{\langle \mathbf{p} \rangle\} - \frac{1}{V}\int_S (\mathbf{p}\otimes\mathbf{v})\mathbf{n}\,dA, \tag{10.5.25}$$

and
$$\langle \operatorname{div}\{\mathbf{p}\otimes\mathbf{p}\} \rangle = \operatorname{div}\{\langle \mathbf{p}\otimes\mathbf{p} \rangle\} + \frac{1}{V}\int_S (\mathbf{p}\otimes\mathbf{p})\mathbf{n}\,dA. \tag{10.5.26}$$

There remains only the averaging of the constant field $\rho_0\,\mathbf{g}$. We have

$$\begin{aligned}\langle \rho_0\,\mathbf{g}\rangle(\mathbf{x}) &:= \int_{\mathcal{E}} \rho_0\,\mathbf{g}\,w(\mathbf{y}-\mathbf{x})dV_{\mathbf{y}} = \rho_0\,\mathbf{g}\int_R w(\mathbf{y}-\mathbf{x})dV_{\mathbf{y}} \\ &= \left(\frac{\rho_0\,\mathbf{g}}{V}\right)\int_{R\cap C(\mathbf{x})} 1\,dV_{\mathbf{y}} \\ &= \left(\frac{\rho_0\,\mathbf{g}}{V}\right)\times \operatorname{vol}(R\cap C(\mathbf{x})).\end{aligned} \tag{10.5.27}$$

Here [recall (4.4.2)] $\operatorname{vol}(R\cap C(\mathbf{x}))$ is the volume of the pore space to be found within cell $C(\mathbf{x})$; for brevity subscript $\mathcal{U}$ has been suppressed. Accordingly,

$$\frac{\operatorname{vol}(R\cap C(\mathbf{x}))}{V} = \nu(\mathbf{x}), \tag{10.5.28}$$

the porosity at $\mathbf{x}$ appropriate to the choice of REV. Hence, from (10.5.27) and (10.5.28),

$$\langle \rho_0\,\mathbf{g}\rangle = \rho_0\,\nu\mathbf{g}. \tag{10.5.29}$$

From (10.5.25), (10.5.26), and (10.5.23), and noting that $\mathbf{v}\,.\,\mathbf{n} = 0$ on S (Why?),

$$\langle \frac{\partial \mathbf{p}}{\partial t} + \frac{1}{\rho_0}\operatorname{div}\{\mathbf{p}\otimes\mathbf{p}\} \rangle = \frac{\partial}{\partial t}\{\langle \mathbf{p} \rangle\} + \frac{1}{\rho_0}\operatorname{div}\{\langle \mathbf{p}\otimes\mathbf{p} \rangle\}. \tag{10.5.30}$$

It follows from (10.5.17), (10.5.24), (10.5.29), and (10.5.30) that averaging (10.5.22) yields

$$\begin{aligned}-\nabla\{\langle P \rangle\} + \left(\frac{\mu}{\rho_0}\right)\Delta\{\langle \mathbf{p} \rangle\} + \rho_0\,\nu\mathbf{g} + \frac{1}{V}\left(\int_S \left(\frac{\mu}{\rho_0}\nabla\mathbf{p} - P\mathbf{1}\right)\mathbf{n}\,dA + \frac{\mu}{\rho_0}\operatorname{div}\left\{\int_S \mathbf{p}\otimes\mathbf{n}\,dA\right\}\right) \\ = \frac{\partial}{\partial t}\{\langle \mathbf{p} \rangle\} + \frac{1}{\rho_0}\operatorname{div}\{\langle \mathbf{p}\otimes\mathbf{p} \rangle\}.\end{aligned} \tag{10.5.31}$$

Remark 10.5.3. Explicit dependence upon porous structure is evident in (10.5.31) via term $\rho_0\,\nu\mathbf{g}$ together with terms

$$\mathbf{f}^{fp}(\mathbf{x}) := \frac{1}{V}\int_{S(\mathbf{x})} \left(\frac{\mu}{\rho_0}\nabla\mathbf{p} - P\mathbf{1}\right)\mathbf{n}\,dA \tag{10.5.32}$$

and [via (10.5.8) and (10.5.23)]

$$\mathbf{F}^{fp}(\mathbf{x}) := \left(\frac{\mu}{V}\right)\operatorname{div}_{\mathbf{x}}\left\{\int_{S(\mathbf{x})} \mathbf{v}^p\otimes\mathbf{n}\,dA\right\}. \tag{10.5.33}$$

In order to obtain a relation which resembles (10.5.22) as closely as possible we define

$$\mathcal{D} := \rho_0^{-1}\langle \mathbf{p}\otimes\mathbf{p}\rangle - \langle\rho\rangle^{-1}(\langle\mathbf{p}\rangle\otimes\langle\mathbf{p}\rangle), \tag{10.5.34}$$

where we note [cf. (10.5.29)] that

$$\langle\rho\rangle = \rho_0\,\nu. \tag{10.5.35}$$

It follows from (10.5.32), (10.5.33), and (10.5.34) that (10.5.31) may be written as

$$-\nabla\{\langle P\rangle\} + \left(\frac{\mu}{\rho_0}\right)\Delta\{\langle\mathbf{p}\rangle\} + \rho_0\,\nu\mathbf{g} + \mathbf{f}^{fp} + \mathbf{F}^{fp} - \operatorname{div}\mathcal{D}$$
$$= \frac{\partial}{\partial t}\{\langle\mathbf{p}\rangle\} + \operatorname{div}\{\langle\rho\rangle^{-1}(\langle\mathbf{p}\rangle\otimes\langle\mathbf{p}\rangle)\}. \tag{10.5.36}$$

Exercise 10.5.3. Verify (10.5.36).

Remark 10.5.4. Force density $\mathbf{F}^{fp}$ is associated with the change in pore space associated with the rate of deformation of the porous body, in view of the appearance of $\mathbf{v}^p$ in (10.5.33). Clearly (10.5.33) requires that

$$\mathbf{F}^{fp} = \mathbf{0} \quad \textit{if } \mathbf{v}^p = \mathbf{0} \qquad \textit{everywhere.} \tag{10.5.37}$$

In particular, $\mathbf{F}^{fp} = \mathbf{0}$ if the porous body does not move and hence, in particular, does not deform.

Remark 10.5.5. Relation (10.5.36) may be written in several equivalent ways upon introducing two candidate velocity fields $\mathbf{V}$ and $\mathbf{Q}$ via [cf. (4.2.15]

$$\mathbf{V} := \frac{\langle\mathbf{p}\rangle}{\langle\rho\rangle} \tag{10.5.38}$$

and

$$\mathbf{Q} := \frac{\langle\mathbf{p}\rangle}{\rho_0}. \tag{10.5.39}$$

Of course, from (10.5.35), (10.5.38), and (10.5.39),

$$\mathbf{Q} = \nu\mathbf{V}. \tag{10.5.40}$$

Velocity $\mathbf{V}$ is the natural choice, upon recalling (4.2.15), and results in (10.5.36) taking the form

$$-\nabla\{\langle P\rangle\} + \mu\Delta\{\nu\mathbf{V}\} + \rho_0\,\nu\mathbf{g} + \mathbf{f}^{fp} + \mathbf{F}^{fp} - \operatorname{div}\mathcal{D}$$
$$= \frac{\partial}{\partial t}\{\langle\rho\rangle\mathbf{V}\} + \operatorname{div}\{\langle\rho\rangle\mathbf{V}\otimes\mathbf{V}\}. \tag{10.5.41}$$

The equivalent formulation in terms of $\mathbf{Q}$ is

$$-\nabla\{\langle P\rangle\} + \mu\Delta\mathbf{Q} + \rho_0\,\nu\mathbf{g} + \mathbf{f}^{fp} + \mathbf{F}^{fp} - \operatorname{div}\mathcal{D}$$
$$= \frac{\partial}{\partial t}\{\rho_0\,\mathbf{Q}\} + \operatorname{div}\left\{\left(\frac{\rho_0}{\nu}\right)\mathbf{Q}\otimes\mathbf{Q}\right\}. \tag{10.5.42}$$

Relations (10.5.41) and (10.5.42) may be compared with (10.5.7). To obtain the analogue of incompressibility criterion (10.5.5), note that from (10.5.19) and (10.5.8)

$$0 = \langle\operatorname{div}\mathbf{v}\rangle = \operatorname{div}\{\langle\mathbf{v}\rangle\} + \frac{1}{V}\int_S \mathbf{v}^p.\mathbf{v}\,dV. \tag{10.5.43}$$

However, via (10.5.23), and (10.5.39),

$$\operatorname{div}\{\langle \mathbf{v}\rangle\} = \operatorname{div}\left\{\frac{\langle \rho_0 \mathbf{v}\rangle}{\rho_0}\right\} = \operatorname{div}\left\{\frac{\langle \mathbf{p}\rangle}{\rho_0}\right\} = \operatorname{div}\mathbf{Q}. \tag{10.5.44}$$

Accordingly, from (10.5.43) and (10.5.44),

$$\operatorname{div}\mathbf{Q} + \frac{1}{V}\int_S \mathbf{v}^P.\mathbf{n}\,dA = 0. \tag{10.5.45}$$

Further, from (10.5.40),

$$\operatorname{div}\mathbf{Q} = \nu\operatorname{div}\mathbf{V} + \nabla\nu\,.\,\mathbf{V}. \tag{10.5.46}$$

Remark 10.5.6. If

$$\mathbf{H}(\mathbf{x}) := \frac{1}{V}\int_{S(\mathbf{x})} \mathbf{v}^p \otimes \mathbf{n}\,dA, \tag{10.5.47}$$

then from (10.5.33)

$$\mathbf{F}^{fp} = \mu\operatorname{div}\mathbf{H}, \tag{10.5.48}$$

and (4.5.45) may be written as

$$\operatorname{div}\mathbf{Q} + \operatorname{tr}\mathbf{H} = 0. \tag{10.5.49}$$

In what follows, only flows through *non-deforming stationary* porous bodies will be considered. Accordingly, since $\mathbf{v}^p = \mathbf{0}$ everywhere,

$$\mathbf{H} = \mathbf{O} \tag{10.5.50}$$

and hence, from (10.5.48) and (10.5.49),

$$\mathbf{F}^{fp} = \mathbf{0} \tag{10.5.51}$$

and

$$\operatorname{div}\mathbf{Q} = 0. \tag{10.5.52}$$

In such case (10.5.42) becomes

$$-\nabla\{\langle P\rangle\} + \mu\Delta\mathbf{Q} + \rho_0\,\nu\mathbf{g} + \mathbf{f}^{fp} - \operatorname{div}\boldsymbol{\mathcal{D}} = \frac{\partial}{\partial t}\{\rho_0\,\mathbf{Q}\} + \operatorname{div}\left\{\frac{\rho_0}{\nu}\mathbf{Q}\otimes\mathbf{Q}\right\}. \tag{10.5.53}$$

Remark 10.5.7. Notice that (10.5.53) may be written as

$$\operatorname{div}\mathbf{T} + \rho_0\,\nu\mathbf{g} + \mathbf{f}^{fp} = \frac{\partial}{\partial t}\{\rho_0\,\mathbf{Q}\} + \operatorname{div}\left\{\frac{\rho_0}{\nu}\mathbf{Q}\otimes\mathbf{Q}\right\}, \tag{10.5.54}$$

where [cf. (10.5.6)]

$$\mathbf{T} := -\langle P\rangle\mathbf{1} + \mu(\nabla\mathbf{Q} + (\nabla\mathbf{Q})^T) - \boldsymbol{\mathcal{D}}. \tag{10.5.55}$$

This is a consequence of the identities

$$\operatorname{div}\{(\nabla\mathbf{Q})^T\} = \nabla\{\operatorname{div}\mathbf{Q}\}, \qquad \text{and} \qquad \operatorname{div}\{\langle P\rangle\mathbf{1}\} = \nabla\{\langle P\rangle\}, \tag{10.5.56}$$

and (10.5.52).

In the case of *creeping flow* (10.5.7) is approximated by

$$-\nabla P + \mu\Delta\mathbf{v} + \rho_0\,\mathbf{g} = \frac{\partial}{\partial t}\{\rho_0\,\mathbf{v}\}, \tag{10.5.57}$$

corresponding to small values of **v** and consequence neglect of the last term in (10.5.7). The relevant forms of (10.5.41) and (10.5.42) are thus

$$-\nabla\{\langle P\rangle\}+\mu\Delta\{\nu\mathbf{V}\}+\rho_0\,\nu\mathbf{g}+\mathbf{f}^{fp}=\frac{\partial}{\partial t}\{\langle\rho\rangle\mathbf{V}\} \tag{10.5.58}$$

and

$$-\nabla\{\langle P\rangle\}+\mu\Delta\mathbf{Q}+\rho_0\,\nu\mathbf{g}+\mathbf{f}^{fp}=\frac{\partial}{\partial t}\{\rho_0\,\mathbf{Q}\}, \tag{10.5.59}$$

respectively.

Exercise 10.5.4. Verify relations (10.5.58) and (10.5.59), taking account of (10.2.4), (10.5.23), and (10.5.38).

In view of (10.5.52), choice **Q** (rather than **V**) results in the closest formal match, at least for creeping flow, with the appropriate form of the Navier–Stokes equations [namely (10.5.57) with (10.5.5)].

Remark 10.5.8. The velocity field **Q** has a very important physical interpretation in the present context, namely flow of an incompressible liquid through a rigid stationary body that it saturates. To see this it is instructive to consider mass conservation of *any* fluid which saturates a stationary rigid body. In such case the (ϵ_1-scale) continuity equation [see (4.2.14)]

$$\frac{\partial\rho}{\partial t}+\operatorname{div}\mathbf{p}=0 \tag{10.5.60}$$

holds throughout all pore space, irrespective of whether the fluid is liquid, gaseous, or a mixture of the two separated by interfaces. Averaging (10.5.60) using (10.5.19) with $\mathbf{u}=\mathbf{p}$ and noting the analogue of (10.5.21) for scalar fields (and hence for ρ in particular) we obtain

$$\frac{\partial}{\partial t}\{\langle\rho\rangle\}+\operatorname{div}\{\langle\mathbf{p}\rangle\}=0. \tag{10.5.61}$$

Exercise 10.5.5. Write down the analogue of (10.5.21) for a scalar field φ [notice that this follows from (10.5.21) upon setting $\mathbf{u}=\varphi\mathbf{k}$, where $\mathbf{k}\neq\mathbf{0}$ is any fixed vector], and verify (10.5.61).

Integrating (10.5.61) over any fixed region $\mathcal{P}$ yields

$$\frac{\partial}{\partial t}\left\{\int_{\mathcal{P}}\langle\rho\rangle dV\right\}=-\int_{\partial\mathcal{P}}\langle\mathbf{p}\rangle\,.\,\mathbf{n}\,dA \tag{10.5.62}$$

on using the divergence theorem. Since from (10.5.35)

$$\begin{aligned}\int_{\mathcal{P}}\langle\rho\rangle dV=\int_{\mathcal{P}}\rho_0\,\nu\,dV&=\rho_0\times\text{ pore volume within }\mathcal{P}\\&=\text{mass of fluid within }\mathcal{P},\end{aligned} \tag{10.5.63}$$

it follows from (10.5.62) that

$$-\int_{\partial\mathcal{P}}\langle\mathbf{p}\rangle\,.\,\mathbf{n}\,dA=\text{ rate of change of mass of fluid within }\mathcal{P}. \tag{10.5.64}$$

Accordingly $\langle\mathbf{p}\rangle\,.\,\mathbf{n}$ has the interpretation of fluid mass flux across $\partial\mathcal{P}$ in the direction of the outward unit normal **n** to $\partial\mathcal{P}$.

Exercise 10.5.6. Convince yourself of the preceding statement. (It is the direct analogue of the interpretation of $\rho\mathbf{v}.\mathbf{n}$ when the continuity equation is integrated over $\mathcal{P}$.)

Accordingly $\langle\mathbf{p}\rangle.\mathbf{n}\,\Delta A$ is interpreted, for small areas ΔA, as delivering the mass flow per unit time across a planar surface of area ΔA to which $\mathbf{n}$ is a normal, in the direction of $\mathbf{n}$. From (10.5.39) this flow rate is $\rho_0\,\mathbf{Q}.\mathbf{n}\,\Delta A$ in the context of incompressible flow. Thus the *volume* of fluid crossing 'ΔA' per unit time in the $\mathbf{n}$ direction is $\mathbf{Q}.\mathbf{n}\,\Delta A$. For this reason

$\mathbf{Q}$ is termed the *volumetric flux vector.*

Now consider the term $\mathbf{f}^{fp}$. This represents a force density associated with the resistance to fluid flow from the porous body: in the event that $\mathbf{F}^{fp}=\mathbf{0}$ it is the sole such density. From (10.5.32) we may write

$$\mathbf{f}^{fp}=\hat{\mathbf{f}}^{fp}+\mu\check{\mathbf{f}}^{fp}, \tag{10.5.65}$$

where

$$\hat{\mathbf{f}}^{fp}:=-\frac{1}{V}\int_S P\mathbf{n}\,dA \tag{10.5.66}$$

and [recall (10.5.23)]

$$\check{\mathbf{f}}^{fp}:=\frac{1}{V}\int_S(\nabla\mathbf{v})\mathbf{n}w\,dA. \tag{10.5.67}$$

The REV *intrinsic* average pressure $\bar{P}$ (i.e., the REV-scale average pressure computed only over pore space) is given by

$$\begin{aligned}\bar{P}(\mathbf{x}) &:= \frac{1}{\mathrm{vol}(R\cap C(\mathbf{x}))}\int_R P(\mathbf{y})d\mathbf{y}\\ &= \frac{1}{\nu(\mathbf{x})V}\int_R P(\mathbf{y})d\mathbf{y}=\frac{1}{\nu(\mathbf{x})}\langle P\rangle(\mathbf{x}),\end{aligned} \tag{10.5.68}$$

via (10.5.28). That is,

$$\bar{P}=\frac{1}{\nu}\langle P\rangle. \tag{10.5.69}$$

Writing

$$\hat{\mathbf{f}}^{fp}(\mathbf{x})=-\frac{1}{V}\int_{S(\mathbf{x})}\{\bar{P}(\mathbf{x})+[P(\mathbf{y})-\bar{P}(\mathbf{x})]\}\mathbf{n}(\mathbf{y})dA_{\mathbf{y}} \tag{10.5.70}$$

yields

$$\hat{\mathbf{f}}^{fp}(\mathbf{x})=-\frac{\bar{P}(\mathbf{x})}{V}\int_{S(\mathbf{x})}\mathbf{n}(\mathbf{y})dA_{\mathbf{y}}+\tilde{\mathbf{f}}^{fp}(\mathbf{x}), \tag{10.5.71}$$

where

$$\tilde{\mathbf{f}}^{fp}(\mathbf{x}):=\int_{S(\mathbf{x})}(P(\mathbf{y})-\bar{P}(\mathbf{x}))\mathbf{n}(\mathbf{y})dA_{\mathbf{y}}. \tag{10.5.72}$$

The first integral in (10.5.71) may be simplified by noting the following:

Result 10.5.2.

$$\frac{1}{V}\int_{S(\mathbf{x})}\mathbf{n}(\mathbf{y})dA_{\mathbf{y}}=\int_{S(\mathbf{x})}w(\mathbf{y}-\mathbf{x})\mathbf{n}(\mathbf{y})dA_{\mathbf{y}}=-\nabla\nu(\mathbf{x}). \tag{10.5.73}$$

Proof: Consider [see Section 10.1, Remark 10.2.1 and $(10.2.23)_2$]

$$R^+:={}^pB^{\epsilon_2}-{}^pB^{\epsilon_1}. \tag{10.5.74}$$

This region consists of all pore space together with an 'enveloping' region between the ϵ_2-scale porous body boundary $\partial^p B^{\epsilon_2}$ and that part, Σ say, of the ϵ_1-scale boundary which does not constitute pore walls. We assume that R^+ is a regular region.[5] Accordingly (here $\mathbf{n}^+$ denotes the outward unit normal to ∂R^+),

$$\int_{R^+} \text{div}\{w(\mathbf{y}-\mathbf{x})\mathbf{1}\}dV_{\mathbf{y}} = \int_{\partial R^+} w(\mathbf{y}-\mathbf{x})\mathbf{n}^+(\mathbf{y})dA_{\mathbf{y}}$$
$$= \int_{\partial R^+ \cap C(\mathbf{x})} w(\mathbf{y}-\mathbf{x})\mathbf{n}^+(\mathbf{y})dA_{\mathbf{y}}. \tag{10.5.75}$$

The last step here is a consequence of the vanishing of $w(\mathbf{y}-\mathbf{x})$ outside $C(\mathbf{x})$. Notice that

$$\partial R^+ = \partial^p B^{\epsilon_2} \cup \partial R \cup \Sigma. \tag{10.5.76}$$

Thus if
$$C(\mathbf{x}) \cap R^+ = C(\mathbf{x}) \cap R, \tag{10.5.77}$$

then
$$\partial R^+ \cap C(\mathbf{x}) = \partial R \cap C(\mathbf{x}) = S(\mathbf{x}). \tag{10.5.78}$$

That is, if $\mathbf{x}$ lies sufficiently within ${}^pB^{\epsilon_2}$ to ensure (10.5.77), then (10.5.75) becomes, from (10.5.78), and noting that $\mathbf{n}^+ = \mathbf{n}$ on ∂R and hence on $S(\mathbf{x})$,

$$\int_{R^+} \text{div}\{w(\mathbf{y}-\mathbf{x})\mathbf{1}\}dV_{\mathbf{y}} = \int_{S(\mathbf{x})} w(\mathbf{y}-\mathbf{x})\mathbf{n}(\mathbf{y})dA_{\mathbf{y}} = \frac{1}{V}\int_{S(\mathbf{x})} \mathbf{n}(\mathbf{y})dA_{\mathbf{y}}. \tag{10.5.79}$$

Further,
$$\int_{R^+} \text{div}_{\mathbf{y}}\{w(\mathbf{y}-\mathbf{x})\mathbf{1}\}dV_{\mathbf{y}} = \int_{R^+} \nabla_{\mathbf{y}} w(\mathbf{y}-\mathbf{x})dV_{\mathbf{y}}$$
$$= -\int_{R^+} \nabla_{\mathbf{x}} w(\mathbf{y}-\mathbf{x})dV_{\mathbf{y}} = -\nabla_{\mathbf{x}}\left\{\int_{R^+} w(\mathbf{y}-\mathbf{x})dV_{\mathbf{y}}\right\}$$
$$= -\nabla_{\mathbf{x}}\left\{\int_{R} w(\mathbf{y}-\mathbf{x})dV_{\mathbf{y}}\right\} = -\nabla\nu(\mathbf{x}). \tag{10.5.80}$$

The result follows from (10.5.79) and (10.5.80).

Remark 10.5.9. Region R^+ was introduced in order to be able to invoke the divergence theorem: notice that pore space R does not constitute a regular region since its boundary is not closed.

From (10.5.71) and (10.5.73),

$$\hat{\mathbf{f}}^{fp} = \bar{P}\nabla\nu + \tilde{\mathbf{f}}^{fp}. \tag{10.5.81}$$

Accordingly, with the aim of re-writing (10.5.58) and (10.5.59), we note that from (10.5.65), (10.5.69), and (10.5.81),

$$-\nabla\{\langle P\rangle\} + \mathbf{f}^{fp} = -\nabla\{\langle P\rangle\} + \hat{\mathbf{f}}^{fp} + \mu\check{\mathbf{f}}^{fp}$$
$$= -\nabla\{\nu\bar{P}\} + \bar{P}\nabla\nu + \tilde{\mathbf{f}}^{fp} + \mu\check{\mathbf{f}}^{fp}$$
$$= -\nu\nabla\bar{P} + \tilde{\mathbf{f}}^{fp} + \mu\check{\mathbf{f}}^{fp}. \tag{10.5.82}$$

[5] This is a region for which the divergence theorem holds. See Kellogg [21].

At this point (10.5.58) and (10.5.59) may be expressed as

$$-\nu\nabla\bar{P}+\tilde{\mathbf{f}}^{fp}+\mu\check{\mathbf{f}}^{fp}+\mu\Delta\{\nu\mathbf{V}\}+\rho_0\,\nu\mathbf{g}=\frac{\partial}{\partial t}\{\langle\rho\rangle\mathbf{V}\} \tag{10.5.83}$$

and

$$-\nu\nabla\bar{P}+\tilde{\mathbf{f}}^{fp}+\mu\check{\mathbf{f}}^{fp}+\mu\Delta\mathbf{Q}+\rho_0\,\nu\mathbf{g}=\frac{\partial}{\partial t}\{\rho_0\,\mathbf{Q}\}. \tag{10.5.84}$$

If $\mathbf{V}$ does not change with time (but may change with location), the flow is termed *steady*. In such case, from (10.5.40) $\mathbf{Q}$ does not change with time: recall that we are presently considering a stationary rigid body, so that ν does not change with time. Thus, for steady flows, the right-hand sides of (10.5.83) and (10.5.84) vanish, yielding

$$-\nu\nabla\bar{P}+\tilde{\mathbf{f}}^{fp}+\mu\check{\mathbf{f}}^{fp}+\mu\Delta\{\nu\mathbf{V}\}+\rho_0\,\nu\mathbf{g}=\mathbf{0} \tag{10.5.85}$$

and

$$-\nu\nabla\bar{P}+\tilde{\mathbf{f}}^{fp}+\mu\check{\mathbf{f}}^{fp}+\mu\Delta\mathbf{Q}+\rho_0\,\nu\mathbf{g}=\mathbf{0}. \tag{10.5.86}$$

Term $\mu\check{\mathbf{f}}^{fp}$ [see (10.5.67)] constitutes a force density associated with viscous drag exerted by pore walls, and is modelled in terms of a permeability tensor. Specifically, we make the following

Modelling Assumption: There exists an invertible tensor field $\mathbf{K}$ (the *permeability tensor*) such that

$$\check{\mathbf{f}}^{fp}=-\mathbf{K}^{-1}\nu^2(\mathbf{V}-\mathbf{V}_B), \tag{10.5.87}$$

where $\mathbf{V}_B$ denotes the ϵ_2-scale (i.e., REV-scale) porous body velocity.

Thus if the body is stationary

$$\check{\mathbf{f}}^{fp}=-\mathbf{K}^{-1}\nu\mathbf{Q}, \tag{10.5.88}$$

and (10.5.86) may be written as

$$-\nabla\bar{P}+\frac{1}{\nu}\tilde{\mathbf{f}}^{fp}-\mu\mathbf{K}^{-1}\mathbf{Q}+\frac{\mu}{\nu}\Delta\mathbf{Q}+\rho_0\,\mathbf{g}=\mathbf{0}. \tag{10.5.89}$$

In the event that $\tilde{\mathbf{f}}^{fp}$ is negligible, then (10.5.89) becomes what is usually termed the *Brinkman equation*, with the addition of a gravitational term:

$$-\nabla\bar{P}-\mu\mathbf{K}^{-1}\mathbf{Q}+\frac{\mu}{\nu}\Delta\mathbf{Q}+\rho_0\,\mathbf{g}=\mathbf{0}. \tag{10.5.90}$$

If $\mu/\nu\,\Delta\mathbf{Q}$ and $\rho_0\,\mathbf{g}$ are negligible in comparison with other terms, then (10.5.90) reduces to

$$\nabla\bar{P}=-\mu\mathbf{K}^{-1}\mathbf{Q}. \tag{10.5.91}$$

If the porous body is isotropic at REV scale, then, for some real-valued function k,

$$\mathbf{K}=k\mathbf{1}, \tag{10.5.92}$$

and (10.5.91) becomes *Darcy's 'law'*:

$$\nabla\bar{P}=-\frac{\mu}{k}\mathbf{Q}. \tag{10.5.93}$$

Remark 10.5.10. Relations equivalent to (10.5.90) appear in the literature in alternate forms as a consequence of differing definitions of the average pressure and the permeability tensor. For example, writing

$$\hat{\mathbf{K}} := \frac{1}{\nu}\mathbf{K} \tag{10.5.94}$$

enables (10.5.85) to be written as

$$-\nabla\bar{P} + \frac{1}{\nu}\tilde{\mathbf{f}}^{fp} - \mu\hat{\mathbf{K}}^{-1}(\mathbf{V}-\mathbf{V}_B) + \frac{\mu}{\nu}\Delta\{\nu\mathbf{V}\} + \rho_0\,\mathbf{g} = \mathbf{0}. \tag{10.5.95}$$

If the body is stationary, $\tilde{\mathbf{f}}^{fp}$ is negligible, and *if the porosity is constant*, then (10.5.95) becomes

$$-\nabla\bar{P} - \mu\hat{\mathbf{K}}^{-1}\mathbf{V} + \mu\Delta\mathbf{V} + \rho_0\,\mathbf{g} = \mathbf{0}. \tag{10.5.96}$$

This is a form of the Brinkman equation which holds in the case of constant porosity. Further, in such case (10.5.91) may be written as

$$\nabla\{\langle P\rangle\} = -\mu\hat{\mathbf{K}}^{-1}\mathbf{V} \tag{10.5.97}$$

which in the case of isotropy takes the form

$$\nabla\{\langle P\rangle\} = -\frac{\mu}{\hat{k}}\mathbf{V}, \tag{10.5.98}$$

where here

$$\hat{\mathbf{K}} = \hat{k}\mathbf{1}. \tag{10.5.99}$$

Exercise 10.5.7. Derive (10.9.97) from (10.5.69), (10.5.40), and (10.5.91).

Remark 10.5.11. The formal similarity between the Brinkman equations [(10.5.90) and (10.5.96)] and the simplifications [(10.5.91) and (10.5.98)] make it essential that the physical interpretations of the pressure average and velocity field be precise when using such relations. Specifically, (10.5.90) and (10.5.96) may be expressed as

$$-\nabla P_{av} - \mu\mathcal{K}^{-1}\mathbf{v} + \alpha\Delta\mathbf{v} + \rho_0\,\mathbf{g} = \mathbf{0}, \tag{10.5.100}$$

where P_{av} denotes average pressure, $\mathcal{K}$ is a permeability tensor, and $\mathbf{v}$ is a velocity. Constant α is the key to interpreting just which average and velocity are intended. If $\alpha = \mu/\nu$, then we must have, from (10.5.90), $P_{av} = \bar{P}$ and $\mathbf{v} = \mathbf{Q}$, while if $\alpha = \mu$, then, from (10.5.96), $P_{av} = \bar{P}$ and $\mathbf{v} = \mathbf{V}$. In the case of Darcy relations (10.5.93), and (10.5.98), and their generalisations (10.5.91) and (10.5.97), we have forms

$$\nabla P_{av} = -\frac{\mu}{k}\mathbf{v} \qquad \text{and} \qquad \nabla P_{av} = -\mu\mathcal{K}^{-1}\mathbf{v}. \tag{10.5.101}$$

Here it is essential to note that if P_{av} is selected to be $\bar{P}$, then $\mathbf{v}$ is to be identified as $\mathbf{Q}$, while choice $P_{av} = \langle P\rangle$ requires that $\mathbf{v}$ be identified with $\mathbf{V}$.

Remark 10.5.12. Form (10.5.93) may be regarded as more practical than form (10.5.98), based on the following considerations. Since $\bar{P}$ is an average computed with respect to fluid-filled pore space within an (ϵ_2-scale) REV, its value just inside the ϵ_2-scale boundary of the body must be expected to differ little from the *external* pressure value at the corresponding part of this boundary. Now consider unidirectional REV-scale steady creeping flow of incompressible fluid through an isotropic cylindrical porous body that it saturates. If the cylinder is of length L and cross-sectional area A, then $\nabla\bar{P}$ can be approximated by $\Delta P/L$, where ΔP denotes the difference in external pressure at end faces. Further, if volume V_0 of liquid is discharged from one end in time Δt, then the creep speed Q is given by $QA = V_0/\Delta t$. (See the discussion of **Q** following Exercise 10.5.6.) Thus the global analogue of (10.5.93) is

$$\frac{\Delta P}{L} = -\frac{\mu}{k} \cdot \frac{V_0}{A\Delta t}. \qquad (10.5.102)$$

Here μ is the known viscosity of the fluid, and ΔP, L, V_0,A, and Δt are measurable quantities. This relation thus delivers a global permeability value k for the uniaxial flow direction. The validity of the model can be tested by applying different pressure differences and checking the corresponding mass discharge rate values $\rho_0 V_0/\Delta t$.

11 Linkage of Microscopic and Macroscopic Descriptions of Material Behaviour via Cellular Averaging

11.1 Preamble

An alternative strategy for relating continuum field values to their microscopic origins is outlined. Spatial averaging of additive molecular quantities is effected in terms of 'cells'. In contrast to weighting function methodology, linear momentum balance for macroscopic regions is established before deducing the local form. The existence of a traction field on the boundary of such regions is derived via the assumption of short-range molecular interactions. The corresponding interaction stress tensor is obtained in the standard manner of continuum mechanics. Unlike balances obtained in terms of weighting functions, which hold for *any* given pair (ϵ, Δ) of length–time scales, fields obtained as cellular averages exist only if their values are somewhat insensitive to changes in ϵ, Δ, and cell shape.

11.2 Cellular Averaging

Recall Subsection 4.4.1 in which selection of an appropriate weighting function [namely a mollified version of relations (4.4.4)] delivered spatial cellular averages. The analyses of Sections 4.2, 5.2 through 5.6, 6.2 and 6.3, 7.2 through 7.5, 8.4 through 8.6, 8.9 and 9.2 through 9.7 apply to such a choice. Thus, in particular, the standard relations [which express mass conservation (4.2.16), momentum balance (2.7.30) with $\mathbf{T} = \mathbf{T}_w$, where $\mathbf{T}_w$ is given by (5.5.20), and energy balance (6.2.75)] hold with field values defined in terms of cellular molecular averages. The associated (i.e., cell-based) notion of material point, and motion thereof, derives from the corresponding velocity field $\mathbf{v}_w$ defined in Section 5.2.

Remark 11.2.1. Of course, the distinguished ϵ-scale choice w_ϵ of weighting-function (see Section 4.3) corresponds to a very special form of cell, namely a spherical region of radius ϵ.

Remark 11.2.2. Relations obtained via weighting function methodology are mathematically precise and hold for any choice of weighting function w which satisfies W.F. 1 through 4 of Section 4.2. However, the physical interpretation of fields is a priori critically dependent upon choice w. In particular, field values must be expected to be sensitive to the length scale embodied in w. However, in situations of practical interest, it may be that field values are somewhat insensitive both to the form of w *and* to a range of length scales. For example, the mass density values at location $\mathbf{x}$ and time t may vary negligibly when computed at time t as (molecular mass content)/volume averages taken over spheres, or cubes of arbitrary orientation, centred at $\mathbf{x}$, having radii or side length ϵ, where ϵ can vary over some interval, (ϵ_1, ϵ_2) say. Indeed, the assumption of such insensitivity is often regarded to be a basis for continuum modelling (cf., e.g., Paterson [5], p. 33, fig. III.1). If such a viewpoint is adopted, it proves possible to elaborate upon the interpretation and evaluation of field values.

Bearing in mind Remark 11.2.2, we define an *ϵ-cell centred at* $\mathbf{x}$ to be a simply connected region of Euclidean space with a piece-wise smooth boundary which satisfies $\epsilon/2 < \|\mathbf{y}-\mathbf{x}\| < 3\epsilon/2$: here $\mathbf{x}$ denotes the centroid of the cell, and $\mathbf{y}$ is any point within the cell. In order to exclude pathological boundaries it is also assumed that nowhere on any subsurface should curvatures be too large, nor should there be too many component subsurfaces.

Now suppose that ψ_i denotes an additive quantity associated with point mass P_i (modelling a molecule or atom), such as its mass m_i, momentum $m_i\mathbf{v}_i$, or kinetic energy $\frac{1}{2}m_i\mathbf{v}_i^2$. Consider $\sum_i{}' \psi_i/V_\epsilon$, where the superposed prime indicates that the sum is taken only over those point masses within an ϵ-cell of volume V_ϵ. If the cell is centred at $\mathbf{x}$ and this volumetric average is $\psi(\mathbf{x})(1+\beta)$, with[1] $|\beta| << 1$, over a range of values macroscopically small yet microscopically large (say $\epsilon_1 < \epsilon < \epsilon_2$), quite independently of the cell shape (modulo the definition of an ϵ-cell), then $\psi(\mathbf{x})$ is termed an *ϵ-representative* of the volumetric average at $\mathbf{x}$. Now suppose that there exists such a representative at every point in a regular region $\mathcal{R}$ which can be decomposed into very many mutually disjoint ϵ-cells. If we make the *modelling assumption* that there exists an ϵ-representative *field* ψ which is spatially continuous on $\mathcal{R}$, then the sum of ψ_i taken over all point masses within $\mathcal{R}$ may be expressed as a Riemann sum (see Appendix B.6) which approximates the (Riemann) integral of ψ over $\mathcal{R}$. More precisely, let $\mathcal{R}$ be partitioned by a set $\mathcal{C}_a$ of ϵ-cells. Here $a \in A$, where A is an indexing set, and

$$\mathcal{R} = \bigcup_{a\in A} \mathcal{C}_a, \qquad \text{where } \mathcal{C}_a \cap \mathcal{C}_{a'} = \phi \text{ if } a \neq a'. \tag{11.2.1}$$

Then the sum

$$\sum_{P_1\in\mathcal{R}} \psi_i = \sum_{a\in A}\sum_i{}^a \psi_i = \sum_{a\in A}(\sum_i{}^a \psi_i/V_a)V_a = \sum_{a\in A}\psi(\mathbf{x}_a)V_a. \tag{11.2.2}$$

Here $\sum_i{}^a$ denotes a sum over point masses P_i in $\mathcal{C}_a$, $V_a := \text{vol}\{\mathcal{C}_a\}$ and $\mathbf{x}_a$ denotes the centroid of $\mathcal{C}_a$. The final expression is a Riemann sum which approximates the integral of ψ over $\mathcal{R}$. We make the identification

$$\int_{\mathcal{R}} \psi(\mathbf{x})dV_{\mathbf{x}} \leftrightarrow \sum_{a\in A}\psi(\mathbf{x}_a)V_a = \sum_{P_i\in\mathcal{R}} \psi_i. \tag{11.2.3}$$

[1] Of course, β will depend upon the quantity in question, $\mathbf{x}$, and the cell concerned. What is intended here is that, for the range of cell shapes and ϵ values stated, $|\beta|$ should be uniformly small.

Remark 11.2.3. For notational reasons it is helpful to write

$$\psi(\mathbf{x}) =: \lim_{\epsilon}\left\{\sum_i{}' \psi_i/V_\epsilon\right\} \tag{11.2.4}$$

and term this the 'ϵ-limit' of the sum/volume ratio. The superposed prime attached to the summation symbol is used to indicate that the sum is taken over an ϵ-cell (understood to be centred at $\mathbf{x}$).

Choices $\psi_i = m_i$ and $\psi_i = m_i\mathbf{v}_i$ are assumed to yield ϵ-representative smooth fields of mass and momentum densities ρ and $\mathbf{p}$. In particular,

$$\rho(\mathbf{x},t) := \lim_{\epsilon}\left\{\sum_i{}' m_i/V_\epsilon\right\} \tag{11.2.5}$$

and

$$\mathbf{p}(\mathbf{x},t) := \lim_{\epsilon}\left\{\sum_i{}' m_i\mathbf{v}_i(t)/V_\epsilon\right\}. \tag{11.2.6}$$

The associated velocity field $\mathbf{v}$ is defined by

$$\mathbf{v} := \mathbf{p}/\rho \tag{11.2.7}$$

[cf. (4.2.15)]. The motion corresponding to the situation at time t_0 is the solution $\boldsymbol{\chi}_{t_0}$ to the initial value problem [cf. (5.2.3) and (5.2.4)]

$$\dot{\boldsymbol{\chi}}_{t_0}(\hat{\mathbf{x}},t) = \mathbf{v}(\boldsymbol{\chi}_{t_0}(\hat{\mathbf{x}},t),t) \tag{11.2.8}$$

with

$$\boldsymbol{\chi}_{t_0}(\hat{\mathbf{x}},t_0) = \hat{\mathbf{x}}. \tag{11.2.9}$$

Now consider the mass within a simply connected region of macroscopic proportion which deforms with motion $\boldsymbol{\chi}_{t_0}$. Accordingly, at any instant t, any point $\mathbf{x}$ on the boundary of this region is moving with the velocity of the mass centre of those molecules which at this instant lie within an ϵ-cell centred at $\mathbf{x}$. [This follows from (11.2.5), (11.2.6), and (11.2.7); see also Remark 4.3.1.] Such a motion can be regarded as minimising net mass exchange between matter within the region and its exterior and motivates the *assumption* that such a mass is (within the approximative scheme of ϵ-cell methodology) conserved. If R denotes the region at time t_0, and

$$R_t := \boldsymbol{\chi}_{t_0}(R,t), \tag{11.2.10}$$

then it follows that [cf. (2.5.5)]

$$\int_R \rho(\hat{\mathbf{x}},t_0)dV_{\hat{\mathbf{x}}} = \int_{R_t} \rho(\mathbf{x},t)dV_{\mathbf{x}}. \tag{11.2.11}$$

The further assumption that $\boldsymbol{\chi}_{t_0}(.,t){:}R \to R_t$ is a class C^1 bijection yields the continuity equation (2.5.16) precisely via the argument presented in Section 2.5.

In respect of linear momentum balance, consider

$$\boldsymbol{\psi}_i = \sum_{j\neq i}\mathbf{f}_{ij}, \tag{11.2.12}$$

where, as before, $\mathbf{f}_{ij}$ denotes the force exerted by P_j upon P_i, and the sum is taken over all molecules of the system in question other than P_i. Suppose [cf. (11.2.4)] that

$$\mathbf{f}(\mathbf{x},t) := \lim_{\epsilon}\left\{\sum_i{}' \boldsymbol{\psi}_i(t)/V_\epsilon\right\} = \lim_{\epsilon}\left\{\sum_i{}' \sum_{j\neq i}\mathbf{f}_{ij}(t)/V_\epsilon\right\} \tag{11.2.13}$$

makes sense. Now

$$\sum_i{}' \sum_{j\neq i}\mathbf{f}_{ij} = \sum_{i\neq j}{}'\sum{}' \mathbf{f}_{ij} + \sum_i{}' \sum_j{}'' \mathbf{f}_{ij}. \tag{11.2.14}$$

A primed sum indicates addition over molecules in a specific ϵ-cell (here centred at $\mathbf{x}$); a double-primed sum is used to indicate a sum over all molecules *not* in the specific ϵ-cell. Since

$$\sum_{i\neq j}{}'\sum{}' \mathbf{f}_{ij} = \frac{1}{2}\sum_{i\neq j}{}'\sum{}' (\mathbf{f}_{ij}+\mathbf{f}_{ji}) \tag{11.2.15}$$

it follows that such a sum vanishes if interactions are pairwise balanced [cf. (5.6.6)]. However, it suffices for our purposes (of obtaining a statement of linear momentum balance and establishing the existence of an interaction stress tensor) to make the following two mild assumptions concerning the nature of interactions:

I.1. *The net self-force associated with particles in any microscopically large*[2] *region is negligible.*

I.2. $\sum_\ell \mathbf{f}_{i\ell} = \sum_n \mathbf{f}_{in}$, *where* $P_iP_n < \delta << \epsilon$.

Exercise 11.2.1. Let

$$\mathcal{F}(R_1,R_2) := \sum_{P_i\in R_1}\sum_{P_j\in R_2}\mathbf{f}_{ij} \tag{11.2.16}$$

denote the force exerted by molecules in region R_2 upon those in region R_1. Suppose now that R_1 and R_2 are mutually disjoint, microscopically large regions. Show that if $R := R_1 \cup R_2$, then

$$\mathcal{F}(R,R) = \mathcal{F}(R_1,R_1) + \mathcal{F}(R_1,R_2) + \mathcal{F}(R_2,R_1) + \mathcal{F}(R_2,R_2). \tag{11.2.17}$$

Deduce from assumption I.1 that, modulo negligible contributions,

$$\mathcal{F}(R_2,R_1) = -\mathcal{F}(R_1,R_2). \tag{11.2.18}$$

Remark 11.2.4. Result (11.2.18) shows that assumption I.1 implies, at the level of net interactions between matter in ϵ-cells or larger regions, that action and reaction are equal and opposite.

Remark 11.2.5. Assumption I.2., which requires that interactions be *formally* of short range, is much weaker than the *actual* short-range stipulation

$$\mathbf{f}_{i\ell} = \mathbf{0} \qquad \text{if } P_iP_\ell > \delta$$

[2] That is, any region which contains very many molecules.

since *individual* interactions may be admitted which have much greater effective range. The motivation here stems from charged particle interactions: these may have no finite range yet effectively result in assumption I.2, within our approximative scheme, by judicious book-keeping in balancing the forces on P_i due to collections of particles of different charges and polarities located at distances in excess of δ from P_i. Assumption I.2 is consistent with what is known about molecular interactions: these are termed very long range if $\delta \sim 10^{-7}\,\mathrm{m}(= 10^3\,\text{Å})$. (It is helpful here to note that typical nearest-neighbour molecular separations in condensed matter are less than 10 Å.)

Let R denote any region of macroscopic dimensions within that region in which ρ and $\mathbf{p}$ are defined. Here, and in what follows, it also will be assumed that ∂R is piecewise-smooth and has an outward unit normal which varies negligibly at the ϵ-length scale on the smooth component subsurfaces of ∂R. Considering the resultant force on matter in R, we have, as a consequence of assumptions I.1 and I.2,

$$\sum_{P_i \,\text{in}\, R} \sum_{j \neq i} \mathbf{f}_{ij} = \sum_{\substack{P_i, P_j \,\text{in}\, R \\ j \neq i}} \sum \mathbf{f}_{ij} + \sum_{P_i \,\text{in}\, R} \sum_{P_j \,\text{not in}\, R} \mathbf{f}_{ij}$$

$$= \mathbf{0} + \sum_i \sum_j \mathbf{f}_{ij}, \tag{11.2.19}$$

where the final sum is over those molecules P_i in R and P_j outside R *which both lie within a distance* δ *from* ∂R. Consequently, the only non-vanishing contribution derives from molecules which lie between inner and outer 'δ-envelopes' of ∂R; that is, surfaces ($\partial R^{-\delta}$ and $\partial R^{+\delta}$, respectively, say) parallel to ∂R but distant δ therefrom. The region between $\partial R^{-\delta}$ and $\partial R^{+\delta}$ may be covered by ϵ-cells whose centres lie on ∂R. Such cells give rise to what we call *ϵ-subsurfaces* and *ϵ-surface cells* (based upon ∂R). Specifically, an ϵ-subsurface based upon ∂R and centred at $\mathbf{x} \in \partial R$ is the intersection with ∂R of an ϵ-cell centred at $\mathbf{x}$. Consider the boundary of such a subsurface: this is that curve, Γ say, formed by the intersection of the boundary of the ϵ-cell with ∂R. Normals to ∂R drawn through points of Γ generate a 'walled' surface. The intersection of this 'wall' with ∂R and $\partial R^{-\delta}$ delineates a thin region of thickness δ which we term an *ϵ-surface cell centred at* $\mathbf{x}$ *and based upon* ∂R. (See Figure 11.1.)

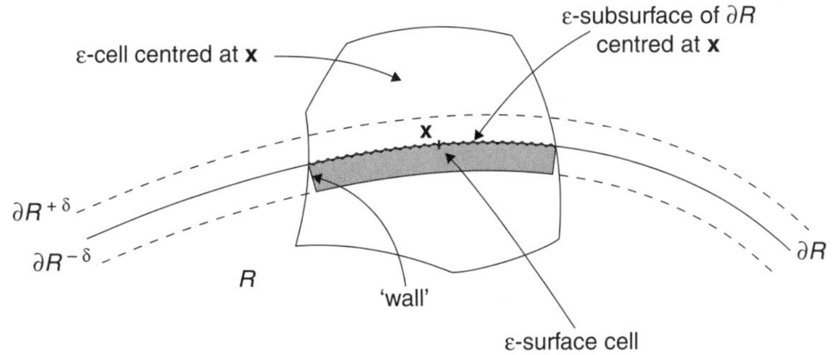

Figure 11.1. An ε-surface cell based upon ∂R and centred at $\mathbf{x}$.

For a system of disjoint ϵ-surface cells based upon ∂R, and whose union is ∂R, the final sum in (11.2.19) may be written as[3]

$$\sum_i \sum_j \mathbf{f}_{ij} = \sum_{\substack{\text{surface} \\ \text{cells}}} \left(\sum_i{}' \sum_j \mathbf{f}_{ij} \right), \tag{11.2.20}$$

where the superposed prime denotes summation over molecules within an ϵ-surface cell. If A_ϵ denotes the area of that portion of ∂R associated with a surface cell, then [cf. (11.2.2)]

$$\sum_{\substack{\text{surface} \\ \text{cells}}} \left(\sum_i{}' \sum_j \mathbf{f}_{ij} \right) = \sum_{\substack{\text{surface} \\ \text{cells}}} \left(\sum_i{}' \sum_j \mathbf{f}_{ij}/A_\epsilon \right) A_\epsilon. \tag{11.2.21}$$

If ratio $\sum_i{}' \sum_j \mathbf{f}_{ij}/A_\epsilon$ is insensitive to changes in the choice of ϵ-cell (centred at $\mathbf{x} \in \partial R$) from which the surface cell is constructed, then this value is denoted by $\lim_\epsilon \left\{ \sum_i{}' \sum_j \mathbf{f}_{ij}/A_\epsilon \right\}$. Writing

$$\mathbf{t}(\mathbf{x},t) := \lim_\epsilon \left\{ \sum_i{}' \sum_j \mathbf{f}_{ij}(t)/A_\epsilon \right\}, \tag{11.2.22}$$

it follows from (11.2.12), (11.2.13), (11.2.19), (11.2.20), (11.2.21), and (11.2.22) that we may make the identification [recall (11.2.3)]

$$\int_R \mathbf{f}(\mathbf{x},t)dV_\mathbf{x} \leftrightarrow \sum_{P_i \in R} \sum_{j \neq i} \mathbf{f}_{ij} = \sum_{\substack{\text{surface} \\ \text{cells}}} \mathbf{t}(\mathbf{x},t)A_\epsilon. \tag{11.2.23}$$

Assuming that ϵ-representative $\mathbf{t}$ is continuous, the final sum constitutes a Riemann sum which approximates the (Riemann) surface integral of $\mathbf{t}$ over ∂R. That is,

$$\int_R \mathbf{f}(\mathbf{x},t)dV_\mathbf{x} \leftrightarrow \int_{\partial R} \mathbf{t}(\mathbf{x},t)dA_\mathbf{x}. \tag{11.2.24}$$

If the integrals in (11.2.24) are assumed to differ negligibly, then

$$\int_R \mathbf{f}(\mathbf{x},t)dV_\mathbf{x} = \int_{\partial R} \mathbf{t}(\mathbf{x},t)dA_\mathbf{x} \tag{11.2.25}$$

for regions R of macroscopic proportions.

Remark 11.2.6. The definitions of $\mathbf{f}$ and $\mathbf{t}$ as ϵ-limits restrict their physical interpretations. For example, $\mathbf{f}(\mathbf{x},t)V_\epsilon$ only represents the force on matter in an ϵ-cell centred at $\mathbf{x}$ at time t due to molecular interactions provided that ϵ lies in some range (ϵ_1, ϵ_2) of values. However, the assumed continuity of $\mathbf{f}$ and $\mathbf{t}$, a further assumption of continuous dependence of $\mathbf{t}$ upon boundary orientation (i.e., upon a choice

[3] Note that here, and later in (11.2.21) and (11.2.22), that the 'j' sum is over molecules P_j which lie outside R.

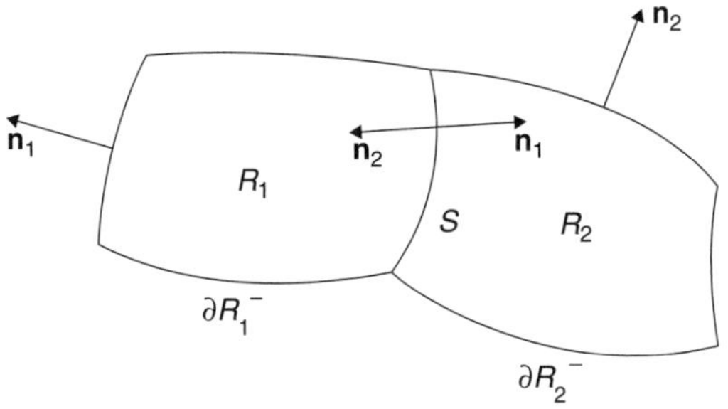

Figure 11.2. Two mutually disjoint but contiguous macroscopic regions with shared sub-boundary $S = \partial R_1 \cap \partial R_2$

of unit normal $\mathbf{n}$), together with the additivity of integrals over disjoint domains, imply that

$$(11.2.25) \text{ holds for arbitrarily small regions } R. \tag{11.2.26}$$

Proof of this assertion is outlined in Aside 11.2.1 below. Its *consequence* is, via Theorem 2.7.1, the existence of an *interaction stress tensor* $\mathbf{T}^-$ such that

$$\int_{\partial R} \mathbf{t}(\mathbf{x},t;\mathbf{n})dA_{\mathbf{x}} = \int_{\partial R} \mathbf{T}^-(\mathbf{x},t)\mathbf{n}(\mathbf{x})dA_{\mathbf{x}}. \tag{11.2.27}$$

Here $\mathbf{n}$ denotes a choice of unit normal field on ∂R which is conventionally chosen to be directed out of region R.

Aside 11.2.1. Let R_1 and R_2 denote two mutually disjoint but contiguous macroscopic regions with $S := \partial R_1 \cap \partial R_2$ (see Figure 11.2).

Let
$$R := R_1 \cup R_2. \tag{11.2.28}$$

Then the boundary of $R_\alpha\,(\alpha = 1,2)$ consists of S together with a subsurface, ∂R_α^- say, of ∂R. That is,

$$\partial R_\alpha = \partial R_\alpha^- \cup S \tag{11.2.29}$$

and

$$\partial R = \partial R_1^- \cup \partial R_2^-. \tag{11.2.30}$$

If $\mathbf{t}$ is assumed to depend continuously upon surface orientation, then from (11.2.25)

$$\int_R \mathbf{f}\,dV = \int_{\partial R} \mathbf{t}(\mathbf{n})dA \qquad \text{and} \qquad \int_{R_\alpha} \mathbf{f}\,dV = \int_{\partial R_\alpha} \mathbf{t}(\mathbf{n}_\alpha)dA. \tag{11.2.31}$$

Here $\mathbf{n}(\mathbf{n}_\alpha)$ denotes the outward unit normal to $R(R_\alpha)$. Now

$$\int_R \mathbf{f}\,dV = \int_{R_1 \cup R_2} \mathbf{f}\,dV = \int_{R_1} \mathbf{f}\,dV + \int_{R_2} \mathbf{f}\,dV. \tag{11.2.32}$$

Accordingly, from (11.2.25),

$$\int_{\partial R} \mathbf{t}(\mathbf{n})dA = \int_{\partial R_1} \mathbf{t}(\mathbf{n}_1)dA + \int_{\partial R_2} \mathbf{t}(\mathbf{n}_2)dA, \tag{11.2.33}$$

so

$$\int_{\partial R_1^-} \mathbf{t}(\mathbf{n})dA + \int_{\partial R_2^-} \mathbf{t}(\mathbf{n})dA = \int_{\partial R} \mathbf{t}(\mathbf{n})dA = \int_{\partial R_1^- \cup S} \mathbf{t}(\mathbf{n}_1)dA + \int_{\partial R_2^- \cup S} \mathbf{t}(\mathbf{n}_2)dA$$

$$= \int_{\partial R_1^-} \mathbf{t}(\mathbf{n})dA + \int_S \mathbf{t}(\mathbf{n}_1)dA + \int_{\partial R_2^-} \mathbf{t}(\mathbf{n})dA + \int_S \mathbf{t}(\mathbf{n}_2)dA. \tag{11.2.34}$$

Hence, noting that $\mathbf{n}_2 = -\mathbf{n}_1$ on S (Why?),

$$\int_S \mathbf{t}(\mathbf{n}_1) + \mathbf{t}(-\mathbf{n}_1)dA = \mathbf{0}. \tag{11.2.35}$$

However, subsurface S can be arbitrarily small (by choosing suitable macroscopic regions R_1 and R_2), so continuity of $\mathbf{t}$ (together with assumed continuous orientation dependence) implies that on any surface S with unit normal fields $\pm\mathbf{n}_1$,

$$\mathbf{t}(-\mathbf{n}_1) = -\mathbf{t}(\mathbf{n}_1). \tag{11.2.36}$$

Proof of (11.2.26) is a consequence of the following simple considerations.

Exercise 11.2.2. If a rectangular region has sides of macroscopic scale lengths x, y and z, show that a rectangular region of sides $\Delta x, \Delta y$ and Δz can be obtained in terms of sums and differences of eight such regions each of which has sides of length no less than $\min\{x,y,z\}$. [*Hint*:

$$\begin{aligned} \Delta x \Delta y \Delta z &\equiv ((x+\Delta x)-x)((y+\Delta y)-y)((z+\Delta z)-z) \\ &\equiv \{(x+\Delta x)(y+\Delta y)(z+\Delta z)-(x+\Delta x)(y+\Delta y)z\} \\ &- \{(x+\Delta x)y(z+\Delta z)-(x+\Delta x)yz\}+\cdots] \end{aligned}$$

Exercise 11.2.3. Pairing boxes as indicated via $\{\cdot\}$ terms in the preceding hint, consider boxes B and B_1 of macroscopic proportions $(a+\Delta a)\times b\times c$ and $a\times b\times c$, respectively, with $\Delta a > 0$ arbitrary. Then (11.2.25) yields

$$\int_B \mathbf{f}\,dV = \int_{\partial B} \mathbf{t}\,dA \quad \text{and} \quad \int_{B_1} \mathbf{f}\,dV = \int_{\partial B_1} \mathbf{t}\,dA. \tag{11.2.37}$$

Let $B_2 := B - B_1$ denote the box of dimension $(\Delta a)\times b\times c$. Thus

$$S := \partial B_1 \cap \partial B_2 \tag{11.2.38}$$

is a plane rectangular surface of area bc. If $(\alpha = 1,2)$

$$\partial B_\alpha^- := \partial B_\alpha - S, \tag{11.2.39}$$

note that

$$\partial B = \partial B_1^- \cup \partial B_2^-. \tag{11.2.40}$$

Show that from (11.2.37)

$$\int_{\partial B} \mathbf{t}\,dA = \int_{B_1} \mathbf{f}\,dV + \int_{B_2} \mathbf{f}\,dV = \int_{\partial B_1} \mathbf{t}\,dA + \int_{B_2} \mathbf{f}\,dV. \tag{11.2.41}$$

Note also that

$$\int_{\partial B} \mathbf{t}\,dA = \int_{\partial B_1^- \cup \partial B_2^-} \mathbf{t}\,dA = \int_{\partial B_1^-} \mathbf{t}\,dA + \int_{\partial B_2^-} \mathbf{t}\,dA. \tag{11.2.42}$$

Using (11.2.36), show that

$$\int_{\partial B_1^-} \mathbf{t}\, dA + \int_{\partial B_2^-} \mathbf{t}\, dA = \int_{\partial B_1} \mathbf{t}\, dA + \int_{\partial B_2} \mathbf{t}\, dA. \tag{11.2.43}$$

Deduce from (11.2.41), (11.2.42), and (11.2.43) that

$$\int_{B_2} \mathbf{f}\, dV = \int_{\partial B_2} \mathbf{t}\, dA. \tag{11.2.44}$$

Using the hint from Exercise 11.2.2, deduce that the integral of **f** over an arbitrarily small rectangular region of dimensions $(\Delta x) \times (\Delta y) \times (\Delta z)$ can be expressed in terms of eight surface integrals which in toto reduce to the integral of **t** over the boundary of this region as a consequence of repeated use of (11.2.36) and arguments of the kind used to obtain (11.2.44). (Non-trivial!)

Remark 11.2.7. A 'wrinkle': motivation for time averaging. In motivating (11.2.25), field **t** has been assumed implicitly to have macroscopic character. In particular, the pseudo-limit in definition (11.2.22) has been assumed to make sense. However, statistical mechanical analyses suggest that this may not be the case (cf. Alblas [43], p. 281). Specifically, the values of **t** might be expected to be subject to non-negligible erratic changes over sub-macroscopic time intervals; that is, **t** values may *fluctuate*. Such fluctuations can be smoothed by time averaging in the manner of Section 8.3. *Accordingly we revise our cellular averaging procedure by implementing an additional temporal smoothing.*

Let ψ_i denote an additive molecular quantity as before, and write

$$s_\psi(\mathbf{y}, \tau; \epsilon) := \sum_i{}' \psi_i(\tau), \tag{11.2.45}$$

where the sum is taken at time τ over all molecules P_i which lie in an ϵ-cell centred at **y**. If the cell (and hence also its boundary) deforms with the motion prescribed by **v** [see (11.2.7)], then the corresponding space–time average at point **x** and time t is

$$\hat{\psi}(\mathbf{x}, t; \epsilon, \Delta) := \frac{1}{\Delta} \int_{t-\Delta}^{t} \frac{s_\psi(\hat{\mathbf{x}}(\tau), \tau; \epsilon)}{V_\epsilon(\tau)}\, d\tau. \tag{11.2.46}$$

Here $\hat{\mathbf{x}}(\tau)$ denotes the location at time τ of the ϵ-scale material point which coincides with **x** at instant t, and $V_\epsilon(\tau)$ is the volume of the cell at time t. If $\hat{\psi}(\mathbf{x}, t; \epsilon, \Delta)$ varies negligibly with ϵ, cell shape, and Δ, when ϵ and Δ are macroscopically small but microscopically large, then for such a range of space–time scales $\hat{\psi}$ essentially may be regarded as a function ψ of **x** and t. This is symbolised by writing

$$\psi(\mathbf{x}, t) := \lim_{\epsilon, \Delta} \{\hat{\psi}(\mathbf{x}, t; \epsilon, \Delta)\} \tag{11.2.47}$$

or, in abbreviated form,

$$\psi(\mathbf{x}, t) := \lim_{\epsilon} \left\{ \overline{\sum_i{}' \psi_i / V_\epsilon} \right\}. \tag{11.2.48}$$

Remark 11.2.8. The foregoing employed velocity field $\mathbf{v}$ [see (11.2.7)] which may itself be subject to fluctuation. Such fluctuation may not be as pronounced as in the case of $\mathbf{t}$, but it could be argued that fluctuations in $\mathbf{t}$ and acceleration $\mathbf{a}$ are commensurate, and so $\mathbf{v}$, as a temporal smoothing of $\mathbf{a}$ (via integration), is less 'lively'. Alternatively, an iterative procedure might be implemented to deliver an appropriate macroscopic field as follows. Compute Δ-time averages of ρ and $\mathbf{p}$ as per (11.2.45) and (11.2.46) with $\psi_i = m_i$ and $m_i\mathbf{v}_i$, respectively, *but keep the cell fixed* (rather than having it deform with the motion prescribed by $\mathbf{v}$). Terming the resulting fields $\bar{\mathbf{p}}_0$ and $\bar{\rho}_0$, define

$$\mathbf{v}_0 := \bar{\mathbf{p}}_0/\bar{\rho}_0. \tag{11.2.49}$$

Now re-compute time averages of ρ and $\mathbf{p}$ [given by (11.2.5) and (11.2.6)] but *require the cell to deform with the motion prescribed by* $\mathbf{v}_0$. Denoting the resulting fields by $\bar{\rho}_1$ and $\bar{\mathbf{p}}_1$ yields a corresponding velocity field

$$\mathbf{v}_1 := \bar{\mathbf{p}}_1/\bar{\rho}_1. \tag{11.2.50}$$

The procedure may be repeated indefinitely. Intuitively (at least to the author!), the sequence of fields $\mathbf{v}_0, \mathbf{v}_1, \ldots$ should get closer and closer to a field $\bar{\mathbf{v}}$ which gives rise to a motion for which the associated deforming cell time-averaged fields $\bar{\rho}$ and $\bar{\mathbf{p}}$ satisfy

$$\bar{\mathbf{p}} = \bar{\rho}\bar{\mathbf{v}}. \tag{11.2.51}$$

In such case the procedure for obtaining time averages of the form $\psi(\mathbf{x},t)$ in (11.2.48) may be reformulated: in such time averaging cells are now considered to deform with the motion prescribed by $\bar{\mathbf{v}}$.

Henceforth in this section it will be assumed that $\bar{\mathbf{v}}$ *exists.* That is, there exists a velocity field $\bar{\mathbf{v}}$, together with a corresponding motion field $\bar{\chi}_{t_0}$ [given by (11.2.8) and (11.2.9) with $\mathbf{v}$ replaced by $\bar{\mathbf{v}}$] in terms of which limits of form (11.2.48) exist for $\psi_i = m_i$ and $\psi_i = m_i\mathbf{v}_i$ when cells deform with motion $\bar{\chi}_{t_0}$. Further, these limits ($\bar{\rho}$ and $\bar{\mathbf{p}}$, say) satisfy (11.2.51). Revisiting the discussion of mass conservation, the choice $\bar{\mathbf{v}}$ and consequent motion $\bar{\chi}_{t_0}$ are seen to be the appropriate choices for minimising net time-averaged mass exchange between a deforming region and its exterior. Specifically, the assumption

$$d/dt\left\{\int_{R_t} \bar{\rho}(\mathbf{x},t)dV_{\mathbf{x}}\right\} = 0 \tag{11.2.52}$$

has been motivated, where [cf. (11.2.10)]

$$R_t := \bar{\chi}_{t_0}(R,t). \tag{11.2.53}$$

Continuity equation

$$\partial\bar{\rho}/\partial t + \operatorname{div}\{\bar{\rho}\bar{\mathbf{v}}\} = 0 \tag{11.2.54}$$

follows via the argument of Section 2.5.

The region B_{t_0} 'occupied' by the system of molecules at time t_0 is [cf. (2.2.1)] defined in terms of the relevant mass density function. Specifically,

$$B_{t_0} := \{\mathbf{x} \in \mathcal{E} : \bar{\rho}(\mathbf{x},t_0) > 0\}. \tag{11.2.55}$$

Any macroscopic region $R_0 \subset B_{t_0}$ may be partitioned into very many ϵ-cells [see (11.2.1)]. Then

$$R_\tau := \bar{\chi}_{t_0}(R_0, \tau) \tag{11.2.56}$$

is a macroscopic region with corresponding partition

$$R_\tau = \sum_{a \in A} C_a(\tau), \tag{11.2.57}$$

where

$$C_a(\tau) := \bar{\chi}_{t_0}(\mathcal{C}_a, \tau). \tag{11.2.58}$$

In order to analyse linear momentum balance, we consider the motion of each molecule P_i in an inertial frame, sum over all molecules in R_τ at time t, and then effect a Δ-time average [cf. (5.4.10)]. Thus the starting point is the equation of motion [cf. (5.4.4) and (5.4.5)]

$$\sum_{j \neq i} \mathbf{f}_{ij} + \mathbf{b}_i = \frac{d}{dt}\{m_i \mathbf{v}_i\}. \tag{11.2.59}$$

The first term gives rise to the time-averaged sum

$$\begin{aligned} &\frac{1}{\Delta}\int_{t-\Delta}^{t} \sum_{a \in A} \sum_{P_i \in C_a(\tau)}{}^{'}_{\tau} \sum_{j \neq i} \mathbf{f}_{ij}(\tau) d\tau \\ &= \sum_{a \in A} \left\{ \frac{1}{\Delta}\int_{t-\Delta}^{t} \sum_{P_i \in C_a(\tau)}{}^{'}_{\tau} \sum_{j \neq i} (\mathbf{f}_{ij}(\tau)/V_\epsilon(\tau)) V_\epsilon(\tau) d\tau \right\}. \end{aligned} \tag{11.2.60}$$

Here symbol

$$\sum_{P_i \in C_a(\tau)}{}^{'}_{\tau}$$

indicates a sum taken over all P_i in cell $C_a(\tau)$ at time τ.

Since cell boundaries deform with the macroscopic motion $\bar{\chi}_{t_0}$ and Δ is macroscopically small, cell volumes $V_\epsilon(\tau)$ will change negligibly over the interval $t - \Delta \leq \tau \leq t$ [$V_\epsilon(\tau) \sim V_\epsilon$, say]. Thus if [cf. (11.2.47)]

$$\bar{\mathbf{f}}(\mathbf{x}, t) := \lim_{\epsilon, \Delta} \left\{ \frac{1}{\Delta}\int_{t-\Delta}^{t} \sum_{P_i \in C_a(\tau)}{}^{'}_{\tau} \sum_{j \neq i} \mathbf{f}_{ij}(\tau)/V_\epsilon \right\} \tag{11.2.61}$$

makes sense, where $\mathbf{x}$ denotes the centroid of $C_a(t)$, then this first term of (11.2.59) gives rise, via (11.2.60), to a Riemann sum. Accordingly, this first term is identified with the corresponding Riemann integral. That is,

$$\sum_{a \in A} \bar{\mathbf{f}}(\mathbf{x}, t) V_\epsilon \leftrightarrow \int_R \bar{\mathbf{f}}(\mathbf{x}, t) dV_{\mathbf{x}}. \tag{11.2.62}$$

Here the region R can be identified with R_t [see (11.2.56)] or R_0 since these differ negligibly: R_τ varies negligibly over time interval $[t - \Delta, t]$. (Why?)

With an eye on (11.2.23) et seq., the argument which preceded (11.2.19) can be repeated. Thus, modulo interaction assumptions I.1 and I.2, at time τ,

$$\sum_{P_i \,\mathrm{in}\, R_\tau} \sum_{j \neq i} \mathbf{f}_{ij}(\tau) = \sum_{P_i \in R_\tau} \sum_{P_j \notin R_\tau} \mathbf{f}_{ij}(\tau), \tag{11.2.63}$$

where P_i and P_j lie within a distance δ of ∂R_τ. At time $t - \Delta$ the region bounded by inner and outer 'δ-envelopes' of $R_{\tau-\Delta}$ may be partitioned into ϵ-surface cells based upon $\partial R_{t-\Delta}$. Further, motion $\bar{\boldsymbol{\chi}}_{t_0}$ mandates a corresponding partition for each τ in $(t - \Delta, t]$: cf. (11.2.58). Accordingly [cf. (11.2.20)]

$$\frac{1}{\Delta}\int_{t-\Delta}^{t} \sum_{P_i \,\mathrm{in}\, R_\tau} \sum_{j \neq i} \mathbf{f}_{ij}(\tau) d\tau = \frac{1}{\Delta}\int_{t-\Delta}^{t} \sum_{\substack{\text{surface}\\ \text{cells}}} \left\{ {\sum_i}'_\tau \sum_{P_j \notin R_\tau} \mathbf{f}_{ij}(\tau)/A_\epsilon \right\} A_\epsilon \, d\tau. \tag{11.2.64}$$

Here ${\sum_i}'$ denotes a sum at time τ over all molecules within an ϵ-surface cell in the partition, and A_ϵ is the area of the corresponding ϵ-subsurface of ∂R_τ. (Notice that such molecules P_i lie *within* R_τ.)

Noting that changes in A_ϵ are prescribed by $\bar{\boldsymbol{\chi}}_{t_0}$ and hence are negligible for $t - \Delta \leq \tau \leq t$, and that the left-hand side of (11.2.64) has been shown to be given by the Riemann sum in (11.2.62), it follows that

$$\sum_a \bar{\mathbf{f}}(\mathbf{x}, t) V_\epsilon = \sum_{\substack{\text{surface}\\ \text{cells}}} \left\{ \frac{1}{\Delta}\int_{t-\Delta}^{t} {\sum_i}'_\tau \sum_{P_j \notin R_\tau} \mathbf{f}_{ij}(\tau) d\tau / A_\epsilon \right\}. \tag{11.2.65}$$

If

$$\bar{\mathbf{t}}(\mathbf{x}, t) := \lim_{\epsilon, \Delta} \left\{ \frac{1}{A_\epsilon \Delta}\int_{t-\Delta}^{t} {\sum_i}'_\tau \sum_{P_j \notin R_\tau} \mathbf{f}_{ij}(\tau) d\tau \right\} \tag{11.2.66}$$

makes sense, where $\mathbf{x}$ denotes the location of the surface cell centre at time t [i.e., the time-integrated double sum taken over a deforming ϵ-subsurface of ∂R_τ is insensitive to changes in A_ϵ and Δ in the sense of (11.2.47)], then from (11.2.65)

$$\sum_a \bar{\mathbf{f}}(\mathbf{x}, t) V_\epsilon = \sum_{\substack{\text{surface}\\ \text{cells}}} \bar{\mathbf{t}}(\mathbf{x}, t) A_\epsilon. \tag{11.2.67}$$

Upon identifying the Riemann sum on the right-hand side of (11.2.67) with the corresponding surface integral, we have, from (11.2.62) and (11.2.67),

$$\int_R \bar{\mathbf{f}}(\mathbf{x}, t) dV_{\mathbf{x}} \leftrightarrow \sum_{a \in A} \bar{\mathbf{f}}(\mathbf{x}, t) V_\epsilon = \sum_{\substack{\text{surface}\\ \text{cells}}} \bar{\mathbf{t}}(\mathbf{x}, t) A_\epsilon \leftrightarrow \int_{\partial R} \bar{\mathbf{t}}(\mathbf{x}, t) dA_{\mathbf{x}}. \tag{11.2.68}$$

The second term in (11.2.59) yields [cf. (11.2.60) et seq.] the time-averaged sum

$$\frac{1}{\Delta}\int_{t-\Delta}^{t} \sum_{a \in A} {\sum_i}'_\tau \mathbf{b}_i(\tau) d\tau = \sum_{a \in A} \left\{ \frac{1}{\Delta}\int_{t-\Delta}^{t} {\sum_i}'_\tau \mathbf{b}_i(\tau) d\tau / V_\epsilon \right\} V_\epsilon. \tag{11.2.69}$$

Thus if

$$\bar{\mathbf{b}}(\mathbf{x}, t) := \lim_{\epsilon, \Delta} \left\{ \frac{1}{V_\epsilon \Delta}\int_{t-\Delta}^{t} {\sum_i}'_\tau \mathbf{b}_i(\tau) d\tau \right\} \tag{11.2.70}$$

makes sense [cf. (11.2.61)], then the right-hand side of (11.2.69) is a Riemann sum which we identify with the corresponding integral. That is,

$$\sum_{a\in A} \bar{\mathbf{b}}(\mathbf{x},t)V_\epsilon \leftrightarrow \int_R \bar{\mathbf{b}}(\mathbf{x},t)dV_{\mathbf{x}}. \tag{11.2.71}$$

Finally, the right-hand side of (11.2.59) results in the time-averaged sum (computed at time τ: this is indicated by subscript τ attached to the summation sign)

$$\mathbf{i}(t;\Delta) := \frac{1}{\Delta}\int_{t-\Delta}^{t} {\sum_{P_i\in R_\tau}}'_{\tau} m_i\dot{\mathbf{v}}_i(\tau)d\tau. \tag{11.2.72}$$

Here the argument of Section 8.7 (which involves consideration of all possibilities of molecules leaving and entering the deforming region R_τ during the time interval $t-\Delta \le \tau \le t$) can be employed to yield

$$\begin{aligned}\mathbf{i}(t;\Delta) = \frac{1}{\Delta}\left\{\sum_{P_i\in R_t}{}_{t}\, m_i\mathbf{v}_i(t) - \sum_{P_i\in R_{t-\Delta}}{}_{t-\Delta}\, m_i\mathbf{v}_i(t-\Delta)\right\} \\ -\frac{1}{\Delta}\sum_i\sum_{t_{i_e}} m_i\mathbf{v}_i(t_{i_e}) + \frac{1}{\Delta}\sum_i\sum_{t_{i_\ell}} m_i\mathbf{v}_i(t_{i_\ell}).\end{aligned} \tag{11.2.73}$$

The sums in the penultimate term are taken over all P_i which enter the deforming region at times τ between $t-\Delta$ and t: each 'entry' gives rise to contribution $m_i\mathbf{v}_i(t_{i_e})$ at the time t_{i_e} of *e*ntry. Similarly, the last term involves sums, one for each '*l*eaving' (at time t_{i_ℓ}) of the deforming region. Such crossings of the deforming region take place at its boundary. Accordingly book-keeping can be effected in terms of an ϵ-subsurface partition of the boundary: given such a partition at time $t-\Delta$, the motion $\bar{\chi}_{t_0}$ induces a partition at subsequent times [cf. (11.2.64)]. Thus we may write

$$\frac{1}{\Delta}\left\{\sum_i\sum_{t_{i_\ell}} m_i\mathbf{v}_i(t_{i_\ell}) - \sum_i\sum_{t_{i_e}} m_i\mathbf{v}_i(t_{i_e})\right\} = \sum_{\substack{\epsilon\text{-subsurface}\\ \text{partition}}} \mathbf{d}_{\epsilon,\Delta}(\mathbf{x},t)A_\epsilon, \tag{11.2.74}$$

where $$\mathbf{d}_{\epsilon,\Delta}(\mathbf{x},t) := \frac{1}{A_\epsilon\Delta}\left\{{\sum_i}'\sum_{t_{i_\ell}} m_i\mathbf{v}_i(t_{i_\ell}) - {\sum_i}'\sum_{t_{i_e}} m_i\mathbf{v}_i(t_{i_e})\right\}. \tag{11.2.75}$$

Here the primed sums indicate that only molecules which cross an ϵ-subsurface of area A_ϵ, centred at $\mathbf{x}$ at time t, are involved. If

$$\bar{\mathbf{d}}(\mathbf{x},t) := \lim_{\epsilon,\Delta}\{\mathbf{d}_{\epsilon,\Delta}(\mathbf{x},t)\} \tag{11.2.76}$$

makes sense, then (11.2.73), (11.2.74), and (11.2.76) yield

$$\begin{aligned}\mathbf{i}(t;\Delta) = \frac{1}{\Delta}\left\{\int_{R_t}\rho(\mathbf{x},t)\mathbf{v}(\mathbf{x},t)dV_{\mathbf{x}} - \int_{R_{t-\Delta}}\rho(\mathbf{x},t-\Delta)\mathbf{v}(\mathbf{x},t-\Delta)dV_{\mathbf{x}}\right\} \\ +\int_{\partial R_t}\bar{\mathbf{d}}(\mathbf{x},t)dA_{\mathbf{x}}.\end{aligned} \tag{11.2.77}$$

Of course, here we have identified the Riemann (surface) sum in (11.2.74) with the last term, and invoked identification (11.2.3) with $\psi_i = m_i\mathbf{v}_i$, together with (11.2.6) and (11.2.7).

At this point the time averaging of (11.2.59) over molecules in a deforming region (R_τ at time τ, with $R := R_t$) has resulted, via (11.2.62), (11.2.71), and (11.2.72) with (11.2.77), in

$$\int_R \bar{\mathbf{f}}(\mathbf{x},t) + \bar{\mathbf{b}}(\mathbf{x},t)dV_\mathbf{x} = \frac{1}{\Delta}\left\{\int_{R_t} \rho(\mathbf{x},t)\mathbf{v}(\mathbf{x},t)dV_\mathbf{x} - \int_{R_{t-\Delta}} \rho(\mathbf{x},t-\Delta)\mathbf{v}(\mathbf{x},t-\Delta)dV_\mathbf{x}\right\} + \int_{\partial R} \bar{\mathbf{d}}(\mathbf{x},t)dA_\mathbf{x}. \tag{11.2.78}$$

Further, from (11.2.68)

$$\int_R \bar{\mathbf{f}}(\mathbf{x},t)dV_\mathbf{x} = \int_{\partial R} \bar{\mathbf{t}}(\mathbf{x},t)dA_\mathbf{x}. \tag{11.2.79}$$

If $\bar{\mathbf{t}}$ satisfies the conditions of Theorem 2.7.1 (namely that its dependence upon location and the orientation of ∂R be continuous), then there exists an interaction stress tensor $\bar{\mathbf{T}}^-$ such that if $\mathbf{n}$ denotes the outward unit normal on ∂R, then

$$\bar{\mathbf{t}}(\mathbf{x},t) = \bar{\mathbf{T}}^-(\mathbf{x},t)\mathbf{n}(\mathbf{x}). \tag{11.2.80}$$

Further, at any time τ [and noting (11.2.51)],

$$\int_{R_\tau} \rho(\mathbf{x},\tau)\mathbf{v}(\mathbf{x},\tau)dV_\mathbf{x} = \int_{R_\tau} \bar{\rho}(\mathbf{x},\tau)\bar{\mathbf{v}}(\mathbf{x},\tau)dV_\mathbf{x} + \int_{R_\tau} (\rho\mathbf{v})(\mathbf{x},\tau) - \overline{\rho\mathbf{v}}(\mathbf{x},\tau)dV_\mathbf{x}. \tag{11.2.81}$$

Term $(\rho\mathbf{v} - \overline{\rho\mathbf{v}})(\mathbf{x},\tau)$ denotes the fluctuation in the momentum field at location $\mathbf{x}$ and time, τ and must be expected to vary, at any fixed time τ, in random manner. Accordingly it would appear reasonable that a spatial average, taken over any macroscopic region such as R_τ, should yield a negligible quantity. In such case [cf. (11.2.78)]

$$\frac{1}{\Delta}\left\{\int_{R_t} (\rho\mathbf{v})(\mathbf{x},t)dV_\mathbf{x} - \int_{R_{t-\Delta}} (\rho\mathbf{v})(\mathbf{x},t-\Delta)dV_\mathbf{x}\right\} = \frac{1}{\Delta}\left\{\int_{R_t} \overline{\rho\mathbf{v}}(\mathbf{x},t)dV_\mathbf{x} - \int_{R_{t-\Delta}} \overline{\rho\mathbf{v}}(\mathbf{x},t-\Delta)dV_\mathbf{x}\right\}, \tag{11.2.82}$$

which we identify with [recall (11.2.51)]

$$d/d\tau\left\{\int_{R_\tau} (\overline{\rho\mathbf{v}})(\mathbf{x},\tau)dV_\mathbf{x}\right\}\Bigg|_{\tau=t} =: \frac{d}{dt}\left\{\int_{R_t} \bar{\rho}\bar{\mathbf{v}}(\mathbf{x},t)dV_\mathbf{x}\right\}. \tag{11.2.83}$$

Since R_τ deforms with the motion $\bar{\chi}_{t_0}$ prescribed by $\bar{\mathbf{v}}$, it follows as a consequence of a standard argument [see (2.5.24) et seq.] that

$$\frac{d}{dt}\left\{\int_{R_t} \bar{\rho}\bar{\mathbf{v}}\,dV\right\} = \int_{R_t} \bar{\rho}\bar{\mathbf{a}}\,dV, \tag{11.2.84}$$

where the acceleration field

$$\bar{\mathbf{a}} := \dot{\bar{\mathbf{v}}} := \frac{\partial \bar{\mathbf{v}}}{\partial t} + (\nabla \bar{\mathbf{v}})\bar{\mathbf{v}}. \tag{11.2.85}$$

At this stage linear momentum balance takes the form [from (11.2.78), (11.2.79), (11.2.80), (11.2.82), (11.2.83), and (11.2.84); recall also that $R := R_t$]

$$\int_{\partial R} \mathbf{T}^{-}\mathbf{n}\, dA + \int_R \bar{\mathbf{b}}\, dV = \int_R \bar{\rho}\bar{\mathbf{a}}\, dV + \int_{\partial R} \bar{\mathbf{d}}\, dA. \tag{11.2.86}$$

If $\mathbf{T}^{-}$ is continuously differentiable, then use of the divergence theorem yields

$$\int_R \{\mathrm{div}\,\mathbf{T}^{-} + \bar{\mathbf{b}} - \bar{\rho}\bar{\mathbf{a}}\} dV = \int_{\partial R} \bar{\mathbf{d}}\, dA. \tag{11.2.87}$$

Provided that field $\mathbf{d}$ depends continuously upon location and orientation $\mathbf{n}$ on ∂R then Theorem 2.7.1 furnishes the existence of the *diffusive stress tensor* $\bar{\mathcal{D}}$ for which

$$\bar{\mathbf{d}} = \bar{\mathcal{D}}\mathbf{n}. \tag{11.2.88}$$

Linear momentum balance now may be written, via (11.8.86) and (11.8.88), in the form

$$\int_{\partial R} \bar{\mathbf{T}}\mathbf{n}\, dA + \int_R \bar{\mathbf{b}}\, dV = \int_R \bar{\rho}\bar{\mathbf{a}}\, dV, \tag{11.2.89}$$

where the stress tensor

$$\bar{\mathbf{T}} := \bar{\mathbf{T}}^{-} - \bar{\mathcal{D}}. \tag{11.2.90}$$

If $\bar{\mathcal{D}}$ is continuously differentiable, then (11.2.89) may be written in the form

$$\int_R \mathrm{div}\,\bar{\mathbf{T}} + \bar{\mathbf{b}} - \bar{\rho}\bar{\mathbf{a}}\, dV = \mathbf{0} \tag{11.2.91}$$

for any macroscopically large region R. Additivity of integrals over disjoint regions yields, via the observation made in Exercise 11.2.2, satisfaction of (11.2.91) for rectangular box-like regions R of arbitrarily small dimensions. Continuity of all terms in the integrand then implies that

$$\mathrm{div}\,\bar{\mathbf{T}} + \bar{\mathbf{b}} = \bar{\rho}\bar{\mathbf{a}}. \tag{11.2.92}$$

11.3 Concluding Remarks

The procedure introduced here was based upon an assumed insensitivity of volume and surface densities of additive molecular quantities to modest changes in geometry and scales of length and time. Such insensitivity motivated the notions of pseudo-limits '$\underset{\epsilon}{\mathrm{lim}}$' and '$\underset{\epsilon,\Delta}{\mathrm{lim}}$' in terms of which fields were defined with clear physical interpretation. Statements of mass conservation and linear momentum balance were established for regions of macroscopic proportion. These (global) statements are the postulates usually adopted in continuum mechanics at the outset. Corresponding local relations followed in standard manner. This ϵ-cell approach was developed by the author before he was made aware of weighting function methodology by Dick Bedeaux, and was used to explore forms of (tensor-valued) moment of momentum balance, energy balance, relations for continua with microstructure, and mixture theory (cf. Murdoch [44], Morro & Murdoch [40], Murdoch & Morro [41], Murdoch [27].) However, the precision, and absence of assumptions of scale insensitivity, associated with weighting function analysis make the latter approach both simpler and more general. In this respect the microstructural considerations and results in Murdoch [27] may serve as motivation for a weighting function approach: see Chapter 15.

12 Modelling the Behaviour of Specific Materials: Constitutive Relations and Objectivity

12.1 Preamble

Continuum mechanics is the study of material behaviour as manifest at macroscopic scales of length and time. Irrespective of molecular constitution, or whether the material is gaseous, liquid, or solid, mathematical descriptions of such behaviour have a common foundation. This foundation has been developed in Chapters 3 through 8, and is codified in terms of the system of balance relations for mass, linear and rotational momentum, and energy, and the physical descriptors (i.e., fields) which appear therein. Such relations, which serve as evolution equations for mass etc., involve terms directly related to molecular behaviour, but take the same form regardless of the specific and explicit scales of length and time associated with the averaging procedure.

In order to model a particular material system, with the aim of predicting its behaviour under prescribed circumstances, it is necessary

(i) to distinguish this behaviour from that of other materials subjected to the same prescribed circumstances, and
(ii) to make precise just what is intended by 'prescribed circumstances'.

In respect of prescription (i), the nature (or 'constitution') of an individual system is identified in terms of *constitutive relations* which specify the 'response' of the material to the consequences of prescribed circumstances. Prescription (ii) involves specification of agencies external to the system (including the effect of gravity and of contact between the system and the exterior world across its boundary) together with initial information. The balance relations, together with constitutive relations and specification of external agencies and initial information, give rise to a system of coupled partial differential equations with associated boundary conditions and initial values. Solutions of this initial-/boundary-value problem then can be compared with actual material behaviour.

In attempting to describe specific material behaviour an observer/modeller O must choose appropriate constitutive relations. Those which have proved to be most useful have been relations which link stress $\mathbf{T}$, heat flux $\mathbf{q}$, and internal energy e to macroscopic motion and temperature distribution. Such constitutive relations should be capable of being understood by any other observer O^*, for otherwise no consensus could be achieved, and the resulting model would be meaningless. To arrive at a consensus it is necessary that the molecular contributions to the stress $\mathbf{T}^*$, heat flux $\mathbf{q}^*$, and internal energy e^* for O^* should be the same as those to $\mathbf{T}, \mathbf{q}$, and e for O *and* that O^* and O should be able to agree on both the nature of the material and its response (i.e., values of $\mathbf{T}, \mathbf{q}$, and e) to all possible manifestations of the specific behaviour of interest.

In Section 12.2 the manner in which O and O^* view space is formalised. Assumed agreement on time lapses, and on distances between simultaneous events, gives rise to an instant-by-instant isometry between space $\mathcal{E}$ as perceived by O and[1] space $\mathcal{E}^*$ as perceived by O^*. Relations between local field values, computed by each observer at the same spatial scale, are obtained. It is then shown how time averaging can be implemented in such a way that field values for pairs of observers correspond to local space–time averages computed instant by instant over the same molecules. (This resolves, in a more satisfactory manner, the problem discussed in Remark 8.9.10.) In this procedure an essential role is played by inertial observers. The local space–time averages $\rho, \mathbf{p}, \mathbf{f}, \mathcal{D}, \mathbf{T}$ and $\mathbf{q}$ are shown to be objective; that is, roughly speaking, $\rho^* = \rho$, $\mathbf{p}^* = \mathbf{Q}\mathbf{p}$, $\mathbf{f}^* = \mathbf{Q}\mathbf{f}$, $\mathbf{q}^* = \mathbf{Q}\mathbf{q}$, $\mathcal{D}^* = \mathbf{Q}\mathcal{D}\mathbf{Q}^T$ and $\mathbf{T}^* = \mathbf{Q}\mathbf{T}\mathbf{Q}^T$. (Here corresponding fields for O and O^* are distinguished by the absence or presence of a superposed asterisk, respectively. Orthogonal map $\mathbf{Q}$ relates displacements as viewed by O to those regarded by O^*.) Such statements of objectivity are derived directly from molecular considerations. Attention is drawn to situations in which stress may be expected to be sensitive to the local spin of a body relative to an (any) inertial observer.

In Section 12.3 the fundamental nature of human communication and consensus is reviewed, together with the consequent notion of objectivity in science. The specific nature of what constitutes objectivity in deterministic continuum mechanics is codified in terms of five distinct aspects of agreement. The implications of such objectivity are derived for the stress in elastica, materials with memory, and viscous fluids. In contrast with standard approaches (cf., e.g., Gurtin [1] and Chadwick [3]), observers are not assumed to employ the same response functions: rather, a specific relation is derived between the pair of response functions chosen by two different observers to describe any particular aspect of behaviour. This relation is then used to derive restrictions on each individual response function.

Section 12.4 contains a review of the long-standing controversy over statements and interpretations of material frame-indifference, invariance under superposed rigid body motions, and objectivity, together with a personal statement of involvement in the issue.

12.2 Microscopic Considerations and the Key Role Played by Inertial Observers

In principle any observer O can monitor what (if anything) is happening at a given location $\mathbf{x}$ and time t in terms of Euclidean space $\mathcal{E}$ ('space' as perceived by O) and a clock. Here $\mathbf{x} \in \mathcal{E}$, and $t \in \mathbb{R}$ is measured from a particular 'happening' chosen by O (so that this happening corresponds to $t = 0$). Another observer O^* will identify any 'event' occurring at $(\mathbf{x}, t)$ for O as taking place at $(\mathbf{x}^*, t^*)$. Here $\mathbf{x}^* \in \mathcal{E}^*$ ('space' as perceived by O^*), and $t^* \in \mathbb{R}^*$ (time measured from a particular happening chosen

[1] Following Noll [45], a conceptual distinction is made between $\mathcal{E}$ and $\mathcal{E}^*$ in order to clarify the differing perspectives of individual observers. Of course, in the conventional approach, a single, universal 'space' is employed.

by O^*): $\mathcal{E}^*$ and $\mathbb{R}^*$ are copies of Euclidean space and the real numbers and are mathematically identical to, but conceptually distinct from, $\mathcal{E}$ and $\mathbb{R}$. In order to relate observations made by O and O^* it is necessary to codify the relationship between $(\mathbf{x},t)$ and $(\mathbf{x}^*,t^*)$. In classical (i.e., Newtonian) physics it is possible to talk of 'simultaneous' events and to agree[2] upon both time lapses between events and distances between simultaneous events. Simultaneity and agreement on time lapses between events allow observers in communication with each other to agree on a reference 'happening' and units of measurement for time lapses. Accordingly, such observers can choose clocks which agree on time, and hence $t^* = t$ for two such observers. More generally there will be a relationship of the form $t^* = \alpha t + \beta$, where $\alpha > 0$. Here α corresponds to ratios of time lapses due to different units of time measurement, and β is the time lapse between the reference happening for O^* and that for O. For simplicity we choose $\alpha = 1$ and $\beta = 0$. Similarly, in respect of distances we assume that O and O^* use the same length units. Thus agreement on the distance between simultaneous events means that at any time t, and for any $\mathbf{x},\mathbf{y} \in \mathcal{E}$,

$$\|\mathbf{y}^* - \mathbf{x}^*\|^* = \|\mathbf{y} - \mathbf{x}\|, \tag{12.2.1}$$

where $(\mathbf{y}^*,t)$ and $(\mathbf{x}^*,t)$ are associated with events for O^* which O regards as occurring at $(\mathbf{y},t)$ and $(\mathbf{x},t)$, respectively. Relation (12.2.1) is a statement that, at any time t, there is an isometry between $\mathcal{E}$ and $\mathcal{E}^*$, say

$$\mathbf{i}_t : \mathcal{E} \to \mathcal{E}^*, \tag{12.2.2}$$

where

$$\mathbf{y}^* = \mathbf{i}_t(\mathbf{y}) \quad \text{and} \quad \mathbf{x}^* = \mathbf{i}_t(\mathbf{x}). \tag{12.2.3}$$

Remark 12.2.1. Term $\|\mathbf{y}^* - \mathbf{x}^*\|^*$ in (12.2.1) involves the norm $\|\cdot\|^*$ on the space $\mathcal{V}^*$ of displacements in $\mathcal{E}^*$.

Maintaining the conceptual distinction between $\mathcal{E}^*$ and $\mathcal{E}$, it is possible to apply the form of argument used in Appendix B.3.2 to show that [cf. (B.3.2) with (12.2.1)]

$$\mathbf{i}_t(\mathbf{y}) - \mathbf{i}_t(\mathbf{x}) = \mathbf{Q}_t(\mathbf{y} - \mathbf{x}), \tag{12.2.4}$$

where $\mathbf{Q}_t$ is an orthogonal linear transformation[3] from $\mathcal{V}$ onto $\mathcal{V}^*$. If $\mathbf{x}_0 \in \mathcal{E}$ and

$$\mathbf{c}(t) := i_t(\mathbf{x}_0), \tag{12.2.5}$$

[2] Upon taking account of the possible different choices of units of length and time.
[3] See Maclane & Birkhoff [46]. Specifically, for any $\mathbf{u},\mathbf{v} \in \mathcal{V}$ and any $\alpha,\beta \in \mathbb{R}$,

$$\mathbf{Q}_t(\alpha\mathbf{u} + \beta\mathbf{v}) = \alpha\mathbf{Q}_t\mathbf{u} + \beta\mathbf{Q}_t\mathbf{v}$$

and (here '.*' denotes the inner product on $\mathcal{V}^*$)

$$\mathbf{Q}_t\mathbf{u}\,.^*\,\mathbf{Q}_t\mathbf{v} = \mathbf{u}\,.\,\mathbf{v}.$$

It follows that

$$\mathbf{Q}_t^T\mathbf{Q}_t = \mathbf{1} \quad \text{and} \quad \mathbf{Q}_t\mathbf{Q}_t^T = \mathbf{1}^*.$$

Here $\mathbf{1}$ and $\mathbf{1}^*$ denote the identity maps on $\mathcal{V}$ and $\mathcal{V}^*$, and

$$\mathbf{Q}_t^T : \mathcal{V}^* \to \mathcal{V}$$

is that linear map for which [cf. (A.8.1)]

$$\mathbf{Q}_t^T\mathbf{v}^*\,.\,\mathbf{u} = \mathbf{v}^*\,.^*\,\mathbf{Q}_t\mathbf{u}$$

for all $\mathbf{u} \in \mathcal{V}$ and all $\mathbf{v}^* \in \mathcal{V}^*$.

then (12.2.4) with $\mathbf{y} = \mathbf{x}_0$ [cf. (B.3.24)] yields

$$\mathbf{x}^* := \mathbf{i}_t(\mathbf{x}) = \mathbf{c}(t) + \mathbf{Q}_t(\mathbf{x} - \mathbf{x}_0). \tag{12.2.6}$$

Of course, $\mathbf{c}(t)$ denotes the location at time t for O^* of any event which occurs at $(\mathbf{x}_0, t)$ for O.

Now consider the motion of a point mass P_i whose location at time t for $O\,(O^*)$ is $\mathbf{x}_i(t)(\mathbf{x}_i^*(t))$. Accordingly, from (12.2.6),

$$\mathbf{x}_i^*(t) = \mathbf{c}(t) + \mathbf{Q}_t(\mathbf{x}_i(t) - \mathbf{x}_0). \tag{12.2.7}$$

The corresponding velocities $\mathbf{v}_i^* := \dot{\mathbf{x}}_i^*$ and $\mathbf{v}_i := \dot{\mathbf{x}}_i$ are thus related by

$$\mathbf{v}_i^*(t) = \dot{\mathbf{c}}(t) + \dot{\mathbf{Q}}_t(\mathbf{x}_i(t) - \mathbf{x}_0) + \mathbf{Q}_t\mathbf{v}_i(t). \tag{12.2.8}$$

If O and O^* monitor the motions of a system of point masses $P_i (i = 1, 2, \ldots, N)$, then the corresponding mass density fields ρ^* and ρ are given by [sums are over $i : 1 \to N$: see (4.2.1)]

$$\rho(\mathbf{x}, t) := \sum_i m_i\, w(\mathbf{x}_i(t) - \mathbf{x}) \tag{12.2.9}$$

and

$$\rho^*(\mathbf{x}^*, t) := \sum_i m_i\, w^*(\mathbf{x}_i^*(t) - \mathbf{x}^*), \tag{12.2.10}$$

where $\mathbf{x}^*$ and $\mathbf{x}$ are related by (12.2.6) and $\mathbf{x}_i^*(t)$ and $\mathbf{x}_i(t)$ by (12.2.7). If the localisation functions w^* and w chosen by O^* and O depend only on distances $\|\mathbf{x}_i^* - \mathbf{x}^*\|^*$ and $\|\mathbf{x}_i - \mathbf{x}\|$, respectively, and involve the same form and scale[4] (so that $w^*(\mathbf{u}^*) = w(\mathbf{u})$ if $\|\mathbf{u}^*\|^* = \|\mathbf{u}\|$), then

$$w^*(\mathbf{x}_i^*(t) - \mathbf{x}^*) = w(\mathbf{x}_i(t) - \mathbf{x}). \tag{12.2.11}$$

It follows from (12.2.9) and (12.2.10) that

$$\rho^*(\mathbf{x}^*, t) = \rho(\mathbf{x}, t). \tag{12.2.12}$$

The momentum density fields for O and O^* are [cf. (4.2.13)]

$$\mathbf{p}(\mathbf{x}, t) := \sum_i m_i \mathbf{v}_i(t) w(\mathbf{x}_i(t) - \mathbf{x}) \tag{12.2.13}$$

and

$$\mathbf{p}^*(\mathbf{x}^*, t) := \sum_i m_i \mathbf{v}_i^*(t) w^*(\mathbf{x}_i^*(t) - \mathbf{x}^*). \tag{12.2.14}$$

From (12.2.8) and (12.2.11)

$$\mathbf{p}^*(\mathbf{x}^*, t) = \sum_i m_i(\dot{\mathbf{c}}(t) + \dot{\mathbf{Q}}_t(\mathbf{x}_i(t) - \mathbf{x}_0) + \mathbf{Q}_t\mathbf{v}_i(t)) w(\mathbf{x}_i(t) - \mathbf{x}).$$

Thus

$$\mathbf{p}^*(\mathbf{x}^*, t) = \rho(\mathbf{x}, t)\dot{\mathbf{c}}(t) + \dot{\mathbf{Q}}_t \sum_i m_i(\mathbf{x}_i(t) - \mathbf{x}_0) w(\mathbf{x}_i(t) - \mathbf{x}) + \mathbf{Q}_t\mathbf{p}(\mathbf{x}, t). \tag{12.2.15}$$

[4] For example, choices (4.3.2) and (4.4.25).

Now [abbreviating and noting (7.3.1)]

$$\sum_i m_i(\mathbf{x}_i - \mathbf{x}_0)w = \sum_i m_i((\mathbf{x}_i - \mathbf{x}) + (\mathbf{x} - \mathbf{x}_0))w = \rho(\mathbf{d} + (\mathbf{x} - \mathbf{x}_0)). \tag{12.2.16}$$

Hence $$\mathbf{p}^*(\mathbf{x}^*,t) = \rho(\mathbf{x},t)\{\dot{\mathbf{c}}(t) + \dot{\mathbf{Q}}_t(\mathbf{d}(\mathbf{x},t) + (\mathbf{x} - \mathbf{x}_0))\} + \mathbf{Q}_t\mathbf{p}(\mathbf{x},t). \tag{12.2.17}$$

The corresponding velocity fields [cf. (4.2.15)]

$$\mathbf{v}^* := \mathbf{p}^*/\rho^* \qquad \text{and} \qquad \mathbf{v} := \mathbf{p}/\rho \tag{12.2.18}$$

are thus related, via (12.2.12), by

$$\mathbf{v}^*(\mathbf{x}^*,t) = \dot{\mathbf{c}}(t) + \mathbf{Q}_t\mathbf{v}(\mathbf{x},t) + \dot{\mathbf{Q}}_t((\mathbf{x} - \mathbf{x}_0) + \mathbf{d}(\mathbf{x},t)). \tag{12.2.19}$$

From (12.2.6) and footnote 3

$$\mathbf{x} - \mathbf{x}_0 = \mathbf{Q}_t^{-1}(\mathbf{x}^* - \mathbf{c}(t)) = \mathbf{Q}_t^T(\mathbf{x}^* - \mathbf{c}(t)), \tag{12.2.20}$$

so (12.2.19) may be written as

$$\mathbf{v}^*(\mathbf{x}^*,t) = \dot{\mathbf{c}}(t) + \mathbf{Q}_t\mathbf{v}(\mathbf{x},t) + \mathbf{W}_t(\mathbf{x}^* - \mathbf{c}(t)) + \dot{\mathbf{Q}}_t\mathbf{d}(\mathbf{x},t), \tag{12.2.21}$$

where[5] $$\mathbf{W}_t := \dot{\mathbf{Q}}_t\mathbf{Q}_t^T. \tag{12.2.22}$$

Exercise 12.2.1. Show that, on suppressing arguments,

$$\mathbf{a}^* := \dot{\mathbf{v}}^* = \ddot{\mathbf{c}} + \mathbf{Q}\mathbf{a} + (\dot{\mathbf{W}} - \mathbf{W}^2)(\mathbf{x}^* - \mathbf{c}) + 2\mathbf{W}(\mathbf{v}^* - \dot{\mathbf{c}}) + \dot{\mathbf{W}}\mathbf{Q}\mathbf{d} + \mathbf{W}\mathbf{Q}\dot{\mathbf{d}}. \tag{12.2.23}$$

(Notice that $\dot{\mathbf{Q}} = \mathbf{W}\mathbf{Q}$ and hence $\ddot{\mathbf{Q}} = \dot{\mathbf{W}}\mathbf{Q} + \mathbf{W}\dot{\mathbf{Q}}$. Note also that, on writing $\mathbf{d}^* := \mathbf{Q}\mathbf{d}$, the last two terms may be expressed as $(\dot{\mathbf{W}} - \mathbf{W}^2)\mathbf{d}^* + \mathbf{W}\widehat{\mathbf{d}^*}$).
Compare this with the point mass acceleration $\mathbf{a}_i^*$ derived from (12.2.8), namely

$$\mathbf{a}_i^* = \dot{\mathbf{v}}_i^* = \ddot{\mathbf{c}} + \mathbf{Q}\mathbf{a} + (\dot{\mathbf{W}} - \mathbf{W}^2)(\mathbf{x}_i^* - \mathbf{c}) + 2\mathbf{W}(\mathbf{v}_i^* - \dot{\mathbf{c}}). \tag{12.2.24}$$

Remark 12.2.2. Notice that the *forms* of (12.2.8) and (12.2.19) differ due to the presence of inhomogeneity measure $\mathbf{d}$ acted upon by $\dot{\mathbf{Q}}$. This is a consequence of (12.2.8) being a strictly local relation, while (12.2.19) involves terms defined using spatial averaging. However, standard treatments of changes of observer also omit term $\dot{\mathbf{Q}}\mathbf{d}$ [cf. (17.8) of Truesdell & Noll [2] and (20.13) of Gurtin, Fried & Anand [23]]. Since $\mathbf{d}$

[5] $\mathbf{W}$ is a skew-symmetric linear transformation on $\mathcal{V}^*$. To see this consider $\mathbf{Q}\mathbf{Q}^T = \mathbf{1}^*$. Time differentiation yields $\dot{\mathbf{Q}}\mathbf{Q}^T + \mathbf{Q}\widehat{\mathbf{Q}^T} = \mathbf{O}^*$. The result follows from noting that $(\dot{\mathbf{Q}})^T = \widehat{\mathbf{Q}^T}$ and that if $\mathbf{A}(\mathbf{B})$ is a linear transformation from $\mathcal{V}(\mathcal{V}^*)$ into $\mathcal{V}^*(\mathcal{V})$ then $(\mathbf{AB})^T = \mathbf{B}^T\mathbf{A}^T$. To see the latter, note that for any $\mathbf{u}^*, \mathbf{v}^* \in \mathcal{V}^*$, $(\mathbf{AB})^T\mathbf{u}^* .^* \mathbf{v}^* = \mathbf{u}^* .^* \mathbf{AB}\mathbf{v}^* = \mathbf{A}^T\mathbf{u}^* . \mathbf{B}\mathbf{v}^* = \mathbf{B}^T\mathbf{A}^T\mathbf{u}^* .^* \mathbf{v}^*$. The former result follows by noting that for any $\mathbf{u}^* \in \mathcal{V}^*$ and $\mathbf{v} \in \mathcal{V}$, $\mathbf{Q}^T\mathbf{u}^* . \mathbf{v} = \mathbf{u}^* .^* \mathbf{Q}\mathbf{v}$. Hence $\widehat{\mathbf{Q}^T}\mathbf{u}^* . \mathbf{v} = \mathbf{u}^* .^* \dot{\mathbf{Q}}\mathbf{v} = (\dot{\mathbf{Q}})^T\mathbf{u}^* . \mathbf{v}$.

is a 'fine-scale' measure, its appearance here gives no cause for concern if fine structure is neglected. However, it is the first indication of a problem concerning relative rotation of observers (measured via $\dot{\mathbf{Q}}$) that becomes significant when attempting time averaging, as we now show.

In relating field values derived in terms of space–time averages of molecular quantities by an observer O to those derived by another observer O^*, it is possible to say that such values represent exactly the same description of material behaviour if and only if such averaging is undertaken instant by instant over the same molecules. In order to accomplish this O and O^* must employ weighting functions which satisfy (12.2.11) and undertake time averaging by following the spatially averaged motion prescribed by the velocity field[6]; that is, O must follow the motion prescribed by $\mathbf{v}$ and O^* that prescribed by $\mathbf{v}^*$.

Suppose that $\hat{\mathbf{x}}(\tau)$ satisfies [for O: see (12.2.18), (12.2.9), and (12.2.13)]

$$\dot{\hat{\mathbf{x}}}(\tau) = \mathbf{v}(\hat{\mathbf{x}}(\tau),\tau) \qquad \text{with } \hat{\mathbf{x}}(t) = \mathbf{x}. \tag{12.2.25}$$

The corresponding *intrinsic* Δ-time average F_Δ of any field F is [cf. (8.4.6) with $\alpha = 1$]

$$F_\Delta(\mathbf{x},t) := \frac{1}{\Delta}\int_{t-\Delta}^{t} F(\hat{\mathbf{x}}(\tau),\tau)d\tau. \tag{12.2.26}$$

The analogues of (12.2.25) and (12.2.26) for O^* are

$$\dot{\hat{\mathbf{x}}}^*(\tau) = \mathbf{v}^*(\hat{\mathbf{x}}^*(\tau),\tau) \qquad \text{with } \hat{\mathbf{x}}^*(t) = \mathbf{x}^* \tag{12.2.27}$$

and

$$F^*_\Delta(\mathbf{x}^*,t) := \frac{1}{\Delta}\int_{t-\Delta}^{t} F^*(\hat{\mathbf{x}}^*(\tau),\tau)d\tau. \tag{12.2.28}$$

If $F = \rho$, then, from (12.2.12),

$$\rho^*_\Delta(\mathbf{x}^*,t) := \frac{1}{\Delta}\int_{t-\Delta}^{t} \rho^*(\hat{\mathbf{x}}^*(\tau),\tau)d\tau = \frac{1}{\Delta}\int_{t-\Delta}^{t} \rho(\hat{\mathbf{x}}(\tau),\tau)d\tau. \tag{12.2.29}$$

That is,

$$\rho^*_\Delta(\mathbf{x}^*,t) = \rho_\Delta(\mathbf{x},t). \tag{12.2.30}$$

Now suppose that $F = \mathbf{f}$: that is [see (5.5.2)]

$$F(\mathbf{x},t) = \mathbf{f}(\mathbf{x},t) := \sum_{i \neq j}\sum \mathbf{f}_{ij}(t)w(\mathbf{x}_i(t) - \mathbf{x}). \tag{12.2.31}$$

Since forces can be represented by displacements (see Appendix A.4), the interaction $\mathbf{f}_{ij}(\tau)$ for O is regarded by O^* as $\mathbf{f}^*_{ij}(\tau)$, where

$$\mathbf{f}^*_{ij}(\tau) = \mathbf{Q}_\tau \mathbf{f}_{ij}(\tau). \tag{12.2.32}$$

Thus [see (12.2.26)]

$$\mathbf{f}_\Delta(\mathbf{x},t) := \frac{1}{\Delta}\sum_{i \neq j}\sum \int_{t-\Delta}^{t} \mathbf{f}_{ij}(\tau)w(\mathbf{x}_i(\tau) - \hat{\mathbf{x}}(\tau))d\tau \tag{12.2.33}$$

[6] In the event that $w = w_\epsilon$, the velocity field $\mathbf{v}$ corresponds to a local ϵ-scale mass centre velocity (recall Remark 4.3.1). Such choice would seem to be natural for any given scale ϵ.

and [see (12.2.28)]

$$\mathbf{f}^*_\Delta(\mathbf{x}^*,t) := \frac{1}{\Delta}\sum_{i\neq j}\sum \int_{t-\Delta}^{t} \mathbf{f}^*_{ij}(\tau)w(\mathbf{x}^*_i(\tau)-\hat{\mathbf{x}}^*(\tau))d\tau \tag{12.2.34}$$

$$= \frac{1}{\Delta}\sum_{i\neq j}\sum \int_{t-\Delta}^{t} \mathbf{Q}_\tau \mathbf{f}_{ij}(\tau)w(\mathbf{x}_i(\tau)-\hat{\mathbf{x}}(\tau))d\tau. \tag{12.2.35}$$

Remark 12.2.3. Since $\mathbf{Q}_\tau$ will in general vary during the time interval $(t-\Delta,t)$ due to the relative rotation of O and O^*, then from (12.2.35)

$$\mathbf{f}^*_\Delta(\mathbf{x}^*,t) \neq \mathbf{Q}_t \mathbf{f}_\Delta(\mathbf{x},t) \tag{12.2.36}$$

in general. At first sight this is a major problem, since (12.2.32) holds for all observer pairs. There is also a problem concerning (12.2.25) and (12.2.27) since

$$\hat{\mathbf{x}}^*(\tau) = \mathbf{c}(\tau) + \mathbf{Q}_\tau(\hat{\mathbf{x}}(\tau)-\mathbf{x}_0) \tag{12.2.37}$$

yields
$$\mathbf{v}^*(\hat{\mathbf{x}}^*(\tau),\tau) = \dot{\hat{\mathbf{x}}}^*(\tau) = \dot{\mathbf{c}}(\tau) + \dot{\mathbf{Q}}_\tau(\hat{\mathbf{x}}(\tau)-\mathbf{x}_0) + \mathbf{Q}_\tau\dot{\hat{\mathbf{x}}}(\tau). \tag{12.2.38}$$

That is, in contrast to (12.2.19), evaluation of (12.2.38) at time t yields

$$\mathbf{v}^*(\mathbf{x}^*,t) = \dot{\mathbf{c}}(t) + \dot{\mathbf{Q}}_t(\mathbf{x}-\mathbf{x}_0) + \mathbf{Q}_t\mathbf{v}(\mathbf{x},t). \tag{12.2.39}$$

Fortunately there is a very simple but profound solution to these difficulties, both of which derive from the time dependence of $\mathbf{Q}$. The solution is for any observer O to note how an inertial observer O_{in} of choice would regard molecular motions and compute averages at the same scales of length and time. Values of such averages can then be identified with field values couched in terms of O-related fields. It is thus necessary at this stage to review what characterises an inertial observer.

In classical physics there is assumed to be a distinguished class of 'inertial' observers for any pair O_{in} and O^*_{in} of which (12.2.6) takes the simplified form

$$\mathbf{x}^* = \mathbf{c}(t) + \mathbf{Q}_0(\mathbf{x}-\mathbf{x}_0) \tag{12.2.40}$$

with
$$\dot{\mathbf{c}}(t) = \mathbf{u}_0. \tag{12.2.41}$$

Here $\mathbf{Q}_0$ is a *time-independent* orthogonal map from $\mathcal{V}_{\text{in}}$ into $\mathcal{V}^*_{\text{in}}$, and $\mathbf{u}_0$ is a fixed element of $\mathcal{V}^*_{\text{in}}$. ($\mathcal{V}^*_{\text{in}}$ and $\mathcal{V}_{\text{in}}$ are the vector spaces associated with O^*_{in} and O_{in}.) For practical purposes an *inertial observer* is one who moves at constant velocity, and without rotating, with respect to a frame of reference based upon certain 'fixed' star systems[7]. From (12.2.40) and (12.2.41) relation (12.2.19) takes the unequivocal form

$$\mathbf{v}^*(\mathbf{x}^*,t) = \mathbf{u}_0 + \mathbf{Q}_0\,\mathbf{v}(\mathbf{x},t), \tag{12.2.42}$$

while (12.2.35) simplifies to

$$\mathbf{f}^*_\Delta(\mathbf{x}^*,t) = \mathbf{Q}_0\,\mathbf{f}_\Delta(\mathbf{x},t). \tag{12.2.43}$$

[7] The naming of various celestial arrangements such as 'Sagittarius' or the 'Plough' centuries ago, and their observation today, indicate that for our purposes such star systems exist.

Remark 12.2.4. That inertial observers are seen to play a central role in our discussion should, upon reflection, come as no surprise. The whole foundation of classical physics is constructed in terms of the postulation of such observers O_{in} and O^*_{in} for whom moving mass points have locations, velocities, and accelerations related at any time t by [see (12.2.40) and (12.2.41) together with (12.2.6) and (12.2.8)]

$$\mathbf{x}^*_i(t) = \mathbf{c}(t) + \mathbf{Q}_0(\mathbf{x}_i(t) - \mathbf{x}_0), \tag{12.2.44}$$

$$\mathbf{v}^*_i(t) = \mathbf{u}_0 + \mathbf{Q}_0\mathbf{v}_i(t), \tag{12.2.45}$$

and

$$\mathbf{a}^*_i(t) = \mathbf{Q}_0\mathbf{a}_i(t). \tag{12.2.46}$$

Any inertial observer monitoring such a motion regards this to be the consequence of a 'force' acting upon the point mass which is defined to be its mass m_i multiplied by its instantaneous acceleration. Thus the force computed by O_{in} is

$$\mathbf{F}_i(t) := m_i\mathbf{a}_i(t), \tag{12.2.47}$$

while the force computed by O^*_{in} is

$$\mathbf{F}^*_i(t) := m_i\mathbf{a}^*_i(t). \tag{12.2.48}$$

Accordingly, from (12.2.46),

$$\mathbf{F}^*_i(t) = \mathbf{Q}_0\mathbf{F}_i(t). \tag{12.2.49}$$

Relation (12.2.47) was the basis of the dynamical considerations of preceding chapters (see Section 5.4). Notice that an implicit assumption has been made that inertial observers agree upon mass.

Exercise 12.2.2. Show from (5.5.3) that with $F = \mathbf{b}$ in (12.2.26)

$$\mathbf{b}^*_\Delta(\mathbf{x}^*, t) = \mathbf{Q}_0\mathbf{b}_\Delta(\mathbf{x}, t). \tag{12.2.50}$$

Recall (5.5.11):

$$\mathcal{D}(\mathbf{x},t) := \sum_i m_i(\mathbf{v}_i(t) - \mathbf{v}(\mathbf{x},t)) \otimes (\mathbf{v}_i(t) - \mathbf{v}(\mathbf{x},t))w(\mathbf{x}_i(t) - \mathbf{x}). \tag{12.2.51}$$

The counterpart relation for O^*_{in} is

$$\mathcal{D}^*(\mathbf{x}^*,t) := \sum_i m_i(\mathbf{v}^*_i(t) - \mathbf{v}^*(\mathbf{x}^*,t)) \otimes (\mathbf{v}^*_i(t) - \mathbf{v}^*(\mathbf{x}^*,t))w^*(\mathbf{x}^*_i(t) - \mathbf{x}^*). \tag{12.2.52}$$

Thus, from (12.2.42) and (12.2.45),

$$\mathcal{D}^*(\mathbf{x}^*,t) = \sum_i m_i\mathbf{Q}_0(\mathbf{v}_i(t) - \mathbf{v}(\mathbf{x},t)) \otimes \mathbf{Q}_0(\mathbf{v}_i(t) - \mathbf{v}(\mathbf{x},t))w(\mathbf{x}_i(t) - \mathbf{x}),$$

so [Show this, recalling (A.8.12).]

$$\mathcal{D}^*(\mathbf{x}^*,t) = \mathbf{Q}_0\mathcal{D}(\mathbf{x},t)\mathbf{Q}_0^T. \tag{12.2.53}$$

It follows from (12.2.28) with $F = \mathcal{D}$ that

$$\mathcal{D}^*_\Delta(\mathbf{x}^*,t) := \frac{1}{\Delta}\int_{t-\Delta}^{t} \mathcal{D}^*(\hat{\mathbf{x}}^*(\tau),\tau)d\tau = \frac{1}{\Delta}\int_{t-\Delta}^{t} \mathbf{Q}_0\mathcal{D}(\hat{\mathbf{x}}(\tau),\tau)\mathbf{Q}_0^T d\tau.$$

Thus, from (12.2.26),

$$\mathcal{D}_\Delta^*(\mathbf{x}^*,t) = \mathbf{Q}_0\mathcal{D}_\Delta(\mathbf{x},t)\mathbf{Q}_0^T. \tag{12.2.54}$$

Accordingly [see (B.10.22)]

$$(\mathrm{div}^*\mathcal{D}_\Delta^*)(\mathbf{x}^*,t) = \mathbf{Q}_0((\mathrm{div}\,\mathcal{D}_\Delta)(\mathbf{x},t)). \tag{12.2.55}$$

At this point the intrinsic time-averaged version of linear momentum balance (5.5.15) for observer O_{in} is, on suppressing subscript w,

$$-\mathrm{div}\,\mathcal{D}_\Delta + \mathbf{f}_\Delta + \mathbf{b}_\Delta = (\rho\mathbf{a})_\Delta, \tag{12.2.56}$$

while for observer O_{in}^* it is

$$-\mathrm{div}^*\mathcal{D}_w^* + \mathbf{f}_\Delta^* + \mathbf{b}_\Delta^* = (\rho^*\mathbf{a}^*)_\Delta. \tag{12.2.57}$$

Recalling (12.2.55), (12.2.43), and (12.2.50), namely

$$\mathrm{div}^*\mathcal{D}_\Delta^* = \mathbf{Q}_0\,\mathrm{div}\,\mathcal{D}_\Delta, \qquad \mathbf{f}_\Delta^* = \mathbf{Q}_0\mathbf{f}_\Delta, \qquad \text{and} \qquad \mathbf{b}_\Delta^* = \mathbf{Q}_0\mathbf{b}_\Delta, \tag{12.2.58}$$

it follows that

$$(\rho^*\mathbf{a}^*)_\Delta = \mathbf{Q}_0(\rho\mathbf{a})_\Delta. \tag{12.2.59}$$

The question now is how inertial observers relate interaction stress tensors [see (5.5.18)].

Exercise 12.2.3. Show from (5.7.2) that

$$\mathbf{a}_i^*(\mathbf{x}^*) := \hat{a}(\|\mathbf{x}_i^* - \mathbf{x}^*\|)(\mathbf{x}_i^* - \mathbf{x}^*)$$

satisfies $\mathbf{a}_i^*(\mathbf{x}^*) = \mathbf{Q}_0\,\mathbf{a}_i(\mathbf{x})$. Deduce from (5.6.11) that ${}_S\mathbf{b}_{ij}^*(\mathbf{x}^*) = \mathbf{Q}_0\,{}_S\mathbf{b}_{ij}(\mathbf{x})$. Show similarly from (5.6.13) and (5.7.25) that $((\hat{\mathbf{b}}_{ij}^H)^*(\mathbf{x}^*))(\mathbf{x}_j^* - \mathbf{x}_i^*) = \mathbf{Q}_0((\hat{b}_{ij}^H)(\mathbf{x}))(\mathbf{x}_j - \mathbf{x}_i)$ and from (5.7.30) that $(\mathbf{b}_{ij}^N)^*(\mathbf{x}^*) = \mathbf{Q}_0\mathbf{b}_{ij}^N(\mathbf{x})$. Deduce from (5.7.19), (5.7.23), (5.7.25), and (5.7.29) that [via (12.2.32) with $\mathbf{Q}_\tau = \mathbf{Q}_0$]

$$(\mathbf{T}_{w_\epsilon}^-)^*(\mathbf{x}^*,t) = \mathbf{Q}_0(\mathbf{T}_{w_\epsilon}^-(\mathbf{x},t))\mathbf{Q}_0^T \tag{12.2.60}$$

for all three choices ${}_S\mathbf{T}_{w_\epsilon}^-$, ${}_H\mathbf{T}_{w_\epsilon}^-$ and ${}_N\mathbf{T}_{w_\epsilon}^-$ of $\mathbf{T}_{w_\epsilon}^-$.

It follows from (12.2.60) and (12.2.26) with (12.2.28) and (12.2.54) that the Cauchy stresses

$$\mathbf{T}_\Delta := \mathbf{T}_\Delta^- - \mathcal{D}_\Delta \quad \text{for } O_{\mathrm{in}} \qquad \text{and} \qquad \mathbf{T}_\Delta^* := (\mathbf{T}_\Delta^-)^* - \mathcal{D}_\Delta^* \quad \text{for } O_{\mathrm{in}}^* \tag{12.2.61}$$

are related by

$$\mathbf{T}_\Delta^* = \mathbf{Q}_0\mathbf{T}_\Delta\mathbf{Q}_0^T \tag{12.2.62}$$

for all candidate stress tensors associated with choice w_ϵ of weighting function.

Exercise 12.2.4. Show via (12.2.61), (12.2.55), (5.5.18), and (12.2.43) that

$$\mathrm{div}^*\mathbf{T}_\Delta^* = \mathbf{Q}_0\,\mathrm{div}\,\mathbf{T}_\Delta. \tag{12.2.63}$$

From the simplified forms of (12.2.8) and (12.2.19) appropriate to inertial observers,

$$\hat{\mathbf{v}}_i^*(t;\mathbf{x}^*) := \mathbf{v}_i^*(t) - \mathbf{v}_w^*(\mathbf{x}^*,t) = \mathbf{Q}_0(\mathbf{v}_i(t) - \mathbf{v}_w(\mathbf{x},t)) = \mathbf{Q}_0\hat{\mathbf{v}}_i(t;\mathbf{x}). \tag{12.2.64}$$

It follows from the results for $\mathbf{a}_i^*$ and $\mathbf{b}_{ij}^*$ in Exercise 12.2.3 and definitions (6.2.15), (6.2.27), (6.2.41), and (6.2.62) that

$$\kappa_w^* = \mathbf{Q}_0\kappa_w,\ ({}_s\mathbf{q}_w^-)^* = \mathbf{Q}_0\,{}_s\mathbf{q}_w^-,\ ({}_b\mathbf{q}_w^-)^* = \mathbf{Q}_0\,{}_b\mathbf{q}_w^-,\ \text{and}\ ({}_2\mathbf{q}_w)^* = \mathbf{Q}_0\,{}_2\mathbf{q}_w. \tag{12.2.65}$$

Accordingly, from (6.2.30) and (6.2.65),

$$\mathbf{q}^*(\mathbf{x}^*,t) = \mathbf{Q}_0\,\mathbf{q}(\mathbf{x},t) \tag{12.2.66}$$

for either choice ${}_s\mathbf{q}_w$ or ${}_b\mathbf{q}_w^+$ of $\mathbf{q}$. (Recall here that $w = w_\epsilon$.) Thus [see (12.2.26) and (12.2.28)]

$$\mathbf{q}_\Delta^* = \mathbf{Q}_0\,\mathbf{q}_\Delta. \tag{12.2.67}$$

From (B.10.13)

$$\mathrm{div}^*\mathbf{q}_\Delta^* = \mathrm{div}\,\mathbf{q}_\Delta. \tag{12.2.68}$$

Remark 12.2.5. Summarising, relations (12.2.62) and (12.2.67) link the values of the Cauchy stress and heat flux as computed by any pair of inertial observers. Such computations are performed as space–time averages over the same (instant by instant) molecules at *any* chosen scales of length ϵ and time Δ. Similar analyses can be made of couple stress and internal energy density but are not here undertaken.

It is now appropriate to review the special nature of fields such as ρ, $\mathrm{div}\,\mathbf{q}$, $\mathbf{q}$, $\mathrm{div}\,\mathbf{T}$ and $\mathbf{T}$. Once these fields are known for a single inertial observer O_{in} then this observer knows precisely how they must appear to any other inertial observer O_{in}^*. [Of course, O_{in} must identify O_{in}^*, via knowledge of $\mathbf{c}$ and $\mathbf{Q}_0$: see (12.2.40).] Said differently, O_{in} and O_{in}^* would be able to agree upon the values of these fields if they were in communication with each other. Accordingly such fields represent aspects of material behaviour that are independent of inertial observers, (and hence are of fundamental importance). These fields are said to be *objective*, terminology which is now generalised, firstly for inertial observers, and then for observers in general.

A scalar field manifest as ϕ to inertial observer O_{in} and as ϕ^* to another observer O_{in}^* is termed *objective* if

$$\phi^*(\mathbf{x}^*,t) = \phi(\mathbf{x},t), \tag{12.2.69}$$

where $\mathbf{x}^*$ and $\mathbf{x}$ are related by (12.2.40). Similarly, a vector field regarded as $\mathbf{u}(\mathbf{u}^*)$ by $O_{\text{in}}(O_{\text{in}}^*)$ is termed *objective* if

$$\mathbf{u}^*(\mathbf{x}^*,t) = \mathbf{Q}_0\,\mathbf{u}(\mathbf{x},t), \tag{12.2.70}$$

while a linear transformation field denoted by $\mathbf{A}(\mathbf{A}^*)$ for $O_{\text{in}}(O_{\text{in}}^*)$ is said to be *objective* if

$$\mathbf{A}^*(\mathbf{x}^*,t) = \mathbf{Q}_0\,\mathbf{A}(\mathbf{x},t)\mathbf{Q}_0^T. \tag{12.2.71}$$

Thus an objective field is one about which inertial observers can agree. Said differently, any inertial observer O_{in} monitoring an objective field can appreciate how this field is regarded by any other inertial observer O_{in}^*, modulo knowledge of their relative motion, prescribed by (12.2.40) and (12.2.41).

Remark 12.2.6. The local molecular averages $\rho_\Delta, \mathbf{p}_\Delta, \mathbf{f}_\Delta$, $\mathcal{D}_\Delta, \mathbf{T}_\Delta^-, \mathbf{T}_\Delta$, and $\mathbf{q}_\Delta$, computed at specific length and time scales, have been shown to be objective.

While it is natural (indeed essential) to focus upon inertial observers, in view of their central role in classical physics, one can examine both how a general observer O is to regard any objective field, and the nature of the relationship between objective fields as manifest to two general observers. To this end, consider general observers O and O^*, and suppose that $O\,(O^*)$ monitors physical behaviour by noting how this behaviour is witnessed by an inertial observer $O_{\text{in}}\,(O^*_{\text{in}})$. At any time t there are isometries between $\mathcal{E}$ and $\mathcal{E}_{\text{in}}$, $\mathcal{E}_{\text{in}}$ and $\mathcal{E}^*_{\text{in}}$, and $\mathcal{E}^*$ and $\mathcal{E}^*_{\text{in}}$, where $\mathcal{E}, \mathcal{E}_{\text{in}}, \mathcal{E}^*$, and $\mathcal{E}^*_{\text{in}}$ are models of 'space' as perceived by O, O_{in}, O^*, and O^*_{in}, respectively: denote these by

$$_{\text{in}}\mathbf{i}_t : \mathcal{E} \to \mathcal{E}_{\text{in}}, \qquad \boldsymbol{\alpha}_t : \mathcal{E}_{\text{in}} \to \mathcal{E}^*_{\text{in}} \qquad \text{and} \qquad {}_{\text{in}}\mathbf{i}^*_t : \mathcal{E}^* \to \mathcal{E}^*_{\text{in}}. \tag{12.2.72}$$

It follows that

$$\mathbf{i}_t := ({}_{\text{in}}\mathbf{i}^*_t)^{-1} \circ \boldsymbol{\alpha}_t \circ {}_{\text{in}}\mathbf{i}_t : \mathcal{E} \to \mathcal{E}^* \tag{12.2.73}$$

is an isometry and thus has form [see (12.2.6)]

$$\mathbf{x}^* := \mathbf{i}_t(\mathbf{x}) = \mathbf{c}(t) + \mathbf{Q}_t(\mathbf{x} - \mathbf{x}_0). \tag{12.2.74}$$

Here

$$\mathbf{c}(t) := \mathbf{i}_t(\mathbf{x}_0), \tag{12.2.75}$$

and $\mathbf{Q}_t$ is an orthogonal linear transformation from $\mathcal{V}$ into $\mathcal{V}^*$ (the spaces of vectors associated with $\mathcal{E}$ and $\mathcal{E}^*$) given by

$$\mathbf{Q}_t := ({}_{\text{in}}\mathbf{Q}^*_t)^T \mathbf{Q}_0 ({}_{\text{in}}\mathbf{Q}_t). \tag{12.2.76}$$

Here ${}_{\text{in}}\mathbf{Q}_t : \mathcal{V} \to \mathcal{V}_{\text{in}}$, $\mathbf{Q}_0 : \mathcal{V}_{\text{in}} \to \mathcal{V}^*_{\text{in}}$, and ${}_{\text{in}}\mathbf{Q}^*_t : \mathcal{V}^* \to \mathcal{V}^*_{\text{in}}$ are the orthogonal linear transformations which correspond to ${}_{\text{in}}\mathbf{i}_t, \boldsymbol{\alpha}_t$ and ${}_{\text{in}}\mathbf{i}^*_t$, respectively. (See Figure 12.1.)

Exercise 12.2.5. Prove (12.2.76).

Any displacement $\mathbf{d}$ between a pair of points in $\mathcal{E}$ at time t is manifest to O_{in} as ${}_{\text{in}}\mathbf{Q}_t\,\mathbf{d}$, which, in turn, appears to O^*_{in} as $\mathbf{Q}_0({}_{\text{in}}\mathbf{Q}_t\,\mathbf{d})$ and to O^* as $({}_{\text{in}}\mathbf{Q}^*_t)^{-1}\mathbf{Q}_0({}_{\text{in}}\mathbf{Q}_t\,\mathbf{d})$, that is, $\mathbf{Q}_t\,\mathbf{d}$. Now consider an objective vector field, denoted by $\mathbf{u}_{\text{in}}$ for O_{in} [and $\mathbf{u}^*_{\text{in}}$ for O^*_{in}, where $\mathbf{u}^*_{\text{in}} = \mathbf{Q}_0\mathbf{u}_{\text{in}}$: see (12.2.70)]. Since vectorial quantities are representable

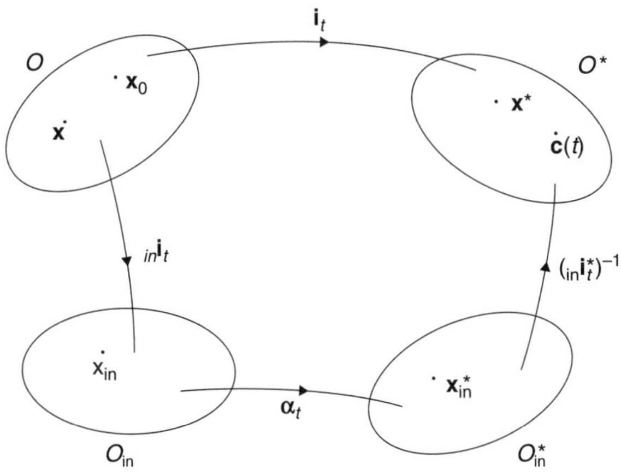

Figure 12.1. Isometry $\boldsymbol{i}_t := ({}_{\text{in}}\boldsymbol{i}^*_t)^{-1} \circ \boldsymbol{\alpha}_t \circ_{\text{in}} \boldsymbol{i}_t \boldsymbol{x}_{\text{in}} = {}_{\text{in}}\,\boldsymbol{i}_t(\boldsymbol{x})$, and points $\boldsymbol{x}^*_{\text{in}} = \boldsymbol{\alpha}_t(\boldsymbol{x}_{\text{in}})$, $\boldsymbol{x}^* = ({}_{\text{in}}\boldsymbol{i}^*_t)^{-1}(\boldsymbol{x}^*_{\text{in}})$, $\boldsymbol{c}(\text{t}) = \boldsymbol{i}_t(\boldsymbol{x}_0)$.

in terms of displacements (see Appendix A.4), O regards this as field $\mathbf{u} := {}_{\rm in}\mathbf{Q}_t^T\mathbf{u}_{\rm in}$ and O^* as $\mathbf{u}^* = ({}_{\rm in}\mathbf{Q}_t^*)^T\mathbf{u}_{\rm in}^*$. Thus

$$\mathbf{u}^* = ({}_{\rm in}\mathbf{Q}_t^*)^T\mathbf{Q}_0\mathbf{u}_{\rm in} = ({}_{\rm in}\mathbf{Q}_T^*)^T\mathbf{Q}_0({}_{\rm in}\mathbf{Q}_t)\mathbf{u} = \mathbf{Q}_t\,\mathbf{u}. \tag{12.2.77}$$

More precisely,

$$\mathbf{u}^*(\mathbf{x}^*,t) = \mathbf{Q}_t\,\mathbf{u}(\mathbf{x},t). \tag{12.2.78}$$

In the same way, an objective linear transformation field $\mathbf{A}_{\rm in}$ for $O_{\rm in}$ [see (12.2.71)] is witnessed by O as $({}_{\rm in}\mathbf{Q}_t)^T\mathbf{A}_{\rm in}({}_{\rm in}\mathbf{Q}_t)$, as $\mathbf{A}_{\rm in}^* = \mathbf{Q}_0\mathbf{A}_{\rm in}\mathbf{Q}_0^T$ by $O_{\rm in}^*$, and as $({}_{\rm in}\mathbf{Q}_t^*)^T\mathbf{A}_{\rm in}^*({}_{\rm in}\mathbf{Q}_t^*)$ by O^*. Writing

$$\mathbf{A} := ({}_{\rm in}\mathbf{Q}_t^T)\mathbf{A}_{\rm in}({}_{\rm in}\mathbf{Q}_t) \qquad \text{and} \qquad \mathbf{A}^* := ({}_{\rm in}\mathbf{Q}_t^*)^T\mathbf{A}_{\rm in}^*({}_{\rm in}\mathbf{Q}_t^*) \tag{12.2.79}$$

yields

$$\mathbf{A}^*(\mathbf{x}^*,t) = \mathbf{Q}_t\mathbf{A}(\mathbf{x},t)\mathbf{Q}_t^T. \tag{12.2.80}$$

Exercise 12.2.6. Prove (12.2.80).

Rather trivially, objective scalar fields ψ satisfy [cf. (12.2.69)]

$$\psi^*(\mathbf{x}^*,t) = \psi(\mathbf{x},t). \tag{12.2.81}$$

Remark 12.2.7. The general relations (12.2.81), (12.2.78), and (12.2.80), which link fields as perceived by *any* pair of observers, are *formally* the same as those for inertial observers [cf. (12.2.69), (12.2.70), and (12.2.71)]. The only difference is the time-dependent nature of the orthogonal map $\mathbf{Q}_t$ as compared with $\mathbf{Q}_0$. Any fields which transform in this general manner give rise to fields for inertial observers which are objective. Accordingly, fields which satisfy the *general* transformation relations are said to be *objective.*

Remark 12.2.8. It follows from Remark 2.2.7 that fields $\rho_\Delta, \mathbf{p}_\Delta, \mathbf{f}_\Delta, \mathbf{b}_\Delta, \mathcal{D}_\Delta, \mathbf{T}_\Delta$, and $\mathbf{q}_\Delta$ give rise to corresponding fields for general observers which are objective. *It is of the utmost importance to notice that meaningful intrinsic time averaging is implemented by inertial observers.* The value of any local space–time average so computed can then be related to an instantaneous value for a general observer. *Time averaging is not implemented in a general frame.*

Remark 12.2.9. In order to record the outcome of any measurement/observation an observer must choose a means of cataloguing locations. This is done by selection of a reference system based upon a number of locations regarded by the observer as fixed: a choice of four non-coplanar points suffices. A co-ordinate system can be constructed on the basis of these locations. Any such choice of locations is termed a *frame of reference* for the observer.

Exercise 12.2.7. Consider how it is possible to construct a co-ordinate system from a choice of non-coplanar points $\mathbf{x}_1, \mathbf{x}_2, \mathbf{x}_3, \mathbf{x}_4 \in \mathcal{E}$ by noting that displacements $\mathbf{u}_i := \mathbf{x}_i - \mathbf{x}_4$ $(i = 1,2,3)$ are linearly independent, and hence any point $\mathbf{x}$ may be written as

$$\mathbf{x} = \mathbf{x}_4 + (\mathbf{x} - \mathbf{x}_4) = \mathbf{x}_4 + \alpha_i(\mathbf{x})\mathbf{u}_i. \tag{12.2.82}$$

[Of course, this constitutes an oblique co-ordinate system with origin $\mathbf{x}_4$, and axes parallel to $\mathbf{u}_i$, in which $\mathbf{x}$ has co-ordinates $\alpha_i(\mathbf{x})$.]

Remark 12.2.10. Consider a body composed of molecules each of which undergoes an erratic localised high-speed motion about a slowly moving node of a lattice-like structure. (This is a simple model of a crystalline solid.) If the lattice rotates with respect to an inertial observer, and if time averaging is implemented at a scale Δ during which this rotation is detectable, then time-averaged fields must a priori be expected to be sensitive to such rotation. In particular, the stress and heat flux fields may depend upon the local rate of rotation (i.e., the spin) of the body relative to inertial observers. Similar remarks apply to any body in which molecules, or groups of molecules, spin in such a way as to act as miniature gyroscopes. It may be that such behaviour is encountered via microscopic vortex motion in superfluid helium (cf. Roberts & Donnelly [47] and Hills & Roberts [48]). Vortex motion at scales less than those associated with a continuum description (namely ε and Δ) is encountered in fluid turbulence and in the same way may give rise to constitutive dependence upon (inertial) spin (cf. Piquet [49]).

Remark 12.2.11. Any observer who regards the Earth as motionless is a good approximation to an inertial observer when modelling behaviour over small time intervals. The limitations of such an approximation become evident when observing motion of the plane of oscillation of a pendulum over several hours. (Originally demonstrated by Foucault: cf. Ohanian [32], p. 93. The effect is most noticeable at either pole but absent on the equator.) In modelling motions of terrestrial fluid (e.g., ocean currents and atmospheric weather systems) at scales in excess of $\varepsilon = 1$ km and $\Delta = 1$ h, one must expect the stress field to be sensitive to spin (computed at these scales) relative to a truly inertial (non-terrestrial) observer.

Remark 12.2.12. The possibility that stress may depend upon spin/vorticity [see (B.5.26) and Remark B.5.3 with (B.5.30)] has called into the question the validity of the *principle of material frame indifference*[8], which is often interpreted in a way that rules out spin dependence (cf., e.g., Gurtin, Fried & Anand [23], p. 251–3.) As will be argued in what follows, it is the understanding of the nature of observer agreement (which is codified in any statement of mfi) that lies behind a long-lived controversy among continuum mechanicians on the issue. For the record, the two most influential figures behind the debate, Walter Noll and Ingo Müller, while disagreeing upon the implications of mfi, do agree on the possibility of spin dependence for stress. *It is most important that researchers should not be dissuaded from modelling behaviour in terms of (inertial) spin-dependent fields of stress and heat flux when this appears to be required: no fundamental physical law is thereby violated.*

12.3 Objectivity

12.3.1 Objectivity in General

Agreement between human beings is essential for the development of knowledge and its dissemination. Our means of communication have evolved historically in terms of structured sounds to which are attributed common meaning (words) and a shared way of measuring sets of individual items (counting). Indeed, each of us replicates such

[8] Abbreviated to mfi.

historical development as we grow up and learn from others. In a scientific context, information is gathered by observation and experimentation. The results of such activity, insofar as these can be agreed upon, form the basis of established scientific knowledge and give rise to the notion of a 'reality' that is independent of observer; that is, to the concept of an 'objective' reality. In the context of material behaviour, knowledge is codified according to the material in question, the specific behaviour of interest, and the length–time scales at which this behaviour is monitored. General mathematical theories provide the framework within which obervations and measurements can be interpreted and catalogued, and accordingly form the basis of scientific agreement. The main disciplines which address physical behaviour are those of classical, relativistic, and quantum physics. Each of these disciplines incorporates fundamental assumptions concerning just what observers can agree upon, and hence upon a particular, relevant, concept of 'objectivity'. Within classical physics, macroscopic material behaviour is most commonly treated in terms of deterministic continuum mechanics, and aspects of individual material systems are modelled in terms of 'ideal' materials. Here it is necessary that observers should agree upon the nature of each class of ideal material. Further, when attempting to interpret and predict aspects of the behaviour of a particular material system, observers should agree on the relevant class to be employed. For any given observer an ideal material is characterised by one or more constitutive relations, each of which involves a 'response' function which links an *effect* (a particular aspect of behaviour) with its *cause*. A candidate response function which helps to characterise an ideal material for a given observer must be rather special in order to comply with considerations of observer agreement/objectivity. At this point it is necessary to delineate precisely the nature of objectivity as it applies to the modelling of reproducible macroscopic behaviour in classical physics.

12.3.2 Objectivity in Deterministic Continuum Mechanics[9]

All observers, whenever in communication, are assumed to agree upon

- O.1. mathematical concepts and results, and relevant physical notions,
- O.2. which observers are inertial,
- O.3. time lapses between events, and distances between simultaneous events,
- O.4. the nature of any given ideal material, and
- O.5. all possible responses of an ideal material, no matter what be the relative motions of observers.

Assumptions O.1 through O.3 have either have been made implicitly in previous chapters or have been discussed explicitly in Section 12.2. *It is assumptions O.4 and O.5 that constitute a statement of mfi.*

Remark 12.3.1. In classical physics it is assumed that *in principle* material behaviour can be monitored in a manner which does not change this behaviour (in contrast to quantum mechanics, in which knowledge of a system is gained via 'observations', each of which usually changes the 'state' of the system). In practice, experiments/observations yield information (via measurements made with specific

[9] See Murdoch [50].

instruments of known accuracy) about the system in question at particular scales of length and time. Deterministic continuum mechanical descriptions/models apply to behaviour witnessed at macroscopic scales of length and time and which can be reproduced by repeated identical experiments. (Here 'reproduced' and 'identical' mean 'the same, modulo accuracy of monitoring devices and an agreed tolerance of error'.) Accordingly this behaviour is not described exactly: what *is* described is the behaviour of an ideal material which mimics what is actually monitored/observed. The success of any choice of ideal material depends upon both the accuracy with which it describes observed behaviour and whether it can accurately predict behaviour previously unwitnessed or unexplained.

12.3.3 Elastic Behaviour

An observer O regards behaviour of a material to be elastic if, in any motion, the stress depends upon the gradient of its deformation from a known configuration. More precisely, O can choose a past time t_0 and consider the motion of the material/body χ_{t_0} corresponding to the situation at time t_0 [see Section 2.3, in particular (2.3.5)]. The body is elastic if, for every $\hat{\mathbf{x}} \in B_{t_0}$, the stress at $(\mathbf{x},t)$ is (here $t \geq t_0$)

$$\mathbf{T}(\mathbf{x},t) = \mathbf{T}_{t_0}(\mathbf{F}(\hat{\mathbf{x}},t),\hat{\mathbf{x}}), \tag{12.3.1}$$

where $\mathbf{x} = \chi_{t_0}(\hat{\mathbf{x}},t)$ and $\mathbf{F}(\hat{\mathbf{x}},t) := \nabla\chi_{t_0}(\hat{\mathbf{x}},t)$. (12.3.2)

Any other observer O^* will, *as a consequence of assumption* O.4, regard the body to be elastic and thus, upon choosing a reference time, t_0^* say, regard stress at $(\mathbf{x}^*,t)$ to be given by

$$\mathbf{T}^*(\mathbf{x}^*,t) = \mathbf{T}^*_{t_0^*}(\mathbf{F}^*(\bar{\mathbf{x}},t),\bar{\mathbf{x}}), \tag{12.3.3}$$

where $\mathbf{x}^* = \chi^*_{t_0^*}(\bar{\mathbf{x}},t)$ and $\mathbf{F}^*(\mathbf{x}^*,t) := \nabla\chi^*_{t_0^*}(\bar{\mathbf{x}},t)$. (12.3.4)

The problem now is to link these relations. Such linkage is based on two considerations:

(i) The objective nature of the stress tensor was established via molecular considerations (see Remark 12.2.7). Accordingly,

$$\mathbf{T}^*(\mathbf{x}^*,t) = \mathbf{Q}_t\mathbf{T}(\mathbf{x},t)\mathbf{Q}_t^T. \tag{12.3.5}$$

Here $\mathbf{Q}_t$ is an orthogonal map from $\mathcal{V}$ into $\mathcal{V}^*$ associated at time t with the isometry $\mathbf{i}_t : \mathcal{E} \to \mathcal{E}^*$ which identifies points viewed by O with those regarded by O^* [see (12.2.73)]. Thus, in (12.3.3),

$$\mathbf{x}^* = \mathbf{i}_t(\mathbf{x}). \tag{12.3.6}$$

(ii) It is possible to link $\mathbf{F}^*$ with $\mathbf{F}$ as follows. From $(12.3.4)_1$, (12.3.6), and $(12.3.2)_1$,

$$\chi^*_{t_0^*}(\bar{\mathbf{x}},t) = \mathbf{i}_t(\chi_{t_0}(\hat{\mathbf{x}},t)). \tag{12.3.7}$$

Recalling (2.3.2), the definitions of $\chi^*_{t_0^*}$ and χ_{t_0} yield

$$\chi^*_{t_0^*}(\bar{\mathbf{x}},t_0^*) = \bar{\mathbf{x}} \quad \text{and} \quad \chi_{t_0}(\hat{\mathbf{x}},t_0) = \hat{\mathbf{x}}. \tag{12.3.8}$$

Thus, at time $t = t_0^*$, relation (12.3.7) implies that

$$\bar{\mathbf{x}} = \chi^*_{t_0^*}(\bar{\mathbf{x}}, t_0^*) = \mathbf{i}_t(\chi_{t_0}(\hat{\mathbf{x}}, t_0^*)) =: \boldsymbol{\lambda}(\hat{\mathbf{x}}). \tag{12.3.9}$$

Here $\boldsymbol{\lambda}$ is an identification of points of[10] B_{t_0} with those of $B^*_{t_0^*}$ which is independent of t. (In fact $\boldsymbol{\lambda}$ is an identification of material points associated with the observer pair.) It follows that (12.3.7) may be written as

$$\chi^*_{t_0^*}(\boldsymbol{\lambda}(\hat{\mathbf{x}}), t) = \mathbf{i}_t(\chi_{t_0}(\hat{\mathbf{x}}, t)). \tag{12.3.10}$$

That is, suppressing arguments, and for fixed time t,

$$\chi^*_{t_0^*} \circ \boldsymbol{\lambda} = \mathbf{i}_t \circ \chi_{t_0}. \tag{12.3.11}$$

Taking the spatial gradient (with respect to location $\hat{\mathbf{x}}$) yields

$$\mathbf{F}^* \nabla \boldsymbol{\lambda} = \mathbf{Q}_t \mathbf{F}. \tag{12.3.12}$$

More precisely,

$$(\mathbf{F}^*(\boldsymbol{\lambda}(\hat{\mathbf{x}}), t))(\nabla \boldsymbol{\lambda}(\hat{\mathbf{x}})) = \mathbf{Q}_t \mathbf{F}(\hat{\mathbf{x}}, t). \tag{12.3.13}$$

Equivalently, writing

$$\mathbf{H} := \nabla \boldsymbol{\lambda}, \tag{12.3.14}$$

$$\mathbf{F}^*(\bar{\mathbf{x}}, t) = \mathbf{Q}_t \mathbf{F}(\hat{\mathbf{x}}, t) \mathbf{H}^{-1}(\hat{\mathbf{x}}). \tag{12.3.15}$$

From (12.3.1), (12.3.3), and (12.3.5),

$$\mathbf{T}^*_{t_0^*}(\mathbf{F}^*(\bar{\mathbf{x}}, t), \bar{\mathbf{x}}) = \mathbf{Q}_t \mathbf{T}_{t_0}(\mathbf{F}(\hat{\mathbf{x}}, t), \hat{\mathbf{x}}) \mathbf{Q}_t^T. \tag{12.3.16}$$

Thus, from (12.3.15),

$$\mathbf{T}^*_{t_0^*}(\mathbf{F}^*(\bar{\mathbf{x}}, t), \bar{\mathbf{x}}) = \mathbf{Q}_t \mathbf{T}_{t_0}(\mathbf{Q}_t^T \mathbf{F}^*(\bar{\mathbf{x}}, t) \mathbf{H}(\boldsymbol{\lambda}^{-1}(\bar{\mathbf{x}}))) \mathbf{Q}_t^T. \tag{12.3.17}$$

Since observers agree upon all possible responses (see the first half of assumption O.5) this relation holds for all possible deformation gradients $\mathbf{F}^*(\bar{\mathbf{x}}, t)$ at time t. Thus the response functions $\mathbf{T}^*_{t_0^*}$ and $\mathbf{T}_{t_0}$, for the same material point [corresponding to $\bar{\mathbf{x}}$ for O^* and $\boldsymbol{\lambda}^{-1}(\bar{\mathbf{x}})$ for O], are linked at any time t by

$$\mathbf{T}^*_{t_0^*}(\mathbf{F}^*) = \mathbf{Q}_t \mathbf{T}_{t_0}(\mathbf{Q}_t^T \mathbf{F}^* \mathbf{H}) \mathbf{Q}_t^T \tag{12.3.18}$$

for any $\mathbf{F}^* \in \mathrm{Invlin}^+ \mathcal{V}^*$ [cf. (A.12.33) and (B.5.11)].

Remark 12.3.2. Standard treatments of mfi do not distinguish between response functions appropriate to different observers. Just why or how different observers should employ precisely the same response function is not clear. Although (12.3.18) was established on the basis of observer agreement, observer O can obtain this relation as that which delineates $\mathbf{T}^*_{t_0^*}$ for any hypothetical observer O^* who chooses t_0^* as the referential time. The problem now is to use (12.3.18) to establish a restriction on $\mathbf{T}_{t_0}$. Such restriction is a consequence of the second half of assumption O.5.

[10] B_{t_0} ($B^*_{t_0^*}$) denotes the domain of $\chi_{t_0}(\cdot, t_0)$ ($\chi^*_{t_0^*}(\cdot, t_0^*)$): cf. (2.5.2).

Consider a motion of the body for which the deformation gradient at a given material point and time t is $\mathbf{F}^*$ for O^*. Consider further two possible relative motions of O and O^* at this time for which $\mathbf{Q}_t$ takes values $\mathbf{Q}_1$ and $\mathbf{Q}_2$. It follows from (12.3.18) that

$$\mathbf{Q}_1\mathbf{T}_{t_0}(\mathbf{Q}_1^T\mathbf{F}^*\mathbf{H})\mathbf{Q}_1^T = \mathbf{T}^*_{t_0^*}(\mathbf{F}^*) = \mathbf{Q}_2\mathbf{T}_{t_0}(\mathbf{Q}_2^T\mathbf{F}^*\mathbf{H})\mathbf{Q}_2^T. \tag{12.3.19}$$

Thus

$$\mathbf{Q}_2^T\mathbf{Q}_1\mathbf{T}_{t_0}(\mathbf{Q}_1^T\mathbf{F}^*\mathbf{H})\mathbf{Q}_1^T\mathbf{Q}_2 = \mathbf{T}_{t_0}(\mathbf{Q}_2^T\mathbf{Q}_1\mathbf{Q}_1^T\mathbf{F}^*\mathbf{H}). \tag{12.3.20}$$

Writing

$$\mathbf{F} := \mathbf{Q}_1^T\mathbf{F}^*\mathbf{H} \quad \text{and} \quad \mathbf{Q} := \mathbf{Q}_2^T\mathbf{Q}_1 \tag{12.3.21}$$

yields

$$\mathbf{T}_{t_0}(\mathbf{QF}) = \mathbf{Q}\mathbf{T}_{t_0}(\mathbf{F})\mathbf{Q}^T. \tag{12.3.22}$$

Remark 12.3.3. Here $\mathbf{Q}$ is[11] an orthogonal map on $\mathcal{V}$. Since from assumption O.5 relation (12.3.18) holds for all relative motions of O and O^*, and all responses to motions of the body, (12.3.22) must hold for all deformation gradients $\mathbf{F}$ and 'all' $\mathbf{Q}$. Since deformation gradient $\mathbf{F}$ takes values in $\text{Invlin}^+\mathcal{V}$, the domain of $\mathbf{T}_{t_0}$ is $\text{Invlin}^+\mathcal{V}$. Accordingly, $\mathbf{QF}$ must lie in $\text{Invlin}^+\mathcal{V}$; that is, $\det(\mathbf{QF}) > 0$. Thus, since $\det(\mathbf{QF}) = (\det\mathbf{Q})(\det\mathbf{F})$ and $\det\mathbf{F} > 0$, it follows that $\det\mathbf{Q} > 0$; that is, $\mathbf{Q}$ should be proper orthogonal. This is indeed the case, as we now argue.

Remark 12.3.4. Consider how observer O regards a triple of orthonormal vectors $\mathbf{e}_1^*, \mathbf{e}_2^*, \mathbf{e}_3^*$ used by O^* to help define a Cartesian co-ordinate system for $\mathcal{E}^*$. At time t these will be regarded by O in the two relative motions as $\mathbf{e}_i' := \mathbf{Q}_1^T\mathbf{e}_i^*$ and $\mathbf{e}_i'' := \mathbf{Q}_2^T\mathbf{e}_i^*$, so $\mathbf{e}_i'' = \mathbf{Q}\mathbf{e}_i$ [see $(12.3.21)_2$]. By virtue of assumption O.1, observers are aware of alternating trilinear forms, and hence can distinguish two categories of Cartesian co-ordinate systems. To do so observer O merely selects any such non-zero form, ω say, and computes $\omega(\mathbf{e}_1, \mathbf{e}_2, \mathbf{e}_3)$ for choice $\{\mathbf{e}_i\}$. The co-ordinate systems fall into those for which $\omega(\mathbf{e}_1, \mathbf{e}_2, \mathbf{e}_3) > 0$ and those for which $\omega(\mathbf{e}_1, \mathbf{e}_2, \mathbf{e}_3) < 0$: term these categories ${}^+\mathcal{C}_\omega$ and ${}^-\mathcal{C}_\omega$, respectively. Observer O^* does likewise, using form $\omega^* \neq \mathbf{0}$, and notes categories ${}^+\mathcal{C}^*_{\omega^*}$ and ${}^-\mathcal{C}^*_{\omega^*}$. Of course, if $\omega(\mathbf{e}_1, \mathbf{e}_2, \mathbf{e}_3) := \mathbf{e}_1 \times \mathbf{e}_2 . \mathbf{e}_3$, then ${}^+\mathcal{C}_\omega$ and ${}^-\mathcal{C}_\omega$ correspond to right- and left-handed co-ordinate systems. However, observers not in communication cannot be expected to understand 'right-handedness', but certainly each appreciates the *difference* between the categories[12] (via choice of alternating form). Specifically, whether $\{\mathbf{e}_i^*\}$ corresponds to $\mathcal{C}_\omega^+$ or $\mathcal{C}_\omega^-$, both $\{\mathbf{e}_i'\}$ and $\{\mathbf{e}_i''\}$ correspond to $\mathcal{C}_\omega^+$ or both to $\mathcal{C}_\omega^-$, $\omega(\mathbf{e}_1', \mathbf{e}_2', \mathbf{e}_3')$ and $\omega(\mathbf{e}_1'', \mathbf{e}_2'', \mathbf{e}_3'')$ have the same sign. Thus $\omega(\mathbf{e}_1'', \mathbf{e}_2'', \mathbf{e}_3'')\omega(\mathbf{e}_1', \mathbf{e}_2', \mathbf{e}_3') > 0$, and so $\omega(\mathbf{Q}\mathbf{e}_1', \mathbf{Q}\mathbf{e}_2', \mathbf{Q}\mathbf{e}_3')\omega(\mathbf{e}_1', \mathbf{e}_2', \mathbf{e}_3') > 0$. Hence [recall (A.12.21)]

$$(\det\mathbf{Q})(\omega(\mathbf{e}_1', \mathbf{e}_2', \mathbf{e}_3'))^2 > 0$$

and so

$$\det\mathbf{Q} > 0. \tag{12.3.23}$$

[11] If $\mathbf{u}, \mathbf{v} \in \mathcal{V}$, then, noting that $\mathbf{Q}_2^T$ is an orthogonal map from $\mathcal{V}^*$ onto $\mathcal{V}$ (see footnote 3),

$$\mathbf{Q}_2^T\mathbf{Q}_1\mathbf{u} . \mathbf{Q}_2^T\mathbf{Q}_1\mathbf{v} = \mathbf{Q}_1\mathbf{u} .^* \mathbf{Q}_1\mathbf{v} = \mathbf{u}.\mathbf{v}.$$

Thus $\mathbf{Q} : \mathcal{V} \to \mathcal{V}$ is orthogonal.

[12] If this were not the case then an observer could not distinguish between optically active fluids which rotate polarised light in different senses.

Remark 12.3.5. The foregoing argument, which resulted in the standard result (12.3.22) for all $\mathbf{F} \in \text{Invlin}^+\mathcal{V}$ and all proper orthogonal $\mathbf{Q}$, was based upon consideration of a single deformation process for the body as witnessed by O together with this same process (which is, of course, completely independent of its observation) as would have been witnessed by O if O were to have undergone a rigidly related superposed motion. Viewed in this way, relation (12.3.22) is transparent and corresponds to O monitoring the situation from a different viewpoint: in so doing, the body, its stress, and the deformation gradient all appear to rotate in appropriate fashion. Said differently, (12.3.22) expresses how the stress in an elastic body changes *under a superposed rigid motion of the* **observer**. This interpretation differs profoundly from what is often cited as a statement of mfi, namely 'invariance under superposed rigid body motions' (denoted by 'isrbm') in the sense of the observer monitoring two distinct deformation processes which are rigidly related. The isrbm viewpoint mandates how the stress changes under a *superposed rigid motion of the* **body**. At first glance there might seem to be no difference between these viewpoints, since what is involved is a rigid *relative* motion of body and observer. The distinction derives from assumption O.2. A motion of the body together with a superposed motion of this body constitute two distinct motions with respect to any inertial frame, one motion being a rotating version of the other. Since material response may depend upon rotation relative to an (any) inertial frame (see Remarks 12.2.10 and 12.2.11), such consideration is physically quite different from an observer witnessing the same physical behaviour from two different perspectives. As will be shown later, standard results mandated by various statements of mfi/isrbm follow from assumptions O.1 through O.5. Further, appropriate restrictions are obtained for response functions sensitive to rotation relative to inertial frames. By way of contrast, mechanicians who subscribe to isrbm rule out the possibility of such spin dependence.

12.3.4 Simple Materials

An observer deems material behaviour to be that of a simple material if the current value of the stress depends upon not only the current value, but also upon all past values of the deformation gradient, computed with respect to a known configuration (see Truesdell & Noll [2], Section 28). Thus, having chosen a past time t_0 (which serves to establish a known configuration), an observer O describes the stress at $(\mathbf{x}, t)$ in a simple material by a relationship of form

$$\mathbf{T}(\mathbf{x}, t) = \mathbf{T}_{t_0}({}^h\mathbf{F}_t(\hat{\mathbf{x}}), \hat{\mathbf{x}}). \qquad (12.3.24)$$

Here $\hat{\mathbf{x}} \in B_{t_0}$, and

$${}^h\mathbf{F}_t(\hat{\mathbf{x}}) : [0, \infty) \to \text{Invlin}^+\mathcal{V} \qquad (12.3.25)$$

with [see (12.3.2)]

$$({}^h\mathbf{F}_t(\hat{\mathbf{x}}))(s) := \mathbf{F}(\hat{\mathbf{x}}, t - s). \qquad (12.3.26)$$

Function ${}^h\mathbf{F}_t(\hat{\mathbf{x}})$ is termed the h*istory of the deformation gradient for* $\hat{\mathbf{x}}$ *at time* t.

Any other observer O^*, on the basis of assumption O.4, will note that the behaviour of the material is described by a stress dependence upon history of the gradient of the deformation from a known configuration (the situation at time t_0^* say, where t_0^* is chosen by O^*). Thus [cf. (12.3.3)]

$$\mathbf{T}^*(\mathbf{x}^*,t) = \mathbf{T}^*_{t_0^*}({}^h\mathbf{F}_t^*(\bar{\mathbf{x}}),\bar{\mathbf{x}}). \tag{12.3.27}$$

The analysis of subsection 12.3.3 yields

$$\mathbf{T}^*(\mathbf{x}^*,t) = \mathbf{Q}_t\mathbf{T}(\mathbf{x},t)\mathbf{Q}_t^T, \tag{12.3.28}$$

and from (12.3.13), for any $s \geq 0$,

$$(\mathbf{F}^*(\boldsymbol{\lambda}(\hat{\mathbf{x}}),t-s))(\nabla\boldsymbol{\lambda}(\hat{\mathbf{x}})) = \mathbf{Q}_{t-s}\mathbf{F}(\hat{\mathbf{x}},t-s). \tag{12.3.29}$$

That is [cf. (12.3.14)]

$${}^h\mathbf{F}_t^*(\bar{\mathbf{x}})\mathbf{H}(\hat{\mathbf{x}}) = {}^h\mathbf{Q}_t\,{}^h\mathbf{F}_t(\hat{\mathbf{x}}), \tag{12.3.30}$$

where

$${}^h\mathbf{Q}_t(s) := \mathbf{Q}(t-s). \tag{12.3.31}$$

It follows from (12.3.24), (12.3.27), (12.3.28), and (12.3.30) that

$$\mathbf{T}^*_{t_0^*}({}^h\mathbf{F}_t^*) = \mathbf{Q}_t\,\mathbf{T}_{t_0}({}^h\mathbf{Q}_t^T\,{}^h\mathbf{F}_t^*\mathbf{H})\mathbf{Q}_t^T, \tag{12.3.32}$$

on omitting explicit dependence on $\hat{\mathbf{x}},\bar{\mathbf{x}}$ [cf. (12.3.17) and (12.3.18)]. This holds for all deformation gradient histories ${}^h\mathbf{F}_t^*$ for O^*; that is, for all functions ${}^h\mathbf{F}_t^* : [0,\infty) \to \text{Invlin}^+\,\mathcal{V}^*$. Upon considering two possible relative motions of O with respect to O^* for which $\mathbf{Q}_t$ takes values $\mathbf{Q}_1$ and $\mathbf{Q}_2$, it follows from (12.3.32) that

$$\mathbf{Q}_1\,\mathbf{T}_{t_0}({}^h(\mathbf{Q}_1)_t^T\,{}^h\mathbf{F}_t^*\,\mathbf{H}^{-1})\mathbf{Q}_1^T = \mathbf{T}^*_{t_0^*}({}^h\mathbf{F}_t^*) = \mathbf{Q}_2\,\mathbf{T}_{t_0}({}^h(\mathbf{Q}_2)_t^T\,{}^h\mathbf{F}_t^*\,\mathbf{H}^{-1})\mathbf{Q}_2^T. \tag{12.3.33}$$

Accordingly, writing

$$\mathbf{Q} := \mathbf{Q}_2^T\mathbf{Q}_1 \qquad \text{and} \qquad {}^h\mathbf{F} := {}^h(\mathbf{Q}_1)_t^T\,{}^h\mathbf{F}_t^*\mathbf{H}^{-1}, \tag{12.3.34}$$

(12.3.33) yields

$$\mathbf{T}_{t_0}({}^h\mathbf{Q}^T\,{}^h\mathbf{F}) = \mathbf{Q}\mathbf{T}_{t_0}({}^h\mathbf{F})\mathbf{Q}^T. \tag{12.3.35}$$

Here

$${}^h\mathbf{Q}(s) := {}^h\mathbf{Q}_2^T(s)\,{}^h\mathbf{Q}_1(s) \tag{12.3.36}$$

denotes the history of any function on $[0,\infty)$, with values which are proper orthogonal maps on $\mathcal{V}$, which takes value $\mathbf{Q}$ at time t. Relation (12.3.35) must hold for all histories ${}^h\mathbf{F}$ with values in Invlin$^+$. The arbitrary natures of ${}^h\mathbf{Q}$ and ${}^h\mathbf{F}$ are consequences of assumption O.5.

Remark 12.3.6. Notice that an elastic material is a special case of simple material.

12.3.5 Viscous Fluids

An observer regards viscous fluid behaviour to be that governed by dependence of the current value of the stress at any given location only upon the current values of density and velocity gradient at this location. Thus for O

$$\mathbf{T}(\mathbf{x},t) = \mathbf{T}_O(\rho(\mathbf{x},t),\mathbf{L}(\mathbf{x},t)), \tag{12.3.37}$$

while for O^*

$$\mathbf{T}^*(\mathbf{x}^*,t) = \mathbf{T}^*_{O^*}(\rho^*(\mathbf{x}^*,t),\,\mathbf{L}^*(\mathbf{x}^*,t)). \tag{12.3.38}$$

Remark 12.3.6. Notice that here the response function $\mathbf{T}_0$ does not explicitly depend upon which 'part' of the fluid is concerned, in contrast to response functions for elastic or simple materials.

Recall (12.2.19): on suppressing time dependence this is

$$\mathbf{v}^*(\mathbf{x}^*) = \dot{\mathbf{c}} + \mathbf{Q}_t\,\mathbf{v}(\mathbf{x}) + \dot{\mathbf{Q}}_t(\mathbf{x} - \mathbf{x}_0 + \mathbf{d}(\mathbf{x})). \tag{12.3.39}$$

Replacing $\mathbf{x}$ by $\mathbf{x} + \mathbf{h}$ (where $\mathbf{h} \in \mathcal{V}$) results in a change in $\mathbf{x}^*$ to $\mathbf{i}_t(\mathbf{x}+\mathbf{h}) = \mathbf{i}_t(\mathbf{x}) + (\mathbf{i}_t(\mathbf{x}+\mathbf{h}) - \mathbf{i}_t(\mathbf{x})) = \mathbf{x}^* + \mathbf{h}^*$, where [cf. (12.2.4) with $\mathbf{y} = \mathbf{x}+\mathbf{h}$]

$$\mathbf{h}^* := \mathbf{Q}_t\,\mathbf{h}. \tag{12.3.40}$$

Accordingly

$$\mathbf{v}^*(\mathbf{x}^* + \mathbf{h}^*) = \dot{\mathbf{c}} + \mathbf{Q}_t\,\mathbf{v}(\mathbf{x}+\mathbf{h}) + \dot{\mathbf{Q}}_t(\mathbf{x} + \mathbf{h} - \mathbf{x}_0 + \mathbf{d}(\mathbf{x}+\mathbf{h})). \tag{12.3.41}$$

Subtracting (12.3.39) from (12.3.41) yields

$$\nabla^*\mathbf{v}^*(\mathbf{x}^*)\mathbf{h}^* + o(\mathbf{h}^*) = \mathbf{Q}_t\nabla\mathbf{v}(\mathbf{x})\mathbf{h} + \dot{\mathbf{Q}}_t(\mathbf{1} + \nabla\mathbf{d}(\mathbf{x}))\mathbf{h} + o(\mathbf{h}). \tag{12.3.42}$$

Noting that from (12.3.40) $\|\mathbf{h}^*\|^* = \|\mathbf{h}\|$ (Why?), writing $\mathbf{h} = s\hat{\mathbf{u}}$ with $s \in \mathbb{R}$ and $\|\hat{\mathbf{u}}\| = 1$, dividing by s and taking the limit as $s \to 0$, we obtain

$$\nabla^*\mathbf{v}^*(\mathbf{x}^*)\mathbf{Q}_t\hat{\mathbf{u}} = \mathbf{Q}_t\nabla\mathbf{v}(\mathbf{x})\hat{\mathbf{u}} + \dot{\mathbf{Q}}_t(\mathbf{1} + \nabla\mathbf{d}(\mathbf{x}))\hat{\mathbf{u}}. \tag{12.3.43}$$

It follows by linearity that (12.3.43) holds for *any* $\hat{\mathbf{u}} \in \mathcal{V}$, whence

$$\nabla^*\mathbf{v}^*(\mathbf{x}^*)\mathbf{Q}_t = \mathbf{Q}_t\,\nabla\mathbf{v}(\mathbf{x}) + \dot{\mathbf{Q}}_t(\mathbf{1} + \nabla\mathbf{d}(\mathbf{x})). \tag{12.3.44}$$

Equivalently, suppressing arguments $\mathbf{x}^*$ and $\mathbf{x}$,

$$\mathbf{L}^* = \mathbf{Q}_t\{\mathbf{L} + (\mathbf{Q}_t^T\dot{\mathbf{Q}}_t)(\mathbf{1} + \nabla\mathbf{d})\}\mathbf{Q}_t^T, \tag{12.3.45}$$

or, rearranging,

$$\mathbf{L} = -(\mathbf{Q}_t^T\,\dot{\mathbf{Q}})(\mathbf{1} + \nabla\mathbf{d}) + \mathbf{Q}_t^T\,\mathbf{L}^*\mathbf{Q}_t. \tag{12.3.46}$$

Given the objectivity of the stress tensor, (12.3.37), (12.3.38), and (12.3.46) yield

$$\mathbf{T}^*_{O^*}(\rho^*, \mathbf{L}^*) = \mathbf{Q}_t\,\mathbf{T}_0(\rho, \mathbf{Q}_t^T\,\mathbf{L}^*\mathbf{Q}_t - (\mathbf{Q}_t^T\dot{\mathbf{Q}}_t)(\mathbf{1} + \nabla\mathbf{d}))\mathbf{Q}_t^T. \tag{12.3.47}$$

Recall from (12.2.12) that $\rho^* = \rho$. Now consider two possible relative motions of O with respect to O^* for which $\mathbf{Q}_t$ and $\dot{\mathbf{Q}}_t$ take values $\mathbf{Q}_1, \mathbf{Q}_2$ and $\dot{\mathbf{Q}}_1, \dot{\mathbf{Q}}_2$. Then, omitting density dependence, (12.3.47) implies that

$$\begin{aligned}&\mathbf{Q}_1\mathbf{T}_O(\mathbf{Q}_1^T\mathbf{L}^*\mathbf{Q}_1 - (\mathbf{Q}_1^T\dot{\mathbf{Q}}_1)(\mathbf{1} + \nabla\mathbf{d}))\mathbf{Q}_1^T\\ &\quad = \mathbf{T}^*_{O^*}(\mathbf{L}^*) = \mathbf{Q}_2\mathbf{T}_O(\mathbf{Q}_2^T\mathbf{L}^*\mathbf{Q}_2 - (\mathbf{Q}_2^T\dot{\mathbf{Q}}_2)(\mathbf{1} + \nabla\mathbf{d}))\mathbf{Q}_2^T.\end{aligned} \tag{12.3.48}$$

Writing

$$\mathcal{L} := \mathbf{Q}_1^T\mathbf{L}^*\mathbf{Q}_1, \qquad \mathbf{Q} := \mathbf{Q}_2^T\mathbf{Q}_1 \qquad \text{and} \qquad \mathbf{S}_\alpha := \mathbf{Q}_\alpha^T\dot{\mathbf{Q}}_\alpha \qquad (\alpha = 1, 2), \tag{12.3.49}$$

(12.3.48) yields

$$\mathbf{Q}\mathbf{T}_O(\mathcal{L} - \mathbf{S}_1(\mathbf{1} + \nabla\mathbf{d}))\mathbf{Q}^T = \mathbf{T}_O(\mathbf{Q}\mathcal{L}\mathbf{Q}^T - \mathbf{S}_2(\mathbf{1} + \nabla\mathbf{d})). \tag{12.3.50}$$

As a consequence of assumption O.5 this relation must hold for all $\mathbf{Q} \in \mathrm{Orth}^+\mathcal{V}$ [see Remark 12.3.4 and (A.16.9)], all $\boldsymbol{\mathcal{L}} \in \mathrm{Lin}\,\mathcal{V}$, all $\nabla\mathbf{d} \in \mathrm{Lin}\,\mathcal{V}$, and all $\mathbf{S}_1, \mathbf{S}_2 \in \mathrm{Sk}\,\mathcal{V}$ (as follows from time differentiation of $\mathbf{Q}_t^T\mathbf{Q}_t = \mathbf{1}$: prove this!). In the case of $\mathbf{Q} = \mathbf{1}$ and $\mathbf{S}_1 \neq \mathbf{S}_2$ (i.e., two motions of O for which at time t the orientation of the body witnessed by O is the same, but for which O is spinning relative to the body at different rates), (12.3.50) becomes

$$\mathbf{T}_O(\mathbf{A}) = \mathbf{T}_O(\mathbf{A} + (\mathbf{S}_1 - \mathbf{S}_2)(\mathbf{1} + \nabla\mathbf{d})) \tag{12.3.51}$$

on writing

$$\mathbf{A} := \boldsymbol{\mathcal{L}} - \mathbf{S}_1(\mathbf{1} + \nabla\mathbf{d}). \tag{12.3.52}$$

The arbitrary natures of $\boldsymbol{\mathcal{L}}$, $\nabla\mathbf{d}, \mathbf{S}_1$ and $\mathbf{S}_2$ imply that (12.3.51) holds for all $\mathbf{A} \in \mathrm{Lin}\,\mathcal{V}$. Thus, for any velocity gradient $\mathbf{L}$,

$$\mathbf{T}_O(\mathbf{L}) = \mathbf{T}_O(\mathbf{L} + \mathbf{S}(\mathbf{1} + \nabla\mathbf{d})), \tag{12.3.53}$$

where $\mathbf{S} \in \mathrm{Sk}\,\mathcal{V}$ is arbitrary. Noting (B.5.26) and (B.5.27) of Appendix B,

$$\mathbf{L} = \mathbf{D} + \mathbf{W}, \tag{12.3.54}$$

where $\mathbf{D}$ and $\mathbf{W}$ are the stretching and spin values, then choice $\mathbf{S} = -\mathbf{W}$ yields

$$\mathbf{T}_O(\mathbf{L}) = \mathbf{T}_0(\mathbf{D} - \mathbf{W}\nabla\mathbf{d}). \tag{12.3.55}$$

Returning to (12.3.48), consider two relative motions of O with respect to O^* for which $\dot{\mathbf{Q}}_1 = \mathbf{O} = \dot{\mathbf{Q}}_2$. In such case

$$\mathbf{Q}_1\mathbf{T}_O(\mathbf{Q}_1^T\mathbf{L}^*\mathbf{Q}_1)\mathbf{Q}_1^T = \mathbf{Q}_2\mathbf{T}_O(\mathbf{Q}_2^T\mathbf{L}^*\mathbf{Q}_2)\mathbf{Q}_2^T. \tag{12.3.56}$$

Exercise 12.3.1. Writing $\mathbf{L} := \mathbf{Q}_1^T\mathbf{L}^*\mathbf{Q}_1$ and $\mathbf{Q} := \mathbf{Q}_2^T\mathbf{Q}_1$, show that

$$\mathbf{T}_O(\mathbf{Q}\mathbf{L}\mathbf{Q}^T) = \mathbf{Q}\mathbf{T}_O(\mathbf{L})\mathbf{Q}^T \tag{12.3.57}$$

holds for all $\mathbf{L} \in \mathrm{Lin}\,\mathcal{V}$ and all $\mathbf{Q} \in \mathrm{Orth}^+$. (Note the arbitrary natures of $\mathbf{L}^*, \mathbf{Q}_1$ and $\mathbf{Q}_2$, courtesy of assumption O.5.)

It follows from (12.3.55) and (12.3.57) that the response function $\mathbf{T}_O$ for a viscous fluid must satisfy

$$\mathbf{T} = \mathbf{T}_O(\rho, \mathbf{D} - \mathbf{W}\nabla\mathbf{d}), \tag{12.3.58}$$

where

$$\mathbf{T}_O(\rho, \mathbf{Q}\mathbf{D}\mathbf{Q}^T - \mathbf{Q}\mathbf{W}\nabla\mathbf{d}\mathbf{Q}^T) = \mathbf{Q}\mathbf{T}_O(\rho, \mathbf{D} - \mathbf{W}\nabla\mathbf{d})\mathbf{Q}^T. \tag{12.3.59}$$

Remark 12.3.7. If fine structure is neglected (cf. Remark 12.2.2), then terms involving $\nabla\mathbf{d}$ are absent. Accordingly, for viscous fluids [cf. (12.3.37), (12.3.58), and (12.3.59)]

$$\mathbf{T} = \mathbf{T}_O(\rho, \mathbf{L}) = \mathbf{T}_O(\rho, \mathbf{D}), \tag{12.3.60}$$

where

$$\mathbf{T}_O(\rho, \mathbf{Q}\mathbf{D}\mathbf{Q}^T) = \mathbf{Q}\mathbf{T}_O(\rho, \mathbf{D})\mathbf{Q}^T \tag{12.3.61}$$

for all $\mathbf{Q} \in \mathrm{Orth}^+$ and all $\mathbf{D} \in \mathrm{Sym}\,\mathcal{V}$. These are standard relations [cf., e.g., Gurtin, Fried & Anand [23], (45.5) and (45.6)].

Remark 12.3.8. At first glance simplification (12.3.60) would appear to rule out spin dependence for viscous fluids, modulo neglect of fine structure. However, this is so because relation (12.3.37) is taken to hold for *any* observer O. Accordingly, the assumption of behaviour sensitive to $\mathbf{L}(=\mathbf{D}+\mathbf{W})$ means that the stress depends a priori on the stretching $\mathbf{D}$ and local spin $\mathbf{W}$ of the body relative to O. Since O can spin arbitrarily without affecting material behaviour, it is not surprising to find a relation of form (12.3.60). Some question concerning such general spin dependence is raised by the more general result (12.3.58). A bigger issue is raised if one wishes to talk about an absolute, rather than relative, concept of spin. This highlights once again the role of inertial frames, since any pair of *inertial* observers will agree on local spin values. Such consideration leads naturally to examining materials for which

$$\mathbf{T}(\mathbf{x},t)=\mathbf{T}_{O_{\text{in}}}(\rho(\mathbf{x},t),\mathbf{L}_{\text{in}}(\mathbf{x},t)). \tag{12.3.62}$$

Here O_{in} is any inertial observer and $\mathbf{L}_{\text{in}}$ is the corresponding velocity gradient. Recall that if O^*_{in} is another inertial observer, then (12.2.42) and (12.2.40) hold. It follows that

$$\mathbf{L}^*_{\text{in}}(\mathbf{x}^*,t)=\mathbf{Q}_0\mathbf{L}_{\text{in}}(\mathbf{x},t)\mathbf{Q}_0^T. \tag{12.3.63}$$

[See (12.3.45), and note that $\dot{\mathbf{Q}}=\mathbf{O}$ and $\mathbf{Q}_t=\mathbf{Q}_0$.] Of course, observer agreement upon the nature of such material (O.4) means that O^*_{in} has, as the counterpart of (12.3.62),

$$\mathbf{T}^*(\mathbf{x}^*,t)=\mathbf{T}^*_{O^*_{\text{in}}}(\rho^*(\mathbf{x}^*,t),\mathbf{L}^*_{\text{in}}(\mathbf{x}^*,t)). \tag{12.3.64}$$

Given objectivity of the stress tensor, (12.3.62) and (12.3.64) yield

$$\mathbf{T}^*_{O^*_{\text{in}}}(\rho^*,\mathbf{L}^*_{\text{in}})=\mathbf{Q}_0\mathbf{T}_{O_{\text{in}}}(\rho,\mathbf{L}_{\text{in}})\mathbf{Q}_0^T. \tag{12.3.65}$$

Noting that $\rho^*=\rho$ [see (12.2.12)] and suppressing this argument, (12.3.65) and (12.3.63) yield

$$\mathbf{T}^*_{O^*_{\text{in}}}(\mathbf{L}^*_{\text{in}})=\mathbf{Q}_0\mathbf{T}_{O_{\text{in}}}(\mathbf{Q}_0^T\mathbf{L}^*_{\text{in}}\mathbf{Q}_0)\mathbf{Q}_0^T. \tag{12.3.66}$$

Since this is to hold for all responses and all relative motions of observers (via O.5), it follows that $\mathbf{L}^*_{\text{in}}\in \text{Lin}\,\mathcal{V}$ is arbitrary, and $\mathbf{Q}_0$ is an arbitrary orthogonal (time-independent) map from $\mathcal{V}$ into $\mathcal{V}^*$. Thus with $\mathbf{Q}_0=\mathbf{Q}_1$ and $\mathbf{Q}_0=\mathbf{Q}_2$ it follows that

$$\mathbf{T}_{O_{\text{in}}}(\mathbf{Q}\mathbf{L}\mathbf{Q}^T)=\mathbf{Q}\mathbf{T}_{O_{\text{in}}}(\mathbf{L})\mathbf{Q}^T \tag{12.3.67}$$

for all $\mathbf{L}\in\text{Lin}\,\mathcal{V}$ and all $\mathbf{Q}\in\text{Orth}^+\mathcal{V}$.

Exercise 12.3.2. Prove (12.3.67). [*Hint*: Set $\mathbf{Q}_0=\mathbf{Q}_1$ and $\mathbf{Q}_0=\mathbf{Q}_2$ in (12.3.66), and write $\mathbf{Q}:=\mathbf{Q}_2^T\mathbf{Q}_1$ and $\mathbf{L}:=\mathbf{Q}_1^T\mathbf{L}^*_{\text{in}}\mathbf{Q}_1$.]

An analogue of (12.3.67) holds for a general observer O for whom the stress in the fluid described by (12.3.62) takes the form

$$\mathbf{T}=\mathbf{T}_O(\mathbf{L}), \tag{12.3.68}$$

where $\mathbf{L}$ denotes the velocity gradient with respect to the motion in an inertial frame chosen by O. If O_{in} denotes an observer for whom this frame is fixed, then at time t

$$\mathbf{L}=\mathbf{Q}_t^T\mathbf{L}_{\text{in}}\mathbf{Q}_t, \tag{12.3.69}$$

where $\mathbf{L}_{\text{in}}$ is the velocity gradient for O_{in}. The stress $\mathbf{T}_{O_{\text{in}}}$ for O_{in} is correspondingly related to that for O by

$$\mathbf{T}_{O_{\text{in}}} = \mathbf{Q}_t \mathbf{T} \mathbf{Q}_t^T. \tag{12.3.70}$$

In the same way, another general observer O^* can select an inertial frame and associated observer O^*_{in}. The counterparts of (12.3.68), (12.3.69), and (12.3.70) are

$$\mathbf{T}^* = \mathbf{T}^*_{O^*}(\mathbf{L}^*), \tag{12.3.71}$$

$$\mathbf{L}^* = (\mathbf{Q}_t^*)^T \mathbf{L}^*_{\text{in}} \mathbf{Q}_t \qquad \text{and} \qquad \mathbf{T}^*_{O^*_{\text{in}}} = \mathbf{Q}_t^* \mathbf{T}^* (\mathbf{Q}_t^*)^T. \tag{12.3.72}$$

From (12.3.71), (12.3.72), (12.3.66), (12.3.63), (12.3.70), (12.3.68), and (12.3.69),

$$\begin{aligned}
\mathbf{T}^*_{O^*}(\mathbf{L}^*) &= (\mathbf{Q}_t^*)^T \mathbf{T}^*_{O^*_{\text{in}}}(\mathbf{Q}_t^* \mathbf{L}^* (\mathbf{Q}_t^*)^T) \mathbf{Q}_t^* \\
&= (\mathbf{Q}_t^*)^T \mathbf{Q}_0 \mathbf{T}_{O_{\text{in}}}(\mathbf{Q}_0^T \mathbf{Q}_t^* \mathbf{L}^* (\mathbf{Q}_t^*)^T \mathbf{Q}_0) \mathbf{Q}_0^T \mathbf{Q}_t^* \\
&= (\mathbf{Q}_t^*)^T \mathbf{Q}_0 \mathbf{Q}_t \mathbf{T}_O (\mathbf{Q}_t^T \mathbf{Q}_0^T \mathbf{Q}_t^* \mathbf{L}^* (\mathbf{Q}_t^*)^T \mathbf{Q}_0 \mathbf{Q}_t) \mathbf{Q}_t^T \mathbf{Q}_0^T \mathbf{Q}_t^*.
\end{aligned} \tag{12.3.73}$$

That is, writing

$$\mathbf{Q} := (\mathbf{Q}_t^*)^T \mathbf{Q}_0 \mathbf{Q}_t, \tag{12.3.74}$$

$$\mathbf{T}^*_{O^*}(\mathbf{L}^*) = \mathbf{Q} \mathbf{T}_O (\mathbf{Q}^T \mathbf{L}^* \mathbf{Q}) \mathbf{Q}^T. \tag{12.3.75}$$

Choosing two relative motions of O with respect to O^* which at time t yield $\mathbf{Q}$ values of $\mathbf{Q}_1$ and $\mathbf{Q}_2$ implies that for all $\mathbf{L}^* \in \text{Lin}\,\mathcal{V}^*$

$$\mathbf{Q}_1 \mathbf{T}_O (\mathbf{Q}_1^T \mathbf{L}^* \mathbf{Q}_1) \mathbf{Q}_1^T = \mathbf{Q}_2 \mathbf{T}_O (\mathbf{Q}_2^T \mathbf{L}^* \mathbf{Q}_1) \mathbf{Q}_2^T. \tag{12.3.76}$$

Writing

$$\mathbf{L} := \mathbf{Q}_1^T \mathbf{L}^* \mathbf{Q}_1 \qquad \text{and} \qquad \mathbf{Q} := \mathbf{Q}_2^T \mathbf{Q}_1, \tag{12.3.77}$$

it follows that

$$\mathbf{T}_O(\mathbf{Q}\mathbf{L}\mathbf{Q}^T) = \mathbf{Q}\mathbf{T}_O(\mathbf{L})\mathbf{Q}^T. \tag{12.3.78}$$

The arbitrary natures of $\mathbf{Q}_1, \mathbf{Q}_2$, and $\mathbf{L}$ mean that (12.3.78) must hold for all $\mathbf{Q} \in \text{Orth}^+\mathcal{V}$ and all $\mathbf{L} \in \text{Lin}\,\mathcal{V}$. Of course, dependence upon density has been suppressed. Noting that $\mathbf{D}$ and $\mathbf{W}$ are independent complementary contributions to $\mathbf{L}$, (12.3.78) may be written as

$$\hat{\mathbf{T}}_O(\rho, \mathbf{Q}\mathbf{D}\mathbf{Q}^T, \mathbf{Q}\mathbf{W}\mathbf{Q}^T) = \mathbf{Q}\hat{\mathbf{T}}_O(\rho, \mathbf{D}, \mathbf{W})\mathbf{Q}^T \tag{12.3.79}$$

for all $\mathbf{D} \in \text{Sym}\,\mathcal{V}$, all $\mathbf{W} \in \text{Sk}\,\mathcal{V}$, and all $\mathbf{Q} \in \text{Orth}^+\mathcal{V}$. Of course,

$$\hat{\mathbf{T}}_O(\rho, \mathbf{D}, \mathbf{W}) := \mathbf{T}_O(\rho, \mathbf{L}). \tag{12.3.80}$$

12.3.6 Other Materials and Considerations

Remark 12.3.9. The approach used here to obtain restrictions imposed upon response functions by objectivity has been applied to materials for which $\mathbf{T}, \mathbf{q}$ and e depend upon objective field values (Murdoch [50], p. 316), and also to materials for which $\mathbf{T}$ depends upon the history of the motion of the whole body. Specifically (cf. Murdoch [51], p. 92–4),

$$\mathbf{T}(\mathbf{x}, t) = \mathbf{T}_{t_0}({}^h\chi_{t_0,t}(.,.), \hat{\mathbf{x}}). \tag{12.3.81}$$

Here ${}^h\chi_{t_0,t}$ denotes the history of the motion relative to the situation at time t_0; that is, for any $\hat{\mathbf{y}} \in B_{t_0}$ and $s \geq 0$,

$$ {}^h\chi_{t_0,t}(\hat{\mathbf{y}},s) := \chi_{t_0}(\hat{\mathbf{y}},t-s) \tag{12.3.82} $$

and

$$ {}^h\chi_{t_0,t}(\hat{\mathbf{x}},0) = \chi_{t_0}(\hat{\mathbf{x}},t) = \mathbf{x}. \tag{12.3.83} $$

This last category of materials was considered because it would appear to represent the most general stress–motion relation that can be envisaged. Of course, (12.3.1) and (12.3.24) are special cases of such a relation.

Remark 12.3.10. Restrictions upon response functions for heat flux and internal energy have not been addressed since no microscopic definition of temperature has been introduced. Although absolute temperature θ is intimately related to h [see (6.2.13)], such considerations are relevant only in the context of a full thermodynamic description. This requires a statement of the second law (cf., e.g., Gurtin, Fried & Anand [23], p. 186) and consequent involvement of entropy. To date an analysis of the microscopic underpinnings of the second law, in the manner of the molecular approach adopted here, has not been developed[13]. (Comments on irreversibility and the second law are made later in Subsection 15.3.6.) However, the procedure for obtaining restrictions upon postulated response functions for $\mathbf{T},\mathbf{q}$, e, and specific entropy η should be evident from the preceding subsection. Notice that the very use of the terms 'temperature θ' and 'specific entropy η' without qualification indicates an implicit assumption that these scalar fields are objective.

Remark 12.3.11. Any pair of observers not in communication cannot be expected to select the same scales of mass, length and time. In particular, O and O^* may use different units of length. In such case $\mathbf{Q}_t$, used in (12.2.4) to link displacements, must be replaced by $\alpha\mathbf{Q}_t$, where α is the dimensionless factor necessary to convert length units employed by O to those adopted by O^*. Similarly, the link (12.2.32) between forces should involve $\beta\mathbf{Q}_\tau$ rather than $\mathbf{Q}_\tau$: here β converts force units for O to those for O^*. However, restrictions upon response functions involved proper orthogonal maps $\mathbf{Q}$ on the space $\mathcal{V}$ of vectors for any observer O which were obtained as combinations $\mathbf{Q}_2^T\mathbf{Q}_1$ [see $(12.3.21)_2$, $(12.3.34)_1$, (12.3.57), and $(12.3.77)_2$]. Here $\mathbf{Q}_1$ and $\mathbf{Q}_2$ should take forms $\gamma\mathbf{Q}_1$ and $\gamma\mathbf{Q}_2$ (where γ is a context-dependent conversion factor), and $\mathbf{Q}_2^T\mathbf{Q}_1 = \mathbf{Q}_2^{-1}\mathbf{Q}_1$ should be replaced by $(\gamma\mathbf{Q}_2)^{-1}(\gamma\mathbf{Q}_1) = \gamma^{-1}\mathbf{Q}_2^T\gamma\mathbf{Q}_1 = \mathbf{Q}_2^T\mathbf{Q}_1$. Thus *the restrictions on response functions are independent of the choice of units of measurement.*

12.4 Remarks on the mfi/isrbm Controversy

12.4.1 Introduction

In classical physics material behaviour is in principle assumed to be independent of its observation (see Remark 12.3.1). In order to explore the consequences of this assumption, when modelling specific behaviour in terms of deterministic continuum mechanics, it is necessary to be more explicit. A detailed history

[13] In particular, in such an approach the entropy density would be scale-dependent.

of how the assumption has been interpreted is given in Truesdell & Noll [2], Section 19. There are essentially two conceptually different interpretations, those of *material frame-indifference* (mfi) and *invariance under superposed rigid body motions* (isrbm).

12.4.2 Material Frame-Indifference

In Truesdell & Noll [2], pp. 36 and 44, mfi is stated variously as 'the response of a material is the same for all observers' and 'constitutive equations must be invariant under changes of frames of reference'. In eliciting the restrictions imposed upon postulated response functions by mfi, three distinct assumptions are involved. These are

MFI 1. Fields delivered by response functions/constitutive relations (e.g., $\mathbf{T}, \mathbf{q}, e$) are objective.

MFI 2. A change of frame/observer is characterised by a relation of form (12.2.6).

MFI 3. The response function in any constitutive relation is the same for all observers.

Remark 12.4.1. In formulations of mfi no distinction is made between 'space' as perceived by different observers. Thus all considerations are based upon a single Euclidean space $\mathcal{E}$ and a single associated vector space $\mathcal{V}$, and $\mathbf{Q}_t$ in (12.2.6) is taken to be an orthogonal map on $\mathcal{V}$. According to Truesdell & Noll, *all* elements of Orth $\mathcal{V}$ should be involved in restrictions upon response functions which follow from mfi. This viewpoint was challenged by Rivlin [52], who argued that only *proper* orthogonal maps should be admitted, and termed this modification of mfi *invariance under superposed rotation* (isr).[14] It is worthy of note that in Truesdell & Noll [2] MFI 3 was not stated explicitly but inferred by example.

12.4.3 Invariance under Superposed Rigid Body Motions

Formally isrbm is almost identical to mfi in that key features MFI 1 and MFI 3 are shared, and a relationship of form (12.2.6) is postulated in which $\mathbf{Q}_t$ is assumed to be a proper orthogonal map. However, a profoundly different interpretation is accorded to this relation. Rather than interpret (12.2.6) in terms of a change of observer, as in mfi or isr, this relation is regarded to represent a rigid displacement of the *body*. Thus, while mfi and isr involve consideration of specific behaviour by different observers, isrbm corresponds to a single observer witnessing different behaviours, any pair of which are related in terms of a rigid motion of the body.

12.4.4 Comparison of mfi, isr, isrbm, and Objectivity

Remark 12.4.2. Acceptance of isrbm constitutes imposition of an a priori restriction upon Nature. While at first it might seem entirely reasonable that a rigid motion of a body should merely rotate the stress and heat flux fields in corresponding fashion,

[14] Rivlin based his objection upon consideration of a material which exhibits optical activity.

the possible sensitivity of these fields to spin relative to an inertial frame becomes a crucial issue. Indeed, isrbm denies the possibility of such sensitivity, since differing rigid motions would give rise to different spins yet result in the same response. The fact that a great deal of material behaviour is well-described by spin-independent constitutive relations should not obscure the essence of the difference between mfi/isr and isrbm. That spin-dependence of stress and heat flux is to be expected in rarefied gases was demonstrated by Müller [53]. The nature of field values as representing space-time averages of molecular quantities lends support to this viewpoint, as do considerations of superfluid helium and turbulence (see Remarks 12.2.10, 12.2.11, and 12.2.12). Further, irbm gives rise to conceptual difficulties. Consider a bucket of liquid at rest in an inertial frame. (For this example a terrestrial frame serves as inertial.) Now contemplate a rigid motion of this liquid and ask how such a motion might be achieved. From the mfi/isr perspective this same liquid is viewed by all possible observers, each of whom moves rigidly relative to the bucket and liquid (e.g., walking round the bucket!). Of course, to such observers the liquid appears to move rigidly, but nothing is actually happening to it.

Remark 12.4.3. Influential criticism of mfi has come from Müller [53] in respect of MFI 3, which he terms the '*principle of material objectivity*'. Here a simplified version of his example[15] suffices to make his point. Consider a constitutive relation of form

$$\mathbf{T} = \hat{\mathbf{T}}(\rho, \mathbf{D}, \mathbf{W} + \mathbf{S}). \tag{12.4.1}$$

Given MFI 3, an inertial observer O_{in} will, since $\mathbf{S} = \mathbf{O}$, use

$$\mathbf{T}_{in} = \hat{\mathbf{T}}(\rho, \mathbf{D}, \mathbf{W}). \tag{12.4.2}$$

To Müller, (12.4.1) and (12.4.2) violate mfi since he regards these relations as demonstrating a frame dependence of the stress response function. Acceptance of this viewpoint would mean that mfi rules out constitutive dependence on spin relative to an inertial frame. Accordingly mfi would become untenable as a physical principle should inertial spin dependence exist in any single material. However, the case for such spin sensitivity has been made several times here. Indeed, ruling out inertial spin dependence (as previously noted in Remark 12.4.2 in respect of isrbm) imposes a restriction upon behaviour which Nature is under no obligation to honour. Müller's results[16] [53], and consequent criticism of mfi as a physical principle, have been the subject of extensive debate and controversy. Wang [55] and Truesdell [56] clearly accepted Müller's interpretation of MFI 3 in respect of regarding relations of form (12.4.1) and (12.4.2) as an example of frame dependence, since they attempted to deny the validity of his results. Their arguments were, respectively, that there was

[15] In [53] Müller obtained expressions for stress and heat flux in a rarefied gas via kinetic theory. These expressions included, inter alia, constitutive dependence upon the objective fields $\dot{\mathbf{g}} + \mathbf{S}\mathbf{g}$ and $(\mathbf{W} + \mathbf{S})\mathbf{g}$. Here $\mathbf{g}$ denotes the temperature gradient, $\mathbf{W}$ the spin in the observer frame, and $\mathbf{S}$ the spin of the observer frame relative to any inertial frame. See also Murdoch [54].

[16] Edelen & McLennan [59], Söderholm [60] and Woods [61] also presented relations, based upon work of Burnett [62], which they claimed to be incompatible with mfi. However, these authors lacked the general approach of Müller in that their relations are implicitly associated only with inertial frames but which they regarded as holding generally. The point is exemplified by considering (12.4.2) to hold for all observers. Their claim is that Burnett's work implies stress dependence upon spin relative to the observer frame, rather than that with respect to an inertial frame, namely $\mathbf{W} + \mathbf{S}$.

no basis in the fundamental tenets of kinetic theory for violation of mfi, and that approximative relations derived on the basis of kinetic theory cannot be regarded as constitutive relations. Speziale [57] claimed that incompatibility of Müller's results with mfi must derive from the Maxwell and Chapman & Enskog approximation procedures employed. Svendsen & Bertram [58] drew attention to the individual strands MFI 1 through MFI 3, characterising MFI 1 as Euclidean frame indifference and MFI 3 as form invariance, and proved that these [when taken together with (12.2.6) as a statement of orientation-preserving change of observer] implied isrbm. *There is, however, an entirely different way of looking at relations (12.4.1) and (12.4.2).* The mathematical content of the equations involves a function $\hat{\mathbf{T}}$ of three independent variables which take values in $\mathbb{R}^+, \mathrm{Sym}\,\mathcal{V}$ and[17] $\mathrm{Sk}\,\mathcal{V}$. Said differently, $\hat{\mathbf{T}}$ is a function with domain $\mathbb{R}^+ \times \mathrm{Sym}\,\mathcal{V} \times \mathrm{Sk}\,\mathcal{V}$ (with co-domain $\mathrm{Lin}\,\mathcal{V}$). The physical content of the equations is that they pertain to a material in which the stress depends upon the mass density, symmetric part of the velocity gradient, and spin relative to an inertial frame. That is, (12.4.1) and (12.4.2) represent the behaviour of a specific material which involves a *single* function $\hat{\mathbf{T}}$. Any observer feeds into the three argument slots of $\hat{\mathbf{T}}(.,.,.)$ the local values of mass density, symmetric part of velocity gradient, and spin relative to an inertial frame, and in so doing obtains the local stress value. Said differently, what distinguishes (12.4.1) and (12.4.2) is not the response *function* ($\hat{\mathbf{T}}$) but the appropriate way observers fill the third slot: inertial observers need only enter $\mathbf{W}(=\mathbf{W}+\mathbf{O})$. Given the objective natures of $\rho, \mathbf{D}$ and $\mathbf{W}+\mathbf{S}$, the restriction imposed upon function $\hat{\mathbf{T}}$ by mfi/isr is that

$$\hat{\mathbf{T}}(a, \mathbf{QBQ}^T, \mathbf{QCQ}^T) = \mathbf{Q}\hat{\mathbf{T}}(a, \mathbf{B}, \mathbf{C})\mathbf{Q}^T \tag{12.4.3}$$

should hold for all $a \in \mathbb{R}^+, \mathbf{B} \in \mathrm{Sym}\,\mathcal{V}, \mathbf{C} \in \mathrm{Sk}\,\mathcal{V}$ and[18] $\mathbf{Q} \in \mathrm{Orth}^+\mathcal{V}$. (Only $\mathrm{Orth}^+\mathcal{V}$ need be involved here. Why?)

Relation (12.4.1) is a very special case of a constitutive relation having the form

$$\mathbf{T} = \tilde{\mathbf{T}}(\rho, \mathbf{D}, \mathbf{W}, \mathbf{S}), \tag{12.4.4}$$

in which the four constitutive variables are not independent. A material described by (12.4.4) with *independent* $\mathbf{W}$ and $\mathbf{S}$ sensitivity is genuinely frame-dependent and would violate mfi. Indeed, writing (12.4.4) equivalently as $\mathbf{T} = \bar{\mathbf{T}}(\rho, \mathbf{D}, \mathbf{W}+\mathbf{S}, \mathbf{W}-\mathbf{S})$ and invoking mfi result in exclusion of $\mathbf{W}-\mathbf{S}$ dependence and reduction to (12.4.1).

Exercise 12.4.1. Consider the function

$$\hat{\mathbf{q}}(\mathbf{a}, \mathbf{B}, \mathbf{C}) := \alpha\mathbf{a} + \beta\mathbf{B} + \gamma\mathbf{C}, \tag{12.4.5}$$

where $\alpha, \beta, \gamma \in \mathbb{R}$ are constants, $\mathbf{a} \in \mathcal{V}, \mathbf{B} \in \mathrm{Sym}\,\mathcal{V}$ and $\mathbf{C} \in \mathrm{Sk}\,\mathcal{V}$. This is a model for heat conduction if $\mathbf{q}$ is interpreted as the heat flux vector, $\mathbf{a}$ as the temperature gradient $\mathbf{g}, \mathbf{B}$ as the symmetric part $\mathbf{D}$ of the velocity gradient, and $\mathbf{C}$ as the spin relative to an inertial frame, namely $\mathbf{W}+\mathbf{S}$. With these interpretations, write down the relation $\mathbf{q} = \hat{\mathbf{q}}(\mathbf{a}, \mathbf{B}, \mathbf{C})$ as employed by a general observer, and also by an inertial observer. What restriction is imposed upon $\hat{\mathbf{q}}$ in (12.4.5) by mfi? How does this differ from the corresponding restriction associated with isr?

[17] See (A.8.15) and (A.8.16).
[18] See (A.16.8) and (A.16.9).

Remark 12.4.4. Since Müller's relations for stress and heat flux can be regarded as involving the same constitutive functions, once the physical interpretations of arguments are made clear, with this proviso MFI 3 is *not* violated. However, just why MFI 3 should hold is called into question by a simple practical example. Consider an elastic structural support beam. An observer who has monitored its manufacture, and subsequent incorporation into a load-bearing structure (e.g., a steel girder in a bridge framework), may choose as reference configuration a stress-free situation as manifest in the factory. Another observer who has supervised the structural embedding may choose as reference the equilibrium configuration of the beam in situ, in which there will be residual stress, $\mathbf{T}_R$, say. If the response functions are $\mathbf{T}_1$ and $\mathbf{T}_2$, respectively, then these *must* differ since $\mathbf{T}_1(\mathbf{1}) = \mathbf{O}$ and $\mathbf{T}_2(\mathbf{1}) = \mathbf{T}_R \neq \mathbf{O}$.

Remark 12.4.5. Objectivity requirements O.1 through O.5 (see Murdoch [50]) were drafted with the aim of identifying in more detail than usual the nature of observer agreement in deterministic continuum mechanics. As a consequence, several issues have been resolved:

(i) Restrictions upon response functions due to objective considerations should involve only proper orthogonal maps.
(ii) There is no requirement for observers to choose the same response functions (but once such choices have been made, specific relations hold between these functions).
(iii) Material behaviour sensitive to spin relative to inertial frames is admissible (specifically, there is no fundamental tenet of classical physics that rules out such behaviour).

Remark 12.4.6. The effects of inertial spin predicted by kinetic theory are small, and have led many to regard mfi not as a principle but as merely a useful tool in obtaining restrictions on response functions (cf., e.g., Edelen & McLennan [59]). Here, in citing O.1 through O.5, a completely different perspective has been adopted. The effects, apparently anomalous, have led to a re-examination and improved understanding of fundamental assumptions in which inertial spin-dependent stress and heat flux are no longer an issue. Indeed, such dependence is a priori to be expected (see Remarks 12.2.8, 12.2.10, and 12.2.12). *The success of constitutive models which do not incorporate inertial spin dependence merely indicates that such dependence is negligible in the behaviour they address.*

12.4.5 A Personal History

As a first-year graduate student at Carnegie-Mellon University in 1971 I was introduced to material frame-indifference by Mort Gurtin via considerations of observer change. A year later, in lectures by Walter Noll, I came to appreciate the clarity and utility of regarding 'events' and 'space-time' in an observer-free, abstract way. In this approach observers are regarded to have conceptually distinct notions of space: this viewpoint has been adopted in Sections 12.2 and 12.3 since it helps to keep clear relations between the fields employed by different observers. As I understood matters at that time, mfi and isrbm differed only in that constitutive restrictions involved, respectively, all orthogonal maps or only proper orthogonal maps. In respect of elastic bodies mfi required that a response function $\hat{\mathbf{T}}$ should satisfy [see (12.3.22) with $\mathbf{Q} = -\mathbf{1}$]

$$\hat{\mathbf{T}}(-\mathbf{F}) = \hat{\mathbf{T}}(\mathbf{F}). \qquad (12.4.6)$$

This seemed unsatisfactory, given that

(i) a deformation gradient of $-\mathbf{F}$ is physically impossible to achieve, and
(ii) even if one were to conceive such a deformation, relation (12.4.6) would seem to involve a material symmetry, namely that an inversion (see Remark A.16.3) should be undetectable. Quite how one might interpret 'inversion' was not clear (e.g., would this involve complete physical inversion at subatomic level?), and imposition of such restriction upon Nature seemed unwarranted. In 1981 I wrote [63] to show that if observers can distinguish between right- and left-handed screws (but not accord to either any preferred status), then only proper orthogonal maps should appear in (12.3.22), and deformation gradients should have positive determinants. The argument used was similar to that of Subsection 12.3.3. In particular, I had relaxed MFI 3 for elastica by having observers employ reference configurations of their choice[19] (and hence use different response functions).

A year later I learned at first hand, in a summer school at Noto, Sicily, of Ingo Müller's relations for stress and heat flux in a rarefied gas which he claimed to violate mfi. This did not appear to be the case for me (as I informed him at the time), and I subsequently wrote [54] to prove this point. In this article mfi was interpreted via

Assumption 1: The physical quantities which characterise the behaviour of a given body are intrinsic, and
Assumption 2: All observers agree upon the nature of any given material.

The first assumption is precisely MFI 1, and the second is a relaxation of MFI 3 to O.4. That different observers should be expected to employ different response functions was emphasised. The final conclusions drawn from Müller's relations were that mfi is not violated thereby, that isrbm is an incorrect statement of the observer-independent nature of material behaviour[20], and that one should be aware of the possibility of inertial spin-dependent behaviour. Müller never accepted this viewpoint, and regarded my article as merely showing the objective natures of $\mathbf{T}$ and $\mathbf{q}$. After twenty years, during which isrbm seemed to be gaining increasing influence (whether as an absolute statement or as a very good approximation in practice), I wrote [50] to express the viewpoint presented in Subsection 12.3.2.

In choosing the title (Objectivity in Classical Continuum Physics: A Rationale for Discarding the 'Principle of Invariance under Superposed Rigid Body Motions' in Favour of Purely Objective Considerations) I was elaborating on Ericksen's viewpoint (see footnote 19). However, in so doing, I was ignoring earlier advice (in

[19] Here I am greatly indebted to Jerry Ericksen, who, in commenting upon a model of heat conduction which I felt illuminated the mfi/isrbm argument, wrote in November 1980, 'As I interpret the word, objectivity refers to what it is that different observers can agree upon. First, different observers can and do make their own choice of reference configurations'.

[20] Walter Noll agreed and, in a letter of November 1983, wrote, 'As to the matter concerning frame-indifference, I fully agree with your viewpoint, and I told Ingo Müller so when I recently met him at a meeting in Providence, RI'.

respect of [63]) by Truesdell, who wrote in October 1980, 'I hope you can avoid using "objectivity" because that term is often used (e.g. by Müller) to denote "invariance under superposed rigid motions"'. Somewhat naively (as it transpired), I was hoping to change Müller's viewpoint. Instead, publication of [50] was met by an immediate and personal attack by Liu [64], a former student of Müller, in which it was clear from the outset that the author had completely missed O.4 and O.5 as together forming a statement which embodies the essence of material indifference to its observation.

Anticipating a prolonged proliferation of published exchanges, the editor-in-chief of *Continuum Mechanics and Thermodynamics*, Kolumban Hutter, invited Liu and myself to enter into a dialogue to end in mutually agreed statements of our final positions. These appeared as Liu [65] and Murdoch [66]. Subsequently, implications of molecular considerations and the key role played by inertial observers were discussed in Murdoch [51]: these form the basis of Section 12.2. Specifically, ensuring that molecularly defined densities can be agreed upon by all observers has implications for constitutive relations. In particular, such relations should be expressed first in terms of Galilean-invariant functions of the motion relative to an inertial frame and thereafter re-phrased for general observers. This should be no surprise if one considers the inconsistency of expressing linear momentum balance in an inertial frame [so that **b** in (2.7.20) represents a genuine force density alone, free of inertial terms[21]] and then postulating a constitutive relation for **T** in terms of motion witnessed by a general observer.

12.4.6 A Final Remark

The three viewpoints represented by isrbm, mfi, and O.4 with O.5, may, very roughly, be summarised as mandating restrictions upon response functions via considerations of

(i) one observer, two motions of the body (isrbm)
(ii) two observers, one motion of the body, same response functions (mfi)
(iii) one motion of the body viewed in terms of two relative motions of one observer. (O.4, O.5)

Implementation of O.4 with O.5 (as here advocated) requires that the constitutive relations adopted by an observer should be invariant under superposed rigid motions of the *observer*. This is what is intended in (iii): see Remark 12.3.5.

[21] That is, terms involving $\ddot{\mathbf{c}}, \dot{\mathbf{W}}$ and $\mathbf{W}$ [see (12.2.23)].

13 Comments on Non-Local Balance Relations

13.1 Preamble

Here we consider two approaches to non-local balance relations which aim in some way to generalise standard forms of balance and in so doing provide novel alternative perspectives in continuum modelling. These are the non-local field theories of Edelen [67] and so-called 'peridynamics' introduced by Silling [68]. Comments are made on these approaches in the light of the developments of earlier chapters.

13.2 Edelen's Non-Local Field Theories

Edelen's approach is to postulate generally accepted *global* balances for a body [cf., e.g., (2.6.4) and (2.6.13)] but admit the possibility of fields which provide zero net input into such balances yet contribute non-trivially to the same balances taken only over *parts* of the body.

In respect of mass conservation, Edelen postulates that [cf. (2.5.2)]

$$\frac{d}{dt}\left\{\int_{B_t} \rho(\mathbf{x},t)dV_{\mathbf{x}}\right\} = 0, \tag{13.2.1}$$

and hence [cf. (2.5.4) et seq., with $R_t = B_t$]

$$0 = \frac{d}{dt}\left\{\int_{B_{t_0}} \rho(\boldsymbol{\chi}_{t_0}(\hat{\mathbf{x}},t))J(\hat{\mathbf{x}},t)d\hat{V}_{\hat{\mathbf{x}}}\right\} = \int_{B_{t_0}} \frac{\partial}{\partial t}\{\rho(\boldsymbol{\chi}_{t_0}(\hat{\mathbf{x}},t))J(\hat{\mathbf{x}},t)\}d\hat{V}_{\hat{\mathbf{x}}}. \tag{13.2.2}$$

If $\hat{\mu}$ denotes any integrable field on B_{t_0} with physical dimension $ML^{-3}T^{-1}$ and such that

$$\int_{B_{t_0}} \hat{\mu}(\hat{\mathbf{x}})d\hat{V}_{\hat{\mathbf{x}}} = 0, \tag{13.2.3}$$

then, from (13.2.2) and abbreviating,

$$\int_{B_{t_0}} \left\{\frac{\partial}{\partial t}\{(\rho \circ \boldsymbol{\chi}_{t_0})J\} + \hat{\mu}\right\} d\hat{V} = 0. \tag{13.2.4}$$

For Edelen the most general point-wise local form of mass balance must take the form

$$\frac{\partial}{\partial t}\{(\rho \circ \boldsymbol{\chi}_{t_0})J\} + \hat{\mu} = 0. \tag{13.2.5}$$

Integrating over an arbitrary part R_{t_0} of B_{t_0} yields

$$\frac{d}{dt}\left\{\int_{R_{t_0}}(\rho\circ\chi_{t_0})J\,d\hat{V}\right\}=\int_{R_{t_0}}\frac{\partial}{\partial t}\{\rho\circ\chi_{t_0}J\}d\hat{V}=-\int_{R_{t_0}}\hat{\mu}\,d\hat{V}. \tag{13.2.6}$$

Equivalently, $$\frac{d}{dt}\left\{\int_{R_t}\rho dV\right\}=-\int_{R_t}\mu\,dV, \tag{13.2.7}$$

where $$\mu(\mathbf{x},t):=\hat{\mu}(\hat{\mathbf{x}})(J(\hat{\mathbf{x}},t))^{-1}\quad\text{and}\quad R_t:=\chi_{t_0}(R_{t_0},t). \tag{13.2.8}$$

Since $$\int_{B_t}\mu\,dV=0=\int_{R_t}\mu\,dV+\int_{B_t-R_t}\mu\,dV, \tag{13.2.9}$$

relation (13.2.7) may be written as

$$\frac{d}{dt}\left\{\int_{R_t}\rho\,dV\right\}=\int_{B_t-R_t}\mu\,dV. \tag{13.2.10}$$

Edelen terms $\hat{\mu}$ a *localisation residual* for the global conservation of mass.

Remark 13.2.1. Recourse to a referential description in the foregoing would seem to be motivated by requiring a time-independent residual $\hat{\mu}$ which depends only upon the chosen reference situation. The alternative would be to assume a family of such residuals parametrised by time t, $\{\mu_t\}$ say, for each member of which

$$\int_{B_t}\mu_t\,dV=0. \tag{13.2.11}$$

Such alternative procedure would be equivalent to that adopted if $\mu_t(\mathbf{x}):=\mu(\mathbf{x},t)$ were to satisfy $(13.2.8)_1$.

Exercise 13.2.1. Show directly from (13.2.5), (13.2.8), and (B.5.36) that

$$\dot{\rho}+\rho\,\mathrm{div}\,\mathbf{v}+\mu=0, \tag{13.2.12}$$

or, equivalently, $$\frac{\partial\rho}{\partial t}+\mathrm{div}\{\rho\mathbf{v}\}+\mu=0. \tag{13.2.13}$$

Remark 13.2.2. Relation (13.2.7) corresponds to a body with an 'internal' mass supply (or depletion) rate but whose total mass remains constant. This is clearly a special case of (8.9.12), which may be written, upon suppressing the subscripts (which refer to the averaging scales of length and time), as

$$\frac{\partial\rho}{\partial t}+\mathrm{div}\{\rho\mathbf{v}\}-\mathcal{G}=0. \tag{13.2.14}$$

Of course, in systems with changing material content (see Section 8.9)

$$\int_{B_t}\mathcal{G}\,dV\neq 0 \tag{13.2.15}$$

in general.

Remark 13.2.3. While the local form (13.2.12) of (13.2.7) admits the simple interpretation of an internal mass supply, the alternative formulation (13.2.10) would not appear to have a physical interpretation: it would correspond to a *non-local* volumetric transfer of matter in $B_t - R_t$ into R_t.

Edelen postulates global linear momentum balance in the standard form (2.6.4) [see (1.4.1) of [67]], namely

$$\int_{\partial B_t} \mathbf{t}\, dS + \int_{B_t} \mathbf{b}\, dV = \frac{d}{dt}\left\{\int_{B_t} \rho\mathbf{v}\, dV\right\}. \tag{13.2.16}$$

Re-stating this in referential form, Edelen considers a class of candidate Piola–Kirchhoff stress tensor fields on B_{t_0} which deliver the required traction $\mathbf{t}$ when referred to ∂B_{t_0} *and* which are continuously differentiable in B_{t_0}. When re-expressed in current situation format these comprise an equivalence class (in the sense of traction delivery and regularity in B_t) of candidate Cauchy stress tensors $\mathbf{T}$. Further, noting that global linear momentum balance is unaffected by the addition of any integrable field $\hat{\boldsymbol{\phi}}$ which has the physical dimension of a force density (namely $ML^{-2}T^{-2}$) and for which $\boldsymbol{\phi} := \hat{\boldsymbol{\phi}} J^{-1}$ satisfies

$$\int_{B_t} \boldsymbol{\phi}\, dV = \mathbf{0}, \tag{13.2.17}$$

Edelen takes the local form of (13.2.16) to be [here (13.2.12) has been used in simplifying the right-hand side]

$$\operatorname{div}\mathbf{T} + \mathbf{b} + \boldsymbol{\phi} = \rho\dot{\mathbf{v}} - \mu\mathbf{v}. \tag{13.2.18}$$

Accordingly, integrating (13.2.18) over any part $R_t \subset B_t$ which convects with the motion of the body, and invoking (13.2.12),

$$\int_{\partial R_t} \mathbf{T}\mathbf{n}\, dS + \int_{R_t} (\mathbf{b} + \boldsymbol{\phi})dV = \frac{d}{dt}\left\{\int_{R_t} \rho\mathbf{v}\, dV\right\}. \tag{13.2.19}$$

From (13.2.17)

$$\int_{R_t} \boldsymbol{\phi}\, dV + \int_{B_t - R_t} \boldsymbol{\phi}\, dV = \mathbf{0}, \tag{13.2.20}$$

so (13.2.19) is expressible in the form

$$\int_{\partial R_t} \mathbf{T}\mathbf{n}\, dS + \int_{R_t} \mathbf{b}\, dV - \int_{B_t - R_t} \boldsymbol{\phi}\, dV = \frac{d}{dt}\left\{\int_{R_t} \rho\mathbf{v}\, dV\right\}. \tag{13.2.21}$$

Remark 13.2.4. In comparing (13.2.18) and (13.2.21) with their counterparts derived from a molecular perspective there are two distinct considerations: (i) non-uniqueness of $\mathbf{T}$, and (ii) interpretation of $\boldsymbol{\phi}$.

(i) In Chapters 5 and 8 the Cauchy stress [cf. (5.5.20) and (8.9.48)]

$$\mathbf{T} := \mathbf{T}^- - \mathcal{D}, \tag{13.2.22}$$

where $\mathcal{D}$ emerged naturally as a thermokinetic field, and

$$\operatorname{div}\mathbf{T}^- = \mathbf{f}, \tag{13.2.23}$$

where $\mathbf{f}$ was a uniquely defined scale-dependent internal force density field [cf. (5.5.2), (8.9.41) and (8.9.44)]. Non-uniqueness of $\mathbf{T}^-$ corresponded to the different solutions $\mathbf{a}_i$ and $\mathbf{b}_{ij}$ to (5.6.2) and (5.6.8), and the consequent interpretations have been discussed in Section 5.8.

(ii) Form (13.2.18) can be compared with (8.9.51):

$$\text{div}\,\mathbf{T} + {}^{\text{ext}}\mathbf{f} + \mathbf{b} + \mathbf{P} = \rho\dot{\mathbf{v}} + \mathcal{G}\mathbf{v}. \tag{8.9.51}$$

As seen from (13.2.12) and (13.2.14), in the context of a system $\mathcal{M}^+$ with time-dependent material content, $\mu = -\mathcal{G}$. Further, since $\mathbf{b}$ in both (2.7.20) and (8.9.51) represents the net volumetric effect of agencies *external* to the body, we are led to

$$\textit{identify } -\boldsymbol{\phi} \qquad \textit{with} \qquad {}^{\text{ext}}\mathbf{f} + \mathbf{P}. \tag{13.2.24}$$

Recall that ${}^{\text{ext}}\mathbf{f}$ represents a force density for $\mathcal{M}^+$ which derives from interactions with the complementary system $\mathcal{M}^-$ and that $\mathbf{P}$ is the net volumetric supply rate of momentum which accompanies the internal net volumetric mass supply to $\mathcal{M}^+$ from $\mathcal{M}^-$.

Remark 13.2.5. At this point one might wonder where non-locality enters into the reckoning. Edelen's viewpoint is that $-\boldsymbol{\phi}$ in (13.2.21) represents a force density whose integral over $B_t - R_t$ represents the net force exerted upon that part of the body within R_t by that part outside R_t. In such case $-\boldsymbol{\phi}$ is a non-local interaction density. Since the only macroscopically long-range interactions are gravitational, it would seem to be clear that $-\boldsymbol{\phi}(\mathbf{x},t)\Delta V$ approximates the gravitational force exerted upon that part of the body within R_t by that part of the body located in a small region of volume ΔV which contains $\mathbf{x} \in B_t - R_t$ at time t. Of course, the resultant gravitational force exerted on a body by itself vanishes, and hence letting $\text{vol}(R_t) \to 0$ yields (13.2.17).

13.3 Peridynamics

In order to model the effects of defects and inhomogeneities in solids, in particular those associated with cracks and fracture, Silling [68] introduced a referential form of linear momentum balance in which the divergence of the stress tensor was replaced by an integral of a force double-density field[1] $\mathcal{F}_S$. Specifically, taking the reference configuration to be the situation at time[2] t_0, this balance is

$$\rho_0(\hat{\mathbf{x}})\ddot{\mathbf{u}}(\hat{\mathbf{x}},t) = \int_{\mathcal{E}} \mathcal{F}_S(\mathbf{u}(\hat{\mathbf{y}},t) - \mathbf{u}(\hat{\mathbf{x}},t), \hat{\mathbf{y}} - \hat{\mathbf{x}}, \hat{\mathbf{x}})dV_{\hat{\mathbf{y}}} + \mathbf{b}_0(\hat{\mathbf{x}},t), \tag{13.3.1}$$

where the displacement field $\mathbf{u} : B_{t_0} \to \mathcal{V}$ is given by

$$\mathbf{u}(\hat{\mathbf{z}},t) := \chi_{t_0}(\hat{\mathbf{z}},t) - \hat{\mathbf{z}}, \tag{13.3.2}$$

and
$$\rho_0(\hat{\mathbf{x}}) := \rho(\hat{\mathbf{x}},t_0), \qquad \mathbf{b}_0(\hat{\mathbf{x}},t) := \mathbf{b}(\chi_{t_0}(\hat{\mathbf{x}},t),t)J(\hat{\mathbf{x}},t). \tag{13.3.3}$$

The comparable form of the general linear momentum balances derived from molecular considerations [namely (5.5.15) in respect of purely spatial averaging, and (8.9.81) with $\mathbf{f}^{\text{ext}}, \mathbf{P}$, and $\mathcal{G}$ absent in respect of space–time averaging] is

$$\rho_0(\hat{\mathbf{x}})\ddot{\mathbf{u}}(\hat{\mathbf{x}},t) = -(\text{div}\,\mathcal{D}_0)(\hat{\mathbf{x}},t) + \mathbf{f}_0(\hat{\mathbf{x}},t) + \mathbf{b}_0(\hat{\mathbf{x}},t), \tag{13.3.4}$$

[1] Notation has been changed for ease of comparison with the molecular considerations of earlier chapters.

[2] See Section 2.5.

where $\mathcal{D}_0$ and $\mathbf{f}_0$ denote the referential counterparts of $\mathcal{D}$ and $\mathbf{f}$. Thus [cf. (8.5.15)]

$$\mathcal{D}_0 := (\det \mathbf{F})(\mathcal{D} \circ \chi_{t_0})\mathbf{F}^{-T} \qquad \text{and} \qquad \mathbf{f}_0 := (\mathbf{f} \circ \chi_{t_0})J. \tag{13.3.5}$$

Silling seems initially to have been unaware of the thermokinetic contribution to the balance represented by the term $-\text{div}\,\mathcal{D}_0$: this omission was rectified by Lehoucq and Sears [69]. Modulo such omission, comparison of (13.3.4) with (13.3.1) leads to the identification

$$\int_{\mathcal{E}} \mathcal{F}_S(\mathbf{u}(\hat{\mathbf{y}},t) - \mathbf{u}(\hat{\mathbf{x}},t), \hat{\mathbf{y}} - \hat{\mathbf{x}}, \hat{\mathbf{x}})dV_{\hat{\mathbf{y}}} \leftrightarrow \mathbf{f}_0(\hat{\mathbf{x}},t). \tag{13.3.6}$$

The constitutive nature of the material is embodied in the choice of $\mathcal{F}_S$. Interpreting value $\mathcal{F}_S(\mathbf{u}(\hat{\mathbf{y}},t) - \mathbf{u}(\hat{\mathbf{x}},t), \hat{\mathbf{y}} - \hat{\mathbf{x}}, \hat{\mathbf{x}})$ to be[3] 'the force vector (per unit volume squared) that the particle $\hat{\mathbf{y}}$ exerts on the particle $\hat{\mathbf{x}}$', and recognising that interactions are localised, Silling assumed that $\mathcal{F}_S$ should vanish whenever $\|\hat{\mathbf{y}} - \hat{\mathbf{x}}\| > \delta$ for some $\delta > 0$. Accordingly (13.3.6) becomes the identification[4]

$$\int_{S_\delta(\hat{\mathbf{x}})} \mathcal{F}_S(\mathbf{u}(\hat{\mathbf{y}},t) - \mathbf{u}(\hat{\mathbf{x}},t), \hat{\mathbf{y}} - \hat{\mathbf{x}}, \hat{\mathbf{x}})dV_{\hat{\mathbf{y}}} \leftrightarrow \mathbf{f}_0(\hat{\mathbf{x}},t). \tag{13.3.7}$$

For (13.3.1) to make sense it is necessary that $\delta > \epsilon$, where ϵ denotes the length scale associated with the definitions of $\rho, \mathbf{v}, \chi_{t_0}$, and $\mathbf{f}$ (and hence of $\rho_0, \mathbf{u}$, and $\mathbf{f}_0$). Thus the peridynamic approach is non-local. Indeed, Silling (in ref. [68], p. 176) states the following:

Statement S1: 'Particles separated by a finite distance can interact with each other.'

Such a viewpoint is emphasised by the phrase 'long-range forces' in the title of [68]. This statement prompts the following fundamental query:

Question Q1: What does Silling mean by a particle?

It would appear here that 'particle' is not synonymous with 'atom' or 'molecule' since the peridynamic approach is a *continuum* theory. Thus one infers that 'particle' is used in the sense of 'material point'. However, as pointed out in Section 2.3, and formalised in Section 5.2, the notion of material point is a scale-dependent artefact whose utility lies in tracing the motion at a length scale ϵ of choice. What *does* make sense at any given scale is the force $\mathbf{F}_\epsilon(\mathbf{x},\mathbf{y},t)$ exerted at any time on molecules in $S_\epsilon(\mathbf{x})$ by molecules in $S_\epsilon(\mathbf{y})$ for any pair of (geometrical) points $\mathbf{x},\mathbf{y} \in \mathcal{E}$. That is (see Section 5.4),

$$\mathbf{F}_\epsilon(\mathbf{x},\mathbf{y},t) := \sum_{i,\mathbf{x}}{}^{t} \sum_{j,\mathbf{y}}{}^{t} \mathbf{f}_{ij}(t), \tag{13.3.8}$$

where $\sum_{i,\mathbf{x}}^{t} \left(\sum_{j,\mathbf{y}}^{t} \right)$ denotes a sum over those point masses $P_i(P_j)$ in $S_\epsilon(\mathbf{x})$ $(S_\epsilon(\mathbf{y}))$ at time t. The relevant force double-density field is (here the sums are taken over *all*

[3] Cf. Silling et al. [70], p. 152.
[4] Recall that $S_\delta(\hat{\mathbf{x}})$ denotes that spherical region of radius δ centred at $\hat{\mathbf{x}}$.

molecules[5])

$$\mathbf{G}_\epsilon(\mathbf{x},\mathbf{y},t) := \sum\sum_{i\neq j} \mathbf{f}_{ij}(t) w_\epsilon(\mathbf{x}_i(t)-\mathbf{x}) w_\epsilon(\mathbf{x}_j(t)-\mathbf{y}). \tag{13.3.9}$$

Notice that the normalisation property (4.2.4) for any weighting function ensures that

$$\int_{\mathcal{E}} \mathbf{G}_\epsilon(\mathbf{x},\mathbf{y},t) dV_{\mathbf{y}} = \sum\sum_{i\neq j} \mathbf{f}_{ij}(t) w_\epsilon(\mathbf{x}_i(t)-\mathbf{x}) = \mathbf{f}(\mathbf{x},t). \tag{13.3.10}$$

This makes precise, in molecular terms, what Silling appears to intend, although he makes no mention of scale dependence. Equation (13.3.10) may be regarded as giving the current situation counterpart of the referentially based identification (13.3.6).

Exercise 13.3.1. Show that if $\mathbf{f}_{ji} = -\mathbf{f}_{ij}$, then

$$\mathbf{G}_\epsilon(\mathbf{y},\mathbf{x},t) = -\mathbf{G}_\epsilon(\mathbf{x},\mathbf{y},t). \tag{13.3.11}$$

Now recall Statement S1 and regard the interaction between Silling's 'particles' to be given by (13.3.8). In such case $\mathbf{G}_\epsilon(\mathbf{x},\mathbf{y},t)$ is the force double-density associated (at scale ϵ and time t) with the action of a particle/material point at $\mathbf{y}$ upon another particle/material point at $\mathbf{x}$. From $(13.3.5)_2$ and (13.3.10)

$$\begin{aligned}\mathbf{f}_0(\hat{\mathbf{x}},t) &= \mathbf{f}(\chi_{t_0}(\hat{\mathbf{x}},t),t) J(\hat{\mathbf{x}},t)\\ &= \left(\int_{\mathcal{E}} \mathbf{G}_\epsilon(\chi_{t_0}(\hat{\mathbf{x}},t),\mathbf{y},t) dV_{\mathbf{y}}\right) J(\hat{\mathbf{x}},t)\\ &= \left(\int_{\mathcal{E}} \mathbf{G}_\epsilon(\chi_{t_0}(\hat{\mathbf{x}},t),\chi_{t_0}(\hat{\mathbf{y}},t),t) J(\hat{\mathbf{y}},t) dV_{\hat{\mathbf{y}}}\right) J(\hat{\mathbf{x}},t).\end{aligned} \tag{13.3.12}$$

Thus
$$\mathbf{f}_0(\hat{\mathbf{x}},t) = \int_{\mathcal{E}} \mathcal{F}_\epsilon(\chi_{t_0}(\hat{\mathbf{x}},t),\chi_{t_0}(\hat{\mathbf{y}},t),t,\hat{\mathbf{x}},\hat{\mathbf{y}}) dV_{\hat{\mathbf{y}}}, \tag{13.3.13}$$

where
$$\mathcal{F}_\epsilon(\chi_{t_0}(\hat{\mathbf{x}},t),\chi_{t_0}(\hat{\mathbf{y}},t),t,\hat{\mathbf{x}},\hat{\mathbf{y}}) := \mathbf{G}_\epsilon(\chi_{t_0}(\hat{\mathbf{x}},t),\chi_{t_0}(\hat{\mathbf{y}},t),t) J(\hat{\mathbf{x}},t) J(\hat{\mathbf{y}},t). \tag{13.3.14}$$

Recalling (13.3.9) thus yields

$$\mathcal{F}_\epsilon(\mathbf{x}(t),\mathbf{y}(t),t,\hat{\mathbf{x}},\hat{\mathbf{y}}) = \sum\sum_{i\neq j} \mathbf{f}_{ij}(t) w_\epsilon(\mathbf{x}_i(t)-\mathbf{x}(t)) w_\epsilon(\mathbf{x}_j(t)-\mathbf{x}(t)) J(\hat{\mathbf{x}},t) J(\hat{\mathbf{y}},t), \tag{13.3.15}$$

where
$$\mathbf{x}(t) := \chi_{t_0}(\hat{\mathbf{x}},t) \quad \text{and} \quad \mathbf{y}(t) := \chi_{t_0}(\hat{\mathbf{y}},t). \tag{13.3.16}$$

Comparison of the two expressions for $\mathbf{f}_0(\hat{\mathbf{x}},t)$ given by (13.3.13) and identification (13.3.6) indicates that Silling's choice $\mathcal{F}_S$ is to be viewed as a special (constitutive) case of the general force double-density function $\mathcal{F}_\epsilon$. To simplify comparison of $\mathcal{F}_S$ with $\mathcal{F}_\epsilon$, note that the former may be regarded to be a function $\mathcal{F}_S$

[5] Of course, the localised selection is achieved by the presence of the weighting functions. Recall also that w_ϵ values have physical dimension L^{-3}.

of $\chi_{t_0}(\hat{\mathbf{y}},t) - \chi_{t_0}(\hat{\mathbf{x}},t)$, $\hat{\mathbf{x}}$, and $\hat{\mathbf{y}}$: knowledge of these quantities is equivalent to that of $\mathbf{u}(\hat{\mathbf{y}},t) - \mathbf{u}(\hat{\mathbf{x}},t), \hat{\mathbf{y}} - \hat{\mathbf{x}}$, and $\hat{\mathbf{y}}$.

Exercise 13.3.2. Prove the last assertion, recalling (13.3.2).

Silling's constitutive function takes values $\tilde{\mathcal{F}}_S(\mathbf{y}(t) - \mathbf{x}(t), \hat{\mathbf{x}}, \hat{\mathbf{y}})$, while the general ϵ-scale function takes values $\mathcal{F}_\epsilon(\mathbf{x}(t), \mathbf{y}(t), t, \hat{\mathbf{x}}, \hat{\mathbf{y}})$, defined by (13.3.15) in terms of molecular locations, together with ϵ-scale localisation effected in terms of weighting function w_ϵ. In particular, unlike $\mathcal{F}_\epsilon$, $\tilde{\mathcal{F}}_S$ embodies no explicit time dependence. The source of such time dependence for $\mathcal{F}_\epsilon$ is evident from (3.13.15): the molecular content of $S_\epsilon(\mathbf{x}(t))$ and $S_\epsilon(\mathbf{y}(t))$, which determine which interactions contribute to $\mathcal{F}_\epsilon$ at time t, may change with time.

Having placed the notion of 'particle' on a secure, scale-dependent basis, it is possible to discuss Silling's assumption S.1. in terms of the general form (13.3.15) of force double density $\mathcal{F}_\epsilon$, of which $\mathcal{F}_S$ has been shown to be a special case. In particular, it is of crucial interest to explore the basis of the assumption that $\mathcal{F}_S$ should vanish if $\|\hat{\mathbf{y}} - \hat{\mathbf{x}}\| > \delta$. The origin of such a cut-off separation δ lies in the nature of intermolecular forces $\mathbf{f}_{ij}$.

Remark 13.3.1. Intermolecular forces are complex: some indication of the nature of $\mathbf{f}_{ij}$ as a time-averaged resultant force between the subatomic constituents of molecules was given in Section 5.4. More specifically, interactions between subatomic discrete entities (nuclei and electrons) are composed of electromagnetic and gravitational contributions, of which the former are overwhelmingly dominant. Although individual interactions are essentially delivered by inverse square (of separation) relations, *net* interactions between the pair of *assemblies* of nuclei and electrons that constitute two electrically neutral molecules decay much more rapidly with separation. While detailed study of intermolecular forces is the province of quantum mechanics, simple modelling of interactions between 'multipoles' provides an indication of asymptotic separation dependence of interactions between pairs of molecules and ions:[6] only in the ion–ion case is inverse-square dependence a possibility. The most general long-range contribution to molecule–molecule interactions is named for van der Waals and decays as the inverse seventh power of separation. As stated in Israelachvili [22]: 'They are long-range forces, effective from very large distances[7] (>50 nm) down to interatomic separations'. Now recall that the values of $\mathcal{F}_\epsilon$ correspond to resultant interactions between pairs of molecular assemblies (each of which includes all molecules in a spherical region of radius ϵ). Notwithstanding the perturbations of pairwise interactions due to neighbouring molecules, and even in the case of net electrically neutral assemblies of *ions*, it is difficult to avoid the conclusion that $\mathcal{F}_\epsilon$ values decay rapidly with separation $\|\mathbf{x}(t) - \mathbf{y}(t)\|$ and are negligible if $\|\mathbf{x}(t) - \mathbf{y}(t)\| > 10^3$ Å.

The conclusion to be drawn from Remark 13.3.1 is that Silling's assertion (S.1) is physically suspect. Said differently, it is necessary to be clear about context when using the term *long-range force*: to a molecular physicist this means a force which can extend to 1,000 Å, but at the continuum level (with associated spatial-averaging length scale ϵ) a force significantly in excess of ϵ is intended.

[6] Cf., e.g., Table 1 in Stone [71] and Fig. 1 of Israelachvili [22].
[7] 1 nm = 10^{-9} m = 10 Å.

Of course, if, say, $\epsilon \sim 10$ Å and $\delta \sim 10^3$ Å, then peridynamic theory would make sense. However, at this level molecular dynamical simulations would appear to be more appropriate since interatomic separations in solids are about 3 Å. Further, only microcracks and micro-inhomogeneities could be modelled at such scales, in contrast with claims associated with the peridynamic approach. One possible situation with longer length scales would be a multiply fractured dielectric with surface distribution of (electronic) charge.

Remark 13.3.2. Given that the aim of peridynamics would seem to be to provide a continuum model of the effects of small-scale cracks and inhomogeneities without precise knowledge of such features, there is a clear parallel with porous body modelling at scales in excess of those at which porosity becomes evident. The problem now is one of how to incorporate surface and edge effects at microscale voids, rather than the flow of fluid through pores discussed in Chapter 10. As with porous media, there are three distinct scale-dependent viewpoints involved (cf., e.g., Murdoch & Hassanizadeh [15]). These are the molecular (discrete) perspective at scale $\epsilon_0 \sim 2$ Å, the scale ϵ_1 at which microcracks are considered (delineated by the boundary at scale[8] ϵ_1), and the scale ϵ_2 at which microcracks are no longer evident. The aim of such an approach is to inform the ϵ_2-scale relations with such aspects of the ϵ_0 and ϵ_1 level knowledge as is available.

[8] See subsection 4.3.2. Here the triangulated geometric boundary (4.3.14) would seem to be most appropriate.

14 Elements of Classical Statistical Mechanics

14.1 Preamble

Statistical physics[1] addresses the microscopic basis of macroscopic behaviour from a probabilistic viewpoint, and is of profound importance (inter alia) in its approach to, and development of, thermodynamics, and in its analysis of fluctuations in observable behaviour. For this reason it is instructive to review basic aspects of classical[2] statistical mechanics and to explore the statistical approach to balance relations.

14.2 Basic Concepts in Classical Statistical Mechanics

14.2.1 Time Evolution in Phase Space of a System of Interacting Point Masses

The *microscopic state* (or *microstate*) of a system $\mathcal{M}$ of point masses $P_i (i = 1, 2, \ldots, N)$ with respect to an inertial frame is defined to be the ordered list

$$\mathbf{X} := (\mathbf{x}_1, \mathbf{x}_2, \ldots, \mathbf{x}_N; \mathbf{p}_1, \mathbf{p}_2, \ldots, \mathbf{p}_N), \tag{14.2.1}$$

where $\mathbf{x}_i$ and $\mathbf{p}_i$ denote the location and momentum of P_i in the frame. The collection of all possible lists is termed *phase space* $\mathbb{P}$. Thus

$$\mathbb{P} = \mathbb{C} \times \mathbb{M}, \tag{14.2.2}$$

where

$$\mathbb{C} := \mathcal{E}^N \quad \text{and} \quad \mathbb{M} := \mathcal{V}^N. \tag{14.2.3}$$

$\mathbb{C}$ is termed *configuration space* and $\mathbb{M}$ *momentum space.*

If a Cartesian co-ordinate system[3] $C(\mathbf{x}_0; \mathbf{e}_1, \mathbf{e}_2, \mathbf{e}_3)$ is chosen for Euclidean space $\mathcal{E}$, then both $\mathbb{C}$ and $\mathbb{M}$ can be identified with $\mathbb{R}^{3N}$. Thus any microstate [see (14.2.1)] may be identified with an ordered list of $3N$ numbers, each having the physical dimension of length, followed by an ordered list of $3N$ numbers,

[1] Cf., e.g., Landau & Lifschitz [72].
[2] Here microscopic dynamics is governed by Newtonian (as distinct from quantum) mechanics.
[3] See Appendix B.2.

each having the physical dimension of momentum. Indeed, we may make the identification[4]

$$\mathbf{X} \leftrightarrow (X_1, X_2, \ldots, X_{6N}), \tag{14.2.4}$$

where the co-ordinates of $\mathbf{x}_i$ are $(X_{3i-2}, X_{3i-1}, X_{3i})$ and the components of $\mathbf{p}_i$ are $(X_{3N+3i-2}, X_{3N+3i-1}, X_{3N+3i})$.

Of course, particle locations $\mathbf{x}_i$ and momenta $\mathbf{p}_i$ will in general change with time, and thus so too will $\mathbf{X}$. To express this time dependence we write

$$\mathbf{X}(t) := (\mathbf{x}_1(t), \mathbf{x}_2(t), \ldots; \mathbf{p}_1(t), \mathbf{p}_2(t), \ldots, \mathbf{p}_N(t)), \tag{14.2.5}$$

or, when using the Cartesian co-ordinate formulation (14.2.4) (see footnote 4),

$$\mathbf{X}(t) := (X_1(t), X_2(t), \ldots, X_{6N}(t)). \tag{14.2.6}$$

As time evolves, the *point* $\mathbf{X}(t)$ in $\mathbb{P}$ given by (14.2.5) [equivalently *point* $\mathbf{X}(t)$ in $\mathbb{P}$, as identified with $\mathbb{R}^{6N}$ in (14.2.6)] traces out a *curve* in $\mathbb{P}$. If $\mathbf{F}_i(t)$ denotes the net force on P_i at time t (a composite of interactions with other point masses in $\mathcal{M}$, together with forces 'external' to $\mathcal{M}$, at this time), then, since the frame is inertial,

$$\dot{\mathbf{p}}_i(t) = \mathbf{F}_i(t). \tag{14.2.7}$$

Hence, provided that $\mathbf{F}_i$ depends only upon the current locations and momenta (and thus velocities) of (possibly all) other point masses in $\mathcal{M}$ and time,[5] the time evolution of $\mathbf{X}$ will be governed by an equation of the form

$$\dot{\mathbf{X}}(t) = \mathcal{F}(\mathbf{X}(t), t). \tag{14.2.8}$$

If $\mathbf{X}$ is known at some time t_0, so that for some $\hat{\mathbf{X}} \in \mathbb{P}$

$$\mathbf{X}(t_0) = \hat{\mathbf{X}}, \tag{14.2.9}$$

then a solution to the initial-value problem (14.2.8) with (14.2.9) is assumed to exist and to be unique. *Thus at the microscopic level material behaviour is considered to be deterministic.*

14.2.2 Ensembles, Probability Density Functions, and Ensemble Averaging

In practice complete knowledge of system $\mathcal{M}$ at any given time t_0 (embodied in $\hat{\mathbf{X}}$) is unobtainable, given the number and nature of the point masses considered.[6] Nevertheless, at least some information about $\mathcal{M}$ is usually available concerning its spatial location, mass, momentum, and energy. All microscopic states compatible with information available about $\mathcal{M}$ up to and including time t_0 are said to be *admissible* at time t_0. It follows that subsequent behaviour of the system may correspond to

[4] No notational distinction will be made between $\mathbf{X}$ given by (14.2.1) and the corresponding relation (14.2.4) when $\mathbb{P}$ is identified with $\mathbb{R}^{6N}$ via choice of co-ordinates. It will be evident from context which interpretation is intended.

[5] Here any explicit time dependence is considered to derive from agencies external to $\mathcal{M}$, such as, for example, forces which confine $\mathcal{M}$ to a time-dependent region. It also has been noted that $\dot{\mathbf{x}}_i = \mathbf{p}_i/m_i$, where m_i denotes the mass of P_i.

[6] Point masses can be interpreted to be fundamental discrete units of any kind (e.g., molecules, atoms, nuclei, or electrons), but it is usual to consider only molecular entities.

any one of the solutions to (14.2.8) with (14.2.9) for which $\hat{\mathbf{X}}$ is admissible at time t_0. *The collection $\mathbb{D}_0$ of admissible states at time t_0, together with the evolution equation (14.2.8), constitute what is known as the* **ensemble** *appropriate to knowledge of $\mathcal{M}$ up to and including time t_0*. Imprecision in knowledge of $\mathbf{X}(t_0)$ is modelled probabilistically: a fundamental postulate of classical statistical mechanics is that there exists a *probability density function*

$$P_0 : \mathbb{P} \to \mathbb{R}^+ \tag{14.2.10}$$

such that[7]

$$\int_{\mathbb{S}} P_0(\hat{\mathbf{X}}) dV_{\hat{\mathbf{X}}} \; (=: \int_{\mathbb{S}} P_0 \, dV) \tag{14.2.11}$$

represents the probability that the system is in a state which lies in a subset $\mathbb{S}$ of $\mathbb{P}$. Accordingly,

$$\int_{\mathbb{D}_0} P_0 \, dV = 1 \qquad \text{and} \qquad \int_{\mathbb{P}-\mathbb{D}_0} P_0 \, dV = 0. \tag{14.2.12}$$

Remark 14.2.1. It is often assumed that any two microstates which lie in $\mathbb{D}_0$ at time t_0 are equally likely; that is, if $\mathbf{X} \in \mathbb{D}_0$, then $P_0(\mathbf{X})$ is the reciprocal of the phase space volume of $\mathbb{D}_0$. Such postulation of 'uniformity of ignorance' constitutes a working hypothesis, and no more, at this level of generality.

Let

$$\boldsymbol{\phi}_0(.,t) : \mathbb{D}_0 \to \mathbb{P} \tag{14.2.13}$$

denote the solution to (14.2.8) with (14.2.9); that is, $\boldsymbol{\phi}_0(\hat{\mathbf{X}},t)$ denotes the state at time t to which the system has evolved if it was in state $\hat{\mathbf{X}}$ at time t_0. Function $\boldsymbol{\phi}_0$ is termed the *motion map with respect to the situation at time* t_0. It is assumed that $\boldsymbol{\phi}_0$ is one-to-one; that is, any state $\mathbf{X}$ which lies in $\boldsymbol{\phi}_0(\mathbb{D}_0,t)$ at time t corresponds to one and only one state $\hat{\mathbf{X}}$ at time t_0 that has evolved to $\mathbf{X}$ at time t. In particular, for any volume-integrable subset $\mathbb{S}$ of $\mathbb{D}_0$, $\boldsymbol{\phi}_0(\hat{\mathbf{X}},t) \in \boldsymbol{\phi}_0(\mathbb{S},t)$ if and *only* if $\hat{\mathbf{X}} \in \mathbb{S}$ at time t_0. Accordingly, if there exists a probability density function $P(.,t)$ associated with states at time t, then

$$\int_{\mathbb{S}} P_0(\hat{\mathbf{X}}) dV_{\hat{\mathbf{X}}} = \int_{\boldsymbol{\phi}_0(\mathbb{S},t)} P(\mathbf{X},t) dV_{\mathbf{X}}. \tag{14.2.14}$$

Here the second integral delivers the probability that the state of the system at time t lies in $\boldsymbol{\phi}_0(\mathbb{S},t)$ given that it lies in $\mathbb{S} \subset \mathbb{P}$ at time t_0. Relation (14.2.14) is, modulo the difference in dimension of the spaces involved, mathematically identical to (2.5.5): $\mathbb{S} \subset \mathbb{P}$ corresponds to $R \subset \mathcal{E}$, P_0 to $\rho(.,t_0)$, $P(.,t)$ to $\rho(.,t)$, and $\boldsymbol{\phi}_0$ to $\boldsymbol{\chi}_{t_0}$. The analogue of (2.5.16) is

$$\frac{\partial P}{\partial t} + \mathrm{div}_{\mathbb{P}}\{P\mathbf{V}\} = 0, \tag{14.2.15}$$

where [cf. (2.3.3)]

$$\mathbf{V}(\mathbf{X},t) := \dot{\mathbf{X}}(t), \qquad \text{with} \quad \mathbf{X}(\tau) := \phi_0(\hat{\mathbf{X}},\tau) \qquad \text{and} \quad \phi_0(\hat{\mathbf{X}},t) = \mathbf{X}. \tag{14.2.16}$$

[7] See Appendix B.11.3 for a discussion of integration in $\mathbb{P}$.

The derivation of (14.2.15) is given in Subsection B.11.3 of Appendix B. If $\mathbf{U}$ is a vector field on an open subset $\mathbb{D}$ of $\mathbb{P}$ (that is, $\mathbf{U} : \mathbb{D} \to \mathcal{V}^{2N}$), then

$$\mathrm{div}_{\mathbb{P}}\mathbf{U} := \sum_{p=1}^{6N} \frac{\partial U_p}{\partial X_p} =: U_{p,p}, \tag{14.2.17}$$

where

$$\mathbf{U}(\mathbf{X}) = (U_1(X_1,\dots,X_{6N}), U_2(X_1,\dots,X_{6N}),\dots, U_{6N}(X_1,\dots,X_{6N})). \tag{14.2.18}$$

Remark 14.2.2. If $\mathbf{F}_i$ [see (14.2.7)] depends only upon locations of (possibly *all*) point masses, then (see Exercise 14.2.1)

$$\mathrm{div}_{\mathbb{P}}\,\mathbf{V} = 0. \tag{14.2.19}$$

Exercise 14.2.1. Note that from (14.2.5) and (14.2.16)

$$\mathbf{V} = (\mathbf{p}_1/m_1,\dots \mathbf{p}_N/m_N; \mathbf{F}_1,\dots,\mathbf{F}_N). \tag{14.2.20}$$

Re-writing this in terms of components, show that

$$V_{1,1} = V_{2,2} = \cdots = V_{3N,3N} = 0. \tag{14.2.21}$$

Deduce (14.2.19) when $\mathbf{F}_1,\dots,\mathbf{F}_N$ depend only upon $X_1,\dots,X_{3N}$.

Exercise 14.2.2. If $f : \mathbb{D} \to \mathbb{R}$ and $\mathbf{U} : \mathbb{D} \to \mathcal{V}^{2N}$ denote scalar and vector fields on an open subset $\mathbb{D}$ of $\mathbb{P}$, show that

$$\mathrm{div}_{\mathbb{P}}\{f\mathbf{U}\} = f\,\mathrm{div}_{\mathbb{P}}\mathbf{U} + \nabla_{\mathbb{P}} f\,.\,\mathbf{U}, \tag{14.2.22}$$

where $\nabla_{\mathbb{P}} f$ denotes that vector field with components $\partial f/\partial X_p =: f_{,p}$ $(p = 1,2,\dots,6N)$ and the inner product of two vector fields $\mathbf{U}$ and $\mathbf{U}'$ with components U_p and U'_q $(p,q = 1,\dots,6N)$ is

$$\mathbf{U}\,.\,\mathbf{U}' := \sum_{p=1}^{6N} U_p U'_p =: U_p U'_p. \tag{14.2.23}$$

Exercise 14.2.3. Show that if (14.2.19) holds, then (14.2.15) simplifies to the *Liouville equation*

$$\frac{\partial P}{\partial t} = -\mathbf{V}\,.\,\nabla_{\mathbb{P}}\, P. \tag{14.2.24}$$

Consider any function of form

$$\alpha(t) := \hat{\alpha}(\mathbf{X}(t)), \tag{14.2.25}$$

where $\hat{\alpha}$ has domain $\mathbb{P}$. Function α of time t is termed a *dynamic variable*[8] whose *expectation* at time t is

$$\langle \alpha; t\rangle := \int_{\mathbb{P}} \hat{\alpha}(\mathbf{X}) P(\mathbf{X},t)\, dV_{\mathbf{X}}. \tag{14.2.26}$$

[8] For example, the mass or momentum associated with point masses which lie within some designated region at time t.

If
$$\mathbb{D}_t := \boldsymbol{\phi}_0(\mathbb{D}_0, t), \tag{14.2.27}$$

then from $(14.2.12)_1$ and (14.2.14) with $\mathbb{S} = \mathbb{D}_0$ it follows that

$$\int_{\mathbb{D}_t} P(\mathbf{X}, t) dV_{\mathbf{X}} = 1. \tag{14.2.28}$$

Both (14.2.27) and (14.2.28) express the fact that at time t the state $\mathbf{X}(t)$ of the system [see (14.2.1)] lies in $\mathbb{D}_t$. Thus (14.2.26) is equivalent to

$$\langle \alpha; t \rangle = \int_{\mathbb{D}_t} \hat{\alpha}(\mathbf{X}) P(\mathbf{X}, t) dV_{\mathbf{X}}. \tag{14.2.29}$$

Remark 14.2.3. Quantity $\langle \alpha; t \rangle$ is the *ensemble average* at time t and represents the average of α values at time t that corresponds to what is known about the system [namely the ensemble-defining information embodied in knowledge of $\mathbb{D}_0, P_0$, and system dynamics determined by (14.2.8)].

The time evolution of expectations may be calculated by invoking (14.2.15): modulo sufficient regularity to justify differentiating under the integral sign,

$$\begin{aligned}\frac{d}{dt}\{\langle \alpha; t \rangle\} &= \int_{\mathbb{P}} \hat{\alpha} \frac{\partial P}{\partial t} dV = -\int_{\mathbb{P}} \hat{\alpha}\, \mathrm{div}_{\mathbb{P}}\{P\mathbf{V}\} dV \\ &= \int_{\mathbb{P}} (\nabla_{\mathbb{P}} \hat{\alpha} \,.\, P\mathbf{V} - \mathrm{div}_{\mathbb{P}}\{\hat{\alpha} P\mathbf{V}\}) dV \\ &= \int_{\mathbb{D}_t} \nabla_{\mathbb{P}} \hat{\alpha} \,.\, P\mathbf{V}\, dV - \int_{\mathbb{D}_t^+} \mathrm{div}_{\mathbb{P}}\{\hat{\alpha}\, P\mathbf{V}\} dV. \end{aligned} \tag{14.2.30}$$

Here the appropriate analogue of identity (B.7.28) in Appendix B.7 has been employed, and the vanishing of P in $\mathbb{P} - \mathbb{D}_t$ [see (14.2.28)] has been noted: $\mathbb{D}_t^+$ denotes any subset of $\mathbb{P}$ which properly contains $\mathbb{D}_t$. If $\mathbb{D}_t$ is a bounded subset of $\mathbb{P}$ (in the sense of the natural metric on $\mathbb{R}^{6N}$, namely $\|\mathbf{X} - \mathbf{Y}\| := \left[\sum_{p=1}^{6N} (X_p - Y_p)^2 \right]^{1/2}$), then $\mathbb{D}_t^+$ can be chosen to be bounded, and P will vanish on its boundary $\partial \mathbb{D}_t^+$. It follows (see Appendix B.11) that the second integral in (14.2.30), which reduces to an integral over $\partial \mathbb{D}_t^+$, vanishes.

Remark 14.2.4. If system $\mathcal{M}$ is confined to a bounded region by conservative forces which depend on point mass locations, all other external agencies have this nature, and point mass interactions are governed by separation-dependent pair potentials which are bounded below, then the system dynamics is conservative, and, for any given total energy, $\mathbb{D}_t$ will be uniformly bounded.

Noll [16] has discussed general sufficient conditions for vanishing of the integral of $\mathrm{div}_{\mathbb{P}}\{\hat{\alpha} P\mathbf{V}\}$ over $\mathbb{P}$ for the dynamical variables α discussed here. We shall assume hereafter that such conditions are met and hence that (14.2.30) reduces to

$$\frac{d}{dt}\{\langle \alpha; t \rangle\} = \int_{\mathbb{D}_t} \nabla_{\mathbb{P}} \hat{\alpha} \,.\, P\mathbf{V}\, dV \left(= \int_{\mathbb{P}} \nabla_{\mathbb{P}} \hat{\alpha} \,.\, P\mathbf{V}\, dV \right) \tag{14.2.31}$$

for subsequent choices of α.

Remark 14.2.5. Noting that from (14.2.25)

$$\dot{\alpha}(t) = \nabla_{\mathbb{P}}\hat{\alpha}(\mathbf{X}(t)).\dot{\mathbf{X}}(t) = \nabla_{\mathbb{P}}\hat{\alpha}(\mathbf{X}(t)).\mathbf{V}(t), \tag{14.2.32}$$

multiplication by $P(\mathbf{X},t)$ followed by integration over $\mathbb{P}$ yields

$$\frac{d}{dt}\{\langle\alpha\rangle\} = \langle\dot{\alpha}\rangle. \tag{14.2.33}$$

The probability, per unit volume of $\mathcal{E}$, of finding point mass P_i at point $\mathbf{x}$ at time t, irrespective of its momentum and the locations and momenta of all other point masses in $\mathcal{M}$, is

$$w_i(\mathbf{x},t) := \int_{\mathbb{P}_i(\mathbf{x})} P(\mathbf{X},t)dV_{\mathbf{X}_i^-}, \tag{14.2.34}$$

where
$$\mathbb{P}_i(\mathbf{x}) := \{\mathbf{X} \in \mathbb{P} : \mathbf{x}_i = \mathbf{x}\}. \tag{14.2.35}$$

Here '$\mathbf{x}_i = \mathbf{x}$' indicates that the entries X_{3i-2}, X_{3i-1}, and X_{3i} (which correspond to the co-ordinates of the location $\mathbf{x}_i$ of P_i) are held fixed at the co-ordinates of the point $\mathbf{x}$. The notation $dV_{\mathbf{X}_i^-}$ is used to indicate that integration is taken over all phase space variables except X_{3i-2}, X_{3i-1}, and X_{3i-2}. Alternative notation, for integration over $\mathbb{P}_i(\mathbf{x})$ of any function f of $\mathbf{X}$ and t, is

$$\int_{\mathbb{P}} f(\mathbf{X},t)\delta(\mathbf{x}_i - \mathbf{x})dV_{\mathbf{X}} := \int_{\mathbb{P}_i(\mathbf{x})} f(\mathbf{X},t)dV_{\mathbf{X}_i^-}. \tag{14.2.36}$$

Remark 14.2.6. The physical dimension of $\mathbb{P}_i(\mathbf{x})$ differs from that of $\mathbb{P}$ by a factor L^{-3}, and hence w_i takes values of physical dimension L^{-3}.

The *ensemble mass density* ρ and *ensemble momentum density* $\mathbf{p}$ are those fields defined by (here m_i denotes the mass of P_i)

$$\rho(\mathbf{x},t) := \sum_{i=1}^{N} m_i w_i(\mathbf{x},t) \quad \text{and} \quad \mathbf{p}(\mathbf{x},t) := \sum_{i=1}^{N} \int_{\mathbb{P}_i(\mathbf{x})} \mathbf{p}_i P(\mathbf{X},t)dV_{\mathbf{X}_i^-}, \tag{14.2.37}$$

where the momentum of P_i in phase space format is

$$\mathbf{p}_i = X_{3N+3i-2}\mathbf{e}_1 + X_{3N+3i-1}\mathbf{e}_2 + X_{3N+3i}\mathbf{e}_3. \tag{14.2.38}$$

The *ensemble velocity field* [cf. (4.2.15)] is

$$\mathbf{v} := \rho^{-1}\mathbf{p}. \tag{14.2.39}$$

Noting (14.2.36), definitions (14.2.37) may be written as

$$\rho(\mathbf{x},t) := \int_{\mathbb{P}} \sum_{i=1}^{N} m_i \delta(\mathbf{x}_i(t) - \mathbf{x}) P(\mathbf{X},t)dV_{\mathbf{X}} \tag{14.2.40}$$

and
$$\mathbf{p}(\mathbf{x},t) := \int_{\mathbb{P}} \sum_{i=1}^{3N} \mathbf{p}_i \delta(\mathbf{x}_i(t) - \mathbf{x}) P(\mathbf{X},t)dV_{\mathbf{X}}. \tag{14.2.41}$$

Writing

$$\rho_{\text{mic}}(\mathbf{x},t) := \sum_{i=1}^{N} m_i \delta(\mathbf{x}_i(t) - \mathbf{x}) \qquad \text{and} \qquad \mathbf{p}_{\text{mic}}(\mathbf{x},t) = \sum_{i=1}^{N} \mathbf{p}_i(t)\delta(\mathbf{x}_i(t) - \mathbf{x}) \tag{14.2.42}$$

for the microscopic distributions of mass and linear momentum, ρ and $\mathbf{p}$ are seen to be the expectations of ρ_{mic} and $\mathbf{p}_{\text{mic}}$, respectively [see (14.2.26)].

14.3 Mass Conservation and Linear Momentum Balance

From $(14.2.37)_1$, (14.2.34), and (14.2.15),

$$\begin{aligned} \frac{\partial \rho}{\partial t}(\mathbf{x},t) &= \sum_{i=1}^{N} m_i \frac{\partial}{\partial t}\left\{\int_{\mathbb{P}_i(\mathbf{x})} P(\mathbf{X},t)dV_{\mathbf{X}_i^-}\right\} \\ &= \sum_{i=1}^{N} m_i \int_{\mathbb{P}_i(\mathbf{x})} \frac{\partial P}{\partial t} dV_{\mathbf{X}_i^-} = -\sum_{i=1}^{N} m_i \int_{\mathbb{P}_i(\mathbf{x})} \text{div}_{\mathbb{P}}\{P\mathbf{V}\}dV_{\mathbf{X}_i^-}. \end{aligned} \tag{14.3.1}$$

Exercise 14.3.1. Show from (14.2.17) that

$$\begin{aligned} \text{div}_{\mathbb{P}}\{P\mathbf{V}\} &= \sum_{p=1}^{6N} \frac{\partial}{\partial X_p}\{PV_p\} \\ &= \sum_{j=1}^{2N}\left(\frac{\partial}{\partial X_{3j-2}}\{PV_{3j-2}\} + \frac{\partial}{\partial X_{3j-1}}\{PV_{3j-1}\} + \frac{\partial}{\partial X_{3j}}\{PV_{3j}\}\right). \end{aligned} \tag{14.3.2}$$

Show further that

$$\int_{\mathbb{P}_i(\mathbf{x})} \text{div}_{\mathbb{P}}\{P\mathbf{V}\}d\mathbf{X}_i^- = \sum_{j=1}^{N}(a_{ij}(\mathbf{x}) + b_{ij}(\mathbf{x})), \tag{14.3.3}$$

where

$$a_{ij}(\mathbf{x}) := \int_{\mathbb{P}_i(\mathbf{x})} \frac{\partial}{\partial X_{3j-2}}\{PX_{3N+3j-2}\} + \frac{\partial}{\partial X_{3j-1}}\{PX_{3N+3j-1}\} + \frac{\partial}{\partial X_{3j}}\{PX_{3N+3j}\}d\mathbf{V}_{\mathbf{X}_i^-} \tag{14.3.4}$$

and

$$\begin{aligned} b_{ij}(\mathbf{x}) :&= \int_{\mathbb{P}_i(\mathbf{x})} \frac{\partial}{\partial X_{3N+3j-2}}\{PV_{3N+3j-2}\} + \frac{\partial}{\partial X_{3N+3j-1}}\{PV_{3N+3j-1}\} \\ &\quad + \frac{\partial}{\partial X_{3N+3j}}\{PV_{3N+3j}\}dV_{\mathbf{X}_i^-}. \end{aligned} \tag{14.3.5}$$

Evaluation of integrals over $\mathbb{P}_i(\mathbf{x})$ involves $(6N-3)$-fold repeated integrals over $\mathbb{R}$ with integration in turn over all phase space variables except X_{3i-2}, X_{3i-1}, and X_{3i}. Since order of integration is unimportant (Why?), consider (here $j \neq i$)

$$\int_{\mathbb{R}} \frac{\partial}{\partial X_{3j-2}}\{PX_{3N+3j-2}\}dX_{3j-2} = [PX_{3N+3j-2}]_{-\infty}^{+\infty}. \tag{14.3.6}$$

The limits here are for X_{3j-2}, where [see (14.3.3) and (14.3.4)] $j=1,2,\ldots,N$. Accordingly X_{3j-2} corresponds to the $\mathbf{e}_1$ co-ordinate of the location of P_j. If all point masses lie in a bounded region, then probability P must vanish outside this region, and so the value of (14.3.6) is zero. Thus the integral of $(\partial/\partial X_{3j-2})\{PX_{3N+3j-1}\}$ over $\mathbb{P}_i(\mathbf{x})$ must vanish.

Exercise 14.3.2. Deduce, by considering the two remaining contributions to $a_{ij}(\mathbf{x})$ in (14.3.4), that *if all point masses lie in a bounded region, then*

$$a_{ij}(\mathbf{x})=0 \qquad \text{if} \quad j\neq i. \tag{14.3.7}$$

Turning to the evaluation of $m_i b_{ij}(\mathbf{x})$, consider

$$\int_{\mathbb{R}} \frac{\partial}{\partial X_{3N+3j-2}}\{PV_{3N+3j-2}\}dX_{3N+3j-2}=[PV_{3N+3j-2}]_{-\infty}^{+\infty}. \tag{14.3.8}$$

Here $V_{3N+3j-2}$ corresponds to the $\mathbf{e}_1$ component of the force on P_j and $X_{3N+3j-2}$ to the $\mathbf{e}_1$ component of the momentum of P_j. Thus, if the momenta of all point masses are uniformly bounded, then P vanishes for sufficiently large momentum components, and the value of (14.3.8) is zero. Thus integration of the integrand in (14.3.8) over all $\mathbb{P}_i(\mathbf{x})$ is zero. In similar fashion the remaining contributions to $b_{ij}(\mathbf{x})$ vanish, whence *if all momenta are uniformly bounded, then*

$$b_{ij}(\mathbf{x})=0 \qquad \text{for all} \quad j. \tag{14.3.9}$$

From (14.3.1), (14.3.3), (14.3.7), and (14.3.9),

$$\frac{\partial\rho}{\partial t}=-\sum_{i=1}^{N} m_i a_{ii}. \tag{14.3.10}$$

Further, from (14.3.4)

$$a_{ii}=\int_{\mathbb{P}_i(\mathbf{x})} \frac{\partial}{\partial X_{3i-2}}\{PX_{3N+3i-2}\}+\frac{\partial}{\partial X_{3i-1}}\{PX_{3N+i-1}\}+\frac{\partial}{\partial X_{3i}}\{PX_{3N+3i}\}dV_{\mathbf{X}_i^-}. \tag{14.3.11}$$

However, since X_{3i-2}, X_{3i-1}, and X_{3i} are not variables of integration and are held fixed at, respectively, the $\mathbf{e}_1,\mathbf{e}_2$, and $\mathbf{e}_3$ co-ordinates of location $\mathbf{x}$, namely x_1,x_2, and x_3,

$$\begin{aligned} m_i a_{ii} &= \frac{\partial}{\partial x_1}\left\{\int_{\mathbb{P}_i(\mathbf{x})} m_i PX_{3N+3i-2}dV_{\mathbf{X}_i^-}\right\}+\frac{\partial}{\partial x_2}\left\{\int_{\mathbb{P}_i(\mathbf{x})} m_i PX_{3N+3i-1}dV_{\mathbf{X}_i^-}\right\} \\ &\quad + \frac{\partial}{\partial x_3}\left\{\int_{\mathbb{P}_i(\mathbf{x})} m_i PX_{3N+3i}\,dV_{\mathbf{X}_i^-}\right\}. \end{aligned} \tag{14.3.12}$$

Exercise 14.3.3. Recalling (14.2.38), deduce that

$$m_i a_{ii}(\mathbf{x})=\operatorname{div}_{\mathbb{P}}\left\{\int_{\mathbb{P}_i(\mathbf{x})} P\mathbf{p}_i\,dV_{\mathbf{X}_i^-}\right\}. \tag{14.3.13}$$

Deduce from (14.3.10), (14.3.13), and $(14.2.37)_2$ that

$$\frac{\partial\rho}{\partial t}+\operatorname{div}\mathbf{p}=0. \tag{14.3.14}$$

Accordingly, *provided that point mass locations and momenta are bounded*, from (14.2.39)

$$\frac{\partial \rho}{\partial t} + \mathrm{div}\{\rho \mathbf{v}\} = 0. \tag{14.3.15}$$

(Recall Remark 14.2.4, which detailed sufficient conditions for such bounded behaviour.)

From $(14.2.37)_2$, (14.2.15), and (14.2.17), and suppressing time dependence (recall P depends upon time),

$$\begin{aligned}\frac{\partial \mathbf{p}}{\partial t}(\mathbf{x}) &= \sum_{i=1}^{N} \int_{\mathbb{P}_i(\mathbf{x})} \mathbf{p}_i \frac{\partial P}{\partial t} dV_{\mathbf{X}_i^-} = -\sum_{i=1}^{N} \int_{\mathbb{P}_i(\mathbf{x})} \mathbf{p}_i \,\mathrm{div}_{\mathbb{P}}\{P\mathbf{V}\} dV_{\mathbf{X}_i^-} \\ &= -\sum_{i=1}^{N}\sum_{p=1}^{6N} \int_{\mathbb{P}_i(\mathbf{x})} \mathbf{p}_i \frac{\partial}{\partial X_p}\{PV_p\} dV_{\mathbf{X}_i^-} \\ &= \mathbf{A}(\mathbf{x}) + \mathbf{B}(\mathbf{x}),\end{aligned} \tag{14.3.16}$$

where

$$\mathbf{A}(\mathbf{x}) := -\sum_{i=1}^{N}\sum_{p=1}^{6N} \int_{\mathbb{P}_i(\mathbf{x})} \frac{\partial}{\partial X_p}\{\mathbf{p}_i PV_p\} dV_{\mathbf{X}_i^-} \tag{14.3.17}$$

and

$$\mathbf{B}(\mathbf{x}) := \sum_{i=1}^{N}\sum_{p=1}^{6N} \int_{\mathbb{P}_i(\mathbf{x})} \frac{\partial}{\partial X_p}\{\mathbf{p}_i\} PV_p \, dV_{\mathbf{X}_i^-}. \tag{14.3.18}$$

Exercise 14.3.4. If X_p is a variable of integration (so that $p \neq 3i-2, 3i-1$, or $3i$), then, on first integrating with respect to this variable, show that

$$\int_{\mathbb{P}_i(\mathbf{x})} \frac{\partial}{\partial X_p}\{\mathbf{p}_i PV_p\} dV_{\mathbf{X}_i^-} = \mathbf{0} \tag{14.3.19}$$

if all point mass locations and momenta are bounded.

From (14.3.17) and (14.3.19)

$$\begin{aligned}\mathbf{A}(\mathbf{x}) &= -\sum_{i=1}^{N} \int_{\mathbb{P}_i(\mathbf{x})} \frac{\partial}{\partial X_{3i-2}}\{\mathbf{p}_i PV_{3i-2}\} + \frac{\partial}{\partial X_{3i-1}}\{\mathbf{p}_i PV_{3i-1}\} + \frac{\partial}{\partial X_{3i}}\{\mathbf{p}_i PV_{3i}\} dV_{\mathbf{X}_i^-} \\ &= -\sum_{i=1}^{N} \int_{\mathbb{P}_i(\mathbf{x})} m_i \mathbf{v}_i \otimes \mathbf{v}_i \left(\frac{\partial P}{\partial x_1}\mathbf{e}_1 + \frac{\partial P}{\partial x_2}\mathbf{e}_2 + \frac{\partial P}{\partial x_3}\mathbf{e}_3 \right) dV_{\mathbf{X}_i^-},\end{aligned} \tag{14.3.20}$$

on noting that $X_{3i-2} = x_1, X_{3i-1} = x_2, X_{3i} = x_3$, $V_{3i-2} = \mathbf{v}_i.\mathbf{e}_1$, $V_{3i-1} = \mathbf{v}_i.\mathbf{e}_2$, $V_{3i} = \mathbf{v}_i.\mathbf{e}_3$, and $\mathbf{v}_i$ is independent of x_1,x_2, and x_3. Accordingly

$$\begin{aligned}\mathbf{A}(\mathbf{x}) &= -\sum_{i=1}^{N} \int_{\mathbb{P}_i(\mathbf{x})} \mathrm{div}_{\mathbf{x}}\{m_i \mathbf{v}_i \otimes \mathbf{v}_i P\} dV_{\mathbf{X}_i^-} \\ &= -\mathrm{div}_{\mathbf{x}} \left\{ \sum_{i=1}^{N} \int_{\mathbb{P}_i(\mathbf{x})} m_i \mathbf{v}_i \otimes \mathbf{v}_i P \, dV_{\mathbf{X}_i^-} \right\},\end{aligned} \tag{14.3.21}$$

on noting that $\mathbf{x}$ co-ordinates do not appear in variable list $\mathbf{X}_i^-$.

Defining the *ensemble thermal velocity* [cf. (5.5.25)] by

$$\tilde{\mathbf{v}}_i(t) := \mathbf{v}_i(t) - \mathbf{v}(\mathbf{x}_i(t),t), \tag{14.3.22}$$

then, from $(14.2.37)_2$, $(14.2.37)_1$, and (14.2.39),

$$\sum_{i=1}^{N}\int_{\mathbb{P}_i(\mathbf{x})} m_i\tilde{\mathbf{v}}_i(t)P(\mathbf{X},t)dV_{\mathbf{X}_i^-} = \sum_{i=1}^{N}\int_{\mathbb{P}_i(\mathbf{x})} (m_i\mathbf{v}_i(t) - m_i\mathbf{v}(\mathbf{x},t))P(\mathbf{X},t)dV_{\mathbf{X}_i^-}$$
$$= \mathbf{p}(\mathbf{x},t) - \rho(\mathbf{x},t)\mathbf{v}(\mathbf{x},t) = \mathbf{0}. \tag{14.3.23}$$

Accordingly, from (14.3.21) and (14.3.23),

$$\mathbf{A}(\mathbf{x},t) = -\mathrm{div}_{\mathbf{x}}\left\{\sum_{i=1}^{N}\int_{\mathbb{P}_i(\mathbf{x})} m_i(\tilde{\mathbf{v}}_i(t)+\mathbf{v}(\mathbf{x},t))\otimes(\tilde{\mathbf{v}}_i(t)+\mathbf{v}(\mathbf{x},t))P(\mathbf{X},t)dV_{\mathbf{X}_i^-}\right\}$$
$$= -(\mathrm{div}\mathcal{D} + \rho\mathbf{v}\otimes\mathbf{v})(\mathbf{x},t), \tag{14.3.24}$$

where [cf. (5.5.11)]
$$\mathcal{D}(\mathbf{x},t) := \sum_{i=1}^{N}\int_{\mathbb{P}_i(\mathbf{x})} m_i\tilde{\mathbf{v}}_i(t)\otimes\tilde{\mathbf{v}}_i(t)P(\mathbf{X},t)dV_{\mathbf{X}_i^-}. \tag{14.3.25}$$

It remains to consider $\mathbf{B}(\mathbf{x})$ given by (14.3.18). Noting that from (14.2.38)

$$\frac{\partial}{\partial X_p}\{\mathbf{p}_i\} = \mathbf{0} \qquad \text{unless} \quad p = 3N+3i-2, 3N+3i-1 \ \text{or} \ 3N+3i, \tag{14.3.26}$$

and, for $j = 2,1,0$,

$$\frac{\partial}{\partial X_{3N+3i-j}}\{\mathbf{p}_i\} = \mathbf{e}_{3-j}, \tag{14.3.27}$$

$$\mathbf{B}(\mathbf{x},t) = \sum_{i=1}^{N}\int_{\mathbb{P}_i(\mathbf{x})} P(\mathbf{X},t)(V_{3N+3i-2}\,\mathbf{e}_1 + V_{3N+3i-1}\,\mathbf{e}_2 + V_{3N+3i}\,\mathbf{e}_3)dV_{\mathbf{X}_i^-}. \tag{14.3.28}$$

However, from (14.2.20)

$$V_{3N+3i-2}\,\mathbf{e}_1 + V_{3N+3i-1}\,\mathbf{e}_2 + V_{3N+3i}\,\mathbf{e}_3 = \mathbf{F}_i, \tag{14.3.29}$$

the resultant force on point mass P_i. Hence

$$\mathbf{B}(\mathbf{x},t) = \sum_{i=1}^{N}\int_{\mathbb{P}_i(\mathbf{x})} P(\mathbf{X},t)\mathbf{F}_i\,dV_{\mathbf{X}_i^-}. \tag{14.3.30}$$

If
$$\mathbf{F}_i = \mathbf{f}_i + \mathbf{b}_i, \tag{14.3.31}$$

where $\mathbf{f}_i$ and $\mathbf{b}_i$ denote the resultant force on P_i from $\mathcal{M}$ point mass interactions and external agencies, respectively, then

$$\mathbf{B} = \mathbf{f} + \mathbf{b}, \tag{14.3.32}$$

where
$$\mathbf{f}(\mathbf{x},t) := \sum_{i=1}^{N} \int_{\mathbb{P}_i(\mathbf{x})} P(\mathbf{X},t)\mathbf{f}_i \, dV_{\mathbf{X}_i^-} \tag{14.3.33}$$

and
$$\mathbf{b}(\mathbf{x},t) := \sum_{i=1}^{N} \int_{\mathbb{P}_i(\mathbf{x})} P(\mathbf{X},t)\mathbf{b}_i \, dV_{\mathbf{X}_i^-}. \tag{14.3.34}$$

Thus, from (14.3.16), (14.3.24), and (14.3.30), linear momentum balance takes the form [cf. (5.5.13)]

$$\frac{\partial \mathbf{p}}{\partial t} = -\operatorname{div}\{\mathcal{D} + \rho\mathbf{v} \otimes \mathbf{v}\} + \mathbf{f} + \mathbf{b}. \tag{14.3.35}$$

It follows from the continuity equation (14.3.14), using exactly the same analysis as in Exercise 5.5.3 et seq., that

$$-\operatorname{div}\mathcal{D} + \mathbf{f} + \mathbf{b} = \rho\mathbf{a}, \tag{14.3.36}$$

where the acceleration field [cf. (5.5.16)]

$$\mathbf{a} := \frac{\partial \mathbf{v}}{\partial t} + (\nabla\mathbf{v})\mathbf{v}. \tag{14.3.37}$$

Balance (14.3.36) has the form of (5.5.15). In order to obtain a balance of standard form (2.7.20) it is necessary to seek a solution $\mathbf{T}^-$ to [cf. (5.5.18)]

$$\operatorname{div}\mathbf{T}^- = \mathbf{f}. \tag{14.3.38}$$

In such case a candidate Cauchy stress tensor is [cf. (5.5.20)]

$$\mathbf{T} := \mathbf{T}^- - \mathcal{D}. \tag{14.3.39}$$

Remark 14.3.1. The approach adopted here was detailed in the seminal paper of Irving & Kirkwood [73]. In deriving a solution to (14.3.38), use was made of δ-function formalism, and a clear physical interpretation was given in their appendix concerning the contribution of individual molecular interactions to integrals of $\mathbf{T}^-\mathbf{n}$ over 'infinitesimal' surfaces with oriented area $d\mathbf{S} = dS\mathbf{n}$. Noll [16] showed how the extensive manipulation of δ-functions used in Irving & Kirkwood [73] could be avoided using simple multivariable calculus. The latter approach is now sketched.

For any $\mathbb{P}$-integrable function f,

$$\int_{\mathbb{P}_i(\mathbf{x})} f(\mathbf{X}) dV_{\mathbf{X}_i^-} = \int_{\mathbb{P}_{ij}(\mathbf{x})} \int_{-\infty}^{+\infty} \int_{-\infty}^{+\infty} \int_{-\infty}^{+\infty} f(\mathbf{X}) dX_{3j-2} dX_{3j-1} dX_{3j} dV_{\mathbf{X}_{ij}^-}, \tag{14.3.40}$$

where $\mathbb{P}_{ij}(\mathbf{x})$ is the set of elements of $\mathbb{P}$ for which location $\mathbf{x}_i$ co-ordinates are held fixed at those of point $\mathbf{x}$ and those for location $\mathbf{x}_j$ are omitted: $dV_{\mathbf{X}_{ij}^-}$ indicates integration over such elements. The inner triple integral in (14.3.40) may be written, upon relabelling location $\mathbf{x}_j$ [with co-ordinates $(X_{3j-2}, X_{3j-1}, X_{3j})$] as $\mathbf{y}$, in the form of an integral over $\mathcal{E}$. Thus, for any given $j \neq i$,

$$\int_{\mathbb{P}_i(\mathbf{x})} f(\mathbf{X}) dV_{\mathbf{X}_i^-} = \int_{\mathbb{P}_{ij}(\mathbf{x})} \int_{\mathcal{E}} f(\mathbf{X})\Big|_{\mathbf{x}_j=\mathbf{y}} dV_{\mathbf{y}} \, dV_{\mathbf{X}_{ij}^-} = \int_{\mathcal{E}} \int_{\mathbb{P}_{ij}(\mathbf{x},\mathbf{y})} f(\mathbf{X}) dV_{\mathbf{X}_{ij}^-} dV_{\mathbf{y}}. \tag{14.3.41}$$

Here $$\mathbb{P}_{ij}(\mathbf{x},\mathbf{y}) := \{\mathbf{X} \in \mathbb{P} : \mathbf{x}_i = \mathbf{x} \quad \text{and} \quad \mathbf{x}_j = \mathbf{y}\}. \tag{14.3.42}$$

Accordingly, from (14.3.33), for any choice of $j \neq i$

$$\mathbf{f}(\mathbf{x},t) = \sum_{i=1}^{N} \int_{\mathbb{P}_i(\mathbf{x})} P(\mathbf{X},t)\mathbf{f}_i \, dV_{\mathbf{X}_i^-} = \sum_{i=1}^{N} \int_{\mathcal{E}} \int_{\mathbb{P}_{ij}(\mathbf{x},\mathbf{y})} P(\mathbf{X},t)\mathbf{f}_i \, dV_{\mathbf{X}_{ij}^-} dV_{\mathbf{y}}. \tag{14.3.43}$$

If $\mathbf{f}_i$ is expressible as the sum of individual forces $\mathbf{f}_{ij}$ on P_i due to P_j, so that [cf. (5.4.4)]

$$\mathbf{f}_i = \sum_{j \neq i} \mathbf{f}_{ij}, \tag{14.3.44}$$

then $$\mathbf{f}(\mathbf{x},t) = \int_{\mathcal{E}} \mathbf{g}(\mathbf{x},\mathbf{y},t) dV_{\mathbf{y}}, \tag{14.3.45}$$

where $$\mathbf{g}(\mathbf{x},\mathbf{y},t) := \sum_{i \neq j}\sum \int_{\mathbb{P}_{ij}(\mathbf{x},\mathbf{y})} P(\mathbf{X},t)\mathbf{f}_{ij} \, dV_{\mathbf{X}_{ij}^-}. \tag{14.3.46}$$

That is, $\mathbf{g}$ is a force double density whose integral over all space with respect to location variable $\mathbf{y}$ yields the body force density $\mathbf{f}$. Now

$$\mathbf{g}(\mathbf{y},\mathbf{x},t) = \sum_{i \neq j}\sum \int_{\mathbb{P}_{ij}(\mathbf{y},\mathbf{x})} P(\mathbf{X},t)\mathbf{f}_{ij} \, dV_{\mathbf{X}_{ij}^-} = \sum_{j \neq i}\sum \int_{\mathbb{P}_{ji}(\mathbf{y},\mathbf{x})} P(\mathbf{X},t)\mathbf{f}_{ji} \, dV_{\mathbf{X}_{ij}^-}, \tag{14.3.47}$$

on relabelling (i,j) as (j,i). (Why is $dV_{\mathbf{X}_{ji}^-} = dV_{\mathbf{X}_{ij}^-}$?). Since

$$\mathbb{P}_{ij}(\mathbf{x},\mathbf{y}) = \mathbb{P}_{ji}(\mathbf{y},\mathbf{x}), \tag{14.3.48}$$

it follows that if interactions are pairwise-balanced [cf. (5.6.6)], then

$$\mathbf{g}(\mathbf{y},\mathbf{x},t) = -\mathbf{g}(\mathbf{x},\mathbf{y},t). \tag{14.3.49}$$

Noll [16] has shown, modulo regularity conditions, that [via (14.3.49)] relation (14.3.38) holds with

$$\mathbf{T}^-(\mathbf{x},t) := -\frac{1}{2}\int_{\mathcal{V}}\int_0^1 \mathbf{g}(\mathbf{x}+\alpha\mathbf{u},\mathbf{x}-(1-\alpha)\mathbf{u},t) \otimes \mathbf{u} \, d\alpha dV_{\mathbf{u}}. \tag{14.3.50}$$

With Remark 14.3.1 in mind, it is helpful to note the following:

Lemma 14.3.1.[9] If $\mathbf{n} \in \mathcal{V}$ with $\mathbf{n}\,.\,\mathbf{n} = 1$, then (omitting time dependence)

$$\frac{1}{2}\int_{\mathcal{V}}\int_0^1 \mathbf{g}(\mathbf{x}+\alpha\mathbf{u},\mathbf{x}-(1-\alpha)\mathbf{u}) \otimes \mathbf{u} d\alpha dV_{\mathbf{u}} = \int_{\mathcal{V}_{\mathbf{n}}}\int_0^1 \mathbf{g}(\mathbf{x}+\alpha\mathbf{u},\mathbf{x}-(1-\alpha)\mathbf{u}) \otimes \mathbf{u} d\alpha dV_{\mathbf{u}}, \tag{14.3.51}$$

where $$\mathcal{V}_{\mathbf{n}} := \{\mathbf{u} \in \mathcal{V} : \mathbf{u}\,.\,\mathbf{n} > 0\}. \tag{14.3.52}$$

[9] This is proved in Noll [16]. See also Murdoch & Bedeaux [77], Appendix A.

Exercise 14.3.5. Show from (14.3.50), Lemma 14.3.1, and (14.3.46) that

$$\mathbf{T}^-(\mathbf{x},t)\mathbf{n} = -\int_{\mathcal{V}_\mathbf{n}}\int_0^1 \sum\sum_{i\neq j}\int_{\mathbb{P}_{ij}(\mathbf{x}+\alpha\mathbf{u},\mathbf{x}-(1-\alpha)\mathbf{u})} P(\mathbf{X},t)\mathbf{f}_{ij}(\mathbf{u}\,.\,\mathbf{n})dV_{\mathbf{X}_{ij}^-}d\alpha dV_\mathbf{u}. \tag{14.3.53}$$

In (14.3.53), $\mathbf{x}+\alpha\mathbf{u}\in\mathcal{E}_\mathbf{n}^+(\mathbf{x})$ and $\mathbf{x}-(1-\alpha)\mathbf{u}\in\mathcal{E}_\mathbf{n}^-(\mathbf{x})$, where [cf. (5.9.15)]

$$\mathcal{E}_\mathbf{n}^+(\mathbf{x}) := \{\mathbf{z}\in\mathcal{E} : (\mathbf{z}-\mathbf{x})\,.\,\mathbf{n}\} > 0 \tag{14.3.54}$$

and

$$\mathcal{E}_\mathbf{n}^-(\mathbf{x}) := \{\mathbf{y}\in\mathcal{E} : (\mathbf{y}-\mathbf{x})\,.\,\mathbf{n}\} < 0 \tag{14.3.55}$$

denote those half-spaces into which the plane $\Pi_\mathbf{n}(\mathbf{x})$ [see (5.8.3)] divides $\mathcal{E}$. Further, the $\mathbb{P}_{ij}$ integral in (14.3.53) involves $\mathbf{x}_i = \mathbf{x}+\alpha\mathbf{u}$ and $\mathbf{x}_j = \mathbf{x}-(1-\alpha)\mathbf{u}$, so that $\mathbf{x}_i\in\mathcal{E}_n^+(\mathbf{x}), \mathbf{x}_j\in\mathcal{E}_\mathbf{n}^-(\mathbf{x})$, and $\mathbf{x}_i$ and $\mathbf{x}_j$ lie on that line through $\mathbf{x}$ with $\mathbf{x}_j-\mathbf{x}_i = \mathbf{u}$. In particular, requirement $\mathbf{u}\,.\,\mathbf{n} > 0$ means $(\mathbf{x}_j-\mathbf{x}_i)\,.\,\mathbf{n} > 0$.

It follows that $\mathbf{T}^-(\mathbf{x},t)\mathbf{n}$ is an ensemble average taken over particle pairs, one of which lies in $\mathcal{E}_n^+(\mathbf{x})$ and the other in $\mathcal{E}_\mathbf{n}^-(\mathbf{x})$, and the line between their locations passes through $\mathbf{x}$. It is a simple matter to check that the physical dimensions of $\mathbf{T}^-(\mathbf{x},t)\mathbf{n}$ are those of force per unit area: notice that the factor $-\mathbf{f}_{ij}$ indicates that this force (per surface area) density is associated with the effect of matter in $\mathcal{E}_\mathbf{n}^+(\mathbf{x})$ upon that in $\mathcal{E}_\mathbf{n}^-(\mathbf{x})$. (Convince yourself!)

The result of integrating $\mathbf{T}^-(\mathbf{x},t)\mathbf{n}$ over a subsurface S of plane $\Pi_\mathbf{n}(\mathbf{x}_0)$ through $\mathbf{x}_0\in\mathcal{E}$ with unit normal $\mathbf{n}$ is given by the following:

Lemma 14.3.2.[10]

$$\int_S\left\{\frac{1}{2}\int_\mathcal{V}\int_0^1 \mathbf{g}(\mathbf{x}+\alpha\mathbf{u},\mathbf{x}-(1-\alpha)\mathbf{u})\otimes\mathbf{u}\,d\alpha d\mathbf{u}\right\}\mathbf{n}\,dS_\mathbf{x}$$
$$= \int_{\mathcal{E}_\mathbf{n}^+(\mathbf{x}_0)}\int_{\mathcal{R}^-(S;\mathbf{z})}\sum\sum_{i\neq j}\int_{\mathbb{P}_{ij}(\mathbf{y},\mathbf{z})} P(\mathbf{X},t)\mathbf{f}_{ji}dV_{\mathbf{X}_{ij}^-}dV_\mathbf{y}dV_\mathbf{z}, \tag{14.3.56}$$

where

$$\mathcal{R}^-(S;\mathbf{z}) := \{\mathbf{y}\in\mathcal{E}_\mathbf{n}^-(\mathbf{x}_0) : \ell(\mathbf{y},\mathbf{z}) \text{ passes through } S\}, \tag{14.3.57}$$

and $\ell(\mathbf{y},\mathbf{z})$ denotes that straight line through points $\mathbf{y}$ and $\mathbf{z}$.

From (14.3.50) and (14.3.56)

$$\int_S \mathbf{T}^-(\mathbf{x},t)\mathbf{n}\,dS_\mathbf{x} = \sum\sum_{i\neq j}\int_{\mathcal{E}^+_{\mathbf{n}(\mathbf{x}_0)}}\int_{\mathcal{R}^-(S;\mathbf{z})}\mathcal{F}_{ij}(\mathbf{y},\mathbf{z},t)dV_\mathbf{y}dV_\mathbf{z}, \tag{14.3.58}$$

where

$$\mathcal{F}_{ij}(\mathbf{y},\mathbf{z},t) := \int_{\mathbb{P}_{ij}(\mathbf{y},\mathbf{z})} P(\mathbf{X},t)\mathbf{f}_{ij}\,dV_{\mathbf{X}_{ij}^-}. \tag{14.3.59}$$

Thus a contribution from P_i and P_j occurs only if $\mathbf{x}_j\in\mathcal{E}_\mathbf{n}^+(\mathbf{x}_0)$, $\mathbf{x}_i\in\mathcal{E}_\mathbf{n}^-(\mathbf{x}_0)$, and the line through $\mathbf{x}_i$ and $\mathbf{x}_j$ intersects S: its value is given by the double force density

[10] This is proved in Noll [16]. See also Murdoch & Bedeaux [77], Appendix A.

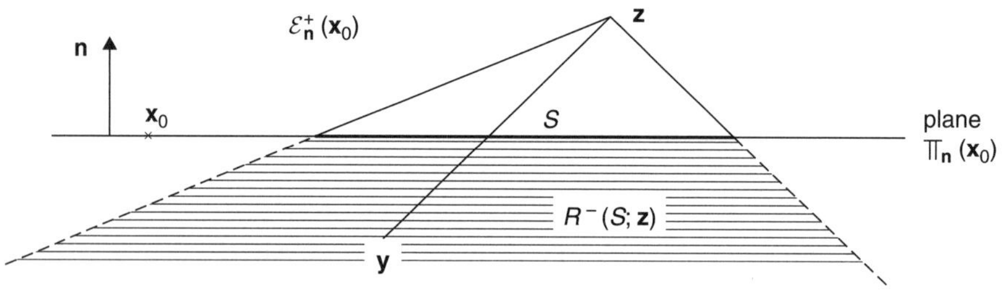

Figure 14.1. Given any point $\mathbf{z}$ 'above' plane $\Pi_{\mathbf{n}}(\mathbf{x}_0)$ through $\mathbf{x}_0$ with normal $\mathbf{n}$, shaded region $R^{-}(S;\mathbf{z})$ consists of those points $\mathbf{y}$ 'below' $\Pi_{\mathbf{n}}(\mathbf{x}_0)$ reached by rays from $\mathbf{z}$ which pass through plane surface region $S \subset \Pi_{\mathbf{n}}(\mathbf{x}_0)$.

ensemble average $\mathcal{F}_{ij}(\mathbf{y},\mathbf{z},t)$ which delivers the expectation of $\mathbf{f}_{ij}$ associated with P_i being located at $\mathbf{y}$ and P_j at $\mathbf{z}$ (at time t). Said differently, given any point $\mathbf{z} \in \mathcal{E}_{\mathbf{n}}^{+}(\mathbf{x}_0)$ and pair of point masses P_i and P_j, force double density $\mathcal{F}_{ij}(\mathbf{y},\mathbf{z},t)$ is defined for all points $\mathbf{y}$ which both lie in $\mathcal{E}_{\mathbf{n}}^{-}(\mathbf{x}_0)$ and can be reached from $\mathbf{z}$ by a line which intersects S. (See Figure 14.1.) Roughly speaking, for small regions $R(\mathbf{z}) \subset \mathcal{E}_{\mathbf{n}}^{+}(\mathbf{x}_0)$ and $R(\mathbf{y}) \subset \mathcal{E}_{\mathbf{n}}^{-}(\mathbf{x}_0)$ surrounding $\mathbf{z}$ and $\mathbf{y}$ (of volumes $\Delta V_{\mathbf{z}}$ and $\Delta V_{\mathbf{y}}$, say), expression $\mathcal{F}_{ij}(\mathbf{y},\mathbf{z},t)\Delta V_{\mathbf{y}}\Delta V_{\mathbf{z}}$ represents the ensemble average at time t of $\mathbf{f}_{ij}$ associated with P_j being located in $R(\mathbf{z})$ and P_i in $R(\mathbf{y})$.

Remark 14.3.2. If interactions $\mathbf{f}_{ij}$ have cut-off range δ (i.e., $\mathbf{f}_{ij} = \mathbf{0}$ if $\|\mathbf{x}_i - \mathbf{x}_j\| > \delta$), then contributions to the integral over S in (14.3.58) derive only from points $\mathbf{y}$ and $\mathbf{z}$ within a distance δ from S.

Recapitulating, Noll's solution to (14.3.38), namely (14.3.50), together with (14.3.46), enables (14.3.36) to be expresses in the standard form [cf. (2.7.20)]

$$\operatorname{div}\mathbf{T} + \mathbf{b} = \rho\mathbf{a}, \tag{14.3.60}$$

where $\mathbf{T}$ is given by (14.3.39).

Remark 14.3.3. Irving & Kirkwood [73] considered systems for which external forces are conservative and interactions are governed by separation-dependent pair potentials. The time evolution of the expectation of the local energy density (the sum of potential energies of external and interaction forces together with kinetic energy) was shown to yield a local balance of form (2.8.24) with $r = 0$. Noll [16] derived the same balance using standard multivariable calculus.

14.4 Generalisation of Irving and Kirkwood/Noll Results

In the derivation of linear momentum balance (14.3.35) minimal assumptions were made concerning the resultant force $\mathbf{F}_i$ experienced by P_i. This force was assumed to have distinguishable contributions from interactions and external agencies, namely $\mathbf{f}_i$ and $\mathbf{b}_i$, but these represented *resultant* effects. Specifically, pairwise interactions [cf. (14.3.44)] were *not* invoked. Accordingly $\mathbf{f}_i$ could depend upon the locations and *momenta* of *all* point masses. In particular, (14.2.19) was not assumed to hold[11]; only

[11] Relation (14.2.19) holds if $\mathbf{f}_i$ and $\mathbf{b}_i$ separately depend only upon the locations of (possibly all) point masses (see Exercise 14.2.1).

(4.2.15) was invoked. However, balance (14.3.35) does not involve an interaction stress tensor. Thus it is natural to ask the following:

Question Q.1: Can the standard balance (14.3.60), obtained by Irving & Kirkwood/Noll on the basis of interactions $\mathbf{f}_{ij}$ governed by separation-dependent pair potentials, be derived under more general forms of interaction?

Answer A.1: Generalisation to velocity-dependent interactions.

Pitteri [74] considered interactions $\mathbf{f}_{ij}$ which depend upon both the locations and *velocities* of the point mass pair P_i and P_j concerned. Specifically, Pitteri considered interactions of the form [cf. (6.2.51)]

$$\mathbf{f}_{ij} = -\nabla_{\mathbf{x}_j}\hat{\phi}_{ij}(r_{ij}) + \mathbf{F}_{ij}(\mathbf{x}_i, \mathbf{x}_j, \mathbf{v}_i, \mathbf{v}_j, t), \tag{14.4.1}$$

where [cf. (6.2.53)]

$$\hat{\phi}_{ji} = \hat{\phi}_{ij}. \tag{14.4.2}$$

Employing Noll's methodology, Pitteri [74] showed that

$$\mathbf{f} = \operatorname{div}\mathbf{T}^- + \mathbf{F}, \tag{14.4.3}$$

where interaction force density $\mathbf{F}$ incorporates the effect of contributions $\mathbf{F}_{ij}$ to $\mathbf{f}_{ij}$, and $\mathbf{T}^-$ is precisely Noll's interaction stress tensor. The corresponding form of linear momentum balance[12] is

$$\operatorname{div}\mathbf{T} + \mathbf{F} + \mathbf{b} = \rho\mathbf{a}, \tag{14.4.4}$$

where $\mathbf{T}$ is given by (14.3.39).

Answer A.2: Generalisation to multibody potentials.

A function $\hat{\psi}$ which depends upon the location of all point masses in a system is said to be a *multibody potential* if the resultant internal force on any point mass P_i is given by

$$\mathbf{f}_i = -\nabla_{\mathbf{x}_i}\hat{\psi}(\mathbf{x}_1, \ldots, \mathbf{x}_N). \tag{14.4.5}$$

Exercise 14.4.1. Recall that conservative pairwise interactions give rise to separation-dependent pair potentials. Specifically [cf. (6.2.51)],

$$\mathbf{f}_{ij} = \nabla_{\mathbf{x}_j}\hat{\phi}_{ij}(r_{ij}) = \hat{\phi}'_{ij}(r_{ij})\frac{(\mathbf{x}_j - \mathbf{x}_i)}{r_{ij}} = -\nabla_{\mathbf{x}_i}\hat{\phi}_{ij}(r_{ij}). \tag{14.4.6}$$

Further, the work done in assembling the point masses in their current locations from infinitely remote locations is[13] (cf. 6.2.54))

$$W = \frac{1}{2}\sum\sum_{i \neq j}\hat{\phi}_{ij}(r_{ij}). \tag{14.4.7}$$

[12] Pitteri also derived an energy balance in the manner of Noll in which the term r [cf. (2.8.24): this term was absent in Noll's analysis] is not solely associated with external influence but includes the effect of power expended in thermal motions associated with interaction contributions $\mathbf{F}_{ij}$.

[13] Provided potentials vanish for infinite separations: see Exercise 6.2.3, specifically (6.2.53).

By regarding W as a function of the N independent locations $\mathbf{x}_1,\ldots,\mathbf{x}_N$, namely

$$W = \hat{W}(\mathbf{x}_1,\ldots,\mathbf{x}_N) := \frac{1}{2}\sum\sum_{i\neq j}\hat{\phi}_{ij}(\|\mathbf{x}_i - \mathbf{x}_j\|), \tag{14.4.8}$$

show that

$$-\nabla_{\mathbf{x}_i}\hat{W}(\mathbf{x}_1,\ldots,\mathbf{x}_N) = \sum_{j\neq i}\mathbf{f}_{ij} = \mathbf{f}_i. \tag{14.4.9}$$

Recently Admal & Tadmor [75] have made a comprehensive study of microscopic interpretations of the stress tensor with the aim of unifying the distinctly different perspectives offered by classical statistical mechanics and weighting function methodology. In particular, these authors study potential functions of form (14.4.5). Exercise 14.4.1 shows that such functions *exist* (at least in the context of conservative pairwise interactions). If ψ is to represent a total binding energy in some way, then its value at time t, is for an observer O,

$$\psi(t) = \hat{\psi}(\mathbf{x}_1(t),\ldots,\mathbf{x}_N(t)). \tag{14.4.10}$$

If O were to view the system from a different perspective at time t, then locations would appear to have undergone (for O) a rigid body deformation (see Remark B.3.1 of Appendix B), yet $\psi(t)$, an observer-independent scalar quantity, would not change value. It follows that ψ effectively depends upon less than the $3N$ scalars necessary to define all N locations. Notice that knowledge of $(\mathbf{x}_1,\ldots,\mathbf{x}_N)$ is equivalent to that of $(\mathbf{x}_1;\mathbf{x}_2 - \mathbf{x}_1,\mathbf{x}_3 - \mathbf{x}_1,\ldots,\mathbf{x}_N - \mathbf{x}_1)$, which is, in turn, equivalent to knowledge of $(\mathbf{x}_1 : r_{12},\hat{\mathbf{u}}_{12};r_{13},\hat{\mathbf{u}}_{13};\ldots,r_{1N},\hat{\mathbf{u}}_{1N})$, where $r_{1j} := \|\mathbf{x}_j - \mathbf{x}_1\|,\hat{\mathbf{u}}_{1j} := (\mathbf{x}_j - \mathbf{x}_1)/r_{1j}$, and $j = 2,3,\ldots,N$. A rigid body deformation can locate $\mathbf{x}_1$ anywhere, orient $\mathbf{x}_2 - \mathbf{x}_1$ in any direction, and then (having chosen the three co-ordinates of $\mathbf{x}_1$ and polar angles θ and ϕ necessary to define the direction of $\mathbf{x}_2 - \mathbf{x}_1$) rotate the system about that line through $\mathbf{x}_1$, parallel to $\mathbf{x}_2 - \mathbf{x}_1$, through any angle. (The foregoing considerations are essentially those which establish Euler angles in rigid-body dynamics: see Goldstein et al. [7], pp.150–4.) It follows that only $3N - 6$ scalars are necessary to delineate the value of ψ, a result stated by Admal & Tadmor without proof. These authors proceed to consider

$$S := \{(r_{12},r_{13},\ldots,r_{1N},r_{23},\ldots,r_{(N-1)N}) : r_{ij} := \|\mathbf{x}_i - \mathbf{x}_j\|,\ \mathbf{x}_i \neq \mathbf{x}_j\}. \tag{14.4.11}$$

That is, S denotes the set of all $N(N-1)/2$ distances between N *distinct* and *distinguishable* points in $\mathcal{E}$, ordered in the manner indicated. (Recall that $\mathbf{x}_i$ is the location of P_i, and point masses are considered to be distinguishable. Thus the first number in an element of S represents the distance between P_1 and P_2, etc.) Set S is clearly a proper subset of $\mathbb{R}^{N(N-1)/2}$, since its elements are ordered lists of positive numbers which are constrained by their interpretation. (For example, $\|\mathbf{x}_i - \mathbf{x}_j\| < \|\mathbf{x}_i - \mathbf{x}_k\| + \|\mathbf{x}_k - \mathbf{x}_j\|$ if $\mathbf{x}_i,\mathbf{x}_j$, and $\mathbf{x}_k$ are non-collinear: in such case, $r_{ij} < r_{ik} + r_{kj}$.) Set S is identified with the set, E say, of equivalence classes of lists $(\mathbf{x}_1,\ldots,\mathbf{x}_N)$ of ordered N-tuples of distinct locations in $\mathcal{E}$, where two such lists are defined to be equivalent if they are related by a rigid deformation. Thus ψ values are determined by elements of S; that is, there exist functions $\tilde{\psi}$ and $\overline{\psi}$ such that

$$\hat{\psi}(\mathbf{x}_1,\ldots,\mathbf{x}_N) = \tilde{\psi}(\{\mathbf{x}_1,\ldots,\mathbf{x}_N)\}) = \overline{\psi}(r_{12},r_{13},\ldots,r_{N(N-1)}), \tag{14.4.12}$$

where $(r_{12}, r_{13}, \dots, r_{(N-1)N}) \in S \subset \mathbb{R}^{N(N-1)/2}$, and $\{(\mathbf{x}_1, \dots, \mathbf{x}_N)\}$ denotes the equivalence class of sets of N ordered distinct locations, each of which is the result of a rigid deformation applied to locations $(\mathbf{x}_1, \dots, \mathbf{x}_N)$. In understanding (14.4.12) it is necessary to note that while knowledge of $\mathbf{x}_1, \dots, \mathbf{x}_N$ suffices to determine the value of $\overline{\psi}$, knowledge of $r_{12}, \dots, r_{(N-1)N}$ does not yield unique locations $\mathbf{x}_1, \dots, \mathbf{x}_N$, but rather a set of isometric images of such locations. Admal & Tadmor assume that $\overline{\psi}$ has an extension to $\mathbb{R}^{(N(N-1)/2}$ which is continuously differentiable and, for such extension, $\check{\psi}$ say, make the (*formal*) identification[14]

$$\nabla_{\mathbf{x}_i} \hat{\psi}(\mathbf{x}_1, \dots, \mathbf{x}_N) \leftrightarrow \nabla_{\mathbf{x}_i} \check{\psi}(r_{12}, \dots, r_{N(N-1)}). \tag{14.4.13}$$

Thus from (14.4.5) follows the identification

$$\mathbf{f}_i \leftrightarrow -\nabla_{\mathbf{x}_i} \check{\psi}(r_{12}, \dots, r_{(N-1)N}). \tag{14.4.14}$$

Exercise 14.4.2. Show that

$$\begin{aligned} &\nabla_{\mathbf{x}_i} \check{\psi}(\|\mathbf{x}_2 - \mathbf{x}_1\|, \|\mathbf{x}_3 - \mathbf{x}_1\|, \dots, \|\mathbf{x}_N - \mathbf{x}_{N-1}\|) \\ &= \sum_{i<k} \frac{\partial \check{\psi}}{\partial r_{ik}} \frac{(\mathbf{x}_i - \mathbf{x}_k)}{r_{ik}} + \sum_{\ell<i} \frac{\partial \check{\psi}}{\partial r_{\ell i}} \frac{(\mathbf{x}_i - \mathbf{x}_\ell)}{r_{\ell i}} \\ &= \sum_{i \neq j} \check{\mathbf{f}}_{ij}, \end{aligned} \tag{14.4.15}$$

where
$$\check{\mathbf{f}}_{ij} := \frac{\partial \check{\psi}}{\partial r_{ij}} \frac{(\mathbf{x}_i - \mathbf{x}_j)}{r_{ij}} \quad \text{if} \quad i<j \tag{14.4.16}$$

and
$$\check{\mathbf{f}}_{ij} := \frac{\partial \check{\psi}}{\partial r_{ji}} \frac{(\mathbf{x}_i - \mathbf{x}_j)}{r_{ji}} \quad \text{if} \quad i>j. \tag{14.4.17}$$

Deduce that
$$\check{\mathbf{f}}_{ji} = -\check{\mathbf{f}}_{ij}. \tag{14.4.18}$$

(Hint: Consider $\check{\mathbf{f}}_{12}$ and $\check{\mathbf{f}}_{21}$.)

Terms $\check{\mathbf{f}}_{ij}$ defined by (14.4.16) and (14.4.17) are candidate interaction forces associated with point mass pairs. Accordingly the analysis following (14.3.44) establishes an interaction stress tensor $\check{\mathbf{T}}^-$ given by (14.3.50) with [see (14.3.46)]

$$\mathbf{g}(\mathbf{x} + \alpha\mathbf{u}, \mathbf{x} - (1-\alpha)\mathbf{u}, t) := \sum\sum_{i \neq j} \int_{\mathbb{P}_{ij}(\mathbf{y},\mathbf{z})} P(\mathbf{X}, t) \check{\mathbf{f}}_{ij} \, dV_{\mathbf{X}_{ij}^-}, \tag{14.4.19}$$

where $\mathbf{y} = \mathbf{x} + \alpha\mathbf{u}$ and $\mathbf{z} = \mathbf{x} - (1-\alpha)\mathbf{u}$. In the integral, $\mathbf{x}_i$ is located at $\mathbf{y} = \mathbf{x} + \alpha\mathbf{u}$ and $\mathbf{x}_j$ at $\mathbf{z} = \mathbf{x} - (1-\alpha)\mathbf{u}$, so $\mathbf{x}_i - \mathbf{x}_j = \mathbf{u}$. Since $\check{\mathbf{f}}_{ij}$ is parallel to $\mathbf{u}$ for all particle pairs [see (4.4.16) and (4.4.17)],

$$\mathbf{g}(\mathbf{x} + \alpha\mathbf{u}, \mathbf{x} - (1-\alpha)\mathbf{u}, t) \quad \textit{is parallel to} \quad \mathbf{u}, \tag{14.4.20}$$

[14] Here $r_{12}, \dots, r_{(N-1)N}$ each lie in $\mathbb{R}$ and are subject to no constraints.

and hence the integral which defines $\check{\mathbf{T}}^-(\mathbf{x},t)$ [see (14.3.50)] has an integrand which is a scalar multiple of the symmetric linear transformation $\mathbf{u}\otimes\mathbf{u}$. Accordingly $\check{\mathbf{T}}^-(\mathbf{x},t)$ is symmetric. Admal & Tadmor [75] argue that the central force decomposition given by (14.4.16) and (14.4.17) is the only physically meaningful partition of $\mathbf{f}_i$ into pairwise terms $\check{\mathbf{f}}_{ij}$. However, since the extension of $\overline{\psi}$ to $\check{\psi}$ is non-unique, such non-uniqueness is inherited by $\check{\mathbf{T}}$.

Remark 14.4.1. Admal & Tadmor [75] conclude that multibody potentials that admit continuously differentiable extensions (as described earlier) give rise to symmetric interaction stress tensors. [Of course, since $\mathcal{D}$ is symmetric, it follows that the stress $\mathbf{T}$ in (14.3.39) is symmetric]. The significance of such a conclusion depends crucially upon whether extensions have any physical basis. The following two remarks address this issue together with not-so-obvious assumptions implicit in the notion of a multibody potential.

Remark 14.4.2. Recall the discussion of Section 5.4 in which the molecular interaction $\mathbf{f}_{ij}$ was related to the subatomic structures of P_i, P_j, and neighbouring molecules. This very general viewpoint furnished an unequivocal and unique physical interpretation of $\mathbf{f}_{ij}$, together with a rationale for expecting pairwise balance $\mathbf{f}_{ji} = -\mathbf{f}_{ij}$. *The non-uniqueness of Admal & Tadmor's candidate interactions is accordingly unphysical.* Of course, this non-uniqueness derives from the non-unique extensions of $\overline{\psi}$ to $\check{\psi}$: each such extension produces a collection of pseudo-interactions $\check{\mathbf{f}}_{ij}$ as artefacts. *It follows that the claim that multibody potentials give rise to symmetric interaction stress tensors is questionable.*

Remark 14.4.3. (General observation on multibody potentials) For a system of interacting point masses in which interactions are governed by separation-dependent pair potentials it is meaningful to talk of the energy of system assembly [see (6.2.51) et seq.] Specifically, the work done in assembling the system molecules in their current locations, starting from a state of infinitely remote dispersal, is independent of the manner or order of assembly. Now consider assembly of a system for which there exists a multibody potential $\hat{\psi}$. From (14.4.5) the work done in moving P_i from location $\mathbf{x}$ to location $\mathbf{x}'$ in the presence of P_j located at $\mathbf{x}_j$ $(j \neq i; j = 1,2,\ldots,N)$ is[15]

$$\int_{\mathbf{x}_i:\mathbf{x}\to\mathbf{x}'} -\mathbf{f}_i \,.\, d\mathbf{r} = [\hat{\psi}(\mathbf{x}_1,\ldots,\mathbf{x}_N)]_{\mathbf{x}_i=\mathbf{x}}^{\mathbf{x}_i=\mathbf{x}'}. \tag{14.4.21}$$

Consistent with the nature of molecular interactions decaying with separation is the assumption that *if* a subset of $\mathbf{x}_1,\ldots,\mathbf{x}_N$ corresponds to locations infinitely removed from each other and from those of the remaining point masses, *then* $\hat{\psi}$ values depend only upon the locations of these remaining point masses. Defining $\hat{\psi}_{12\ldots p}(\mathbf{x}_1,\mathbf{x}_2,\ldots,\mathbf{x}_p)$ to be the value of $\hat{\psi}$ when $P_{p+1},\ldots,P_N$ are remotely dispersed [here $p = 1,2,\ldots(N-1)$], then this quantity represents the work done in locating P_p at $\mathbf{x}_p$ from remote dispersal in the presence of $P_1,\ldots,P_{p-1}$ at $\mathbf{x}_1,\ldots,\mathbf{x}_{p-1}$. Accordingly the work done in assembling the point masses at locations $\mathbf{x}_1,\ldots\mathbf{x}_N$ (from a state of

[15] The notation should be self-evident: the value of the integral is path-independent. Notice that the value of the integral is what might be described as work 'against' the force $\mathbf{f}_i$.

remote dispersal) in the order P_1 at $\mathbf{x}_1$, then P_2 at $\mathbf{x}_2$, then P_3 at $\mathbf{x}_3$, etc. is[16]

$$W_{12\ldots N} := \hat{\psi}_{12}(\mathbf{x}_1,\mathbf{x}_2) + \hat{\psi}_{123}(\mathbf{x}_1,\mathbf{x}_2,\mathbf{x}_3) + \cdots + \hat{\psi}_{12\ldots N}(\mathbf{x}_1,\ldots,\mathbf{x}_N). \tag{14.4.22}$$

Of course, the order of assembly can be performed in $N!$ ways. Thus, for $\hat{\psi}(\mathbf{x}_1,\ldots,\mathbf{x}_N)$ to make sense as an energy of assembly, it is necessary that, for all permutations $\pi(1,2,\ldots,N)$ of $\{1,2,\ldots,N\}$, the work $W_{\pi(1,2,\ldots,N)}$, defined by obvious analogy with $W_{12\ldots N}$, should be the same.

14.5 Selection of a Probability Density Function: Projection Operator Methodology

This section outlines an approach and procedure that were established and developed in Murdoch & Bedeaux [76–79] following the seminal work of Zwanzig [80, 81]. While the Irving & Kirkwood/Noll approach furnishes balance relations which are *formally* those employed in deterministic continuum mechanics,[17] it is necessary to address two inter-related fundamental issues.

Issue 1: How are Irving & Kirkwood/Noll field values to be linked with measurable quantities?
Issue 2: How is the probability density function (upon which all fields depend) to be chosen?

In [73] Irving & Kirkwood identified the field values of deterministic continuum mechanics with statistical averages of local measurement values conducted in oft-repeated experiments. Since each local measurement value reflects a space-time average of molecular behaviour, it follows that field values were identified with statistical averages of local molecular space-time averages. In [73] it was assumed that such averages can be identified with space-time averages of the ensemble averages which appear in their balance relations. However, Irving & Kirkwood did not implement this additional space-time averaging.[18]

There is an alternative viewpoint and methodology which bears directly upon Issue 2. Recall that P is determined by P_0 and microscopic dynamics [see (14.2.10) et seq.], where P_0 is associated with initial ($t = t_0$) information about the system. Since initial information requires measurement/observation, there are associated scales ϵ_0 and Δ_0 of length and time. Now consider macroscopic behaviour that is *continuously* reproducible at space-time scales (ϵ_0, Δ_0) in the sense that if the macroscopic situation is monitored at these scales at any given time $t \geq t_0$, then subsequent behaviour of the system at these scales is uniquely determined (modulo a possible 'toleration of error'). Accordingly, any two microstates $\mathbf{X}(t)$ and $\mathbf{X}'(t)$ which give rise to the known (ϵ_0, Δ_0) situation at time t must have evolved [via (14.2.8)], at any subsequent time t_1, to microstates $\mathbf{X}(t_1)$ and $\mathbf{X}'(t_1)$ which give rise to the same (ϵ_0, Δ_0) situation (modulo

16 No work is assumed to be necessary in locating P_1 at $\mathbf{x}_1$ when all other point masses are infinitely dispersed.

17 Recall Section 3.8: deterministic (as opposed to 'statistical'; cf. Kröner [82]) continuum mechanics pertains to *reproducible* behaviour.

18 It has been suggested that this constitutes too much averaging (private communication with Oliver Penrose, 1994) and that only a further spatial averaging is required; that is, in some sense the ensemble averaging is equivalent to time averaging. This viewpoint was adopted by Admal & Tadmor [75].

the tolerance of variation) at this time. At this stage the term '(ϵ_0, Δ_0) situation' has not been made precise. The projection operator approach addresses the selection of a probability density function and delineation of scale-dependent macroscopic situations [*in the context of continuously reproducible behaviour at prescribed scales* (ϵ_0, Δ_0)] on the basis of the following assumptions.

P.O.[19]*1.* It is possible to identify a list of $n\,(<< 6N)$ scalar quantities which, at any given time, characterise the system at scale ϵ_0. Such a list is termed an ϵ_0-*scale macrostate*, $\mathbf{A}_{\epsilon_0}$ say. Specifically, there exists a map

$$\mathbf{a}_{\epsilon_0} : \mathbb{P}_N^- \longrightarrow \mathbb{A}_{\epsilon_0} \tag{14.5.1}$$

which identifies any element in the physically relevant subset[20] $\mathbb{P}_N^-$ of phase space $\mathbb{P}_N$ with an element in the set $\mathbb{A}_{\epsilon_0} := \mathbf{a}_{\epsilon_0}(\mathbb{P}_N^-) \subset \mathbb{R}^n$ of ϵ_0-scale macrostates.

P.O.2. It is possible to define a differentiable change of variables in which n microstate variables are replaced by macrostate variables. That is, there exists a differentiable bijection[21]

$$\mathbf{a}_{\epsilon_0}^+ : \mathbb{P}_N^- \longrightarrow \mathbb{A}_{\epsilon_0} \times \mathbb{P}_{N,n}^- \tag{14.5.2}$$

where

$$\mathbf{a}_{\epsilon_0}^+(\mathbf{X}) = (\mathbf{a}_{\epsilon_0}(\mathbf{X}), \mathbf{X}^-). \tag{14.5.3}$$

P.O.3. (*Local equilibrium hypothesis*) Any two microstates which give rise to the same ϵ_0-scale macrostate are equally likely.

P.O.4. (*Dynamic ergodicity at scales* ϵ_0, Δ_0) If f denotes any function of microstate that represents an ϵ_0-scale spatial average, then the average of f, computed over all microstates that give rise to the same ϵ_0-scale macrostate (with equal weighting accorded to all such microstates), yields the Δ_0-time average f_{Δ_0} [cf. (5.4.10)] of f.

Remark 14.5.1. Since $n << 6N$, $\mathbf{a}_{\epsilon_0}$ is termed the *reduction map*.

Remark 14.5.2. In view of P.O.3, if at time t the system is in an ϵ_0-scale macrostate $\mathbf{A} \in \mathbb{A}_{\epsilon_0}$, and if $\mathbf{X}, \mathbf{Y}$ denote microstates for which $\mathbf{a}_{\epsilon_0}(\mathbf{X}) = \mathbf{A} = \mathbf{a}_{\epsilon_0}(\mathbf{Y})$, then $P(\mathbf{X}, t) = P(\mathbf{Y}, t)$.

Remark 14.5.3. If $\mathbb{D} \subset \mathbb{P}_N^-$ is volume-measurable (see Appendix B.11.3), then from P.O.2, if f is continuous on $\mathbb{D}$,

$$\int_{\mathbb{D}} f(\mathbf{Y})dV_{\mathbf{Y}} = \int_{\mathbf{a}_{\epsilon_0}^+(\mathbb{D})} \tilde{f}(\mathbf{A}, \mathbf{Y}^-)J(\mathbf{A}, \mathbf{Y}^-)dV_{\mathbf{A}}dV_{\mathbf{Y}^-}. \tag{14.5.4}$$

[19] *P*rojection *O*perator.

[20] For example, the system may be confined in some way, so that not all of configuration space is physically relevant.

[21] Here $\mathbb{P}_{N,n}^-$ denotes the set of sublists $\mathbf{X}^-$ of $\mathbb{P}_N^-$ in which the n replaceable microstate variables have been omitted.

Here $$(\mathbf{A},\mathbf{Y}^-) = \mathbf{a}^+_{\epsilon 0}(\mathbf{Y}), \qquad \tilde{f}(\mathbf{A},\mathbf{Y}^-) := f((\mathbf{a}^+_{\epsilon 0})^{-1}(\mathbf{A},\mathbf{Y}^-)), \tag{14.5.5}$$

and J denotes the Jacobian associated with map $(\mathbf{a}^+_{\epsilon 0})^{-1}$. In view of P.O.3 it is instructive to consider integrals over subsets of $\mathbb{P}^-_N$ whose elements correspond to the same macrostate, $\mathbf{A}$ say. Relation (14.5.4) indicates how this may be accomplished, for if $\mathbf{A} \in \mathbb{A}_{\epsilon 0}$ and

$$\mathbb{D}_\mathbf{A} := \{\mathbf{X} \in \mathbb{D} : \mathbf{a}_{\epsilon 0}(\mathbf{X}) = \mathbf{A}\}, \tag{14.5.6}$$

then $$\int_{\mathbb{D}_\mathbf{A}} f(\mathbf{Y}) dV_\mathbf{Y} := \int_{\mathbf{a}^+_{\epsilon 0}(\mathbb{D}_\mathbf{A})} \tilde{f}(\mathbf{A},\mathbf{Y}^-) J(\mathbf{A},\mathbf{Y}^-) dV_{\mathbf{Y}^-}. \tag{14.5.7}$$

Notice that the result of integration is a function of $\mathbf{A}$. It is convenient to adopt the suggestive *notation*[22]

$$\int_\mathbb{D} f(Y)\delta(\mathbf{a}(\mathbf{Y}) - \mathbf{A}) dV_\mathbf{Y} := \int_{\mathbb{D}_\mathbf{A}} f(\mathbf{Y}) dV_\mathbf{Y}. \tag{14.5.8}$$

Now consider any function f of microstate that represents an ϵ_0-scale spatial average in a continuously reproducible (ϵ_0, Δ_0) scale process. According to P.O.4 the corresponding (ϵ_0, Δ_0) average is related to

$$f_{\Delta_0}(\mathbf{X}) := \int_{\mathbb{P}^-_N} f(\mathbf{Y})\delta(\mathbf{a}(\mathbf{Y}) - \mathbf{a}(\mathbf{X})) dV_\mathbf{Y}. \tag{14.5.9}$$

The function f_{Δ_0} of microstate $\mathbf{X}$ is clearly piecewise constant on microstates which share the same ϵ_0 macrostate (Why?). However, a requirement of any sensible averaging is that the average of a constant function should leave this function unchanged. If $f = k$, then (14.5.9) yields

$$k = k_{\Delta_0} = k \int_{\mathbb{P}^-_N} \delta(\mathbf{a}(\mathbf{Y}) - \mathbf{a}(\mathbf{X})) dV_\mathbf{Y}. \tag{14.5.10}$$

Accordingly (14.5.9) is modified via the introduction of

$$\mu(\mathbf{a}(\mathbf{X})) := \int_{\mathbb{P}^-_N} \delta(\mathbf{a}(\mathbf{Y}) - \mathbf{a}(\mathbf{X})) dV_\mathbf{Y}, \tag{14.5.11}$$

and the (averaging) *projection operator* $\mathcal{P}$ is defined on any function f as above by

$$\mathcal{P}f(\mathbf{X}) := \frac{1}{\mu(\mathbf{a}(\mathbf{X}))} \int_{\mathbb{P}^-_N} f(\mathbf{Y})\delta(\mathbf{a}(\mathbf{Y}) - \mathbf{a}(\mathbf{X})) dV_\mathbf{Y}. \tag{14.5.12}$$

Exercise 14.5.1. Verify that

(i) if f is a constant k, then $\mathcal{P}f = f$, and
(ii) if $\mathbf{a}(\mathbf{X}) = \mathbf{a}(\mathbf{Y})$, then $\mathcal{P}f(\mathbf{X}) = \mathcal{P}f(\mathbf{Y})$.

It follows from (ii) above that if f is any function of microstate, then $\mathcal{P}f$ induces a function $\widehat{\mathcal{P}f}$ on $\mathbb{A}_{\epsilon 0}$ for which

$$\widehat{\mathcal{P}f} \circ \mathbf{a} = \mathcal{P}f. \tag{14.5.13}$$

[22] Subscript ϵ_0 is supressed from now on for brevity.

That is, if $\mathbf{X} \in \mathbb{P}_N^-$ and $\mathbf{a}(\mathbf{X}) = \mathbf{A}$, then

$$\widehat{\mathcal{P}f}(\mathbf{A}) = \mathcal{P}f(\mathbf{X}). \tag{14.5.14}$$

Definition (14.5.12) of $\mathcal{P}$ leads to two simple results:

$\mathcal{P}$.1.
$$\mathcal{P}^2 = \mathcal{P} \tag{14.5.15}$$

$\mathcal{P}$.2.
$$\text{If } \mathcal{Q}f := f - \mathcal{P}f =: (1 - \mathcal{P})f, \tag{14.5.16}$$

then
$$\mathcal{Q}^2 = \mathcal{Q} \quad \text{and} \quad \mathcal{P}\mathcal{Q} = 0 = \mathcal{Q}\mathcal{P}. \tag{14.5.17}$$

Exercise 14.5.2. Prove 14.5.15 and (14.5.17).

Remark 14.5.4. Property $\mathcal{P}$*.1* establishes that $\mathcal{P}$ is a projection in the space of integrable functions defined on $\mathbb{P}_N^-$, while $\mathcal{P}$*.2* introduces the complementary projection associated via *P.O.4* with fluctuations in ϵ_0-scale spatial averages from the corresponding (ϵ_0, Δ_0) space-*time* averages. In this connection it is a straightforward matter, using (14.5.13) and (14.5.12), to obtain properties

$\mathcal{P}$*.3*. If f and g are real-valued integrable functions defined on $\mathbb{P}_N^-$, then

$$\mathcal{P}((\mathcal{P}f)(\mathcal{P}g)) = (\mathcal{P}f)(\mathcal{P}g) \tag{14.5.18}$$

and
$$\mathcal{P}((\mathcal{Q}f)(\mathcal{P}g)) = 0. \tag{14.5.19}$$

More suggestively, writing

$$\bar{f} := \mathcal{P}f \quad \text{and} \quad f' := f - \bar{f} = \mathcal{Q}f, \tag{14.5.20}$$

(14.5.18) and (14.5.19) become

$$\overline{\bar{f}\bar{g}} = \bar{f}\bar{g} \quad \text{and} \quad \overline{f'\bar{g}} = 0. \tag{14.5.21}$$

Exercise 14.5.3. Show that

$$\overline{fg} = \bar{f}\bar{g} + \overline{f'g'}. \tag{14.5.22}$$

Properties (14.5.21), and hence result (14.5.22), do not always hold for averages: recall the *assumption* in (8.6.4) of such a result (as a plausible approximation).

Further properties are

$\mathcal{P}$*.4*.
$$\mathcal{P}P = P \tag{14.5.23}$$

and

$\mathcal{P}$*.5*. If f denotes an ϵ_0-scale function of microstate, then

$$\mathcal{P}(fP) = (\mathcal{P}f)P. \tag{14.5.24}$$

Property $\mathcal{P}.4$ is an immediate consequence of *P.O.3*, and $\mathcal{P}.5$ follows from *P.O.3*, (14.5.13), and (14.5.12).

Remark 14.5.5. Matters are greatly simplified if

(i) the system is confined to a closed and bounded region, and
(ii) forces on point masses depend only upon their locations[23].

Both conditions will be assumed to hold in what follows. In particular, from (ii) and Remark 14.2.2,

$$\mathrm{div}_{\mathbb{P}}\,\mathbf{V} = 0. \tag{14.5.25}$$

Further, $\mathcal{P}$ is formally 'symmetric' as a linear operator on a function space relevant to condition (i). Specifically, let $\mathcal{F}_0(\mathbb{P}_N^-,\mathbb{R})$ denote the space of real-valued volume-integrable functions on $\mathbb{P}_N^-$ with compact support.[24] If $f,g \in \mathcal{F}_0(\mathbb{P}_N^-,\mathbb{R})$, then

$$(f,g) := \int_{\mathbb{P}_N^-} f(\mathbf{X})g(\mathbf{X})dV_{\mathbf{X}} \tag{14.5.26}$$

is an inner product on this space, and $\mathcal{P}$ satisfies (cf. Appendix A.8.)

$\mathcal{P}.6.$ $$(\mathcal{P}f,g) = (f,\mathcal{P}g). \tag{14.5.27}$$

The main purpose of projection operator methodology is to compare the time evolution of ϵ_0-scale functions of microstate [in particular, macrostate $\mathbf{A}_{\epsilon_0}$: see *P.O.1*] with their (ϵ_0,Δ_0)-scale counterparts. The former fluctuate in time, while the latter do not [recall that the context is that of (ϵ_0,Δ_0) reproducible behaviour].

If f represents an ϵ_0-scale function of microstate, then

$$\frac{d}{dt}\{f(\mathbf{X}(t))\} = \nabla_{\mathbb{P}} f(\mathbf{X}(t)).\mathbf{V}(\mathbf{X}(t)) = -(Lf)(\mathbf{X}(t)), \tag{14.5.28}$$

where $$Lf := -\nabla_{\mathbb{P}} f.\mathbf{V} \tag{14.5.29}$$

defines the *Liouville operator* L. [Here $\mathbf{V}$ depends only upon $\mathbf{X}(t)$, in contrast to (14.2.16), in view of condition (ii) and Remark 14.5.5.]

Exercise 14.5.4. Show that the time evolution of P given by (14.2.15) may, in view of (14.5.25), be written as

$$\frac{\partial P}{\partial t} = LP. \tag{14.5.30}$$

Noting that (14.5.28) is expressible as

$$\dot{\widehat{f \circ \mathbf{X}}} = -(Lf)\circ\mathbf{X}, \tag{14.5.31}$$

setting $g = -Lf$ yields

$$\dot{\widehat{g \circ \mathbf{X}}} = -L(-Lf)\circ\mathbf{X} = L^2 f\circ\mathbf{X},$$

23 That is, both point mass interactions and forces which confine the system depend only upon point mass locations.

24 That is, the support $\{\mathbf{X} \in \mathbb{P}_N^- : f(\mathbf{X}) \neq 0\}$ of f is closed and bounded.

so that
$$\widehat{f \circ \ddot{\mathbf{X}}} = L^2 f \circ \mathbf{X}. \tag{14.5.32}$$

Repeating the process yields the nth order time derivative of $f \circ \mathbf{X}$ as $(-L)^n f \circ \mathbf{X}$. If

$$F := f \circ \mathbf{X} \tag{14.5.33}$$

is analytic at $t = t_0$ and $\mathbf{X}(t_0) = \mathbf{X}_0$, then

$$F(t) = \sum_{n=0}^{\infty} F^{(n)}(t_0) \frac{(t - t_0)^n}{n!} \tag{14.5.34}$$

$$= \sum_{n=0}^{\infty} (-L)^n f(\mathbf{X}_0) \frac{(t - t_0)^n}{n!}. \tag{14.5.35}$$

Formally, (14.5.33) and (14.5.35) yield

$$f(\mathbf{X}(t)) = (e^{-L(t-t_0)}\{f\})(\mathbf{X}(t_0)). \tag{14.5.36}$$

Remark 14.5.6. While the exponential $e^{\mathcal{L}}$ makes sense for any bounded linear transformation $\mathcal{L}$ on a finite-dimensional normed vector space,[25] L is an *unbounded* linear operator on an infinite-dimensional space of functions. Nevertheless, relation (14.5.36) can be given a rigorous mathematical interpretation within the theory of *semigroups* (cf., e.g., Belleni-Morante [83]). Notice that (14.5.36) indicates that $e^{-L(t-t_0)}$ is an operator which 'updates' the value of f, at any point $\mathbf{X}_0 \in \mathbb{P}_N^-$ at time t_0, to its value at the corresponding point $\mathbf{X}(t)$ after a time lapse $(t - t_0)$. The nature of the microscopic dynamics considered [see (14.2.8) with $\mathcal{F}$ dependent only upon $\mathbf{X}(t)$ and $\hat{\mathbf{X}} = \mathbf{X}_0$ in (14.2.9)] means that updating over a time lapse τ_1, followed by a further lapse τ_2 updating, yields the same result as a single updating over a time lapse $\tau_1 + \tau_2$. Thus, formally,

$$e^{-L\tau_2} e^{-L\tau_1} = e^{-L(\tau_1 + \tau_2)}. \tag{14.5.37}$$

Such pairwise composition is associative; that is,

$$e^{-L\tau_1}(e^{-L\tau_2} e^{-L\tau_3}) = (e^{-L\tau_1} e^{-L\tau_2}) e^{-L\tau_3} \tag{14.5.38}$$

since both sides are equivalent to $e^{-L(\tau_1+\tau_2+\tau_3)}$. Property (14.5.38) establishes the composition as that of a semigroup (see Jacobson [84]).

From (14.5.30)

$$\frac{\partial^2 P}{\partial t^2} = \frac{\partial}{\partial t}\{LP\} = L\frac{\partial P}{\partial t} = L^2 P, \tag{14.5.39}$$

and similarly, for any integer $n \geq 3$,

$$\frac{\partial^n P}{\partial t^n} = L^n P. \tag{14.5.40}$$

[25] Cf., e.g., Goertzel & Tralli [89], Appendix 1B.

If $P(\mathbf{X},t)$ is analytic in t at $t = t_0$, then

$$P(\mathbf{X},t) = \sum_{n=0}^{\infty} \frac{\partial^n P}{\partial t^n}(\mathbf{X},t_0)\frac{(t-t_0)^n}{n!},$$

and formally [cf. (14.5.36)]

$$P(\mathbf{X},t) = e^{L(t-t_0)}P(\mathbf{X},t_0). \tag{14.5.41}$$

Any microscopic process $\mathbf{X}(t)$ induces an ϵ_0-scale macroscopic process [see (14.5.1)] via

$$\mathbf{A}(t) := \mathbf{a}_{\epsilon_0}(\mathbf{X}(t)). \tag{14.5.42}$$

The probability density function for such macrostates is

$$\mathbf{W}(\mathbf{A},t) := \int_{\mathbb{P}_N^-} P(\mathbf{Y},t)\delta(\mathbf{a}_{\epsilon_0}(\mathbf{Y}) - \mathbf{A})dV_{\mathbf{Y}} \tag{14.5.43}$$

$$= \widehat{\mathcal{P}P}(\mathbf{A},t)\mu(\mathbf{A}), \tag{14.5.44}$$

noting *P.O.3* and (14.5.13). After considerable formal manipulation (see Murdoch & Bedeaux [79]) the *master equation*

$$\frac{\partial W}{\partial t} = -\mathrm{div}_{\mathbb{A}}(W\mathbf{V}_{\mathbb{A}}) + \int_{\mathbf{a}_{\epsilon_0}(\mathbb{P}_N^-)} (T(\mathbf{A},\mathbf{A}')W(\mathbf{A}',t) - T(\mathbf{A}',\mathbf{A})W(\mathbf{A},t))dV_{\mathbf{A}'} \tag{14.5.45}$$

is obtained via a 'fading memory' hypothesis. Vector field $\mathbf{V}_{\mathbb{A}}$ in $\mathbb{A}_{\epsilon_0}$ space denotes the average rate of change of ϵ_0-scale macrostate taken over all microscopic processes which instantaneously yield macrostate $\mathbf{A}$. Term $T(\mathbf{A},\mathbf{A}')W(\mathbf{A}',t)$ denotes the $\mathbb{A}_{\epsilon_0}$-space density which delivers the rate of increase in probability of macrostate $\mathbf{A}$ due to macrostate $\mathbf{A}'$ evolving into $\mathbf{A}$, and $T(\mathbf{A}',\mathbf{A})W(\mathbf{A},t)$ denotes the probability decrease rate due to macrostate $\mathbf{A}$ evolving into $\mathbf{A}'$. Assuming that the integrand is dominated by transitions $\mathbf{A}'$ close to $\mathbf{A}$ leads, via truncation of the Taylor series in $\mathbf{A}' - \mathbf{A}$ at second-order terms, to the corresponding *Fokker-Planck equation*

$$\frac{\partial W}{\partial t} + \mathrm{div}_{\mathbb{A}}\{W(\mathbf{V}_{\mathbb{A}} + \boldsymbol{\mathcal{A}}_1) - \frac{1}{2}\mathrm{div}_{\mathbb{A}}(W\boldsymbol{\mathcal{A}}_2)\} = 0. \tag{14.5.46}$$

Here $\boldsymbol{\mathcal{A}}_1(\mathbf{A}) \in \mathbb{A}_{\epsilon_0}$ and $\boldsymbol{\mathcal{A}}_2(\mathbf{A})$ is a linear transformation on $\mathbb{A}_{\epsilon_0}$. The physical natures of $\boldsymbol{\mathcal{A}}_1$ and $\boldsymbol{\mathcal{A}}_2$ are given by

$$\frac{d}{dt}\{\langle\mathbf{A}\rangle\} = \langle\mathbf{V}_{\mathbb{A}} + \boldsymbol{\mathcal{A}}_1\rangle \tag{14.5.47}$$

and

$$\frac{d}{dt}\{\langle(\mathbf{A} - \langle\mathbf{A}\rangle)\otimes(\mathbf{A} - \langle\mathbf{A}\rangle)\rangle\} = \langle\boldsymbol{\mathcal{A}}_2\rangle. \tag{14.5.48}$$

Here $\langle\cdot\rangle$ denotes the expectation of any function of $\mathbf{A}$, computed over $\mathbb{A}_{\epsilon_0}$ with respect to the appropriate probability density W. Fields $\boldsymbol{\mathcal{A}}_1$ and $\boldsymbol{\mathcal{A}}_2$ are related by

$$\boldsymbol{\mathcal{A}}_1 = \frac{1}{2W_0}\mathrm{div}_{\mathbb{A}}\{W_0\boldsymbol{\mathcal{A}}_2\}, \tag{14.5.49}$$

where W_0 is the macroscopic probability density which corresponds [see (14.5.44)] to the equilibrium probability density P_0 on $\mathbb{P}_N^-$. Relation (4.5.49) is the relevant *fluctuation-dissipation* equation.

Remark 14.5.7. The analysis leading to the fundamental relation (14.5.45) is extensive and complex. Furthermore, it depends crucially upon the following operator identity:

$$\mathcal{Q}e^{Lt} = \mathcal{Q}e^{\mathcal{Q}L\mathcal{Q}t}\mathcal{Q} + \int_0^t \mathcal{Q}e^{\mathcal{Q}L\mathcal{Q}(t-t')}\mathcal{Q}L\mathcal{P}e^{Lt'}dt'. \tag{14.5.50}$$

While this can be proved *formally* by integration by parts (cf. Murdoch & Bedeaux [79], p. 932), its rigorous derivation would appear to be an open problem (cf. Lamb, Murdoch & Stewart [85]) despite providing the theoretical basis for specific applications of projection operators (cf. Grabert [86]).

Remark 14.5.8. Returning to *P.O.1* and (14.5.1), it is clear that physical interpretations depend upon the choice of reduction map $\mathbf{a}_{\epsilon_0}$. If the system is contained within a rectangular box [see (i) in Remark 14.5.5] of dimensions $2L_1 \times 2L_2 \times 2L_3$, then, choosing a Cartesian co-ordinate system with origin at the centre of the box and axes parallel to its edges, formal Fourier series of the microscopic distributions (see Remark 4.4.5) of mass, momentum, and energy can be computed, and terms corresponding to wavelengths smaller than ϵ_0 then discarded. Such truncated series can be obtained by using the weighting function given by (4.4.40) and (4.4.41) for each of the five scalar-valued microscopic fields of mass, momentum components, and energy. The microscopic fields of mass and momentum were given in (14.2.42) while that of energy is, on suppressing time dependence,

$$e_{\text{mic}}(\mathbf{x}) := \sum_{i=1}^{N}\left\{\frac{1}{2}m_i\mathbf{v}_i^2 + \frac{1}{2}\sum_{j\neq i}\phi_{ij}(r_{ij}) + \sum_{\alpha=1}^{6}\psi_{i\alpha}(d_{i\alpha})\right\}\delta(\mathbf{x}_i - \mathbf{x}). \tag{14.5.51}$$

Here interactions have been assumed to be governed by separation-dependent pair potentials ϕ_{ij} and confinement by potentials $\psi_{i\alpha}$ associated with each face of the box and dependent upon the distance $d_{i\alpha}$ of P_i from face 'α'. Each scalar field involves $(2N_1+1)(2N_2+1)(2N_3+1)$ Fourier coefficients, where N_i is the integral part of $2\epsilon_0^{-1}L_i$ [see (4.4.40) and (4.4.41)]. These may be ordered, and in total form a list in $\mathbb{R}^n$, where $n = 5(2N_1+1)(2N_2+1)(2N_3+1)$. Since the total mass and energy remain unchanged, and these appear as Fourier coefficients, there are effectively $n-2$ variables. If the reduction map has its range in $\mathbb{R}^n$, then the *underlying assumption behind its adoption is that $n-2$ of the $6N$ phase space variables can be replaced by Fourier coefficients.* These $(n-2)$ coefficients, together with the remaining $(6N-n+2)$ phase space variables, represent an independent set of variables which completely characterise microstates.

Remark 14.5.9. The reason for choosing the reduction map $\mathbf{a}_\epsilon$ of Remark 14.5.8, which singles out Fourier coefficients associated with microscopic distributions of mass, momentum, and energy, is that the continuity equation and balances of momentum and energy are precisely evolution equations for these quantities. However, effecting the change of variables from those of phase space to a mix of reduced space variables [the $(n-2)$ Fourier coefficients] and complementary phase state variables is problematic. However, an alternative is to focus solely upon those Fourier coefficients associated with microscopic distributions of mass and momentum. In this respect, Murdoch & Bedeaux [87] showed rigorously that the phase space description of a system of N point masses is precisely equivalent to knowledge of a minimal set of $6N$ complex Fourier coefficients associated with the microscopic distributions of matter and momentum. This minimal set of coefficients defines two truncated Fourier series which involve only wavelengths in excess of a certain value ϵ_{min}. Accordingly complete knowledge of both truncated series is equivalent, at any time, to complete knowledge of the system microstate and vice versa. Continuum descriptions of mass and momentum distributions *at any scale* $\epsilon_0 > \epsilon_{min}$ are obtained by further truncation to exclude wavelengths smaller than ϵ_0.

15 Summary and Suggestions for Further Study

15.1 Preamble

In drawing attention to the microscopic, scale-dependent basis of continuum concepts, field values, and balance relations, the intention has been to complement and inform formal, axiomatic approaches to continuum mechanics. Here the methodology and main aspects of the study are summarised in the form of concluding remarks. Since the discussion has not been exhaustive, suggestions are made for possible extension of the procedures here employed to interfacial phenomena and boundary conditions, generalised and structured continua, configurational forces, reacting mixtures, electromagnetic effects, and irreversibility.

15.2 Summary

Remark 15.2.1. In Chapter 3 the manifest scale dependence of the boundary of a solid body was shown to imply scale dependence of the associated mass density field ρ. Further, spatial continuity of ρ, together with interpretation of the integral of ρ over a region R as yielding the mass within R, were shown to be incompatible with the fundamentally discrete nature of matter. Upon modelling fundamental discrete entities (i.e., molecules, atoms, or ions) as point masses, these issues were resolved in Chapter 4 via local spatial averaging in terms of scale-dependent weighting functions w. The continuity equation followed directly from temporal differentiation of the w-based mass density function ρ_w upon defining the velocity field $\mathbf{v}_w$ as $\rho_w^{-1}\mathbf{p}_w$, where $\mathbf{p}_w$ denotes the corresponding momentum density. The spatial smoothness of $\rho_w, \mathbf{p}_w$, and $\mathbf{v}_w$ is determined by the smoothness of w, and temporal smoothness is governed by the smoothness of w together with that of point mass motions.

Remark 15.2.2. The starting point for kinematical considerations was $\mathbf{v}_w$ (identified, for the simplest choice of w, with a local mass centre velocity associated with point masses). The concept of (scale-dependent) motion $\boldsymbol{\chi}_w$ was *derived* from that of $\mathbf{v}_w$ [via (5.2.7) and (5.2.8)], as was the notion of material point. Accordingly, material points are seen to be scale-dependent mathematical artefacts whose utility lies in their delineation of local scale-dependent distortion and mass transport.

Remark 15.2.3. Linear momentum balance was shown to take the form [cf. (5.5.15)] $-\text{div}\,\mathcal{D}_w + \mathbf{f}_w + \mathbf{b}_w = \rho_w \mathbf{a}_w$, on assuming point masses interact. Here $\mathbf{f}_w$ and $\mathbf{b}_w$ are interaction and external force densities, and $\mathcal{D}_w$ is a symmetric tensor of thermokinetic character. (Any quantity is regarded to be of thermal nature if its definition involves the velocity of individual point masses relative to the local continuum velocity $\mathbf{v}_w$: this is the essence of the kinetic theory of heat.) The standard form of balance was derived by obtaining solutions $\mathbf{T}_w^-$ to $\text{div}\,\mathbf{T}_w^- = \mathbf{f}_w$ and identifying the stress tensor with $\mathbf{T}_w := \mathbf{T}_w^- - \mathcal{D}_w$. The non-uniqueness of interaction stress tensor $\mathbf{T}_w^-$ was explored, together with interpretations of integrals of $\mathbf{T}_w^-\mathbf{n}$ over planes with unit normal $\mathbf{n}$. In particular, the integral of $\mathbf{T}_w\mathbf{n}$ over such a plane does not have the conventional interpretation of yielding the force exerted by one part of a body on the other, where the parts are separated by the plane. Rather, there is a contribution from $-\mathcal{D}_w\mathbf{n}$, which corresponds to a rate of momentum exchange associated with microscopic thermal motions, together with a contribution from $\mathbf{T}_w^-\mathbf{n}$ which derives from microscopic interactions weighted in such a way that de-emphasises those associated with point masses near the plane.

Remark 15.2.4. Different local forms of energy balance were obtained which depended upon assumptions about interactions. If only the resultant force $\mathbf{F}_i$ on any individual molecule P_i is employed, then balances (6.2.24) and (6.2.32) are appropriate and constitute evolution equations for the heat energy density per unit mass h. If interactions $\mathbf{f}_{ij}$ between molecules P_i and P_j are identifiable and balanced (that is, $\mathbf{f}_{ji} = -\mathbf{f}_{ij}$), but may nevertheless depend upon other molecules, then balance (6.2.75) followed. Here internal energy per unit mass $e = \beta + h$, where β is a local measure of stored energy. If interactions are governed by separation-dependent pair potentials, then balance (6.2.69) was derived. This is formally equivalent to (6.2.75), but β is now identified as an energy of assembly (sometimes termed *binding energy*) density per unit mass.

Remark 15.2.5. In Chapter 7 the time evolution of localised moments of molecular masses and momenta were shown to result in local balance relations which delineate the evolution of a measure $\mathbf{d}$ of inhomogeneity and of generalised (i.e., second-order tensor-valued) ‘internal’ moment of momentum. The skew part of the latter balance involves couple stress, body couple of external origin, and internal moment of momentum. Since no molecular microstructure was featured in the derivation, it follows that couple stress, etc. are to be expected quite generally. However, attention was drawn to the fine-scale nature of the averaging procedure (see Remark 15.3.2).

Remark 15.2.6. The necessity for, and implications of, time averaging were examined in Chapter 8, and time-averaged versions of the continuity equation and balances of linear momentum and of energy were established. Field values therein were identified in terms of molecular averages both in space *and* time. A novel means of analysing material systems whose molecular content varies with time was introduced via the notion of a membership function. Applications to rocket and jet propulsion were followed by derivations of local balances of mass, momentum, and energy in which full account was taken of mass exchange between a system and its material complement.

Remark 15.2.7. The microscopic basis of the continuum theory of mixtures was established in Chapter 9, and a paradox concerning the interpretation of partial stress was resolved.

Remark 15.2.8. In Chapter 10 fluid flow through a porous medium was examined at two scales, ϵ_1 and ϵ_2. The former is the scale at which pore boundary is to be considered, and the latter is a scale large compared with pore size and porous body structure. The nature of porosity as dependent on both scales was emphasised and followed by derivations of linear momentum balance at any scale. A hierarchy of relations, culminating in that named for Darcy, was established for flow of incompressible, linearly viscous fluid through a porous body it saturates by averaging ϵ_1-scale relations at scale ϵ_2.

Remark 15.2.9. An alternative form of spatial averaging was outlined in Chapter 11, less mathematically rigorous than weighting function methodology, but which provides additional insight into molecular averaging by the use of *cells*. Such approach has yielded results which might motivate scrutiny from the weighting function perspective (see Subsection 15.3.2 to follow).

Remark 15.2.10. The necessity of observer agreement in establishing scientific knowledge was discussed in Chapter 12. The nature and consequences of such objectivity in the context of deterministic continuum mechanics was explored from both molecular and macroscopic viewpoints. The manner in which restrictions upon constitutive relations follow from objectivity was indicated in the cases of elastica, simple materials, and viscous fluids. Particular attention was paid to a long-standing controversy concerning possible constitutive dependence upon spin relative to an inertial frame. Such dependence, to be expected in certain circumstances, was shown to violate no fundamental tenet of physics, but rather to indicate the shortcomings of two postulated 'principles' which purport to reflect the indifference of material behaviour to its observation. One such principle denies to Nature the possibility of behaviour which is spin-dependent, while subscribers to the other require observers to employ the same response functions.

Remark 15.2.11. Comments on non-local considerations were made in Chapter 13, and attention was drawn to the similarity between peridynamics and scale ϵ_2 modelling of porous media (see Remark 15.2.8).

Remark 15.2.12. Since the microscopic foundation of macroscopic behaviour is the raison d'être of statistical physics, two approaches to classical statistical mechanics were outlined in Chapter 14 for comparison.

15.3 Suggestions for Further Study

15.3.1 Interfacial Phenomena and Boundary Conditions

From a continuum viewpoint surface and interfacial effects are modelled in terms of bidimensional continua. That is, the material systems which give rise to these effects are considered to 'occupy', at any instant, a geometrical surface (cf., e.g., Moeckel [92] for fluids and Gurtin & Murdoch [93] for solids). For example, consider the

Laplace relation [cf., e.g., Landau & Lifschitz [4], (60.3)]

$$\Delta p = 2\kappa\sigma \tag{15.3.1}$$

which links the equilibrium pressure difference Δp between two immiscible fluids for which the interfacial tension is σ, and κ is the mean curvature of the surface which models the interfacial region. This relation is actually the normal component of linear momentum balance for the interface in the absence of external body force (per unit area): the tangential component of this balance requires that σ be constant. More generally, in equilibrium the surface stress T on the interface between two bulk continua in which the stress tensors are $\mathbf{T}_1$ and $\mathbf{T}_2$ satisfies (again in the absence of external body force per unit area)

$$(\mathbf{T}_1 - \mathbf{T}_2)\mathbf{n}.\mathbf{n} = \mathsf{T}\cdot\mathsf{L}. \tag{15.3.2}$$

Here L denotes the curvature tensor associated with the choice $\mathbf{n}$ of unit normal directed from bulk medium 1 into medium 2 (cf. Gurtin & Murdoch [93], Theorem 5.3). Accordingly, if either $\mathsf{T} = \mathsf{O}$ (surface stress absent or negligible) or $\mathsf{L} = \mathsf{O}$ (interfacial surface locally planar), then

$$\mathbf{T}_1\mathbf{n}.\mathbf{n} = \mathbf{T}_2\mathbf{n}.\mathbf{n}. \tag{15.3.3}$$

Indeed, the standard *boundary condition*

$$\mathbf{T}_1\mathbf{n} = \mathbf{T}_2\mathbf{n} \tag{15.3.4}$$

for contiguous bulk continua is a consequence of interfacial linear momentum balance upon neglect of all interfacial effects.

Given the insight into continuum modelling provided by study of its microscopic basis, it is natural to investigate the link between bidimensional modelling of interfacial behaviour and its microscopic origin.

Remark 15.3.1. Surface effects in solid bodies derive from molecules located at the boundary[1] together with those which lie within a few molecular spacings therefrom. Under most circumstances, liquid–liquid and liquid–vapour interfaces have similar thickness,[2] of order $10\,\text{Å}\,(= 10^{-9}\,\text{m})$. Thus interfacial regions are characterised by very large density gradients. For example, in liquid–vapour systems the vapour density is typically only 0.1% that of the liquid. Accordingly, crossing a water-water vapour interface of thickness $10\,\text{Å}$ involves a density change of order $10^3\,\text{kg}\cdot\text{m}^{-3}$ at STP and hence a notional density gradient magnitude of $10^{12}\,\text{kg}\cdot\text{m}^{-4}$!

The preceding remark suggests that interfacial regions should be delineated via mass density considerations. Consider variation of ρ_ϵ [see (4.3.4)] across such a region. If $\epsilon \geq 10\,\text{Å}$, then the thickness of this region, as sensed by change in ρ_ϵ, must be expected to be of order 2ϵ. This is illustrated in Figure 15.1 in the case of a planar lattice boundary. In particular, a choice of $\epsilon \sim 10^{-6}\,\text{m}$ (see Exercise 3.6.2) results in a model interfacial region of thickness much greater than that associated with molecules whose behaviour gives rise to interfacial effects.

[1] More precisely, those molecules which define the geometric boundary of the body at a scale slightly larger than nearest-neighbour separations [cf. (4.3.13) et seq.].

[2] Cf., e.g., Rusanov [94] and Rowlinson & Widom [95].

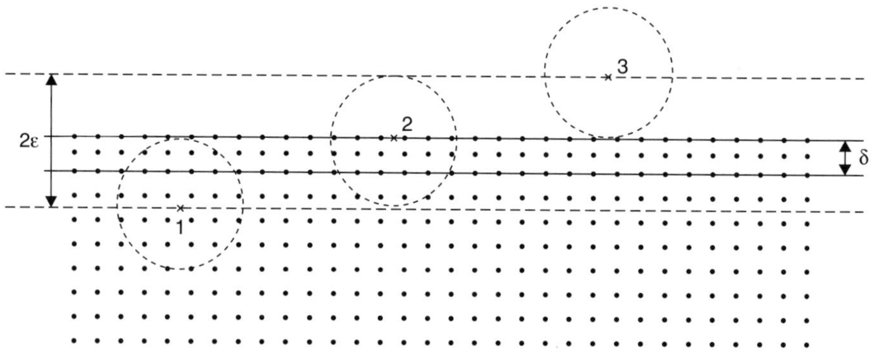

Figure 15.1. The density ρ_ϵ changes from its bulk value (ρ_0 say) to zero over a distance 2ϵ as a planar lattice boundary is crossed. At locations 1, 2, and 3 the values of ρ_ϵ are approximately ρ_0, $\frac{1}{2}\rho_0$, and 0, respectively. Interfacial effects derive only from molecules on or near the boundary which comprise a layer of thickness $\delta \approx 10\,\text{Å}$.

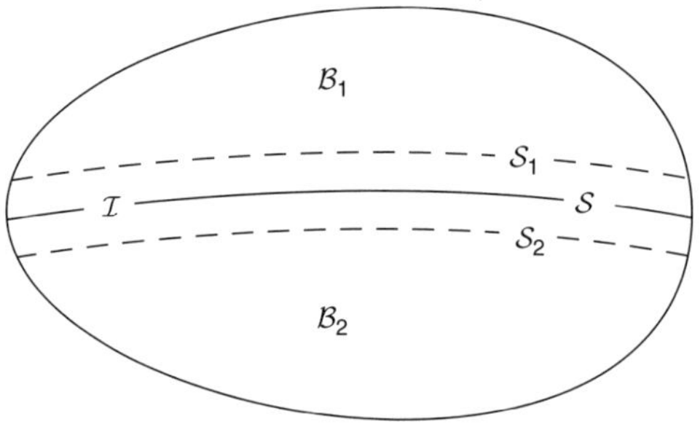

Figure 15.2. Interfacial region I between bulk regions $\mathcal{B}_1$ and $\mathcal{B}_2$ is bounded by surfaces S_1 and S_2. Model surface S lies between S_1 and S_2.

The conventional way to obtain balance relations consists, roughly speaking, of

(i) selecting a model surface $\mathcal{S}$ within the interfacial region $\mathcal{I}$, and
(ii) obtaining surface balance relations by integrating bulk balance relations across $\mathcal{I}$ normal to $\mathcal{S}$.

This procedure results in an overall model in which bulk balance relations hold in regions $\mathcal{B}_1$ and $\mathcal{B}_2$ which are separated by $\mathcal{I}$, and interfacial balance relations hold on $\mathcal{S}$. Accordingly no relations are assigned to the regions between $\mathcal{S}$ and $\mathcal{S}_\alpha (\alpha = 1,2)$, where $\mathcal{S}_\alpha := \partial\mathcal{B}_\alpha \cap \partial\mathcal{I}$ (see Figure 15.2). This modelling 'gap' may be 'filled' by extending bulk relations from $\mathcal{S}_\alpha$ to $\mathcal{S}$ along normals to $\mathcal{S}$, and then modifying interfacial balances by subtraction of integrated versions of these extended relations (along normals to $\mathcal{S}$). This results in modified interfacial balances which involve *surface excess* fields (cf., e.g., Murdoch [96]).

The foregoing constitutes an established approach to interfacial modelling and is informed by molecular considerations insofar as the analysis is based upon bulk fields which are identifiable in terms of scale-dependent volumetric molecular averages. However, an alternative, more direct approach might be possible, as is now indicated.

Interfacial molecules for a solid body might be identified as those which constitute boundary molecules (see Section 4.3) at a scale ϵ_1 slightly in excess of nearest-neighbour separations, together with all other molecules distant less than ϵ_1 therefrom. Such molecules constitute, instant by instant, a well-defined material system: there is a clear membership function in the sense of Section 8.7. For a liquid–liquid interface the same procedure can be adopted by selecting a particular molecular constituent. The problem is now to define a model surface $\mathcal{S}$. The first choice could be the triangulated geometric boundary at scale ϵ_1 [see (4.3.14)]. However, given erratic molecular motions, such a surface would fluctuate. Characterisation of interfacial molecules for single-component liquid–vapour interfaces would appear to be less obvious. Having delineated the interfacial material system (and hence choice of membership function) and defined a model surface (in general time-dependent), it is necessary to define scale-dependent surface fields. A first attempt could be to define the surface mass density at scale ϵ_2 ($\epsilon_2 \geq \epsilon_1$) by

$$\rho_{\mathcal{S}}(\mathbf{x},t) := \sum_i m_i e_i(t) \,.\, \frac{4\epsilon_2}{3} \,.\, w_{\epsilon_2}(\mathbf{x}_i(t) - \mathbf{x}). \tag{15.3.5}$$

Here a molecule P_i contributes if and only if

(i) P_i is in the interface at time t $(e_i(t) = 1)$, and
(ii) P_i lies within a distance ϵ_2 of $\mathbf{x} \in \mathcal{S}$.

The weighting attached to m_i is thus $(\pi\epsilon_2^2)^{-1}$ or 0. An alternative approach could be to delineate interfaces in terms of inhomogeneity measure **d**. (Note the behaviour of **d** in the three locations depicted in Figure 15.1.)

15.3.2 Generalised and Structured Continua

A *generalised continuum*[3] is any continuum model which incorporates, within its balance relations, terms additional to those which appear in the continuity equation (2.5.16), linear momentum balance (2.7.20) with symmetric stress tensor, and energy balance (2.8.24). Thus balance of mass (8.9.12) for a time-dependent system, balance of generalised moment of momentum (7.2.32), and balance of energy (7.4.45), all correspond to generalised continua. The analysis of Chapter 7 indicates that *all* material systems are to be expected to manifest generalised characteristics upon adoption of a fine-scale viewpoint. The success of standard forms of balance (coupled with constitutive relations) attests to the negligible nature of extra terms in most situations of interest. However, couple-stress effects have been detected (cf. Truesdell & Noll [2], §98) and linked to material inhomogeneity and/or microstructure (cf., e.g., Mindlin [26] and Toupin [25]).

In respect of inhomogeneity, it might be of interest to exploit the apparently new moment of mass relation (7.3.7) by the inclusion of inhomogeneity measure **d** as a constitutive variable.

Perhaps the simplest and most successful theory addressing microstructure is that which models nematic liquid crystalline phases, wherein long molecules tend to align with their neighbours (cf. de Gennes [24]). This co-operative behaviour is manifest at the macroscopic level and modelled in terms of a kinematic variable (termed a *director*) which represents the local molecular alignment (cf. Ericksen [97] and Leslie

[3] Cf., e.g., Kröner [30].

[98]). Similar situations which involve large rigid molecules and large deformable molecules are modelled by, respectively, the micropolar and micromorphic theories of Eringen and co-workers (cf. Kröner, [30]). Balance relations for materials which consist of large molecules may be motivated via cellular averaging, in the manner of Chapter 11, on the basis of physically plausible and intelligible assumptions concerning the natures of molecular motions, atomic motions within molecules, and interactions between atoms (cf. Murdoch [27]). It might be instructive to revisit the arguments of this work using weighting function methodology.

15.3.3 Configurational Forces

Continuum modelling of the time evolution of cracks, inhomogeneities, and phase interfaces may be effected by introducing the notion of so-called configurational forces (cf., e.g., Maugin [99] and Gurtin [100]). The viewpoints of these authors are somewhat different but are both strictly grounded in continuum/macroscopic formalism. A simple example illustrates a *microscopic* perspective. Consider an atomic lattice in which there is a single 'vacancy' which may migrate with time: see Figure 15.3. Such motion is illusory: a vacancy (which consists of nothing!) cannot of itself be traced. Of course, what characterises the vacancy is the location of nearby lattice atoms, the pattern of which is broken by an absentee. The collection of near-neighbours which signals the vacancy at any instant is a time-dependent material system which changes as a consequence of atomic interactions with other lattice atoms and matter external to the lattice. Similarly, a crack or cavity in a body is delineated at any instant (and given length scale) by boundary molecules of the body which form a time-dependent material system. It is forces on this system which drive crack or cavity evolution. Together with the interfacial considerations of subsection 15.3.1, these examples give classical microscopic interpretations to the motion of a lattice vacancy, crack, or interface, and forces thereon, via the methodology of Chapter 8 applied to the appropriate material system at length and time scales of choice. The difficulty in implementing the foregoing is incorporation of microscopic mechanisms (which drive changing material content of the relevant systems) into a continuum description.

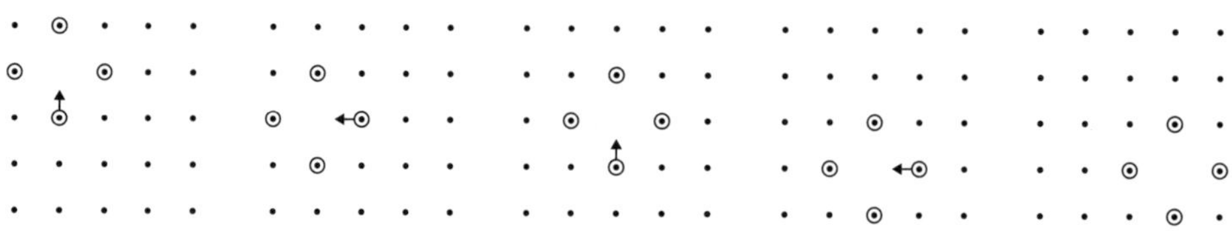

Figure 15.3. The 'motion' of a vacancy in an atomic lattice, by way of single 'steps', is signalled by changes in its near neighbours (indicated by ⊙). The arrows indicate which atom actually 'jumps' in each 'step'.

15.3.4 Reacting Mixtures

The methodology of Section 9.7 might be extended by considering molecules to be assemblies of nuclei and electrons, and taking complete classical account of detailed individual reactions involving ionic constituents and energy exchange.

15.3.5 Electromagnetic Effects

As discussed in Section 5.4, molecular interactions derive from forces between atomic nuclei and electrons. Such fundamental discrete entities are characterised not only by mass but also *electric charge*. The charge associated with a nucleus is an integral multiple of that of an electron but has opposite sign. It is the forces between nuclei and electrons which derive from their *charges*[4] that, upon suitable temporal averaging, yield intermolecular force: interactions associated with mass (i.e., gravitational interactions) are insignificant at molecular length scales. Given a system of identifiable discrete entities with net charge,[5] then, if each is modelled as a point mass P_i of mass m_i and charge q_i, the *charge density* field σ_ϵ at scale ϵ may be defined via

$$\sigma_\epsilon(\mathbf{x},t) := \sum_{i=1}^{N} q_i\, w_\epsilon(\mathbf{x}_i(t) - \mathbf{x}). \tag{15.3.6}$$

It follows, precisely as in derivation of the continuity equation in Section 4.2, that

$$\frac{\partial \sigma_\epsilon}{\partial t} + \operatorname{div} \mathbf{j}_\epsilon = 0, \tag{15.3.7}$$

where
$$\mathbf{j}_\epsilon(\mathbf{x},t) := \sum_{i=1}^{N} q_i\, \mathbf{v}_i(t) w_\epsilon(\mathbf{x}_i(t) - \mathbf{x}). \tag{15.3.8}$$

Field $\mathbf{j}_\epsilon$ is known as the (*electric*) *current density* (at scale ϵ) associated with the given system of charged discrete entities. While study of matter at the atomic level is the province of quantum mechanics, the derivation of such a simply understood relation as (15.3.7) suggests that a classical approach may be of some utility in the interpretation of field values and the relations they satisfy.[6] A starting point could be to model a conductor as a binary mixture of electrons and positively-charged ions. The deformation of the mixture as a whole would be determined by the scale-dependent, mass-based, velocity field $\mathbf{v}_\epsilon$ [see (5.2.7) and (5.2.8) with $w = w_\epsilon$]. The electric current, governed by motions of the free electrons, would have an associated *electrical* velocity field $\mathbf{u}_\epsilon := \mathbf{j}_\epsilon / \sigma_\epsilon$ (where $\mathbf{j}_\epsilon$ and σ_ϵ are computed solely for the set of electrons) and give rise to an electronic motion map at scale ϵ defined by strict analogy with (5.2.3) and (5.2.4). (Notice that the electrical velocity field coincides with the velocity field for the system of electrons, since all electrons have the same charge and mass. (Convince yourself!) The main problem is to characterise, and distinguish between, electric and magnetic fields. Here it is to be expected that individual account be taken of all electrons and nuclei. However, any resulting description will suffer from the absence, in point mass modelling, of electron *spin* (cf., e.g., Coulson [101], footnote on p. 7).

15.3.6 Irreversibility

It is possible to contemplate the time-reversed counterpart of any given material behaviour. However, it is a matter of experience that such counterparts of observed

[4] These are termed *electromagnetic* forces.

[5] Any such entity is termed an *ion* and may be an electron, a nucleus, or an assembly of nuclei and electrons with non-zero net charge.

[6] The relations are known collectively as *Maxwell's equations*, and (15.3.7) is termed *conservation of charge*.

reproducible (time-varying) macroscopic behaviour are not often witnessed[7]. For example, after water in a stationary glass has been stirred, its subsequent swirling motion dies down. Time reversal in such case would involve essentially motionless water spontaneously developing swirling characteristics: such behaviour is not observed. The decay of the swirling motion is understood in terms of *dissipation*: the loss of macroscopic kinetic energy (modelled by density $\rho\mathbf{v}^2/2$) is accompanied by a gain in microscopic kinetic energy (represented by density ρh) which is identified[8] as 'heat'. In order to appreciate and model the 'one-way' nature of macroscopic processes (that is, their innate 'irreversibility'), it is natural to explore how this might be related to microscopic considerations.

Consider the behaviour of a material system[9] over a time interval $0 \le t \le T$. The location of the mass centre of molecule P_i in the time-reversed motion, time t units after its initiation, is given by

$$\mathbf{x}_i^{\mathrm{rev}}(t) := \mathbf{x}_i(T-t). \tag{15.3.9}$$

That is, this location is that of P_i in the *original* motion after $T-t$ time units. Differentiation with respect to time yields

$$\mathbf{v}_i^{\mathrm{rev}}(t) = -\mathbf{v}_i(T-t), \tag{15.3.10}$$

where $\mathbf{v}_i^{\mathrm{rev}}$ denotes the velocity of P_i in the time-reversed motion.

The mass density (corresponding to choice w of weighting function) at location $\mathbf{x}$ and time t in the time-reversed motion is, from (4.2.1) and (15.3.9),

$$\rho_w^{\mathrm{rev}}(\mathbf{x},t) := \sum_i m_i w(\mathbf{x}_i^{\mathrm{rev}}(t) - \mathbf{x}) = \sum_i m_i w(\mathbf{x}_i(T-t) - \mathbf{x}). \tag{15.3.11}$$

That is,

$$\rho_w^{\mathrm{rev}}(\mathbf{x},t) = \rho_w(\mathbf{x}, T-t). \tag{15.3.12}$$

Similarly, the time-reversed momentum density [see (4.2.13)] is

$$\mathbf{p}_w^{\mathrm{rev}}(\mathbf{x},t) := \sum_i m_i \mathbf{v}_i^{\mathrm{rev}}(t) w(\mathbf{x}_i^{\mathrm{rev}}(t) - \mathbf{x}) = -\sum_i m_i \mathbf{v}_i(T-t) w(\mathbf{x}_i(T-t) - \mathbf{x}),$$

so that

$$\mathbf{p}_w^{\mathrm{rev}}(\mathbf{x},t) = -\mathbf{p}_w(\mathbf{x}, T-t). \tag{15.3.13}$$

Noting that

$$\frac{\partial \rho_w^{\mathrm{rev}}}{\partial t}(\mathbf{x},t) = -\frac{\partial \rho_w}{\partial t}(\mathbf{x}, T-t) \tag{15.3.14}$$

and

$$(\mathrm{div}\,\mathbf{p}_w^{\mathrm{rev}})(\mathbf{x},t) = -(\mathrm{div}\,\mathbf{p}_w)(\mathbf{x}, T-t), \tag{15.3.15}$$

it follows that

$$\left(\frac{\partial \rho_w^{\mathrm{rev}}}{\partial t} + \mathrm{div}\,\mathbf{p}_w^{\mathrm{rev}}\right)(\mathbf{x},t) = -\left(\frac{\partial \rho_w}{\partial t} + \mathrm{div}\,\mathbf{p}_w\right)(\mathbf{x}, T-t) = 0. \tag{15.3.16}$$

[7] Motion of the mass centre of a rigid body in vacuo, under the sole influence of a constant gravitational field, *is* reversible: in watching a film of this motion, run both forwards and backwards, a viewer would be unable to decide which version depicts the actual motion.

[8] See Subsection 5.5.2.

[9] See Section 4.2.

That is, *the continuity equation holds for time-reversed motions*. Of course, the time-reversed velocity field is [cf. (4.2.15)]

$$\mathbf{v}_w^{\text{rev}} := \mathbf{p}_w^{\text{rev}} / \rho_w^{\text{rev}}, \tag{15.3.17}$$

so that, from (15.3.12) and (15.3.13),

$$\mathbf{v}_w^{\text{rev}}(\mathbf{x}, t) = -\mathbf{v}_w(\mathbf{x}, T - t). \tag{15.3.18}$$

Turning next to balances of linear momentum, generalised moment of momentum, and energy, recall that these relations were derived on the basis of the equations which govern molecular mass centre motions in an inertial frame. In particular, for molecule P_i (see (5.4.4))

$$\sum_{j \neq i} \mathbf{f}_{ij} + \mathbf{b}_i = m_i \ddot{\mathbf{x}}_i. \tag{15.3.19}$$

Suppose that all initial locations $\mathbf{x}_i(0)$ and velocities $\dot{\mathbf{x}}_i(0)$ are known. If $\mathbf{f}_{ij}(t)$ is a known function of $\mathbf{x}_i(t), \mathbf{x}_j(t), \dot{\mathbf{x}}_i(t)$ and $\dot{\mathbf{x}}_j(t)$, and if $\mathbf{b}_i(t)$ is a known function of $\mathbf{x}_i(t), \dot{\mathbf{x}}_i(t)$, and possibly[10] t, then the system of equations (15.3.19), together with the initial information (and any explicit time dependence of $\mathbf{b}_i$), mandate the time evolution of molecular trajectories $\mathbf{x}_i$. Accordingly, the time evolution of all fields which appear in the balance relations are prescribed over the time interval $[0, T]$ and satisfy these relations. (Convince yourself by looking at the definitions of $\mathbf{f}_w, \mathbf{b}_w, \mathcal{D}_w, \mathbf{a}_w$ and $\mathbf{T}_w^-$ in respect of linear momentum balance: cf. (5.5.2), (5.5.3), (5.5.11) and (5.5.8), (5.5.16) and (5.6.5) or (5.6.10).)

Now consider the time-reversed version of the foregoing, namely the time evolution of[11] $\mathbf{x}_i^r$ prescribed by

$$\sum_{j \neq i} \mathbf{f}_{ij}^r + \mathbf{b}_i^r = m_i \ddot{\widehat{\mathbf{x}_i^r}}, \tag{15.3.20}$$

together with

$$\mathbf{x}_i^r(0) := \mathbf{x}_i(T), \qquad \dot{\widehat{\mathbf{x}_i^r}}(0) := -\dot{\mathbf{x}}_i(T) \tag{15.3.21}$$

and

$$\mathbf{b}_i^r(t) := \mathbf{b}_i^{\text{rev}}(t) := \mathbf{b}_i(T - t). \tag{15.3.22}$$

That is, we consider the time evolution of molecular trajectories beginning at the end of the process described in (15.3.19) et seq and subject to the external influence (represented by $\mathbf{b}_i$) in reverse. *If*

$$\mathbf{f}_{ij}^r(t) = \mathbf{f}_{ij}^{\text{rev}}(t) := \mathbf{f}_{ij}(T - t), \tag{15.3.23}$$

then it is a simple matter to show that [see (15.3.9)]

$$\mathbf{x}_i^r = \mathbf{x}_i^{\text{rev}}, \tag{15.3.24}$$

and individual molecules retrace their original trajectories. In such case the macroscopic motion (prescribed by $\mathbf{v}_w^{\text{rev}}$: see (15.3.18) and Section 5.2) is similarly reversed,

[10] External force $\mathbf{b}_i$ will include the effect of gravity and of any confinement of the system: such confinement may be time dependent as, for example, gas in an enclosure whose volume is governed by a moveable piston.

[11] At this stage it is unclear whether or not $\mathbf{x}_i^r = \mathbf{x}_i^{\text{rev}}$.

and field values are simply related to those in the 'forward' process. Given any macroscopic field $\mathcal{F}_w$, let $\mathcal{F}_w^{\text{rev}}(\mathbf{x},t)$ denote its value at location $\mathbf{x}$ in the reversed motion time t after its initiation. It follows directly from the definitions of fields that

$$\mathcal{F}_w^{\text{rev}}(\mathbf{x},t) = (-1)^p \mathcal{F}_w(\mathbf{x}, T-t), \tag{15.3.25}$$

where

$$p = 0 \quad \text{for} \quad \mathcal{F} = \mathbf{f}, \mathcal{D}, \mathbf{T}^-, \mathbf{T}, \mathbf{b}, \rho, \mathbf{a}, \mathbf{c}, \mathbf{J}, \mathbf{M}, \hat{\mathbf{M}}, \mathbf{C}, \text{ or } \mathbf{d}, \tag{15.3.26}$$

and

$$p = 1 \quad \text{for} \quad \mathcal{F} = \mathbf{L}, Q, r, \kappa, \mathbf{q}, A, \text{ or } \mathbf{B}. \tag{15.3.27}$$

Remark 15.3.2. The foregoing has shown that macroscopic processes are reversible if both (15.3.22) *and* (15.3.23) hold. Satisfaction of (15.3.22) requires that the effect $\mathbf{b}_i$ of external influences on each and every molecule must be replicated in reverse for the duration of the (reverse) process. This is a most significant observation, because $\mathbf{b}_i$ incorporates the effect of any system confinement (see Footnote 10) together with external radiation and conduction across any contiguous boundary[12]. Reversal of such external influences cannot be replicated in general. If, however, $\mathbf{b}_i(t)$ depends only on $\mathbf{x}_i(t)$, then (15.3.22) *is* satisfied[13]. Thus for thermally isolated systems confined to fixed regions, the possibility of time reversal of macroscopic behaviour is governed by satisfaction (or otherwise) of (15.2.23). Recall that, in discussing (15.3.19), interactions $\mathbf{f}_{ij}$ could depend upon velocities $\dot{\mathbf{x}}_i$ and $\dot{\mathbf{x}}_j$. In such case (15.2.23) cannot be expected to hold. For example, suppose that [cf. (6.2.51)]

$$\mathbf{f}_{ij} = \alpha(r_{ij})(\mathbf{x}_j - \mathbf{x}_i) + \beta(r_{ij})(\dot{\mathbf{x}}_j - \dot{\mathbf{x}}_i). \tag{15.3.28}$$

In the case of reversed molecular trajectories the corresponding interaction would be

$$\mathbf{f}_{ij}^{\text{rev}} := \alpha(r_{ij}^{\text{rev}})(\mathbf{x}_j^{\text{rev}} - \mathbf{x}_i^{\text{rev}}) + \beta(r_{ij}^{\text{rev}})(\dot{\widehat{\mathbf{x}_j^{\text{rev}}}} - \dot{\widehat{\mathbf{x}_i^{\text{rev}}}}). \tag{15.3.29}$$

Thus, from (15.3.9) and (15.3.10),

$$\begin{aligned} \mathbf{f}_{ij}^{\text{rev}}(t) &= \alpha(r_{ij}(T-t))(\mathbf{x}_j(T-t) - \mathbf{x}_i(T-t)) - \beta(r_{ij}(T-t))(\dot{\mathbf{x}}_j(T-t) - \dot{\mathbf{x}}_i(T-t)) \\ &\neq \mathbf{f}_{ij}(T-t) \;\; \textit{unless } \beta \textit{ is identically zero.} \end{aligned} \tag{15.3.30}$$

The conclusion to be drawn from Remark 15.3.2 is that for thermally isolated systems confined to fixed regions, and for which interactions depend only upon current molecular locations, reversible behaviour is to be expected at both microscopic and macroscopic levels[14]. However, if interactions are sensitive to molecular velocities, then irreversible behaviour at all scales is to be expected.

[12] Recall (6.2.6) in respect of radiation. The molecules which comprise any contiguous boundary system ($\mathcal{M}^-$, say) may be added to those of the system in question ($\mathcal{M}^+$, say) to form a composite system as described in Chapter 8. Then ${}^{\text{ext}}Q$ [see (8.9.63)] represents the rate of heat per unit volume supplied to $\mathcal{M}^+$ from $\mathcal{M}^-$: in (8.9.63) term $\mathbf{f}_{ij}$ represents the force on P_i in $\mathcal{M}^+$ due to P_j in $\mathcal{M}^-$, and is hence a force external to $\mathcal{M}^+$ which must be regarded as contributing to $\mathbf{b}_i$ in (15.3.19).

[13] This would be the case if the system were to be confined to a fixed region, thermally isolated, and were $\mathbf{b}_i(t)$ to be the sum of the effect $m_i\mathbf{g}$ of gravity together with a function of the distance of $\mathbf{x}_i(t)$ from the boundary of the region.

[14] The temporal averaging implemented in Chapter 8 does not affect this conclusion: the counterpart of (15.3.25) is

$$\mathcal{F}_{w,\Delta}^{\text{rev}}(\mathbf{x},t) = (-1)^p \mathcal{F}_{w,\Delta}(\mathbf{x}, T-t+\Delta).$$

Remark 15.3.3. Summation of relations (15.3.28) over all molecules $P_j (j \neq i)$ yields from the second terms a composite force on P_i which depends on its motion relative to an averaged motion of other molecules weighted according to their proximity to P_i. Such a composite force has the formal attributes of a viscous 'drag' on a body (here P_i) moving through a fluid (here molecules P_j).

Remark 15.3.4. Recalling the subatomic considerations of Subsections 5.4 and 6.3, one is led to question whether or not interactions between electrons and[15] nuclei are invariant under time reversal: such invariance, or its absence, would a priori be expected to be inherited by molecular interactions. Discussion of such interactions is the essence of quantum electrodynamics, and is far beyond our simplistic considerations. However, two observations are in order.

Observation 1. The force on a charge q moving with velocity $\mathbf{v}$ in an electromagnetic field is

$$\mathbf{F} = q(\mathbf{E} + \frac{1}{c}\mathbf{v} \times \mathbf{B}), \tag{15.3.31}$$

where $\mathbf{E}$ and $\mathbf{B}$ denote the values of the electrostatic and magnetic induction fields at the location of the charge. Time reversal leaves $\mathbf{E}$ unchanged and $\mathbf{v}$ becomes $-\mathbf{v}$. Thus $\mathbf{F}$ is invariant if and only if $\mathbf{B}$ reverses sign. Here we note the need for further study outlined in Subsection 15.3.5. In this context time reversal requires not only that electron orbits be reversed but also that electron spins must change sign.

Observation 2. Influence of one electron on another is not instantaneous, but is transmitted at the speed of light[16]. (Of course, a proper discussion of this observation requires a relativistic approach.)

The foregoing analysis was undertaken from a classical mechanical view of microscopic dynamics. Specifically, this involved complete and exact knowledge of initial molecular mass centre locations, together with precise details both of the nature of molecular interactions and the effect of the environment. Such information sufficed to delineate exact molecular trajectories and thereby yield the time evolution of the macroscopic fields (at any space-time scales of choice) which appear in, and (automatically) satisfy, the balance relations. Of course such exhaustive knowledge of a material system is not available for both physical and practical reasons. Eliciting such knowledge requires observation/interaction with the system which thereby changes: this is at the heart of quantum mechanics. Further, given the vast numbers of molecules involved in systems of macroscopic interest[17], even limited information about each and every molecule is impossible to obtain. There are two conceptually different approaches to overcoming such difficulties, namely those of continuum thermodynamics and of statistical mechanics. Both approaches involve introduction of two scalar quantities, temperature and entropy. The entropy has, for thermomechanically isolated systems, the property of increasing with time. Statements which express the time evolution of entropy are formulations of the so-called 'second law' of thermodynamics.[18]

[15] That is, electron-electron, electron-nucleus, and nucleus-nucleus interactions.

[16] Recall Remark 5.4.1. See also Landau & Lifschitz [72], pp.32-33, for quantum mechanical aspects of irreversibility.

[17] E.g., there are of order 3×10^{19} molecules in 1 ml of air at sea level.

[18] The first such 'law' is any formulation of energy balance.

In continuum mechanics the most commonly accepted formulation of the second law is the Clausius-Duhem inequality (cf., e.g., Gurtin *et al* [23], p.187) in which entropy is regarded to be a primitive concept.[19] Together with constitutive relations and requirements of objectivity (recall Chapter 12), the Clausius-Duhem inequality has proved to be very successful in mandating restrictions upon, and relations between, response functions. This inequality involves fields $\mathbf{q}, r$, and those of entropy and temperature. Since fields $\mathbf{q}$ and r have here been established in terms of spatial (and spatio-temporal) averages of molecular quantities, the obvious question is whether (and if so, how) it is possible to establish the existence of a scale-dependent entropy field on the basis of microscopic considerations, and thereby elucidate assumptions implicit in the Clausius-Duhem inequality. To this end, and for comparison, we conclude by sketching a statistical mechanical approach to entropy.

Consider continuously reproducible macroscopic behaviour at spacetime scales (ϵ, Δ) beginning at time t_0: see Subsection 14.5. This means that whatever macroscopic information [at scales (ϵ, Δ)] is available at this time (identified as its 'macrostate' $\mathbf{A}(t_0)$), the system will always evolve to yield a unique macrostate $\mathbf{A}(t)$ at any subsequent time t. Of course, to any given macrostate $\mathbf{A}$ corresponds a subset of phase space $\mathbb{P}$ each of whose elements ('microstates') give rise to $\mathbf{A}$. Suppose that $\mathbb{D}_{t_0}$ and $\mathbb{D}_t$ denote those subsets of $\mathbb{P}$ which give rise to $\mathbf{A}(t_0)$ and $\mathbf{A}(t)$, respectively. 'Reproducibility' means that if $\hat{\mathbf{X}} \in \mathbb{D}_{t_0}$ then $\mathbf{X}(t) \in \mathbb{D}_t$, where $\mathbf{X}(t)$ is prescribed by the underlying dynamics in $\mathbb{P}$: see (14.2.8) and (14.2.9). However, while every microstate which at time t_0 lies in $\mathbb{D}_{t_0}$ has evolved so as to lie in $\mathbb{D}_t$ at time t, it is *not* the case that if $\mathbf{Y} \in \mathbb{D}_t$ then there must be an element $\mathbf{Y}_0 \in \mathbb{D}_{t_0}$ which has evolved to $\mathbf{Y}$ at time[20] t. Said differently, $\mathbb{D}_t$ is in some sense larger than $\mathbb{D}_{t_0}$ if (ϵ, Δ)-scale behaviour is irreversible. Quantifying this notion requires a measure of size associated with subsets of $\mathbb{P}$, the most natural of which is that of volume: see Appendix B, Subsection B.11.1. Accordingly any monotonic increasing function of phase space volume satisfies

$$S(\mathrm{vol}(\mathbb{D}_t)) \geq S(\mathrm{vol}(\mathbb{D}_{t_0})), \tag{15.3.32}$$

and is thus a candidate (global) entropy function for the behaviour considered.

15.4 A Final Remark

There have been three key ingredients in the averaging procedures discussed in this book: the use of spatial weighting functions, temporal averaging, and delineation of time-dependent systems in terms of membership functions. In combination these procedures may be applied widely to mathematical models which embody discrete or continuous, and possibly time-dependent, features.

[19] Further, entropy enters into the inequality in terms of an entropy density *field* so that, in particular, the entropy of the whole system is the integral of this field over the region it occupies. Day [104] developed a comparable theory in which entropy was a *derived* entity.

[20] Consider water swirling in a stationary glass at time t_0. This dies down and, at any later time t, the reduced swirling motion is consistent with a much greater variety of initial swirling than that actually witnessed at time t_0. Although the motion may be macroscopically reproducible, if the swirling is monitored only at time t, then it is impossible to infer the macroscopic situation at time t_0.

Appendix A: Vectors, Vector Spaces, and Linear Algebra

Preamble

The functions of space and time (i.e., *fields*) used to model material behaviour take values which may be real numbers, vectors, or higher-order *tensors*.[1] Formal manipulations of tensors (i.e., *tensor algebra*) are best understood in terms of *vector spaces*.[2] Here basic concepts and results are reviewed for completeness and for establishing familiarity with the notation employed. Vectorial entities (i.e., entities which have both direction and magnitude and combine like displacements) are modelled in terms of a three-dimensional inner-product vector space $\mathcal{V}$, and higher-order tensorial entities are described in terms of algebraic constructs of $\mathcal{V}$.

Simple considerations of rectilinear changes of position (i.e., displacements) and the notion of perpendicularity are used to establish the three-dimensional inner product vector space $\mathcal{V}$ used to model vectorial quantities, irrespective of their physical dimensions of mass, length, and time, and units of measurement. Linear transformations on $\mathcal{V}$ are defined and shown to have algebraic features in common with $\mathcal{V}$, so motivating the definition of a general abstract vector space. The transpose of a linear transformation $\mathbf{L}$ on $\mathcal{V}$ and the tensor product of two vectors are defined without recourse to basis-dependent representations: such representations are derived upon selecting an orthonormal basis for $\mathcal{V}$. Criteria which establish the invertibility or otherwise of a linear transformation $\mathbf{L}$ on $\mathcal{V}$ are identified, and the principal invariants and characteristic equation of $\mathbf{L}$ are analysed using alternating trilinear forms on $\mathcal{V}$. Skew and orthogonal linear transformations on $\mathcal{V}$ are characterised: the former are identified with (axial) vectors and the latter related to rotations. Symmetric linear transformations are shown to have orthonormal bases of eigenvectors, the square root of a positive-definite symmetric transformation is defined, and the

[1] Tensor analysis categorises the differing natures of field values: scalars and vectors are, respectively, tensors of order 0 and 1. Measures of stress and strain, and spatial derivatives of velocity fields, fall into the category of second-order tensors (often merely termed *tensors* if no higher-order category is required), while in linear elasticity a fourth-order tensor is encountered. In discussing couple stress it proves necessary to introduce third-order tensors.

[2] In the literature tensors are often treated solely in terms of co-ordinate systems and the manner in which their representations change upon change of co-ordinates. Such manipulations of symbols with superscripts and subscripts tend to mask both the underlying physics *and* algebra. Here the approach is *direct* (i.e., free of co-ordinate considerations). This allows us to work in a natural way with intrinsic quantities rather than their representations.

latter is used in establishing the polar decomposition of linear transformations with positive determinants. Third-order tensors are introduced as linear maps from $\mathcal{V}$ into the space of linear transformations on $\mathcal{V}$, and their representations and combinations with vectors and linear transformations are discussed. Finally, comparisons are made between direct (i.e., component-free) notation for quantities having the nature of vectors, linear transformations or third-order tensors, representations of these in terms of components with respect to any orthonormal basis, and Cartesian tensor notation.

A.1 The Algebra of Displacements

Our concept of vectorial entities and the notion of an abstract vector space[3] derive from intuitive ideas concerning locations and of the relation between pairs of locations. Specifically, we can conceive of idealised locations in space (*points*), and given any two points, we can agree on their separation (the distance between these points) and the direction of one from the other. Accordingly we can formalise the notion of *displacement*: that is, the rectilinear change in location necessary to arrive at one point having started from another. Any two displacements are regarded as equivalent if they involve the same separation and the same direction (implicit here is the notion of parallelism). Roughly speaking, one can regard a displacement as an order: move a given distance in a given direction. Any two displacements can be combined in a natural way: merely follow orders! We denote by $\mathbf{d}_1 + \mathbf{d}_2$ that displacement achieved by first displacing from any point via displacement $\mathbf{d}_1$ and then by a further displacement $\mathbf{d}_2$. Here the symbol $+$ is employed because the formal properties of such displacement combination are precisely those of addition of real numbers. Specifically, if we write $\mathbf{0}$ for the displacement corresponding to 'stay where you are', then for any displacements $\mathbf{d}_1, \mathbf{d}_2, \mathbf{d}_3$, and $\mathbf{d}$,

D.1. $$\mathbf{d}_1 + \mathbf{d}_2 = \mathbf{d}_2 + \mathbf{d}_1 \tag{A.1.1}$$

D.2. $$\mathbf{d}_1 + (\mathbf{d}_2 + \mathbf{d}_3) = (\mathbf{d}_1 + \mathbf{d}_2) + \mathbf{d}_3 \tag{A.1.2}$$

and

D.3. $$\mathbf{0} + \mathbf{d} = \mathbf{d} = \mathbf{d} + \mathbf{0}. \tag{A.1.3}$$

We term $\mathbf{0}$ the *zero displacement.*

Property (D.2) implies that brackets are unnecessary when combining any number of displacements, and (D.1) implies that the order of terms in any such combinations is unimportant.

From the definition of $\mathbf{0}$, if $\mathbf{d}_1 + \mathbf{d}_2 = \mathbf{0}$, then displacement $\mathbf{d}_2$ must be that unique displacement in the direction opposite to that of $\mathbf{d}_1$ but with the same associated distance. That is, for each displacement $\mathbf{d}_1$ there exists a displacement [which we write as $(-\mathbf{d}_1)$, again mimicking usage in $\mathbb{R}$] for which

D.4. $$\mathbf{d}_1 + (-\mathbf{d}_1) = \mathbf{0} = (-\mathbf{d}_1) + \mathbf{d}_1. \tag{A.1.4}$$

[3] See, for example, Halmos [88].

We write

D.5. $$\mathbf{d}_2 - \mathbf{d}_1 := \mathbf{d}_2 + (-\mathbf{d}_1). \qquad \text{(A.1.5)}$$

Exercise A.1.1. Show that $-(\mathbf{d}_1 - \mathbf{d}_2) = \mathbf{d}_2 - \mathbf{d}_1$.

Given $\alpha \in \mathbb{R}$ ($\alpha \neq 0$) and any displacement $\mathbf{d}$, we write $\alpha\mathbf{d}$ to represent that displacement which has associated distance $|\alpha|$ times that of $\mathbf{d}$ and which is (i) in the same direction as $\mathbf{d}$ if $\alpha > 0$ and (ii) in the opposite direction if $\alpha < 0$. Also, by convention, we write

D.6. $$0\mathbf{d} = \mathbf{0} \qquad \text{(A.1.6)}$$

and

D.7. $$\alpha\mathbf{0} = \mathbf{0} \qquad \text{(A.1.7)}$$

for any $\alpha \in \mathbb{R}$. It follows that, for all $\alpha, \beta \in \mathbb{R}$ and all displacements $\mathbf{d}$,

D.8. $$\alpha(\beta\mathbf{d}) = (\alpha\beta)\mathbf{d} \qquad \text{(A.1.8)}$$

and

D.9. $$(\alpha + \beta)\mathbf{d} = \alpha\mathbf{d} + \beta\mathbf{d}. \qquad \text{(A.1.9)}$$

Further, for any displacements $\mathbf{d}_1$ and $\mathbf{d}_2$ (essentially via consideration of similar triangles),

D.10. $$\alpha(\mathbf{d}_1 + \mathbf{d}_2) = \alpha\mathbf{d}_1 + \alpha\mathbf{d}_2. \qquad \text{(A.1.10)}$$

Exercise A.1.2. Show that if we had assumed D.9, then D.6 follows on setting $\alpha + \beta = 0$. Show further that D.7 follows from D.10 and D.8 on setting $\mathbf{d}_2 = -\mathbf{d}_1$.

A.2 Dimensionality

Two non-zero displacements $\mathbf{d}_1$ and $\mathbf{d}_2$ are said to be *parallel* if $\mathbf{d}_2 = \alpha\mathbf{d}_1$ for some $\alpha \in \mathbb{R}$; that is, the direction associated with $\mathbf{d}_1$ is the same as, or opposite to, that of $\mathbf{d}_2$. If $\mathbf{d}_1$ and $\mathbf{d}_2$ are non-zero non-parallel displacements then these displacements, taken from a point P, define points Q and R which together with P form a triangle Δ. This triangle defines a unique plane, Π say. Now consider any displacement $\mathbf{d}_3$ taken from P. If this takes us to a point S which does not lie in Π, then we say that $\mathbf{d}_1, \mathbf{d}_2$, and $\mathbf{d}_3$ are *linearly independent displacements*. If S lies in Π, then $\mathbf{d}_1, \mathbf{d}_2$, and $\mathbf{d}_3$ are said to be *linearly dependent*.

Exercise A.2.1. If S lies in Π, show that there exist unique $\alpha, \beta \in \mathbb{R}$ such that $\mathbf{d}_3 = \alpha\mathbf{d}_1 + \beta\mathbf{d}_2$. (*Hint*: Draw the line through S parallel to $\mathbf{d}_2$, and label the point it which this line meets PQ as Q'. Then $\mathbf{d}_3$ is the combination of the displacement of P to Q' followed by that of Q' to S.)

Result A.2.1. If $\mathbf{d}_1, \mathbf{d}_2$, and $\mathbf{d}_3$ are linearly independent displacements, then any displacement $\mathbf{d}$ may be written as

D.11. $$\mathbf{d} = \alpha_1 \mathbf{d}_1 + \alpha_2 \mathbf{d}_2 + \alpha_3 \mathbf{d}_3, \tag{A.2.1}$$

where α_1, α_2, and α_3 are unique real numbers.

Numbers α_1, α_2, and α_3 are termed the *components* of $\mathbf{d}$ with respect to choice $\mathbf{d}_1, \mathbf{d}_2$, and $\mathbf{d}_3$ of displacements.

[Proof of the result follows from drawing that line through S parallel to $\mathbf{d}_3$. This line meets Π in a unique point, L say. Now draw that line through L parallel to $\mathbf{d}_2$. This line meets the line joining P and Q in a unique point, M say. Then that displacement taking P to S (namely $\mathbf{d}$) is the combination of displacements P to M, M to L, and then L to S.]

Corollary to Result A.2.1. If $\mathbf{d}_1, \mathbf{d}_2$, and $\mathbf{d}_3$ are linearly independent displacements, and if, for some $\alpha_1, \alpha_2, \alpha_3 \in \mathbb{R}$,

$$\alpha_1 \mathbf{d}_1 + \alpha_2 \mathbf{d}_2 + \alpha_3 \mathbf{d}_3 = \mathbf{0},$$

then $$\alpha_1 = \alpha_2 = \alpha_3 = 0. \tag{A.2.2}$$

[The proof follows on noting that $\alpha_1 = \alpha_2 = \alpha_3 = 0$ satisfies the displacement equation $(\text{A.2.2})_1$ and that this is the *only* solution as a consequence of the uniqueness incorporated in Result A.2.1.]

Since any three linearly independent displacements form a foundation in terms of which any displacement can be constructed, any such displacement triad is termed a *basis* for the set of all displacements. Each member of such a triad embodies a degree of freedom in moving in space. The three degrees of such freedom is why we term space *three-dimensional*. The set of all displacements will be denoted by $\mathcal{D}$.

A.3 Angles, Magnitudes, and Euclidean Structure

Consider the triangle $\triangle$ of Section A.2. The angle $\angle RPQ =: \theta$ is termed the *angle between* $\mathbf{d}_1$ *and* $\mathbf{d}_2$. Note that $0 \leq \theta \leq \pi : \theta = 0$ and $\theta = \pi$ correspond to the degenerate cases in which $\mathbf{d}_1$ and $\mathbf{d}_2$ are parallel (with, respectively, the same or opposite directions). If $\theta = \pi/2$ (so that sides PQ and PR are perpendicular) the displacements are said to be *orthogonal*.

In specifying a displacement, one needs to know what distance is involved, and being asked to move a certain distance (in a given direction) requires communication of a length scale. Once such a scale ℓ is adopted, then the distance associated with a displacement $\mathbf{d}$ is a dimensionless number, d say, of ℓ-length units. Of course, different choices of ℓ can be made (e.g., meter, foot, ångstrom), but this causes no problem as long as the choice is made explicit: it is possible to relate one choice to another by the appropriate conversion factor (e.g., x metres $= kx$ feet, where $k \simeq 3.28$). Accordingly, once selection of a scale ℓ is adopted, to each $\mathbf{d} \in \mathcal{D}$ can be assigned the number of ℓ units of length involved in such displacement: we write this number as $\|\mathbf{d}\|_\ell$, term it the *magnitude* of $\mathbf{d}$, and write $\mathcal{D}_\ell$ rather than $\mathcal{D}$ to make clear that a choice of scale has been made. It follows that if $\alpha \in \mathbb{R}$, then, for all $\mathbf{d} \in \mathcal{D}_\ell$,

D.12. $$\|\alpha \mathbf{d}\|_\ell = |\alpha| \|\mathbf{d}\|_\ell. \quad \text{(A.3.1)}$$

(Convince yourself of this!)

Given any three non-collinear points P, Q, and R, the distance of R from P is less than the sum of the distances of Q from P and R from Q. If P, Q, and R are collinear, then one has equality if and only if R lies between P and Q. These statements are a consequence of empirical record. If these displacements are labelled $\mathbf{d}_1$ (P to Q) and $\mathbf{d}_2$ (Q to R), then we have

D.13. $$\|\mathbf{d}_1 + \mathbf{d}_2\|_\ell \leq \|\mathbf{d}_1\|_\ell + \|\mathbf{d}_2\|_\ell \quad \text{(A.3.2)}$$

with equality holding if and only if $\mathbf{d}_1$ and $\mathbf{d}_2$ are in the same direction. Further, if PQ and QR are perpendicular (so that $\mathbf{d}_1$ and $\mathbf{d}_2$ are orthogonal displacements), then the Pythagorean theorem yields

$$PR^2 = PQ^2 + QR^2.$$

That is

D.14. $$\|\mathbf{d}_1 + \mathbf{d}_2\|_\ell^2 = \|\mathbf{d}_1\|_\ell^2 + \|\mathbf{d}_2\|_\ell^2 \text{ if } \mathbf{d}_1 \text{ and } \mathbf{d}_2 \text{ are orthogonal.} \quad \text{(A.3.3)}$$

Further, if P, Q, R, and S are four non-coplanar points for which PQ, QR, and RS are mutually perpendicular, then by twofold use of the Pythagorean theorem (and noting that PR is perpendicular to RS),

$$PS^2 = PR^2 + RS^2 = (PQ^2 + QR^2) + RS^2.$$

Accordingly, labelling the displacement from R to S as $\mathbf{d}_3$,

D.15. $$\|\mathbf{d}_1 + \mathbf{d}_2 + \mathbf{d}_3\|_\ell^2 = \|\mathbf{d}_1\|_\ell^2 + \|\mathbf{d}_2\|_\ell^2 + \|\mathbf{d}_3\|_\ell^2 \quad \text{(A.3.4)}$$

if $\mathbf{d}_1, \mathbf{d}_2$, and $\mathbf{d}_3$ are mutually orthogonal displacements.

The additional structure on $\mathcal{D}_\ell$ provided by properties D.12 and D.13 is said to endow a *norm* $\|\cdot\|_\ell$ on $\mathcal{D}_\ell$. The notion of angle between pairs of displacements, and the properties D.14 and D.15 associated with orthogonal displacements, are said to provide $\mathcal{D}_\ell$ with Euclidean structure.

A.4 Vectorial Entities and the Fundamental Space $\mathcal{V}$

The formal rules of manipulation associated with displacements are shared by many physical entities. In particular, velocities, accelerations, momenta, and forces all combine in essentially the same way as displacements. More specifically, each of these sets of physical descriptors can be *represented* by displacements.[4] For example, a force $\mathbf{F}$ of magnitude F force units can be represented by a displacement in the same

[4] This is not a trivial statement but requires case-by-case justification. In fact, it follows from the definitions of velocity and acceleration as time derivatives of displacements, of momentum as mass multiplied by velocity, and of force defined in terms of the time derivative of momentum via Newton's second law.

direction whose associated distance is cF length units. Then the sum of two forces $\mathbf{F}_1$ and $\mathbf{F}_2$ is represented by the combination of the two displacements which separately represent $\mathbf{F}_1$ and $\mathbf{F}_2$: if this is $\mathbf{d}$, then the resultant force is in the direction of $\mathbf{d}$ and has as magnitude the product of c^{-1} with the distance associated with $\mathbf{d}$. The choice $c\,(> 0)$ of conversion factor has dimensions of length per unit force but is otherwise arbitrary. *However, if we choose specific measures of mass, length, and time, then it is highly convenient to choose* $c = 1$. *Indeed, if we do this for every vectorial entity (i.e., choose for each such entity a conversion factor numerically one – of course this factor would have a case-by-case physical dimensionality), then we can work entirely in terms of a single three-dimensional space* $\mathcal{V}$ *when modelling vectorial entities.* Space $\mathcal{V}$ is essentially a copy of $\mathcal{D}_\ell$ with ℓ taken as the unit of length measure. This procedure of abstraction is the direct analogue of how scalar entities are represented by real numbers. [Given a choice of mass, length, and time measures, a given $x \in \mathbb{R}(x > 0)$ can, for example, represent a volume of measure x (length)^3 units or a mass density x mass units per unit volume.] For future reference we now list properties of $\mathcal{V}$ and introduce an inner product which formalises the Euclidean structure found in $\mathcal{D}_\ell$.

The structure and properties of $\mathcal{V}$, whose elements are termed *vectors*, are given as follows.

V.1. To each $\mathbf{v} \in \mathcal{V}$ corresponds a non-negative number $\|\mathbf{v}\|$ (termed the *norm* of $\mathbf{v}$) and, if $\|\mathbf{v}\| \neq 0$, a unique direction in space. (A.4.1)

V.2. Two vectors $\mathbf{v}_1$ and $\mathbf{v}_2$ are said to be equal if they have the same direction and $\|\mathbf{v}_1\| = \|\mathbf{v}_2\|$. (A.4.2)

V.3. To each ordered pair of vectors $\mathbf{v}_1, \mathbf{v}_2$ corresponds another element of $\mathcal{V}$ written as $\mathbf{v}_1 + \mathbf{v}_2$. (A.4.3)

This pairwise combination of vectors satisfies

V.4.
$$\mathbf{v}_2 + \mathbf{v}_1 = \mathbf{v}_1 + \mathbf{v}_2 \tag{A.4.4}$$

and

V.5.
$$\mathbf{v}_1 + (\mathbf{v}_2 + \mathbf{v}_3) = (\mathbf{v}_1 + \mathbf{v}_2) + \mathbf{v}_3. \tag{A.4.5}$$

V.6. There exists a distinguished vector $\mathbf{0}$ (termed the *zero vector*) for which $\|\mathbf{0}\| = 0$ and, for each $\mathbf{v} \in \mathcal{V}$,

$$\mathbf{v} + \mathbf{0} = \mathbf{v}. \tag{A.4.6}$$

V.7. For each $\mathbf{v} \in \mathcal{V}$ there exists a vector $\mathbf{v}^-$ such that

$$\mathbf{v} + \mathbf{v}^- = \mathbf{0}. \tag{A.4.7}$$

Vector $\mathbf{v}^-$ is denoted by $(-\mathbf{v})$ and

DV.7.
$$\mathbf{v}_1 - \mathbf{v}_2 := \mathbf{v}_1 + (-\mathbf{v}_2). \tag{A.4.8}$$

V.8. To each $\alpha \in \mathbb{R}$ and $\mathbf{v} \in \mathcal{V}$ corresponds an element of $\mathcal{V}$ denoted by $\alpha\mathbf{v}$.

This pairwise combination of real numbers with vectors satisfies (here $\mathbf{v}, \mathbf{v}_1, \mathbf{v}_2 \in \mathcal{V}$, and $\alpha, \beta \in \mathbb{R}$ are arbitrary choices)

V.9. $$1\mathbf{v} = \mathbf{v} \tag{A.4.9}$$

V.10. $$\alpha(\beta\mathbf{v}) = (\alpha\beta)\mathbf{v} \tag{A.4.10}$$

V.11. $$(\alpha+\beta)\mathbf{v} = \alpha\mathbf{v} + \beta\mathbf{v} \tag{A.4.11}$$

V.12. $$\alpha(\mathbf{v}_1 + \mathbf{v}_2) = \alpha\mathbf{v}_1 + \alpha\mathbf{v}_2 \tag{A.4.12}$$

V.13. $$\|\alpha\mathbf{v}\| = |\alpha|\,\|\mathbf{v}\| \tag{A.4.13}$$

V.14. $$\|\mathbf{v}\| = 0 \quad \text{if and only if} \quad \mathbf{v} = \mathbf{0} \tag{A.4.14}$$

V.15. $$\|\mathbf{v}_1 + \mathbf{v}_2\| \leq \|\mathbf{v}_1\| + \|\mathbf{v}_2\| \tag{A.4.15}$$

V.16. There exist vector triples $\mathbf{b}_1, \mathbf{b}_2$, and $\mathbf{b}_3$ for which

$$\alpha_1\mathbf{b}_1 + \alpha_2\mathbf{b}_2 + \alpha_3\mathbf{b}_3 = \mathbf{0}$$

implies $\alpha_1 = \alpha_2 = \alpha_3 = 0$, *and* any $\mathbf{v} \in \mathcal{V}$ is expressible as

$$\mathbf{v} = v_1\mathbf{b}_1 + v_2\mathbf{b}_2 + v_3\mathbf{b}_3. \tag{A.4.16}$$

Numbers v_1, v_2, and v_3 are termed the *components* of $\mathbf{v}$ with respect to choice of triple $\mathbf{b}_1, \mathbf{b}_2$, and $\mathbf{b}_3$.

Remark A.4.1. Properties V.3 through V.15 are those shared in general by any normed vector space[5] over $\mathbb{R}$. The link with the physical world stems from V.16 (which establishes $\mathcal{V}$ as three-dimensional), the association in V.1 of every non-zero vector with a direction in space, and the satisfaction by the set of displacements of all properties of $\mathcal{V}$. In particular, our discussion of $\mathcal{D}_\ell$ establishes the *existence* of a space $\mathcal{V}$ which can represent vectorial entities.

Remark A.4.2. Space $\mathcal{V}$ will be employed hereafter to model displacements, velocities, accelerations, forces, and couples: $\mathbf{v} \in \mathcal{V}$ will be used to model a value of any one of these. In so doing $\|\mathbf{v}\|$, a number (v say) will represent the *magnitude* of the vectorial quantity, understood to be v units of measurement of this particular quantity (e.g., v ms^{-1}, v ms^{-2}, v N, or v Nm), while the direction of $\mathbf{v}$ will be that of the relevant quantity. In order to emphasise, and distinguish between, the physical natures of entities modelled by 'copies' of $\mathcal{V}$, we could use notation such as $\mathcal{V}_{(a,b,c)}$, where the vectorial entities in question have physical dimension $M^aL^bT^c$. (Here M, L, and T denote *m*ass, *l*ength, and *t*ime, and a, b, and c are integers.) However, it should be clear from context just how $\mathcal{V}$ is used in what follows.

A.5 Products in $\mathcal{V}$ (Products of Physical Descriptors)

(i) *Multiplication by a scalar*

Given $\alpha \in \mathbb{R}$ and $\mathbf{v} \in \mathcal{V}$, then if α is a pure number (i.e., has no physical dimension), $\alpha\mathbf{v} \in \mathcal{V}$ [see V.8]. If, however, α represents the value of a scalar entity (i.e., has

[5] These properties define an abstract normed vector space over $\mathbb{R}$.

physical dimension), then $\alpha\mathbf{v}$, strictly speaking, lies in another 'copy' of $\mathcal{V}$. We do not make such distinction but note that the physical dimensions of $\alpha, \mathbf{v}$, and $\alpha\mathbf{v}$ in general will all differ, and thus so too will their units of measure (e.g., α kg, $\mathbf{v}$ ms^{-1}, and $\alpha\mathbf{v}$ kg ms^{-1}).

(ii) *Scalar multiplication of vectors*
If $\mathbf{v}_1, \mathbf{v}_2 \in \mathcal{V}$, then their *scalar product*

SP.1.
$$\mathbf{v}_1 . \mathbf{v}_2 = \|\mathbf{v}_1\| \|\mathbf{v}_2\| \cos\theta. \tag{A.5.1}$$

Here $\theta (0 \leq \theta \leq \pi)$ denotes the angle between the directions defined by $\mathbf{v}_1$ and $\mathbf{v}_2$. It follows that:

SP.2.
$$\mathbf{v}_1 . \mathbf{v}_2 = \mathbf{v}_2 . \mathbf{v}_1 \tag{A.5.2}$$

SP.3.
$$\mathbf{v} . \mathbf{v} = \|\mathbf{v}\|^2 \tag{A.5.3}$$

SP.4.
$$(\alpha\mathbf{v}_1) . \mathbf{v}_2 = \alpha(\mathbf{v}_1 . \mathbf{v}_2) = \mathbf{v}_1 . (\alpha\mathbf{v}_2) \tag{A.5.4}$$

Property SP.4 follows from SP.1 and V.13 if $\alpha > 0$ (so $|\alpha| = \alpha$) and is trivial if $\alpha = 0$ [see D.6]. If $\alpha < 0$, then $|\alpha| = -\alpha$ and $\alpha\mathbf{v}_1 = -|\alpha|\mathbf{v}_1 = |\alpha|(-\mathbf{v}_1)$, a vector which makes an angle $\pi - \theta$ with $\mathbf{v}_2$. Thus

$$(\alpha\mathbf{v}_1) . \mathbf{v}_2 = |\alpha| \| -\mathbf{v}_1\| \|\mathbf{v}_2\| \cos(\pi - \theta) = -|\alpha| \|\mathbf{v}_1\| \|\mathbf{v}_2\| \cos\theta = \alpha(\mathbf{v}_1 . \mathbf{v}_2).$$

Further, if $\mathbf{v}_1, \mathbf{v}_2, \mathbf{v}_3 \in \mathcal{V}$, then

SP.5.
$$\mathbf{v}_1 . (\mathbf{v}_2 + \mathbf{v}_3) = \mathbf{v}_1 . \mathbf{v}_2 + \mathbf{v}_1 . \mathbf{v}_3. \tag{A.5.5}$$

This non-trivial property can be proved by appeal to $\mathcal{D}$ and the notion of perpendicular projections: the perpendicular projection of displacement $\mathbf{d}_2$ on the direction defined by $\mathbf{d}_1$ is $\|\mathbf{d}_2\| \cos\theta$, where θ is the angle between $\mathbf{d}_1$ and $\mathbf{d}_2$. A sketch reveals that the projection of $(\mathbf{d}_2 + \mathbf{d}_3)$ on the direction of $\mathbf{d}_1$ is the sum of the separate projections of $\mathbf{d}_2$ and $\mathbf{d}_3$ on this direction.

Properties SP.2, SP.4, and SP.5 ensure that the rules of manipulation associated with operations $+$ and $.$ are formally the same as those in $\mathbb{R}$ for addition and multiplication. Notice, however, $(\mathbf{v}_1 . \mathbf{v}_2)\mathbf{v}_3 \neq (\mathbf{v}_2 . \mathbf{v}_3)\mathbf{v}_1$ (Why?) and $(\mathbf{v}_1 . \mathbf{v}_2) . \mathbf{v}_3$ is undefined (Why?).

Two vectors $\mathbf{v}_1$ and $\mathbf{v}_2$ are said to be *orthogonal* if

$$\mathbf{v}_1 . \mathbf{v}_2 = 0. \tag{A.5.6}$$

From SP.1 this implies that $\|\mathbf{v}_1\| = 0$ (so $\mathbf{v}_1 = \mathbf{0}$) or $\|\mathbf{v}_2\| = 0$ (so $\mathbf{v}_2 = \mathbf{0}$) *or* $\cos\theta = 0$ (so $\theta = \pi/2$, and the directions associated with $\mathbf{v}_1$ and $\mathbf{v}_2$ are perpendicular). (A.5.7)

Exercise A.5.1. Suppose that $\mathbf{u}$ is a given vector and

$$\mathbf{u} . \mathbf{v} = 0 \qquad \text{for all} \quad \mathbf{v} \in \mathcal{V}.$$

Show that $\mathbf{u} = \mathbf{0}$ [recall (V.14) and use (SP.3)]. Deduce that if

$$\mathbf{u}_1 . \mathbf{v} = \mathbf{u}_2 . \mathbf{v} \qquad \text{for all} \quad \mathbf{v} \in \mathcal{V},$$

then $\mathbf{u}_1 = \mathbf{u}_2$.

Remark A.5.1. The notion of scalar product arises naturally in defining work and the rate of working associated with forces and motions. If a constant force $\mathbf{F}$ acts on a moving location (e.g., a particular point on the boundary of a moving body), then the work done when this location has undergone a displacement $\mathbf{d}$ is defined to be $\mathbf{F}.\mathbf{d}$. If any force (possibly time-dependent) $\mathbf{F}$ acts on a location which moves with velocity $\mathbf{v}$, then $\mathbf{F}.\mathbf{v}$ is termed the *rate of working* of $\mathbf{F}$, and the work done by $\mathbf{F}$ over a time interval $t_1 \leq t \leq t_2$ is

$$w := \int_{t_1}^{t_2} \mathbf{F}(t).\mathbf{v}(t)d\tau.$$

Remark A.5.2. Clearly, different 'copies' of $\mathcal{V}$ are involved in computation of $\mathbf{F}.\mathbf{d}$ and $\mathbf{F}.\mathbf{v}$, in the sense of the differing physical dimensions. For example, in SP.1 with $\mathbf{v}_1 = \mathbf{F}$ and $\mathbf{v}_2 = \mathbf{d}$, $\|\mathbf{v}_1\|$ has dimension MLT^{-2} and $\|\mathbf{v}_2\|$ dimension L.

(iii) *Vector multiplication*
If $\mathbf{v}_1, \mathbf{v}_2 \in \mathcal{V}$, then their *vector product*

VP.1. $$\mathbf{v}_1 \times \mathbf{v}_2 := \|\mathbf{v}_1\|\|\mathbf{v}_2\| \sin\theta\, \mathbf{n} \tag{A.5.8}$$

Here vector $\mathbf{n}$ has norm $\|\mathbf{n}\| = 1$. The direction of $\mathbf{n}$ is that in which a right-handed screw would move (if embedded in a motionless body) if its axis were initially perpendicular to the directions of $\mathbf{v}_1$ and $\mathbf{v}_2$, with groove/slot parallel to $\mathbf{v}_1$, and then the groove/slot rotated through angle θ (so that the slot is now parallel to $\mathbf{v}_2$).[6] Vector multiplication satisfies, for any $\mathbf{v}, \mathbf{v}_1, \mathbf{v}_2 \in \mathcal{V}$ and any $\alpha \in \mathbb{R}$,

VP.2. $$\mathbf{v}_2 \times \mathbf{v}_1 = -(\mathbf{v}_1 \times \mathbf{v}_2) \tag{A.5.9}$$

VP.3. $$\mathbf{v} \times \mathbf{v} = \mathbf{0} \tag{A.5.10}$$

VP.4. $$(\alpha\mathbf{v}_1) \times \mathbf{v}_2 = \alpha(\mathbf{v}_1 \times \mathbf{v}_2) = \mathbf{v}_1 \times (\alpha\mathbf{v}_2) \tag{A.5.11}$$

These properties follow from definition VP.1 and, for VP.4, separate examination of cases $\alpha > 0, \alpha = 0$, and $\alpha < 0$. Further, non-trivially [see Rutherford [102] for details: the geometrical argument involves perpendicular projections. Cf. SP.5 et seq.],

VP.5 $$\mathbf{v}_1 \times (\mathbf{v}_2 + \mathbf{v}_3) = \mathbf{v}_1 \times \mathbf{v}_2 + \mathbf{v}_1 \times \mathbf{v}_3. \tag{A.5.12}$$

Properties VP.2, VP.4, and VP.5 ensure that the rules of manipulation for operations $+$ and *twofold* products $\times$ are formally the same as those in $\mathbb{R}$ for addition and multiplication *provided that the order of factors is maintained.*

Exercise A.5.2. Show that if $\mathbf{v}_1 \times \mathbf{v}_2 = \mathbf{0}$, with $\mathbf{v}_1$ and $\mathbf{v}_2$ both non-zero, then $\mathbf{v}_1$ and $\mathbf{v}_2$ must be parallel. Is this the same as saying $\mathbf{v}_1$ and $\mathbf{v}_2$ have the same associated direction?

Remark A.5.3. Vector multiplication is usually motivated physically by the notion of the moment of a force, $\mathbf{F}$ say, about a point, P say. If $\mathbf{d}$ denotes the displacement

[6] Notice that the screw could be in one of two possible orientations, and the prescribed rotation could involve either 'screwing' or 'unscrewing'. Convince yourself that the direction of motion of the screw would be the same for all these possibilities.

from P of any point Q on the line of action of $\mathbf{F}$, then this moment is $\mathbf{d} \times \mathbf{F}$. Notice that $\mathbf{d}, \mathbf{F}$, and $\mathbf{d} \times \mathbf{F}$ reside in different 'copies' of $\mathcal{V}$ (Why?) (cf. Remark A.5.2).

Exercise A.5.3. Prove that $\mathbf{d} \times \mathbf{F}$ is independent of the choice of Q.

(iv) *Triple products*
Given $\mathbf{v}_1, \mathbf{v}_2, \mathbf{v}_3 \in \mathcal{V}$, we may compute $\mathbf{v}_1 . (\mathbf{v}_2 \times \mathbf{v}_3)$ and $\mathbf{v}_1 \times (\mathbf{v}_2 \times \mathbf{v}_3)$, termed the *triple scalar product* and *triple vector product*, respectively. The *t*riple *s*calar *p*roduct satisfies

TSP.1. $$\mathbf{v}_1 . (\mathbf{v}_2 \times \mathbf{v}_3) = \mathbf{v}_2 . (\mathbf{v}_3 \times \mathbf{v}_1) = \mathbf{v}_3 . (\mathbf{v}_1 \times \mathbf{v}_2). \quad \text{(A.5.13)}$$

Equality follows from noting that each represents the volume of a parallelipiped of sides $\|\mathbf{v}_1\|$, $\|\mathbf{v}_2\|$ and $\|\mathbf{v}_3\|$ parallel to $\mathbf{v}_1$, $\mathbf{v}_2$ and $\mathbf{v}_3$, respectively, or each represents the negative of this volume. Such observation yields the following result:

TSP.2. $\mathbf{v}_1, \mathbf{v}_2$, and $\mathbf{v}_3$ are coplanar if and only if $\mathbf{v}_1 . (\mathbf{v}_2 \times \mathbf{v}_3) = 0$. (A.5.14)

(*Coplanar* means displacements $\mathbf{v}_1$, $\mathbf{v}_2$, and $\mathbf{v}_3$ length units from any given point P_0 yield points P_1, P_2, and P_3 for which P_0, P_1, P_2, and P_3 are coplanar.)

Remark A.5.4. Recalling the origin of our notion of linear independence of three vectors as precisely their non-coplanar nature (see Section A.2), we have

TSP.3. $\mathbf{v}_1, \mathbf{v}_2$, and $\mathbf{v}_3$ are linearly independent if and only if $\mathbf{v}_1 . (\mathbf{v}_2 \times \mathbf{v}_3) \neq 0$.(A.5.15)

Triple vector products satisfy

TVP.1. $$\mathbf{v}_1 \times (\mathbf{v}_2 \times \mathbf{v}_3) = (\mathbf{v}_1 . \mathbf{v}_3)\mathbf{v}_2 - (\mathbf{v}_1 . \mathbf{v}_2)\mathbf{v}_3. \quad \text{(A.5.16)}$$

Exercise A.5.4. Deduce that

TVP.2. $$(\mathbf{v}_1 \times \mathbf{v}_2) \times \mathbf{v}_3 = (\mathbf{v}_1 . \mathbf{v}_3)\mathbf{v}_2 - (\mathbf{v}_3 . \mathbf{v}_2)\mathbf{v}_1. \quad \text{(A.5.17)}$$

[*Hint*: Write $(\mathbf{v}_1 \times \mathbf{v}_2) \times \mathbf{v}_3 = -[\mathbf{v}_3 \times (\mathbf{v}_1 \times \mathbf{v}_2)]$.]

Thus, in particular,

$$\mathbf{v}_1 \times (\mathbf{v}_2 \times \mathbf{v}_3) \neq (\mathbf{v}_1 \times \mathbf{v}_2) \times \mathbf{v}_3. \quad \text{(A.5.18)}$$

(Why?)

Proof of TVP.1 is most simply obtained by employing a basis of mutually orthogonal vectors, each of magnitude 1.

A.6 Unit Vectors, Orthonormal Bases, and Related Components

Any vector $\mathbf{e} \in \mathcal{V}$ for which $\|\mathbf{e}\| = 1$ is termed a *unit* vector. Recall from Section A.2. that any three non-coplanar displacements are linearly independent and form a basis for $\mathcal{D}$. Accordingly, any three vectors in $\mathcal{V}$ form a basis if their directions, taken from any given point, define three non-coplanar lines. In particular, choose unit vectors $\mathbf{e}_1$, $\mathbf{e}_2$, and $\mathbf{e}_3$ in mutually perpendicular directions. Thus

$$\mathbf{e}_1 . \mathbf{e}_1 = \mathbf{e}_2 . \mathbf{e}_2 = \mathbf{e}_3 . \mathbf{e}_3 = 1 \quad \text{(A.6.1)}$$

and $$\mathbf{e}_1 . \mathbf{e}_2 = \mathbf{e}_2 . \mathbf{e}_3 = \mathbf{e}_3 . \mathbf{e}_1 = 0. \quad \text{(A.6.2)}$$

(Why?) These results can be abbreviated to

$$\mathbf{e}_i \,.\, \mathbf{e}_j = \delta_{ij}, \tag{A.6.3}$$

where the Kronecker delta symbol δ_{ij} is defined by

$$\delta_{ij} = 1 \quad \text{if} \quad i = j, \qquad \delta_{ij} = 0 \quad \text{if} \quad i \neq j. \tag{A.6.4}$$

(Here $i,j = 1,2,3$.) Set $\{\mathbf{e}_1, \mathbf{e}_2, \mathbf{e}_3\}$ is said to be an *orthonormal basis*.

Writing $\mathbf{v} \in \mathcal{V}$ in terms of basis $\mathbf{e}_1$, $\mathbf{e}_2$, and $\mathbf{e}_3$ yields

$$\mathbf{v} = v_1\mathbf{e}_1 + v_2\mathbf{e}_2 + v_3\mathbf{e}_3 \tag{A.6.5}$$

for unique numbers v_1, v_2, and v_3 [the components of $\mathbf{v}$ with respect to this basis: see (A.4.16)]. Scalar multiplication by $\mathbf{e}_1$ yields

$$\mathbf{v}\,.\,\mathbf{e}_1 = (v_1\mathbf{e}_1 + v_2\mathbf{e}_2 + v_3\mathbf{e}_3)\,.\,\mathbf{e}_1 = v_1\mathbf{e}_1\,.\,\mathbf{e}_1 + v_2\mathbf{e}_2\,.\,\mathbf{e}_1 + v_3\mathbf{e}_3\,.\,\mathbf{e}_1 = v_1.$$

Similarly, $v_2 = \mathbf{v}\,.\,\mathbf{e}_2$ and $v_3 = \mathbf{v}\,.\,\mathbf{e}_3$, and hence

$$\mathbf{v} = (\mathbf{v}\,.\,\mathbf{e}_1)\mathbf{e}_1 + (\mathbf{v}\,.\,\mathbf{e}_2)\mathbf{e}_2 = (\mathbf{v}\,.\,\mathbf{e}_3)\mathbf{e}_3 =: (\mathbf{v}\,.\,\mathbf{e}_i)\mathbf{e}_i. \tag{A.6.6}$$

Here we have used for the first time the *summation convention* in which a repeated suffix (here i) indicates that summation over this suffix, for values 1, 2, and 3, is intended. This convention will be used henceforth. (Occasionally we encounter repeated suffices where summation is not intended, but in such case this will be stated explicitly.) In particular, (A.6.6) may be written as

$$\mathbf{v} = v_i\mathbf{e}_i. \tag{A.6.7}$$

The scalar product in terms of components is given by

$$\mathbf{u}\,.\,\mathbf{v} = (u_1\mathbf{e}_1 + u_2\mathbf{e}_2 + u_3\mathbf{e}_3).(v_1\mathbf{e}_1 + v_2\mathbf{e}_2 + v_3\mathbf{e}_3).$$

That is, on invoking (A.5.5) and (A.6.3),

$$\mathbf{u}\,.\,\mathbf{v} = u_1v_1 + u_2v_2 + u_3v_3 = u_iv_i. \tag{A.6.8}$$

Thus

$$\|\mathbf{v}\|^2 = \mathbf{v}\,.\,\mathbf{v} = v_1^2 + v_2^2 + v_3^2, \tag{A.6.9}$$

and if θ denotes the angle between $\mathbf{u}$ and $\mathbf{v}$, then from (A.5.1)

$$\cos\theta = (u_1v_1 + u_2v_2 + u_3v_3)/(u_1^2 + u_2^2 + u_3^2)^{1/2}(v_1^2 + v_2^2 + v_3^2)^{1/2}. \tag{A.6.10}$$

If $\mathbf{e}_1$, $\mathbf{e}_2$, and $\mathbf{e}_3$ are chosen so that $\mathbf{e}_3 = \mathbf{e}_1 \times \mathbf{e}_2$, then basis $\{\mathbf{e}_1, \mathbf{e}_2, \mathbf{e}_3\}$ is said to be *right-handed*. [Recall that definition (A.5.8) involved the notion of a right-handed screw.] It follows that

$$\mathbf{e}_1 = \mathbf{e}_2 \times \mathbf{e}_3 \qquad \text{and} \qquad \mathbf{e}_2 = \mathbf{e}_3 \times \mathbf{e}_1. \tag{A.6.11}$$

Exercise A.6.1. Convince yourself of results (A.6.11).

Exercise A.6.2. For a right-handed orthonormal basis $\{\mathbf{e}_1, \mathbf{e}_2, \mathbf{e}_3\}$, show that

(i) $\mathbf{u}\times\mathbf{v} = (u_2v_3 - u_3v_2)\mathbf{e}_1 + (u_3v_1 - u_1v_3)\mathbf{e}_2 + (u_1v_2 - u_2v_1)\mathbf{e}_3,$ (A.6.12)

(ii) $(\mathbf{u}\times\mathbf{v}).\mathbf{w} = (u_2v_3 - u_3v_2)w_1 + (u_3v_1 - u_1v_3)w_2 + (u_1v_2 - u_2v_1)w_3$ (A.6.13)

$$= \begin{vmatrix} u_1 & u_2 & u_3 \\ v_1 & v_2 & v_3 \\ w_1 & w_2 & w_3 \end{vmatrix},$$

and [see (A.5.16)]

(iii) $(\mathbf{u}\times\mathbf{v})\times\mathbf{w} = (\mathbf{u}.\mathbf{w})\mathbf{v} - (\mathbf{v}.\mathbf{w})\mathbf{u}.$

Remark A.6.1. Product (i) above can be written as

$$\mathbf{u}\times\mathbf{v} = \epsilon_{ijk}u_jv_k\mathbf{e}_i,$$

where ϵ_{ijk} denotes the *permutation factor*: $\epsilon_{ijk} = 1(-1)$ if ordered triple (i,j,k) is an even (odd) permutation of $(1, 2, 3)$, and $\epsilon_{ijk} = 0$ otherwise.

A.7 Linear Transformations on $\mathcal{V}$ and the General Definition of a Vector Space over $\mathbb{R}$

A map[7] $\mathbf{L}:\mathcal{V}\to\mathcal{V}$ which preserves sums and multiplication by real numbers is termed a *linear transformation on* $\mathcal{V}$. That is, for all $\mathbf{v}_1,\mathbf{v}_2 \in \mathcal{V}$ and all $\alpha_1,\alpha_2 \in \mathbb{R}$,

$$\mathbf{L}(\alpha_1\mathbf{v}_1 + \alpha_2\mathbf{v}_2) = \alpha_1\mathbf{L}\mathbf{v}_1 + \alpha_2\mathbf{L}\mathbf{v}_2, \tag{A.7.1}$$

where[8] $\mathbf{Lv}$ denotes the *image* of $\mathbf{v}$ under $\mathbf{L}$.

Result A.7.1. A linear transformation is uniquely determined by its action on a basis[9] for $\mathcal{V}$. (That is, if $\mathbf{b}_1$, $\mathbf{b}_2$, and $\mathbf{b}_3$ form a basis for $\mathcal{V}$, then knowledge of $\mathbf{Lb}_1$, $\mathbf{Lb}_2$, and $\mathbf{Lb}_3$ uniquely determines $\mathbf{L}$. Thus $\mathbf{Lv}$ is uniquely determined by $\mathbf{Lb}_1, \mathbf{Lb}_2$ and $\mathbf{Lb}_3$ for any $\mathbf{v}\in\mathcal{V}$.)

Proof. If $\mathbf{v}\in\mathcal{V}$, then since $\mathbf{b}_1$, $\mathbf{b}_2$, and $\mathbf{b}_3$ form a basis, $\mathbf{v}$ is expressible as a linear combination of $\mathbf{b}_1$, $\mathbf{b}_2$, and $\mathbf{b}_3$; that is, there exist unique numbers v_1, v_2, and v_3 such that $\mathbf{v} = v_1\mathbf{b}_1 + v_2\mathbf{b}_2 + v_3\mathbf{b}_3$. [See (A.2.1).] Thus, since $\mathbf{L}$ is linear,

$$\begin{aligned}\mathbf{Lv} &= \mathbf{L}(v_1\mathbf{b}_1 + v_2\mathbf{b}_2 + v_3\mathbf{b}_3) = \mathbf{L}(v_1\mathbf{b}_1) + \mathbf{L}(v_2\mathbf{b}_2) + \mathbf{L}(v_3\mathbf{b}_3)\\ &= v_1(\mathbf{Lb}_1) + v_2(\mathbf{Lb}_2) + v_3(\mathbf{Lb}_3).\end{aligned}$$

The set $\mathrm{Lin}\,\mathcal{V}$ of all linear transformations on $\mathcal{V}$ has a natural structure based upon that of $\mathcal{V}$. In particular, if $\mathbf{L}\in\mathrm{Lin}\,\mathcal{V}$ and $\alpha\in\mathbb{R}$, then $\alpha\mathbf{L}\in\mathrm{Lin}\,\mathcal{V}$ is defined, for any $\mathbf{v}\in\mathcal{V}$, by

$$(\alpha\mathbf{L})\mathbf{v} := \alpha(\mathbf{Lv}). \tag{A.7.2}$$

Further, if $\mathbf{L}_1,\mathbf{L}_2\in\mathrm{Lin}\,\mathcal{V}$, then $\mathbf{L}_1+\mathbf{L}_2\in\mathrm{Lin}\,\mathcal{V}$ is defined, for any $\mathbf{v}\in\mathcal{V}$, by

$$(\mathbf{L}_1+\mathbf{L}_2)\mathbf{v} := (\mathbf{L}_1\mathbf{v}) + (\mathbf{L}_2\mathbf{v}). \tag{A.7.3}$$

[7] That is, a function which is defined on $\mathcal{V}$ and takes values in $\mathcal{V}$.

[8] That is, $\mathbf{Lv}$ is the value of function $\mathbf{L}$ at $\mathbf{v}\in\mathcal{V}$.

[9] Recall from the last paragraph of Section A.2 that a basis for $\mathcal{V}$ is a set of three linearly independent vectors [see (A.2.2)].

Remark A.7.1. Notice that in (A.7.2) $\alpha\mathbf{L}$ denotes multiplication of a linear transformation by a number, while $\alpha(\mathbf{Lv})$ denotes multiplication of a vector (namely $\mathbf{Lv}$) by the same number. Notice also that in (A.7.3) the $+$ sign on the left-hand side (the *sum* of two linear transformations) is defined by the right-hand side (the known sum of two vectors $\mathbf{Lv}_1$ and $\mathbf{Lv}_2$). Thus the symbol $+$ is being used in two different senses. Strictly speaking, we should distinguish the two summation signs from each other (and from summation of real numbers). However, this would be somewhat tedious. In practice there should be no confusion since, whenever a symbol $\mathcal{A}+\mathcal{B}$ is used, the summands $\mathcal{A}$ and $\mathcal{B}$ should be of the same nature (namely, both real numbers or both vectors or both linear transformations on $\mathcal{V}$) *and have the same physical dimension*!

Exercise A.7.1. Prove that $\alpha\mathbf{L}$ and $\mathbf{L}_1+\mathbf{L}_2$ are linear transformation on $\mathcal{V}$.

We define the zero linear transformation $\mathbf{O}$ on $\mathcal{V}$ by

$$\mathbf{Ov}=\mathbf{0} \tag{A.7.4}$$

and the additive inverse $(-\mathbf{L})$ of $\mathbf{L}$ by

$$(-\mathbf{L})+\mathbf{L}=\mathbf{O}. \tag{A.7.5}$$

Thus, for any $\mathbf{v}\in\mathcal{V}$,

$$(-\mathbf{L})\mathbf{v}+\mathbf{Lv}=((-\mathbf{L})+\mathbf{L})\mathbf{v}=\mathbf{Ov}=\mathbf{0}$$

and hence (Show this!)

$$(-\mathbf{L})\mathbf{v}=-(\mathbf{Lv}). \tag{A.7.6}$$

It follows that the set Lin $\mathcal{V}$ of linear transformations formally satisfies V.3 through V.12 in Section A.4, provided that *vector* is replaced by *linear transformation*, $\mathbf{0}$ by $\mathbf{O}$, and pairwise combinations of $\mathbb{R}$ and Lin $\mathcal{V}$ are defined by (A.7.2) and (A.7.3). This leads us to generalise the notion of the physical *vectorial* space $\mathcal{V}$ as follows.

A *vector* (or *linear*) *space* $\mathcal{U}$ *over* $\mathbb{R}$ is a set, together with pairwise combinations (written as $\mathbf{u}_1+\mathbf{u}_2$ and $\alpha\mathbf{u}$ for any $\mathbf{u}_1,\mathbf{u}_2,\mathbf{u}\in\mathcal{U}$ and $\alpha\in\mathbb{R}$) which lie in $\mathcal{U}$, and for which V.3 through V.12 are satisfied. A subset $\{\mathbf{u}_1,\mathbf{u}_2,\ldots,\mathbf{u}_\ell\}$ of $\mathcal{U}$ is termed *linearly independent* if

$$\alpha_1\mathbf{u}_1+\alpha_2\mathbf{u}_2+\cdots+\alpha_\ell\mathbf{u}_\ell=\mathbf{0}$$

for some $\alpha_1,\alpha_2,\cdots,\alpha_\ell\in\mathbb{R}$ implies that $\alpha_1=\alpha_2=\ldots=\alpha_\ell=0$. A subset $\{\mathbf{u}'_1,\mathbf{u}'_2,\ldots,\mathbf{u}'_m\}$ of $\mathcal{U}$ is said to span $\mathcal{U}$ if every $\mathbf{u}\in\mathcal{U}$ is expressible as a *linear combination* of $\mathbf{u}'_1,\ldots,\mathbf{u}'_m$; that is, there exist numbers $\alpha_1,\ldots,\alpha_m$ such that

$$\mathbf{u}=\alpha_1\mathbf{u}'_1+\alpha_2\mathbf{u}'_2+\ldots+\alpha_m\mathbf{u}'_m.$$

A subset $\{\mathbf{b}_1,\mathbf{b}_2,\ldots,\mathbf{b}_n\}$ of $\mathcal{U}$ which both spans $\mathcal{U}$ and is linearly independent is termed a *basis* for $\mathcal{U}$. If a basis exists, then any other basis can be shown to have the same number n of elements. In such case $\mathcal{U}$ is said to be of *dimension n* (or *n-dimensional*). (See, for example, Halmos [88].)

It follows that Lin $\mathcal{V}$ has the structure of a vector space *and more.* Indeed, if $\mathbf{L}_1,\mathbf{L}_2\in$ Lin $\mathcal{V}$, then their composition $\mathbf{L}_1\mathbf{L}_2$, defined for any $\mathbf{v}\in\mathcal{V}$ by

$$(\mathbf{L}_1\mathbf{L}_2)\mathbf{v}:=\mathbf{L}_1(\mathbf{L}_2\mathbf{v}), \tag{A.7.7}$$

also belongs to Lin $\mathcal{V}$. Further, Lin $\mathcal{V}$ contains an identity element $\mathbf{1}$ defined for all $\mathbf{v} \in \mathcal{V}$ by

$$\mathbf{1v} := \mathbf{v}. \tag{A.7.8}$$

Exercise A.7.2. Show that $\mathbf{L}_1\mathbf{L}_2$ and $\mathbf{1}$ are elements of Lin $\mathcal{V}$ and that

$$\mathbf{1L} = \mathbf{L} = \mathbf{L1} \tag{A.7.9}$$

for all $\mathbf{L} \in \text{Lin}\,\mathcal{V}$.

A.8 The Transpose of a Linear Transformation on $\mathcal{V}$ and Tensor Products of Vectors

If $\mathbf{L} \in \text{Lin}\,\mathcal{V}$, consider $\mathbf{L}^T : \mathcal{V} \to \mathcal{V}$, where

$$\mathbf{L}^T\mathbf{u} \,.\, \mathbf{v} := \mathbf{u} \,.\, \mathbf{Lv} \tag{A.8.1}$$

for any $\mathbf{u}, \mathbf{v} \in \mathcal{V}$. Thus, for any $\alpha_1, \alpha_2 \in \mathbb{R}$ and $\mathbf{u}_1, \mathbf{u}_2, \mathbf{v} \in \mathcal{V}$,

$$\begin{aligned}
\mathbf{L}^T(\alpha_1\mathbf{u}_1 + \alpha_2\mathbf{u}_2) \,.\, \mathbf{v} :&= (\alpha_1\mathbf{u}_1 + \alpha_2\mathbf{u}_2) \,.\, \mathbf{Lv} \\
&= \alpha_1(\mathbf{u}_1 \,.\, \mathbf{Lv}) + \alpha_2(\mathbf{u}_2 \,.\, \mathbf{Lv}) \\
&= \alpha_1(\mathbf{L}^T\mathbf{u}_1) \,.\, \mathbf{v} + \alpha_2(\mathbf{L}^T\mathbf{u}_2) \,.\, \mathbf{v} \\
&= (\alpha_1(\mathbf{L}^T\mathbf{u}_1) + \alpha_2(\mathbf{L}^T\mathbf{u}_2)) \,.\, \mathbf{v}.
\end{aligned} \tag{A.8.2}$$

Since $\mathbf{v}$ is arbitrary, from Exercise A.5.1

$$\mathbf{L}^T(\alpha_1\mathbf{u}_1 + \alpha_2\mathbf{u}_2) = \alpha_1\mathbf{L}^T\mathbf{u}_1 + \alpha_2\mathbf{L}^T\mathbf{u}_2,$$

and hence [see (A.7.1)] $\mathbf{L}^T \in \text{Lin}\,\mathcal{V}$. $\mathbf{L}^T$ is termed the *transpose* of $\mathbf{L}$.

Exercise A.8.1. Show that

$$\mathbf{1}^T = \mathbf{1}, \tag{A.8.3}$$

$$(\alpha\mathbf{L})^T = \alpha\mathbf{L}^T, \tag{A.8.4}$$

$$(\mathbf{L}_1 + \mathbf{L}_2)^T = \mathbf{L}_1^T + \mathbf{L}_2^T, \quad \text{and} \tag{A.8.5}$$

$$(\mathbf{L}^T)^T = \mathbf{L}. \tag{A.8.6}$$

(*Hint*: In each case let the left-hand side act on an arbitrary vector $\mathbf{u}$ and form the scalar product with an arbitrary vector $\mathbf{v}$.)

Consider, for $\mathbf{L}_1, \mathbf{L}_2 \in \text{Lin}\,\mathcal{V}$ and any choice $\mathbf{u}, \mathbf{v} \in \mathcal{V}$,

$$(\mathbf{L}_1\mathbf{L}_2)^T\mathbf{u} \,.\, \mathbf{v} = \mathbf{u} \,.\, (\mathbf{L}_1\mathbf{L}_2)\mathbf{v} = \mathbf{u} \,.\, \mathbf{L}_1(\mathbf{L}_2\mathbf{v}) = \mathbf{L}_1^T\mathbf{u} \,.\, \mathbf{L}_2\mathbf{v} = \mathbf{L}_2^T\mathbf{L}_1^T\mathbf{u} \,.\, \mathbf{v}.$$

Thus, from Exercise A.5.1,

$$(\mathbf{L}_1\mathbf{L}_2)^T\mathbf{u} = \mathbf{L}_2^T\mathbf{L}_1^T\mathbf{u}$$

and hence

$$(\mathbf{L}_1\mathbf{L}_2)^T = \mathbf{L}_2^T\mathbf{L}_1^T. \tag{A.8.7}$$

The *tensor product* of any pair of vectors $\mathbf{a},\mathbf{b}$ is denoted by $\mathbf{a}\otimes\mathbf{b}$ and defined for any $\mathbf{v}\in\mathcal{V}$ by

$$(\mathbf{a}\otimes\mathbf{b})\mathbf{v} := (\mathbf{b}\,.\,\mathbf{v})\mathbf{a}. \tag{A.8.8}$$

Since
$$\begin{aligned}(\mathbf{a}\otimes\mathbf{b})(\alpha_1\mathbf{v}_1+\alpha_2\mathbf{v}_2) &= (\mathbf{b}\,.\,(\alpha_1\mathbf{v}_1+\alpha_2\mathbf{v}_2))\mathbf{a}\\ &= \alpha_1(\mathbf{b}\,.\,\mathbf{v}_1)\mathbf{a}+\alpha_2(\mathbf{b}\,.\,\mathbf{v}_2)\mathbf{a}\\ &= \alpha_1(\mathbf{a}\otimes\mathbf{b})\mathbf{v}_1+\alpha_2(\mathbf{a}\otimes\mathbf{b})\mathbf{v}_2,\end{aligned}$$

we have
$$\mathbf{a}\otimes\mathbf{b}\in \operatorname{Lin}\mathcal{V}. \tag{A.8.9}$$

Further, notice that for all $\mathbf{u},\mathbf{v}\in\mathcal{V}$

$$(\mathbf{b}\otimes\mathbf{a})\mathbf{u}\,.\,\mathbf{v} = (\mathbf{a}\,.\,\mathbf{u})(\mathbf{b}.\mathbf{v}) = (\mathbf{b}\,.\,\mathbf{v})(\mathbf{a}\,.\,\mathbf{u}) = \mathbf{u}\,.\,(\mathbf{a}\otimes\mathbf{b})\mathbf{v},$$

while, by definition,

$$(\mathbf{a}\otimes\mathbf{b})^T\mathbf{u}\,.\,\mathbf{v} = \mathbf{u}\,.\,(\mathbf{a}\otimes\mathbf{b})\mathbf{v}.$$

Thus
$$[(\mathbf{a}\otimes\mathbf{b})^T-(\mathbf{b}\otimes\mathbf{a})]\mathbf{u}\,.\,\mathbf{v}=0,$$

and so, from Exercise A.5.1,

$$((\mathbf{a}\otimes\mathbf{b})^T-\mathbf{b}\otimes\mathbf{a})\mathbf{u}=\mathbf{0}$$

for all $\mathbf{u}\in\mathcal{V}$. Hence

$$(\mathbf{a}\otimes\mathbf{b})^T=\mathbf{b}\otimes\mathbf{a}. \tag{A.8.10}$$

Exercise A.8.2. Show that

$$(\mathbf{a}\otimes\mathbf{b})(\mathbf{c}\otimes\mathbf{d}) = (\mathbf{b}\,.\,\mathbf{c})(\mathbf{a}\otimes\mathbf{d}). \tag{A.8.11}$$

[*Hint*: Consider $(\mathbf{a}\otimes\mathbf{b})(\mathbf{c}\otimes\mathbf{d})\mathbf{v} = (\mathbf{a}\otimes\mathbf{b})((\mathbf{c}\otimes\mathbf{d})\mathbf{v})$.]

If $\mathbf{L}_1,\mathbf{L}_2\in\operatorname{Lin}\mathcal{V}$, and $\mathbf{a},\mathbf{b}\in\mathcal{V}$, then for any $\mathbf{v}\in\mathcal{V}$ we have

$$\begin{aligned}(\mathbf{L}_1(\mathbf{a}\otimes\mathbf{b})\mathbf{L}_2)\mathbf{v} &= \mathbf{L}_1(\mathbf{a}\otimes\mathbf{b})(\mathbf{L}_2\mathbf{v})\\ &= \mathbf{L}_1((\mathbf{b}\,.\,\mathbf{L}_2\mathbf{v})\mathbf{a}) = (\mathbf{b}\,.\,\mathbf{L}_2\mathbf{v})\mathbf{L}_1\mathbf{a}\\ &= (\mathbf{L}_2^T\mathbf{b}\,.\,\mathbf{v})\mathbf{L}_1\mathbf{a} = (\mathbf{L}_1\mathbf{a})\otimes(\mathbf{L}_2^T\mathbf{b})\mathbf{v}.\end{aligned}$$

Thus
$$\mathbf{L}_1(\mathbf{a}\otimes\mathbf{b})\mathbf{L}_2 = (\mathbf{L}_1\mathbf{a})\otimes(\mathbf{L}_2^T\mathbf{b}). \tag{A.8.12}$$

Remark A.8.1. The definitions of transpose and tensor product (also known as a *dyadic*) are direct in the sense that no appeal to bases has been made. The next section will detail matrix *representations* of vectors and linear transformations which follow from selection of appropriate (orthonormal) bases. In particular it will turn out that a (square) matrix representation of $\mathbf{L}^T$ is the transpose of the corresponding matrix representation of $\mathbf{L}$.

A linear transformation is said to be *symmetric* if

$$\mathbf{L}^T=\mathbf{L} \tag{A.8.13}$$

and *skew(-symmetric)* if

$$\mathbf{L}^T = -\mathbf{L}. \tag{A.8.14}$$

Exercise A.8.2. Show that

(i) scalar multiples of symmetric (skew) linear transformations are symmetric (skew),
(ii) sums of symmetric (skew) linear transformations are symmetric (skew), and
(iii) $\mathbf{O} \in \operatorname{Lin} \mathcal{V}$ is both symmetric *and* skew.

Results (i), (ii), and (iii) suffice to establish the sets

$$\operatorname{Sym} \mathcal{V} := \{\mathbf{L} \in \operatorname{Lin} \mathcal{V} : \mathbf{L}^T = \mathbf{L}\} \tag{A.8.15}$$

and

$$\operatorname{Sk} \mathcal{V} := \{\mathbf{L} \in \operatorname{Lin} \mathcal{V} : \mathbf{L}^T = -\mathbf{L}\} \tag{A.8.16}$$

as vector spaces in their own right. These subsets of Lin $\mathcal{V}$ are termed *subspaces* (of Lin $\mathcal{V}$).

Exercise A.8.3. Show that if $\mathbf{L} \in \operatorname{Lin} \mathcal{V}$, then

$$\mathbf{L} + \mathbf{L}^T \in \operatorname{Sym} \mathcal{V}, \qquad \mathbf{L} - \mathbf{L}^T \in \operatorname{Sk} \mathcal{V} \tag{A.8.17}$$

and

$$\mathbf{L} \in (\operatorname{Sym} \mathcal{V}) \cap (\operatorname{Sk} \mathcal{V}) \qquad \text{implies } \mathbf{L} = \mathbf{O}. \tag{A.8.18}$$

Since

$$\mathbf{L} = \frac{1}{2}(\mathbf{L} + \mathbf{L}^T) + \frac{1}{2}(\mathbf{L} - \mathbf{L}^T), \tag{A.8.19}$$

every linear transformation $\mathbf{L}$ can be written as the sum of a symmetric element of Lin $\mathcal{V}$ [the *symmetric part* $\frac{1}{2}(\mathbf{L} + \mathbf{L}^T)$ of $\mathbf{L}$] with a skew element of Lin $\mathcal{V}$ [the *skew part* $\frac{1}{2}(\mathbf{L} - \mathbf{L}^T)$ of $\mathbf{L}$]. Of course, from (A.8.3)

$$\mathbf{1} \in \operatorname{Sym} \mathcal{V} \tag{A.8.20}$$

and the corresponding decomposition (A.8.19) is $\mathbf{1} = \mathbf{1} + \mathbf{O}$.

The *wedge product* of $\mathbf{a}, \mathbf{b}$ is defined by

$$\mathbf{a} \wedge \mathbf{b} := \mathbf{a} \otimes \mathbf{b} - \mathbf{b} \otimes \mathbf{a}. \tag{A.8.21}$$

Exercise A.8.4. Show that $\mathbf{a} \wedge \mathbf{b} \in \operatorname{Sk} \mathcal{V}$ via (A.8.10) and [see (A.5.17)]

$$(\mathbf{a} \wedge \mathbf{b})\mathbf{v} = -(\mathbf{a} \times \mathbf{b}) \times \mathbf{v}. \tag{A.8.22}$$

A.9 Orthonormal Bases and Matrix Representation of Vectors and Linear Transformations

Suppose that $\{\mathbf{e}_1, \mathbf{e}_2, \mathbf{e}_3\}$ is an orthonormal basis [see (A.6.3)] and $\mathbf{L} \in \operatorname{Lin} \mathcal{V}$. Then (see Result A.7.1) $\mathbf{L}$ is uniquely determined by its action on this basis, namely the vectors $\mathbf{L}\mathbf{e}_1$, $\mathbf{L}\mathbf{e}_2$, and $\mathbf{L}\mathbf{e}_3$. Each of these three vectors is determined by its components. For example [see (A.6.6) with $\mathbf{v} = \mathbf{L}\mathbf{e}_1$],

$$\mathbf{L}\mathbf{e}_1 = (\mathbf{e}_1 . \mathbf{L}\mathbf{e}_1)\mathbf{e}_1 + (\mathbf{e}_2 . \mathbf{L}\mathbf{e}_1)\mathbf{e}_2 + (\mathbf{e}_3 . \mathbf{L}\mathbf{e}_1)\mathbf{e}_3.$$

Thus $\mathbf{L}$ is uniquely characterised by the nine numbers $(i,j = 1,2,3)$

$$L_{ij} := \mathbf{e}_i \,.\, \mathbf{L}\mathbf{e}_j. \tag{A.9.1}$$

If $\mathbf{v} \in \mathcal{V}$, then the $\mathbf{e}_i$ component of $\mathbf{Lv}$ is [see (A.6.5)]

$$\begin{aligned} \mathbf{e}_i \,.\, \mathbf{Lv} &= \mathbf{e}_i \,.\, \{\mathbf{L}(v_j \mathbf{e}_j)\} \\ &= v_j(\mathbf{e}_i \,.\, \mathbf{L}\mathbf{e}_j) = v_j L_{ij} = L_{ij} v_j. \end{aligned} \tag{A.9.2}$$

[Of course, we are here using the summation convention introduced in (A.6.7) et seq.] That is,

$$\mathbf{Lv} = (\mathbf{e}_i \,.\, \mathbf{Lv})\mathbf{e}_i = L_{ij} v_j \mathbf{e}_i. \tag{A.9.3}$$

Thus, if we represent $\mathbf{e}_1$, $\mathbf{e}_2$, $\mathbf{e}_3$, and $\mathbf{v}$ by column matrices

$$\begin{bmatrix} 1 \\ 0 \\ 0 \end{bmatrix}, \begin{bmatrix} 0 \\ 1 \\ 0 \end{bmatrix}, \begin{bmatrix} 0 \\ 0 \\ 1 \end{bmatrix}, \quad \text{and} \quad \begin{bmatrix} v_1 \\ v_2 \\ v_3 \end{bmatrix},$$

respectively, then the column matrix which represents $\mathbf{Lv}$ is, from (A.9.3),

$$\begin{bmatrix} \mathbf{Lv} \,.\, \mathbf{e}_1 \\ \mathbf{Lv} \,.\, \mathbf{e}_2 \\ \mathbf{Lv} \,.\, \mathbf{e}_3 \end{bmatrix} = \begin{bmatrix} L_{11}v_1 + L_{12}v_2 + L_{13}v_3 \\ L_{21}v_1 + L_{22}v_2 + L_{23}v_3 \\ L_{31}v_1 + L_{32}v_2 + L_{33}v_3 \end{bmatrix} = \begin{bmatrix} L_{11} & L_{12} & L_{13} \\ L_{21} & L_{22} & L_{23} \\ L_{31} & L_{32} & L_{33} \end{bmatrix} \begin{bmatrix} v_1 \\ v_2 \\ v_3 \end{bmatrix}. \tag{A.9.4}$$

Accordingly we note that

$$\text{the square matrix} \begin{bmatrix} L_{11} & L_{12} & L_{13} \\ L_{21} & L_{22} & L_{23} \\ L_{31} & L_{32} & L_{33} \end{bmatrix} \text{represents } \mathbf{L}. \tag{A.9.5}$$

Thus the action of $\mathbf{L}$ upon $\mathbf{v}$ which yields $\mathbf{Lv}$ is mirrored by the product of matrix $[L_{ij}]$ with column matrix $[v_1 \; v_2 \; v_3]^T$ (the transpose of row *matrix* $[v_1 \; v_2 \; v_3]$)). It is a simple matter to see that $[\alpha v_1 \; \alpha v_2 \; \alpha v_3]^T$ represents $\alpha\mathbf{v}$ and $[\alpha L_{ij}]$ represents $\alpha\mathbf{L}$ for any $\alpha \in \mathbb{R}$. Likewise, if $\mathbf{L}, \mathbf{M} \in \operatorname{Lin}\mathcal{V}$, then the matrix which represents $\mathbf{L} + \mathbf{M}$ is $[L_{ij} + M_{ij}] = [L_{ij}] + [M_{ij}]$, the sum of the matrices which represent $\mathbf{L}$ and $\mathbf{M}$.

From (A.9.1) and (A.8.1),

$$(L^T)_{ij} := \mathbf{e}_i \,.\, \mathbf{L}^T \mathbf{e}_j = \mathbf{L}\mathbf{e}_i \,.\, \mathbf{e}_j = \mathbf{e}_j \,.\, \mathbf{L}\mathbf{e}_i = L_{ji}. \tag{A.9.6}$$

That is, *the matrix which represents the transpose of* $\mathbf{L} \in \operatorname{Lin}\mathcal{V}$ *is the transpose of the matrix which represents* $\mathbf{L}$, and the two senses in which '*transpose*' is used have been shown to be naturally related.

Exercise A.9.1. Show that the matrix which represents a symmetric (skew) linear transformation is a symmetric (skew) matrix.

Remark A.9.1. While matrix representations are useful, it is important to note that changing from one orthonormal basis to another will change the matrix representation of $\mathbf{L} \in \operatorname{Lin}\mathcal{V}$ in general and that, since there is no natural choice of orthonormal basis, there is no natural choice of matrix. Accordingly, in dealing with linear transformations such as stress tensor values, it is more natural to work directly with these values rather than representations. Said differently, use of representations introduces considerations not directly related to physical quantities which have the character of vectors or linear transformations.

Consider the linear transformation $L_{pq}\mathbf{e}_p \otimes \mathbf{e}_q$. (Notice in our use of the summation convention that symbols p and q are dummy suffices and could carry any pair of different labels.) Now

$$\mathbf{e}_i \,.\, (L_{pq}\mathbf{e}_p \otimes \mathbf{e}_q)\mathbf{e}_j = L_{pq}(\mathbf{e}_i \,.\, \mathbf{e}_p)(\mathbf{e}_q \,.\, \mathbf{e}_j)$$
$$= L_{pq}\delta_{ip}\delta_{qj} = L_{ij} = \mathbf{e}_i \,.\, \mathbf{L}\mathbf{e}_j. \tag{A.9.7}$$

It follows that

$$\mathbf{L} = L_{pq}\,\mathbf{e}_p \otimes \mathbf{e}_q, \tag{A.9.8}$$

and hence that any $\mathbf{L} \in \text{Lin}\,\mathcal{V}$ is expressible in terms of $\{\mathbf{e}_p \otimes \mathbf{e}_q\}$, where $p,q = 1,2,3$. That is, the set $\{\mathbf{e}_p \otimes \mathbf{e}_q\}$ *spans* Lin $\mathcal{V}$. In fact, $\{\mathbf{e}_p \otimes \mathbf{e}_q\}$ are linearly independent. To see this consider

$$\alpha_{pq}\mathbf{e}_p \otimes \mathbf{e}_q = \mathbf{O}. \tag{A.9.9}$$

$$\therefore \quad \mathbf{e}_i \,.\, (\alpha_{pq}\mathbf{e}_p \otimes \mathbf{e}_q)\mathbf{e}_j = \mathbf{e}_i \,.\, \mathbf{O}\mathbf{e}_j = \mathbf{e}_i \,.\, \mathbf{0} = 0,$$

for any $i,j = 1,2,3$. That is,

$$\alpha_{pq}(\mathbf{e}_i \,.\, \mathbf{e}_p)(\mathbf{e}_q \,.\, \mathbf{e}_j) = 0,$$

whence

$$0 = \alpha_{pq}\,\delta_{ip}\,\delta_{qj} = \alpha_{ij} \tag{A.9.10}$$

and so $\{\mathbf{e}_p \otimes \mathbf{e}_q\}$ are linearly independent. Accordingly, from (A.9.8), (A.9.9), and (A.9.10),

$$\{\mathbf{e}_p \otimes \mathbf{e}_q\} \qquad \text{is a basis for} \qquad \text{Lin}\,\mathcal{V}, \tag{A.9.11}$$

and

$$\text{the dimension of Lin}\,\mathcal{V} \text{ is } 9. \tag{A.9.12}$$

If $\mathbf{L},\mathbf{M} \in \text{Lin}\,\mathcal{V}$, then

$$\mathbf{LM} = (\mathbf{LM})_{ij}\,\mathbf{e}_i \otimes \mathbf{e}_j. \tag{A.9.13}$$

On the other hand,

$$\mathbf{LM} = (L_{ip}\,\mathbf{e}_i \otimes \mathbf{e}_p)(M_{qj}\,\mathbf{e}_q \otimes \mathbf{e}_j)$$
$$= L_{ip}M_{qj}(\mathbf{e}_i \otimes \mathbf{e}_p)(\mathbf{e}_q \otimes \mathbf{e}_j) = L_{ip}M_{qj}(\mathbf{e}_p \,.\, \mathbf{e}_q)\mathbf{e}_i \otimes \mathbf{e}_j = L_{ip}M_{qj}\delta_{pq}\mathbf{e}_i \otimes \mathbf{e}_j. \tag{A.9.14}$$

Since $\{\mathbf{e}_i \otimes \mathbf{e}_j\}$ is a basis for Lin $\mathcal{V}$, the components of $\mathbf{LM}$ with respect to this basis are unique. Thus, from (A.9.13) and (A.9.14),

$$(\mathbf{LM})_{ij} = L_{ip}M_{pj}. \tag{A.9.15}$$

That is, the matrix which represents the composition of two linear transformations (with respect to any given orthonormal basis) is the product of the representations of the individual transformations (in the same order).

Exercise A.9.2. What are the matrix representations of $\mathbf{e}_1 \otimes \mathbf{e}_1, \mathbf{e}_1 \otimes \mathbf{e}_2$, etc. with respect to basis $\{\mathbf{e}_p \otimes \mathbf{e}_q\}$? Show that

$$\mathbf{e}_1 \otimes \mathbf{e}_1 + \mathbf{e}_2 \otimes \mathbf{e}_2 + \mathbf{e}_3 \otimes \mathbf{e}_3 = \mathbf{1}. \tag{A.9.16}$$

Note that for any $\mathbf{v} \in \mathcal{V}$,

$$(\mathbf{e}_1 \otimes \mathbf{e}_1)\mathbf{v} = v_1 \mathbf{e}_1. \tag{A.9.17}$$

That is, $\mathbf{e}_1 \otimes \mathbf{e}_1$ singles out what might be called the *vector component of* $\mathbf{v}$ *associated with the direction defined by* $\mathbf{e}_1$. In fact,

$$\mathbf{P}_1 := \mathbf{e}_1 \otimes \mathbf{e}_1 \tag{A.9.18}$$

is termed the *perpendicular projection of* $\mathcal{V}$ *upon the space of vectors spanned by* $\mathbf{e}_1$ (i.e., vectors of form $\alpha \mathbf{e}_1$ with $\alpha \in \mathbb{R}$) and similarly for $\mathbf{P}_2$ and $\mathbf{P}_3$.

Exercise A.9.3. Show that

$$\mathbf{P}_i^2 = \mathbf{P}_i \qquad \text{and} \qquad \mathbf{P}_i \mathbf{P}_j = \mathbf{O} \qquad \text{if} \quad i \neq j. \tag{A.9.19}$$

(Here $\mathbf{P}_i^2$ denotes $\mathbf{P}_i \mathbf{P}_i$.) Show further that

$$(\mathbf{P}_1 + \mathbf{P}_2)^2 = \mathbf{P}_1 + \mathbf{P}_2. \tag{A.9.20}$$

Describe the result of $\mathbf{P}_1 + \mathbf{P}_2$ acting upon $\mathbf{v} \in \mathcal{V}$.

Exercise A.9.4. Recalling (A.8.11), show that if $\{\mathbf{e}_1, \mathbf{e}_2, \mathbf{e}_3\}$ is an orthonormal basis and $\mathbf{L}_1 := \mathbf{e}_1 \otimes \mathbf{e}_2$, $\mathbf{L}_2 := \mathbf{e}_2 \otimes \mathbf{e}_3$, then

$$\mathbf{L}_1 \mathbf{L}_2 = \mathbf{e}_1 \otimes \mathbf{e}_3 \qquad \text{and} \qquad \mathbf{L}_2 \mathbf{L}_1 = \mathbf{O}.$$

(This shows not only that $\mathbf{L}_1$ and $\mathbf{L}_2$ do not commute but also that $\mathbf{L}_2 \mathbf{L}_1 = \mathbf{O}$, although $\mathbf{L}_2 \neq \mathbf{O}$ and $\mathbf{L}_1 \neq \mathbf{O}$.)

A.10 Invertibility

A linear transformation $\mathbf{L}$ is termed *invertible* if it is bijective (i.e., 1:1) and surjective (i.e., 'onto'). Specifically,

(i) if $\mathbf{L}\mathbf{u}_1 = \mathbf{L}\mathbf{u}_2$, then

$$\mathbf{u}_1 = \mathbf{u}_2 \qquad \text{(bijectivity) and ,} \tag{A.10.1}$$

(ii) if $\mathbf{v} \in \mathcal{V}$, then there exists a vector $\mathbf{u} \in \mathcal{V}$ such that

$$\mathbf{L}\mathbf{u} = \mathbf{v} \qquad \text{(surjectivity).} \tag{A.10.2}$$

In fact only one of (i) and (ii) is necessary: each implies the other. Indeed, each of the following is sufficient to ensure that $\mathbf{L} \in \operatorname{Lin} \mathcal{V}$ is invertible:

Inv 1. $\mathbf{L}$ is bijective. (A.10.3)

Inv 2. $\mathbf{L}\mathbf{u} = \mathbf{0}$ implies $\mathbf{u} = \mathbf{0}$. (A.10.4)

Inv 3. $\mathbf{L}$ preserves bases. (A.10.5)

Inv 4. $\mathbf{L}$ is surjective. (A.10.6)

Proofs: If $\mathbf{L}$ is bijective, then $\mathbf{Lu} = \mathbf{0} = \mathbf{L0}$ implies that $\mathbf{u} = \mathbf{0}$, and so Inv 1 $\Rightarrow$ Inv 2. Now suppose that $\{\mathbf{u}_1, \mathbf{u}_2, \mathbf{u}_3\}$ is a basis for $\mathcal{V}$, and consider $\mathbf{Lu}_1, \mathbf{Lu}_2, \mathbf{Lu}_3$. To check whether this is a basis, we consider $\alpha_1 \mathbf{Lu}_1 + \alpha_2 \mathbf{Lu}_2 + \alpha_3 \mathbf{Lu}_3 = \mathbf{0}$ (and hope to show that this implies $\alpha_1 = \alpha_2 = \alpha_3 = 0$). Equivalently, $\mathbf{L}(\alpha_1 \mathbf{u}_1 + \alpha_2 \mathbf{u}_2 + \alpha_3 \mathbf{u}_3) = \mathbf{0}$, which by Inv 2 implies that $\alpha_1 \mathbf{u}_1 + \alpha_2 \mathbf{u}_2 + \alpha_3 \mathbf{u}_3 = \mathbf{0}$ and hence $\alpha_1 = \alpha_2 = \alpha_3 = 0$ since $\mathbf{u}_1$, $\mathbf{u}_2$, and $\mathbf{u}_3$ are linearly independent. Thus $\{\mathbf{Lu}_1, \mathbf{Lu}_2, \mathbf{Lu}_3\}$ is a basis for $\mathcal{V}$, and hence Inv 1 $\Rightarrow$ Inv 3. Now suppose that $\{\mathbf{u}_1, \mathbf{u}_2, \mathbf{u}_3\}$ is a basis, and Inv 3 holds. Thus $\{\mathbf{Lu}_1, \mathbf{Lu}_2, \mathbf{Lu}_3\}$ is a basis, and hence any $\mathbf{v} \in \mathcal{V}$ can be expressed in the form

$$\mathbf{v} = \beta_1 \mathbf{Lu}_1 + \beta_2 \mathbf{Lu}_2 + \beta_3 \mathbf{Lu}_3 = \mathbf{L}(\beta_1 \mathbf{u}_1 + \beta_2 \mathbf{u}_2 + \beta_3 \mathbf{u}_3) = \mathbf{Lu},$$

where $\mathbf{u} := \beta_1 \mathbf{u}_1 + \beta_2 \mathbf{u}_2 + \beta_3 \mathbf{u}_3$. Thus $\mathbf{L}$ is surjective, and so Inv 3 $\Rightarrow$ Inv 4. At this point we have proved Inv 1 $\Rightarrow$ Inv 2 $\Rightarrow$ Inv 3 $\Rightarrow$ Inv 4. Since Inv 2 $\Rightarrow$ Inv 1 [with $\mathbf{u} := \mathbf{u}_1 - \mathbf{u}_2$, Inv 2 yields $\mathbf{L}(\mathbf{u}_1 - \mathbf{u}_2) = \mathbf{0}$ implies $\mathbf{u}_1 - \mathbf{u}_2 = \mathbf{0}$, and hence $\mathbf{Lu}_1 - \mathbf{Lu}_2 = \mathbf{0}$ implies $\mathbf{u}_1 = \mathbf{u}_2$], it suffices to show Inv 4 $\Rightarrow$ Inv 2. If $\{\mathbf{v}_1, \mathbf{v}_2, \mathbf{v}_3\}$ is a basis for $\mathcal{V}$, then Inv 4 yields the existence of vectors $\mathbf{u}_1$, $\mathbf{u}_2$, and $\mathbf{u}_3$ for which $\mathbf{Lu}_i = \mathbf{v}_i$ $(i = 1,2,3)$. If $\alpha_1 \mathbf{u}_1 + \alpha_2 \mathbf{u}_2 + \alpha_3 \mathbf{u}_3 = \mathbf{0}$, then $\mathbf{L}(\alpha_1 \mathbf{u}_1 + \alpha_2 \mathbf{u}_2 + \alpha_3 \mathbf{u}_3) = \mathbf{L0} = \mathbf{0}$ so that $\alpha_1 \mathbf{Lu}_1 + \alpha_2 \mathbf{Lu}_2 + \alpha_3 \mathbf{Lu}_3 = \mathbf{0}$, namely $\alpha_1 \mathbf{v}_1 + \alpha_2 \mathbf{v}_2 + \alpha_3 \mathbf{v}_3 = \mathbf{0}$. Since $\{\mathbf{v}_i\}$ is a basis, $\alpha_1 = \alpha_2 = \alpha_3 = 0$, and hence $\mathbf{u}_1$, $\mathbf{u}_2$, and $\mathbf{u}_3$ are linearly independent and thus form a basis. Now suppose that $\mathbf{Lu} = \mathbf{0}$. Since $\{\mathbf{u}_i\}$ is a basis, $\mathbf{u} = \beta_1 \mathbf{u}_1 + \beta_2 \mathbf{u}_2 + \beta_3 \mathbf{u}_3$ for some numbers β_1, β_2, and β_3. Thus $\mathbf{0} = \mathbf{Lu} = \mathbf{L}(\beta_1 \mathbf{u}_1 + \beta_2 \mathbf{u}_2 + \beta_3 \mathbf{u}_3) = \beta_1 \mathbf{v}_1 + \beta_2 \mathbf{v}_2 + \beta_3 \mathbf{v}_3$. However, $\{\mathbf{v}_i\}$ is a basis, and hence $\beta_1 = \beta_2 = \beta_3 = 0$. Thus $\mathbf{u} = \mathbf{0}$, and accordingly Inv 4 $\Rightarrow$ Inv 2.

Remark A.10.1. From Inv 2 we can deduce that *if* $\mathbf{L}$ *is not invertible, then there exists a non-zero vector* $\mathbf{u}$ *such that* $\mathbf{Lu} = \mathbf{0}$.

While any one of Inv 1 through Inv 4 can be used to decide whether or not $\mathbf{L} \in \operatorname{Lin} \mathcal{V}$ is invertible, none is particularly practicable for this purpose.

Problem: Is there a simple criterion to decide whether or not $\mathbf{L} \in \operatorname{Lin} \mathcal{V}$ is invertible?

Solution: Recall (A.5.15), namely that three vectors $\mathbf{u}, \mathbf{v}$, and $\mathbf{w}$ are linearly independent if and only if $\mathbf{u} \times \mathbf{v} . \mathbf{w} \neq 0$. Now suppose that $\mathbf{L}$ is invertible. By Inv 3 it follows that if $\mathbf{u}$, $\mathbf{v}$, and $\mathbf{w}$ are linearly independent, then so are $\mathbf{Lu}, \mathbf{Lv}$, and $\mathbf{Lw}$. Hence $\mathbf{Lu} \times \mathbf{Lv} . \mathbf{Lw} \neq 0$. Choosing a right-handed orthonormal basis $\mathbf{u} = \mathbf{e}_1, \mathbf{v} = \mathbf{e}_2$ and $\mathbf{w} = \mathbf{e}_3$, and recalling (A.6.13), this translates [noting that $(\mathbf{Lu})_1 = (\mathbf{Le}_1)_1 = \mathbf{Le}_1 . \mathbf{e}_1 = L_{11}$, etc.] into

$$\mathbf{Le}_1 \times \mathbf{Le}_2 . \mathbf{Le}_3 = \begin{vmatrix} \mathbf{Le}_1 . \mathbf{e}_1 & \mathbf{Le}_1 . \mathbf{e}_2 & \mathbf{Le}_1 . \mathbf{e}_3 \\ \mathbf{Le}_2 . \mathbf{e}_1 & \mathbf{Le}_2 . \mathbf{e}_2 & \mathbf{Le}_2 . \mathbf{e}_3 \\ \mathbf{Le}_3 . \mathbf{e}_1 & \mathbf{Le}_3 . \mathbf{e}_2 & \mathbf{Le}_3 . \mathbf{e}_3 \end{vmatrix} = \begin{vmatrix} L_{11} & L_{12} & L_{13} \\ L_{21} & L_{22} & L_{23} \\ L_{31} & L_{32} & L_{33} \end{vmatrix} \neq 0. \tag{A.10.7}$$

That is, if $\mathbf{L}$ is invertible, then the determinant det $[L_{ij}]$ of any matrix representation $[L_{ij}]$ of $\mathbf{L}$ is non-zero. On the other hand, if $\mathbf{L}$ is *not* invertible, then by Inv 2 there must be a vector $\mathbf{u} \neq \mathbf{0}$ such that $\mathbf{Lu} = \mathbf{0}$. Writing $\mathbf{u}$ in terms of the orthonormal basis $\{\mathbf{e}_1, \mathbf{e}_2, \mathbf{e}_3\}$ yields $\mathbf{L}(u_1 \mathbf{e}_1 + u_2 \mathbf{e}_2 + u_3 \mathbf{e}_3) = \mathbf{L}u_i \mathbf{e}_i = u_i \mathbf{Le}_i = \mathbf{0}$, where not all u_1, u_2, and u_3 are zero. Thus the foregoing final equality indicates that $\mathbf{Le}_1, \mathbf{Le}_2$, and $\mathbf{Le}_3$ are linearly dependent, and hence $\mathbf{Le}_1 \times \mathbf{Le}_2 . \mathbf{Le}_3 = 0$. However, as above, $\mathbf{Le}_1 \times \mathbf{Le}_2 . \mathbf{Le}_3 = \det [L_{ij}]$, so non-invertibility of $\mathbf{L}$ implies det $[L_{ij}] = 0$. Accordingly we have

Inv 5. **L** *is invertible if and only if the determinant of any matrix representation* $[L_{ij}]$ *of* **L** *is non-zero.* (A.10.8)

If $\{\mathbf{e}_1, \mathbf{e}_2, \mathbf{e}_3\}$ is a right-handed orthonormal basis (i.e., $\mathbf{e}_1 \times \mathbf{e}_2 = \mathbf{e}_3$), then from (A.10.7), noting $\mathbf{e}_1 \times \mathbf{e}_2 . \mathbf{e}_3 = 1$,

$$\det[L_{ij}] = \frac{(\mathbf{Le}_1 \times \mathbf{Le}_2) . \mathbf{Le}_3}{\mathbf{e}_1 \times \mathbf{e}_2 . \mathbf{e}_3}. \tag{A.10.9}$$

Remark A.10.2 At this point the number $\det[L_{ij}]$ might be expected to depend upon the choice of orthonormal basis. In the next section we shall *define* the determinant $\det \mathbf{L}$ of $\mathbf{L} \in \operatorname{Lin} \mathcal{V}$, quite independently of the choice of basis, in terms of alternating trilinear forms (of which the triple scalar product is an example). In so doing it will prove possible to define two other scalar invariants[10] associated with **L** which characterise its structure.

If **L** is invertible and

$$\mathbf{Lu} = \mathbf{v}, \tag{A.10.10}$$

we write

$$\mathbf{v} = \mathbf{L}^{-1}\mathbf{u}, \tag{A.10.11}$$

and define

$$\operatorname{Invlin} \mathcal{V} := \{\mathbf{L} : \mathbf{L} \in \operatorname{Lin} \mathcal{V},\ \mathbf{L} \text{ is invertible}\}. \tag{A.10.12}$$

Some properties of $\operatorname{Invlin} \mathcal{V}$ are

Inv 6. $\mathbf{L}^{-1} \in \operatorname{Lin} \mathcal{V}$, (A.10.13)

Inv 7. $\mathbf{L}^{-1}\mathbf{L} = \mathbf{1} = \mathbf{L}\mathbf{L}^{-1}$, (A.10.14)

Inv 8. $\mathbf{1}^{-1} = \mathbf{1}$ and $(\mathbf{L}^{-1})^{-1} = \mathbf{L}$, (A.10.15)

Inv 9. $(\alpha\mathbf{L})^{-1} = \alpha^{-1}\mathbf{L}^{-1}$ if $\alpha \neq 0$, (A.10.16)

and

Inv 10. *If* $\mathbf{L}_1, \mathbf{L}_2 \in \operatorname{Invlin} \mathcal{V}$, then $\mathbf{L}_1\mathbf{L}_2 \in \operatorname{Invlin} \mathcal{V}$ and $(\mathbf{L}_1\mathbf{L}_2)^{-1} = \mathbf{L}_2^{-1}\mathbf{L}_1^{-1}$. (A.10.13)

Proofs: We first observe that by definition $\mathbf{L}^{-1}$ is unique: since $\mathbf{L} : \mathcal{V} \to \mathcal{V}$ must be 1:1 and 'onto', the same is true of $\mathbf{L}^{-1}$.

Inv 6: If $\mathbf{v}_1, \mathbf{v}_2 \in \mathcal{V}$ and $\mathbf{Lu}_1 = \mathbf{v}_1$, $\mathbf{Lu}_2 = \mathbf{v}_2$, then

$$\mathbf{L}(\alpha_1\mathbf{u}_1 + \alpha_2\mathbf{u}_2) = \alpha_1\mathbf{Lu}_1 + \alpha_2\mathbf{Lu}_2 = \alpha_2\mathbf{v}_1 + \alpha_2\mathbf{v}_2.$$

[10] Invariant in the sense of basis-independent.

This implies that $\mathbf{L}^{-1}(\alpha_1\mathbf{v}_1 + \alpha_2\mathbf{v}_2) = \alpha_1\mathbf{u}_1 + \alpha_2\mathbf{u}_2 = \alpha_1\mathbf{L}^{-1}\mathbf{v}_1 + \alpha_2\mathbf{L}^{-1}\mathbf{v}_2$, and so $\mathbf{L}^{-1} \in \mathrm{Invlin}\,\mathcal{V}$.

Inv 7: For any $\mathbf{u} \in \mathcal{V}, \mathbf{v} := \mathbf{L}\mathbf{u}$ implies that $\mathbf{u} = \mathbf{L}^{-1}\mathbf{v}$. Thus

$$\mathbf{L}^{-1}\mathbf{L}\mathbf{u} = \mathbf{L}^{-1}\mathbf{v} = \mathbf{u} = \mathbf{1}\mathbf{u}, \qquad \text{so} \quad \mathbf{L}^{-1}\mathbf{L} = \mathbf{1}.$$

Also, for any $\mathbf{v} \in \mathcal{V}$ there exists a $\mathbf{u} \in \mathcal{V}$ such that

$$\mathbf{L}^{-1}\mathbf{v} = \mathbf{u} \qquad \text{and hence} \qquad \mathbf{L}\mathbf{u} = \mathbf{v}.$$

Thus

$$\mathbf{L}\mathbf{L}^{-1}\mathbf{v} = \mathbf{L}\mathbf{u} = \mathbf{v} = \mathbf{1}\mathbf{v}, \qquad \text{so} \quad \mathbf{L}\mathbf{L}^{-1} = \mathbf{1}.$$

$(\text{Inv } 8)_1$ is trivial and $(\text{Inv } 8)_2$ follows from Inv 7. Property Inv 9 holds since

$$(\alpha^{-1}\mathbf{L}^{-1})(\alpha\mathbf{L})\mathbf{u} = \alpha^{-1}\alpha\mathbf{L}^{-1}\mathbf{L}\mathbf{u} = \mathbf{u} = \mathbf{1}\mathbf{u} \;\text{ for all }\; \mathbf{u} \in \mathcal{V}.$$

Similarly, Inv 10 follows by noting that for all $\mathbf{u} \in \mathcal{V}$

$$\begin{aligned}(\mathbf{L}_2^{-1}\mathbf{L}_1^{-1})(\mathbf{L}_1\mathbf{L}_2\mathbf{u}) &= \mathbf{L}_2^{-1}(\mathbf{L}_1^{-1}\mathbf{L}_1)\mathbf{L}_2\mathbf{u} = \mathbf{L}_2^{-1}\mathbf{1}\mathbf{L}_2\mathbf{u} \\ &= \mathbf{L}_2^{-1}\mathbf{L}_2\mathbf{u} = \mathbf{u} = \mathbf{1}\mathbf{u}.\end{aligned}$$

Remark A.10.3. If $\mathbf{L}_1, \mathbf{L}_2 \in \mathrm{Lin}\,\mathcal{V}$, then *in general* (cf. Exercise A.9.4)

$$\mathbf{L}_1\mathbf{L}_2 \neq \mathbf{L}_2\mathbf{L}_1. \tag{A.10.18}$$

(If $\mathbf{L}_1\mathbf{L}_2 = \mathbf{L}_2\mathbf{L}_1$, then $\mathbf{L}_1$ and $\mathbf{L}_2$ are said to *commute*.)

Exercise A.10.1. Use $\mathbf{L}\mathbf{L}^{-1} = \mathbf{1}$, (A.8.3), and (A.8.7) to show that

$$(\mathbf{L}^{-1})^T = (\mathbf{L}^T)^{-1}. \tag{A.10.19}$$

We thus denote either side of (A.10.19) by $\mathbf{L}^{-T}$ without ambiguity.

Properties $\mathbf{1} \in \mathrm{Invlin}\,\mathcal{V}$, Inv 6, and the observation that $\mathbf{L}_1\mathbf{L}_2 \in \mathrm{Invlin}\,\mathcal{V}$ whenever $\mathbf{L}_1, \mathbf{L}_2 \in \mathrm{Invlin}\,\mathcal{V}$ [see (A.10.13), (A.10.15), and (A.10.17)] establish (cf., e.g., Jacobson [84]) that

Inv 11. $\qquad$ Invlin $\mathcal{V}$ has the structure of a group. $\qquad$ (A.10.20)

(The group operation is that of composition of linear maps.)

A.11 Alternating Trilinear Forms on $\mathcal{V}$

A map

$$\omega : \mathcal{V} \times \mathcal{V} \times \mathcal{V} \longrightarrow \mathbb{R} \tag{A.11.1}$$

which satisfies (for all $\alpha_1, \alpha_2 \in \mathbb{R}$ and all $\mathbf{u}_1, \mathbf{u}_2, \mathbf{v}, \mathbf{w} \in \mathcal{V}$)

$$\omega(\alpha, \mathbf{u}_1 + \alpha_2\mathbf{u}_2, \mathbf{v}, \mathbf{w}) = \alpha_1\omega(\mathbf{u}_1, \mathbf{v}, \mathbf{w}) + \alpha_2\omega(\mathbf{u}_2, \mathbf{v}, \mathbf{w}) \tag{A.11.2}$$

and changes sign whenever two arguments are transposed, so that

$$\begin{aligned}\omega(\mathbf{u}, \mathbf{v}, \mathbf{w}) &= -\omega(\mathbf{v}, \mathbf{u}, \mathbf{w}) = \omega(\mathbf{v}, \mathbf{w}, \mathbf{u}) = -\omega(\mathbf{w}, \mathbf{v}, \mathbf{u}) \\ &= \omega(\mathbf{w}, \mathbf{u}, \mathbf{v}) = -\omega(\mathbf{u}, \mathbf{w}, \mathbf{v}),\end{aligned} \tag{A.11.3}$$

is termed an *alternating trilinear form on* $\mathcal{V}$. Here *alternating* refers to property (A.11.3), *form* to values lying in $\mathbb{R}$, and *trilinear* to property (A.11.2) which holds for each of the three argument slots. That is,

$$\omega(\mathbf{u},\alpha_1\mathbf{v}_1+\alpha_2\mathbf{v}_2,\mathbf{w})=\alpha_1\omega(\mathbf{u},\mathbf{v}_1,\mathbf{w})+\alpha_2\omega(\mathbf{u},\mathbf{v}_2,\mathbf{w}) \tag{A.11.4}$$

and

$$\omega(\mathbf{u},\mathbf{v},\alpha_1\mathbf{w}_1+\alpha_2\mathbf{w}_2)=\alpha_1\omega(\mathbf{u},\mathbf{v},\mathbf{w}_1)+\alpha_2\omega(\mathbf{u},\mathbf{v},\mathbf{w}_2). \tag{A.11.5}$$

Exercise A.11.1. Show that (A.11.2) and (A.11.3) together imply (A.11.4) and (A.11.5).

Exercise A.11.2. Show that

$$\omega(\mathbf{u},\mathbf{v},\mathbf{w}) := \mathbf{u}\times\mathbf{v}\,.\,\mathbf{w}$$

is an alternating trilinear form on $\mathcal{V}$.

The three-dimensionality of $\mathcal{V}$ together with trilinearity result in there being essentially only one non-zero alternating trilinear form, modulo multiplication by a non-zero real number. Specifically, we have

ATF 1. If $\{\mathbf{b}_1,\mathbf{b}_2,\mathbf{b}_3\}$ is a basis then ω is uniquely determined by the value of $\omega(\mathbf{b}_1,\mathbf{b}_2,\mathbf{b}_3)$, and

ATF 2. Given any two non-zero[11] forms ω_1 and ω_2, then $\boldsymbol{\omega}_1 = k\boldsymbol{\omega}_2$ for some $k \neq 0$.

To see ATF 1, suppose that $\{\mathbf{b}_1,\mathbf{b}_2,\mathbf{b}_3\}$ is a basis. Then

$$\omega(\mathbf{u},\mathbf{v},\mathbf{w})=\omega(u_i\mathbf{b}_i,v_j\mathbf{b}_j,w_k\mathbf{b}_k), \tag{A.11.6}$$

where summation convention is intended in each argument, and u_i, v_j, and w_k are the components of $\mathbf{u}$, $\mathbf{v}$, and $\mathbf{w}$ with respect to the chosen basis. Trilinearity [(A.11.2), (A.11.4), and (A.11.5)] implies from (A.11.6) that

$$\omega(\mathbf{u},\mathbf{v},\mathbf{w})=u_iv_jw_k\,\omega(\mathbf{b}_i,\mathbf{b}_j,\mathbf{b}_k). \tag{A.11.7}$$

The only non-zero terms in this triple sum correspond to terms for which (i,j,k) is a permutation of (1, 2, 3), since if any two of $\mathbf{b}_i$, and $\mathbf{b}_j$ are the same in any term of this sum, then the value of ω is zero [e.g., $\omega(\mathbf{b}_1,\mathbf{b}_1,\mathbf{b}_2)=0$ because interchanging the two $\mathbf{b}_1$'s changes the value of ω by factor -1 yet also leaves it unchanged!]. Accordingly

$$\omega(\mathbf{u},\mathbf{v},\mathbf{w})=u_iv_jw_k\epsilon_{ijk}\,\omega(\mathbf{b}_1,\mathbf{b}_2,\mathbf{b}_3). \tag{A.11.8}$$

Here $\epsilon_{ijk}=1$ or -1 according to whether (i,j,k) is an even or odd permutation of (1, 2, 3), respectively. Thus ATF 1 holds as a consequence of (A.11.8) and the uniqueness of components u_i, v_j, and w_k.

Given ω_α with $\omega_\alpha(\mathbf{b}_1,\mathbf{b}_2,\mathbf{b}_3)\neq 0$ $(\alpha=1,2)$, then clearly

$$\omega_1(\mathbf{b}_1,\mathbf{b}_2,\mathbf{b}_3)=k\omega_2(\mathbf{b}_1,\mathbf{b}_2,\mathbf{b}_3) \tag{A.11.9}$$

[11] ω is zero if $\omega(\mathbf{u},\mathbf{v},\mathbf{w})=0$ for all $\mathbf{u},\mathbf{v},\mathbf{w}\in\mathcal{V}$.

defines a number $k \neq 0$. (Why?) It follows that

$$\begin{aligned}\omega_1(\mathbf{u},\mathbf{v},\mathbf{w}) &= u_i v_j w_k \epsilon_{ijk}\,\omega_1(\mathbf{b}_1,\mathbf{b}_2,\mathbf{b}_3)\\ &= k u_i v_j w_k \epsilon_{ijk}\omega_2(\mathbf{b}_1,\mathbf{b}_2,\mathbf{b}_3)\\ &= k u_i v_j w_k\,\omega_2(\mathbf{b}_i,\mathbf{b}_j,\mathbf{b}_k)\\ &= k\omega_2(u_i\mathbf{b}_i, v_j\mathbf{b}_j, w_k\mathbf{b}_k) = k\omega_2(\mathbf{u},\mathbf{v},\mathbf{w}),\end{aligned} \tag{A.11.10}$$

and ATF 2. holds.

Further, we have

ATF 3. If ω is a non-zero trilinear alternating form in $\mathcal{V}$, then $\mathbf{u},\mathbf{v},\mathbf{w} \in \mathcal{V}$ are linearly independent if and only if $\omega(\mathbf{u},\mathbf{v},\mathbf{w}) \neq 0$. (A.11.11)

To see this, suppose that $\mathbf{u},\mathbf{v}$, and $\mathbf{w}$ are linearly dependent. Then, for some $\alpha,\beta,\gamma \in \mathbb{R}$, not all zero, $\alpha\mathbf{u}+\beta\mathbf{v}+\gamma\mathbf{w} = \mathbf{0}$, and so one of $\mathbf{u},\mathbf{v}$, or $\mathbf{w}$ is expressible in terms of the other two. Suppose without loss of generality that $\mathbf{w} = a\mathbf{u}+b\mathbf{v}$. Then

$$\begin{aligned}\omega(\mathbf{u},\mathbf{v},\mathbf{w}) &= \omega(\mathbf{u},\mathbf{v},a\mathbf{u}+b\mathbf{v}) = a\omega(\mathbf{u},\mathbf{v},\mathbf{u})+b\omega(\mathbf{u},\mathbf{v},\mathbf{v})\\ &= 0.\end{aligned}$$

Now suppose that $\mathbf{u},\mathbf{v}$, and $\mathbf{w}$ are linearly independent (and hence form a basis). Then, from ATF 1, ω is uniquely defined by the value of $\omega(\mathbf{u},\mathbf{v},\mathbf{w})$. If $\omega(\mathbf{u},\mathbf{v},\mathbf{w}) = 0$, then this implies that $\omega = 0$, a contradiction (since we assumed $\omega \neq 0$). Thus $\omega(\mathbf{u},\mathbf{v},\mathbf{w}) \neq 0$.

Remark A.11.1. Properties ATF 2 and ATF 3 are the basis of the remainder of our discussion of linear algebra associated with $\mathcal{V}$.

A.12 Principal Invariants of $\mathbf{L} \in \operatorname{Lin}\mathcal{V}$

A.12.1 The First Principal Invariant: $I_1(\mathbf{L}) = \operatorname{tr}\mathbf{L}$

If $\mathbf{L} \in \operatorname{Lin}\mathcal{V}$, consider, for any $\omega \neq 0$,

$$\omega_{\mathbf{L}}^{I}(\mathbf{u},\mathbf{v},\mathbf{w}) := \omega(\mathbf{L}\mathbf{u},\mathbf{v},\mathbf{w})+\omega(\mathbf{u},\mathbf{L}\mathbf{v},\mathbf{w})+\omega(\mathbf{u},\mathbf{v},\mathbf{L}\mathbf{w}). \tag{A.12.1}$$

It is a simple matter to verify that $\omega_{\mathbf{L}}^{I}$ is a trilinear alternating form. (Prove this assertion!). Accordingly, from ATF 2 there is a number $I_1(\mathbf{L})$ such that

$$\omega_{\mathbf{L}}^{I} = I_1(\mathbf{L})\omega. \tag{A.12.2}$$

$I_1(\mathbf{L})$ is termed the *first principal invariant* of $\mathbf{L}$ and is also known as the *trace* of $\mathbf{L}$, written as $\operatorname{tr}\mathbf{L}$. Thus

$$\omega_{\mathbf{L}}^{I} =: (\operatorname{tr}\mathbf{L})\omega. \tag{A.12.3}$$

Notice that this definition is independent of the choice of ω: had another non-zero selection ω' been made, then [by ATF 2] for some $k \in \mathbb{R}$, $\omega' = k\omega$, and multiplication of (A.12.1) by k would yield the same value of $I_1(\mathbf{L})$ via (A.12.2) when ω is replaced by ω'.

The trace function on $\mathcal{V}$ has simple properties. In particular, we have:

Tr 1. $$\operatorname{tr}\mathbf{1} = 3 \tag{A.12.4}$$

Tr 2. $$\operatorname{tr}(\alpha\mathbf{L}) = \alpha \operatorname{tr}\mathbf{L} \tag{A.12.5}$$

Tr 3. $$\operatorname{tr}(\mathbf{L}_1 + \mathbf{L}_2) = \operatorname{tr}\mathbf{L}_1 + \operatorname{tr}\mathbf{L}_2 \tag{A.12.6}$$

Exercise A.12.1. Prove Tr 1 through Tr 3 from (A.12.1) and properties of ω.

Further, with choice of orthonormal basis $\{\mathbf{e}_1, \mathbf{e}_2, \mathbf{e}_3\}$, we compute $\operatorname{tr}(\mathbf{a} \otimes \mathbf{b})$ by noting

$$\begin{aligned} \operatorname{tr}(\mathbf{a} \otimes \mathbf{b})\omega(\mathbf{e}_1, \mathbf{e}_2, \mathbf{e}_3) &= \omega((\mathbf{a} \otimes \mathbf{b})\mathbf{e}_1, \mathbf{e}_2, \mathbf{e}_3) \\ &\quad + \omega(\mathbf{e}_1, (\mathbf{a} \otimes \mathbf{b})\mathbf{e}_2, \mathbf{e}_3) + \omega(\mathbf{e}_1, \mathbf{e}_2, (\mathbf{a} \otimes \mathbf{b})\mathbf{e}_3) \\ &= (\mathbf{b} \,.\, \mathbf{e}_1)\omega(\mathbf{a}, \mathbf{e}_2, \mathbf{e}_3) + (\mathbf{b} \,.\, \mathbf{e}_2)\omega(\mathbf{e}_1, \mathbf{a}, \mathbf{e}_3) \\ &\quad + (\mathbf{b} \,.\, \mathbf{e}_3)\omega(\mathbf{e}_1, \mathbf{e}_2, \mathbf{a}). \end{aligned} \tag{A.12.7}$$

Now
$$\begin{aligned} \omega(\mathbf{a}, \mathbf{e}_2, \mathbf{e}_3) &= \omega(a_1\mathbf{e}_1 + a_2\mathbf{e}_2 + a_3\mathbf{e}_3, \mathbf{e}_2, \mathbf{e}_3) \\ &= a_1\omega(\mathbf{e}_1, \mathbf{e}_2, \mathbf{e}_3) + a_2\omega(\mathbf{e}_2, \mathbf{e}_2, \mathbf{e}_3) + a_3\omega(\mathbf{e}_3, \mathbf{e}_2, \mathbf{e}_3) \\ &= a_1\omega(\mathbf{e}_1, \mathbf{e}_2, \mathbf{e}_3). \qquad \text{(Why?)} \end{aligned} \tag{A.12.8}$$

Similarly $\omega(\mathbf{e}_1, \mathbf{a}, \mathbf{e}_3) = a_2\omega(\mathbf{e}_1, \mathbf{e}_2, \mathbf{e}_3)$ and $\omega(\mathbf{e}_1, \mathbf{e}_2, \mathbf{a}) = a_3\omega(\mathbf{e}_1, \mathbf{e}_2, \mathbf{e}_3)$. Thus, from (A.12.7), (A.12.8) et seq.,

$$\operatorname{tr}(\mathbf{a} \otimes \mathbf{b})\omega(\mathbf{e}_1, \mathbf{e}_2, \mathbf{e}_3) = (b_1a_1 + b_2a_2 + b_3a_3)\omega(\mathbf{e}_1, \mathbf{e}_2, \mathbf{e}_3),$$

and hence we have [see (A.6.8)]

Tr 4. $$\operatorname{tr}(\mathbf{a} \otimes \mathbf{b}) = \mathbf{a} \,.\, \mathbf{b} \;\; (= \operatorname{tr}(\mathbf{b} \otimes \mathbf{a})). \tag{A.12.9}$$

Linearity properties Tr 2 and Tr 3 together with (A.9.8) and Tr 4 imply that, for any $\mathbf{L} \in \operatorname{Lin}\mathcal{V}$,

$$\operatorname{tr}\mathbf{L} = \operatorname{tr}(L_{pq}\mathbf{e}_p \otimes \mathbf{e}_q) = L_{pq}\operatorname{tr}(\mathbf{e}_p \otimes \mathbf{e}_q) = L_{pq}\mathbf{e}_p \,.\, \mathbf{e}_q = L_{pq}\delta_{pq}. \tag{A.12.10}$$

Thus, for any matrix representation $[L_{ij}]$ of $\mathbf{L}$,

Tr 5. $$\operatorname{tr}\mathbf{L} = L_{11} + L_{22} + L_{33}, \tag{A.12.11}$$

and from (A.9.6),

Tr 6. $$\operatorname{tr}\mathbf{L}^T = \operatorname{tr}\mathbf{L}. \tag{A.12.12}$$

In particular, *the trace of a linear transformation is the sum of the diagonal elements of any matrix representation of* $\mathbf{L}$ (i.e., the trace of any such matrix).

Further, for any $\mathbf{a}, \mathbf{b}, \mathbf{c}, \mathbf{d} \in \mathcal{V}$,

$$\begin{aligned} \operatorname{tr}((\mathbf{a} \otimes \mathbf{b})(\mathbf{c} \otimes \mathbf{d})) &= \operatorname{tr}((\mathbf{b}\,.\,\mathbf{c})(\mathbf{a} \otimes \mathbf{d}) \\ &= (\mathbf{b}\,.\,\mathbf{c}) \operatorname{tr}(\mathbf{a} \otimes \mathbf{d}) = (\mathbf{b}\,.\,\mathbf{c})(\mathbf{a}\,.\,\mathbf{d}) \\ &= (\mathbf{a}\,.\,\mathbf{d}) \operatorname{tr}(\mathbf{c} \otimes \mathbf{b}) = \operatorname{tr}((\mathbf{a}\,.\,\mathbf{d})(\mathbf{c} \otimes \mathbf{b}) \\ &= \operatorname{tr}((\mathbf{c} \otimes \mathbf{d})(\mathbf{a} \otimes \mathbf{b})). \end{aligned} \tag{A.12.13}$$

Thus, if $\mathbf{A}, \mathbf{B} \in \operatorname{Lin} \mathcal{V}$, then, via Tr 2, Tr 3 and (A.9.8),

$$\begin{aligned} \operatorname{tr}(\mathbf{AB}) &= \operatorname{tr}(A_{pq}(\mathbf{e}_p \otimes \mathbf{e}_q) B_{rs}(\mathbf{e}_r \otimes \mathbf{e}_s)) \\ &= A_{pq} B_{rs} \operatorname{tr}((\mathbf{e}_p \otimes \mathbf{e}_q)(\mathbf{e}_r \otimes \mathbf{e}_s)) \\ &= A_{pq} B_{rs} \operatorname{tr}((\mathbf{e}_r \otimes \mathbf{e}_s)(\mathbf{e}_p \otimes \mathbf{e}_q) \qquad \text{(via (A.12.13))} \\ &= \operatorname{tr}(B_{rs}(\mathbf{e}_r \otimes \mathbf{e}_s) A_{pq}(\mathbf{e}_p \otimes \mathbf{e}_q)) = \operatorname{tr}(\mathbf{BA}). \end{aligned}$$

That is, we have

Tr 7. $$\operatorname{tr}(\mathbf{AB}) = \operatorname{tr}(\mathbf{BA}). \tag{A.12.14}$$

Exercise A.12.2. Prove Tr 7 directly from (A.12.11), recalling (A.9.15).

A.12.2. The Second Principal Invariant $I_2(\mathbf{L})$

Consider, for $\mathbf{L} \in \operatorname{Lin} \mathcal{V}$,

$$\omega_{\mathbf{L}}^{II}(\mathbf{u}, \mathbf{v}, \mathbf{w}) := \omega(\mathbf{u}, \mathbf{Lv}, \mathbf{Lw}) + \omega(\mathbf{Lu}, \mathbf{v}, \mathbf{Lw}) + \omega(\mathbf{Lu}, \mathbf{Lv}, \mathbf{w}). \tag{A.12.15}$$

Exercise A.12.3. Show that $\omega_{\mathbf{L}}^{II}$ is an alternating trilinear form.

From ATF 2 it follows that there is a number $I_2(\mathbf{L})$, termed the *second principal invariant* of $\mathbf{L}$, independent of the choice of non-zero form ω, such that

$$\omega_{\mathbf{L}}^{II} = I_2(\mathbf{L})\omega. \tag{A.12.16}$$

It turns out that $I_2(\mathbf{L})$ can be expressed in terms of trace operations. To this end consider, for any $\mathbf{u}, \mathbf{v}, \mathbf{w} \in \mathcal{V}$,

$$\begin{aligned} (\operatorname{tr}\mathbf{L})^2 \omega(\mathbf{u}, \mathbf{v}, \mathbf{w}) &= (\operatorname{tr}\mathbf{L})(\operatorname{tr}\mathbf{L})\omega(\mathbf{u}, \mathbf{v}, \mathbf{w}) \\ &= \operatorname{tr}\mathbf{L}\{\omega(\mathbf{Lu}, \mathbf{v}, \mathbf{w}) + \omega(\mathbf{u}, \mathbf{Lv}, \mathbf{w}) + \omega(\mathbf{u}, \mathbf{v}, \mathbf{Lw})\} \\ &= \omega(\mathbf{L}^2\mathbf{u}, \mathbf{v}, \mathbf{w}) + \omega(\mathbf{Lu}, \mathbf{Lv}, \mathbf{w}) + \omega(\mathbf{Lu}, \mathbf{v}, \mathbf{Lw}) \\ &\quad + \omega(\mathbf{Lu}, \mathbf{Lv}, \mathbf{w}) + \omega(\mathbf{u}, \mathbf{L}^2\mathbf{v}, \mathbf{w}) + \omega(\mathbf{u}, \mathbf{Lv}, \mathbf{Lw}) \\ &\quad + \omega(\mathbf{Lu}, \mathbf{v}, \mathbf{Lw}) + \omega(\mathbf{u}, \mathbf{Lv}, \mathbf{Lw}) + \omega(\mathbf{u}, \mathbf{v}, \mathbf{L}^2\mathbf{w}) \\ &= \omega(\mathbf{L}^2\mathbf{u}, \mathbf{v}, \mathbf{w}) + \omega(\mathbf{u}, \mathbf{L}^2\mathbf{v}, \mathbf{w}) + \omega(\mathbf{u}, \mathbf{v}, \mathbf{L}^2\mathbf{w}) + 2I_2(\mathbf{L})\omega(\mathbf{u}, \mathbf{v}, \mathbf{w}) \\ &= \{\operatorname{tr}(\mathbf{L}^2) + 2I_2(\mathbf{L})\}\omega(\mathbf{u}, \mathbf{v}, \mathbf{w}). \end{aligned}$$

Accordingly $$I_2(\mathbf{L}) = \frac{1}{2}\{(\operatorname{tr}\mathbf{L})^2 - \operatorname{tr}(\mathbf{L}^2)\}. \tag{A.12.17}$$

Exercise A.12.4. Using (A.12.12) and (A.8.7), show that

$$I_2(\mathbf{L}^T) = I_2(\mathbf{L}). \tag{A.12.18}$$

A.12.3. The Third Principal Invariant: $I_3(\mathbf{L}) = \det \mathbf{L}$

If $\mathbf{L} \in \operatorname{Lin} \mathcal{V}$, define

$$\omega_{\mathbf{L}}^{III}(\mathbf{u}, \mathbf{v}, \mathbf{w}) := \omega(\mathbf{Lu}, \mathbf{Lv}, \mathbf{Lw}). \tag{A.12.19}$$

Exercise A.12.5. Show that $\omega_{\mathbf{L}}^{III}$ is an alternating trilinear form.

It follows from ATF 2 that

$$\omega_{\mathbf{L}}^{III} = \mathbf{I}_3(\mathbf{L})\omega \tag{A.12.20}$$

for some number $I_3(\mathbf{L})$ which is independent of the choice of form ω. This number is termed the *third principal invariant*, or *determinant* det $\mathbf{L}$, of $\mathbf{L}$. That is,

$$\omega_{\mathbf{L}}^{III}(\mathbf{u}, \mathbf{v}, \mathbf{w}) = \omega(\mathbf{Lu}, \mathbf{Lv}, \mathbf{Lw}) = (\det \mathbf{L})\omega(\mathbf{u}, \mathbf{v}, \mathbf{w}). \tag{A.12.21}$$

Remark A.12.1. At this point one should realise that here the use of *determinant* refers directly to a linear transformation $\mathbf{L}$ and not to the determinant of a (square) matrix *representation* of $\mathbf{L}$ [see (A.9.5)]. As in the case of the trace operation, it is necessary to justify the dual use of this term.

Exercise A.12.6. Show from (A.12.21) that

Det 1. $$\det \mathbf{1} = 1, \tag{A.12.22}$$

and

Det 2. $$\det(\alpha \mathbf{1}) = \alpha^3 \qquad \text{for any } \alpha \in \mathbb{R}. \tag{A.12.23}$$

If $\{\mathbf{u}, \mathbf{v}, \mathbf{w}\}$ is a basis and $\mathbf{L}$ is invertible, then, from Inv 3 [see (A.10.5)], $\{\mathbf{Lu}, \mathbf{Lv}, \mathbf{Lw}\}$ is a basis. Thus, from ATF 1, for any $\omega \neq 0, \omega(\mathbf{u}, \mathbf{v}, \mathbf{w})$ and $\omega(\mathbf{Lu}, \mathbf{Lv}, \mathbf{Lw})$ must be non-zero. Hence, from (A.12.21), $\det \mathbf{L} \neq 0$. Conversely, if $\det \mathbf{L} \neq 0$, then from (A.12.21) $\omega(\mathbf{Lu}, \mathbf{Lv}, \mathbf{Lw}) \neq 0$, so $\mathbf{Lu}, \mathbf{Lv}$, and $\mathbf{Lw}$ must be linearly independent and hence span $\mathcal{V}$. Thus $\mathbf{L}$ is surjective and hence invertible via Inv 4. Accordingly we have [cf. (A.10.8)]

Det 3. $\mathbf{L} \in \operatorname{Lin} \mathcal{V}$ is invertible if and only if det $\mathbf{L} \neq 0$.

Equivalently,

Det 3′. $\mathbf{L} \in \operatorname{Lin} \mathcal{V}$ fails to be invertible if and only if det $\mathbf{L} = 0$.

Noting that $$\omega(\mathbf{u}, \mathbf{v}, \mathbf{w}) := \mathbf{u} \times \mathbf{v} \,.\, \mathbf{w} \tag{A.12.24}$$

is an alternating trilinear form (Convince yourself of this!), it follows from (A.12.21) that

Det 4. If $\mathbf{u}, \mathbf{v}$, and $\mathbf{w}$ are linearly independent, then

$$\det \mathbf{L} = \frac{\mathbf{Lu} \times \mathbf{Lv} \,.\, \mathbf{Lw}}{\mathbf{u} \times \mathbf{v} \,.\, \mathbf{w}}. \tag{A.12.25}$$

Comparison with (A.10.9) yields

Det 5. $$\det \mathbf{L} = \det[L_{ij}]. \tag{A.12.26}$$

That is, all matrix representatives of $\mathbf{L}$ have the same determinantal value, namely det $\mathbf{L}$.

Now consider the determinant of the composition of two linear transformations on $\mathcal{V}$. For any basis $\{\mathbf{u}, \mathbf{v}, \mathbf{w}\}$ and any $\omega \neq 0$,

$$\begin{aligned}\omega(\mathbf{L}_1\mathbf{L}_2\mathbf{u}, \mathbf{L}_1\mathbf{L}_2\mathbf{v}, \mathbf{L}_1\mathbf{L}_2\mathbf{w}) &= \omega(\mathbf{L}_1(\mathbf{L}_2\mathbf{u}), \mathbf{L}_1(\mathbf{L}_2\mathbf{v}), \mathbf{L}_1(\mathbf{L}_2\mathbf{w})) \\ &= (\det \mathbf{L}_1)\omega(\mathbf{L}_2\mathbf{u}, \mathbf{L}_2\mathbf{v}, \mathbf{L}_2\mathbf{w}) \\ &= (\det \mathbf{L}_1)(\det \mathbf{L}_2)\omega(\mathbf{u}, \mathbf{v}, \mathbf{w}).\end{aligned} \tag{A.12.27}$$

On the other hand,

$$\begin{aligned}\omega(\mathbf{L}_1\mathbf{L}_2\mathbf{u}, \mathbf{L}_1\mathbf{L}_2\mathbf{v}, \mathbf{L}_1\mathbf{L}_2\mathbf{w}) &= \omega((\mathbf{L}_1\mathbf{L}_2)\mathbf{u}, (\mathbf{L}_1\mathbf{L}_2)\mathbf{v}, (\mathbf{L}_1\mathbf{L}_2)\mathbf{w}) \\ &= \det(\mathbf{L}_1\mathbf{L}_2)\omega(\mathbf{u}, \mathbf{v}, \mathbf{w}).\end{aligned} \tag{A.12.28}$$

Comparison of (A.12.27) and (A.12.28) yields

Det 6. $$\det(\mathbf{L}_1\mathbf{L}_2) = (\det \mathbf{L}_1)(\det \mathbf{L}_2). \tag{A.12.29}$$

If $\mathbf{L}$ is invertible, then by Det 1 and Det 6,

$$1 = \det \mathbf{1} = \det(\mathbf{L}\mathbf{L}^{-1}) = (\det \mathbf{L})(\det \mathbf{L}^{-1}).$$

Hence

Det 7. $$\det(\mathbf{L}^{-1}) = (\det \mathbf{L})^{-1}. \tag{A.12.30}$$

Considering, in the manner of the derivation of (A.12.17), $(\operatorname{tr}\mathbf{L})^3\omega(\mathbf{u}, \mathbf{v}, \mathbf{w})$ and $(\operatorname{tr}\mathbf{L})(\operatorname{tr}\mathbf{L}^2)\omega(\mathbf{u}, \mathbf{v}, \mathbf{w})$, it can be shown that

Det 8. $$\det \mathbf{L} = \frac{1}{6}\{(\operatorname{tr}\mathbf{L})^3 - 3(\operatorname{tr}\mathbf{L})(\operatorname{tr}\mathbf{L}^2) + 2\operatorname{tr}\mathbf{L}^3\}. \tag{A.12.31}$$

Exercise A.12.7. Prove (A.12.31).

Exercise A.12.8. Using (A.12.12) and (A.8.7), show that

$$\operatorname{tr}\mathbf{L}^2 = \operatorname{tr}(\mathbf{L}^T)^2 \qquad \text{and} \qquad \operatorname{tr}\mathbf{L}^3 = \operatorname{tr}(\mathbf{L}^T)^3.$$

Deduce from (A.12.31) that

Det 9. $$\det \mathbf{L}^T = \det \mathbf{L}. \tag{A.12.32}$$

Remark A.12.2. From (A.12.12), (A.12.18), and (A.12.32), *the principal invariants of* $\mathbf{L}$ *and* $\mathbf{L}^T$ *are the same.*

We define

$$\mathrm{Invlin}^{+}\mathcal{V} := \{\mathbf{L} \in \mathrm{Lin}\,\mathcal{V} : \det \mathbf{L} > 0\}. \tag{A.12.33}$$

Exercise A.12.9. Show that

(i) if $\mathbf{L}_1, \mathbf{L}_2 \in \mathrm{Invlin}^{+}\mathcal{V}$, then $\mathbf{L}_1\mathbf{L}_2 \in \mathrm{Invlin}^{+}\mathcal{V}$, and
(ii) if $\mathbf{L} \in \mathrm{Invlin}^{+}\mathcal{V}$, then $\mathbf{L}^{-1} \in \mathrm{Invlin}^{+}\mathcal{V}$.

Since $\mathbf{1} \in \mathrm{Invlin}^{+}\mathcal{V}$ (Why?) it follows from Exercise A.12.9 that $\mathrm{Invlin}^{+}\mathcal{V}$ is a group. Since $\mathrm{Invlin}^{+}\mathcal{V} \subset \mathrm{Invlin}\,\mathcal{V}$ [see (A.10.12)], $\mathrm{Invlin}^{+}\mathcal{V}$ is termed a (*proper*) *subgroup* of $\mathrm{Invlin}\,\mathcal{V}$. (Here '*proper*' indicates that $\mathrm{Invlin}^{+}\mathcal{V} \neq \mathrm{Invlin}\,\mathcal{V}$. This is seen by noting that $-\mathbf{1} \in \mathrm{Invlin}\,\mathcal{V}$ but $-\mathbf{1} \notin \mathrm{Invlin}^{+}\mathcal{V}$.)

A.13 Eigenvectors, Eigenvalues, and the Characteristic Equation for a Linear Transformation

Given $\mathbf{L} \in \mathrm{Lin}\,\mathcal{V}$, then $\mathbf{v} \neq \mathbf{0} \in \mathcal{V}$ is termed an *eigenvector* of $\mathbf{L}$ (with corresponding *eigenvalue* λ) if, for some $\lambda \in \mathbb{R}$,

$$\mathbf{L}\mathbf{v} = \lambda\mathbf{v}. \tag{A.13.1}$$

Accordingly,

$$(\mathbf{L} - \lambda\mathbf{1})\mathbf{v} = \mathbf{0} \tag{A.13.2}$$

for some $\mathbf{v} \neq \mathbf{0}$, and hence [see (A.10.4)] $\mathbf{L} - \lambda\mathbf{1}$ is not invertible. Thus, from Det 3′,

$$\det(\mathbf{L} - \lambda\mathbf{1}) = 0. \tag{A.13.3}$$

Hence, if ω is a non-zero trilinear alternating form and $\mathbf{u}, \mathbf{v}$, and $\mathbf{w}$ are linearly independent, then from (A.12.21)

$$\omega((\mathbf{L} - \lambda\mathbf{1})\mathbf{u},\, (\mathbf{L} - \lambda\mathbf{1})\mathbf{v},\, (\mathbf{L} - \lambda\mathbf{1})\mathbf{w}) = 0. \tag{A.13.4}$$

It follows from the trilinearity of ω that

$$\begin{aligned}
&\omega(\mathbf{Lu}, \mathbf{Lv}, \mathbf{Lw}) - \lambda\{\omega(\mathbf{u}, \mathbf{Lv}, \mathbf{Lw}) + \omega(\mathbf{Lu}, \mathbf{v}, \mathbf{Lw}) + \omega(\mathbf{Lu}, \mathbf{Lv}, \mathbf{w})\} \\
&\quad + \lambda^2\{\omega(\mathbf{Lu}, \mathbf{v}, \mathbf{w}) + \omega(\mathbf{u}, \mathbf{Lv}, \mathbf{w}) + \omega(\mathbf{u}, \mathbf{v}, \mathbf{Lw})\} \\
&\quad - \lambda^3\omega(\mathbf{u}, \mathbf{v}, \mathbf{w}) = 0.
\end{aligned}$$

Hence

$$(\lambda^3 - I_1(\mathbf{L})\lambda^2 + I_2(\mathbf{L})\lambda - I_3(\mathbf{L}))\omega(\mathbf{u}, \mathbf{v}, \mathbf{w}) = 0.$$

Since $\omega(\mathbf{u}, \mathbf{v}, \mathbf{w}) \neq 0$ (Why?), we have the *characteristic equation* for $\mathbf{L}$:

$$C(\mathbf{L}; \lambda) := \lambda^3 - I_1(\mathbf{L})\lambda^2 + I_2(\mathbf{L})\lambda - I_3 = 0 \tag{A.13.5}$$

which may be written as

$$C(\mathbf{L}; \lambda) \equiv \lambda^3 - (\mathrm{tr}\,\mathbf{L})\lambda^2 + \frac{1}{2}\{(\mathrm{tr}\,\mathbf{L})^2 - \mathrm{tr}\,\mathbf{L}^2\}\lambda - \det \mathbf{L} = 0. \tag{A.13.6}$$

Now suppose that (A.13.3) holds for some $\lambda \in \mathbb{R}$. This means by Det 3′ that $(\mathbf{L} - \lambda\mathbf{1})$ is not invertible and hence by Remark A.10.1 that there exists a non-zero vector $\mathbf{u}$ such that $(\mathbf{L}-\lambda\mathbf{1})\mathbf{u}=\mathbf{0}$. Accordingly, there *exists* an eigenvector $\mathbf{u}$ of $\mathbf{L}$ with eigenvalue λ. Hence, in conjunction with (A.13.3), we have

E.1. $\mathbf{L}$ *has an eigenvector* $\mathbf{v}$ *with eigenvalue* λ *if and only if* $\det(\mathbf{L}-\lambda\mathbf{1})=0$; that is, if and only if λ is a solution of $C(\mathbf{L};\lambda)=0$. (A.13.8)

Cubic $C(\mathbf{L};\lambda)$ has real coefficients and so has zeros which are real or occur in complex conjugate pairs. Thus $C(\mathbf{L};\lambda)=0$ has either a single real zero (and two complex conjugate zeros) or three real zeros (which may involve repetitions). Thus we have the following:

E.2. $\mathbf{L}$ has either a single real eigenvalue λ_1, or three real eigenvalues λ_1,λ_2, and λ_3 which may include repetition.

Further simple observations are:

E.3. If $\mathbf{v}$ is an eigenvector of $\mathbf{L}$ with eigenvalue λ, then so too is $\alpha\mathbf{v}$, for any $\alpha\neq 0$.

E.4. The eigenvalues of $\mathbf{L}^T$ are the same as those of $\mathbf{L}$. (Why?)

Notice that E.4. says nothing about associated eigen*vectors* of $\mathbf{L}$ and $\mathbf{L}^T$: in general these will differ.

A.14 A Natural Inner Product for Lin $\mathcal{V}$

Recall that Lin $\mathcal{V}$ has the structure of a nine-dimensional vector space [see Section A.7 and (A.9.12)]. Indeed, Lin $\mathcal{V}$ has extra structure in the form of composition of linear transformations and the existence of a distinguished element $\mathbf{1}$. It also proves possible to furnish Lin $\mathcal{V}$ with an *inner product* which generalises the notion of the scalar product in $\mathcal{V}$. Recall the definition of a general vector/linear space $\mathcal{U}$ in Section A.7. An inner product on $\mathcal{U}$ is a map which assigns to each ordered pair $(\mathbf{u}_1,\mathbf{u}_2)$ of elements of $\mathcal{U}$ a real number, denoted by $\mathbf{u}_1\,.\,\mathbf{u}_2$, for which

IP 1.
$$\mathbf{u}_1\,.\,\mathbf{u}_2=\mathbf{u}_2\,.\,\mathbf{u}_1,$$

IP 2.
$$\mathbf{u}\,.\,\mathbf{u}\geq 0,\ \text{with}\ \mathbf{u}\,.\,\mathbf{u}=0 \quad \text{if and only if} \quad \mathbf{u}=\mathbf{0}, \tag{A.14.2}$$

and

IP 3.
$$(\alpha_1\mathbf{u}_1+\alpha_2\mathbf{u}_2)\,.\,\mathbf{u}_3=\alpha_1\mathbf{u}_1\,.\,\mathbf{u}_3+\alpha_2\mathbf{u}_2\,.\,\mathbf{u}_3.$$

Given an inner product on $\mathcal{U}$, the associated *norm* is defined by

$$\|\mathbf{u}\|:=(\mathbf{u}\,.\,\mathbf{u})^{1/2}. \tag{A.14.3}$$

Exercise A.14.1. Show that

N.1. $$\|\mathbf{u}\| = 0 \quad \text{if and only if} \quad \mathbf{u} = \mathbf{0} \tag{A.14.5}$$

and

N.2. $$\|\alpha\mathbf{u}\| = |\alpha|\,\|\mathbf{u}\| \quad \text{for any} \quad \alpha \in \mathbb{R}. \tag{A.14.6}$$

Further, we have

N.3. $$\mathbf{u}_1 . \mathbf{u}_2 \leq |\mathbf{u}_1 . \mathbf{u}_2| \leq \|\mathbf{u}_1\|\,\|\mathbf{u}_2\|. \tag{A.14.7}$$

Proof. The result is trivial if $\mathbf{u}_1 = \mathbf{0}$ or $\mathbf{u}_2 = \mathbf{0}$ or $\mathbf{u}_1 = \mathbf{0} = \mathbf{u}_2$. If $\mathbf{u}_1 \neq \mathbf{0}$ and $\mathbf{u}_2 \neq \mathbf{0}$, then for any $t \in \mathbb{R}$

$$\begin{aligned} 0 \leq \|\mathbf{u}_1 + t\mathbf{u}_2\|^2 &= (\mathbf{u}_1 + t\mathbf{u}_2).(\mathbf{u}_1 + t\mathbf{u}_2) = \|\mathbf{u}_1\|^2 + 2t\mathbf{u}_1 . \mathbf{u}_2 + t^2\|\mathbf{u}_2\|^2 \\ &= \|\mathbf{u}_1\|^2 + \left[\left\{\frac{\mathbf{u}_1 . \mathbf{u}_2}{\|\mathbf{u}_2\|\cdot} + t\|\mathbf{u}_2\|\right\}^2 - \frac{(\mathbf{u}_1 . \mathbf{u}_2)^2}{\|\mathbf{u}_2\|^2}\right]. \end{aligned}$$

Setting $t = -\mathbf{u}_1 . \mathbf{u}_2 / \|\mathbf{u}_2\|^2$ yields

$$\|\mathbf{u}_1\|^2 - \frac{(\mathbf{u}_1 . \mathbf{u}_2)^2}{\|\mathbf{u}_2\|^2} \geq 0$$

so that $\|\mathbf{u}_1\|^2\|\mathbf{u}_2\|^2 \geq (\mathbf{u}_1 . \mathbf{u}_2)^2$. The result follows upon taking positive square roots.

It follows that

$$\begin{aligned} \|\mathbf{u}_1 + \mathbf{u}_2\|^2 &= (\mathbf{u}_1 + \mathbf{u}_2).(\mathbf{u}_1 + \mathbf{u}_2) \\ &= \|\mathbf{u}_1\|^2 + 2\mathbf{u}_1 . \mathbf{u}_2 + \|\mathbf{u}_2\|^2 \leq \|\mathbf{u}_1\|^2 + 2\|\mathbf{u}_1\|\,\|\mathbf{u}_2\| + \|\mathbf{u}_2\|^2 = (\|\mathbf{u}_1\| + \|\mathbf{u}_2\|)^2. \end{aligned}$$

Taking positive square roots yields

N.4. $$\|\mathbf{u}_1 + \mathbf{u}_2\| \leq \|\mathbf{u}_1\| + \|\mathbf{u}_2\|. \tag{A.14.8}$$

Exercise A.14.2. If $\mathbf{u}_1 \neq \mathbf{0}, \mathbf{u}_2 \neq \mathbf{0}$, and $\mathbf{u}_2 \neq \alpha\mathbf{u}_1$ for any $\alpha \in \mathbb{R}$, show that $\mathbf{u}_1 . \mathbf{u}_2 < \|\mathbf{u}_1\|\,\|\mathbf{u}_2\|$. If $\mathbf{u}_1 \neq \mathbf{0}, \mathbf{u}_2 \neq \mathbf{0}$, and $\mathbf{u}_2 = \alpha\mathbf{u}_1$ for some $\alpha \in \mathbb{R}$, show that $|\mathbf{u}_1 . \mathbf{u}_2| = \|\mathbf{u}_1\|\,\|\mathbf{u}_2\|$; if $\alpha > 0$, note that $\mathbf{u}_1 . \mathbf{u}_2 = \|\mathbf{u}_1\|\,\|\mathbf{u}_2\|$, and deduce that in this case $\|\mathbf{u}_1 + \mathbf{u}_2\| = \|\mathbf{u}_1\| + \|\mathbf{u}_2\|$.

If $\mathbf{L}_1, \mathbf{L}_2 \in \text{Lin}\,\mathcal{V}$, then

Lin IP 1. $$\mathbf{L}_1 \cdot \mathbf{L}_2 := \text{tr}(\mathbf{L}_1^T\mathbf{L}_2). \tag{A.14.9}$$

To show that this is an inner product on the (vector/linear) space Lin $\mathcal{V}$, notice that from (A.12.12) with $\mathbf{L} = \mathbf{L}_2^T\mathbf{L}_1$, (A.8.7) and (A.8.6),

$$\mathbf{L}_2 \cdot \mathbf{L}_1 := \text{tr}(\mathbf{L}_2^T\mathbf{L}_1) = \text{tr}((\mathbf{L}_2^T\mathbf{L}_1)^T) = \text{tr}(\mathbf{L}_1^T\mathbf{L}_2) = \mathbf{L}_1 . \mathbf{L}_2.$$

Thus IP 1 is satisfied. Further, using (A.8.4) and (A.8.5) and with $\mathbf{L}_3 \in \operatorname{Lin}\mathcal{V}$,

$$\begin{aligned}(\alpha_1\mathbf{L}_1 + \alpha_2\mathbf{L}_2)\cdot\mathbf{L}_3 &= \operatorname{tr}((\alpha_1\mathbf{L}_1 + \alpha_2\mathbf{L}_2)^T\mathbf{L}_3)\\ &= \operatorname{tr}(\alpha_1\mathbf{L}_1^T\mathbf{L}_3 + \alpha_2\mathbf{L}_2^T\mathbf{L}_3)\\ &= \alpha_1\operatorname{tr}(\mathbf{L}_1^T\mathbf{L}_3) + \alpha_2\operatorname{tr}(\mathbf{L}_2^T\mathbf{L}_3) = \alpha_2\mathbf{L}_1\cdot\mathbf{L}_3 + \alpha_2\mathbf{L}_2\cdot\mathbf{L}_3,\end{aligned}$$

so guaranteeing IP 3. To establish IP 2 requires the following observation:

Remark A.14.1. Given $\mathbf{L},\mathbf{M} \in \operatorname{Lin}\mathcal{V}$, then [using matrix representations with respect to an orthonormal basis $\{\mathbf{e}_1,\mathbf{e}_2,\mathbf{e}_3\}$: see (A.9.8)]

$$\begin{aligned}\mathbf{L}\cdot\mathbf{M} = \operatorname{tr}(\mathbf{L}^T\mathbf{M}) &= \operatorname{tr}\{(L_{pq}\mathbf{e}_p\otimes\mathbf{e}_q)^T M_{rs}\mathbf{e}_r\otimes\mathbf{e}_s\}\\ &= \operatorname{tr}\{L_{pq}M_{rs}(\mathbf{e}_q\otimes\mathbf{e}_p)(\mathbf{e}_r\otimes\mathbf{e}_s)\}\\ &= L_{pq}M_{rs}\operatorname{tr}\{(\mathbf{e}_p\,.\,\mathbf{e}_r)(\mathbf{e}_q\otimes\mathbf{e}_s)\} = L_{pq}M_{rs}(\mathbf{e}_p\,.\,\mathbf{e}_r)(\mathbf{e}_q\,.\,\mathbf{e}_s)\\ &= L_{pq}M_{rs}\delta_{pr}\delta_{qs} = L_{pq}M_{pq}.\end{aligned}$$

[Here we have used (A.8.4), (A.8.5), (A.8.10), and (A.8.11).] That is,

$$\mathbf{L}\cdot\mathbf{M} = L_{pq}M_{pq}, \tag{A.14.10}$$

so $\mathbf{L}\cdot\mathbf{M}$ *is the sum of products of corresponding elements in matrix representations of* $\mathbf{L}$ *and* $\mathbf{M}$ *with respect to any orthonormal basis.*

Setting $\mathbf{M} = \mathbf{L}$ yields

$$\mathbf{L}\cdot\mathbf{L} = \sum_{p=1}^{3}\sum_{q=1}^{3} L_{pq}^2. \tag{A.14.11}$$

Accordingly $\mathbf{L}\cdot\mathbf{L} \geq 0$ and $\mathbf{L}\cdot\mathbf{L} = 0$ if and only if each matrix element $L_{pq} = 0$ and hence $\mathbf{L} = \mathbf{O}$. Thus IP 2 is satisfied, and having already shown that IP 1 and IP 3 hold, it follows that (A.14.9) *defines an inner product on* Lin $\mathcal{V}$.

Remark A.14.2. Since $\mathbf{L}_1\cdot\mathbf{L}_2$ is defined in terms of $\mathbf{L}_1^T$ and $\mathbf{L}_2$, and the definition of $\mathbf{L}_1^T$ derives from the scalar/inner product on $\mathcal{V}$ [see (A.8.1)], it follows that the inner product (A.14.9) on Lin $\mathcal{V}$ is *natural.* Said differently, inner product (A.14.9) is *induced* by the inner product on $\mathcal{V}$.

Exercise A.14.3. Note that $\mathbf{1}\cdot\mathbf{L} = \operatorname{tr}\mathbf{L}$, and deduce that $\|\mathbf{1}\| = \sqrt{3}$.

Exercise A.14.4. Show that if $\mathbf{W} \in \operatorname{Sk}\mathcal{V}$ and $\mathbf{S} \in \operatorname{Sym}\mathcal{V}$, then

Lin IP 2.
$$\mathbf{W}\cdot\mathbf{S} = 0. \tag{A.14.12}$$

Since N.1 through N.4 [see (A.14.5) through (A.14.8)] hold for any norm on any vector space $\mathcal{U}$, the norm on vector space Lin $\mathcal{V}$ satisfies

Lin N.0. $$\|\mathbf{L}\| := (\mathbf{L}\cdot\mathbf{L})^{1/2} = (\mathrm{tr}(\mathbf{L}^T\mathbf{L}))^{1/2} = \left(\sum_{p=1}^{3}\sum_{q=1}^{3} L_{pq}^2\right)^{1/2} \qquad (A.14.13)$$

for any $\mathbf{L} \in$ Lin $\mathcal{V}$ and any matrix representation $[L_{pq}]$.

Lin N.1. $$\|\mathbf{L}\| = 0 \quad \text{if and only if} \quad \mathbf{L} = \mathbf{O}. \qquad (A.14.14)$$

Lin N.2. $$\|\alpha\mathbf{L}\| = |\alpha|\|\mathbf{L}\| \quad \text{for any} \quad \alpha \in \mathbb{R}. \qquad (A.14.15)$$

Also, for any $\mathbf{L}_1, \mathbf{L}_2 \in$ Lin $\mathcal{V}$,

Lin N.3. $$\mathbf{L}_1\cdot\mathbf{L}_2 \leq |\mathbf{L}_1\cdot\mathbf{L}_2| \leq \|\mathbf{L}_1\|\|\mathbf{L}_2\| \qquad (A.14.16)$$

and

Lin N.4. $$\|\mathbf{L}_1+\mathbf{L}_2\| \leq \|\mathbf{L}_1\| + \|\mathbf{L}_2\|. \qquad (A.14.17)$$

Exercise A.14.5. Prove, using (A.12.14) with $\mathbf{A} = \mathbf{L}$ and $\mathbf{B} = \mathbf{L}^T$,

Lin N.5. $$\|\mathbf{L}^T\| = \|\mathbf{L}\|. \qquad (A.14.18)$$

We now list four additional useful properties of Lin $\mathcal{V}$.

Lin N.6. If $\{\mathbf{e}_1, \mathbf{e}_2, \mathbf{e}_3\}$ is an orthonormal basis for $\mathcal{V}$, then $\{\mathbf{e}_p \otimes \mathbf{e}_q\}$ is an orthonormal basis for Lin $\mathcal{V}$ $(p, q = 1, 2, 3)$. (A.14.19)

Proof. Recall that $\{\mathbf{e}_p \otimes \mathbf{e}_q\}$ is a basis for Lin $\mathcal{V}$ [see (A.9.11)]. To check orthonormality, consider

$$\begin{aligned}(\mathbf{e}_p \otimes \mathbf{e}_q)\cdot(\mathbf{e}_r \otimes \mathbf{e}_s) &= \mathrm{tr}\{(\mathbf{e}_q \otimes \mathbf{e}_p)(\mathbf{e}_r \otimes \mathbf{e}_s)\}\\ &= \mathrm{tr}\{(\mathbf{e}_p\,.\,\mathbf{e}_r)(\mathbf{e}_q \otimes \mathbf{e}_s)\} = (\mathbf{e}_p\,.\,\mathbf{e}_r)(\mathbf{e}_q\,.\,\mathbf{e}_s) = \delta_{pr}\delta_{qs}\\ &= 0 \qquad \text{unless both } r = p \textit{ and } s = q,\end{aligned}$$

while if $r = p$ *and* $s = q$ the value is 1.

Lin N.7. If $\mathbf{v} \in \mathcal{V}$, then[12] $\|\mathbf{v}\otimes\mathbf{v}\| = \|\mathbf{v}\|^2$. (A.14.20)

Proof. $$\begin{aligned}\|\mathbf{v}\otimes\mathbf{v}\|^2 &= \mathrm{tr}((\mathbf{v}\otimes\mathbf{v})^T(\mathbf{v}\otimes\mathbf{v})) = \mathrm{tr}((\mathbf{v}\otimes\mathbf{v})(\mathbf{v}\otimes\mathbf{v}))\\ &= \mathrm{tr}((\mathbf{v}\,.\,\mathbf{v})(\mathbf{v}\otimes\mathbf{v})) = (\mathbf{v}\,.\,\mathbf{v})\mathrm{tr}(\mathbf{v}\otimes\mathbf{v})\\ &= (\mathbf{v}\,.\,\mathbf{v})(\mathbf{v}\,.\,\mathbf{v}) = \|\mathbf{v}\|^4.\end{aligned}$$

[12] Notice that two norms are involved here, the first in Lin $\mathcal{V}$ and the second in $\mathcal{V}$.

The result follows on taking positive square roots.

Lin N.8. If $\mathbf{L}_1, \mathbf{L}_2 \in \operatorname{Lin}\mathcal{V}$, then $\|\mathbf{L}_1\mathbf{L}_2\| \le \|\mathbf{L}_1\|\|\mathbf{L}_2\|$. (A.14.21)

This result is proved at the end of Section A.17 [see (A.17.24)].

Lin N.9. If $\mathbf{v} \in \mathcal{V}$ and $\mathbf{L} \in \operatorname{Lin}\mathcal{V}$, then (see footnote 12)

$$\|\mathbf{L}\mathbf{v}\| \le \|\mathbf{L}\|\|\mathbf{v}\|. \tag{A.14.22}$$

Proof.
$$\begin{aligned}\|\mathbf{L}\mathbf{v}\|^2 &= \mathbf{L}\mathbf{v}\,.\,\mathbf{L}\mathbf{v} = \operatorname{tr}(\mathbf{L}\mathbf{v}\otimes\mathbf{L}\mathbf{v}) = \operatorname{tr}(\mathbf{L}(\mathbf{v}\otimes\mathbf{v})\mathbf{L}^T)\\ &= \operatorname{tr}(\mathbf{L}^T\mathbf{L}(\mathbf{v}\otimes\mathbf{v})) = \mathbf{L}^T\mathbf{L}\cdot(\mathbf{v}\otimes\mathbf{v}) \le \|\mathbf{L}^T\mathbf{L}\|\|\mathbf{v}\otimes\mathbf{v}\|\\ &\le \|\mathbf{L}^T\|\|\mathbf{L}\|\|\mathbf{v}\|^2 = \|\mathbf{L}\|^2\|\mathbf{v}\|^2.\end{aligned} \tag{A.14.23}$$

The result follows on taking positive square roots. Justification of the steps comes from (A.12.9) ($\mathbf{a} = \mathbf{b} = \mathbf{L}\mathbf{v}$), (A.8.12) ($\mathbf{a} = \mathbf{b} = \mathbf{v}$, $\mathbf{L}_1 = \mathbf{L}$, $\mathbf{L}_2 = \mathbf{L}^T$), (A.12.14) ($\mathbf{A} = \mathbf{L}(\mathbf{v}\otimes\mathbf{v}), \mathbf{B} = \mathbf{L}^T$), (A.14.7) for $\mathcal{U} = \operatorname{Lin}\mathcal{V}$ ($\mathbf{u}_1 = \mathbf{L}^T\mathbf{L}$, $\mathbf{u}_2 = \mathbf{v}\otimes\mathbf{v}$), and (A.14.21) and (A.14.18).

A.15 Skew Linear Transformations and Axial Vectors

If $\mathbf{W}$ is skew and non-zero, so that

$$\mathbf{W}^T = -\mathbf{W} \quad \text{and} \quad \mathbf{W} \neq \mathbf{O}, \tag{A.15.1}$$

then

$$\text{(i) } \operatorname{tr}\mathbf{W} = 0 \quad \text{and} \quad \text{(ii) } \det\mathbf{W} = 0. \tag{A.15.2}$$

Exercise A.15.1. Prove (i) by taking the trace of both sides of $(A.15.1)_1$ and noting that $\operatorname{tr}\mathbf{W}^T = \operatorname{tr}\mathbf{W}$ [see (A.12.12)]. Prove (ii) by taking the determinant of both sides of $(A.15.1)_1$ and noting that $\det\mathbf{W}^T = \det\mathbf{W}$ [see (A.12.32)] and $\det(-\mathbf{W}) = (-1)^3\det\mathbf{W}$ [see (A.12.23)].

It follows, from (A.15.2) and (A.13.8) with (A.13.7), that the eigenvalues λ of $\mathbf{W}$ are given by

$$\lambda^3 - \left(\frac{1}{2}\operatorname{tr}\mathbf{W}^2\right)\lambda = 0. \tag{A.15.3}$$

(Check this!) Equivalently [since $-\operatorname{tr}\mathbf{W}^2 = \operatorname{tr}(-\mathbf{W}\mathbf{W}) = \operatorname{tr}(\mathbf{W}^T\mathbf{W}) = \mathbf{W}\cdot\mathbf{W}$],

$$\lambda(\lambda^2 + \frac{1}{2}\|\mathbf{W}\|^2) = 0. \tag{A.15.4}$$

Hence $\mathbf{W}$ has a single real eigenvalue 0, and there exists a non-zero vector $\mathbf{w}$ such that

$$\mathbf{W}\mathbf{w} = \mathbf{0}. \tag{A.15.5}$$

It follows that for any vector $\mathbf{v} \in \mathcal{V}$,

$$\mathbf{W}\mathbf{v}\,.\,\mathbf{w} = \mathbf{v}\,.\,\mathbf{W}^T\mathbf{w} = -\mathbf{v}\,.\,\mathbf{W}\mathbf{w} = 0. \tag{A.15.6}$$

Since $\mathbf{W}\mathbf{v}\,.\,\mathbf{v} = \mathbf{v}\,.\,\mathbf{W}^T\mathbf{v} = -\mathbf{v}\,.\,\mathbf{W}\mathbf{v}$,

we also have $\mathbf{W}\mathbf{v}\,.\,\mathbf{v} = 0$. (A.15.7)

Relations (A.15.6) and (A.15.7) indicate that, for all $\mathbf{v} \in \mathcal{V}, \mathbf{W}\mathbf{v}$ is orthogonal to both $\mathbf{w}$ and $\mathbf{v}$. Accordingly, for some $k \in \mathbb{R}$,

$$\mathbf{W}\mathbf{v} = k\mathbf{w} \times \mathbf{v}. \tag{A.15.8}$$

Since eigenvectors are determined only to within a non-zero scalar multiple (see E.3 of Section A.1.13) we can choose $\mathbf{w}$ so that

$$\mathbf{W}\mathbf{v} = \mathbf{w} \times \mathbf{v}. \tag{A.15.9}$$

In this case $\mathbf{w}$ is termed the *axial vector* corresponding to W.

Remark A.15.1. Note that the result (A.8.22) of Exercise A.8.4 yields $-\mathbf{a} \times \mathbf{b}$ as the axial vector of $\mathbf{a} \wedge \mathbf{b} \in \mathrm{Sk}\,\mathcal{V}$.

Given a choice of orthonormal basis $\{\mathbf{e}_1, \mathbf{e}_2, \mathbf{e}_3\}$ with $\mathbf{e}_3 = \mathbf{e}_2 \times \mathbf{e}_2$ and choosing $\mathbf{v} = \mathbf{e}_j$ in (A.15.9), we have

$$W_{ij} := \mathbf{e}_i \,.\, \mathbf{W}\mathbf{e}_j = \mathbf{e}_i \,.\, (\mathbf{w} \times \mathbf{e}_j) = \mathbf{w} \,.\, (\mathbf{e}_j \times \mathbf{e}_i), \tag{A.15.10}$$

on invoking (A.5.13). Hence with $(i,j) = (1, 2), (1, 3)$, and $(2, 3)$,

$$\left.\begin{array}{l} W_{12} = \mathbf{w} \,.\, (\mathbf{e}_2 \times \mathbf{e}_1) = -\mathbf{w} \,.\, \mathbf{e}_3 = -w_3 \\ W_{13} = \mathbf{w} \,.\, (\mathbf{e}_3 \times \mathbf{e}_1) = \mathbf{w} \,.\, \mathbf{e}_2 = w_2 \\ W_{23} = \mathbf{w} \,.\, (\mathbf{e}_3 \times \mathbf{e}_2) = -\mathbf{w} \,.\, \mathbf{e}_1 = -w_1 \end{array}\right\}. \tag{A.15.11}$$

Thus the matrix representation of $\mathbf{W}$ and components of $\mathbf{w}$ are related by

$$[\mathbf{W}] = \begin{bmatrix} 0 & -w_3 & w_2 \\ w_3 & 0 & -w_1 \\ -w_2 & w_1 & 0 \end{bmatrix}, \qquad \text{where } \mathbf{w} = w_i \mathbf{e}_i. \tag{A.15.12}$$

Exercise A.15.2. Verify the matrix representation of (A.15.9); that is, show that the result of pre-multiplying column matrix $[v_1 \;\; v_2 \;\; v_3]^T$ by $[\mathbf{W}]$ is the column matrix which represents vector $\mathbf{w} \times \mathbf{v}$. (Of course, here $\mathbf{v} = v_i \mathbf{e}_i$.)

Now suppose that $\mathbf{W}_1$ and $\mathbf{W}_2$ are skew linear transformations with corresponding axial vectors $\mathbf{w}_1$ and $\mathbf{w}_2$. Then

$$\begin{aligned} \mathbf{W}_1 \cdot \mathbf{W}_2 &:= \mathrm{tr}(\mathbf{W}_1^T \mathbf{W}_2) = W_1^T \mathbf{W}_2 \mathbf{e}_i \,.\, \mathbf{e}_i && [(\text{A.14.9}), (\text{A.12.11})] \\ &= \mathbf{W}_2 \mathbf{e}_i \,.\, \mathbf{W}_1 \mathbf{e}_i = (\mathbf{w}_2 \times \mathbf{e}_i) \,.\, (\mathbf{w}_1 \times \mathbf{e}_i) && [(\text{A.8.1}), (\text{A.15.9})] \\ &= (\mathbf{e}_i \times (\mathbf{w}_1 \times \mathbf{e}_i)) \,.\, \mathbf{w}_2 && [(\text{A.5.13})] \\ &= (\mathbf{w}_1 (\mathbf{e}_i \,.\, \mathbf{e}_i) - (\mathbf{e}_i \,.\, \mathbf{w}_1) \mathbf{e}_i) \,.\, \mathbf{w}_2 && [(\text{A.5.16})] \\ &= (3\mathbf{w}_1 - \mathbf{w}_1) \,.\, \mathbf{w}_2 = 2\mathbf{w}_1 \,.\, \mathbf{w}_2. && [(\text{A.6.6})] \end{aligned}$$

That is, $$\mathbf{W}_1 \cdot \mathbf{W}_2 = 2\mathbf{w}_1 \,.\, \mathbf{w}_2. \tag{A.15.13}$$

In particular, with $\mathbf{W}_1 = \mathbf{W}_2 = \mathbf{W}$ and $\mathbf{w}_1 = \mathbf{w}_2 = \mathbf{w}$,

$$\|\mathbf{W}\|^2 = 2\|\mathbf{w}\|^2. \tag{A.15.14}$$

Remark A.15.2. Definition (A.5.8) of the vector product required appeal to the notion of a right-handed screw and thus is somewhat artificial: we could have chosen an essentially equivalent product using the notion of a *left*-handed screw. A choice of such a so-called orientation is not natural: there is no compelling physical reason to choose one orientation in preference to another. In fact vector products of pairs of vectors (i.e., elements of $\mathcal{V}$) and axial vectors associated with skew linear transformations can be distinguished from 'genuine' vectors by considering component representations with respect to *any* orthonormal basis $\{\mathbf{e}_1, \mathbf{e}_2, \mathbf{e}_3\}$. If $\mathbf{v} \in \mathcal{V}$, then

$$\mathbf{v} = v_1\mathbf{e}_1 + v_2\mathbf{e}_2 + \mathbf{v}_3\mathbf{e}_3,$$

where $v_i := \mathbf{v}\,.\,\mathbf{e}_i$, $i = 1,2,3$. On choosing instead the orthonormal basis $\{\mathbf{e}'_1, \mathbf{e}'_2, \mathbf{e}'_3\}$, where $\mathbf{e}'_i := -\mathbf{e}_i$, we have

$$\mathbf{v} = v'_1\mathbf{e}'_1 + v'_2\mathbf{e}'_2 + v'_3\mathbf{e}'_3,$$

where $v'_i := \mathbf{v}\,.\,\mathbf{e}'_i = \mathbf{v}\,.\,-\mathbf{e}_i = -\mathbf{v}\,.\,\mathbf{e}_i = -v_i$.

That is, the components of any $\mathbf{v} \in \mathcal{V}$ change sign when the orthonormal basis vectors are all reversed. However, we can see from (A.6.12) that reversal of basis vectors leaves the components of $\mathbf{u} \times \mathbf{v}$ unchanged.

Noting that elements of a square matrix remain unchanged under basis reversal [see (A.9.1)], it follows that the components of $\mathbf{Wv}$ reverse sign if $\mathbf{v} \in \mathcal{V}$. From this and (A.15.9) it follows that the components of $\mathbf{w} \times \mathbf{v}$ reverse sign, and hence the components of $\mathbf{w}$ remain unchanged (Why?). We term expressions of form $\mathbf{u} \times \mathbf{v}$ ($\mathbf{u}, \mathbf{v} \in \mathcal{V}$) and $\mathbf{w}$ [given by (A.15.9) for some skew linear transformation $\mathbf{W}$] *pseudo-vectors* (see Goldstein, Poole, & Safko [7]). In similar fashion, any triple scalar product $\mathbf{u} \times \mathbf{v}\,.\,\mathbf{w}$ with $\mathbf{u}, \mathbf{v}, \mathbf{w} \in \mathcal{V}$ is a *pseudo-scalar* since basis reversal in any component formulation results in a change of sign. (Of course, a 'genuine' scalar has a value completely independent of any basis.) In fact there is no need to introduce either the vector *or* triple scalar product: one can work with the wedge product (A.8.21) and alternating trilinear forms and in so doing obviate any need to appeal to choice of orientation.

A.16 Orthogonal Transformations and Their Characterisation

Let $\mathbf{q} : \mathcal{V} \to \mathcal{V}$ preserve inner products. That is, suppose that, for all $\mathbf{u}, \mathbf{v} \in \mathcal{V}$,

$$\mathbf{q}(\mathbf{u})\,.\,\mathbf{q}(\mathbf{v}) = \mathbf{u}\,.\,\mathbf{v}. \tag{A.16.1}$$

We term $\mathbf{q}$ an *orthogonal map*. It follows that

(i) $\mathbf{q}$ preserves orthonormal bases and thus is surjective ('onto'), and
(ii) $\mathbf{q}(\mathbf{u}) = \mathbf{0}$ implies $\mathbf{u} = \mathbf{0}$.

To see (i), let $\{\mathbf{e}_1, \mathbf{e}_2, \mathbf{e}_3\}$ be an orthonormal basis, and consider

$$\alpha_1\mathbf{q}(\mathbf{e}_1) + \alpha_2\mathbf{q}(\mathbf{e}_2) + \alpha_3\mathbf{q}(\mathbf{e}_3) = \mathbf{0}.$$

Scalar multiplication by $\mathbf{q}(\mathbf{e}_1)$ and use of (A.16.1) yield

$$\alpha_1\mathbf{e}_1\,.\,\mathbf{e}_1 + \alpha_2\mathbf{e}_2\,.\,\mathbf{e}_1 + \alpha_3\mathbf{e}_3\,.\,\mathbf{e}_1 = 0$$

and hence $\alpha_1 = 0$. Similarly, $\alpha_2 = 0 = \alpha_3$, and hence $\{\mathbf{q}(\mathbf{e}_1), \mathbf{q}(\mathbf{e}_2), \mathbf{q}(\mathbf{e}_3)\}$ is a basis. In fact this basis is orthonormal since $\mathbf{q}(\mathbf{e}_i) . \mathbf{q}(\mathbf{e}_j) = \mathbf{e}_i . \mathbf{e}_j = \delta_{ij}$. To show surjectivity, suppose that $\mathbf{v} \in \mathcal{V}$. We must exhibit a vector $\mathbf{u}$ such that $\mathbf{q}(\mathbf{u}) = \mathbf{v}$. Now $\mathbf{v} = \alpha_i \mathbf{q}(\mathbf{e}_i)$ for some $\alpha_1, \alpha_2, \alpha_3$ since $\{\mathbf{q}(\mathbf{e}_i)\}$ is a basis. Consider $\mathbf{u} := \alpha_i \mathbf{e}_i$ and note that $\mathbf{q}(\mathbf{u}) . \mathbf{q}(\mathbf{e}_j) = \mathbf{u} . \mathbf{e}_j = \alpha_j$. Hence

$$\mathbf{q}(\mathbf{u}) = (\mathbf{q}(\mathbf{u}) . \mathbf{q}(\mathbf{e}_j))\mathbf{q}(\mathbf{e}_j) = (\mathbf{u} . \mathbf{e}_j)\mathbf{q}(\mathbf{e}_j) = \alpha_j \mathbf{q}(\mathbf{e}_j) = \mathbf{v}.$$

To prove (ii) it is necessary only to set $\mathbf{v} = \mathbf{u}$ in (A.16.1): if $\mathbf{q}(\mathbf{u}) = \mathbf{0}$, then

$$\mathbf{u} . \mathbf{u} = \mathbf{q}(\mathbf{u}) . \mathbf{q}(\mathbf{u}) = \mathbf{0} . \mathbf{0} = 0, \qquad \text{whence} \quad \mathbf{u} = \mathbf{0}.$$

Further, $\mathbf{q}$ *is linear* since (A.16.1) implies that

$$\begin{aligned}(\mathbf{q}(\alpha_1\mathbf{u}_1 + \alpha_2\mathbf{u}_2) - \alpha_1\mathbf{q}(\mathbf{u}_1) - \alpha_2\mathbf{q}(\mathbf{u}_2)) . \mathbf{q}(\mathbf{v}) \\ = (\alpha_1\mathbf{u}_1 + \alpha_2\mathbf{u}_2) . \mathbf{v} - \alpha_1\mathbf{u}_1 . \mathbf{v} - \alpha_2\mathbf{u}_2 . \mathbf{v} = 0\end{aligned}$$

for every $\mathbf{v} \in \mathcal{V}$. However, since $\mathbf{q}$ is surjective, it follows that

$$\mathbf{q}(\alpha_1\mathbf{u}_1 + \alpha_2\mathbf{u}_2) - \alpha_1\mathbf{q}(\mathbf{u}_1) - \alpha_2\mathbf{q}(\mathbf{u}_2) = \mathbf{0}.$$

Thus $\mathbf{q}$ *is linear and* [see (A.10.4)] *invertible.* We write $\mathbf{q}(\mathbf{u})$ as $\mathbf{Qu}$ and note that (A.16.1) becomes

Orth. 1. $$\mathbf{Qu} . \mathbf{Qv} = \mathbf{u} . \mathbf{v} \tag{A.16.2}$$

for all $\mathbf{u}, \mathbf{v} \in \mathcal{V}$. Hence, for all $\mathbf{v} \in \mathcal{V}$,

$$\mathbf{Q}^T\mathbf{Qu} . \mathbf{v} = \mathbf{u} . \mathbf{v}$$

and so $\mathbf{Q}^T\mathbf{Qu} = \mathbf{u}$ for all $\mathbf{u} \in \mathcal{V}$. Thus

$$\mathbf{Q}^T\mathbf{Q} = \mathbf{1}. \tag{A.16.3}$$

Also, *if* $\mathbf{Q}^T\mathbf{Q} = \mathbf{1}$, then $\mathbf{Qu} . \mathbf{Qv} = \mathbf{Q}^T\mathbf{Qu} . \mathbf{v} = \mathbf{u} . \mathbf{v}$. Accordingly we have

Orth. 2. $$\mathbf{Q} \quad \text{is orthogonal if and only if} \quad \mathbf{Q}^T = \mathbf{Q}^{-1}. \tag{A.16.4}$$

From (A.12.22), (A.16.3), (A.12.29), and (A.12.30)

$$1 = \det \mathbf{1} = \det(\mathbf{Q}^T\mathbf{Q}) = (\det \mathbf{Q}^T)(\det \mathbf{Q}) = (\det \mathbf{Q})^2.$$

Thus we have

Orth. 3. $$\det \mathbf{Q} \pm 1. \tag{A.16.5}$$

Notice

Orth. 4. $$\mathbf{1} \text{ is orthogonal,} \tag{A.16.6}$$

and

Orth. 5. $$\text{If } \mathbf{Q}_1 \text{ and } \mathbf{Q}_2 \text{ are orthogonal, then } \mathbf{Q}_1\mathbf{Q}_2 \text{ is orthogonal.} \tag{A.16.7}$$

To see this, note that for all $\mathbf{u}, \mathbf{v} \in \mathcal{V}$

$$(\mathbf{Q}_1\mathbf{Q}_2)\mathbf{u} . (\mathbf{Q}_1\mathbf{Q}_2)\mathbf{v} = \mathbf{Q}_1(\mathbf{Q}_2\mathbf{u}) . \mathbf{Q}_1(\mathbf{Q}_2\mathbf{v}) = \mathbf{Q}_2\mathbf{u} . \mathbf{Q}_2\mathbf{v} = \mathbf{u} . \mathbf{v}.$$

Remark A.16.1. The set

$$\mathrm{Orth}\,\mathcal{V} := \{\mathbf{Q} \in \mathrm{Lin}\,\mathcal{V} : \mathbf{Q} \text{ is orthogonal}\} \tag{A.16.8}$$

has the structure of a group by virtue of (A.16.6), (A.16.7), and (A.16.4) (cf. (A.10.20)]. It is a subgroup of Invlin $\mathcal{V}$ and is termed the *orthogonal group* (on $\mathcal{V}$).

Exercise A.16.1. Given

$$\mathrm{Orth}^{+}\mathcal{V} := \{\mathbf{Q} \in \mathrm{Orth}\,\mathcal{V} : \det\mathbf{Q} = 1\}, \tag{A.16.9}$$

show that $\mathbf{1} \in \mathrm{Orth}^{+}\mathcal{V}$, $\mathbf{Q}^{-1} \in \mathrm{Orth}^{+}\mathcal{V}$ if $\mathbf{Q} \in \mathrm{Orth}^{+}\mathcal{V}$, and if $\mathbf{Q}_1, \mathbf{Q}_2 \in \mathrm{Orth}^{+}\mathcal{V}$, then $\mathbf{Q}_1\mathbf{Q}_2 \in \mathrm{Orth}^{+}\mathcal{V}$.

It follows that $\mathrm{Orth}^{+}\mathcal{V}$ is a group (a subgroup of Orth$\mathcal{V}$, termed the *proper orthogonal group on* $\mathcal{V}$). Elements of $\mathrm{Orth}^{+}\mathcal{V}$ are termed *proper orthogonal* linear transformations.

If $\mathbf{e}$ is an eigenvector of $\mathbf{Q} \in \mathrm{Orth}\,\mathcal{V}$, then its eigenvalue λ must satisfy

$$\mathbf{e}\,.\,\mathbf{e} = \mathbf{Qe}\,.\,\mathbf{Qe} = (\lambda\mathbf{e})\,.\,(\lambda\mathbf{e}) = \lambda^2\mathbf{e}\,.\,\mathbf{e}.$$

Accordingly we have

Orth. 6. The only eigenvalues of $\mathbf{Q} \in \mathrm{Orth}\,\mathcal{V}$ are ± 1. (A.16.10)

Now consider the characteristic equation for a *proper* orthogonal linear transformation, namely [recall (A.13.7)]

$$C(\mathbf{Q};\lambda) := \lambda^3 - (\mathrm{tr}\,\mathbf{Q})\lambda^2 + \frac{1}{2}((\mathrm{tr}\,\mathbf{Q})^2 - \mathrm{tr}(\mathbf{Q}^2))\lambda - 1 = 0. \tag{A.16.11}$$

Since C is a cubic in λ, can only have zeros at $\lambda = \pm 1$, and $C(\mathbf{Q};0) = -1$, there are only three possibilities: (i) a single zero at $\lambda = 1$, (ii) a double zero at $\lambda = -1$ and single zero at $\lambda = 1$, and (iii) a triple zero at $\lambda = 1$.

Exercise A.16.2. Convince yourself that (i) through (iii) are the only possibilities.

Case (i). Let $\mathbf{e}_1$ be an eigenvector corresponding to $\lambda = 1$. Take $\|\mathbf{e}_1\| = 1$ without loss of generality (Why?), and let $\mathbf{e}_2, \mathbf{e}_3$ be chosen so that $\{\mathbf{e}_1, \mathbf{e}_2, \mathbf{e}_3\}$ is an orthonormal basis with $\mathbf{e}_1 \times \mathbf{e}_2 = \mathbf{e}_3$. We can calculate the corresponding matrix representation of $\mathbf{Q}$. Indeed, $Q_{11} := \mathbf{e}_1\,.\,\mathbf{Qe}_1 = \mathbf{e}_1\,.\,\mathbf{e}_1 = 1$, $Q_{12} := \mathbf{e}_1\,.\,\mathbf{Qe}_2 = \mathbf{Q}^T\mathbf{e}_1\,.\,\mathbf{e}_2 = \mathbf{Q}^{-1}\mathbf{e}_1\,.\,\mathbf{e}_2 = \mathbf{e}_1\,.\,\mathbf{e}_2 = 0$, $Q_{21} := \mathbf{e}_2\,.\,\mathbf{Qe}_1 = \mathbf{e}_2\,.\,\mathbf{e}_1 = 0$, and similarly, $Q_{13} = 0 = Q_{31}$. Let $Q_{22} := \mathbf{e}_2\,.\,\mathbf{Qe}_2 =: a$, $Q_{23} := \mathbf{e}_2\,.\,\mathbf{Qe}_3 =: b$, $Q_{32} := \mathbf{e}_3\,.\,\mathbf{Qe}_2 =: c$, and $Q_{33} := \mathbf{e}_3\,.\,\mathbf{Qe}_3 =: d$. Thus

$$\begin{aligned}\mathbf{Qe}_2 &= ((\mathbf{Qe}_2)\,.\,\mathbf{e}_1)\mathbf{e}_1 + ((\mathbf{Qe}_2)\,.\,\mathbf{e}_2)\mathbf{e}_2 + ((\mathbf{Qe}_2\,.\,\mathbf{e}_3)\mathbf{e}_3,\\ &= 0\mathbf{e}_1 + a\mathbf{e}_2 + c\mathbf{e}_3.\end{aligned} \tag{A.16.12}$$

Similarly, $$\mathbf{Qe}_3 = 0\mathbf{e}_2 + b\mathbf{e}_2 + d\mathbf{e}_3. \tag{A.16.13}$$

However, $\mathbf{Qe}_2$ and $\mathbf{Qe}_3$ are unit vectors (Why?). Thus

$$a^2 + c^2 = 1 \qquad \text{and} \qquad b^2 + d^2 = 1. \tag{A.16.14}$$

Further, $$\mathbf{Qe}_2\,.\,\mathbf{Qe}_3 = \mathbf{e}_2\,.\,\mathbf{e}_3 = 0. \tag{A.16.15}$$

Hence $$ab + cd = 0. \tag{A.16.16}$$

Given (A.16.14), we may write, without loss of generality,

$$a = \cos\alpha, \qquad c = \sin\alpha \qquad \text{and} \qquad b = \cos\beta, \qquad d = \sin\beta.$$

Thus relations (A.16.14) are automatically satisfied, and (A.16.16) yields

$$\cos(\alpha - \beta) = 0. \tag{A.16.17}$$

Finally, since we are here considering only *proper* orthogonal transformations, it follows that [see (A.16.9)]

$$1 = \det\mathbf{Q} = ad - bc = -\sin(\alpha - \beta). \tag{A.16.18}$$

(Prove this!) From (A.16.17) and (A.16.18) we have

$$\alpha - \beta = 3\pi/2, \tag{A.16.19}$$

whence

$$\cos\beta = \cos(\alpha - 3\pi/2) = -\sin\alpha \qquad \text{and} \qquad \sin\beta = \sin(\alpha - 3\pi/2) = \cos\alpha. \tag{A.16.20}$$

Thus (Show this!)

$$[\mathbf{Q}] = \begin{bmatrix} 1 & 0 & 0 \\ 0 & \cos\alpha & -\sin\alpha \\ 0 & \sin\alpha & \cos\alpha \end{bmatrix}. \tag{A.16.21}$$

Case (ii). With eigenvectors $\mathbf{e}_1, \mathbf{e}_2$ such that $\mathbf{Qe}_1 = \mathbf{e}_1$ and $\mathbf{Qe}_2 = -\mathbf{e}_2$ ($\|\mathbf{e}_1\| = 1 = \|\mathbf{e}_2\|$ without loss of generality), let $\mathbf{e}_3 := \mathbf{e}_1 \times \mathbf{e}_2$. Noting that

$$\mathbf{e}_1\,.\,\mathbf{e}_2 = \mathbf{Qe}_1\,.\,\mathbf{Qe}_2 = \mathbf{e}_1\,.\,-\mathbf{e}_2 = -\mathbf{e}_1\,.\,\mathbf{e}_2,$$

we deduce that $\mathbf{e}_1\,.\,\mathbf{e}_2 = 0$, so $\{\mathbf{e}_1, \mathbf{e}_2, \mathbf{e}_3\}$ is an orthonormal (right-handed) basis. The corresponding matrix representative of $\mathbf{Q}$ has $Q_{11} = \mathbf{e}_1\,.\,\mathbf{Qe}_1 = \mathbf{e}_1\,.\,\mathbf{e}_1 = 1$, $Q_{12} = \mathbf{e}_1\,.\,\mathbf{Qe}_2 = \mathbf{e}_1\,.\,-\mathbf{e}_2 = 0$, $Q_{21} = \mathbf{e}_2\,.\,\mathbf{Qe}_1 = \mathbf{e}_2\,.\,\mathbf{e}_1 = 0$, $Q_{22} = \mathbf{e}_2\,.\,\mathbf{Qe}_2 = \mathbf{e}_2\,.\,-\mathbf{e}_2 = -1$, $Q_{13} = \mathbf{e}_1.\mathbf{Qe}_3 = \mathbf{Q}^T\mathbf{e}_1\,.\,\mathbf{e}_3 = \mathbf{Q}^{-1}\mathbf{e}_1\,.\,\mathbf{e}_3 = \mathbf{e}_1\,.\,\mathbf{e}_3 = 0$, $Q_{31} = \mathbf{e}_3\,.\,\mathbf{Qe}_1 = \mathbf{e}_3\,.\,\mathbf{e}_1 = 0$, $Q_{23} = \mathbf{e}_2\,.\,\mathbf{Qe}_3 = \mathbf{Q}^T\mathbf{e}_2\,.\,\mathbf{e}_3 = \mathbf{Q}^{-1}\mathbf{e}_2\,.\,\mathbf{e}_3 = -\mathbf{e}_2\,.\,\mathbf{e}_3 = 0$, $Q_{32} = \mathbf{e}_3\,.\,\mathbf{Qe}_2 = -\mathbf{e}_3\,.\,\mathbf{e}_2 = 0$, and

$$\begin{aligned} Q_{33} = \mathbf{e}_3\,.\,\mathbf{Qe}_3 &= (\mathbf{e}_1 \times \mathbf{e}_2)\,.\,\mathbf{Qe}_3 = (\mathbf{Qe}_1 \times -\mathbf{Qe}_2)\,.\,\mathbf{Qe}_3 \\ &= -(\det\mathbf{Q})(\mathbf{e}_1 \times \mathbf{e}_2\,.\,\mathbf{e}_3) = -\det\mathbf{Q} = -1. \end{aligned}$$

Thus $$[\mathbf{Q}] = \begin{bmatrix} 1 & 0 & 0 \\ 0 & -1 & 0 \\ 0 & 0 & -1 \end{bmatrix}. \tag{A.16.22}$$

Accordingly (A.16.22) is a special case of (A.16.21) corresponding to $\alpha = \pi$.

Case (iii). If $\lambda = 1$ is a triple zero, then

$$C(\mathbf{Q};\lambda) \equiv (\lambda - 1)^3 \equiv \lambda^3 - 3\lambda^2 + 3\lambda - 1. \tag{A.16.23}$$

In particular [cf. (A.13.7)],

$$\operatorname{tr}\mathbf{Q} = 3. \tag{A.16.24}$$

Choose $\mathbf{e}_1$ so that $\mathbf{Q}\mathbf{e}_1 = \mathbf{e}_1$ with $\|\mathbf{e}_1\| = 1$. Proceeding exactly as in case (i),

$$3 = \operatorname{tr}\mathbf{Q} = 1 + a + d. \tag{A.16.25}$$

Since $|a| \leq 1$ and $|d| \leq 1$ (Why?), it follows that

$$a = 1 = d, \tag{A.16.26}$$

and hence [see (A.16.14)] $b = 0 = c$. Thus

$$[\mathbf{Q}] = \begin{bmatrix} 1 & 0 & 0 \\ 0 & 1 & 0 \\ 0 & 0 & 1 \end{bmatrix}, \tag{A.16.27}$$

a special case of (A.16.21) corresponding to $\alpha = 0$. Of course, in this case

$$\mathbf{Q} = \mathbf{1}. \tag{A.16.28}$$

Remark A.16.2. Since (A.16.21) has been shown to exhaust all possibilities, we have, for any $\mathbf{Q} \in \mathrm{Orth}^{+}\mathcal{V}$, the existence of a right-handed orthonormal basis $\{\mathbf{e}_1, \mathbf{e}_2, \mathbf{e}_3\}$ and angle α such that

$$\mathbf{Q}\mathbf{e}_1 = \mathbf{e}_1,\ \mathbf{Q}\mathbf{e}_2 = \cos\alpha\,\mathbf{e}_2 + \sin\alpha\,\mathbf{e}_3 \text{ and } \mathbf{Q}\mathbf{e}_3 = -\sin\alpha\,\mathbf{e}_2 + \cos\alpha\,\mathbf{e}_3. \tag{A.16.29}$$

The effect of $\mathbf{Q}$ can be visualised in terms of a rotation. If we imagine a point $\mathbf{x}$ in a rigid body, and lines through $\mathbf{x}$ parallel to $\mathbf{e}_1, \mathbf{e}_2$, and $\mathbf{e}_3$ fixed in this body, then the action of $\mathbf{Q}$ on $\mathbf{e}_1$, $\mathbf{e}_2$, and $\mathbf{e}_3$ does not change the direction of the line parallel to $\mathbf{e}_1$ but turns the lines parallel to $\mathbf{e}_2$ and $\mathbf{e}_3$ through an angle α in the sense of a right-handed screw with axis parallel to $\mathbf{e}_1$, rotating from $\mathbf{e}_2$ towards $\mathbf{e}_3$.

Exercise A.16.3. Show that

$$\det\mathbf{Q} = -1 \qquad \text{if and only if} \qquad -\mathbf{Q} \in \mathrm{Orth}^{+}\mathcal{V}. \tag{A.16.30}$$

Remark A.16.3. If $\det\mathbf{Q} = -1$, then from (A.16.30) $\mathbf{Q} = (-\mathbf{1})(-\mathbf{Q})$, where $-\mathbf{Q} \in \mathrm{Orth}^{+}\mathcal{V}$. In this case $\mathbf{Q}$ may be visualised in terms of a rotation, as in Remark A.16.2, followed by an *inversion* in which every point $\mathbf{y} \in \mathcal{E}$ is mapped into $\mathbf{y}'$, where $\mathbf{y}' - \mathbf{x} = -(\mathbf{y} - \mathbf{x})$. Of course, no rigid body can undergo such a distortion, which can be considered the result of successive reflections in three mutually perpendicular planes through $\mathbf{x}$. Any two such reflections yield a result achievable by a single rotation through π, but the remaining reflection is physically unachievable. (Convince yourself of this!)

Exercise A.16.4. Notice that from (A.16.29),

$$\mathbf{Q} = \mathbf{e}_1 \otimes \mathbf{e}_1 + \cos\alpha(\mathbf{e}_2 \otimes \mathbf{e}_2 + \mathbf{e}_3 \otimes \mathbf{e}_3) + \sin\alpha\,\mathbf{e}_3 \wedge \mathbf{e}_2. \tag{A.16.31}$$

Writing [see (A.8.21)]

$$\mathbf{W} := \mathbf{e}_3 \wedge \mathbf{e}_2 = \mathbf{e}_3 \otimes \mathbf{e}_2 - \mathbf{e}_2 \otimes \mathbf{e}_3,$$

show from Remark A.15.1 that the axial vector corresponding to $\mathbf{W}$ is $\mathbf{e}_1$. Show further that

$$-\mathbf{W}^2 = \mathbf{e}_2 \otimes \mathbf{e}_2 + \mathbf{e}_3 \otimes \mathbf{e}_3 \tag{A.16.32}$$

and that (A.16.31) may be written as

$$\mathbf{Q} = \mathbf{1} + \sin\alpha\, \mathbf{W} + (1 - \cos\alpha)\mathbf{W}^2. \tag{A.16.33}$$

Verify that

$$\mathbf{W}^3 = -\mathbf{W}. \tag{A.16.34}$$

Defining (cf., e.g., Goertzel & Tralli, Appendix 1B [89]), for $\mathbf{A} \in \mathrm{Lin}\,\mathcal{V}$,

$$\exp \mathbf{A} := \mathbf{1} + \mathbf{A} + \mathbf{A}^2/2! + \mathbf{A}^3/3! + \ldots, \tag{A.16.35}$$

show from (A.16.34) that

$$\mathbf{Q} = \exp(\alpha \mathbf{W}). \tag{A.16.36}$$

A.17 Symmetric and Positive-Definite Linear Transformations

In Section A.13 it was shown that any linear transformation has at least one real eigenvalue. Now suppose that $\mathbf{e}_1$ is an eigenvector of $\mathbf{L} \in \mathrm{Sym}\,\mathcal{V}$ [see (A.1.15)] with eigenvalue $\lambda_1 \in \mathbb{R}$. Without loss of generality $\|\mathbf{e}_1\| = 1$ (if not, $\mathbf{e}' := \mathbf{e}_1/\|\mathbf{e}_1\|$ has this property), and we may choose an orthonormal basis $\{\mathbf{e}_1, \mathbf{e}_2, \mathbf{e}_3\}$. Since $[\mathbf{L}^T] = [\mathbf{L}]^T$ for any $\mathbf{L} \in \mathrm{Lin}\,\mathcal{V}$ [see (A.9.6)], if $\mathbf{L} \in \mathrm{Sym}\,\mathcal{V}$, then $[\mathbf{L}]^T = [\mathbf{L}]$ for *any* matrix representation of $\mathbf{L}$. The matrix representation of $\mathbf{L}$ with respect to $\{\mathbf{e}_1, \mathbf{e}_2, \mathbf{e}_3\}$ thus takes the form

$$[\mathbf{L}] = \begin{bmatrix} \lambda_1 & 0 & 0 \\ 0 & a & b \\ 0 & b & d \end{bmatrix} \tag{A.17.1}$$

for some $a, b, d \in \mathbb{R}$. The characteristic equation for $\mathbf{L}$ is

$$\det(\mathbf{L} - \lambda\mathbf{1}) = \det[\mathbf{L} - \lambda\mathbf{1}] = 0. \tag{A.17.2}$$

Thus

$$(\lambda - \lambda_1)\{(a - \lambda)(d - \lambda) - b^2\} = 0, \tag{A.17.3}$$

and so $\lambda = \lambda_1$ or

$$\left(\lambda - \frac{(a+d)}{2}\right)^2 = \frac{(a-d)^2}{4} + b^2 \geq 0. \tag{A.17.4}$$

Since (A.17.4) has two real (possibly equal) solutions λ (Why?), (A.17.2) *has three real solutions with possible repetitions*. If (A.17.2) has three equal solutions λ_1, then equality must hold in (A.17.4), and hence $a = d, b = 0$, and $(a+d)/2 = \lambda_1$. Thus $a = \lambda_1 = d$, and from (A.17.1)

$$\mathbf{L} = \lambda_1 \mathbf{1} = \lambda_1 \mathbf{e}_i \otimes \mathbf{e}_i. \tag{A.17.5}$$

It may be that (A.17.2) has a repeated solution $\lambda_2 \neq \lambda_1$; that is, equality holds in (A.17.4), so $a = d, b = 0$, and $(a+d)/2 = \lambda_2 \neq \lambda_1$. In such case

$$\mathbf{L} = \lambda_1 \mathbf{e}_1 \otimes \mathbf{e}_1 + \lambda_2(\mathbf{e}_2 \otimes \mathbf{e}_2 + \mathbf{e}_3 \otimes \mathbf{e}_3). \quad (A.17.6)$$

Finally, if $\mathbf{L}$ has three distinct eigenvalues λ_1, λ_2, and λ_3, and $\mathbf{u}_1, \mathbf{u}_2$, and $\mathbf{u}_3$ are corresponding eigenvectors, then

$$\lambda_i \mathbf{u}_i \,.\, \mathbf{u}_j = \mathbf{L}\mathbf{u}_i \,.\, \mathbf{u}_j = \mathbf{u}_i \,.\, \mathbf{L}^T \mathbf{u}_j = \mathbf{u}_i \,.\, \mathbf{L}\mathbf{u}_j = \mathbf{u}_i \,.\, \lambda_j \mathbf{u}_j. \quad (A.17.7)$$

Thus

$$(\lambda_i - \lambda_j)\mathbf{u}_i \,.\, \mathbf{u}_j = 0, \quad (A.17.8)$$

so $\lambda_i \neq \lambda_j$ implies $\mathbf{u}_i \,.\, \mathbf{u}_j = 0$. That is, $\mathbf{u}_1, \mathbf{u}_2$, and $\mathbf{u}_3$ are mutually orthogonal. If $\mathbf{e}_i := \mathbf{u}_i / \|\mathbf{u}_i\|$, then $\{\mathbf{e}_1, \mathbf{e}_2, \mathbf{e}_3\}$ constitutes an orthonormal basis with respect to which

$$[\mathbf{L}] = \mathrm{diag}(\lambda_1, \lambda_2, \lambda_3). \quad (A.17.9)$$

Equivalently,

$$\mathbf{L} = \lambda_1 \mathbf{e}_1 \otimes \mathbf{e}_1 + \lambda_2 \mathbf{e}_2 \otimes \mathbf{e}_2 + \lambda_3 \mathbf{e}_3 \otimes \mathbf{e}_3. \quad (A.17.10)$$

Exercise A.17.1. Show that if $\mathbf{L} \in \mathrm{Sym}$, then via (A.17.9) and (A.12.17)

$$\mathrm{tr}\,\mathbf{L} = \lambda_1 + \lambda_2 + \lambda_3, \qquad I_2(\mathbf{L}) = \lambda_1\lambda_2 + \lambda_2\lambda_3 + \lambda_3\lambda_1 \qquad \text{and } \det \mathbf{L} = \lambda_1\lambda_2\lambda_3.$$

Show also that if $\lambda_1\lambda_2\lambda_3 \neq 0$, then

$$\mathbf{L}^{-1} = \lambda_1^{-1}\mathbf{e}_1 \otimes \mathbf{e}_1 + \lambda_2^{-1}\mathbf{e}_2 \otimes \mathbf{e}_2 + \lambda_3^{-1}\mathbf{e}_3 \otimes \mathbf{e}_3.$$

A linear transformation $\mathbf{L}$ is said to be *positive definite* if

$$\mathbf{L}\mathbf{v} \,.\, \mathbf{v} > 0 \qquad \text{whenever} \quad \mathbf{v} \neq \mathbf{0}. \quad (A.17.11)$$

Accordingly

$$\mathbf{L}\mathbf{v} = \mathbf{0} \qquad \text{implies} \qquad \mathbf{v} = \mathbf{0} \quad (A.17.12)$$

and hence from (A.10.4) $\mathbf{L}$ *is invertible*. Notice that the skew part of $\mathbf{L}$ plays no role in (A.17.11) since

$$\frac{1}{2}(\mathbf{L} - \mathbf{L}^T)\mathbf{v} \,.\, \mathbf{v} = \frac{1}{2}(\mathbf{L}\mathbf{v} \,.\, \mathbf{v} - \mathbf{L}^T\mathbf{v} \,.\, \mathbf{v}) = \frac{1}{2}(\mathbf{L}\mathbf{v} \,.\, \mathbf{v} - \mathbf{v} \,.\, \mathbf{L}\mathbf{v}) = 0. \quad (A.17.13)$$

Now consider *symmetric* positive-definite linear transformations, the set of which is denoted by $\mathrm{Sym}^+\mathcal{V}$; that is,

$$\mathrm{Sym}^+\mathcal{V} := \{\mathbf{L} \in \mathrm{Lin}\,\mathcal{V} : \mathbf{L} = \mathbf{L}^T \text{ and } \mathbf{L}\mathbf{v} \,.\, \mathbf{v} > 0 \text{ whenever } \mathbf{v} \neq \mathbf{0}\}. \quad (A.17.14)$$

Since $\mathbf{L} \in \mathrm{Sym}\,\mathcal{V}$, it is expressible in form (A.17.10). Further, for any eigenvector $\mathbf{e}$,

$$\lambda \|\mathbf{e}\|^2 = \lambda \mathbf{e} \,.\, \mathbf{e} = \mathbf{L}\mathbf{e} \,.\, \mathbf{e} > 0 \quad (A.17.15)$$

whence $\lambda > 0$. That is, *if* $\mathbf{L} \in \mathrm{Sym}^+\mathcal{V}$, *then there exists an orthonormal basis* $\{\mathbf{e}_1, \mathbf{e}_2, \mathbf{e}_3\}$ *of eigenvectors with positive eigenvalues* λ_1, λ_2, and λ_3 *for which (A.17.10) holds.* Further, from Exercise A.17.1,

$$\mathrm{tr}\,\mathbf{L} > 0, \qquad I_2(\mathbf{L}) > 0 \qquad \text{and} \qquad \det \mathbf{L} > 0. \quad (A.17.16)$$

Exercise A.17.2. If $\mathbf{L} \in \mathrm{Lin}\,\mathcal{V}$, show that

$$\mathbf{L}\mathbf{L}^T \in \mathrm{Sym}\mathcal{V} \qquad \text{and} \qquad \mathbf{L}^T\mathbf{L} \in \mathrm{Sym}\mathcal{V} \tag{A.17.17}$$

and that each of these symmetric transformations have non-negative eigenvalues.

We now prove (A.14.21), namely that

$$\textit{if} \qquad \mathbf{L}_1, \mathbf{L}_2 \in \mathrm{Lin}\,\mathcal{V}, \qquad \textit{then} \qquad \|\mathbf{L}_1\mathbf{L}_2\| \leq \|\mathbf{L}_1\|\|\mathbf{L}_2\|. \tag{A.17.18}$$

Firstly, notice that

$$\begin{aligned}\|\mathbf{L}_1\mathbf{L}_2\|^2 := \mathbf{L}_1\mathbf{L}_2 \cdot \mathbf{L}_1\mathbf{L}_2 &:= \mathrm{tr}\{(\mathbf{L}_1\mathbf{L}_2)^T\mathbf{L}_1\mathbf{L}_2\} = \mathrm{tr}\{\mathbf{L}_2^T\mathbf{L}_1^T\mathbf{L}_1\mathbf{L}_2\}\\ &= \mathrm{tr}\{\mathbf{L}_2\mathbf{L}_2^T\mathbf{L}_1^T\mathbf{L}_1\} = \mathrm{tr}\{(\mathbf{L}_2\mathbf{L}_2^T)^T\mathbf{L}_1^T\mathbf{L}_1\}\\ &=: \mathbf{L}_2\mathbf{L}_2^T \cdot \mathbf{L}_1^T\mathbf{L}_1 \leq \|\mathbf{L}_2\mathbf{L}_2^T\|\|\mathbf{L}_1^T\mathbf{L}_1\|.\end{aligned} \tag{A.17.19}$$

Here properties (A.8.7), (A.12.14), and the norm property (A.14.7), valid for any inner product space $\mathcal{U}$, have been invoked for space Lin $\mathcal{V}$. Now, from (A.17.17), $\mathbf{L} := \mathbf{L}_1^T\mathbf{L}_1 \in \mathrm{Sym}\mathcal{V}$ and is expressible in form (A.17.10) with $\lambda_i \geq 0$ $(i = 1,2,3)$. Thus

$$\|\mathbf{L}_1^T\mathbf{L}_1\|^2 = \mathbf{L}_1^T\mathbf{L}_1 \cdot \mathbf{L}_1^T\mathbf{L}_1 = \lambda_1^2 + \lambda_2^2 + \lambda_3^2, \tag{A.17.20}$$

noting that $\{\mathbf{e}_i \otimes \mathbf{e}_j\}$ is an orthonormal basis for Lin $\mathcal{V}$. Further,

$$\|\mathbf{L}_1\|^4 = (\mathbf{L}_1 \cdot \mathbf{L}_1)^2 = [\mathrm{tr}(\mathbf{L}_1^T\mathbf{L}_1)]^2 = (\lambda_1 + \lambda_2 + \lambda_3)^2. \tag{A.17.21}$$

Since each $\lambda_i \geq 0$, $(\lambda_1 + \lambda_2 + \lambda_3)^2 \geq \lambda_1^2 + \lambda_2^2 + \lambda_3^2$, and so

$$\|\mathbf{L}_1^T\mathbf{L}_1\|^2 \leq \|\mathbf{L}_1\|^4. \tag{A.17.22}$$

Taking positive square roots,

$$\|\mathbf{L}_1^T\mathbf{L}_1\| \leq \|\mathbf{L}_1\|^2 \qquad \text{and similarly} \qquad \|\mathbf{L}_2\mathbf{L^T}_2\| \leq \|\mathbf{L}_2\|^2. \tag{A.17.23}$$

From (A.17.19) and (A.17.23), if $\mathbf{L}_1$ and $\mathbf{L}_2 \in \mathrm{Lin}\,\mathcal{V}$ then

$$\|\mathbf{L}_1\mathbf{L}_2\| \leq \|\mathbf{L}_1\|\|\mathbf{L}_2\|. \tag{A.17.24}$$

Of course, if one or both of $\mathbf{L}_1$, and $\mathbf{L}_2$ equals $\mathbf{O}$, then (A.17.18) holds as an equality.

If $\mathbf{L} \in \mathrm{Lin}\,\mathcal{V}$ and for some $\mathbf{A} \in \mathrm{Lin}\,\mathcal{V}$

$$\mathbf{A}^2 = \mathbf{L}, \tag{A.17.25}$$

then $\mathbf{A}$ is said to be *a* square root of $\mathbf{L}$.

Exercise A.17.3. Notice that if $\{\mathbf{e}_1, \mathbf{e}_2, \mathbf{e}_3\}$ is an orthonormal basis for $\mathcal{V}$, then each of the eight linear transformations $\pm\mathbf{e}_1 \otimes \mathbf{e}_1 \pm \mathbf{e}_2 \otimes \mathbf{e}_2 \pm \mathbf{e}_3 \otimes \mathbf{e}_3$ is a square root of $\mathbf{1}$. Write down the corresponding matrices.

If $\mathbf{L} \in \mathrm{Sym}^+\mathcal{V}$, *then there exists a unique positive-definite symmetric square root* $\mathbf{A}$. Indeed, noting that in such case $\mathbf{L}$ is expressible in the form (A.17.10) with $\lambda_i > 0$ $(i = 1,2,3)$, we have

$$\mathbf{A} = \lambda_1^{1/2}\mathbf{e}_1 \otimes \mathbf{e}_1 + \lambda_2^{1/2}\mathbf{e}_2 \otimes \mathbf{e}_2 + \lambda_3^{1/2}\mathbf{e}_3 \otimes \mathbf{e}_3. \tag{A.17.26}$$

Here $\lambda_i^{1/2}$ denotes the positive square root of λ_i. We write

$$\mathbf{A} =: \mathbf{L}^{1/2}. \tag{A.17.27}$$

Exercise A.17.4. Check that $(\mathbf{L}^{1/2})^2 = \mathbf{L}$ and $\mathbf{L}^{1/2} \in \mathrm{Sym}^+\mathcal{V}$.

A.18 The Polar Decomposition Theorem

If $\mathbf{F} \in \mathrm{Invlin}^+\mathcal{V}$ [see (A.12.33)], then there exist unique $\mathbf{U}, \mathbf{V} \in \mathrm{Sym}^+\mathcal{V}$ and unique $\mathbf{R} \in \mathrm{Orth}^+\mathcal{V}$ [see (A.16.9)] such that

$$\mathbf{RU} = \mathbf{F} = \mathbf{VR}. \tag{A.18.1}$$

Proof. If the result holds, then

$$\mathbf{F}^T\mathbf{F} = (\mathbf{RU})^T(\mathbf{RU}) = \mathbf{U}^T\mathbf{R}^T\mathbf{RU} = \mathbf{U1U} = \mathbf{U}^2, \tag{A.18.2}$$

and so $\mathbf{U}^2 \in \mathrm{Sym}^+\mathcal{V}$. (Verify this!) Thus we are led to define [see (A.17.26)]

$$\mathbf{U} := (\mathbf{F}^T\mathbf{F})^{1/2} \tag{A.18.3}$$

and note that $\mathbf{U} \in \mathrm{Sym}^+\mathcal{V}$ (see Exercise A.17.4).

Now define

$$\mathbf{R}_1 := \mathbf{FU}^{-1}. \tag{A.18.4}$$

Clearly, from (A.18.3), (A.12.29), (A.12.30), and $(\text{A.17.16})_3$,

$$\det\mathbf{R}_1 = \det(\mathbf{FU}^{-1}) = (\det\mathbf{F})(\det\mathbf{U}^{-1}) = (\det\mathbf{F})(\det\mathbf{U})^{-1} > 0. \tag{A.18.5}$$

Further,

$$\mathbf{R}_1^T\mathbf{R}_1 = (\mathbf{FU}^{-1})^T\mathbf{FU}^{-1} = (\mathbf{U}^{-1})^T\mathbf{F}^T\mathbf{FU}^{-1} = (\mathbf{U}^T)^{-1}\mathbf{U}^2\mathbf{U}^{-1} = \mathbf{U}^{-1}\mathbf{U}^2\mathbf{U} = \mathbf{1}. \tag{A.18.6}$$

Accordingly $\mathbf{R}_1 \in \mathrm{Orth}\,\mathcal{V}$, and via (A.18.5) and (A.16.9), $\mathbf{R}_1 \in \mathrm{Orth}^+\mathcal{V}$. Similarly, consideration of $(\text{A.18.1})_2$ leads to a candidate $\mathbf{V}$ for which $\mathbf{V}^2 = \mathbf{FF}^T$ (Show this!). Hence we choose

$$\mathbf{V} := (\mathbf{FF}^T)^{1/2}, \tag{A.18.7}$$

on noting $\mathbf{FF}^T \in \mathrm{Sym}^+\mathcal{V}$. With

$$\mathbf{R}_2 := \mathbf{V}^{-1}\mathbf{F} \tag{A.18.8}$$

we have

$$\mathbf{R}_2\mathbf{R}_2^T = (\mathbf{V}^{-1}\mathbf{F})(\mathbf{V}^{-1}\mathbf{F})^T = \mathbf{V}^{-1}\mathbf{FF}^T(\mathbf{V}^{-1})^T = \mathbf{V}^{-1}\mathbf{V}^2(\mathbf{V}^T)^{-1} = \mathbf{V}^{-1}\mathbf{V}^2\mathbf{V}^{-1} = \mathbf{1}. \tag{A.18.9}$$

Since $\det\mathbf{R}_2 > 0$ (Why?), it follows from (A.18.9) that $\mathbf{R}_2 \in \mathrm{Orth}^+\mathcal{V}$.

At this point we have shown that

$$\mathbf{R}_1\mathbf{U} = \mathbf{F} = \mathbf{VR}_2, \tag{A.18.10}$$

where by their construction $\mathbf{U}, \mathbf{V}, \mathbf{R}_1$, and $\mathbf{R}_2$ are *unique* [and given by (A.18.3), (A.18.4), (A.18.7), and (A.18.8)]. To show that $\mathbf{R}_1 = \mathbf{R}_2$, consider

$$\begin{aligned}\mathbf{U} = (\mathbf{F}^T\mathbf{F})^{1/2} &= ((\mathbf{VR}_2)^T\mathbf{VR}_2)^{1/2} = (\mathbf{R}_2^T\mathbf{V}^2\mathbf{R}_2)^{1/2} \\ &= \mathbf{R}_2^T\mathbf{VR}_2.\end{aligned} \tag{A.18.11}$$

Thus

$$\mathbf{R}_2\mathbf{U} = \mathbf{V}\mathbf{R}_2 (= \mathbf{F}). \tag{A.18.12}$$

Comparison with (A.18.10) yields

$$\mathbf{R}_2\mathbf{U} = \mathbf{R}_1\mathbf{U} \tag{A.18.13}$$

and hence $\mathbf{R}_2 = \mathbf{R}_1$.

Remark A.18.1. If $\mathbf{F} \in \text{Invlin}\,\mathcal{V}$, then the foregoing argument again yields unique decompositions $\mathbf{RU} = \mathbf{F} = \mathbf{VR}$ with $\mathbf{U}, \mathbf{V} \in \text{Sym}^+\mathcal{V}$ and $\mathbf{R} \in \text{Orth}\,\mathcal{V}$ (rather than $\mathbf{R} \in \text{Orth}^+\mathcal{V}$). Show this!

A.19 Third-Order Tensors and Elements of Tensor Algebra

In modelling the effect of couples transmitted across surfaces we encounter the notion of a couple-stress field whose values map $\mathcal{V}$ linearly into Lin $\mathcal{V}$. Any such map is termed a *third-order tensor*.[11] The set of all such maps is denoted by $\text{Lin}(\mathcal{V}, \text{Lin}\,\mathcal{V})$. Thus, if $\mathbf{C} \in \text{Lin}(\mathcal{V}, \text{Lin}\,\mathcal{V})$, then, for all $\mathbf{v}, \mathbf{v}_1, \mathbf{v}_2 \in \mathcal{V}$ and $\alpha_1, \alpha_2 \in \mathbb{R}$,

$$\mathbf{Cv} \in \text{Lin}\,\mathcal{V} \qquad \text{and} \qquad \mathbf{C}(\alpha_1\mathbf{v}_1 + \alpha_2\mathbf{v}_2) = \alpha_1\mathbf{C}\mathbf{v}_1 + \alpha_2\mathbf{C}\mathbf{v}_2. \tag{A.19.1}$$

Set $\text{Lin}(\mathcal{V}, \text{Lin}\,\mathcal{V})$ has the structure of a vector space over $\mathbb{R}$ when the natural definitions

$$(\mathbf{C}_1 + \mathbf{C}_2)\mathbf{v} := \mathbf{C}_1\mathbf{v} + \mathbf{C}_2\mathbf{v}, \qquad \text{and} \qquad (\alpha\mathbf{C})\mathbf{v} := \alpha(\mathbf{Cv}) \tag{A.19.2}$$

are introduced. [Here $\mathbf{C}_1, \mathbf{C}_2, \mathbf{C} \in \text{Lin}(\mathcal{V}, \text{Lin}\,\mathcal{V})$, $\mathbf{v} \in \mathcal{V}$, and $\alpha \in \mathbb{R}$ are arbitrary.] The natural 'zero' element $\mathcal{O}$ of $\text{Lin}(\mathcal{V}, \text{Lin}\,\mathcal{V})$ satisfies $\mathcal{O}\mathbf{v} = \mathbf{O} \in \text{Lin}\,\mathcal{V}$.

Exercise A.19.1. Show that $\text{Lin}(\mathcal{V}, \text{Lin}\,\mathcal{V})$ is a vector space. (See Sections A.7 and A.4).

It follows from (A.19.2) that any $\mathbf{C} \in \text{Lin}(\mathcal{V}, \text{Lin}\,\mathcal{V})$ is uniquely determined by its action on a basis for $\mathcal{V}$. In particular, if $\{\mathbf{e}_1, \mathbf{e}_2, \mathbf{e}_3\}$ is an orthonormal basis for $\mathcal{V}$ and $\mathbf{v} = v_k\mathbf{e}_k$, then

$$\mathbf{Cv} = \mathbf{C}v_k\mathbf{e}_k = v_k\mathbf{C}\mathbf{e}_k. \tag{A.19.3}$$

Since $\mathbf{C}\mathbf{e}_k \in \text{Lin}\,\mathcal{V}$, it is completely characterised [see (A.9.1)] by the 9 numbers (recall $i, j = 1, 2, 3$)

$$\mathbf{e}_i \,.\, (\mathbf{C}\mathbf{e}_k)\mathbf{e}_j =: C_{ijk}. \tag{A.19.4}$$

It follows that $\mathbf{C}$ *itself is completely characterised by the 27 numbers* C_{ijk}.

Given $\mathbf{a}, \mathbf{b}, \mathbf{c} \in \mathcal{V}$, we define the *triple tensor product* $\mathbf{a} \otimes \mathbf{b} \otimes \mathbf{c} \in \text{Lin}(\mathcal{V}, \text{Lin}\,\mathcal{V})$ by

$$(\mathbf{a} \otimes \mathbf{b} \otimes \mathbf{c})\mathbf{u} := (\mathbf{c} \cdot \mathbf{u})\mathbf{a} \otimes \mathbf{b} \quad \text{for any} \quad \mathbf{u} \in \mathcal{V}. \tag{A.19.5}$$

[11] Elements of $\mathcal{V}$ and Lin $\mathcal{V}$ are termed, respectively, *first-order* and *second-order tensors*.

Exercise A.19.2. Check that $\mathbf{a}\otimes\mathbf{b}\otimes\mathbf{c}\in \mathrm{Lin}(\mathcal{V},\mathrm{Lin}\,\mathcal{V})$; that is, $(\mathbf{a}\otimes\mathbf{b}\otimes\mathbf{c})\mathbf{u}\in \mathrm{Lin}\,\mathcal{V}$ and $(\mathbf{a}\otimes\mathbf{b}\otimes\mathbf{c})(\alpha_1\mathbf{u}_1+\alpha_2\mathbf{u}_2)=\alpha_1(\mathbf{a}\otimes\mathbf{b}\otimes\mathbf{c})\mathbf{u}_1+\alpha_2(\mathbf{a}\otimes\mathbf{b}\otimes\mathbf{c})\mathbf{u}_2$.

From (A.8.8) and (A.19.5),

$$((\mathbf{a}\otimes\mathbf{b}\otimes\mathbf{c})\mathbf{u})\mathbf{v}=(\mathbf{c}\,.\,\mathbf{u})(\mathbf{b}\,.\,\mathbf{v})\mathbf{a} \tag{A.19.6}$$

and hence $$\mathbf{e}_i\,.\,((\mathbf{a}\otimes\mathbf{b}\otimes\mathbf{c})\mathbf{e}_\mathbf{k})\mathbf{e}_j=(\mathbf{c}\,.\,\mathbf{e}_k)(\mathbf{b}\,.\,\mathbf{e}_j)(\mathbf{e}_i\,.\,\mathbf{a})=c_k\,b_j\,a_i. \tag{A.19.7}$$

Accordingly [see (A.19.4)]

$$(\mathbf{a}\otimes\mathbf{b}\otimes\mathbf{c})_{ijk}=a_i\,b_j\,c_k. \tag{A.19.8}$$

Exercise A.19.3. Show from (A.19.6) that the triple tensor product $\mathbf{a}\otimes\mathbf{b}\otimes\mathbf{c}$ is linear in each argument; that is, $(\alpha_1\mathbf{a}_1+\alpha_2\mathbf{a}_2)\otimes\mathbf{b}\otimes\mathbf{c}=\alpha_1\mathbf{a}_1\otimes\mathbf{b}\otimes\mathbf{c}+\alpha_2\mathbf{a}_2\otimes\mathbf{b}\otimes\mathbf{c}$, etc. Deduce that

$$\mathbf{a}\otimes\mathbf{b}\otimes\mathbf{c}=a_ib_jc_k\mathbf{e}_i\otimes\mathbf{e}_j\otimes\mathbf{e}_k. \tag{A.19.9}$$

Exercise A.19.4. Prove that if $\mathbf{C}\in \mathrm{Lin}(\mathcal{V},\mathrm{Lin}\mathcal{V})$, then

$$\mathbf{C}=C_{pqr}\,\mathbf{e}_p\otimes\mathbf{e}_q\otimes\mathbf{e}_r \tag{A.19.10}$$

by showing that $\mathbf{e}_i.(\mathbf{C}\mathbf{e}_k)\mathbf{e}_j$ coincides with $\mathbf{e}_i.((C_{pqr}\,\mathbf{e}_p\otimes\mathbf{e}_q\otimes\mathbf{e}_r)\mathbf{e}_k)\mathbf{e}_j$ for all i,j,k. [Note that (A.19.8) and (A.19.9) constitute a special case of this result.]

It follows from (A.19.10) that $\{\mathbf{e}_p\otimes\mathbf{e}_q\otimes\mathbf{e}_r\}$, with $p,q,r=1,2,3$, *spans* $\mathrm{Lin}(\mathcal{V},\mathrm{Lin}\mathcal{V})$. In fact this set is linearly independent since, if for 27 numbers α_{pqr}

$$\alpha_{pqr}\,\mathbf{e}_p\otimes\mathbf{e}_q\otimes\mathbf{e}_r=\mathcal{O},$$

then $\alpha_{ijk}=0$ for each and every $i,j,k=1,2,3$. To see this, note that

$$0=\mathbf{e}_i.(\mathcal{O}\mathbf{e}_k)\mathbf{e}_j=\mathbf{e}_i.((\alpha_{pqr}\mathbf{e}_p\otimes\mathbf{e}_q\otimes\mathbf{e}_r)\mathbf{e}_k)\mathbf{e}_j=\alpha_{pqr}\delta_{rk}\delta_{qj}\delta_{ip}=\alpha_{ijk}.$$

Accordingly $\{\mathbf{e}_p\otimes\mathbf{e}_q\otimes\mathbf{e}_r\}$ *is a basis for* $\mathrm{Lin}\{\mathcal{V},\mathrm{Lin}\mathcal{V}\}$, *and hence this space has dimension* 27.

Recall definition (A.8.1) of the transpose $\mathbf{L}^T$ of $\mathbf{L}\in\mathrm{Lin}\,\mathcal{V}$. Specifically, for any $\mathbf{u},\mathbf{v}\in\mathcal{V}$,

$$\mathbf{L}^T\mathbf{u}\,.\,\mathbf{v}:=\mathbf{L}\mathbf{v}\,.\,\mathbf{u}, \tag{A.19.11}$$

and $\mathbf{L}^T$ was shown to be an element of Lin $\mathcal{V}$ via (A.8.2). Further, the transpose operation maps Lin $\mathcal{V}$ linearly into Lin $\mathcal{V}$ [see (A.8.4) and (A.8.5)]. There are two natural analogues of the transpose for elements of $\mathrm{Lin}(\mathcal{V},\mathrm{Lin}\,\mathcal{V})$, namely $\mathbf{C}^T$ and $\mathbf{C}^\sim$, defined for any $\mathbf{u},\mathbf{v},\mathbf{w}\in\mathcal{V}$ by

$$((\mathbf{C}^T\mathbf{u})\mathbf{v}):=(\mathbf{C}\mathbf{v})\mathbf{u} \tag{A.19.12}$$

and $$((\mathbf{C}^\sim\mathbf{u})\mathbf{v}).\mathbf{w}:=((\mathbf{C}\mathbf{w})\mathbf{v})\,.\,\mathbf{w}. \tag{A.19.13}$$

Exercise A.19.5. Show that $\mathbf{C}^T$ and $\mathbf{C}^\sim$ are elements of $\mathrm{Lin}(\mathcal{V}, \mathrm{Lin}\,\mathcal{V})$ and that the operation maps

$$^T : \mathbf{C} \to \mathbf{C}^T \qquad \text{and} \qquad ^\sim : \mathbf{C} \to \mathbf{C}^\sim$$

are linear from $\mathrm{Lin}(\mathcal{V}, \mathrm{Lin}\,\mathcal{V})$ into itself. That is,

$$(\alpha_1\mathbf{C}_1 + \alpha_2\mathbf{C}_2)^T = \alpha_1\mathbf{C}_1^T + \alpha_2\mathbf{C}_2^T \text{ and } (\alpha_1\mathbf{C}_1 + \alpha_2\mathbf{C}_2)^\sim = \alpha_1\mathbf{C}_1^\sim + \alpha_2\mathbf{C}_2^\sim \qquad \text{(A.19.14)}$$

for all $\alpha_1, \alpha_2 \in \mathbb{R}$ and $\mathbf{C}_1, \mathbf{C}_2 \in \mathrm{Lin}(\mathcal{V}, \mathrm{Lin}\,\mathcal{V})$).

Notice that for any $\mathbf{u}, \mathbf{v} \in \mathcal{V}$,

$$\begin{aligned}((\mathbf{a}\otimes\mathbf{b}\otimes\mathbf{c})^T\mathbf{u})\mathbf{v} &= ((\mathbf{a}\otimes\mathbf{b}\otimes\mathbf{c})\mathbf{v})\mathbf{u} = (\mathbf{c}\,.\,\mathbf{v})(\mathbf{b}\,.\,\mathbf{u})\mathbf{a} \\ &= ((\mathbf{a}\otimes\mathbf{c}\otimes\mathbf{b})\mathbf{u})\mathbf{v}.\end{aligned}$$

Accordingly
$$(\mathbf{a}\otimes\mathbf{b}\otimes\mathbf{c})^T = \mathbf{a}\otimes\mathbf{c}\otimes\mathbf{b}. \qquad \text{(A.19.15)}$$

Exercise A.19.6. Show that

$$(\mathbf{a}\otimes\mathbf{b}\otimes\mathbf{c})^\sim = \mathbf{c}\otimes\mathbf{b}\otimes\mathbf{a}. \qquad \text{(A.19.16)}$$

Notice that [cf. (A.19.10)]

$$\mathbf{C}^T = (C^T)_{ijk}\mathbf{e}_i\otimes\mathbf{e}_j\otimes\mathbf{e}_k, \qquad \text{(A.19.17)}$$

and, from $(\text{A.19.14})_1$ and (A.19.15),

$$\begin{aligned}\mathbf{C}^T &= (C_{ijk}\mathbf{e}_i\otimes\mathbf{e}_j\otimes\mathbf{e}_k)^T = C_{ijk}(\mathbf{e}_i\otimes\mathbf{e}_j\otimes\mathbf{e}_k)^T \\ &= C_{ijk}\mathbf{e}_i\otimes\mathbf{e}_k\otimes\mathbf{e}_j = C_{ikj}\mathbf{e}_i\otimes\mathbf{e}_j\otimes\mathbf{e}_k,\end{aligned} \qquad \text{(A.19.18)}$$

where in the last step the dummy suffices k and j have been re-labelled j and k. Comparison of (A.19.17) with (A.19.18), together with the uniqueness of the definition of $(C^T)_{ijk}$, yields

$$(C^T)_{ijk} = C_{ikj}. \qquad \text{(A.19.19)}$$

Exercise A.19.7. Use $(\text{A.19.14})_2$ and (A.19.16) to show that if [see (A.19.10)]

$$\mathbf{C}^\sim = (C^\sim)_{ijk}\mathbf{e}_i\otimes\mathbf{e}_j\otimes\mathbf{e}_k, \qquad \text{(A.19.20)}$$

then
$$(C^\sim)_{ijk} = C_{kji}. \qquad \text{(A.19.21)}$$

At this point it is useful to review the manner in which vectors, linear transformations, and third-order tensors can combine. In particular, if $\mathbf{v} \in \mathcal{V}, \mathbf{L} \in \mathcal{V}$, and $\mathbf{C} \in \mathrm{Lin}(\mathcal{V}, \mathrm{Lin}\,\mathcal{V})$, then $\mathbf{L}\mathbf{v} \in \mathcal{V}, \mathbf{C}\mathbf{v} \in \mathrm{Lin}\,\mathcal{V}, (\mathbf{C}\mathbf{v})\mathbf{L} \in \mathrm{Lin}\,\mathcal{V}, \mathbf{L}(\mathbf{C}\mathbf{v}) \in \mathrm{Lin}\,\mathcal{V}$, and $\mathbf{C}\mathbf{v}\cdot\mathbf{L} \in \mathbb{R}$. We may ask if there are other possibilities, and whether there is a natural inner product on $\mathrm{Lin}(\mathcal{V}, \mathrm{Lin}\,\mathcal{V})$.

The answers are based upon (A.9.8) and (A.19.10). These relations show that general elements of $\mathrm{Lin}\,\mathcal{V}$ and $\mathrm{Lin}(\mathcal{V}, \mathrm{Lin}\,\mathcal{V})$ are expressible as linear combinations of tensor products of vectors. Such products $\mathbf{a}\otimes\mathbf{b}$ and $\mathbf{a}\otimes\mathbf{b}\otimes\mathbf{c}$ [see (A.8.8) and (A.19.5)] are termed *simple tensors* (of orders 2 and 3, respectively). Consider ($\alpha = 2,3$)

$$\otimes^{\alpha}\mathcal{V} := \{\text{linear combinations of order } \alpha \text{ simple tensors}\}. \tag{A.19.22}$$

Addition and multiplication by scalars are simply defined in $\otimes^{\alpha}\mathcal{V}$ by

$$(\alpha_1(\mathbf{a}_1\otimes\mathbf{b}_1)+\alpha_2(\mathbf{a}_2\otimes\mathbf{b}_2))\mathbf{v} := \alpha_1(\mathbf{a}_1\otimes\mathbf{b}_1)\mathbf{v}+\alpha_2(\mathbf{a}_2\otimes\mathbf{b}_2)\mathbf{v} \tag{A.19.23}$$

and

$$(\alpha_1(\mathbf{a}_1\otimes\mathbf{b}_1\otimes\mathbf{c}_1)+\alpha_2(\mathbf{a}_2\otimes\mathbf{b}_2\otimes\mathbf{c}_2))\mathbf{v} := \alpha_1(\mathbf{a}_1\otimes\mathbf{b}_1\otimes\mathbf{c}_1)\mathbf{v}+\alpha_2(\mathbf{a}_2\otimes\mathbf{b}_2\otimes\mathbf{c}_2)\mathbf{v}. \tag{A.19.24}$$

Clearly, any element of $\otimes^2\mathcal{V}$ can be identified with an element of $\mathrm{Lin}\,\mathcal{V}$ and [via (A.9.8)] vice versa. Similarly, $\otimes^3\mathcal{V}$ can be identified with $\mathrm{Lin}(\mathcal{V},\mathrm{Lin}\,\mathcal{V})$ via (A.19.24) and (A.19.10).

Now consider the following combinations between simple tensors:

$$(\mathbf{a}\otimes\mathbf{b}\otimes\mathbf{c}) : (\mathbf{u}\otimes\mathbf{v}\otimes\mathbf{w}) := (\mathbf{c}\,.\,\mathbf{u})(\mathbf{b}\,.\,\mathbf{v})\mathbf{a}\otimes\mathbf{w} \tag{A.19.25}$$

$$(\mathbf{a}\otimes\mathbf{b}\otimes\mathbf{c}) : (\mathbf{u}\otimes\mathbf{v}) := (\mathbf{c}\,.\,\mathbf{u})(\mathbf{b}\,.\,\mathbf{v})\mathbf{a} \tag{A.19.26}$$

$$(\mathbf{a}\otimes\mathbf{b}) : (\mathbf{u}\otimes\mathbf{v}\otimes\mathbf{w}) := (\mathbf{b}\,.\,\mathbf{u})(\mathbf{a}\,.\,\mathbf{v})\mathbf{w} \tag{A.19.27}$$

$$(\mathbf{a}\otimes\mathbf{b}) : (\mathbf{u}\otimes\mathbf{v}) := (\mathbf{b}\,.\,\mathbf{u})(\mathbf{a}\,.\,\mathbf{v}) \tag{A.19.28}$$

and

$$(\mathbf{a}\otimes\mathbf{b}\otimes\mathbf{c}) \,\vdots\, (\mathbf{u}\otimes\mathbf{v}\otimes\mathbf{w}) := (\mathbf{c}\,.\,\mathbf{u})(\mathbf{b}\,.\,\mathbf{v})(\mathbf{a}\,.\,\mathbf{w}). \tag{A.19.29}$$

These definitions generalise immediately to linear combinations of simple tensors (i.e., to $\otimes^2\mathcal{V}$ and $\otimes^3\mathcal{V}$). For example,

$$\begin{aligned}&(\alpha_1(\mathbf{a}_1\otimes\mathbf{b}_1)+\alpha_2(\mathbf{a}_2\otimes\mathbf{b}_2)) : (\mathbf{u}\otimes\mathbf{v}\otimes\mathbf{w})\\ &\quad := (\alpha_1\mathbf{a}_1\otimes\mathbf{b}_1) : (\mathbf{u}\otimes\mathbf{v}\otimes\mathbf{w}) + (\alpha_2\mathbf{a}_2\otimes\mathbf{b}_2) : (\mathbf{u}\otimes\mathbf{v}\otimes\mathbf{w}),\end{aligned} \tag{A.19.30}$$

and

$$\begin{aligned}&(\mathbf{a}\otimes\mathbf{b}) : (\alpha_1(\mathbf{u}_1\otimes\mathbf{v}_1\otimes\mathbf{w}_1)+\alpha_2(\mathbf{u}_2\otimes\mathbf{v}_2\otimes\mathbf{w}_2))\\ &\quad := (\mathbf{a}\otimes\mathbf{b}) : (\alpha_1\mathbf{u}_1\otimes\mathbf{v}_1\otimes\mathbf{w}_1) + (\mathbf{a}\otimes\mathbf{b}) : (\alpha_2\mathbf{u}_2\otimes\mathbf{v}_2\otimes\mathbf{w}_2).\end{aligned} \tag{A.19.31}$$

Thus, if $\mathbf{C},\mathbf{C}' \in \otimes^3\mathcal{V}$ and $\mathbf{L},\mathbf{L}' \in \otimes^2\mathcal{V}$, then definitions (A.19.25) through (A.19.29) yield, respectively, $\mathbf{C} : \mathbf{C}' \in \mathrm{Lin}\,\mathcal{V}, \mathbf{C} : \mathbf{L} \in \mathcal{V}, \mathbf{L} : \mathbf{C} \in \mathcal{V}, \mathbf{L} : \mathbf{L}' \in \mathbb{R}$, and $\mathbf{C}\,\vdots\,\mathbf{C}' \in \mathbb{R}$. In terms of an orthonormal basis $\{\mathbf{e}_1,\mathbf{e}_2,\mathbf{e}_3\}$,

$$\begin{aligned}\mathbf{C}:\mathbf{C}' = C_{pqr}\,\mathbf{e}_p\otimes\mathbf{e}_q\otimes\mathbf{e}_r : C_{ijk}\mathbf{e}_i\otimes\mathbf{e}_j\otimes\mathbf{e}_k &= C_{pqr}C_{ijk}\delta_{ri}\delta_{qj}\mathbf{e}_p\otimes\mathbf{e}_k\\ &= C_{pji}C_{ijk}\mathbf{e}_p\otimes\mathbf{e}_k.\end{aligned} \tag{A.19.32}$$

Equivalently,

$$(\mathbf{C}:\mathbf{C}')_{pk} = C_{pji}\,C'_{ijk}. \tag{A.19.33}$$

Exercise A.19.8. Show similarly that

$$(\mathbf{C}:\mathbf{L})_p = C_{pji}L_{ij}, \tag{A.19.34}$$

$$(\mathbf{L}:\mathbf{C})_r = L_{ij}C_{jir}, \tag{A.19.35}$$

$$\mathbf{L}:\mathbf{L}' = L_{ij}L'_{ji} = \mathbf{L}':\mathbf{L}, \tag{A.19.36}$$

and

$$\mathbf{C}\,\vdots\,\mathbf{C}' = C_{ijk}C'_{kji} = \mathbf{C}'\,\vdots\,\mathbf{C}. \tag{A.19.37}$$

Notice that from (A.14.10) with $\mathbf{M} = \mathbf{L}'$, and (A.19.36),

$$\mathbf{L}\cdot\mathbf{L}' = L_{pq}L'_{pq} = (L^T)_{qp}\,L'_{pq} = \mathbf{L}^T:\mathbf{L}'. \tag{A.19.38}$$

Now define

$$\mathbf{C}\cdot\mathbf{C}' := \mathbf{C}^{\sim}\,\vdots\,\mathbf{C}'. \tag{A.19.39}$$

Thus

$$\mathbf{C}\cdot\mathbf{C}' = (C_{pqr}\mathbf{e}_p\otimes\mathbf{e}_q\otimes\mathbf{e}_r)^{\sim}\,\vdots\,C'_{ijk}\mathbf{e}_i\otimes\mathbf{e}_j\otimes\mathbf{e}_k = C_{rqp}C'_{ijk}\delta_{ri}\delta_{qj}\delta_{pk} = C_{ijk}C'_{ijk}. \tag{A.19.40}$$

Exercise A.19.9. Show from (A.19.40) that (A.19.39) defines an inner product on $\otimes^3\mathcal{V}$.

While (A.19.25) through (A.19.29) introduced combinations which were of lower order than those of the individual tensors involved, higher-order tensors may be constructed. Indeed, $\mathbf{a}\otimes\mathbf{b}\otimes\mathbf{c}$ may be regarded as $(\mathbf{a}\otimes\mathbf{b})\otimes\mathbf{c}$ via $((\mathbf{a}\otimes\mathbf{b})\otimes\mathbf{c})\mathbf{v} = (\mathbf{c}\,.\,\mathbf{v})\mathbf{a}\otimes\mathbf{b}$, so constructing a third-order tensor from second- and first-order tensors. More generally, $\mathbf{L}\otimes\mathbf{a}\in\otimes^3\mathcal{V}$ is defined by

$$(\mathbf{L}\otimes\mathbf{a})\mathbf{v} := (\mathbf{a}\,.\,\mathbf{v})\mathbf{L} \tag{A.19.41}$$

for any $\mathbf{v}\in\mathcal{V}$.

Exercise A.19.10. Show that

$$(\mathbf{L}\otimes\mathbf{a})^T\mathbf{b} = \mathbf{L}\mathbf{b}\otimes\mathbf{a} \tag{A.19.42}$$

by proving this result for $\mathbf{L} = \mathbf{c}\otimes\mathbf{d}$ and then invoking linearity.

Exercise A.19.11. Show that

$$((\mathbf{a}\otimes\mathbf{b}\otimes\mathbf{c}):(\mathbf{u}\otimes\mathbf{v}))\,.\,\mathbf{w} = ((\mathbf{a}\otimes\mathbf{b}\otimes\mathbf{c})^{\sim}\mathbf{w})\cdot(\mathbf{u}\otimes\mathbf{v}).$$

Deduce, by linearity, that

$$(\mathbf{C}:\mathbf{L})\cdot\mathbf{w} = (\mathbf{C}^{\sim}\mathbf{w})\cdot\mathbf{L}. \tag{A.19.43}$$

Exercise A.19.12. Show that

$$(\mathbf{C}^T\mathbf{v})^T = (\mathbf{C}^{\sim})^T\mathbf{v}. \tag{A.19.44}$$

[*Hint*: Consider $(\mathbf{C}^T\mathbf{v})^T\mathbf{k}\,.\,\mathbf{l} = (\mathbf{C}^T\mathbf{v})\mathbf{l}\,.\,\mathbf{k} = (\mathbf{C}\mathbf{l})\mathbf{v}\,.\,\mathbf{k} = (\mathbf{C}^{\sim}\mathbf{k})\mathbf{v}\,.\,\mathbf{l}$.]

Remark A.19.1. In this section only those elements of tensor algebra employed elsewhere (particularly in Chapter 7) have been introduced. In general this subject addresses multilinear maps and, in particular, establishes natural identifications between such maps which have the same tensorial order. (See, for example, Greub [90].) We have essentially seen this with $\otimes^\alpha \mathcal{V}\,(\alpha = 2,3)$. Any element $\mathbf{L} \in \otimes^2\mathcal{V}$ can be interpreted to be an element of $\mathrm{Lin}\,\mathcal{V}$ or a bilinear map from $\mathcal{V}$ into $\mathbb{R}$ [via $(\mathbf{u},\mathbf{v}) \to \mathbf{Lu}\,.\,\mathbf{v}$] or a linear map from $\mathrm{Lin}\,\mathcal{V}$ into $\mathbb{R}$ (via $\mathbf{L}' \to \mathbf{L}'\cdot\mathbf{L}$). Similarly, $\mathbf{C} \in \otimes^3\mathcal{V}$ can be regarded as an element of $\mathrm{Lin}(\mathcal{V},\mathrm{Lin}\,\mathcal{V})$, a linear mapping from $\mathrm{Lin}\,\mathcal{V}$ into $\mathcal{V}$ (via $\mathbf{L} \to \mathbf{C}:\mathbf{L}$), a trilinear map from $\mathcal{V}$ into $\mathbb{R}$ [via $(\mathbf{u},\mathbf{v},\mathbf{w}) \to (\mathbf{Cu})\mathbf{v}\,.\,\mathbf{w}$], and a linear map from $\mathrm{Lin}(\mathcal{V},\mathrm{Lin}\,\mathcal{V})$ into $\mathbb{R}$ (via $\mathbf{C}' \to \mathbf{C}\cdot\mathbf{C}'$). In this context, recall the central role played by (alternating) trilinear forms in establishing principal invariants of linear maps on $\mathcal{V}$ (see Section A.11). Such forms lie in $\otimes^3\mathcal{V}$. Consider

$$\omega := \mathbf{a}\wedge\mathbf{b}\wedge\mathbf{c} := \mathbf{a}\otimes\mathbf{b}\otimes\mathbf{c}+\mathbf{b}\otimes\mathbf{c}\otimes\mathbf{a}+\mathbf{c}\otimes\mathbf{a}\otimes\mathbf{b}-\mathbf{a}\otimes\mathbf{c}\otimes\mathbf{b}-\mathbf{b}\otimes\mathbf{a}\otimes\mathbf{c}-\mathbf{c}\otimes\mathbf{b}\otimes\mathbf{a}. \tag{A.19.45}$$

If $\mathbf{a}\otimes\mathbf{b}\otimes\mathbf{c}$ etc. are identified with trilinear forms [so that $(\mathbf{a}\otimes\mathbf{b}\otimes\mathbf{c})(\mathbf{u},\mathbf{v},\mathbf{w}) := (\mathbf{c}\,.\,\mathbf{u})(\mathbf{b}\,.\,\mathbf{v})(\mathbf{q}\,.\,\mathbf{w})$, etc.], then ω is an alternating trilinear form on $\mathcal{V}$. (Show this!)

Exercise A.19.13. Show that if $\{\mathbf{e}_1,\mathbf{e}_2,\mathbf{e}_3\}$ is an orthonormal basis, then

$$\omega = a_i b_j c_k \mathbf{e}_i\wedge\mathbf{e}_j\wedge\mathbf{e}_k. \tag{A.19.46}$$

Show also that [here ϵ_{ijk} is as in (A.11.8)]

$$\mathbf{e}_i\wedge\mathbf{e}_j\wedge\mathbf{e}_k = \epsilon_{ijk}\mathbf{e}_1\wedge\mathbf{e}_2\wedge\mathbf{e}_3. \tag{A.19.47}$$

Noting that

$$(\mathbf{a}\otimes\mathbf{b}\otimes\mathbf{c})(\mathbf{u},\mathbf{v},\mathbf{w}) = (\mathbf{u}\otimes\mathbf{v}\otimes\mathbf{w})(\mathbf{a},\mathbf{b},\mathbf{c}), \tag{A.19.48}$$

deduce that

$$(\mathbf{a}\wedge\mathbf{b}\wedge\mathbf{c})(\mathbf{u},\mathbf{v},\mathbf{w}) = (\mathbf{u}\wedge\mathbf{v}\wedge\mathbf{w})(\mathbf{a},\mathbf{b},\mathbf{c}). \tag{A.19.49}$$

A.20 Direct, Component, and Cartesian Tensor Notation

Continuum modelling is implemented in terms of physically interpretable fields with the aim of capturing the essence of macroscopic material behaviour. Any such field f is a function of location, in Euclidean space $\mathcal{E}$, and time. If f makes sense at location $\mathbf{x} \in \mathcal{E}$ and time t, then the value of the field is denoted by $f(\mathbf{x},t)$. Such a description is minimal and *direct*. For example, mass density, velocity, stress, and couple stress are described via, respectively, a scalar field ρ, vectorial field $\mathbf{v}$, linear transformation field $\mathbf{T}$, and third-order tensor field $\mathbf{C}$. To address a specific situation, such as fluid flow through a cylindrical pipe, it is natural and sensible to adopt a choice of co-ordinates for $\mathcal{E}$ and related basis for $\mathcal{V}$ which might change with location. However, when considering general situations, the introduction of co-ordinates and components is artificial, adds nothing to understanding, and may obscure and/or complicate matters. For example, $\mathbf{T}(\mathbf{x},t)$ denotes the value of the stress tensor at location $\mathbf{x}$ and time t and has the physical interpretation that for any unit vector $\mathbf{n}$ the (traction) vector $\mathbf{T}(\mathbf{x},t)\mathbf{n}$ is the force per unit area at $(\mathbf{x},t)$, exerted across any smooth surface S through $\mathbf{x}$ with normal $\mathbf{n}$ at this point, by the material on the side of S into which $\mathbf{n}$ is directed upon material on the other side of S. If a Cartesian co-ordinate system is chosen (i.e., a location $\mathbf{x}_0$ is chosen as origin together with an orthonormal basis $\{\mathbf{e}_1,\mathbf{e}_2,\mathbf{e}_3\}$) then $\mathbf{T}(\mathbf{x},t)$ is represented as $T_{ij}(x_1,x_2,x_3,t)\mathbf{e}_i \otimes \mathbf{e}_j$ and the preceding traction vector value as $T_{ij}(x_1,x_2,x_3,t)n_j\mathbf{e}_i$. (Of course, we have here used summation convention and $i,j = 1,2,3$.) The nine functions T_{ij} of four real variables all change with any change of origin and/or orthonormal basis.

If $\mathbf{u},\mathbf{v} \in \mathcal{V}$, $\mathbf{A},\mathbf{B} \in \mathrm{Lin}\,\mathcal{V}$, and $\mathbf{C} \in \mathrm{Lin}(\mathcal{V},\mathrm{Lin}\,\mathcal{V})$, then the combinations $\mathbf{u}.\mathbf{v}$, $\mathbf{u}\otimes\mathbf{v},\mathbf{Au},\mathbf{AB},\mathbf{ABu}$, and $\mathbf{Cu}$ are intelligible, well-defined elements of $\mathbb{R},\mathrm{Lin}\,\mathcal{V},\mathcal{V}$, $\mathrm{Lin}\,\mathcal{V},\mathcal{V}$, and $\mathrm{Lin}\,\mathcal{V}$, respectively. In Cartesian component format these take the forms

$$\mathbf{u}.\mathbf{v} = u_i v_i, \quad \mathbf{u}\otimes\mathbf{v} = u_i v_j \mathbf{e}_i \otimes \mathbf{e}_j, \tag{A.20.1}$$

$$\mathbf{Au} = (A_{ij}\mathbf{e}_i \otimes \mathbf{e}_j)u_k\mathbf{e}_k = A_{ij}u_j\mathbf{e}_i, \tag{A.20.2}$$

$$\mathbf{AB} = (A_{ij}\mathbf{e}_i \otimes \mathbf{e}_j)(B_{k\ell}\mathbf{e}_k \otimes \mathbf{e}_\ell) = A_{ij}B_{j\ell}\mathbf{e}_i \otimes \mathbf{e}_\ell, \tag{A.20.3}$$

$$\mathbf{ABu} = (\mathbf{AB})\mathbf{u} = (A_{ij}B_{j\ell}\mathbf{e}_i \otimes \mathbf{e}_\ell)u_p\mathbf{e}_p = A_{ij}B_{jp}u_p\mathbf{e}_i, \tag{A.20.4}$$

and

$$\mathbf{Cu} = (C_{ijk}\mathbf{e}_i \otimes \mathbf{e}_j \otimes \mathbf{e}_k)u_\ell\mathbf{e}_\ell = C_{ijk}u_k\mathbf{e}_i \otimes \mathbf{e}_j. \tag{A.20.5}$$

To see how the preceding representations change when another orthonormal basis $\{\mathbf{e}'_1,\mathbf{e}'_2,\mathbf{e}'_3\}$ is chosen, consider the linear transformation $\mathbf{Q}$ defined by (here $i = 1,2,3$)

$$\mathbf{Q}\mathbf{e}'_i = \mathbf{e}_i. \tag{A.20.6}$$

(Recall that a linear transformation is uniquely prescribed by its action on a basis: see Result A.7.1.) It follows that

$$\mathbf{Q} = \mathbf{e}_i \otimes \mathbf{e}'_i. \tag{A.20.7}$$

(Verify this!) Accordingly, via (A.8.10) and (A.8.11),

$$\begin{aligned}\mathbf{QQ}^T &= (\mathbf{e}_i \otimes \mathbf{e}'_i)(\mathbf{e}_j \otimes \mathbf{e}'_j)^T \\ &= (\mathbf{e}_i \otimes \mathbf{e}'_i)(\mathbf{e}'_j \otimes \mathbf{e}_j) = (\mathbf{e}'_i.\mathbf{e}'_j)\mathbf{e}_i \otimes \mathbf{e}_j = \delta_{ij}\mathbf{e}_i \otimes \mathbf{e}_j = \mathbf{e}_i \otimes \mathbf{e}_i.\end{aligned}$$

That is, from (A.9.16)

$$\mathbf{QQ}^T = \mathbf{1}, \tag{A.20.8}$$

and hence $\mathbf{Q}$ *is orthogonal.*

Exercise A.20.1. Show that $\mathbf{Q}^T\mathbf{Q} = \mathbf{1}$.

The matrix representation of $\mathbf{Q}$ with respect to $\{\mathbf{e}_1,\mathbf{e}_2,\mathbf{e}_3\}$ is $[Q_{ij}]$, where [see (A.9.1)]

$$Q_{ij} = \mathbf{e}_i.\mathbf{Q}\mathbf{e}_j = \mathbf{e}_i.(\mathbf{e}_k \otimes \mathbf{e}'_k)\mathbf{e}_j = \mathbf{e}_i.(\mathbf{e}'_k.\mathbf{e}_j)\mathbf{e}_k = (\mathbf{e}'_k.\mathbf{e}_j)\delta_{ik}.$$

That is,

$$Q_{ij} = \mathbf{e}'_i.\mathbf{e}_j. \tag{A.20.9}$$

Now consider the matrix representation $[Q'_{ij}]$ of $\mathbf{Q}$ with respect to $\{\mathbf{e}'_1,\mathbf{e}'_2,\mathbf{e}'_3\}$. This is given by

$$Q'_{ij} = \mathbf{e}'_i \,.\, \mathbf{Q}\mathbf{e}'_j = \mathbf{e}'_i \,.\, (\mathbf{e}_k \otimes \mathbf{e}'_k)\mathbf{e}'_j = \mathbf{e}'_i \,.\, (\mathbf{e}'_k \,.\, \mathbf{e}'_j)\mathbf{e}_k = \delta_{kj}\mathbf{e}'_i \,.\, \mathbf{e}_k.$$

That is,
$$Q'_{ij} = \mathbf{e}'_i \,.\, \mathbf{e}_j. \tag{A.20.10}$$

Thus, somewhat unexpectedly, the two matrix representations are the same: $Q'_{ij} = Q_{ij}$ and

$$Q_{ij}\mathbf{e}_i \otimes \mathbf{e}_j = \mathbf{Q} = Q_{ij}\mathbf{e}'_i \otimes \mathbf{e}'_j. \tag{A.20.11}$$

In order to relate the component forms of $\mathbf{v} \in \mathcal{V}$, $\mathbf{A} \in \mathrm{Lin}\,\mathcal{V}$, and $\mathbf{C} \in \mathrm{Lin}(\mathcal{V},\mathrm{Lin}\,\mathcal{V})$, note that

$$\mathbf{e}_i = (\mathbf{e}_i \,.\, \mathbf{e}'_j)\mathbf{e}'_j = (\mathbf{e}'_j \,.\, \mathbf{e}_i)\mathbf{e}'_j,$$

so, from (A.20.9),
$$\mathbf{e}_i = Q_{ji}\mathbf{e}'_j. \tag{A.20.12}$$

Comparison of the component forms of $\mathbf{v}$ with respect to $\{\mathbf{e}_1,\mathbf{e}_2,\mathbf{e}_3\}$ and $\{\mathbf{e}'_1,\mathbf{e}'_2,\mathbf{e}'_3\}$ yields

$$v_i\mathbf{e}_i = \mathbf{v} = v'_k\mathbf{e}'_k, \tag{A.20.13}$$

so that
$$v'_j = \mathbf{v} \,.\, \mathbf{e}'_j = v_i\mathbf{e}_i \,.\, \mathbf{e}'_j = v_i\mathbf{e}'_j \,.\, \mathbf{e}_i.$$

Thus, from (A.20.9),
$$v'_j = Q_{ji}v_i. \tag{A.20.14}$$

Similarly,
$$A_{pq}\mathbf{e}_p \otimes \mathbf{e}_q = \mathbf{A} = A'_{rs}\mathbf{e}'_r \otimes \mathbf{e}'_s \tag{A.20.15}$$

implies that the matrix representations $[A_{pq}]$ and $[A'_{rs}]$ of $\mathbf{A}$ with respect to the two bases are related via

$$A'_{ij} = \mathbf{e}'_i \,.\, \mathbf{A}\mathbf{e}'_j = \mathbf{e}'_i \,.\, (A_{pq}\mathbf{e}_p \otimes \mathbf{e}_q)\mathbf{e}'_j = (\mathbf{e}'_i \,.\, A_{pq}\mathbf{e}_p)(\mathbf{e}_q \,.\, \mathbf{e}'_j) = (\mathbf{e}'_i \,.\, \mathbf{e}_p)A_{pq}(\mathbf{e}'_j \,.\, \mathbf{e}_q).$$

That is, from (A.20.9),

$$A'_{ij} = Q_{ip}A_{pq}Q_{jq} = Q_{ip}Q_{jq}A_{pq}. \tag{A.20.16}$$

For a third-order tensor $\mathbf{C}$, the component forms with respect to the two bases satisfy [see (A.19.10)]

$$C_{pqr}\mathbf{e}_p \otimes \mathbf{e}_q \otimes \mathbf{e}_r = \mathbf{C} = C'_{\ell mn}\mathbf{e}'_\ell \otimes \mathbf{e}'_m \otimes \mathbf{e}'_n. \tag{A.20.17}$$

Accordingly [see (A.19.4)]

$$\begin{aligned} C'_{ijk} &= \mathbf{e}'_i \,.\, (\mathbf{C}\mathbf{e}'_k)\mathbf{e}'_j = \mathbf{e}'_i \,.\, ((C_{pqr}\mathbf{e}_p \otimes \mathbf{e}_q \otimes \mathbf{e}_r)\mathbf{e}'_k)\mathbf{e}'_j \\ &= C_{pqr}(\mathbf{e}_r \,.\, \mathbf{e}'_k)(\mathbf{e}_q \,.\, \mathbf{e}'_j)(\mathbf{e}'_i \,.\, \mathbf{e}_p). \end{aligned}$$

Thus, from (A.20.9),

$$C'_{ijk} = Q_{ip}Q_{jq}Q_{kr}C_{pqr}. \tag{A.20.18}$$

Cartesian tensor notation corresponds to the preceding component representations with respect to orthonormal bases *but with the basis vectors omitted.* Thus $\mathbf{v}$ is represented by v_i, $\mathbf{T}$ by T_{ij}, and $\mathbf{C}$ by C_{ijk}. Standard treatments of Cartesian tensors define a tensor of first order (or *vector*) to be an ordered triple v_i of real numbers which transforms under a change of basis via (A.20.14), a second-order tensor to be an array A_{pq} $(p,q = 1,2,3)$ which transforms via (A.20.16), and a third-order tensor to be an array C_{pqr} $(p,q,r = 1,2,3)$ which transforms via (A.20.18). Such a viewpoint, in its consideration only of representations, does not lend itself to the clearest view of the entities being represented. This is not to denigrate the use of such notation, but merely to remark that in our emphasis on fundamentals it is unnecessary. Indeed, our discussion serves as an introduction to, and justification for, Cartesian tensor notation.

Appendix B: Calculus in Euclidean Point Space $\mathcal{E}$

Preamble

Here geometric and analytical pre-requisites for continuum modelling are developed and linked to the algebraic considerations of Appendix A. This material has been included in order to emphasise the direct (i.e., co-ordinate-free) approach employed, which may not be familiar to the reader. The aim has been to provide a reasonably self-contained basis for understanding the notation and methodology used in the main body of the work.

Continuum modelling of material behaviour requires, among other things, mathematical prescriptions of

(i) the location and distribution of matter for the physical system (or *body*) of interest at a given time,
(ii) changes in location of a body and any associated distortion,
(iii) spatial and temporal variation of local system descriptors (e.g., mass density or velocity), and
(iv) physical descriptors such as mass, momentum, and kinetic energy, which are additive over disjoint regions. (Such descriptors are termed *extensive*.)

Central to such prescriptions are the notions of *point* and *space*, here formalised in terms of Euclidean space $\mathcal{E}$. Distortion is described via one-to-one mappings of points into points (*deformations*), and local spatial variation of descriptors is treated in terms of generalisations of the derivative of a function of a real variable. Analysis of point (iv) involves relating values of extensive descriptors associated with finite regions to their local densities, and is accomplished via volume integration.

Euclidean point space $\mathcal{E}$ is codified in terms of points, displacements, and their inter-relationships. Any Cartesian co-ordinate system for $\mathcal{E}$ is obtained via selection of a distinguished point (or *origin*) and an orthonormal basis for $\mathcal{V}$ (cf. Appendix A.6). In order to analyse deformations in general, it proves helpful first to consider those which preserve distances between points (*isometries*) and those in which distortion is everywhere the same (*homogeneous deformations*). Any isometry is shown to correspond to a translation followed by a rotation, or vice versa. Homogeneous deformations are characterised (via use of the polar decomposition theorem of Appendix A.18) in terms of combinations of one-dimensional deformations (*simple stretches*), rotations, and translations. Further, such deformations are shown to map planes into planes (and parallel planes into parallel planes). The derivative of a real-valued function of a real variable furnishes a local linear approximation to the function. This notion is generalised to maps from $\mathcal{E}$ into $\mathbb{R}, \mathcal{E}, \mathcal{V}$, and Lin $\mathcal{V}$; that is, to scalar fields, deformations, vector fields, and order-two tensor fields. The values of these derivatives lie in $\mathcal{V}, \mathrm{Lin}\,\mathcal{V}, \mathrm{Lin}\,\mathcal{V}$, and Lin $(\mathcal{V}, \mathrm{Lin}\,\mathcal{V})$, respectively. Cartesian co-ordinate representations of these fields are established and expressed in Cartesian tensor notation. Following remarks on generalisations to higher-order tensor fields, and on second derivatives, useful identities, involving derivatives of products and composition of fields, are listed and proved. The derivative of the determinantal function (a map from Lin $\mathcal{V}$ into $\mathbb{R}$) is defined and evaluated for elements of Invlin $\mathcal{V}$. The local volume change ratio associated with a differentiable deformation $\mathbf{d}$ is analysed in terms of its local (homogeneous) linearisation and identified with the determinant of the derivative of $\mathbf{d}$ (its *Jacobian*). The *motion* of a body with respect to its location at a fixed time is defined, and standard kinematic measures of time rates of change of deformations are established, together with the notion of the material time derivative of any physical descriptor. The question of if, and how, 'volume' can be ascribed to a general bounded region R is addressed in terms of upper and lower estimates of sums of volumes of rectangular boxes which lie within R or intersect its boundary. This leads naturally to a brief discussion of Riemann integration of a field f over R, the existence of integrals for continuous fields, and their evaluation via repeated definite integrals of functions of a single real variable. Two key theorems are discussed: one enables local relations to be established from global results, and the other relates a current spatial integral to a referential integral description. The divergence theorem for vector fields is generalised to higher-order tensor fields, and identities used in the main text are established. Finally, calculations relating to Sections 7.4, 10.5, 12.2, and 14.2 are detailed.

B.1 Euclidean Point Space $\mathcal{E}$

Recalling intuitive notions of locations and displacements discussed in Appendix A.1., the properties of 'space' as we perceive it may be characterised in terms of a set $\mathcal{E}$ of points and the three-dimensional vectorial space $\mathcal{V}$ (see Appendix A.4) of displacements between pairs of points. To any ordered pair $(\mathbf{x}, \mathbf{y})$ of points corresponds a unique element of $\mathcal{V}$ (the displacement necessary to reach $\mathbf{y}$ starting from $\mathbf{x}$). We write this displacement as $\mathbf{y} - \mathbf{x}$. Further, to any pair $(\mathbf{x}, \mathbf{v})$ with $\mathbf{x} \in \mathcal{E}$ and $\mathbf{v} \in \mathcal{V}$ corresponds a unique point (the point reached by undergoing a displacement $\mathbf{v}$ starting from $\mathbf{x}$). We denote this point by $\mathbf{x} + \mathbf{v}$. Accordingly

$$\mathbf{y} - \mathbf{x} \in \mathcal{V} \qquad \text{and} \qquad \mathbf{x} + \mathbf{v} \in \mathcal{E}. \tag{B.1.1}$$

See Figure B.1.

Of course, we are here employing symbols $-$ and $+$ in a different sense to their use in vectorial space $\mathcal{V}$. The interpretations given to the combinations in (B.1.1) means that if $\mathbf{x}, \mathbf{y}, \mathbf{z} \in \mathcal{E}$, and $\mathbf{u}, \mathbf{v} \in \mathcal{V}$, then

$$\mathbf{x} + (\mathbf{y} - \mathbf{x}) = \mathbf{y}, \tag{B.1.2}$$

$$(\mathbf{x} + \mathbf{u}) + \mathbf{v} = \mathbf{x} + (\mathbf{u} + \mathbf{v}), \ \text{and} \tag{B.1.3}$$

$$(\mathbf{z} - \mathbf{y}) + (\mathbf{y} - \mathbf{x}) = \mathbf{z} - \mathbf{x}. \tag{B.1.4}$$

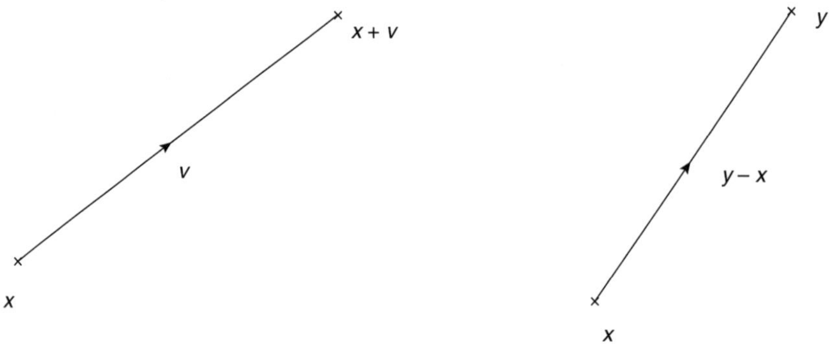

Figure B.1. A displacement v from a point x defines a unique point of $\mathcal{E}$ which is denoted by $x + v$. The displacement (an element of $\mathcal{V}$) necessary to reach point y starting from x is denoted by $y - x$.

Exercise B.1.1. Draw sketches to illustrate (B.1.2) through (B.1.4). In these identities note where + is used in the sense of $(\text{B.1.1})_2$ and where addition in $\mathcal{V}$ is intended. Also, by means of a sketch, note that

$$(\mathbf{y}-\mathbf{x})+\mathbf{v} = (\mathbf{y}+\mathbf{v})-\mathbf{x}, \text{ and} \tag{B.1.5}$$

$$-(\mathbf{y}-\mathbf{x}) = (\mathbf{x}-\mathbf{y}). \tag{B.1.6}$$

Caveat: *Notice that* $\mathbf{x}+\mathbf{y}$ *is not defined*!

The foregoing approach, due to Walter Noll, differs from general models of space in that no special point has been singled out to serve as origin. When such a distinguished point, $\mathbf{x}_0$ say, *is* selected, it is usual for a general point $\mathbf{x}$ to be identified with its displacement from $\mathbf{x}_0$. However, there is no *natural* choice of origin. Hence any model of space which identifies points with displacements from an origin is only a *representation* of our notion of space, since different choices of origin yield different 'position vectors' to identify any given point.

The distance $d(\mathbf{x},\mathbf{y})$ between any pair of points $\mathbf{x},\mathbf{y} \in \mathcal{E}$ is the magnitude of the displacement from $\mathbf{x}$ to $\mathbf{y}$. That is, the distance function or *metric* on $\mathcal{E}$ is

$$d(\mathbf{x},\mathbf{y}) := \|\mathbf{y}-\mathbf{x}\| = ((\mathbf{y}-\mathbf{x}).(\mathbf{y}-\mathbf{x}))^{1/2}. \tag{B.1.7}$$

Clearly [recall (A.4.13) with $\alpha = -1$ and (B.1.6)]

$$d(\mathbf{y},\mathbf{x}) = d(\mathbf{x},\mathbf{y}). \tag{B.1.8}$$

Further [and not surprisingly in view of (A.3.2)!], from (A.4.15), (with $\mathbf{v}_1 = \mathbf{z}-\mathbf{y}, \mathbf{v}_2 = \mathbf{y}-\mathbf{x}$) we have

$$\begin{aligned} d(\mathbf{x},\mathbf{z}) = \|\mathbf{z}-\mathbf{x}\| &= \|(\mathbf{z}-\mathbf{y})+(\mathbf{y}-\mathbf{x})\| \\ &\le \|\mathbf{z}-\mathbf{y}\| + \|\mathbf{y}-\mathbf{x}\| = d(\mathbf{y},\mathbf{z}) + d(\mathbf{x},\mathbf{y}). \end{aligned}$$

That is,

$$d(\mathbf{x},\mathbf{z}) \le d(\mathbf{x},\mathbf{y}) + d(\mathbf{y},\mathbf{z}). \tag{B.1.9}$$

Of course,

$$d(\mathbf{x},\mathbf{y}) \ge 0 \tag{B.1.10}$$

and $$d(\mathbf{x},\mathbf{y}) = 0 \quad \text{implies that points } \mathbf{x} \text{ and } \mathbf{y} \text{ coincide.} \tag{B.1.11}$$

(Why?)

B.2 Cartesian Co-ordinate Systems for $\mathcal{E}$

On selecting a point $\mathbf{x}_0$ and ordered orthonormal basis $\{\mathbf{e}_1,\mathbf{e}_2,\mathbf{e}_3\}$ for $\mathcal{V}$, each $\mathbf{x} \in \mathcal{E}$ can be identified with an ordered triple (x_1,x_2,x_3) of real numbers (i.e., element of $\mathbb{R}^3$) by defining

$$x_i := (\mathbf{x} - \mathbf{x}_0) \,.\, \mathbf{e}_i \qquad (i = 1,2,3). \tag{B.2.1}$$

Representation (x_1,x_2,x_3) depends both upon choice $\mathbf{x}_0$ *and* basis $\{\mathbf{e}_i\}$. Triple (x_1,x_2,x_3) is termed the *set of co-ordinates for point* $\mathbf{x}$ in the Cartesian co-ordinate system $C(\mathbf{x}_0;\mathbf{e}_1,\mathbf{e}_2,\mathbf{e}_3)$.

A change of origin from $\mathbf{x}_0$ to $\mathbf{x}'_0$ yields the co-ordinate system $C(\mathbf{x}'_0;\mathbf{e}_1,\mathbf{e}_2,\mathbf{e}_3)$ in which the co-ordinates (x'_1,x'_2,x'_3) of $\mathbf{x}$ are given by

$$x'_i := (\mathbf{x} - \mathbf{x}'_0) \,.\, \mathbf{e}_i = ((\mathbf{x} - \mathbf{x}_0) + (\mathbf{x}_0 - \mathbf{x}'_0)) \,.\, \mathbf{e}_i,$$

so that $$x'_i = x_i + c_i, \tag{B.2.2}$$

where c_i are the components of the displacement $\mathbf{x}_0 - \mathbf{x}'_0$.

A change of orthonormal basis to $\{\mathbf{e}'_1,\mathbf{e}'_2,\mathbf{e}'_3\}$ yields system $C(\mathbf{x}_0;\mathbf{e}'_1,\mathbf{e}'_2,\mathbf{e}'_3)$ and new co-ordinates (x'_1,x'_2,x'_3) for $\mathbf{x}$ given by [see (A.6.6) with $\mathbf{v} = \mathbf{e}'_i$ and (A.20.9)]

$$\begin{aligned} x'_i &:= (\mathbf{x} - \mathbf{x}_0) \,.\, \mathbf{e}'_i = (\mathbf{x} - \mathbf{x}_0).(\mathbf{e}'_i \,.\, \mathbf{e}_j)\mathbf{e}_j \\ &= Q_{ij}(\mathbf{x} - \mathbf{x}_0) \,.\, \mathbf{e}_j = Q_{ij} x_j. \end{aligned}$$

That is, $$x'_i = Q_{ij} x_j, \tag{B.2.3}$$

where $[Q_{ij}]$ is the matrix representation of $\mathbf{Q}$ given in (A.20.9). Of course, a general change of Cartesian co-ordinates involves both a change of origin *and* change of orthonormal basis, and, in view of (B.2.2) and (B.2.3), takes the form

$$x_i \longrightarrow x'_i, \tag{B.2.4}$$

where $$x'_i = Q_{ij} x_j + c_i. \tag{B.2.5}$$

B.3 Deformations in $\mathcal{E}$

B.3.1 Introduction

Any *continuous, bijective* (i.e., 'one-to-one' and 'onto') map

$$\mathbf{d} : \mathcal{E} \to \mathcal{E}$$

is termed a *deformation in* $\mathcal{E}$. The restriction of a deformation in $\mathcal{E}$ to any region of $\mathcal{E}$ occupied by a material system/body models a possible distortion of this body. It is usual to require that in such context **d** be differentiable in the sense to be described in Subsection B.4.2. There are, however, two special forms of deformation which can be analysed purely algebraically (and interpreted geometrically). These are isometric and homogeneous deformations. (In fact the former turn out to be a special case of the latter, but this is by no means obvious from the outset.)

B.3.2 Isometries and Their Characterisation

A map
$$\mathbf{i}:\mathcal{E}\to\mathcal{E} \tag{B.3.1}$$

which preserves distances between pairs of points is termed an *isometry* in $\mathcal{E}$. Such maps are encountered when describing rigid body changes of position and in relating kinematic phenomena viewed by two different observers who agree upon distances between simultaneous events.

By definition, for every $\mathbf{x},\mathbf{y}\in\mathcal{E}$,

$$\|\mathbf{i}(\mathbf{y})-\mathbf{i}(\mathbf{x})\| = \|\mathbf{y}-\mathbf{x}\|. \tag{B.3.2}$$

Equivalently, for every $\mathbf{x}\in\mathcal{E}$ and $\mathbf{u}\in\mathcal{V}$,

$$\|\mathbf{i}(\mathbf{x}+\mathbf{u})-\mathbf{i}(\mathbf{x})\| = \|\mathbf{u}\|. \tag{B.3.3}$$

'Common sense'/experience suggests that the angle θ between displacements $\mathbf{u}$ and $\mathbf{v}$ from a given point $\mathbf{x}$ to points $\mathbf{x}+\mathbf{u}$ and $\mathbf{x}+\mathbf{v}$ is preserved by an isometry. That is, if θ' is the angle between displacements

$$\mathbf{u}':=\mathbf{i}(\mathbf{x}+\mathbf{u})-\mathbf{i}(\mathbf{x}) \qquad \text{and} \qquad \mathbf{v}':=\mathbf{i}(\mathbf{x}+\mathbf{v})-\mathbf{i}(\mathbf{x}) \tag{B.3.4}$$

from $\mathbf{i}(\mathbf{x})$, then $\theta'=\theta$. To *prove* this on the basis of (B.3.2) we note that from (B.3.3)

$$\|\mathbf{u}'\| = \|\mathbf{i}(\mathbf{x}+\mathbf{u})-\mathbf{i}(\mathbf{x})\| = \|\mathbf{u}\| \qquad \text{and similarly} \qquad \|\mathbf{v}'\|=\|\mathbf{v}\|. \tag{B.3.5}$$

Further, from (B.3.4), (B.3.5), and (B.3.2),

$$\begin{aligned}
2\mathbf{u}'.\mathbf{v}' &= \mathbf{u}'.\mathbf{u}'+\mathbf{v}'.\mathbf{v}'-(\mathbf{u}'-\mathbf{v}').(\mathbf{u}'-\mathbf{v}')\\
&= \|\mathbf{u}'\|^2+\|\mathbf{v}'\|^2-\|(\mathbf{i}(\mathbf{x}+\mathbf{u})-\mathbf{i}(\mathbf{x}))-(\mathbf{i}(\mathbf{x}+\mathbf{v})-\mathbf{i}(\mathbf{x}))\|^2\\
&= \|\mathbf{u}\|^2+\|\mathbf{v}\|^2-\|\mathbf{i}(\mathbf{x}+\mathbf{u})-\mathbf{i}(\mathbf{x}+\mathbf{v})\|^2\\
&= \|\mathbf{u}\|^2+\|\mathbf{v}\|^2-\|(\mathbf{x}+\mathbf{u})-(\mathbf{x}+\mathbf{v})\|^2\\
&= \|\mathbf{u}\|^2+\|\mathbf{v}\|^2-\|\mathbf{u}-\mathbf{v}\|^2\\
&= \mathbf{u}.\mathbf{u}+\mathbf{v}.\mathbf{v}-(\mathbf{u}-\mathbf{v}).(\mathbf{u}-\mathbf{v}) = 2\mathbf{u}.\mathbf{v}.
\end{aligned}$$

Accordingly
$$\mathbf{u}'.\mathbf{v}'=\mathbf{u}.\mathbf{v}, \tag{B.3.6}$$

and so
$$\|\mathbf{u}'\|\,\|\mathbf{v}'\|\cos\theta' = \|\mathbf{u}\|\,\|\mathbf{v}\|\cos\theta. \tag{B.3.7}$$

Hence, from (B.3.5),
$$\cos\theta'=\cos\theta. \tag{B.3.8}$$

However, without loss of generality, θ' and θ can be assumed to lie in $[0,\pi]$ (Why?), whence (B.3.8) yields

$$\theta' = \theta. \tag{B.3.9}$$

Now define, for any $\mathbf{x} \in \mathcal{E}$,

$$\mathbf{q}(\mathbf{x};\mathbf{u}) := \mathbf{i}(\mathbf{x}+\mathbf{u}) - \mathbf{i}(\mathbf{x}). \tag{B.3.10}$$

In particular, note $\mathbf{q}(\mathbf{x};\mathbf{u}) \in \mathcal{V}$. From (B.3.4) and (B.3.6)

$$\mathbf{q}(\mathbf{x};\mathbf{u}) \,.\, \mathbf{q}(\mathbf{x};\mathbf{v}) = \mathbf{u} \,.\, \mathbf{v}. \tag{B.3.11}$$

Thus, for all $\alpha_1,\alpha_2 \in \mathbb{R}$ and $\mathbf{u}_1,\mathbf{u}_2,\mathbf{v} \in \mathcal{V}$,

$$\begin{aligned}&(\mathbf{q}(\mathbf{x};\alpha_1\mathbf{u}_1 + \alpha_2\mathbf{u}_2) - \alpha_1\mathbf{q}(\mathbf{x};\mathbf{u}_1) - \alpha_2\mathbf{q}(\mathbf{x};\mathbf{u}_2)).\mathbf{q}(\mathbf{x};\mathbf{v}) \\ &= (\alpha_1\mathbf{u}_1 + \alpha_2\mathbf{u}_2).\mathbf{v} - \alpha_1\mathbf{u}_1 \,.\, \mathbf{v} - \alpha_2\mathbf{u}_2 \,.\, \mathbf{v} = 0.\end{aligned} \tag{B.3.12}$$

However, we can choose $\mathbf{v} = \mathbf{e}_i$, where $i = 1,2,3$ and $\{\mathbf{e}_1,\mathbf{e}_2,\mathbf{e}_3\}$ is an orthonormal basis for $\mathcal{V}$. It follows from (B.3.11) that $\{\mathbf{q}(\mathbf{x};\mathbf{e}_i)\}$ is also an orthonormal basis for $\mathcal{V}$, and from (B.3.12) we can conclude that

$$\mathbf{q}(\mathbf{x};\alpha_1\mathbf{u}_1 + \alpha_2\mathbf{u}_2) = \alpha_1\mathbf{q}(\mathbf{x};\mathbf{u}_1) + \alpha_2\mathbf{q}(\mathbf{x};\mathbf{u}_2). \tag{B.3.13}$$

That is,

$$\mathbf{q}(\mathbf{x};.) : \mathcal{V} \to \mathcal{V} \tag{B.3.14}$$

is *linear*. We write

$$\mathbf{Q}(\mathbf{x})\mathbf{u} := \mathbf{q}(\mathbf{x};\mathbf{u}) = \mathbf{i}(\mathbf{x}+\mathbf{u}) - \mathbf{i}(\mathbf{x}). \tag{B.3.15}$$

Thus

$$\mathbf{Q}(\mathbf{x}) \in \operatorname{Lin}\mathcal{V} \tag{B.3.16}$$

and, from (B.3.11), for all $\mathbf{u},\mathbf{v} \in \mathcal{V}$

$$\mathbf{Q}(\mathbf{x})\mathbf{u} \,.\, \mathbf{Q}(\mathbf{x})\mathbf{v} = \mathbf{u} \,.\, \mathbf{v}. \tag{B.3.17}$$

Accordingly

$$\mathbf{Q}(\mathbf{x})\mathbf{u} = \mathbf{0} \quad \text{implies} \quad \mathbf{u} = \mathbf{0} \tag{B.3.18}$$

[prove this by setting $\mathbf{v} = \mathbf{u}$ in (B.3.17)], and hence $\mathbf{Q}(\mathbf{x})$ is invertible [see (A.10.4)]. Also, from (B.3.17)

$$\mathbf{Q}(\mathbf{x})^T\mathbf{Q}(\mathbf{x}) = \mathbf{1} \tag{B.3.19}$$

(Prove this!), so

$$\mathbf{Q}(\mathbf{x}) \quad \text{is orthogonal.} \tag{B.3.20}$$

Finally we show that $\mathbf{Q}(\mathbf{x})$ is independent of $\mathbf{x}$. Indeed, for any $\mathbf{x}_1,\mathbf{x}_2 \in \mathcal{E}$ and any $\mathbf{u} \in \mathcal{V}$, from (B.3.15)

$$\begin{aligned}\mathbf{Q}(\mathbf{x}_1)\mathbf{u} &= \mathbf{i}(\mathbf{x}_1 + \mathbf{u}) - \mathbf{i}(\mathbf{x}_1) \\ &= \mathbf{i}(\mathbf{x}_2 + [(\mathbf{x}_1 - \mathbf{x}_2) + \mathbf{u}]) - \mathbf{i}(\mathbf{x}_2 + (\mathbf{x}_1 - \mathbf{x}_2)) \\ &= \mathbf{Q}(\mathbf{x}_2)[(\mathbf{x}_1 - \mathbf{x}_2) + \mathbf{u}] - \mathbf{Q}(\mathbf{x}_2)(\mathbf{x}_1 - \mathbf{x}_2) \\ &= \mathbf{Q}(\mathbf{x}_2)\mathbf{u}.\end{aligned} \tag{B.3.21}$$

Thus

$$\mathbf{Q}(\mathbf{x}_1) = \mathbf{Q}(\mathbf{x}_2) =: \mathbf{Q}. \tag{B.3.22}$$

From (B.3.15) and (B.3.22) we have the following:

Result B.3.1. If $\mathbf{i}$ is an isometry in $\mathcal{E}$, then there exists an orthogonal linear transformation $\mathbf{Q}$ on $\mathcal{V}$ such that, for all $\mathbf{x} \in \mathcal{E}$, and all $\mathbf{u} \in \mathcal{V}$,

$$\mathbf{i}(\mathbf{x}+\mathbf{u}) - \mathbf{i}(\mathbf{x}) = \mathbf{Q}\mathbf{u}. \tag{B.3.23}$$

Equivalently, with $\mathbf{y} := \mathbf{x}+\mathbf{u}$, for all $\mathbf{x}, \mathbf{y} \in \mathcal{E}$

$$\mathbf{i}(\mathbf{y}) - \mathbf{i}(\mathbf{x}) = \mathbf{Q}(\mathbf{y}-\mathbf{x}). \tag{B.3.24}$$

To nail down a specific isometry associated with choice $\mathbf{Q}$ it is necessary (and sufficient) to select the image $\mathbf{c}$ of some selected point $\mathbf{x}_0$. Thus, with $\mathbf{y} = \mathbf{x}_0$ in (B.3.24),

$$\mathbf{Q}(\mathbf{x}_0 - \mathbf{x}) = \mathbf{i}(\mathbf{x}_0) - \mathbf{i}(\mathbf{x}) = \mathbf{c} - \mathbf{i}(\mathbf{x}),$$

and so

$$\mathbf{i}(\mathbf{x}) = \mathbf{c} + \mathbf{Q}(\mathbf{x}-\mathbf{x}_0). \tag{B.3.25}$$

Exercise B.3.1. Check that $\mathbf{i}$ given by (B.3.25) is a deformation (i.e., a bijection) by noting that

(i) if $\mathbf{i}(\mathbf{y}) - \mathbf{i}(\mathbf{x}) = \mathbf{0}$, then $\mathbf{Q}(\mathbf{y}-\mathbf{x}) = \mathbf{0}$ so $\mathbf{y} = \mathbf{x}$ (Why?), and
(ii) any point $\mathbf{y} \in \mathcal{E}$ is the image of point $\mathbf{x}_0 + \mathbf{Q}^{-1}(\mathbf{y}-\mathbf{c})$.

To appreciate the geometrical interpretation of $\mathbf{i}$ it is helpful to regard this map as the composition of two simpler maps as follows.

Exercise B.3.2. Show that if

$$\mathbf{i}_1(\mathbf{x}) := \mathbf{x} + (\mathbf{c}-\mathbf{x}_0), \tag{B.3.26}$$

$$\mathbf{i}_2(\mathbf{x}) := \mathbf{x}_0 + \mathbf{Q}(\mathbf{x}-\mathbf{x}_0), \quad \text{and} \tag{B.3.27}$$

$$\mathbf{i}_3(\mathbf{x}) := \mathbf{x} + \mathbf{Q}^{-1}(\mathbf{c}-\mathbf{x}_0), \tag{B.3.28}$$

then [see (B.3.25)]

$$(\mathbf{i}_1 \circ \mathbf{i}_2)(\mathbf{x}) = \mathbf{i}(\mathbf{x}) = (\mathbf{i}_2 \circ \mathbf{i}_3)(\mathbf{x}). \tag{B.3.29}$$

Any deformation of form (B.3.26) in which every point undergoes the same displacement $\mathbf{u}_0$ (here $\mathbf{u}_0 = \mathbf{c}-\mathbf{x}_0$) is termed a *translation* $\mathbf{u}_0$. Notice that $\mathbf{i}_3$ is also a translation with $\mathbf{u}_0 = \mathbf{Q}^{-1}(\mathbf{c}-\mathbf{x}_0)$.

Now consider deformations of form (B.3.27). Notice that $\mathbf{i}_2(\mathbf{x}_0) = \mathbf{x}_0$ so point $\mathbf{x}_0$ remains unchanged. Since $\mathbf{Q}$ is orthogonal there are two possibilities: $\det \mathbf{Q} = \pm 1$ [see (A.16.5)].

If $\mathbf{Q} = \mathbf{1}$, then $\mathbf{i}_2(\mathbf{x}) = \mathbf{x}$, and nothing changes.

If $\det \mathbf{Q} = 1$, and $\mathbf{Q} \neq \mathbf{1}$, then the analyses preceding (A.16.21) and (A.16.22) yield a unit eigenvector $\mathbf{e}_1$ of $\mathbf{Q}$ with eigenvalue $\lambda = 1$. Consider any point $\mathbf{x}$ on the line through $\mathbf{x}_0$ parallel to $\mathbf{e}_1$: thus $\mathbf{x} = \mathbf{x}_0 + a\mathbf{e}_1$ for some $a \in \mathbb{R}$ (and $|a| = \|\mathbf{x}-\mathbf{x}_0\|$). Hence $\mathbf{i}_2(\mathbf{x}) = \mathbf{x}_0 + \mathbf{Q}(\mathbf{x}-\mathbf{x}_0) = \mathbf{x}_0 + \mathbf{Q}a\mathbf{e}_1 = \mathbf{x}_0 + a\mathbf{Q}\mathbf{e}_1 = \mathbf{x}_0 + a\mathbf{e}_1 = \mathbf{x}$, and such points remain unchanged. Recalling Remark A.16.2, it follows that $\mathbf{i}_2$ represents a rotation about this line: the angle α and sense of rotation were given in this remark.

If $\det \mathbf{Q} = -1$, then $\det(-\mathbf{Q}) = 1$, and

$$\mathbf{i}_2(\mathbf{x}) = (\mathbf{i}_2' \circ \mathbf{i}_2'')(\mathbf{x}) = (\mathbf{i}_2'' \circ \mathbf{i}_2')(\mathbf{x}), \tag{B.3.30}$$

where $\quad \mathbf{i}_2'(\mathbf{x}) := \mathbf{x}_0 + (-\mathbf{Q})(\mathbf{x} - \mathbf{x}_0) \quad$ and $\quad \mathbf{i}_2''(\mathbf{x}) := \mathbf{x}_0 + (-\mathbf{1})(\mathbf{x} - \mathbf{x}_0).$ (B.3.31)

If $\mathbf{Q} = -\mathbf{1}$, then $\mathbf{i}_2 = \mathbf{i}_2''$, and the displacement $\mathbf{i}_2''(\mathbf{x}) - \mathbf{x}_0$ of $\mathbf{i}_2(\mathbf{x})$ from $\mathbf{x}_0$ is equal in magnitude but opposite in direction to the displacement $\mathbf{x} - \mathbf{x}_0$ of $\mathbf{x}$ from $\mathbf{x}_0$. Map $\mathbf{i}_2''$ is termed an *inversion in $\mathcal{E}$ through* $\mathbf{x}_0$. If $\mathbf{Q} \neq -\mathbf{1}$, then $\mathbf{i}_2'$ is a rotation (Why?). See Remark A.16.3, which identifies an inversion with a rotation together with a reflection.

Relations (B.3.25), (B.3.29), and (B.3.30) serve to establish the following:

Result B.3.2. (*Representation theorem for isometries*) Every isometry takes the form (B.3.25) and may be regarded, if $\det \mathbf{Q} > 0$, as the combination of a translation together with a rotation[1], while if $\det \mathbf{Q} < 0$ it may be regarded as the combination of a translation, rotation, and inversion.

Remark B.3.1. Since inversions (and reflections: see Remark A.16.3) are physically impossible to accomplish with material systems, deformations of actual bodies which preserve distance (*rigid-body deformations*) are those of form (B.3.24) with $\det \mathbf{Q} > 0$.

B.3.3 Homogeneous Deformations

Here we consider maps $\mathbf{h} : \mathcal{E} \to \mathcal{E}$ of form

$$\mathbf{h}(\mathbf{x}) := \mathbf{h}_0 + \mathbf{H}(\mathbf{x} - \mathbf{x}_0), \tag{B.3.32}$$

where $\mathbf{h}_0, \mathbf{x}_0 \in \mathcal{E}$ and $\mathbf{H} \in \text{Invlin}\,\mathcal{V}$ are arbitrary.

Exercise B.3.3. Show that if $\mathbf{h}(\mathbf{x}) = \mathbf{h}(\mathbf{y})$, then $\mathbf{H}(\mathbf{y} - \mathbf{x}) = \mathbf{0}$ and hence $\mathbf{x} = \mathbf{y}$ (Why?). Show also that if $\mathbf{y} \in \mathcal{E}$, then $\mathbf{h}(\mathbf{x}) = \mathbf{y}$, where $\mathbf{x} = \mathbf{x}_0 + \mathbf{H}^{-1}(\mathbf{y} - \mathbf{h}_0)$.

This exercise shows that $\mathbf{h}$ is bijective and hence is a deformation. Since $\mathbf{h}_0$ and $\mathbf{H}$ are independent of $\mathbf{x}$, the deformation $\mathbf{h}$ is termed *homogeneous*. Note that for any $\mathbf{x}, \mathbf{y} \in \mathcal{E}$

$$\mathbf{h}(\mathbf{y}) - \mathbf{h}(\mathbf{x}) = \mathbf{H}(\mathbf{y} - \mathbf{x}), \tag{B.3.33}$$

so the displacement of point $\mathbf{h}(\mathbf{y})$ from $\mathbf{h}(\mathbf{x})$ depends only upon the displacement of $\mathbf{y}$ from $\mathbf{x}$ and is linear in this displacement.

Exercise B.3.4.

(i) Show that the identity map $\mathbf{id}(\mathbf{x}) := \mathbf{x}$ is homogeneous with $\mathbf{H} = \mathbf{1}$ by writing $\mathbf{id}(\mathbf{x}) = \mathbf{x}_0 + (\mathbf{x} - \mathbf{x}_0)$.
(ii) Show that from Exercise B.3.3, $\mathbf{h}^{-1}(\mathbf{x}) = \mathbf{x}_0 + \mathbf{H}^{-1}(\mathbf{x} - \mathbf{h}_0)$. Verify that $(\mathbf{h}^{-1} \circ \mathbf{h})(\mathbf{x}) = \mathbf{x} = (\mathbf{h} \circ \mathbf{h}^{-1})(\mathbf{x})$.
(iii) Show that if

$$\mathbf{h}'(\mathbf{x}) := \mathbf{h}_0' + \mathbf{H}'(\mathbf{x} - \mathbf{x}_0') \quad \text{and} \quad \mathbf{h}''(\mathbf{x}) := \mathbf{h}_0'' + \mathbf{H}''(\mathbf{x} - \mathbf{x}_0'')$$

[1] Here 'rotation' is intended to include the trivial situation of $\mathbf{Q} = \mathbf{1}$ in (B.3.27) which yields the identity (or 'leave it alone') map in $\mathcal{E}$.

are two homogeneous deformations (so that $\mathbf{h}_0', \mathbf{x}_0', \mathbf{h}_0'', \mathbf{x}_0'' \in \mathcal{E}$ and $\mathbf{H}', \mathbf{H}'' \in \text{Invlin}\,\mathcal{V}$), then $\mathbf{h} := \mathbf{h}'' \circ \mathbf{h}'$ is a homogeneous deformation (B.3.32) with $\mathbf{h}_0 = \mathbf{h}_0'' + \mathbf{H}''(\mathbf{h}_0' - \mathbf{x}_0'')$ and $\mathbf{H} = \mathbf{H}''\mathbf{H}'$.

Remark B.3.2. Any isometry $\mathbf{i}$ is a homogeneous deformation, as follows from characterisation (B.3.25): here $\mathbf{h}_0 = \mathbf{c}$ and $\mathbf{H} = \mathbf{Q}$.

Exercise B.3.5. Show that

$$\mathbf{h}_1(\mathbf{x}) := \mathbf{x}_0 + \mathbf{H}(\mathbf{x} - \mathbf{x}_0), \mathbf{h}_2(\mathbf{x}) := \mathbf{x} + (\mathbf{h}_0 - \mathbf{x}_0), \text{ and } \mathbf{h}_3(\mathbf{x}) := \mathbf{x} + \mathbf{H}^{-1}(\mathbf{h}_0 - \mathbf{x}_0), \tag{B.3.34}$$

then

$$\mathbf{h}_2 \circ \mathbf{h}_1 = \mathbf{h} = \mathbf{h}_1 \circ \mathbf{h}_3. \tag{B.3.35}$$

Deformation $\mathbf{h}_1$ leaves $\mathbf{x}_0$ unchanged, while $\mathbf{h}_2$ and $\mathbf{h}_3$ are translations $\mathbf{h}_0 - \mathbf{x}_0$ and $\mathbf{H}^{-1}(\mathbf{h}_0 - \mathbf{x}_0)$, respectively. Deformations of form $\mathbf{h}_1$ are termed *homogeneous deformations about* $\mathbf{x}_0$ and may be interpreted geometrically with the aid of the polar decomposition theorem (see Appendix A.18).

Exercise B.3.6. If $\mathbf{RU} = \mathbf{H} = \mathbf{VR}$ are the polar decompositions of $\mathbf{H} \in \text{Invlin}\,\mathcal{V}$ [see (A.18.1): here $\mathbf{F} = \mathbf{H}$], show that

$$\mathbf{h}_4 \circ \mathbf{h}_5 = \mathbf{h}_1 = \mathbf{h}_6 \circ \mathbf{h}_4, \tag{B.3.36}$$

where

$$\mathbf{h}_4(\mathbf{x}) := \mathbf{x}_0 + \mathbf{R}(\mathbf{x} - \mathbf{x}_0), \mathbf{h}_5(\mathbf{x}) := \mathbf{x}_0 + \mathbf{U}(\mathbf{x} - \mathbf{x}_0), \text{ and } \mathbf{h}_6(\mathbf{x}) := \mathbf{x}_0 + \mathbf{V}(\mathbf{x} - \mathbf{x}_0). \tag{B.3.37}$$

Of course, $\mathbf{h}_4$ is an isometry of form (B.3.27), and is either a rotation about a line through $\mathbf{x}_0$ (if $\det \mathbf{R} = 1$) or the combination of a rotation about a line through $\mathbf{x}_0$ with an inversion through $\mathbf{x}_0$ (if $\det \mathbf{R} = -1$: see Result B.3.2). Deformations $\mathbf{h}_5$ and $\mathbf{h}_6$ may be decomposed further into simpler forms, as we now show. Since $\mathbf{U}$ and $\mathbf{V}$ are symmetric and positive-definite, there exists for each an orthonormal basis of eigenvectors. If these are denoted by (here $i = 1,2,3$) $\{\mathbf{u}_i\}$ and $\{\mathbf{v}_i\}$ respectively, and $\{\lambda_i\}$, and $\{\mu_i\}$ denote the corresponding sets of eigenvalues, then [see (B.17.10)]

$$\mathbf{U} = \lambda_i \mathbf{u}_i \otimes \mathbf{u}_i \qquad \text{and} \qquad \mathbf{V} = \mu_i \mathbf{v}_i \otimes \mathbf{v}_i. \tag{B.3.38}$$

Now consider the map (here no summation is implied)

$$\mathbf{s}^i(\mathbf{x}) := \mathbf{x} + (\lambda_i - 1)(\mathbf{u}_i \otimes \mathbf{u}_i)(\mathbf{x} - \mathbf{x}_0) \tag{B.3.39}$$

$$= \mathbf{x} + (\lambda_i - 1)(\mathbf{u}_i \,.\, (\mathbf{x} - \mathbf{x}_0))\mathbf{u}_i. \tag{B.3.40}$$

Exercise B.3.7. Show that

$$(\mathbf{s}^1 \circ \mathbf{s}^2 \circ \mathbf{s}^3)(\mathbf{x}) = \mathbf{h}_5(\mathbf{x}). \tag{B.3.41}$$

Show further that $\mathbf{s}^1, \mathbf{s}^2$, and $\mathbf{s}^3$ commute so that any composition of all three deformations results in $\mathbf{h}_5$. [It is helpful to recall (A.9.16) and (A.9.19).]

Notice that $\mathbf{s}^i(\mathbf{x}_0) = \mathbf{x}_0$. Further, selecting a Cartesian co-ordinate system $C(\mathbf{x}_0; \mathbf{u}_1, \mathbf{u}_2, \mathbf{u}_3)$ (see Section B.2), the co-ordinates of $\mathbf{s}^1(\mathbf{x})$ are given by

$$s_1^1(\mathbf{x}) = x_1 + (\lambda_1 - 1)x_1 = \lambda_1 x_1, \; s_2^1(\mathbf{x}) = x_2 \qquad \text{and} \qquad s_3^1(\mathbf{x}) = x_3. \tag{B.3.42}$$

Accordingly $\mathbf{s}^1$ leaves unchanged points on the plane $x_1 = 0$ (i.e., the plane through $\mathbf{x}_0$ with normal $\mathbf{u}_1$) and moves any point on the plane $x_1 = k$ (the plane through $\mathbf{x}_0 + k\mathbf{u}_1$ with normal $\mathbf{u}_1$) through a displacement $\lambda_1 k \mathbf{u}_1$. If $\lambda_1 > 1$, then this corresponds to a uniaxial stretching (or scaling up) *away* from the plane $x_1 = 0$, and parallel to $\mathbf{u}_1$ with scaling factor λ_1. If $0 < \lambda_1 < 1$, then there is a corresponding downscaling parallel to $\mathbf{u}_1$ *towards* the plane $x_1 = 0$ with scaling factor λ_1. Mapping $\mathbf{s}^1$ accordingly is termed a *simple stretch about* $\mathbf{x}_0$, *parallel to* $\mathbf{u}_1$, *of scale* λ_1. Similarly, choices $i = 2$ and $i = 3$ yield simple stretches about $\mathbf{x}_0$ parallel to $\mathbf{u}_2$ and $\mathbf{u}_3$) of scales λ_2 and λ_3, respectively. Thus Exercise B.3.7. shows that $\mathbf{h}_5$ may be regarded as the composition of three simple stretches about $\mathbf{x}_0$, parallel to the directions defined by the eigenvectors of $\mathbf{U}$, with scales given by the corresponding eigenvalues.

Remark B.3.3. Of course, the analysis can be repeated in respect of $\mathbf{h}_6$, given $(B.3.37)_3$, so this deformation may be regarded as the result of three successive simple stretches about $\mathbf{x}_0$ parallel to $\mathbf{v}_i$ of scale μ_i $(i = 1,2,3)$.

Combining results (B.3.35), (B.3.36), and (B.3.41) with Remark B.3.3 yields the following:

Result B.3.3. (*Characterisation of homogeneous deformations*) Any homogeneous deformation may be regarded as the combination of five successive deformations of simple nature, namely a translation, followed by three simple stretches about the same point, and then a rotation (or rotation plus inversion) about this point: this combination corresponds to $\mathbf{h}_4 \circ \mathbf{s}^1 \circ \mathbf{s}^2 \circ \mathbf{s}^3 \circ \mathbf{h}_3$.

Exercise B.3.8. Describe the nature and order of deformations in the combination $\mathbf{h}_2 \circ \mathbf{h}_4 \circ \mathbf{s}^1 \circ \mathbf{s}^2 \circ \mathbf{s}^3$.

Exercise B.3.9. If [see $(B.3.38)_2$ and compare with (B.3.39)]

$$ {}^i\mathbf{s}(\mathbf{x}) := \mathbf{x} + (\mu_i - 1)(\mathbf{v}_i . (\mathbf{x} - \mathbf{x}_0))\mathbf{v}_i, \tag{B.3.43} $$

show that [see $(B.3.37)_3$]

$$ \mathbf{h}_6 = {}^1\mathbf{s} \circ {}^2\mathbf{s} \circ {}^3\mathbf{s}. \tag{B.3.44} $$

Describe the nature and order of deformations in the combinations $\mathbf{h}_2 \circ {}^1\mathbf{s} \circ {}^2\mathbf{s} \circ {}^3\mathbf{s} \circ \mathbf{h}_4$ and ${}^1\mathbf{s} \circ {}^2\mathbf{s} \circ {}^3\mathbf{s} \circ \mathbf{h}_4 \circ \mathbf{h}_3$.

Remark B.3.4. Since inversions cannot be effected for material systems (see Remark B.3.1) it is necessary to exclude homogeneous deformations for which $\det \mathbf{R} < 0$ (see Exercise B.3.6) as models of material distortion. Equivalently, for such distortion $\det \mathbf{H} > 0$ (Why?). Thus we term homogeneous deformations with $\det \mathbf{H} > 0$ as *physically admissible.* Of course, Result B.3.3 is modified for such deformations by the exclusion of inversions.

Of later interest will be the following:

Result B.3.4. Homogeneous deformations map planes into planes. Further, if two planes are parallel then the image planes under a homogeneous deformation are parallel.

Proof. A plane is characterised by a point, $\mathbf{x}_0'$ say, lying on it together with any normal line. If $\mathbf{n} \in \mathcal{V}$ is parallel to this line, and if $\mathbf{x}$ denotes any point on the plane, then displacement $(\mathbf{x} - \mathbf{x}_0')$ is orthogonal to $\mathbf{n}$, whence

$$(\mathbf{x} - \mathbf{x}_0') . \mathbf{n} = 0. \tag{B.3.45}$$

Any homogeneous deformation $\mathbf{h}$ [see (B.3.32)] may be written in the form

$$\mathbf{h}(\mathbf{x}) = \mathbf{h}_0 + \mathbf{H}(\mathbf{x} - \mathbf{x}_0') + \mathbf{H}(\mathbf{x}_0' - \mathbf{x}_0). \tag{B.3.46}$$

Equivalently,

$$\mathbf{x} - \mathbf{x}_0' = \mathbf{H}^{-1}(\mathbf{h}(\mathbf{x}) - \mathbf{c}_0) \tag{B.3.47}$$

where point

$$\mathbf{c}_0 := \mathbf{h}_0 + \mathbf{H}(\mathbf{x}_0' - \mathbf{x}_0). \tag{B.3.48}$$

Thus, if $\mathbf{x}$ lies on the plane, then (B.3.45) and (B.3.47) imply that

$$\mathbf{H}^{-1}(\mathbf{h}(\mathbf{x}) - \mathbf{c}_0) . \mathbf{n} = 0.$$

That is [see (A.10.19) et seq.],

$$(\mathbf{h}(\mathbf{x}) - \mathbf{c}_0) . \mathbf{H}^{-T}\mathbf{n} = 0. \tag{B.3.49}$$

Thus $\mathbf{h}(\mathbf{x})$ lies on that plane through point $\mathbf{c}_0$ [given by (B.3.48)] with normal parallel to $\mathbf{H}^{-T}\mathbf{n}$.

If two planes are parallel, then they have a common normal line (parallel to $\mathbf{n}$, say), and thus their images are planes, both having a common normal line (parallel to $\mathbf{H}^{-T}\mathbf{n}$).

B.4 Generalisation of the Concept of a Derivative

B.4.1 Preamble

The notion of a derivative is associated with the problem of finding the closest linear approximation to the graph of a real-valued function of a real variable at a given point, namely the tangent line at this point. Consider the function

$$f : (a,b) \longrightarrow \mathbb{R} \tag{B.4.1}$$

defined on the open interval (a,b). If $x \in (a,b)$ and

$$f(x+h) = f(x) + \ell h + o(h) \qquad \text{as} \quad h \to 0 \tag{B.4.2}$$

for some $\ell \in \mathbb{R}$, then f is said to be *differentiable* at x with derivative

$$f'(x) := \ell. \tag{B.4.3}$$

The local linear approximation to f at x is

$$f_{\text{lin}}(x+h) := f(x) + f'(x)h \tag{B.4.4}$$

for *any* $h \in \mathbb{R}$. [Increment $f'(x)h$, a function of both x and h which is linear in h, is termed the *differential* of f: see Apostol [91], p. 105.]

The definition (B.4.2) of $f'(x)$ is often introduced somewhat differently. Rewriting this equation as

$$\frac{f(x+h)-f(x)}{h} = \ell + \frac{o(h)}{h} \tag{B.4.5}$$

and noting that by definition of 'o' order notation

$$\lim_{h\to 0}\left\{\frac{o(h)}{h}\right\} = 0, \tag{B.4.6}$$

it follows that

$$f'(x) = \ell = \lim_{h\to 0}\left\{\frac{f(x+h)-f(x)}{h}\right\}. \tag{B.4.7}$$

While (B.4.2) and (B.4.7) are equivalent in the context of functions of a real variable, it is the former version which generalises: arguments may be vectorial, and hence the division employed in (B.4.5) cannot be effected. Convince yourself as follows.

Exercise B.4.1. Try to define vector division as follows. Given any $\mathbf{a},\mathbf{b} \in \mathcal{V}$ and $\mathbf{c} = \mathbf{a} \div \mathbf{b}$, then we should wish either $\mathbf{b}\times\mathbf{c} = \mathbf{a}$ or $\mathbf{c}\times\mathbf{b} = \mathbf{a}$. Notice that only one such equality is possible unless $\mathbf{c} = \mathbf{0}$ (Why?). Since $\mathbf{b}\times\mathbf{c}$ and $\mathbf{c}\times\mathbf{b}$ are orthogonal to $\mathbf{b}$, such 'division' could be possible only if $\mathbf{a}$ were to be orthogonal to $\mathbf{b}$.

B.4.2 Differentiation of a Scalar Field

Let ϕ denote a scalar field (e.g., the mass density or temperature associated with a continuous body). Specifically, suppose that

$$\phi : \mathcal{D} \to \mathbb{R}, \tag{B.4.8}$$

where $\mathcal{D}$ is an open subset of $\mathcal{E}$. If $\mathbf{x} \in \mathcal{D}$ and

$$\phi(\mathbf{x}+\mathbf{h}) = \phi(\mathbf{x}) + \boldsymbol{\ell}.\mathbf{h} + o(\mathbf{h}) \qquad \text{as } \mathbf{h}\to\mathbf{0} \tag{B.4.9}$$

for some vector $\boldsymbol{\ell} \in \mathcal{V}$, then $\boldsymbol{\ell}$ is termed the *derivative* (or *gradient*) *of* ϕ *evaluated at* $\mathbf{x}$. Of course, here $\mathbf{h} \in \mathcal{V}$ and $o(\mathbf{h})$ represents a quantity that tends to zero faster than $\mathbf{h}$; that is,

$$\text{if} \quad a = o(\mathbf{h}), \quad \text{then} \quad \lim_{\mathbf{h}\to\mathbf{0}}\left\{\frac{a}{\|\mathbf{h}\|}\right\} = 0. \tag{B.4.10}$$

Writing

$$\nabla\phi(\mathbf{x}) := \boldsymbol{\ell} \tag{B.4.11}$$

yields

$$\phi(\mathbf{x}+\mathbf{h}) = \phi(\mathbf{x}) + \nabla\phi(\mathbf{x}).\mathbf{h} + o(\mathbf{h}) \qquad \text{as } \mathbf{h}\to\mathbf{0}. \tag{B.4.12}$$

The local linear approximation to ϕ at $\mathbf{x}$ is

$$\phi_{\text{lin}}(\mathbf{x}+\mathbf{h}) := \phi(\mathbf{x}) + \nabla\phi(\mathbf{x}).\mathbf{h} \tag{B.4.13}$$

for *any* $\mathbf{h} \in \mathcal{V}$.

Let $\mathbf{h} = s\hat{\mathbf{u}}$ in (B.4.12), where $\hat{\mathbf{u}}$ is a fixed unit vector and $s \in \mathbb{R}$. Then (B.4.12) can be written as

$$\phi(\mathbf{x}+s\hat{\mathbf{u}}) - \phi(\mathbf{x}) = s\nabla\phi(\mathbf{x}).\hat{\mathbf{u}} + o(s) \qquad \text{as } s\to 0, \tag{B.4.14}$$

noting $\|\mathbf{h}\| = s$. At this point division throughout by s [note the impossibility of dividing by $\mathbf{h}$ in (B.4.12)] and taking the limit as $s \to 0$ yield

$$\nabla\phi(\mathbf{x}).\hat{\mathbf{u}} = \lim_{s\to 0}\left\{\frac{\phi(\mathbf{x}+s\hat{\mathbf{u}})-\phi(\mathbf{x})}{s}\right\}. \tag{B.4.15}$$

The right-hand side of (B.4.15) delivers the rate at which ϕ changes when moving away from $\mathbf{x}$ in the direction of $\hat{\mathbf{u}}$ and is known as the *directional derivative of ϕ associated with the direction defined by* $\hat{\mathbf{u}}$. Thus $\nabla\phi$ is a vector field whose value at point $\mathbf{x}$ has a component in any given direction which is the directional derivative of ϕ at $\mathbf{x}$ associated with this direction.

Suppose that $C(\mathbf{x}_0;\mathbf{e}_1,\mathbf{e}_2,\mathbf{e}_3)$ is a Cartesian co-ordinate system for $\mathcal{E}$ (see Section B.2). Thus $\mathbf{x} \in \mathcal{E}$ has corresponding co-ordinates (x_1,x_2,x_3), where

$$x_i := (\mathbf{x}-\mathbf{x}_0).\mathbf{e}_i \qquad (i = 1,2,3). \tag{B.4.16}$$

The scalar-valued function ϕ of position now can be represented by $\check{\phi}$, a function defined on an open set $\check{\mathcal{D}}$ in $\mathbb{R}^3$ (which corresponds to the domain $\mathcal{D}$ of ϕ), where

$$\check{\phi}(x_1,x_2,x_3) := \phi(\mathbf{x}). \tag{B.4.17}$$

We now can obtain the co-ordinate *representation* of $\nabla\phi$. Noting that the vector $\nabla\phi(\mathbf{x})$ has components

$$\nabla\phi(\mathbf{x}).\mathbf{e}_i \qquad (i = 1,2,3),$$

from (B.4.15) with $\hat{\mathbf{u}} = \mathbf{e}_1$

$$\begin{aligned}\nabla\phi(\mathbf{x}).\mathbf{e}_1 &= \lim_{s\to 0}\left\{\frac{\phi(\mathbf{x}+s\mathbf{e}_1)-\phi(\mathbf{x})}{s}\right\}\\ &= \lim_{s\to 0}\left\{\frac{\check{\phi}(x_1+s,x_2,x_3)-\check{\phi}(x_1,x_2,x_3)}{s}\right\}.\end{aligned}$$

That is,

$$\nabla\phi(\mathbf{x}).\mathbf{e}_1 = \frac{\partial\check{\phi}}{\partial x_1}(x_1,x_2,x_3). \tag{B.4.18}$$

Notation. If f is a function of n real variables, then the derivative with respect to the rth argument/variable/slot will be denoted by $f_{,r}$. Hence (B.4.18) may be written as

$$\nabla\phi(\mathbf{x}).\mathbf{e}_1 = \check{\phi}_{,1}(x_1,x_2,x_3). \tag{B.4.19}$$

Similarly (Show this!),

$$\nabla\phi(\mathbf{x}).\mathbf{e}_2 = \check{\phi}_{,2}(x_1,x_2,x_3) \qquad \text{and} \qquad \nabla\phi(\mathbf{x}).\mathbf{e}_3 = \check{\phi}_{,3}(x_1,x_2,x_3). \tag{B.4.20}$$

Thus, since [see (A.6.6) with $\mathbf{v} = \nabla\phi(\mathbf{x})$]

$$\nabla\phi(\mathbf{x}) = (\nabla\phi(\mathbf{x}).\mathbf{e}_i)\mathbf{e}_i, \tag{B.4.21}$$

(B.4.19) and (B.4.20) yield

$$\nabla\phi(\mathbf{x}) = \check{\phi}_{,i}(x_1,x_2,x_3)\mathbf{e}_i. \tag{B.4.22}$$

In many texts $\breve{\phi}_{,i}\mathbf{e}_i$ is denoted by grad $\breve{\phi}$, co-ordinates are labelled (x,y,z), and unit orthonormal bases are denoted by $\mathbf{i},\mathbf{j}$, and $\mathbf{k}$, so (B.4.22) is expressed as

$$\operatorname{grad}\breve{\phi} = \frac{\partial\breve{\phi}}{\partial x}\mathbf{i} + \frac{\partial\breve{\phi}}{\partial y}\mathbf{j} + \frac{\partial\breve{\phi}}{\partial z}\mathbf{k}. \tag{B.4.23}$$

In Cartesian tensor notation (see Appendix A.20)

$$\nabla\phi = \phi_{,i}. \tag{B.4.24}$$

Such notation does not distinguish between ϕ and $\breve{\phi}$. (What *is* the difference between ϕ and $\breve{\phi}$?)

B.4.3 Differentiation of Point-Valued Fields

Let

$$\mathbf{d} : \mathcal{D} \to \mathcal{E}, \tag{B.4.25}$$

where $\mathcal{D}$ is an open subset of $\mathcal{E}$. Thus $\mathbf{d}$ maps points in $\mathcal{D}$ into points and could [if bijective: see (B.3.1) et seq.] represent the deformation of a body which initially occupies the region $\mathcal{D}$. We say that $\mathbf{d}$ is differentiable at $\hat{\mathbf{x}} \in \mathcal{D}$ if there exists a linear transformation $\mathbf{L} \in \operatorname{Lin}\mathcal{V}$ such that

$$\mathbf{d}(\hat{\mathbf{x}}+\mathbf{h}) = \mathbf{d}(\hat{\mathbf{x}}) + \mathbf{L}\mathbf{h} + o(\mathbf{h}) \qquad \text{as } \mathbf{h} \to \mathbf{0}. \tag{B.4.26}$$

Transformation $\mathbf{L}$ is termed the *derivative* or *gradient of* $\mathbf{d}$ *evaluated at* $\hat{\mathbf{x}}$. Writing

$$\nabla\mathbf{d}(\hat{\mathbf{x}}) := \mathbf{L} \tag{B.4.27}$$

yields

$$\mathbf{d}(\hat{\mathbf{x}}+\mathbf{h}) = \mathbf{d}(\hat{\mathbf{x}}) + \nabla\mathbf{d}(\hat{\mathbf{x}})\mathbf{h} + o(\mathbf{h}) \qquad \text{as } \mathbf{h} \to \mathbf{0}. \tag{B.4.28}$$

The local linear approximation to $\mathbf{d}$ at $\hat{\mathbf{x}}$ is

$$\mathbf{d}_{\text{lin}}(\hat{\mathbf{x}}+\mathbf{h}) := \mathbf{d}(\hat{\mathbf{x}}) + \nabla\mathbf{d}(\hat{\mathbf{x}})\mathbf{h} \tag{B.4.29}$$

for *any* $\mathbf{h} \in \mathcal{V}$.

Exercise B.4.2. Show, by comparing (B.4.26) and (B.4.27) with $\mathbf{i}$ given by (B.3.25) and $\mathbf{h}$ given by (B.3.32), that $\nabla\mathbf{i}(\mathbf{x}) = \mathbf{Q}$ and $\nabla\mathbf{h}(\mathbf{x}) = \mathbf{H}$. (That is, isometries, and homogeneous deformations in general, are differentiable and have derivatives which are independent of location.)

Remark B.4.1. If $\mathbf{d}$ is a deformation which is differentiable at $\hat{\mathbf{x}}$ and $\det(\nabla\mathbf{d}(\hat{\mathbf{x}})) \neq 0$, then the linear approximation at $\hat{\mathbf{x}}$ [see (B.4.29)] is a homogeneous deformation with $\mathbf{h}_0 = \mathbf{d}(\hat{\mathbf{x}})$ and $\mathbf{H} = \nabla\mathbf{d}(\hat{\mathbf{x}})$.

If $\hat{\mathbf{u}}$ is a fixed unit vector and $s \in \mathbb{R}$, then, with $\mathbf{h} = s\hat{\mathbf{u}}$ in (B.4.28),

$$\mathbf{d}(\hat{\mathbf{x}}+s\hat{\mathbf{u}}) = \mathbf{d}(\hat{\mathbf{x}}) + \nabla\mathbf{d}(\hat{\mathbf{x}})s\hat{\mathbf{u}} + o(s) \qquad \text{as } s \to 0. \tag{B.4.30}$$

Thus [cf. (B.4.15)]

$$\nabla\mathbf{d}(\hat{\mathbf{x}})\hat{\mathbf{u}} = \lim_{s\to 0}\left\{\frac{\mathbf{d}(\hat{\mathbf{x}}+s\hat{\mathbf{u}}) - \mathbf{d}(\hat{\mathbf{x}})}{s}\right\}, \tag{B.4.31}$$

and hence acting upon $\hat{\mathbf{u}}$ with linear transformation $\nabla\mathbf{d}(\hat{\mathbf{x}})$ yields the directional derivative of $\mathbf{d}$ at $\hat{\mathbf{x}}$ associated with the direction defined by $\hat{\mathbf{u}}$.

Question: What are the co-ordinate representations of $\mathbf{d}$ and $\nabla\mathbf{d}$?

Answer: Choice of a Cartesian co-ordinate system $C(\mathbf{x}_0;\mathbf{e}_1,\mathbf{e}_2,\mathbf{e}_3)$ gives a representation $\check{\mathbf{d}}$ of $\mathbf{d}$. Specifically [cf. (B.4.17)],

$$\check{\mathbf{d}} : \check{\mathcal{D}} \subset \mathbb{R}^3 \to \mathcal{E}, \tag{B.4.32}$$

where

$$\check{\mathbf{d}}(\hat{x}_1,\hat{x}_2,\hat{x}_3) = \mathbf{d}(\hat{\mathbf{x}}) =: \mathbf{x}, \tag{B.4.33}$$

with

$$(j=1,2,3) \quad \hat{x}_j := (\hat{\mathbf{x}} - \mathbf{x}_0)\,.\,\mathbf{e}_j. \tag{B.4.34}$$

The co-ordinate representation of $\mathbf{x}$ is given by

$$\begin{aligned} x_i := (\mathbf{x}-\mathbf{x}_0)\,.\,\mathbf{e}_i &= (\mathbf{d}(\hat{\mathbf{x}})-\mathbf{x}_0)\,.\,\mathbf{e}_i \\ &= (\check{\mathbf{d}}(\hat{x}_1,\hat{x}_2,\hat{x}_3)-\mathbf{x}_0)\,.\,\mathbf{e}_i \qquad \text{(B.4.35)} \\ &=: \check{x}_i(\hat{x}_1,\hat{x}_2,\hat{x}_3). \qquad \text{(B.4.36)} \end{aligned}$$

That is, each $\mathbf{x}$ co-ordinate x_i is a function $\check{x}_i$ of the co-ordinates $(\hat{x}_1,\hat{x}_2,\hat{x}_3)$ of $\hat{\mathbf{x}}$.

To obtain the co-ordinate representation of $\nabla\mathbf{d}$, note that, since $\nabla\mathbf{d}(\hat{\mathbf{x}}) \in \operatorname{Lin}\mathcal{V}$, this map may be represented by a 3×3 matrix, namely $[\mathbf{e}_i\,.\,\nabla\mathbf{d}(\hat{\mathbf{x}})\mathbf{e}_j]$ (see Appendix A.9). From (B.4.31) with $\hat{\mathbf{u}} = \mathbf{e}_1$, together with (B.4.35) and (B.4.36),

$$\begin{aligned} \mathbf{e}_i\,.\,\nabla\mathbf{d}(\hat{\mathbf{x}})\mathbf{e}_1 &= \left(\lim_{s\to 0}\left\{\frac{\mathbf{d}(\hat{\mathbf{x}}+s\mathbf{e}_1)-\mathbf{d}(\hat{\mathbf{x}})}{s}\right\}\right).\mathbf{e}_i \\ &= \lim_{s\to 0}\left\{\frac{[(\mathbf{d}(\hat{\mathbf{x}}+s\mathbf{e}_1)-\mathbf{x}_0)-(\mathbf{d}(\hat{\mathbf{x}})-\mathbf{x}_0)]}{s}\,.\,\mathbf{e}_i\right\} \\ &= \lim_{s\to 0}\left\{\frac{\check{x}_i(\hat{x}_1+s,\hat{x}_2,\hat{x}_3)-\check{x}_i(\hat{x}_1,\hat{x}_2,\hat{x}_3)}{s}\right\} \\ &= \frac{\partial \check{x}_i}{\partial x_1}(\hat{x}_1,\hat{x}_2,\hat{x}_3) = \check{x}_{i,1}(\hat{x}_1,\hat{x}_2,\hat{x}_3). \end{aligned} \tag{B.4.37}$$

Similarly, with $\mathbf{h} = s\mathbf{e}_2$ and $\mathbf{h} = s\mathbf{e}_3$, respectively,

$$\mathbf{e}_i\,.\,\nabla\mathbf{d}(\hat{\mathbf{x}})\mathbf{e}_2 = \check{x}_{i,2}(\hat{x}_1,\hat{x}_2,\hat{x}_3) \quad \text{and} \quad \mathbf{e}_i\,.\,\nabla\mathbf{d}(\hat{\mathbf{x}})\mathbf{e}_3 = \check{x}_{i,3}(\hat{x}_1,\hat{x}_2,\hat{x}_3). \tag{B.4.38}$$

(Show this!) Accordingly

$$\mathbf{e}_i\,.\,\nabla\mathbf{d}(\hat{\mathbf{x}})\mathbf{e}_j = \check{x}_{i,j}(\hat{x}_1,\hat{x}_2,\hat{x}_3). \tag{B.4.39}$$

That is, the matrix representative of $\nabla\mathbf{d}(\hat{\mathbf{x}})$ associated with choice $C(\mathbf{x}_0;\mathbf{e}_1,\mathbf{e}_2,\mathbf{e}_3)$ of Cartesian co-ordinate system is

$$\begin{bmatrix} \check{x}_{1,1} & \check{x}_{1,2} & \check{x}_{1,3} \\ \check{x}_{2,1} & \check{x}_{2,2} & \check{x}_{2,3} \\ \check{x}_{3,1} & \check{x}_{3,2} & \check{x}_{3,3} \end{bmatrix} \quad (= [\check{x}_{i,j}]), \tag{B.4.40}$$

where all derivatives are evaluated at $(\hat{x}_1,\hat{x}_2,\hat{x}_3)$.

In Cartesian tensor notation $\nabla\mathbf{d}$ is denoted by $x_{i,j}$.

B.4.4 Differentiation of Vector Fields

Let
$$\mathbf{v} : \mathcal{D} \to \mathcal{V} \tag{B.4.41}$$

denote a vector field on an open subset $\mathcal{D}$ of $\mathcal{E}$. For example, $\mathbf{v}$ could denote the velocity field of a continuous body or the displacement field of such a body from its location at some given time. We say that $\mathbf{v}$ is *differentiable* at $\mathbf{x} \in \mathcal{D}$ if there exists a linear transformation $\mathbf{L}$ such that

$$\mathbf{v}(\mathbf{x}+\mathbf{h}) = \mathbf{v}(\mathbf{x}) + \mathbf{L}\mathbf{h} + o(\mathbf{h}) \qquad \text{as } \mathbf{h} \to \mathbf{0}, \tag{B.4.42}$$

write
$$\nabla\mathbf{v}(\mathbf{x}) := \mathbf{L}, \tag{B.4.43}$$

and term $\nabla\mathbf{v}(\mathbf{x})$ the *derivative* or *gradient* of $\mathbf{v}$ *at* $\mathbf{x}$. The local linear approximation to $\mathbf{v}$ at $\mathbf{x}$ is

$$\mathbf{v}_{\text{lin}}(\mathbf{x}+\mathbf{h}) := \mathbf{v}(\mathbf{x}) + (\nabla\mathbf{v}(\mathbf{x}))\mathbf{h} \tag{B.4.44}$$

for *any* $\mathbf{h} \in \mathcal{V}$.

Exercise B.4.3. Show that if $\hat{\mathbf{u}}$ is any fixed unit vector, then

$$\nabla\mathbf{v}(\mathbf{x})\hat{\mathbf{u}} = \lim_{s\to 0}\left\{\frac{\mathbf{v}(\mathbf{x}+s\hat{\mathbf{u}}) - \mathbf{v}(\mathbf{x})}{s}\right\}. \tag{B.4.45}$$

[*Hint*: Proceed as in (B.4.30) and (B.4.31).]

Given a Cartesian co-ordinate system $C(\mathbf{x}_0;\mathbf{e}_1,\mathbf{e}_2,\mathbf{e}_3)$, $\mathbf{v}$ can be written as

$$\begin{aligned}\mathbf{v}(\mathbf{x}) &= \check{v}_1(x_1,x_2,x_3)\mathbf{e}_1 + \check{v}_2(x_1,x_2,x_3)\mathbf{e}_2 + \check{v}_3(x_1,x_2,x_3)\mathbf{e}_3 \\ &= \check{v}_k(x_1,x_2,x_3)\mathbf{e}_k,\end{aligned} \tag{B.4.46}$$

where $\quad x_i := (\mathbf{x}-\mathbf{x}_0)\,.\,\mathbf{e}_i \quad$ and $\quad \check{v}_k(x_1,x_2,x_3) := v_k(\mathbf{x}) := \mathbf{v}(\mathbf{x})\,.\,\mathbf{e}_k\,. \qquad$ (B.4.47)

That is, x_i are the co-ordinates of $\mathbf{x}$ and v_k are the components of $\mathbf{v}$. The matrix representation of $\nabla\mathbf{v}(\mathbf{x})$ with respect to orthonormal basis $\{\mathbf{e}_1,\mathbf{e}_2,\mathbf{e}_3\}$ is $[L_{ij}]$ where (see Appendix A.9)

$$L_{ij} := \mathbf{e}_i\,.\,(\nabla\mathbf{v}(\mathbf{x}))\mathbf{e}_j. \tag{B.4.48}$$

From (B.4.45) and (B.4.46)

$$\begin{aligned}(\nabla\mathbf{v}(\mathbf{x}))\mathbf{e}_1 &= \lim_{s\to 0}\left\{\frac{\mathbf{v}(\mathbf{x}+s\mathbf{e}_1) - \mathbf{v}(\mathbf{x})}{s}\right\} \\ &= \lim_{s\to 0}\left\{\left[\frac{(\check{v}_k(x_1+s,x_2,x_3) - \check{v}_k(x_1,x_2,x_3))}{s}\right]\mathbf{e}_k\right\} \\ &= \check{v}_{k,1}\,\mathbf{e}_k.\end{aligned}$$

Thus
$$L_{i1} = \mathbf{e}_i\,.\,(\nabla\mathbf{v}(\mathbf{x}))\mathbf{e}_1 = \check{v}_{i,1}. \tag{B.4.49}$$

Similarly (Show this!),

$$L_{i2} = \check{v}_{i,2} \qquad \text{and} \qquad L_{i3} = \check{v}_{i,3}. \tag{B.4.50}$$

Accordingly the matrix representative of $\nabla\mathbf{v}$ associated with choice $C(\mathbf{x}_0;\mathbf{e}_1,\mathbf{e}_2,\mathbf{e}_3)$ of Cartesian co-ordinate system is

$$[L_{ij}] = \begin{bmatrix} \check{v}_{1,1} & \check{v}_{1,2} & \check{v}_{1,3} \\ \check{v}_{2,1} & \check{v}_{2,2} & \check{v}_{2,3} \\ \check{v}_{3,1} & \check{v}_{3,2} & \check{v}_{3,3} \end{bmatrix} = [\check{v}_{i,j}]. \tag{B.4.51}$$

Of course [see (A.9.7), (A.9.8), (B.4.43), and (B.4.48)],

$$\nabla\mathbf{v} = \mathbf{L} = L_{ij}\mathbf{e}_i \otimes \mathbf{e}_j = \check{v}_{i,j}\mathbf{e}_i \otimes \mathbf{e}_j. \tag{B.4.52}$$

In Cartesian tensor notation $\nabla\mathbf{v}$ is denoted by $v_{i,j}$.

B.4.5 Differentiation of Linear Transformation Fields

Let

$$\mathbf{A} : \mathcal{D} \to \operatorname{Lin}\mathcal{V} \tag{B.4.53}$$

denote a linear transformation field on an open subset $\mathcal{D}$ of $\mathcal{E}$. For example, $\mathbf{A}$ could represent the Cauchy stress tensor, velocity gradient, or deformation gradient. We say that $\mathbf{A}$ is *differentiable* at $\mathbf{x} \in \mathcal{D}$ if there exists a linear map $\boldsymbol{\mathcal{L}}$ from $\mathcal{V}$ into $\operatorname{Lin}\mathcal{V}$ (so $\boldsymbol{\mathcal{L}} \in \operatorname{Lin}(\mathcal{V}, \operatorname{Lin}\mathcal{V})$: see Appendix A.19) such that

$$\mathbf{A}(\mathbf{x}+\mathbf{h}) = \mathbf{A}(\mathbf{x}) + \boldsymbol{\mathcal{L}}\mathbf{h} + o(\mathbf{h}) \qquad \text{as } \mathbf{h} \to \mathbf{0}. \tag{B.4.54}$$

We write

$$\nabla\mathbf{A}(\mathbf{x}) := \boldsymbol{\mathcal{L}} \tag{B.4.55}$$

and term $\nabla\mathbf{A}(\mathbf{x})$ the *derivative* or *gradient of* $\mathbf{A}$ *at* $\mathbf{x}$. The local linear approximation to $\mathbf{A}$ at $\mathbf{x}$ is [cf. (B.4.13), (B.4.29), and (B.4.44)]

$$\mathbf{A}_{\text{lin}}(\mathbf{x}+\mathbf{h}) := \mathbf{A}(\mathbf{x}) + \nabla\mathbf{A}(\mathbf{x})\mathbf{h}. \tag{B.4.56}$$

Exercise B.4.4. Show that if $\hat{\mathbf{u}}$ is any fixed unit vector, then

$$\nabla\mathbf{A}(\mathbf{x})\hat{\mathbf{u}} = \lim_{s\to 0}\left\{\frac{\mathbf{A}(\mathbf{x}+s\hat{\mathbf{u}}) - \mathbf{A}(\mathbf{x})}{s}\right\}. \tag{B.4.57}$$

(Cf. Exercise B.4.2.)

In terms of a Cartesian co-ordinate system $C(\mathbf{x}_0;\mathbf{e}_1,\mathbf{e}_2,\mathbf{e}_3)$, field value

$$\mathbf{A}(\mathbf{x}) = A_{ij}(\mathbf{x})\mathbf{e}_i \otimes \mathbf{e}_j = \check{A}_{ij}(x_1,x_2,x_3)\mathbf{e}_i \otimes \mathbf{e}_j, \tag{B.4.58}$$

where [see (A.9.7) and (A.9.8)] $A_{ij} := \mathbf{e}_i \,.\, \mathbf{A}\mathbf{e}_j$.

Exercise B.4.5. Show from (B.4.57) and (B.4.58) that, for $p = 1,2,3$,

$$\nabla\mathbf{A}(\mathbf{x})\mathbf{e}_p = \check{A}_{ij,p}(x_1,x_2,x_3)\mathbf{e}_i \otimes \mathbf{e}_j. \tag{B.4.59}$$

It follows from (A.19.5) that

$$\nabla\mathbf{A}(\mathbf{x}) = \check{A}_{ij,k}(x_1,x_2,x_3)\mathbf{e}_i \otimes \mathbf{e}_j \otimes \mathbf{e}_k. \tag{B.4.60}$$

Exercise B.4.6. Verify (B.4.60).

In Cartesian tensor notation $\nabla\mathbf{A}$ is denoted by $A_{ij,k}$.

B.4.6 Remarks

Remark B.4.2. The generalisation of the foregoing to higher-order tensor fields is straightforward. For example, if $\mathbf{C}$ is third-order, then its *derivative* or *gradient* $\nabla\mathbf{C}$ takes values in[2] $\mathrm{Lin}(\mathcal{V},\mathrm{Lin}(\mathcal{V},\mathrm{Lin}\,\mathcal{V}))$, and with $\mathbf{C} = C_{ijk}\mathbf{e}_i \otimes \mathbf{e}_j \otimes \mathbf{e}_k$,

$$\nabla\mathbf{C} = \check{C}_{ijk,\ell}\mathbf{e}_i \otimes \mathbf{e}_j \otimes \mathbf{e}_k \otimes \mathbf{e}_\ell, \tag{B.4.61}$$

where
$$(\mathbf{e}_i \otimes \mathbf{e}_j \otimes \mathbf{e}_k \otimes \mathbf{e}_\ell)\mathbf{a} := (\mathbf{e}_\ell\,.\,\mathbf{a})\mathbf{e}_i \otimes \mathbf{e}_j \otimes \mathbf{e}_k. \tag{B.4.62}$$

In Cartesian tensor notation $\nabla\mathbf{C}$ is denoted by $C_{ijk,\ell}$.

Remark B.4.3. Notice that the derivative of a tensor field of order n is a tensor field of order $n+1$. Consider here $\phi\,(n=0)$, $\mathbf{v}\,(n=1)$, $\mathbf{L}\,(n=2)$, and $\mathbf{C}\ (n=3)$.

Remark B.4.4. If $\mathbf{a} = \nabla\phi$, then $\nabla\mathbf{a} = \nabla(\nabla\phi) =: \nabla\nabla\phi$ is a second derivative of ϕ. Similarly, $\nabla\mathbf{A}$ is a second derivative if $\mathbf{A} = \nabla\mathbf{a}$. It turns out that second derivatives (if continuous) take symmetric values. In particular, noting that $\nabla\nabla\phi$ takes values in $\mathrm{Lin}\,\mathcal{V}$ and $\nabla\nabla\mathbf{a}$ in $\mathrm{Lin}(\mathcal{V},\mathrm{Lin}\,\mathcal{V})$,

$$(\nabla\nabla\phi)\mathbf{u}\,.\,\mathbf{v} = (\nabla\nabla\phi)\mathbf{v}\,.\,\mathbf{u} \qquad \text{and} \qquad ((\nabla\nabla\mathbf{a})\mathbf{u})\mathbf{v} = ((\nabla\nabla\mathbf{a})\mathbf{v})\mathbf{u}. \tag{B.4.63}$$

Exercise B.4.7. Show that in terms of Cartesian co-ordinates

$$(\nabla\nabla\phi)\mathbf{e}_i\,.\,\mathbf{e}_j = \check{\phi}_{,ij}\mathbf{e}_i. \qquad \text{and} \qquad ((\nabla\nabla\mathbf{a})\mathbf{e}_j)\mathbf{e}_k = \check{a}_{i,jk}, \tag{B.4.64}$$

where ($f = \check{\phi}$ or $\check{a}_i$)

$$f_{,ij} := \partial^2 f/\partial x_i \partial x_j. \tag{B.4.65}$$

B.4.7 Differentiation of Products and Compositions

If ϕ, ψ denote scalar fields, $\mathbf{u},\mathbf{v}$ vector fields, $\mathbf{A},\mathbf{B}$ linear transformation fields, and $\mathbf{C}$ is a field with values in $\mathrm{Lin}(\mathcal{V},\mathrm{Lin}\,\mathcal{V})$, then

(i) $$\nabla\{\phi\psi\} = \phi\nabla\psi + \psi\nabla\phi \tag{B.4.66}$$

(ii) $$\nabla\{\phi\mathbf{v}\} = \phi\nabla\mathbf{v} + \mathbf{v}\otimes\nabla\phi \tag{B.4.67}$$

(iii) $$\nabla\{\mathbf{u}\,.\,\mathbf{v}\} = (\nabla\mathbf{v})^T\mathbf{u} + (\nabla\mathbf{u})^T\mathbf{v} \tag{B.4.68}$$

(iv) $$\nabla\{\mathbf{u}\otimes\mathbf{v}\} = \mathbf{u}\otimes\nabla\mathbf{v} + (\nabla\mathbf{u}\otimes\mathbf{v})^T \tag{B.4.69}$$

(v) $$\nabla\{\phi\mathbf{A}\} = \phi\nabla\mathbf{A} + \mathbf{A}\otimes\nabla\phi \tag{B.4.70}$$

(vi) $$\nabla\{\mathbf{A}\mathbf{u}\} = \mathbf{A}\nabla\mathbf{u} + (\nabla\mathbf{A})^T\mathbf{u} \tag{B.4.71}$$

(vii) $$\nabla\{\mathbf{A}\mathbf{B}\} = \mathbf{A}\nabla\mathbf{B} + ((\nabla\mathbf{A})^T\mathbf{B})^T \tag{B.4.72}$$

(viii) $$\nabla\{\mathbf{C}\mathbf{u}\} = \mathbf{C}\nabla\mathbf{u} + (\nabla\mathbf{C})^T\mathbf{u} \tag{B.4.73}$$

[2] $\mathrm{Lin}(\mathcal{V},\mathrm{Lin}(\mathcal{V},\mathrm{Lin}\,\mathcal{V}))$ denotes the set of linear maps from $\mathcal{V}$ into $\mathrm{Lin}(\mathcal{V},\mathrm{Lin}\,\mathcal{V})$ and has the natural structure of a linear/vector space of dimension 81. Prove this!

Direct proofs of these results are formally very similar. If $\mathcal{A}$ and $\mathcal{B}$ are differentiable tensor fields, then

$$\begin{aligned}(\mathcal{AB})(\mathbf{x}+\mathbf{h}) &= \mathcal{A}(\mathbf{x}+\mathbf{h})\mathcal{B}(\mathbf{x}+\mathbf{h})\\ &= (\mathcal{A}(\mathbf{x}) + \nabla\mathcal{A}(\mathbf{x})\mathbf{h} + o(\mathbf{h}))(\mathcal{B}(\mathbf{x}) + \nabla\mathcal{B}(\mathbf{x})\mathbf{h} + o(\mathbf{h}))\\ &= (\mathcal{AB})(\mathbf{x}) + \mathcal{A}(\mathbf{x})\nabla\mathcal{B}(\mathbf{x})\mathbf{h} + (\nabla\mathcal{A}(\mathbf{x})\mathbf{h})\mathcal{B}(\mathbf{x}) + o(\mathbf{h}). \end{aligned} \tag{B.4.74}$$

Further, by definition,

$$(\mathcal{AB})(\mathbf{x}+\mathbf{h}) = (\mathcal{AB})(\mathbf{x}) + \nabla\{\mathcal{AB}\}(\mathbf{x})\mathbf{h} + o(\mathbf{h}). \tag{B.4.75}$$

Comparing (B.4.74) and (B.4.75) yields

$$\nabla\{\mathcal{AB}\}\mathbf{h} = \mathcal{A}(\nabla\mathcal{B})\mathbf{h} + ((\nabla\mathcal{A})\mathbf{h})\mathcal{B} + o(\mathbf{h}). \tag{B.4.76}$$

Letting $\mathbf{h} = s\mathbf{e}$ for any $\mathbf{e} \in \mathcal{V}$ with $\|\mathbf{e}\| = 1$, dividing (B.4.76) throughout by s, and taking the limit as $s \to 0$, we have

$$\nabla\{\mathcal{AB}\}\mathbf{e} = \mathcal{A}(\nabla\mathcal{B})\mathbf{e} + ((\nabla\mathcal{A})\mathbf{e})\mathcal{B}. \tag{B.4.77}$$

It follows by linearity of all derivatives that for *any* $\mathbf{a} \in \mathcal{V}$

$$(\nabla\{\mathcal{AB}\})\mathbf{a} = \mathcal{A}((\nabla\mathcal{B})\mathbf{a}) + ((\nabla\mathcal{A})\mathbf{a})\mathcal{B}. \tag{B.4.78}$$

(Case $\mathbf{a} = \mathbf{0}$ is trivial: if $\mathbf{a} \neq \mathbf{0}$, then choose $\mathbf{e} := \mathbf{a}/\|\mathbf{a}\|$ in (B.4.77) and multiply throughout by $\|\mathbf{a}\|$.) In interpreting (B.4.78) in the differing cases it is important to note that whatever combination of $\mathcal{A}$ and $\mathcal{B}$ is intended by $\mathcal{AB}$ (e.g., scalar product if $\mathcal{A} = \mathbf{u}, \mathcal{B} = \mathbf{v}$, and $\mathcal{AB} = \mathbf{u}\,.\,\mathbf{v}$) is also that intended for $\mathcal{A}$ with $(\nabla\mathcal{B})\mathbf{a}$ and $(\nabla\mathcal{A})\mathbf{a}$ with $\mathcal{B}$. Further, if $\mathcal{A}, \mathcal{B}$, or $\mathcal{AB}$ is scalar-valued, then (cf. (B.4.9), (B.4.11)) $(\nabla\mathcal{A})\mathbf{a} = \nabla\mathcal{A}\,.\,\mathbf{a}$, $(\nabla\mathcal{B})\mathbf{a} = \nabla\mathcal{B}\,.\,\mathbf{a}$, or $\nabla\{\mathcal{AB}\}\mathbf{a} = (\nabla\{\mathcal{AB}\})\,.\,\mathbf{a}$.

(i) $\mathcal{A} = \phi, \mathbf{B} = \psi$, and from (B.4.78), for any $\mathbf{a} \in \mathcal{V}$

$$\nabla\{\phi\psi\}\,.\,\mathbf{a} = \phi(\nabla\psi\,.\,\mathbf{a}) + ((\nabla\phi)\,.\,\mathbf{a})\psi = (\phi\nabla\psi + \psi\nabla\phi)\,.\,\mathbf{a}.$$

Hence (B.4.66) holds: cf. Exercise B.5.1.

(ii) $\mathcal{A} = \phi, \mathcal{B} = \mathbf{v}$ yields $(\nabla\{\phi\mathbf{v}\})\mathbf{a} = \phi((\nabla\mathbf{v})\mathbf{a}) + (\nabla\phi\,.\,\mathbf{a})\mathbf{v}$, so

$$(\nabla\{\phi\mathbf{v}\} - \phi\nabla\mathbf{v} - \mathbf{v}\otimes\nabla\phi)\mathbf{a} = \mathbf{0} \qquad \text{for all } \mathbf{a} \in \mathcal{V},$$

and thus (B.4.67) holds.

(iii) $\mathcal{A} = \mathbf{u}, \mathcal{B} = \mathbf{v}$ yields $(\nabla\{\mathbf{u}\,.\,\mathbf{v}\})\,.\,\mathbf{a} = \mathbf{u}\,.\,(\nabla\mathbf{v})\mathbf{a} + (\nabla\mathbf{u})\mathbf{a}\,.\,\mathbf{v}$ and so $(\nabla\{\mathbf{u}\,.\,\mathbf{v}\} - (\nabla\mathbf{v})^T\mathbf{u} - (\nabla\mathbf{u})^T\mathbf{v})\,.\,\mathbf{a} = 0$ for all $\mathbf{a} \in \mathcal{V}$ and thus (B.4.68) holds.

(iv) $\mathcal{A} = \mathbf{u}, \mathcal{B} = \mathbf{v}$ yields $(\nabla\{\mathbf{u}\otimes\mathbf{v}\})\mathbf{a} = (\nabla\mathbf{u})\mathbf{a}\otimes\mathbf{v} + \mathbf{u}\otimes(\nabla\mathbf{u})\mathbf{a}$. However, from (A.19.42), with $\mathbf{L} = \nabla\mathbf{u}$, $\mathbf{b} = \mathbf{a}$ and $\mathbf{a} = \mathbf{v}$, we have $(\nabla\mathbf{u})\mathbf{a}\otimes\mathbf{v} = (\nabla\mathbf{u}\otimes\mathbf{v})^T\mathbf{a}$, and hence (B.4.69) holds since $\mathbf{a} \in \mathcal{V}$ is arbitrary.

(v) $\mathcal{A} = \phi, \mathcal{B} = \mathbf{A}$ yields $(\nabla\{\phi\mathbf{A}\})\mathbf{a} = \phi(\nabla\mathbf{A})\mathbf{a} + (\nabla\phi\,.\,\mathbf{a})\mathbf{A} = (\phi\nabla\mathbf{A} + \mathbf{A}\otimes\nabla\phi)\mathbf{a}$, so (B.4.70) holds. [Here (A.19.41) has been invoked, with $\mathbf{L} = \mathbf{A}, \mathbf{a} = \nabla\phi$, and $\mathbf{v} = \mathbf{a}$.]

(vi) $\mathcal{A} = \mathbf{A}, \mathcal{B} = \mathbf{u}$ yields $(\nabla\{\mathbf{A}\mathbf{u}\})\mathbf{a} = ((\nabla\mathbf{A})\mathbf{a})\mathbf{u} + \mathbf{A}(\nabla\mathbf{u})\mathbf{a} = ((\nabla\mathbf{A})^T\mathbf{u} + \mathbf{A}\nabla\mathbf{u})\mathbf{a}$, so (B.4.71) holds.

(vii) $\mathcal{A}=\mathbf{A}, \mathcal{B}=\mathbf{B}$ yields $(\nabla\{\mathbf{AB}\})\mathbf{a}=((\nabla\mathbf{A})\mathbf{a})\mathbf{B}+\mathbf{A}(\nabla\mathbf{B})\mathbf{a}$. Now, for any $\mathbf{b}\in\mathcal{V}$, $((\nabla\mathbf{A})\mathbf{a})\mathbf{Bb}=((\nabla\mathbf{A})^T\mathbf{Bb})\mathbf{a}=(((\nabla\mathbf{A})^T\mathbf{B})\mathbf{b})\mathbf{a}=(((\nabla\mathbf{A})^T\mathbf{B})^T\mathbf{a})\mathbf{b}$, so $((\nabla\mathbf{A})\mathbf{a})\mathbf{B}=((\nabla\mathbf{A})^T\mathbf{B})^T\mathbf{a}$ and (B.4.72) holds.

(viii) $\mathcal{A}=\mathbf{C}, \mathcal{B}=\mathbf{u}$ yields $(\nabla\{\mathbf{Cu}\})\mathbf{a}=((\nabla\mathbf{C})\mathbf{a})\mathbf{u}+\mathbf{C}(\nabla\mathbf{u})\mathbf{a}=:((\nabla\mathbf{C})^T\mathbf{u}+\mathbf{C}\nabla\mathbf{u})\mathbf{a}$, and (B.4.73) holds.

Remark B.4.5. The preceding proofs were derived directly from definitions of the relevant derivatives/gradients. Alternate proofs involving Cartesian co-ordinate representations require only use of the product rule for real-valued functions. Such proofs are given below in order to compare use of direct notation with that of Cartesian tensor notation (see Appendix A.20).

Recall the representations[3] (B.4.22), (B.4.52), (B.4.60), and (B.4.61):

$$\nabla\phi=\phi_{,i}\mathbf{e}_i,\qquad \nabla\mathbf{v}=v_{i,j}\mathbf{e}_i\otimes\mathbf{e}_j, \tag{B.4.79}$$

$$\nabla\mathbf{A}=A_{ij,k}\mathbf{e}_i\otimes\mathbf{e}_j\otimes\mathbf{e}_k,\quad\text{and}\quad \nabla\mathbf{C}=C_{ijk,\ell}\mathbf{e}_i\otimes\mathbf{e}_j\otimes\mathbf{e}_k\otimes\mathbf{e}_\ell. \tag{B.4.80}$$

Thus, with choice $\phi\psi$ of scalar field, $(\text{B.4.79})_1$ becomes

$$\begin{aligned}\nabla\{\phi\psi\}&=(\phi\psi)_{,i}\mathbf{e}_i=(\phi_{,i}\psi+\phi\psi_{,i})\mathbf{e}_i\\&=\psi(\phi_{,i}\mathbf{e}_i)+\phi(\psi_{,i}\mathbf{e}_i)=\psi\nabla\phi+\phi\nabla\psi.\end{aligned}$$

With choice $\phi\mathbf{v}$ of vector field, $(\text{B.4.79})_2$ becomes

$$\begin{aligned}\nabla\{\phi\mathbf{v}\}&=(\phi v_i)_{,j}\mathbf{e}_i\otimes\mathbf{e}_j=(\phi_{,j}v_i+\phi v_{i,j})\mathbf{e}_i\otimes\mathbf{e}_j\\&=v_i\mathbf{e}_i\otimes\phi_{,j}\mathbf{e}_j+\phi(v_{i,j}\mathbf{e}_i\otimes\mathbf{e}_j)=\mathbf{v}\otimes\nabla\phi+\phi\nabla\mathbf{v}.\end{aligned}$$

With $\phi=\mathbf{u}.\mathbf{v}=u_jv_j$, $(\text{B.4.79})_1$ yields

$$\begin{aligned}\nabla\{\mathbf{u}.\mathbf{v}\}&=(u_jv_j)_{,i}\mathbf{e}_i=(u_{j,i}v_j+u_jv_{j,i})\mathbf{e}_i\\&=((\nabla\mathbf{u})^T_{ij}v_j+(\nabla\mathbf{v})^T_{ij}u_j)\mathbf{e}_i=((\nabla\mathbf{u})^T\mathbf{v}+(\nabla\mathbf{v})^T\mathbf{u})_i\mathbf{e}_i=(\nabla\mathbf{u})^T\mathbf{v}+(\nabla\mathbf{v})^T\mathbf{u}.\end{aligned}$$

With $\mathbf{A}=\mathbf{u}\otimes\mathbf{v}=u_iv_j\mathbf{e}_i\otimes\mathbf{e}_j$ (so that $A_{ij}=u_iv_j$), $(\text{B.4.80})_1$ yields

$$\begin{aligned}\nabla\{\mathbf{u}\otimes\mathbf{v}\}&=(u_iv_j)_{,k}\mathbf{e}_i\otimes\mathbf{e}_j\otimes\mathbf{e}_k=(u_{i,k}v_j+u_iv_{j,k})\mathbf{e}_i\otimes\mathbf{e}_j\otimes\mathbf{e}_k\\&=u_{i,k}\mathbf{e}_i\otimes v_j\mathbf{e}_j\otimes\mathbf{e}_k+u_i\mathbf{e}_i\otimes v_{j,k}\mathbf{e}_j\otimes\mathbf{e}_k\\&=(u_{i,k}\mathbf{e}_i\otimes\mathbf{e}_k\otimes\mathbf{v})^T+\mathbf{u}\otimes\nabla\mathbf{v}=(\nabla\mathbf{u}\otimes\mathbf{v})^T+\mathbf{u}\otimes\nabla\mathbf{v}.\end{aligned}$$

With $\phi\mathbf{A}$ replacing $\mathbf{A}$ in $(\text{B.4.80})_1$,

$$\begin{aligned}\nabla\{\phi\mathbf{A}\}&=(\phi A_{ij})_{,k}\mathbf{e}_i\otimes\mathbf{e}_j\otimes\mathbf{e}_k=(\phi_{,k}A_{ij}+\phi A_{ij,k})\mathbf{e}_i\otimes\mathbf{e}_j\otimes\mathbf{e}_k\\&=A_{ij}\mathbf{e}_i\otimes\mathbf{e}_j\otimes\phi_{,k}\mathbf{e}_k+\phi\nabla\mathbf{A}=\mathbf{A}\otimes\nabla\phi+\phi\nabla\mathbf{A}.\end{aligned}$$

With $\mathbf{v}=\mathbf{Au}$ in $(\text{B.4.79})_2$, noting that $v_i=A_{ik}u_k$,

$$\begin{aligned}\nabla\{\mathbf{Au}\}&=(A_{ik}u_k)_{,j}\mathbf{e}_i\otimes\mathbf{e}_j=(A_{ik,j}u_k+A_{ik}u_{k,j})\mathbf{e}_i\otimes\mathbf{e}_j\\&=(((\nabla\mathbf{A})^T)_{ijk}u_k+(\mathbf{A}\nabla\mathbf{u})_{ij})\mathbf{e}_i\otimes\mathbf{e}_j=(\nabla\mathbf{A})^T\mathbf{u}+\mathbf{A}\nabla\mathbf{u}.\end{aligned}$$

[3] Here $\phi_{,i}$, $v_{i,j}$, $A_{ij,k}$, and $C_{ijk,\ell}$ are co-ordinate representations previously denoted by $\check{\phi}_{,i}$, etc. The distinctions between ϕ and $\check{\phi}$ etc. are here omitted for notational simplicity.

With $\mathbf{A}$ replaced by $\mathbf{AB}$ in $(B.4.80)_1$, noting that $(\mathbf{AB})_{ij} = A_{ip}B_{pj}$,

$$\begin{aligned}\nabla\{\mathbf{AB}\} &= (A_{ip}B_{pj})_{,k}\mathbf{e}_i \otimes \mathbf{e}_j \otimes \mathbf{e}_k = (A_{ip,k}B_{pj} + A_{ip}B_{pj,k})\mathbf{e}_i \otimes \mathbf{e}_j \otimes \mathbf{e}_k \\ &= ((\nabla\mathbf{A})^T_{ikp}B_{pj} + (\mathbf{A}\nabla\mathbf{B})_{ijk})\mathbf{e}_i \otimes \mathbf{e}_j \otimes \mathbf{e}_k = ((\nabla\mathbf{A})^T\mathbf{B})^T + \mathbf{A}(\nabla\mathbf{B}))_{ijk}\mathbf{e}_i \otimes \mathbf{e}_j \otimes \mathbf{e}_k \\ &= ((\nabla\mathbf{A})^T\mathbf{B})^T + \mathbf{A}\nabla\mathbf{B}.\end{aligned}$$

Finally, with $\mathbf{A} = \mathbf{Cu}$ in $(B.4.80)_1$, noting that $(\mathbf{Cu})_{ij} = C_{ijp}u_p$,

$$\begin{aligned}\nabla\{\mathbf{Cu}\} &= (C_{ijp}u_p)_{,k}\mathbf{e}_i \otimes \mathbf{e}_j \otimes \mathbf{e}_k = (C_{ijp,k}u_p + C_{ijp}u_{p,k})\mathbf{e}_i \otimes \mathbf{e}_j \otimes \mathbf{e}_k \\ &= ((\nabla\mathbf{C})^T_{ijkp}u_p + (\mathbf{C}\nabla\mathbf{u})_{ijk})\mathbf{e}_i \otimes \mathbf{e}_j \otimes \mathbf{e}_k \\ &= ((\nabla\mathbf{C})^T\mathbf{u} + \mathbf{C}\nabla\mathbf{u})_{ijk}\mathbf{e}_i \otimes \mathbf{e}_j \otimes \mathbf{e}_k = (\nabla\mathbf{C})^T\mathbf{u} + \mathbf{C}\nabla\mathbf{u}.\end{aligned}$$

Notice here the definition of $(\nabla\mathbf{C})^T$ employed in the direct proof of (viii), namely

$$((\nabla\mathbf{C})^T\mathbf{u})\mathbf{a} := ((\nabla\mathbf{C})\mathbf{a})\mathbf{u}. \tag{B.4.81}$$

Exercise B.4.8. Show that $((\nabla\mathbf{C})^T)_{ijk\ell} = (\nabla\mathbf{C})_{ij\ell k}$.

Exercise B.4.9. Show both directly, and using Cartesian co-ordinate representations, that

$$\nabla\{\phi\mathbf{C}\} = \phi\nabla\mathbf{C} + \mathbf{C}\nabla\phi. \tag{B.4.82}$$

Remark B.4.6. Two approaches to tensor algebra and calculus have been presented, the first via a direct, co-ordinate-free formulation, and the second in terms of Cartesian co-ordinate representations. Individual choice is made on the basis of familiarity and taste. The direct viewpoint reflects the absence of any general physically distinguished co-ordinate system, and renders explicit underlying linear and multilinear structure. On the other hand, Cartesian tensor notation is more mathematically accessible.

B.4.8 Differentiation of the Determinant Function

Consider

$$\det : \mathrm{Lin}\,\mathcal{V} \to \mathbb{R}, \tag{B.4.83}$$

and recall the natural inner product (and associated norm) on $\mathcal{V}$ (see Appendix A.14). In order to define what we might mean by the derivative (or gradient) of det evaluated at $\mathbf{A} \in \mathrm{Lin}\,\mathcal{V}$, consider

$$\det(\mathbf{A} + \mathbf{H}) = \det\mathbf{A} + ?\mathbf{H} + o(\mathbf{H}) \tag{B.4.84}$$

as $\mathbf{H} \to \mathbf{O}$. Of course, here $\mathbf{H} \in \mathrm{Lin}\,\mathcal{V}$. Note that both $\det(\mathbf{A} + \mathbf{H})$ and $\det\mathbf{A}$ lie in $\mathbb{R}$. Further, the candidate derivative ? should be linear in $\mathbf{H}$. This leads us to require that $?\mathbf{H}$ take values in $\mathbb{R}$ and be linear in $\mathbf{H}$. We thus define [cf. (B.4.9)] the derivative at $\mathbf{A}$ (if it exists) to be $\mathbf{L} \in \mathrm{Lin}\,\mathcal{V}$, where

$$\det(\mathbf{A} + \mathbf{H}) = \det\mathbf{A} + \mathbf{L}\cdot\mathbf{H} + o(\mathbf{H}) \quad \text{as} \quad \mathbf{H} \to \mathbf{O}. \tag{B.4.85}$$

Let ω denote any non-zero trilinear alternating form and $\{\mathbf{u},\mathbf{v},\mathbf{w}\}$ represent any basis for $\mathcal{V}$. Then (see Appendix A.11 and Appendix A.12)

$$\begin{aligned}
&\omega((\mathbf{A}+\mathbf{H})\mathbf{u},(\mathbf{A}+\mathbf{H})\mathbf{v},(\mathbf{A}+\mathbf{H})\mathbf{w}) \\
&\quad= \omega(\mathbf{Au},\mathbf{Av},\mathbf{Aw}) \\
&\qquad+ \omega(\mathbf{Hu},\mathbf{Av},\mathbf{Aw}) + \omega(\mathbf{Au},\mathbf{Hv},\mathbf{Aw}) + \omega(\mathbf{Au},\mathbf{Av},\mathbf{Hw}) \\
&\qquad+ \omega(\mathbf{Hu},\mathbf{Hv},\mathbf{Aw}) + \omega(\mathbf{Hu},\mathbf{Av},\mathbf{Hw}) + \omega(\mathbf{Au},\mathbf{Hv},\mathbf{Hw}) \\
&\qquad+ \omega(\mathbf{Hu},\mathbf{Hv},\mathbf{Hw}).
\end{aligned} \tag{B.4.86}$$

Further, *if* $\mathbf{A}$ *is invertible*,

$$\begin{aligned}
\omega(\mathbf{Hu},\mathbf{Av},\mathbf{Aw}) &= \omega((\mathbf{HA}^{-1})\mathbf{Au},\mathbf{Av},\mathbf{Aw}), \\
\omega(\mathbf{Au},\mathbf{Hv},\mathbf{Aw}) &= \omega(\mathbf{Au},(\mathbf{HA}^{-1})\mathbf{Av},\mathbf{Aw}), \quad \text{and} \\
\omega(\mathbf{Au},\mathbf{Av},\mathbf{Hw}) &= \omega(\mathbf{Au},\mathbf{Av},(\mathbf{HA}^{-1})\mathbf{Aw}).
\end{aligned} \tag{B.4.87}$$

Accordingly [see (A.12.1) et seq. with $\mathbf{L}=\mathbf{HA}^{-1},\mathbf{u}=\mathbf{Au}$, etc.]

$$\begin{aligned}
&\omega(\mathbf{Hu},\mathbf{Av},\mathbf{Aw}) + \omega(\mathbf{Au},\mathbf{Hv},\mathbf{Aw}) + \omega(\mathbf{Au},\mathbf{Av},\mathbf{Hw}) \\
&\qquad= \operatorname{tr}(\mathbf{HA}^{-1})\omega(\mathbf{Au},\mathbf{Av},\mathbf{Aw}).
\end{aligned} \tag{B.4.88}$$

Similarly,

$$\begin{aligned}
\omega(\mathbf{Hu},\mathbf{Hv},\mathbf{Aw}) &= \omega((\mathbf{HA}^{-1})\mathbf{Au},(\mathbf{HA}^{-1})\mathbf{Av},\mathbf{Aw}), \\
\omega(\mathbf{Hu},\mathbf{Av},\mathbf{Hw}) &= \omega((\mathbf{HA}^{-1})\mathbf{Au},\mathbf{Av},(\mathbf{HA}^{-1})\mathbf{Aw}), \quad \text{and} \\
\omega(\mathbf{Au},\mathbf{Hv},\mathbf{Hw}) &= \omega(\mathbf{Au},(\mathbf{HA}^{-1})\mathbf{Av},(\mathbf{HA}^{-1})\mathbf{Aw}).
\end{aligned} \tag{B.4.89}$$

Accordingly [see (A.12.15) et seq.]

$$\begin{aligned}
&\omega(\mathbf{Hu},\mathbf{Hv},\mathbf{Aw}) + \omega(\mathbf{Hu},\mathbf{Av},\mathbf{Hw}) + \omega(\mathbf{Au},\mathbf{Hv},\mathbf{Hw}) \\
&\qquad= I_2(\mathbf{HA}^{-1})\omega(\mathbf{Au},\mathbf{Av},\mathbf{Aw}).
\end{aligned} \tag{B.4.90}$$

From (B.4.86), (B.4.88), and (B.4.90), together with definition (A.12.21) of the determinant function, we have

$$\begin{aligned}
\det(\mathbf{A}+\mathbf{H})\omega(\mathbf{u},\mathbf{v},\mathbf{w}) &= (\det\mathbf{A})\omega(\mathbf{u},\mathbf{v},\mathbf{w}) + \operatorname{tr}(\mathbf{HA}^{-1})(\det\mathbf{A})\omega(\mathbf{u},\mathbf{v},\mathbf{w}) \\
&\quad+ I_2(\mathbf{HA}^{-1})(\det\mathbf{A})\omega(\mathbf{u},\mathbf{v},\mathbf{w}) + (\det\mathbf{H})\omega(\mathbf{u},\mathbf{v},\mathbf{w}).
\end{aligned} \tag{B.4.91}$$

Since $\omega(\mathbf{u},\mathbf{v},\mathbf{w}) \neq 0$ (Why?) this yields

$$\det(\mathbf{A}+\mathbf{H}) = \det\mathbf{A} + \operatorname{tr}(\mathbf{HA}^{-1})\det\mathbf{A} + I_2(\mathbf{HA}^{-1})\det\mathbf{A} + \det\mathbf{H}. \tag{B.4.92}$$

Now [see (A.12.14) and (A.14.9)]

$$\operatorname{tr}(\mathbf{HA}^{-1}) = \operatorname{tr}(\mathbf{A}^{-1}\mathbf{H}) = \operatorname{tr}((\mathbf{A}^{-T})^T\mathbf{H}) = \mathbf{A}^{-T}\cdot\mathbf{H}, \tag{B.4.93}$$

where [see (A.10.19) et seq.]

$$\mathbf{A}^{-T} := (\mathbf{A}^{-1})^T. \tag{B.4.94}$$

It will be shown in Remark B.4.7 that both $I_2(\mathbf{HA}^{-1})$ and $\det\mathbf{H}$ are of order $o(\mathbf{H})$ as $\mathbf{H}\to\mathbf{O}$. Thus (B.4.92) may be written, via (B.4.93), as

$$\det(\mathbf{A}+\mathbf{H}) = \det\mathbf{A} + (\det\mathbf{A})\mathbf{A}^{-T}\cdot\mathbf{H} + o(\mathbf{H}) \qquad \text{as } \mathbf{H}\to\mathbf{O}. \tag{B.4.95}$$

Accordingly [cf. (B.4.85)] we have the following:

Result B.4.1. (Derivative of the determinant function) The determinant function det on Lin $\mathcal{V}$ is differentiable at every invertible element $\mathbf{A}$ of Lin $\mathcal{V}$ and

$$\nabla(\det)(\mathbf{A}) = (\det\mathbf{A})\mathbf{A}^{-T}. \tag{B.4.96}$$

Remark B.4.7. (*Order of magnitude arguments in* Lin $\mathcal{V}$) Consider the orders of magnitude of $I_2(\mathbf{HA}^{-1})$ and $\det\mathbf{H}$ as $\mathbf{H}\to\mathbf{O}$.

(i) From (A.12.17)

$$I_2(\mathbf{HA}^{-1}) = \frac{1}{2}\{(\mathrm{tr}(\mathbf{HA}^{-1}))^2 - \mathrm{tr}((\mathbf{HA}^{-1})^2)\}. \tag{B.4.97}$$

Hence

$$\begin{aligned} |I_2(\mathbf{HA}^{-1})| &\le \frac{1}{2}\{(\mathrm{tr}(\mathbf{HA}^{-1}))^2 + |\mathrm{tr}((\mathbf{HA}^{-1})^2)|\} \\ &= \frac{1}{2}\{(\mathbf{H}\cdot\mathbf{A}^{-T})^2 + |\mathrm{tr}(\mathbf{HA}^{-1}\mathbf{HA}^{-1})|\} \\ &= \frac{1}{2}\{(\mathbf{H}\cdot\mathbf{A}^{-T})^2 + |\mathbf{H}^T\cdot\mathbf{A}^{-1}\mathbf{HA}^{-1}|\}. \end{aligned} \tag{B.4.98}$$

However, from (A.14.16) and (A.14.18),

$$\mathbf{H}\cdot\mathbf{A}^{-T} \le \|\mathbf{H}\|\|\mathbf{A}^{-T}\| = \|\mathbf{H}\|\|\mathbf{A}^{-1}\| \tag{B.4.99}$$

and, via (A.17.18) and (A.14.18),

$$\begin{aligned} |\mathbf{H}^T\cdot\mathbf{A}^{-1}\mathbf{HA}^{-1}| &\le \|\mathbf{H}^T\|\|\mathbf{A}^{-1}\mathbf{HA}^{-1}\| \\ &\le \|\mathbf{H}\|\|\mathbf{A}^{-1}\|\|\mathbf{H}\|\|\mathbf{A}^{-1}\| \\ &= \|\mathbf{H}\|^2\|\mathbf{A}^{-1}\|^2. \end{aligned} \tag{B.4.100}$$

Thus (B.4.98), (B.4.99), and (B.4.100) yield

$$|\mathbf{I}_2(\mathbf{HA}^{-1})| \le \|\mathbf{A}^{-1}\|^2\|\mathbf{H}\|^2, \tag{B.4.101}$$

so $\qquad I_2(\mathbf{HA}^{-1}) \quad$ is of order $\quad O(\|\mathbf{H}\|^2) \quad$ as $\mathbf{H}\to\mathbf{O}$ (B.4.102)

[and hence of order $o(\|\mathbf{H}\|)$ as $\mathbf{H}\to\mathbf{O}$].

(ii) From (A.12.31)

$$\begin{aligned} \det\mathbf{H} &= \frac{1}{6}\{\mathrm{tr}\,\mathbf{H})^3 - 3(\mathrm{tr}\,\mathbf{H})(\mathrm{tr}(\mathbf{H}^2)) + 2\,\mathrm{tr}(\mathbf{H}^3)\} \\ &= \frac{1}{6}\{(\mathbf{1}\cdot\mathbf{H})^3 - 3(\mathbf{1}\cdot\mathbf{H})(\mathbf{H}^T\cdot\mathbf{H}) + 2\mathbf{H}^T\cdot\mathbf{H}^2\}. \end{aligned} \tag{B.4.103}$$

Thus repeated use of (A.14.16) yields

$$|\det \mathbf{H}| \leq \frac{1}{6}\{(\sqrt{3}\|\mathbf{H}\|)^3 + 3\sqrt{3}\|\mathbf{H}\|\,\|\mathbf{H}^T\|\,\|\mathbf{H}\| + 2\|\mathbf{H}^T\|\,\|\mathbf{H}\|^2\}.$$

Hence, invoking (A.14.18),

$$|\det \mathbf{H}| \leq \frac{1}{3}(2\sqrt{3}+1)\|\mathbf{H}\|^3, \tag{B.4.104}$$

and so $\det \mathbf{H}$ is of order $O(\|\mathbf{H}\|^3)$ as $\mathbf{H} \to \mathbf{O}$ (B.4.105)

[and hence of order $o(\|\mathbf{H}\|)$ as $\mathbf{H} \to \mathbf{O}$].

B.5 Jacobians, Physically Admissible Deformations, and Kinematics

Any deformation $\mathbf{d}$ in $\mathcal{E}$ [see (B.3.1)] prescribes a distortion of regions in $\mathcal{E}$ and in general results in a change in the volumes of the regions. To analyse volume changes in the neighbourhood of a point $\hat{\mathbf{x}}$, consider a parallelipiped[4] Π defined by $\hat{\mathbf{x}}$ and three non-coplanar displacements $\mathbf{u}_1, \mathbf{u}_2$, and $\mathbf{u}_3$, say. [Thus the vertices of Π are the eight points $\hat{\mathbf{x}}, \hat{\mathbf{x}} + \mathbf{u}_i$, $\hat{\mathbf{x}} + \mathbf{u}_i + \mathbf{u}_j$ $(i \neq j)$, and $\hat{\mathbf{x}} + \mathbf{u}_1 + \mathbf{u}_2 + \mathbf{u}_3$, with $i,j = 1,2,3$.]

Exercise B.5.1. Convince yourself that the volume of Π is

$$\text{vol}(\Pi) = |\mathbf{u}_1 \times \mathbf{u}_2 \,.\, \mathbf{u}_3|. \tag{B.5.1}$$

A deformation $\mathbf{d}$ maps Π into a region $\mathbf{d}(\Pi)$ bounded by six (in general curved) surfaces, each of which is the image of a plane face of Π. Region $\mathbf{d}(\Pi)$ is approximated by using the local linear approximation $\mathbf{d}_{\text{lin}}$ to $\mathbf{d}$ at $\hat{\mathbf{x}}$ [see (B.4.29)]. Since $\mathbf{d}_{\text{lin}}$ is a homogeneous deformation (see Exercise B.4.2), and such deformations map parallel planes into parallel planes (see Result B.3.4), $\mathbf{d}_{\text{lin}}(\Pi)$ is a parallelipiped. Further, its volume [see (B.5.1)] is

$$\text{vol}(\mathbf{d}_{\text{lin}}(\Pi)) = |\mathbf{v}_1 \times \mathbf{v}_2 \,.\, \mathbf{v}_3|, \tag{B.5.2}$$

where [here $i = 1,2,3$ and recall (B.4.29)]

$$\mathbf{v}_i := \mathbf{d}_{\text{lin}}(\hat{\mathbf{x}} + \mathbf{u}_i) - \mathbf{d}_{\text{lin}}(\hat{\mathbf{x}}) = \nabla\mathbf{d}(\hat{\mathbf{x}})\mathbf{u}_i. \tag{B.5.3}$$

From (B.5.2), (B.5.3), (A.12.25) with $\mathbf{L} = \nabla\mathbf{d}(\hat{\mathbf{x}})$, and (B.5.1),

$$\text{vol}(\mathbf{d}_{\text{lin}}(\Pi)) = |\det \nabla\mathbf{d}(\hat{\mathbf{x}})|\text{vol}(\Pi). \tag{B.5.4}$$

As $\|\mathbf{u}_1\|, \|\mathbf{u}_2\|$, and $\|\mathbf{u}_3\| \to 0$, map $\mathbf{d}_{\text{lin}}$ approximates $\mathbf{d}$ ever more closely: this is the essence of the definition of differentiability at $\hat{\mathbf{x}}$. Accordingly parallelipiped $\mathbf{d}_{\text{lin}}(\Pi)$ approximates region $\mathbf{d}(\Pi)$ ever more closely, and hence

$$\frac{\text{vol}(\mathbf{d}(\Pi))}{\text{vol}(\Pi)} \longrightarrow |\det \nabla\mathbf{d}(\hat{\mathbf{x}})| \quad \text{as} \quad \max\{\|\mathbf{u}_1\|, \|\mathbf{u}_2\|, \|\mathbf{u}_3\|\} \to 0. \tag{B.5.5}$$

[Equivalently, the limit exists as $(\|\mathbf{u}_1\|^2 + \|\mathbf{u}_2\|^2 + \|\mathbf{u}_3\|^2)^{1/2} \to 0$.]

[4] That is, a region bounded by three pairs of parallel planes.

The function

$$J_{\mathbf{d}} := |\det \nabla \mathbf{d}| \tag{B.5.6}$$

is termed the *Jacobian associated with the differentiable deformation* $\mathbf{d}$ and characterises the *local volume change factor* appropriate to $\mathbf{d}$.

Remark B.5.1. To reflect our experience that matter always occupies 'space' (so that material which 'occupies' a region of volume ΔV cannot be compressed into a region of zero volume), any physically admissible deformation $\mathbf{d}$ which is differentiable should satisfy $J_{\mathbf{d}} \neq 0$. (Said differently, the local volume change factor should never vanish for physically admissible differentiable deformations.) Further, since any differentiable $\mathbf{d}$ has a linear approximation at any point, this approximation is homogeneous (see Remark B.4.1), and inversions cannot be effected in practice (see Remark B.3.5), to make physical sense $\mathbf{d}$ should satisfy $\det \nabla \mathbf{d} > 0$. Thus [cf. (B.5.6)]

$$\textit{if } \mathbf{d} \text{ is physically admissible and differentiable, then } J_{\mathbf{d}} = \det \nabla \mathbf{d}. \tag{B.5.7}$$

Any material system (or *body*) of interest is denoted by $\mathcal{B}$ and considered to occupy an open subset B_t of $\mathcal{E}$ at any given time t. In comparing B_{t_2} with B_{t_1} $(t_2 > t_1)$ there will in general have been a distortion of $\mathcal{B}$. This distortion is modelled as a physically admissible differentiable deformation

$$\mathbf{d}_{t_1,t_2} : B_{t_1} \to B_{t_2}. \tag{B.5.8}$$

In order to describe the distortion of the body as time evolves, a specific time t_0 may be chosen as reference: the family $\{\mathbf{d}_{t_0,t}\}$ of deformations then can be considered as t varies. This family is termed the *motion of* $\mathcal{B}$ *with respect to the situation at time* t_0, and we write

$$\chi_{t_0}(.,t) := \mathbf{d}_{t_0,t}. \tag{B.5.9}$$

Accordingly

$$\chi_{t_0}(.,t) : B_{t_0} \to B_t. \tag{B.5.10}$$

Further,

$$\mathbf{F}(\hat{\mathbf{x}},t) := \nabla \chi_{t_0}(\hat{\mathbf{x}},t) \tag{B.5.11}$$

is termed the *deformation gradient at* $(\hat{\mathbf{x}},t)$ *in motion* χ_{t_0}.

Remark B.5.2. In view of Remark B.5.1,

$$J := \det \mathbf{F} > 0. \tag{B.5.12}$$

The velocity field $\mathbf{v}$ associated with motion χ_{t_0} is given by (see Chapter 2, Section 2.3)

$$\mathbf{v}(\chi_{t_0}(\hat{\mathbf{x}},t),t) = \dot{\chi}_{t_0}(\hat{\mathbf{x}},t). \tag{B.5.13}$$

Suppressing time dependence yields

$$\mathbf{v} \circ \chi_{t_0} = \dot{\chi}_{t_0}. \tag{B.5.14}$$

Suppose that $\mathbf{v}$ is differentiable at

$$\mathbf{x} := \chi_{t_0}(\hat{\mathbf{x}},t). \tag{B.5.15}$$

Then (suppressing time dependence and recalling $\boldsymbol{\chi}_{t_0}(\cdot,t)$ is differentiable)

$$\begin{aligned}\boldsymbol{\chi}_{t_0}(\hat{\mathbf{x}}+\mathbf{h}) &= \boldsymbol{\chi}_{t_0}(\hat{\mathbf{x}}) + \nabla\boldsymbol{\chi}_{t_0}(\hat{\mathbf{x}})\mathbf{h} + o(\mathbf{h}) \quad \text{as} \quad \mathbf{h}\to\mathbf{0} \\ &= \mathbf{x} + \mathbf{F}(\hat{\mathbf{x}})\mathbf{h} + o(\mathbf{h}) \quad \text{as} \quad \mathbf{h}\to\mathbf{0}. \end{aligned} \tag{B.5.16}$$

Further,

$$\begin{aligned}(\mathbf{v}\circ\boldsymbol{\chi}_{t_0})(\hat{\mathbf{x}}+\mathbf{h}) &= \mathbf{v}(\boldsymbol{\chi}_{t_0}(\hat{\mathbf{x}}+\mathbf{h})) \\ &= \mathbf{v}(\mathbf{x}+\mathbf{F}(\hat{\mathbf{x}})\mathbf{h}+o(\mathbf{h})) \\ &= \mathbf{v}(\mathbf{x}) + \nabla\mathbf{v}(\mathbf{x})(\mathbf{F}(\hat{\mathbf{x}})\mathbf{h}+o(\mathbf{h})) + o(\mathbf{h}). \end{aligned} \tag{B.5.17}$$

Thus

$$(\mathbf{v}\circ\boldsymbol{\chi}_{t_0})(\hat{\mathbf{x}}+\mathbf{h}) = (\mathbf{v}\circ\boldsymbol{\chi}_{t_0})(\hat{\mathbf{x}}) + \mathbf{L}(\mathbf{x})\mathbf{F}(\hat{\mathbf{x}})\mathbf{h} + o(\mathbf{h}) \quad \text{as} \quad \mathbf{h}\to\mathbf{0}, \tag{B.5.18}$$

where

$$\mathbf{L} := \nabla\mathbf{v}. \tag{B.5.19}$$

Accordingly vector field $\mathbf{v}\circ\boldsymbol{\chi}_{t_0}$ on B_{t_0} is differentiable at $\hat{\mathbf{x}}$ with derivative

$$\nabla(\mathbf{v}\circ\boldsymbol{\chi}_{t_0})(\hat{\mathbf{x}}) = \mathbf{L}(\mathbf{x})\mathbf{F}(\hat{\mathbf{x}}). \tag{B.5.20}$$

However, from (B.5.13) this derivative is the same as that of $\dot{\boldsymbol{\chi}}_{t_0}$ at $\hat{\mathbf{x}}$, namely $\nabla\dot{\boldsymbol{\chi}}_{t_0}(\hat{\mathbf{x}})$. Assuming that spatial and temporal differentiation can be interchanged (this is the case if all second partial derivatives of representative co-ordinate fields are continuous as functions of position and time), then

$$\nabla\dot{\boldsymbol{\chi}}_{t_0} = \frac{d}{dt}\{\nabla\boldsymbol{\chi}_{t_0}\} = \dot{\mathbf{F}}, \tag{B.5.21}$$

and hence

$$\mathbf{L}\mathbf{F} = \dot{\mathbf{F}}. \tag{B.5.22}$$

Since $J = \det\mathbf{F} > 0$ [see (B.5.12)], linear transformation $\mathbf{F}$ is invertible and hence we have

Result B.5.1.

$$\nabla\mathbf{v} = \mathbf{L} = \dot{\mathbf{F}}\mathbf{F}^{-1}. \tag{B.5.23}$$

The polar decomposition [see (B.18.1)]

$$\mathbf{F} = \mathbf{R}\mathbf{U} \tag{B.5.24}$$

yields

$$\dot{\mathbf{F}}\mathbf{F}^{-1} = (\dot{\mathbf{R}}\mathbf{U} + \mathbf{R}\dot{\mathbf{U}})\mathbf{U}^{-1}\mathbf{R}^{-1}.$$

Thus, noting that $\mathbf{R}$ takes orthogonal values,

$$\mathbf{L} = \dot{\mathbf{R}}\mathbf{R}^T + \mathbf{R}\dot{\mathbf{U}}\mathbf{U}^{-1}\mathbf{R}^T. \tag{B.5.25}$$

The *spin* and *stretching* fields $\mathbf{W}$ and $\mathbf{D}$ are defined by

$$\mathbf{W} := \operatorname{sk}\mathbf{L} = \frac{1}{2}(\mathbf{L} - \mathbf{L}^T) \tag{B.5.26}$$

and

$$\mathbf{D} := \operatorname{sym}\mathbf{L} = \frac{1}{2}(\mathbf{L} + \mathbf{L}^T). \tag{B.5.27}$$

From (B.4.52) the matrix representation of $\mathbf{W}$ is $[W_{ij}]$, where

$$W_{ij} = \frac{1}{2}(\check{v}_{i,j} - \check{v}_{j,i}). \tag{B.5.28}$$

Further, from (A.15.12) the axial vector associated with $\mathbf{W}$ is

$$\mathbf{w} := -W_{23}\mathbf{e}_1 + W_{13}\mathbf{e}_2 - W_{12}\mathbf{e}_3 = \frac{1}{2}(\check{v}_{3,2} - \check{v}_{2,3})\mathbf{e}_1 + \frac{1}{2}(\check{v}_{1,3} - \check{v}_{3,1})\mathbf{e}_2 + \frac{1}{2}(\check{v}_{2,1} - \check{v}_{1,2})\mathbf{e}_3. \tag{B.5.29}$$

Thus

$$\mathbf{w} = \frac{1}{2}\operatorname{curl}\mathbf{v}. \tag{B.5.30}$$

That is, the *axial vector associated with the spin tensor* $\mathbf{W}$ *is* $\frac{1}{2}\operatorname{curl}\mathbf{v}$.

Remark B.5.3. In fluid dynamics curl $\mathbf{v}$ is known as the *vorticity vector field.*

Exercise B.5.3 Show that from (B.5.25), and definitions (B.5.26) and (B.5.27),

$$\mathbf{W} = \dot{\mathbf{R}}\mathbf{R}^T + \frac{1}{2}\mathbf{R}(\dot{\mathbf{U}}\mathbf{U}^{-1} - \mathbf{U}^{-1}\dot{\mathbf{U}})\mathbf{R}^T, \tag{B.5.31}$$

and

$$\mathbf{D} = \frac{1}{2}\mathbf{R}(\dot{\mathbf{U}}\mathbf{U}^{-1} + \mathbf{U}^{-1}\dot{\mathbf{U}})\mathbf{R}^T. \tag{B.5.32}$$

[*Hint*: Setting $\mathbf{Q} = \mathbf{R}$ in Exercise 2.6.3 of Chapter 2 implies that $\dot{\mathbf{R}}\mathbf{R}^T$ takes skew-symmetric values.]

The time rate of change of the local magnification factor $J = \det\mathbf{F}$ is given by the following:

Result B.5.2.

$$\dot{\widehat{\det\mathbf{F}}} = \operatorname{tr}(\dot{\mathbf{F}}\mathbf{F}^{-1})\det\mathbf{F}. \tag{B.5.33}$$

Proof. Since $\mathbf{F}$ is differentiable with respect to time,

$$\mathbf{F}(t+h) = \mathbf{F}(t) + \dot{\mathbf{F}}(t)h + o(h) \quad \text{as} \quad h \to 0. \tag{B.5.34}$$

Accordingly [see (B.4.85) with $\mathbf{A} = \mathbf{F}(t)$ and $\mathbf{H} = \dot{\mathbf{F}}(t)h + o(h)$]

$$\begin{aligned}\det(\mathbf{F}(t+h)) &= \det(\mathbf{F}(t) + \dot{\mathbf{F}}(t)h + o(h))\\ &= \det(\mathbf{F}(t)) + (\nabla(\det)(\mathbf{F}(t)))\cdot(\dot{\mathbf{F}}(t)h + o(h)) + o(h).\end{aligned}$$

Hence, from (B.4.96) with $\mathbf{A} = \mathbf{F}(t)$,

$$\begin{aligned}\det(\mathbf{F}(t+h)) - \det(\mathbf{F}(t)) &= (\det(\mathbf{F}(t)))(\mathbf{F}(t))^{-T}\cdot\dot{\mathbf{F}}(t)h + o(h)\\ &= \det(\mathbf{F}(t))\operatorname{tr}((\mathbf{F}(t))^{-1}\dot{\mathbf{F}}(t))h + o(h).\end{aligned} \tag{B.5.35}$$

It follows that $\det\mathbf{F}$ is time differentiable at any time t, and (B.5.33) holds via (A.12.14).

From (B.5.23) and definition (B.7.1) to follow, we deduce the following:

Corollary B.5.1.

$$\dot{\widehat{\det\mathbf{F}}} = \operatorname{tr}\mathbf{L}\,.\det\mathbf{F} = \operatorname{div}\mathbf{v}\,.\det\mathbf{F}. \tag{B.5.36}$$

Derivation of the continuity equation in Chapter 2 required Corollary 2.5.1 [see (2.5.9) of Chapter 2] together with the following:

Result B.5.3. If $\mathbf{x}$ is a differentiable function of time t, $(\mathbf{x}(t),t) \in \mathcal{T}_B$ [see (2.3.1) of Chapter 2], and f is a scalar-valued field on $\mathcal{T}_B$ which is differentiable both in space and time, then

$$\dot{f}(t) := \frac{d}{dt}\{f(\mathbf{x}(t),t)\} \tag{B.5.37}$$

is given by

$$\dot{f} = (\nabla f).\dot{\mathbf{x}} + \frac{\partial f}{\partial t}. \tag{B.5.38}$$

(Here ∇f is computed with t fixed and $\partial f/\partial t$ with $\mathbf{x}$ fixed.)

Proof. Although the result follows from basic definitions of derivatives, it is simpler to adopt co-ordinate representations and invoke the chain rule for functions of several variables. For a choice of Cartesian co-ordinates, we have the representations

$$\mathbf{x} \leftrightarrow (x_1,x_2,x_3) \qquad \text{and} \qquad f(\mathbf{x}(t),t) \leftrightarrow \check{f}(x_1(t),x_2(t),x_3(t),t).$$

Accordingly, from (B.5.37),

$$\dot{f} \leftrightarrow \frac{\partial \check{f}}{\partial x_1}\dot{x}_1 + \frac{\partial \check{f}}{\partial x_2}\dot{x}_2 + \frac{\partial \check{f}}{\partial x_3}\dot{x}_3 + \frac{\partial \check{f}}{\partial t}. \tag{B.5.39}$$

Since $\partial \check{f}/\partial x_i \, \mathbf{e}_i \leftrightarrow \nabla f$ and $\dot{x}_i\mathbf{e}_i \leftrightarrow \dot{\mathbf{x}}$, the result follows.

Remark B.5.4. The preceding proof holds when f is a tensor field of *any* order, n say: (B.5.39) corresponds to

$$\dot{f} = (\nabla f)\dot{\mathbf{x}} + \frac{\partial f}{\partial t}. \tag{B.5.40}$$

Here ∇f is that tensor field of order $(n+1)$ which maps $\mathbf{e}_j$ into $\partial \check{f}/\partial x_j$ when a Cartesian co-ordinate system representation is adopted: see, for example, $\nabla \mathbf{v}$ in (B.4.49) for order 1 (i.e., vector) fields.

B.6 (Riemann) Integration over Spatial Regions

Here we address fundamental aspects of Riemann integrals over regions in $\mathcal{E}$. The discussion is not comprehensive, but identifies such integrals as limits of sums (a basic ingredient in the cellular averaging approach of Section 11.2) and establishes two central results. The first is that used to obtain the local form of any balance relation, and the second is employed to establish referential forms of relations.

It is a simple matter to calculate the volume of a rectangular box or parallelipiped.[5] However, assigning a volume to a general closed and bounded region R with piecewise smooth boundary ∂R (we term such a region *regular*) requires care. Since R is bounded, it may be enclosed within a rectangular box B. This box may be partitioned into smaller boxes by three systems of planes: each system consists of planes

[5] The volume of a parallelipiped [see (A.5.13) et seq.] is equal to that of an associated rectangular box. Convince yourself! (It may help to recall how the area of a parallelogram can be shown to equal that of an associated rectangle.) See Exercise B.5.1.

parallel to a pair of opposite faces of B and which intersect B. Denoting such a partition by $\mathcal{P}$, consider the sums

$$s(R;\mathcal{P}) := \sum_{R,\mathcal{P}}^{-} \Delta V \qquad \text{and} \qquad S(R;\mathcal{P}) := \sum_{R,\mathcal{P}}^{+} \Delta V. \tag{B.6.1}$$

Here $\sum_{R,\mathcal{P}}^{-}$ denotes the sum of the volumes ΔV of those small boxes associated with $\mathcal{P}$ which lie entirely within R, while $\sum_{R,\mathcal{P}}^{+}$ represents the sum of the volumes of those small boxes which either lie within R or which contain boundary points of R. If the volume vol(R) of R is to make sense, then $s(R;\mathcal{P})$ and $S(R;\mathcal{P})$ constitute lower and upper bounds for vol(R). That is,

$$s(R;\mathcal{P}) \leq \text{vol}(R) \leq S(R;\mathcal{P}). \tag{B.6.2}$$

Partition $\mathcal{P}$ may be 'refined' by choosing more plane sections of B (and hence subdividing at least some small boxes created by $\mathcal{P}$ into smaller boxes). If $\mathcal{P}'$ denotes such a refinement, then

$$s(R;\mathcal{P}) \leq s(R;\mathcal{P}') \leq \text{vol}(R) \leq S(R;\mathcal{P}') \leq S(R;\mathcal{P}). \tag{B.6.3}$$

Exercise B.6.1. Convince yourself of (B.6.3) by considering any box in partition $\mathcal{P}$ that is subdivided in partition $\mathcal{P}'$ into smaller boxes. If such a subdivided box intersects ∂R, then not all the associated smaller boxes in $\mathcal{P}'$ may intersect ∂R. (Draw a sketch to illustrate this.) Any such smaller box must thus lie within R or lie completely outside R. In the former case $s(R;\mathcal{P}') > s(R;\mathcal{P})$, while in the latter case $S(R;\mathcal{P}') < S(R;\mathcal{P})$.

The *fineness* $\phi(R;\mathcal{P}')$ of any partition $\mathcal{P}'$ of R is the maximum *span*[6] of the small boxes associated with $\mathcal{P}'$. Suppose that

$$S(R;\mathcal{P}') - s(R;\mathcal{P}') \to 0 \qquad \text{as} \quad \phi(R;\mathcal{P}') \to 0. \tag{B.6.4}$$

Notice that as partition $\mathcal{P}'$ is refined, the set $\{S(R;\mathcal{P}')\}$ of (positive) numbers is bounded below by $s(R;\mathcal{P})$ and hence has a greatest lower bound,[7] $V(>0)$ say. Similarly, $\{s(R;\mathcal{P}')\}$ is a set of (positive) numbers bounded above by $S(R;\mathcal{P})$ and hence has a least upper bound,[8] $v(>0)$ say. Thus, for any refinement $\mathcal{P}'$ of $\mathcal{P}$,

$$s(R;\mathcal{P}') \leq v \leq V \leq S(R;\mathcal{P}'). \tag{B.6.5}$$

Accordingly, *if (B.6.4) holds*, then from (B.6.3)

$$v = \text{vol}(R) = V. \tag{B.6.6}$$

Remark B.6.1. The existence of vol(R) thus follows from a fundamental property of the real number system $\mathbb{R}$ *if* criterion (B.6.4) is satisfied. Notice that $S(R;\mathcal{P}')-s(R;\mathcal{P}')$ is the sum of the volumes of those small boxes in $\mathcal{P}'$ which intersect ∂R. Fortunately

[6] The span of a subset of $\mathcal{E}$ is the maximum separation of any pair of points in this subset. Thus the span of a rectangular box of dimensions $\ell \times w \times h$ is $(\ell^2 + w^2 + h^2)^{1/2}$.

[7] The existence of such a (unique) bound is an axiomatic property of real numbers (cf., e.g., Apostol [91], Axiom 10, p. 7).

[8] The existence of this (unique) bound is a consequence of the axiomatic property in footnote 7.

a method exists by which volumes can be *calculated* for regular regions, namely integration. It can be shown that (B.6.4) holds for any regular region R. (This is a non-trivial result.)

If f denotes a *non-negative* real-valued function on R, then we can consider the sums

$$s(f;R,\mathcal{P}) := \sum_{\mathcal{P}}^{-} f_{\min}\Delta V \qquad \text{and} \qquad S(f;R;\mathcal{P}) := \sum_{\mathcal{P}}^{+} f_{\max}\Delta V. \tag{B.6.7}$$

Here the sums are taken over the small boxes defined by $\mathcal{P}$ as in (B.6.1), and $f_{\min}$ ($f_{\max}$) denotes the minimum (maximum) value of f in each box. If

$$S(f;R;\mathcal{P}') - s(f;R;\mathcal{P}') \to 0 \qquad \text{as} \quad \phi(R;\mathcal{P}') \to 0 \tag{B.6.8}$$

for refinements $\mathcal{P}'$ of $\mathcal{P}$, then an argument similar to that used to show the existence of vol(R) can be used to show that a common limit (in the sense of increasing refinement) for $S(f;R;\mathcal{P})$ and $s(f;R,\mathcal{P})$ exists. This limit is termed the *value of the Riemann integral of f over R* and is denoted by

$$\int_R f, \qquad \int_R f\,dV, \qquad \text{or} \qquad \int_R f(\mathbf{x})dV_{\mathbf{x}}. \tag{B.6.9}$$

Remark B.6.2. Note that if $f(\mathbf{x}) \equiv 1$, then (B.6.7) reduces to (B.6.1), and

$$\text{vol}(R) = \int_R 1\,dV. \tag{B.6.10}$$

Remark B.6.3. Notice that in (B.6.7) it is assumed that $f_{\min}$ and $f_{\max}$ exist. This is certainly the case if f is continuous on R since this region is closed and bounded, and thus so too will be all small boxes in partitions $\mathcal{P}'$: any continuous function on a closed and bounded set in $\mathcal{E}$ is bounded above and below and attains its bounds (cf., e.g., Apostol [91], Theorem 4–20).

Exercise B.6.2. Show that (B.6.4) implies the existence of a common limit for $S(f;R;\mathcal{P}')$ and $s(f;R;\mathcal{P}')$ as $\phi(R;\mathcal{P}') \to 0$ if f is continuous on R.

Now consider the sum

$$\sigma(f;R;\mathcal{P}') := \sum_{\mathcal{P}'}^{+} f(\mathbf{x})\Delta V_{\mathbf{x}}, \tag{B.6.11}$$

where $\mathbf{x}$ denotes *any* point of R within (or on the boundary of) a small box in partition $\mathcal{P}'$ and $\Delta V_{\mathbf{x}}$ is the volume of this box. Clearly, within each such box $f_{\min} \leq f(\mathbf{x}) \leq f_{\max}$, and hence

$$s(f;R;\mathcal{P}') \leq \sigma(f;R;\mathcal{P}') \leq S(f;R;\mathcal{P}'). \tag{B.6.12}$$

Accordingly, if f is Riemann integrable over R, then from (B.6.8)

$$\sigma(f;R;\mathcal{P}') \to \int_R f\,dV \qquad \text{as} \quad \phi(R;\mathcal{P}') \to 0. \tag{B.6.13}$$

Exercise B.6.3. Prove (B.6.12), noting that f takes non-negative values.

Remark B.6.4. If $f \geq 0, = 0$, and ≤ 0 on regular regions R_1, R_2, respectively, and R_3, and $R := R_1 \cup R_2 \cup R_3$, then

$$\int_R f\, dV := \int_{R_1} f\, dV - \int_{R_3} (-f) dV. \tag{B.6.14}$$

Remark B.6.5. If $\mathbf{f}$ takes values in $\mathcal{V}$, $\{\mathbf{u}_1, \mathbf{u}_2, \mathbf{u}_3\}$ is a basis for $\mathcal{V}$, and if $\mathbf{f} \,.\, \mathbf{u}_i$ is integrable over R for $i = 1,2,3$, then the integral of $\mathbf{f}$ over R is that vector denoted by $\int_R \mathbf{f}\, dV$ for which

$$\left(\int_R \mathbf{f}\, dV\right) . \mathbf{u}_i := \int_R \mathbf{f} \,.\, \mathbf{u}_i\, dV. \tag{B.6.15}$$

Remark B.6.6. The extension to integrals of tensor-valued functions of all orders follows by generalising from Remark B.6.5 one order at a time. In particular, if $\mathbf{F}$ takes values in $\mathrm{Lin}\,\mathcal{V}$ and $\mathbf{F}\mathbf{u}_j$ $(j = 1,2,3)$ is integrable over R in the sense of (B.6.15), then

$$\left(\int_R \mathbf{F}\, dV\right) \mathbf{u}_j := \int_R \mathbf{F}\mathbf{u}_j\, dV. \tag{B.6.16}$$

At this point two leading questions need to be addressed:

Question 1: Is there a simple criterion which ensures satisfaction of (B.6.8)?

Question 2: How are the foregoing integrals to be evaluated?

Answer 1. If $f, \mathbf{f}$, or $\mathbf{F}$ in (B.6.14), (B.6.15), or (B.6.16) is a continuous function of position, then (B.6.8) holds (for $f, \mathbf{f} \,.\, \mathbf{u}_i$, or $\mathbf{F}\mathbf{u}_j \,.\, \mathbf{u}_i$) for any regular region R. The proof requires noting that continuity of a function on a closed, bounded subset of $\mathcal{E}$ implies *uniform* continuity on this set.[9] Thus, given $\epsilon > 0$, for any points $\mathbf{x}, \mathbf{y} \in B$, one can find a number $\delta_1 > 0$ such that

$$|f(\mathbf{y}) - f(\mathbf{x})| < \frac{\epsilon}{2\,\mathrm{vol}(R)} \qquad \text{whenever} \qquad \|\mathbf{y} - \mathbf{x}\| < \delta_1. \tag{B.6.17}$$

Thus if $\phi(R; \mathcal{P}') < \delta_1$, then in each and every small box

$$f_{\max} - f_{\min} < \epsilon / 2\,\mathrm{vol}(R). \tag{B.6.18}$$

Now $$S(f;R;\mathcal{P}') - s(f;R;\mathcal{P}') = \sum_{\mathcal{P}'}^{-} (f_{\max} - f_{\min}) \Delta V + \sum_{\mathcal{P}'}^{'} (f_{\max} - f_{\min}) \Delta V, \tag{B.6.19}$$

where $\sum_{\mathcal{P}'}^{'}$ denotes a sum over small boxes which intersect ∂R. Since f is continuous on the closed and bounded box B which contains R, it is bounded and attains its bounds within this box (cf. Remark B.6.3), and hence there exist non-negative numbers $\underline{f}_R$ and $\bar{f}_R$ such that

$$\sum_{\mathcal{P}'}^{'} (f_{\max} - f_{\min}) \Delta V < (\bar{f}_R - \underline{f}_R) \sum_{\mathcal{P}'}^{'} \Delta V. \tag{B.6.20}$$

[9] This is a standard result (cf., e.g., Apostol [91], Theorem 4–24).

However, since (B.6.4) holds (recall Remark B.6.1 and note that R is regular) and may be written as

$$\sum_{\mathcal{P}'}{}' \Delta V \to 0 \qquad \text{as} \quad \phi(R;\mathcal{P}') \to 0, \tag{B.6.21}$$

we can find a number δ_2 such that

$$\sum_{\mathcal{P}'}{}' \Delta V < \frac{\epsilon}{2(\bar{f}_R - \underline{f}_R)} \qquad \text{if} \quad \phi(R;\mathcal{P}') < \delta_2. \tag{B.6.22}$$

It follows from (B.6.19), (B.6.18), (B.6.20), and (B.6.22) that

$$S(f;R;\mathcal{P}') - s(f;R;\mathcal{P}') < \epsilon \qquad \text{if} \quad \phi(R;\mathcal{P}') < \min(\delta_1,\delta_2). \tag{B.6.23}$$

That is, (B.6.8) holds.

Answer 2: Choose a Cartesian co-ordinate system $C(\mathbf{x}_0;\mathbf{e}_1,\mathbf{e}_2,\mathbf{e}_3)$ (see Appendix B.2), where $\mathbf{e}_1,\mathbf{e}_2$, and $\mathbf{e}_3$ are parallel to box edges. Any point $\mathbf{x}$ in a small box will have co-ordinates (x_1,x_2,x_3), where $\tilde{x}_i \le x_i \le \tilde{x}_i + \Delta x_i$ for some $\tilde{x}_i$ and $\Delta x_i > 0\,(i = 1,2,3)$. Of course, $\Delta V = (\Delta x_1)(\Delta x_2)(\Delta x_3)$ and $\mathbf{x}$ in (B.6.11) can be chosen to be $(\tilde{x}_1,\tilde{x}_2,\tilde{x}_3)$, namely that box vertex $\tilde{\mathbf{x}}$ with smallest co-ordinates. Then [see (B.6.11)]

$$\sigma(f;R;\mathcal{P}') = \sum_{\mathcal{P}'}{}^{+} f(\mathbf{x})\Delta x_1 \Delta x_2 \Delta x_3. \tag{B.6.24}$$

Since [see (B.2.1) which implies $\mathbf{x} = \mathbf{x}_0 + x_i\mathbf{e}_i$]

$$f(\mathbf{x}) = f(\mathbf{x}_0 + x_i\mathbf{e}_i) =: \check{f}(x_1,x_2,x_3), \tag{B.6.25}$$

(B.6.24), with choice $x_i = \tilde{x}_i$ in each small box, yields

$$\begin{aligned}\sigma(f;R;\mathcal{P}') &= \sum_{\mathcal{P}'}{}^{+} \check{f}(\tilde{x}_1,\tilde{x}_2,\tilde{x}_3)\Delta x_1 \Delta x_2 \Delta x_3 \\ &= \sum_3 \left\{ \sum_2 \left\{ \sum_1 \check{f}(\tilde{x}_1,\tilde{x}_2,\tilde{x}_3)\Delta x_1 \right\} \Delta x_2 \right\} \Delta x_3. \end{aligned} \tag{B.6.26}$$

Here $\sum_1$ denotes a sum over all boxes with the same values of $\tilde{x}_2,\tilde{x}_3$ and hence is a function of $\tilde{x}_2,\tilde{x}_3$. Sum $\sum_2$ represents a sum over all boxes with the same value of $\tilde{x}_3$ and thus yields a function of $\tilde{x}_3$. As $\phi(R;\mathcal{P}') \to 0$, so $\Delta x_1, \Delta x_2$, and $\Delta x_3 \to 0$ (Why?). Hence, from (B.6.13) and (B.6.26), if f is continuous on R, then

$$\begin{aligned}\int_R f\,dV &= \lim_{\phi(R;\mathcal{P}')\to 0} \{\sigma(f;R;\mathcal{P}')\} \\ &= \lim_{\Delta x_3 \to 0} \left\{ \sum_3 A(\tilde{x}_3)\Delta x_3 \right\}, \end{aligned} \tag{B.6.27}$$

where $$A(\tilde{x}_3) := \lim_{\Delta x_2 \to 0} \left\{ \sum_2 B(\tilde{x}_2,\tilde{x}_3)\Delta x_2 \right\} \tag{B.6.28}$$

and
$$B(\tilde{x}_2,\tilde{x}_3) := \lim_{\Delta x_1 \to 0} \left\{ \sum_1 \tilde{f}(\tilde{x}_1,\tilde{x}_2,\tilde{x}_3)\Delta x_1 \right\}. \tag{B.6.29}$$

Each of the limit sums (B.6.27), (B.6.28), and (B.6.29) constitutes a definite (Riemann) integral over an interval or union of disjoint intervals in $\mathbb{R}$. Accordingly $\int_R f\,dV$ can be computed in terms of Riemann integrals of functions of a single real variable, each taken over a union of intervals in $\mathbb{R}$. Of course, evaluation of these integrals in terms of ante-derivatives is the content of the fundamental theorem of integral calculus (see Apostol [91], Theorem 9–32).

Remark B.6.7. The sums in (B.6.7) can be generalised by partitioning B into any set of mutually disjoint regular subregions [such regularity ensures that volumes ΔV are meaningful and may be computed via (B.6.10) with R as the relevant subregion] and repeating the analysis with $\phi(R;\mathcal{P})$ now the maximum span of the subregions associated with the partition. It can be shown that for continuous f and regular R the integral (B.6.9) is independent of partition choice. Selection of partitions associated with particular curvilinear co-ordinate systems (e.g., spherical polars) leads to standard results which equate the triple integrals corresponding to Cartesian co-ordinates with those associated with the curvilinear system. The latter involve Jacobians. While we do not give details, the analysis is essentially that outlined in Theorem B.6.2 to follow.

We now consider two key results.

Theorem B.6.1. (*Localisation*) If f is a continuous field on a regular region $\mathcal{R}$ and
$$\int_R f\,dV = 0 \tag{B.6.30}$$
for every regular subregion $R \subset \mathcal{R}$, then
$$f = 0 \quad \text{in} \quad \mathcal{R}. \tag{B.6.31}$$

Proof. Suppose that f takes real values. If $\mathbf{x}$ is any interior point of $\mathcal{R}$, then $f(\mathbf{x}) > 0, f(\mathbf{x}) = 0$, or $f(\mathbf{x}) < 0$. If $f(\mathbf{x}) > 0$, then by continuity there exists a sphere centered at $\mathbf{x}$ which lies within $\mathcal{R}$ and throughout which $f > 0$. Choosing R in (B.6.30) to be this sphere contradicts (B.6.30). (Why?) Thus $f(\mathbf{x}) \not> 0$. Similarly, $f(\mathbf{x}) \not< 0$. (Prove this!) Accordingly $f(\mathbf{x}) = 0$ for all $\mathbf{x}$ not on $\partial\mathcal{R}$. However, by continuity, boundary values of f also must vanish. Note that if f is a vectorial (or higher-order tensor) field, then the result remains valid (see Remarks B.6.5 and B.6.6).

Theorem B.6.2. If $\mathbf{d} : \hat{R} \to R := \mathbf{d}(\hat{R})$ is a physically admissible continuously differentiable deformation, where $\hat{R} \subset \mathcal{E}$ is regular, and f is a continuous function on R, then
$$\int_R f(\mathbf{x})dV_{\mathbf{x}} = \int_{\hat{R}} f(\mathbf{d}(\hat{\mathbf{x}}))J_{\mathbf{d}}(\hat{\mathbf{x}})dV_{\hat{\mathbf{x}}}. \tag{B.6.32}$$

Sketch of proof. Any partition $\hat{\mathcal{P}}$ of $\hat{R}$ is mapped by $\mathbf{d}$ into a partition $\mathcal{P}$ of R. Suppose that $\hat{\mathcal{P}}$ consists of parallelipipeds. If $\Pi_{\hat{\mathbf{x}}}$ denotes a parallelipiped containing $\hat{\mathbf{x}}$, and

$\hat{\mathbf{y}} \in \Pi_{\hat{\mathbf{x}}}$, then a mean value theorem (see Apostol [91], Theorem 6–17) yields the existence of a point $\hat{\mathbf{z}} \in \Pi_{\hat{\mathbf{x}}}$ which lies on the line segment joining $\hat{\mathbf{y}}$ to $\hat{\mathbf{x}}$ such that

$$\mathbf{d}(\hat{\mathbf{y}}) - \mathbf{d}(\hat{\mathbf{x}}) = \nabla\mathbf{d}(\hat{\mathbf{z}})(\hat{\mathbf{y}} - \hat{\mathbf{x}}). \tag{B.6.33}$$

Thus [recalling (A.14.22)]

$$\|\mathbf{d}(\hat{\mathbf{y}}) - \mathbf{d}(\hat{\mathbf{x}}\| = \|\nabla\mathbf{d}(\hat{\mathbf{z}})(\hat{\mathbf{y}} - \hat{\mathbf{x}})\| \leq \|\nabla\mathbf{d}(\hat{\mathbf{z}})\|\|\hat{\mathbf{y}} - \hat{\mathbf{x}}\|. \tag{B.6.34}$$

Since, by hypothesis, $\nabla\mathbf{d}$ is continuous on the closed and bounded set $\hat{R}$, its norm $\|\nabla\mathbf{d}\|$ is uniformly bounded on $\hat{R}$. Thus, for some $K > 0$,

$$\|\mathbf{d}(\hat{\mathbf{y}})\| - \mathbf{d}(\hat{\mathbf{x}})\| \leq K\|\hat{\mathbf{y}} - \hat{\mathbf{x}}\| \quad \text{for all } \hat{\mathbf{y}} \in \Pi_{\hat{\mathbf{x}}}. \tag{B.6.35}$$

Accordingly, if $\phi(\hat{R};\hat{\mathcal{P}}) < \delta$, then $\|\hat{\mathbf{y}} - \hat{\mathbf{x}}\| < \delta$ for any $\Pi_{\hat{\mathbf{x}}}$ in $\hat{\mathcal{P}}$ and so $\|\mathbf{d}(\hat{\mathbf{y}}) - \mathbf{d}(\hat{\mathbf{x}})\| < K\delta$ for any image $\mathbf{d}(\Pi_{\hat{\mathbf{x}}})$ in $\mathcal{P}$. That is, $\phi(R;\mathcal{P}) < K\delta$. Thus if $\Delta V_{\mathbf{x}} := \text{vol}(\mathbf{d}(\Pi_{\hat{\mathbf{x}}}))$, then

$$\begin{aligned}\int_R f(\mathbf{x})dV_{\mathbf{x}} &:= \lim_{\phi(R;\mathcal{P})\to 0}\left\{\sum_{\mathcal{P}}^{+} f(\mathbf{x})\Delta V_{\mathbf{x}}\right\}\\ &= \lim_{\phi(\hat{R};\hat{\mathcal{P}})\to 0}\left\{\sum_{\hat{\mathcal{P}}}^{+} f(\mathbf{d}(\hat{\mathbf{x}}))\text{vol}(\mathbf{d}(\Pi_{\hat{\mathbf{x}}}))\right\}.\end{aligned} \tag{B.6.36}$$

However, from (B.5.5) and (B.5.7),

$$\text{vol}(\mathbf{d}(\Pi_{\hat{\mathbf{x}}})) - J_{\mathbf{d}}\text{vol}(\Pi_{\hat{\mathbf{x}}}) \to 0 \qquad \text{as} \quad \phi(\hat{R};\hat{\mathcal{P}}) \to 0. \tag{B.6.37}$$

Writing $\Delta V_{\hat{\mathbf{x}}} := \text{vol}(\Pi_{\hat{\mathbf{x}}})$, (B.6.36) and (B.6.37) suggest that

$$\begin{aligned}\int_R f(\mathbf{x})dV_{\mathbf{x}} &= \lim_{\phi(\hat{R};\hat{\mathcal{P}})}\left\{\sum_{\mathcal{P}}^{+} f(\mathbf{d}(\hat{\mathbf{x}}))J_{\mathbf{d}}(\hat{\mathbf{x}})\Delta V_{\hat{\mathbf{x}}}\right\}\\ &=: \int_{\hat{R}} f(\mathbf{d}(\hat{\mathbf{x}}))J_{\mathbf{d}}(\hat{\mathbf{x}})dV_{\hat{\mathbf{x}}}.\end{aligned} \tag{B.6.38}$$

Remark B.6.8. Assertion (B.5.5) is intuitively correct but was not proved analytically. Similarly, the equivalence of the sums over $\hat{\mathcal{P}}$ in (B.6.36) and (B.6.38) was not strictly proved. Strict proofs of these results and that of the limit in (B.6.4) for regular regions (see Remark B.6.1) are subtle and not here attempted.

B.7 Divergences and Divergence Theorems

If $\mathbf{v}$ denotes a vectorial field, then

$$\text{div}\,\mathbf{v} := \text{tr}(\nabla\mathbf{v}). \tag{B.7.1}$$

Recalling the Cartesian matrix and co-ordinate representation (B.4.52) of $\mathbf{L} = \nabla\mathbf{v}$ and (A.12.11), we have

$$\text{div}\,\mathbf{v} = \partial\check{v}_i/\partial x_i = \check{v}_{i,i}. \tag{B.7.2}$$

More precisely, with x_i and $\check{v}_i$ given by (B.4.47),

$$(\text{div}\,\mathbf{v})(\mathbf{x}) = \check{v}_{i,i}(x_1,x_2,x_3). \tag{B.7.3}$$

In Cartesian tensor notation div **v** is denoted by $v_{i,i}$.

Theorem B.7.1. (Divergence theorem for vectorial fields)

If R is a regular region in $\mathcal{E}$ and **n** denotes the outward unit normal field on ∂R, then, for any vectorial field **v** with continuous spatial derivative defined on an open set $\mathcal{D}$ containing R,

$$\int_{\partial R} \mathbf{v}\,.\,\mathbf{n}\,dA = \int_R \operatorname{div}\mathbf{v}\,dV. \tag{B.7.4}$$

Remark B.7.1. Proofs can be found in standard advanced calculus textbooks (cf., e.g., Apostol [91], Theorem 11–37). Validity under weaker hypotheses is established in Kellogg [21].

It is possible to define the divergence of a linear transformation field **T** in such a way as to obtain a direct analogue of (B.7.4), namely

$$\int_{\partial R} \mathbf{T}\mathbf{n}\,dA = \int_R \operatorname{div}\mathbf{T}\,dV. \tag{B.7.5}$$

Notice that since the left-hand side of (B.7.5) is a vector (Why?), field div **T** must take values in $\mathcal{V}$. Consider, for any $\mathbf{k} \in \mathcal{V}$,

$$\left(\int_{\partial R} \mathbf{T}\mathbf{n}\,dA\right).\mathbf{k} = \int_{\partial R} \mathbf{T}\mathbf{n}\,.\,\mathbf{k}\,dA = \int_{\partial R} \mathbf{n}\,.\,\mathbf{T}^T\mathbf{k}\,dA = \int_R \operatorname{div}\{\mathbf{T}^T\mathbf{k}\}dV. \tag{B.7.6}$$

If (B.7.5) is to hold, then

$$\left(\int_{\partial R} \mathbf{T}\mathbf{n}\,dA\right).\mathbf{k} = \left(\int_R \operatorname{div}\mathbf{T}\,dV\right).\mathbf{k} = \int_R (\operatorname{div}\mathbf{T})\,.\,\mathbf{k}\,dV. \tag{B.7.7}$$

Comparison of (B.7.6) and (B.7.7) shows that it suffices to define div **T** to be that vector field which, given any $\mathbf{k} \in \mathcal{V}$, satisfies

$$(\operatorname{div}\mathbf{T})\,.\,\mathbf{k} := \operatorname{div}\{\mathbf{T}^T\mathbf{k}\}. \tag{B.7.8}$$

Remark B.7.2. In order for the foregoing to comply with (B.7.4) it suffices for $\mathbf{T}^T\mathbf{k}$ to have a continuous derivative in $\mathcal{D}$ for any $\mathbf{k} \in \mathcal{V}$. Notice the differences in uses of the symbol div in (B.7.8): its application to linear transformation fields is defined in terms of that appropriate to vectorial fields.

To obtain the Cartesian co-ordinate representation of div **T** [cf. (B.7.2)], note that values $\mathbf{T}(\mathbf{x}) \in \operatorname{Lin}\mathcal{V}$ are represented in system $C(\mathbf{x}_0;\mathbf{e}_1,\mathbf{e}_2,\mathbf{e}_3)$ by the 3×3 matrix $[T_{ij}(\mathbf{x})]$, where $T_{ij} := \mathbf{e}_i\,.\,\mathbf{T}\mathbf{e}_j$ [see (A.9.1)]. Further, since **x** is represented by its co-ordinates (x_1,x_2,x_3), then [see (A.9.8)]

$$\mathbf{T}(\mathbf{x}) = \check{T}_{pq}(x_1,x_2,x_3)\mathbf{e}_p \otimes \mathbf{e}_q. \tag{B.7.9}$$

From (B.7.8), with $\mathbf{k} = \mathbf{e}_i$, (A.8.10), and (B.7.2),

$$\begin{aligned}(\operatorname{div}\mathbf{T}).\mathbf{e}_i &= \operatorname{div}\{\mathbf{T}^T\mathbf{e}_i\} = \operatorname{div}\{(\check{T}_{pq}\,\mathbf{e}_q \otimes \mathbf{e}_p)\mathbf{e}_i\} = \operatorname{div}\{\check{T}_{pq}(\mathbf{e}_p\,.\,\mathbf{e}_i)\mathbf{e}_q\}\\ &= \operatorname{div}\{\check{T}_{iq}\,\mathbf{e}_q\} = \frac{\partial}{\partial x_q}\{\check{T}_{iq}\} =: \check{T}_{iq,q}.\end{aligned} \tag{B.7.10}$$

That is, relabelling dummy suffix q and recalling (A.6.6) with $\mathbf{v} = \operatorname{div}\mathbf{T}$,

$$\operatorname{div}\mathbf{T} = ((\operatorname{div}\mathbf{T}).\mathbf{e}_i)\mathbf{e}_i = \check{T}_{ij,j}\,\mathbf{e}_i. \tag{B.7.11}$$

More precisely,
$$(\operatorname{div}\mathbf{T})(\mathbf{x}) = \check{T}_{ij,j}(x_1,x_2,x_3)\mathbf{e}_i. \tag{B.7.12}$$

In Cartesian tensor notation

$$\operatorname{div}\mathbf{T} \qquad \text{is denoted by} \qquad T_{ij,j}. \tag{B.7.13}$$

The process by which (B.7.4) was used to establish (B.7.5), with appropriate definition of divergence, can be repeated to derive divergence theorems for tensor fields of all orders. In particular, if $\mathbf{C}: \mathcal{D} \to \mathrm{Lin}(\mathcal{V}, \mathrm{Lin}\,\mathcal{V})$ is a third-order tensor field, then the direct analogue of (B.7.4) is

$$\int_{\partial R} \mathbf{C}\mathbf{n}\,dA = \int_R \operatorname{div}\mathbf{C}\,dV. \tag{B.7.14}$$

Notice that $\operatorname{div}\mathbf{C}$ must take values in $\mathrm{Lin}\,\mathcal{V}$. (Why?) If (B.7.14) is to hold, then for any $\mathbf{k} \in \mathcal{V}$

$$\begin{aligned}\int_R (\operatorname{div}\mathbf{C})\mathbf{k}\,dV &= \left(\int_R \operatorname{div}\mathbf{C}\,dV\right)\mathbf{k} = \left(\int_{\partial R} \mathbf{C}\mathbf{n}\,dA\right)\mathbf{k} = \int_{\partial R} (\mathbf{C}\mathbf{n})\mathbf{k}\,dA \\ &= \int_{\partial R} (\mathbf{C}^T\mathbf{k})\mathbf{n}\,dA = \int_R \operatorname{div}\{\mathbf{C}^T\mathbf{k}\}dV,\end{aligned} \tag{B.7.15}$$

on recalling (A.9.12) and invoking (B.7.5) with $\mathbf{T} = \mathbf{C}^T\mathbf{k}$. To ensure (B.7.14) it thus suffices to define div $\mathbf{C}$ to be that linear transformation field which, given any $\mathbf{k} \in \mathcal{V}$, satisfies

$$(\operatorname{div}\mathbf{C})\mathbf{k} := \operatorname{div}\{\mathbf{C}^T\mathbf{k}\}. \tag{B.7.16}$$

Remark B.7.3. To comply with (B.7.5) it suffices for $((\mathbf{C}^T)\mathbf{k})\boldsymbol{\ell}$ to have a continuous derivative in $\mathcal{D}$ for any $\mathbf{k}, \boldsymbol{\ell} \in \mathcal{V}$. Notice that, as in (B.7.8), definition (B.7.16) involves two different uses of symbol div.

For the Cartesian co-ordinate representation of div $\mathbf{C}$, note from (A.19.10) that [cf. (B.7.9)]

$$\begin{aligned}\mathbf{C}(\mathbf{x}) &= C_{pqr}(\mathbf{x})\mathbf{e}_p \otimes \mathbf{e}_q \otimes \mathbf{e}_r \\ &= \check{C}_{pqr}(x_1,x_2,x_3)\mathbf{e}_p \otimes \mathbf{e}_q \otimes \mathbf{e}_r.\end{aligned} \tag{B.7.17}$$

Since $(\operatorname{div}\mathbf{C})(\mathbf{x}) \in \mathrm{Lin}\,\mathcal{V}$,

$$\begin{aligned}(\operatorname{div}\mathbf{C})_{ij} &:= \mathbf{e}_i.(\operatorname{div}\mathbf{C})\mathbf{e}_j = \mathbf{e}_i.\operatorname{div}\{\mathbf{C}^T\mathbf{e}_j\} = (\operatorname{div}\{\mathbf{C}^T\mathbf{e}_j\}).\mathbf{e}_i \\ &= \operatorname{div}\{(\mathbf{C}^T\mathbf{e}_j)^T\mathbf{e}_i\}.\end{aligned} \tag{B.7.18}$$

Here (B.7.16) was invoked with $\mathbf{k} = \mathbf{e}_j$ and then (B.7.8) with $\mathbf{T} = \mathbf{C}^T\mathbf{e}_j$ and $\mathbf{k} = \mathbf{e}_i$. From (B.7.17) and (A.19.18),

$$\begin{aligned}(\mathbf{C}^T\mathbf{e}_j)^T\mathbf{e}_i &= ((\check{C}_{pqr}\mathbf{e}_p \otimes \mathbf{e}_r \otimes \mathbf{e}_q)\mathbf{e}_j)^T\mathbf{e}_i = (\check{C}_{pqr}(\mathbf{e}_q.\mathbf{e}_j)\mathbf{e}_p \otimes \mathbf{e}_r)^T\mathbf{e}_i \\ &= (\check{C}_{pjr}\mathbf{e}_r \otimes \mathbf{e}_p)\mathbf{e}_i = \check{C}_{pjr}(\mathbf{e}_p.\mathbf{e}_i)\mathbf{e}_r = \check{C}_{ijr}\,\mathbf{e}_r.\end{aligned} \tag{B.7.19}$$

Thus, from (B.7.18), (B.7.19), and (B.7.2) with $\mathbf{v} = \check{C}_{ijr}\,\mathbf{e}_r$,

$$(\operatorname{div}\mathbf{C})_{ij} = \operatorname{div}\{\check{C}_{ijr}\,\mathbf{e}_r\} = \check{C}_{ijr,r} = \check{C}_{ijk,k}. \tag{B.7.20}$$

That is,

$$\operatorname{div}\mathbf{C} = \check{C}_{ijk,k}\,\mathbf{e}_i \otimes \mathbf{e}_j \tag{B.7.21}$$

or, more precisely,

$$(\operatorname{div}\mathbf{C})(\mathbf{x}) = \check{C}_{ijk,k}(x_1,x_2,x_3)\mathbf{e}_i \otimes \mathbf{e}_j. \tag{B.7.22}$$

In Cartesian tensor notation

$$\operatorname{div}\mathbf{C} \qquad \text{is denoted by} \qquad C_{ijk,k}. \tag{B.7.23}$$

Exercise B.7.1. If (B.7.14) holds, show that if $\mathbf{k},\boldsymbol{\ell} \in \mathcal{V}$, then [recall (A.19.16)]

$$\int_R (\operatorname{div}\mathbf{C})\mathbf{k}\,.\,\boldsymbol{\ell}\,dV = \int_{\partial R} (\mathbf{Cn})\mathbf{k}\,.\,\boldsymbol{\ell}\,dA = \int_{\partial R} (\mathbf{C}^{\sim}\boldsymbol{\ell})\mathbf{k}\,.\,\mathbf{n}\,dA = \int_R \operatorname{div}\{(\mathbf{C}^{\sim}\boldsymbol{\ell})\mathbf{k}\}. \tag{B.7.24}$$

Deduce that (B.7.14) holds if div $\mathbf{C}$ is defined by

$$(\operatorname{div}\mathbf{C})\mathbf{k}\,.\,\boldsymbol{\ell} := \operatorname{div}\{(\mathbf{C}^{\sim}\boldsymbol{\ell})\mathbf{k}\}, \tag{B.7.25}$$

for any $\mathbf{k},\boldsymbol{\ell} \in \mathcal{V}$. Reconcile (B.7.25) with definition (B.7.16). [*Hint*: Note from (B.7.16) that

$$(\operatorname{div}\mathbf{C})\mathbf{k}\,.\,\boldsymbol{\ell} = (\operatorname{div}\{\mathbf{C}^T\mathbf{k}\})\,.\,\boldsymbol{\ell} = \operatorname{div}\{(\mathbf{C}^T\mathbf{k})^T\boldsymbol{\ell}\}, \tag{B.7.26}$$

and show that

$$(\mathbf{C}^T\mathbf{k})^T\boldsymbol{\ell} = ((\mathbf{C}^{\sim})\boldsymbol{\ell})\mathbf{k} \tag{B.7.27}$$

by considering $(\mathbf{C}^T\mathbf{k})^T\boldsymbol{\ell}\,.\,\mathbf{v}$ for any $\mathbf{v} \in \mathcal{V}$.]

The following identities prove useful in manipulations of balance relations. (Here the symbols have the same meanings as in Subsection B.4.6 and $\mathbf{w}$ denotes a vector field.)

(i) $$\operatorname{div}\{\phi\mathbf{v}\} = \phi\operatorname{div}\mathbf{v} + \nabla\phi\,.\,\mathbf{v} \tag{B.7.28}$$

(ii) $$\operatorname{div}\{\mathbf{u}\otimes\mathbf{v}\} = (\operatorname{div}\mathbf{v})\mathbf{u} + (\nabla\mathbf{u})\mathbf{v} \tag{B.7.29}$$

(iii) $$\operatorname{div}\{\phi\mathbf{A}\} = \phi\operatorname{div}\mathbf{A} + \mathbf{A}\nabla\phi \tag{B.7.30}$$

(iv) $$\operatorname{div}\{\mathbf{A}^T\mathbf{v}\} = (\operatorname{div}\mathbf{A})\,.\,\mathbf{v} + \mathbf{A}\cdot\nabla\mathbf{v} \tag{B.7.31}$$

(v) $$\operatorname{div}\{\mathbf{u}\otimes\mathbf{v}\otimes\mathbf{w}\} = (\operatorname{div}\mathbf{w})\mathbf{u}\otimes\mathbf{v} + (\nabla\mathbf{u})\mathbf{w}\otimes\mathbf{v} + \mathbf{u}\otimes(\nabla\mathbf{v})\mathbf{w} \tag{B.7.32}$$

(vi) $$\operatorname{div}\{\mathbf{A}\otimes\mathbf{v}\} = (\operatorname{div}\mathbf{v})\mathbf{A} + (\nabla\mathbf{A})\mathbf{v} \tag{B.7.33}$$

(vii) $$\operatorname{div}\{\phi\mathbf{C}\} = \phi\operatorname{div}\mathbf{C} + \mathbf{C}\nabla\phi \tag{B.7.34}$$

(viii) $$\operatorname{div}\{\mathbf{C}^T\mathbf{v}\} = (\operatorname{div}\mathbf{C})\mathbf{v} + \mathbf{C} : (\nabla\mathbf{v})^T \tag{B.7.35}$$

(ix) $$\operatorname{div}\{\mathbf{C}^{\sim} : \mathbf{A}\} = (\operatorname{div}\mathbf{C})\cdot\mathbf{A} + \mathbf{C}\cdot\nabla\mathbf{A} \tag{B.7.36}$$

Proofs.

(i) Take the trace of (B.4.67) and note that (A.12.9) yields (B.7.28).

(ii) From (B.7.8) with $\mathbf{T} = \mathbf{u} \otimes \mathbf{v}$, using (A.8.10), (B.7.28) with $\phi = \mathbf{u}.\mathbf{k}$, and (B.4.68),

$$(\operatorname{div}\{\mathbf{u} \otimes \mathbf{v}\}).\mathbf{k} = \operatorname{div}\{(\mathbf{v} \otimes \mathbf{u})\mathbf{k}\} = \operatorname{div}\{(\mathbf{u}.\mathbf{k})\mathbf{v}\} = (\mathbf{u}.\mathbf{k})\operatorname{div}\mathbf{v} + (\nabla\mathbf{u})^T\mathbf{k}.\mathbf{v}$$
$$= ((\operatorname{div}\mathbf{v})\mathbf{u} + (\nabla\mathbf{u})\mathbf{v}).\mathbf{k},$$

so the result follows from the arbitrary nature of $\mathbf{k}$ (see Exercise A.5.1).

(iii) From (B.7.8) and (B.7.28) with $\mathbf{v} = \mathbf{A}^T k$,

$$(\operatorname{div}\{\phi\mathbf{A}\}).\mathbf{k} = \operatorname{div}\{\phi\mathbf{A}^T\mathbf{k}\} = \phi\operatorname{div}\{\mathbf{A}^T\mathbf{k}\} + \nabla\phi.\mathbf{A}^T\mathbf{k}$$
$$= (\phi\operatorname{div}\mathbf{A} + \mathbf{A}\nabla\phi).\mathbf{k},$$

whence (B.7.30).

(iv) With $\mathbf{v}$ replaced by $\mathbf{A}^T\mathbf{v}$ (B.7.1) yields, via (B.4.71) with $\mathbf{A}$ replaced by $\mathbf{A}^T$ and $\mathbf{u}$ by $\mathbf{k}$,

$$\operatorname{div}\{\mathbf{A}^T\mathbf{v}\} = \operatorname{tr}\{\nabla(\mathbf{A}^T\mathbf{v})\} = \operatorname{tr}\{\mathbf{A}^T\nabla\mathbf{v} + (\nabla(\mathbf{A}^T))^T\mathbf{v}\}$$
$$= \mathbf{A}\cdot\nabla\mathbf{v} + (\operatorname{div}\mathbf{A}).\mathbf{v}, \qquad \text{(B.7.37)}$$

on recalling (A.14.9) and noting (B.7.37) with $\mathbf{v} = \mathbf{k}$ yields, for any vector $\mathbf{k}, (\operatorname{div}\mathbf{A}).\mathbf{k} = \operatorname{tr}\{(\nabla(\mathbf{A}^T))^T\mathbf{k}\}$. It follows that $(\operatorname{div}\mathbf{A}).\mathbf{v} = \operatorname{tr}\{(\nabla(\mathbf{A}^T))^T\mathbf{v}\}$ since no differentiation of $\mathbf{v}$ is involved.

(v) From (B.7.16) with $\mathbf{C} = \mathbf{u}\otimes\mathbf{v}\otimes\mathbf{w}$ and noting (A.19.15),

$$(\operatorname{div}\{\mathbf{u}\otimes\mathbf{v}\otimes\mathbf{w}\})\mathbf{k} = \operatorname{div}\{(\mathbf{u}\otimes\mathbf{w}\otimes\mathbf{v})\mathbf{k}\} = \operatorname{div}\{(\mathbf{v}.\mathbf{k})\mathbf{u}\otimes\mathbf{w}\}$$
$$= (\mathbf{v}.\mathbf{k})\operatorname{div}\{\mathbf{u}\otimes\mathbf{w}\} + (\mathbf{u}\otimes\mathbf{w})\nabla(\mathbf{v}.\mathbf{k})$$
$$= (((\operatorname{div}\mathbf{w})\mathbf{u} + (\nabla\mathbf{u})\mathbf{w})\otimes\mathbf{v})\mathbf{k} + (\mathbf{w}.(\nabla\mathbf{v})^T\mathbf{k})\mathbf{u}$$
$$= ((\operatorname{div}\mathbf{w})\mathbf{u}\otimes\mathbf{v} + (\nabla\mathbf{u})\mathbf{w}\otimes\mathbf{v} + \mathbf{u}\otimes(\nabla\mathbf{v})\mathbf{w})\mathbf{k},$$

whence (B.7.32). Here use has been made of (B.7.30) with $\phi = \mathbf{v}.\mathbf{k}$ and $\mathbf{A} = \mathbf{u}\otimes\mathbf{w}$, together with (B.7.29).

(vi) From (B.7.16) with $\mathbf{C} = \mathbf{A}\otimes\mathbf{v}$, recalling (A.19.42) with $\mathbf{L} = \mathbf{A}$, $\mathbf{a} = \mathbf{v}$, and $\mathbf{b} = \mathbf{k}$,

$$(\operatorname{div}\{\mathbf{A}\otimes\mathbf{v}\})\mathbf{k} = \operatorname{div}\{(\mathbf{A}\otimes\mathbf{v})^T\mathbf{k}\} = \operatorname{div}\{\mathbf{A}\mathbf{k}\otimes\mathbf{v}\} = (\operatorname{div}\mathbf{v})\mathbf{A}\mathbf{k} + (\nabla\{\mathbf{A}\mathbf{k}\})\mathbf{v}$$
$$= ((\operatorname{div}\mathbf{v})\mathbf{A} + (\nabla\mathbf{A})\mathbf{v})\mathbf{k},$$

whence (B.7.33). Here (B.7.29) has been used, with $\mathbf{u} = \mathbf{A}\mathbf{k}$, and from (B.4.71) with $\mathbf{u} = \mathbf{k}$ it has been noted that $(\nabla\{\mathbf{A}\mathbf{k})\mathbf{v} = ((\nabla\mathbf{A})^T\mathbf{k})\mathbf{v} = ((\nabla\mathbf{A})\mathbf{v})\mathbf{k}$.

(vii) From (B.7.16) and (B.7.30) with $\mathbf{A} = \mathbf{C}^T\mathbf{k}$,

$$(\operatorname{div}\{\phi\mathbf{C}\})\mathbf{k} = \operatorname{div}\{(\phi\mathbf{C})^T\mathbf{k}\} = \operatorname{div}\{\phi(\mathbf{C}^T\mathbf{k})\}$$
$$= \phi\operatorname{div}\{\mathbf{C}^T\mathbf{k}\} + (\mathbf{C}^T\mathbf{k})\nabla\phi = (\phi\operatorname{div}\mathbf{C} + \mathbf{C}\nabla\phi)\mathbf{k},$$

whence (B.7.34).

(viii) From (B.7.8) with $\mathbf{T} = \mathbf{C}^T\mathbf{v}$, and (A.19.44) and (B.7.31) with $\mathbf{A} = (\mathbf{C}^{\sim}\mathbf{k})^T$,

$$(\text{div}\{\mathbf{C}^T\mathbf{v}\}).\mathbf{k} = \text{div}\{(\mathbf{C}^T\mathbf{v})^T\mathbf{k}\} = \text{div}\{((\mathbf{C}^{\sim})^T\mathbf{v})\mathbf{k}\} = \text{div}\{(\mathbf{C}^{\sim}\mathbf{k})\mathbf{v}\}$$
$$= (\text{div}\{(\mathbf{C}^{\sim}\mathbf{k})^T\}).\mathbf{v} + (\mathbf{C}^{\sim}\mathbf{k})^T \cdot \nabla\mathbf{v}.$$

However, for any $\boldsymbol{\ell} \in \mathcal{V}$,

$$(\text{div}\,\mathbf{C})^T\mathbf{k}.\boldsymbol{\ell} = (\text{div}\,\mathbf{C})\boldsymbol{\ell}.\mathbf{k} = (\text{div}\{(\mathbf{C}^T\boldsymbol{\ell})\}).\mathbf{k} = (\text{div}\{(\mathbf{C}^{\sim}\mathbf{k})^T\}).\boldsymbol{\ell},$$

and hence
$$\text{div}\{(\mathbf{C}^{\sim}\mathbf{k})^T\} = (\text{div}\,\mathbf{C})^T\mathbf{k} \tag{B.7.38}$$

on using (B.7.38) with $\mathbf{v} = \boldsymbol{\ell}$.

Further, from (A.19.43) with $\mathbf{w} = \mathbf{k}$ and $\mathbf{L} = (\nabla\mathbf{v})^T$,

$$(\mathbf{C}^{\sim}\mathbf{k})^T \cdot \nabla\mathbf{v} = \mathbf{C}^{\sim}\mathbf{k} \cdot (\nabla\mathbf{v})^T = (\mathbf{C} : (\nabla\mathbf{v})^T).\mathbf{k}. \tag{B.7.39}$$

The result follows from (B.7.38) and (B.7.40), on noting from (B.7.39) that

$$\text{div}\{(\mathbf{C}^{\sim}\mathbf{k})^T\}.\mathbf{v} = (\text{div}\,\mathbf{C})^T\mathbf{k}.\mathbf{v} = (\text{div}\,\mathbf{C})\mathbf{v}.\mathbf{k}.$$

(ix) It suffices to prove the result for simple tensors, say $\mathbf{C} = \mathbf{a}\otimes\mathbf{b}\otimes\mathbf{c}$ and $\mathbf{A} = \mathbf{u}\otimes\mathbf{v}$, via linearity considerations. (Convince yourself!) Here

$$\mathbf{C}^{\sim} : \mathbf{A} = (\mathbf{c}\otimes\mathbf{b}\otimes\mathbf{a}) : (\mathbf{u}\otimes\mathbf{v}) = (\mathbf{a}.\mathbf{u})(\mathbf{b}.\mathbf{v})\mathbf{c}.$$

From (i), with $\phi = (\mathbf{a}.\mathbf{u})(\mathbf{b}.\mathbf{v})$ and $\mathbf{v} = \mathbf{c}$,

$$\begin{aligned}\text{div}\{\mathbf{C}^{\sim} : \mathbf{A}\} &= (\mathbf{a}.\mathbf{u})(\mathbf{b}.\mathbf{v})\text{div}\,\mathbf{c} + \nabla\{(\mathbf{a}.\mathbf{u})(\mathbf{b}.\mathbf{v})\}.\mathbf{c}\\ &= (\mathbf{a}.\mathbf{u})(\mathbf{b}.\mathbf{v})\text{div}\,\mathbf{c} + [(\nabla\mathbf{a})^T\mathbf{u} + (\nabla\mathbf{u})^T\mathbf{a}](\mathbf{b}.\mathbf{v}).\mathbf{c}\\ &\quad + [(\nabla\mathbf{b})^T\mathbf{v} + (\nabla\mathbf{v})^T\mathbf{b}](\mathbf{a}.\mathbf{u}).\mathbf{c}.\end{aligned} \tag{B.7.40}$$

However, from (v),

$$(\text{div}\{\mathbf{a}\otimes\mathbf{b}\otimes\mathbf{c}\}) \cdot (\mathbf{u}\otimes\mathbf{v}) \tag{B.7.41}$$
$$= [(\text{div}\,\mathbf{c})\mathbf{a}\otimes\mathbf{b} + (\nabla\mathbf{a})\mathbf{c}\otimes\mathbf{b} + \mathbf{a}\otimes(\nabla\mathbf{b})\mathbf{c}] \cdot (\mathbf{u}\otimes\mathbf{v})$$
$$= (\mathbf{a}.\mathbf{u})(\mathbf{b}.\mathbf{v})\text{div}\,\mathbf{c} + ((\nabla\mathbf{a})\mathbf{c}.\mathbf{u})(\mathbf{b}.\mathbf{v}) + (\mathbf{a}.\mathbf{u})((\nabla\mathbf{b})\mathbf{c}.\mathbf{v}), \tag{B.7.42}$$

and, from (B.4.69),

$$\begin{aligned}(\mathbf{a}\otimes\mathbf{b}\otimes\mathbf{c}) \cdot \nabla\{\mathbf{u}\otimes\mathbf{v}\} &= (\mathbf{a}\otimes\mathbf{b}\otimes\mathbf{c}) \cdot [\mathbf{u}\otimes\nabla\mathbf{v} + (\nabla\mathbf{u}\otimes\mathbf{v})^T]\\ &= (\mathbf{a}.\mathbf{u})\text{tr}\{(\mathbf{c}\otimes\mathbf{b})\nabla\mathbf{v}\} + (\mathbf{a}\otimes\mathbf{c}\otimes\mathbf{b}) \cdot (\nabla\mathbf{u}\otimes\mathbf{v})\\ &= (\mathbf{a}.\mathbf{u})((\nabla\mathbf{v})^T\mathbf{b}.\mathbf{c}) + (\mathbf{b}.\mathbf{v})\text{tr}\{(\mathbf{c}\otimes\mathbf{a})\nabla\mathbf{u}\}\\ &= (\mathbf{a}.\mathbf{u})(\nabla\mathbf{v})^T\mathbf{b}.\mathbf{c} + (\mathbf{b}.\mathbf{v})((\nabla\mathbf{u})\mathbf{c}.\mathbf{a}).\end{aligned} \tag{B.7.43}$$

Comparison of identities (B.7.41), (B.7.42), and (B.7.43) yields (B.7.36).

Remark B.7.4. Proofs of (B.7.28) through (B.7.36) are generally easier if Cartesian representations are employed. This is particularly the case for (ix). Here $\mathbf{C} = C_{ijk}\,\mathbf{e}_i\otimes\mathbf{e}_j\otimes\mathbf{e}_k$ and $\mathbf{A} = A_{pq}\mathbf{e}_p\otimes\mathbf{e}_q$, so

$$\mathbf{C}^{\sim} : \mathbf{A} = (C_{ijk}\,\mathbf{e}_k\otimes\mathbf{e}_j\otimes\mathbf{e}_i) : (A_{pq}\,\mathbf{e}_p\otimes\mathbf{e}_q) = C_{ijk}A_{pq}\delta_{ip}\delta_{jq}\mathbf{e}_k = C_{ijk}A_{ij}\,\mathbf{e}_k.$$

Thus $\quad \operatorname{div}\{\mathbf{C}^{\sim} : \mathbf{A}\} = (C_{ijk}A_{ij})_{,k} = C_{ijk,k}A_{ij} + C_{ijk}A_{ij,k}$

$$= (\operatorname{div}\mathbf{C})_{ij}A_{ij} + C_{ijk}(\nabla\mathbf{A})_{ijk} = (\operatorname{div}\mathbf{C})\cdot\mathbf{A} + \mathbf{C}\cdot\nabla\mathbf{A}. \qquad \text{(B.7.44)}$$

Exercise B.7.2. Prove (i) through (viii) using Cartesian representations.

B.8 Calculations in Section 7.4

The derivations of (7.4.12), and of (7.4.21) from (7.4.20), follow from

Lemma B.8.1. If $\mathbf{a},\mathbf{b},\mathbf{c} \in \mathcal{V}$ and $\mathbf{L} \in \operatorname{Lin}\mathcal{V}$, then

(i) $$(\mathbf{c}\otimes\mathbf{b})\mathbf{L}\mathbf{a} = (\mathbf{a}\otimes\mathbf{b}\otimes\mathbf{c})^{\sim} : \mathbf{L}^T, \qquad \text{(7.4.14)}$$

(ii) $$\mathbf{a}\otimes(\mathbf{L}\mathbf{b}) = (\mathbf{a}\otimes\mathbf{b})\mathbf{L}^T, \text{ and} \qquad \text{(A.8.12)}'$$

(iii) $$\mathbf{a}\otimes\mathbf{b}\otimes(\mathbf{L}\mathbf{c}) = (\mathbf{a}\otimes\mathbf{b}\otimes\mathbf{c})\mathbf{L}^T. \qquad \text{(B.8.1)}$$

Proofs. (i) It suffices to prove the result with $\mathbf{L}$ a simple tensor. The general result follows by linearity: specifically, $\mathbf{L} \in \operatorname{Lin}\mathcal{V}$ may be expressed as a linear combination of simple tensors, and we note that if $\mathbf{L}_1,\mathbf{L}_2 \in \operatorname{Lin}\mathcal{V}$, then $(\mathbf{c}\otimes\mathbf{b})(\mathbf{L}_1+\mathbf{L}_2)\mathbf{a} = (\mathbf{c}\otimes\mathbf{b})\mathbf{L}_1\mathbf{a} + (\mathbf{c}\otimes\mathbf{b})\mathbf{L}_2\mathbf{a}$. (Prove this!) With $\mathbf{L} = \mathbf{d}\otimes\mathbf{e}$,

$$(\mathbf{c}\otimes\mathbf{b})\mathbf{L}\mathbf{a} = (\mathbf{c}\otimes\mathbf{b})(\mathbf{d}\otimes\mathbf{e})\mathbf{a} = (\mathbf{e}\,.\,\mathbf{a})(\mathbf{c}\otimes\mathbf{b})\mathbf{d} = (\mathbf{e}\,.\,\mathbf{a})(\mathbf{b}\,.\,\mathbf{d})\mathbf{c}$$
$$= (\mathbf{c}\otimes\mathbf{b}\otimes\mathbf{a}) : (\mathbf{e}\otimes\mathbf{d}) = (\mathbf{a}\otimes\mathbf{b}\otimes\mathbf{c})^{\sim} : (\mathbf{d}\otimes\mathbf{e})^T.$$

(ii) Writing $\mathbf{L}_1 = \mathbf{1}$ and $\mathbf{L}_2 = \mathbf{L}^T$ in (A.8.12) yields (A.8.12)$'$.

(iii) The result follows by noting that, for any $\mathbf{k} \in \mathcal{V}$,

$$(\mathbf{a}\otimes\mathbf{b}\otimes\mathbf{L}\mathbf{c})\mathbf{k} = (\mathbf{L}\mathbf{c}\,.\,\mathbf{k})\mathbf{a}\otimes\mathbf{b} = (\mathbf{c}\,.\,\mathbf{L}^T\mathbf{k})\mathbf{a}\otimes\mathbf{b} = ((\mathbf{a}\otimes\mathbf{b}\otimes\mathbf{c})\mathbf{L}^T)\mathbf{k}.$$

In obtaining (7.4.21), note that $2\kappa_w$ is expressed as the sum of six terms in the final equation of (7.4.20). The first is immediately labelled via definition (7.4.22). The remaining terms are sums of scalar multiples of expressions which involve $\mathbf{u} = (\mathbf{x}_i - \mathbf{x}), \mathbf{v} = \check{\mathbf{v}}_i$, and $\mathbf{L} = \mathbf{L}_w$. These five terms simplify as follows to yield the corresponding terms of (7.4.21) via definitions (7.4.23), (7.4.24), and (7.4.25).
Second term:

$$\mathbf{L}\mathbf{v}^2\mathbf{u} = \mathbf{L}(\mathbf{v}\,.\,\mathbf{v})\mathbf{u} = \mathbf{L}(\mathbf{u}\otimes\mathbf{v})\mathbf{1}\mathbf{v} = \mathbf{L}((\mathbf{v}\otimes\mathbf{v}\otimes\mathbf{u})^{\sim} : \mathbf{1}^T) \qquad \text{[via (7.4.14)]}$$
$$= \mathbf{L}((\mathbf{u}\otimes\mathbf{v}\otimes\mathbf{v}) : \mathbf{1}). \qquad \text{(B.8.2)}$$

Third term:

$$(\mathbf{v}\otimes\mathbf{v})\mathbf{L}\mathbf{u} = (\mathbf{u}\otimes\mathbf{v}\otimes\mathbf{v})^{\sim} : \mathbf{L}^T \qquad \text{[via (7.4.14)]}. \qquad \text{(B.8.3)}$$

Fourth term:

$$(\mathbf{v}\,.\,\mathbf{L}\mathbf{u})\mathbf{u} = (\mathbf{L}^T\mathbf{v}\,.\,\mathbf{u})\mathbf{u} = (\mathbf{u}\otimes\mathbf{u})\mathbf{L}^T\mathbf{v} = (\mathbf{v}\otimes\mathbf{u}\otimes\mathbf{u})^{\sim} : (\mathbf{L}^T)^T \qquad \text{(via (7.4.14))}$$
$$= (\mathbf{u}\otimes\mathbf{u}\otimes\mathbf{v}) : \mathbf{L}. \qquad \text{(B.8.4)}$$

Fifth term:

$$(\mathbf{Lu}\,.\,\mathbf{Lu})\mathbf{v} = (\mathbf{v}\otimes\mathbf{Lu})\mathbf{Lu} = ((\mathbf{v}\otimes\mathbf{u})\mathbf{L}^T)\mathbf{Lu} \qquad [\text{via (A.8.12)}'] $$
$$= (\mathbf{v}\otimes\mathbf{u})((\mathbf{L}^T\mathbf{L})\mathbf{u}) = (\mathbf{u}\otimes\mathbf{u}\otimes\mathbf{v})^{\sim} : (\mathbf{L}^T\mathbf{L})^T \qquad [\text{via (7.4.14)}]$$
$$= (\mathbf{u}\otimes\mathbf{u}\otimes\mathbf{v})^{\sim} : \mathbf{L}^T\mathbf{L}. \tag{B.8.5}$$

Sixth term:

$$(\mathbf{Lu}\,.\,\mathbf{Lu})\mathbf{u} = (\mathbf{u}\otimes\mathbf{Lu})\mathbf{Lu} = ((\mathbf{u}\otimes\mathbf{u})\mathbf{L}^T)\mathbf{Lu} \quad (\text{via (B.18.12)}')$$
$$= (\mathbf{u}\otimes\mathbf{u})(\mathbf{L}^T\mathbf{L})\mathbf{u} = (\mathbf{u}\otimes\mathbf{u}\otimes\mathbf{u})^{\sim} : (\mathbf{L}^T\mathbf{L})^T \qquad [\text{via (7.4.14)}]$$
$$= (\mathbf{u}\otimes\mathbf{u}\otimes\mathbf{u}) : \mathbf{L}^T\mathbf{L}. \tag{B.8.6}$$

Similarly, (7.4.31) follows from the final equation of (7.4.30) and definitions. The individual manipulations are now listed.

Second term:

$$\mathbf{u}\otimes\mathbf{v}\otimes\mathbf{Lu} = (\mathbf{u}\otimes\mathbf{v}\otimes\mathbf{u})\mathbf{L}^T \qquad [\text{via (B.8.1)}]$$
$$= (\mathbf{u}\otimes\mathbf{u}\otimes\mathbf{v})^T\mathbf{L}^T. \tag{B.8.7}$$

Third term:
$$\mathbf{u}\otimes\mathbf{v}_w\otimes\mathbf{v} = (\mathbf{u}\otimes\mathbf{v}\otimes\mathbf{v}_w)^T. \tag{B.8.8}$$

Fourth term:

$$\mathbf{u}\otimes\mathbf{v}_w\otimes\mathbf{Lu} = (\mathbf{u}\otimes\mathbf{v}_w\otimes\mathbf{u})\mathbf{L}^T \qquad [\text{via (B.8.1)}]$$
$$= (\mathbf{u}\otimes\mathbf{u}\otimes\mathbf{v}_w)^T\mathbf{L}^T. \tag{B.8.9}$$

Fifth term:

$$\mathbf{u}\otimes\mathbf{Lu}\otimes\mathbf{v} = (\mathbf{u}\otimes\mathbf{v}\otimes\mathbf{Lu})^T = ((\mathbf{u}\otimes\mathbf{v}\otimes\mathbf{u})\mathbf{L}^T)^T \qquad [\text{via (B.8.1)}]$$
$$= ((\mathbf{u}\otimes\mathbf{u}\otimes\mathbf{v})^T\mathbf{L}^T)^T. \tag{B.8.10}$$

Sixth term:

$$\mathbf{u}\otimes\mathbf{Lu}\otimes\mathbf{Lu} = (\mathbf{u}\otimes\mathbf{Lu}\otimes\mathbf{u})\mathbf{L}^T \qquad [\text{via (B.8.1)}]$$
$$= (\mathbf{u}\otimes\mathbf{u}\otimes\mathbf{Lu})^T\mathbf{L}^T = ((\mathbf{u}\otimes\mathbf{u}\otimes\mathbf{u})\mathbf{L}^T)^T\mathbf{L}^T \qquad [\text{via (B.8.1)}].$$

B.9 Proof of Results 10.5.1

In respect of (10.5.18) we have

$$\langle\nabla\mathbf{u}\rangle(\mathbf{x}) := \int_{\mathcal{E}} \nabla\mathbf{u}(\mathbf{y})w(\mathbf{y}-\mathbf{x})dV_{\mathbf{y}} = \int_R \nabla\mathbf{u}(\mathbf{y})w(\mathbf{y}-\mathbf{x})dV_{\mathbf{y}}$$
$$= \int_R \nabla_{\mathbf{y}}\{w\mathbf{u}\} - \mathbf{u}\otimes\nabla_{\mathbf{y}}\,w\,dV_{\mathbf{y}}, \tag{B.9.1}$$

via (B.4.67) with $\phi = w$ and $\mathbf{v} = \mathbf{u}$. To simplify the first integral, note that for any fixed vector $\mathbf{k}$

$$\left(\int_R \nabla_{\mathbf{y}}\{w\mathbf{u}\}dV_{\mathbf{y}}\right)\mathbf{k} = \int_R \nabla_{\mathbf{y}}\{w\mathbf{u}\}\mathbf{k}\,dV_{\mathbf{y}}$$

$$= \int_R \mathrm{div}_{\mathbf{y}}\{w\mathbf{u}\otimes\mathbf{k}\}dV_{\mathbf{y}} = \int_{\partial R}(w\mathbf{u}\otimes\mathbf{k})\mathbf{n}\,dA_{\mathbf{y}}$$
$$= \int_{S(\mathbf{x})}(w\mathbf{u}\otimes\mathbf{n})\mathbf{k}\,dA_{\mathbf{y}} = \left(\int_{S(\mathbf{x})} w\mathbf{u}\otimes\mathbf{n}\,dA_{\mathbf{y}}\right)\mathbf{k}. \quad \text{(B.9.2)}$$

Here (B.7.29) with $\mathbf{u} = w\mathbf{u}, \mathbf{v} = \mathbf{k}$ has been invoked together with the divergence theorem (B.7.5) with $\mathbf{T} = w\mathbf{u}\otimes\mathbf{k}$, and $S(\mathbf{x})$ is given by (10.5.14): the surface integral over ∂R is restricted to $S(\mathbf{x})$ by the presence of factor $w(\mathbf{y}-\mathbf{x})$ in the integrand. Noting that $w = V^{-1}$ on $S(\mathbf{x})$ [see (10.5.11)], and noting the arbitrary nature of $\mathbf{k}$, (B.9.2) implies that

$$\int_R \nabla_{\mathbf{y}}\{w\mathbf{u}\}dV_{\mathbf{y}} = \frac{1}{V}\int_S \mathbf{u}\otimes\mathbf{n}\,dA_{\mathbf{y}}. \quad \text{(B.9.3)}$$

The last term in (B.9.1) is

$$-\int_R \mathbf{u}(\mathbf{y})\otimes\nabla_{\mathbf{y}}\{w(\mathbf{y}-\mathbf{x})\}dV_{\mathbf{y}} = +\int_R \mathbf{u}(\mathbf{y})\otimes\nabla_{\mathbf{x}}\{w(\mathbf{y}-\mathbf{x})\}dV_{\mathbf{y}}$$
$$= \int_R \nabla_{\mathbf{x}}\{w(\mathbf{y}-\mathbf{x})\mathbf{u}(\mathbf{y})\}dV_{\mathbf{y}}$$
$$= \nabla_{\mathbf{x}}\left\{\int_R w(\mathbf{y}-\mathbf{x})\mathbf{u}(\mathbf{y})dV_{\mathbf{y}}\right\} = \nabla\{\langle\mathbf{u}\rangle\}(\mathbf{x}). \quad \text{(B.9.4)}$$

From (B.9.1), (B.9.3), and (B.9.4),

$$\langle\nabla\mathbf{u}\rangle = \nabla\{\langle\mathbf{u}\rangle\} + \frac{1}{V}\int_S \mathbf{u}\otimes\mathbf{n}\,dA, \quad \text{(B.9.5)}$$

namely result (10.5.18).

Taking the trace of relation (B.9.5), recalling (A.12.9) and (B.7.1), and noting that (here $\mathbf{A} = \nabla\mathbf{u}$)

$$\mathrm{tr}\{\langle\mathbf{A}\rangle\} = \mathrm{tr}\left\{\int_R \mathbf{A}w\,dV\right\} = \mathrm{tr}\left\{\int_R A_{ij}\mathbf{e}_i\otimes\mathbf{e}_j\,w\,dV\right\}$$
$$= \mathrm{tr}\left\{\left(\int_R A_{ij}\,w\,dV\right)(\mathbf{e}_i\otimes\mathbf{e}_j)\right\} = \int_R A_{ii}\,w\,dV = \langle\mathrm{tr}\,\mathbf{A}\rangle, \quad \text{(B.9.6)}$$

yield result (10.5.19).

To obtain (10.5.20), note that

$$\langle\mathrm{div}\,\mathbf{B}\rangle(\mathbf{x}) := \int_{\mathcal{E}}((\mathrm{div}\,\mathbf{B})(\mathbf{y}))w(\mathbf{y}-\mathbf{x})dV_{\mathbf{y}} = \int_R((\mathrm{div}\,\mathbf{B})(\mathbf{y}))w(\mathbf{y}-\mathbf{x})dV_{\mathbf{y}}$$
$$= \int_R \mathrm{div}_{\mathbf{y}}\{w\mathbf{B}\} - \mathbf{B}\nabla_{\mathbf{y}}w\,dV_{\mathbf{y}}$$
$$= \int_{\partial R} w\mathbf{B}\mathbf{n}\,dA_{\mathbf{y}} + \int_R \mathbf{B}\nabla_{\mathbf{x}}w\,dV_{\mathbf{y}}$$
$$= \int_{S(\mathbf{x})} w\mathbf{B}\mathbf{n}\,dA_{\mathbf{y}} + \int_R \mathrm{div}_{\mathbf{x}}\{\mathbf{B}w\}dV_{\mathbf{y}}$$

$$= \frac{1}{V}\int_{S(\mathbf{x})} \mathbf{B}\mathbf{n}\, dA_{\mathbf{y}} + \operatorname{div}_{\mathbf{x}}\left\{\int_{R} \mathbf{B}w\, dV_{\mathbf{y}}\right\}$$
$$= \frac{1}{V}\int_{S(\mathbf{x})} \mathbf{B}\mathbf{n}\, dA_{\mathbf{y}} + \operatorname{div}\left\{\int_{\mathcal{E}} \mathbf{B}w\, dV_{\mathbf{y}}\right\}. \tag{B.9.7}$$

Here (B.7.30) has been invoked, with $\phi = w$ and $\mathbf{A} = \mathbf{B}$, together with the divergence theorem (B.7.5), with $\mathbf{T} = w\mathbf{B}$.

The final result (10.5.21) requires care since $R(=R_t)$ depends upon time in general: there is no assumption at this point that pore structure be rigid (or, equivalently, that the porous body be undeformable). In particular,

$$\frac{\partial}{\partial t}\{\langle\mathbf{u}\rangle\}(\mathbf{x}) = \frac{\partial}{\partial t}\left\{\int_{\mathcal{E}} \mathbf{u}(\mathbf{y})w(\mathbf{y}-\mathbf{x})dV_{\mathbf{y}}\right\} = \frac{\partial}{\partial t}\left\{\int_{R_t} \mathbf{u}(\mathbf{y})w(\mathbf{y}-\mathbf{x})dV_{\mathbf{y}}\right\}. \tag{B.9.8}$$

Noting that $\mathbf{u}$ may depend upon time but that w does not, use of the transport theorem (2.5.24) in Chapter 2 yields

$$\frac{\partial}{\partial t}\left\{\int_{R_t} \mathbf{u}w\, dV\right\} = \frac{\partial}{\partial t}\left\{\int_{R_t} \rho\left(\frac{\mathbf{u}w}{\rho}\right) dV\right\} = \int_{R_t} \rho\,\dot{\overline{\left(\frac{\mathbf{u}w}{\rho}\right)}}\, dV. \tag{B.9.9}$$

Exercise B.9.1. Show that

$$\rho\,\dot{\overline{\left(\frac{\mathbf{u}w}{\rho}\right)}} = \frac{\partial}{\partial t}\{\mathbf{u}w\} + (\nabla\{\mathbf{u}w\})\mathbf{v} + (\operatorname{div}\mathbf{v})\mathbf{u}w \tag{B.9.10}$$

by recalling (2.5.28) of Chapter 2 and the continuity equation for the fluid.

Noting that $w(\mathbf{y}-\mathbf{x})$ does not depend upon time,

$$\frac{\partial}{\partial t}\{\mathbf{u}w\} + \nabla\{\mathbf{u}w\}\mathbf{v} + (\operatorname{div}\mathbf{v})\mathbf{u}w = \frac{\partial \mathbf{u}}{\partial t}w + \operatorname{div}\{\mathbf{u}w \otimes \mathbf{v}\} \tag{B.9.11}$$

on use of (B.7.29). From (B.9.8), (B.9.9), (B.9.10), and (B.9.11),

$$\frac{\partial}{\partial t}\{\langle\mathbf{u}\rangle\} = \int_{R_t} \frac{\partial \mathbf{u}}{\partial t} w + \operatorname{div}\{\mathbf{u}w \otimes \mathbf{v}\} dV$$
$$= \int_{\mathcal{E}} \frac{\partial \mathbf{u}}{\partial t} w\, dV + \int_{S} (\mathbf{u}w \otimes \mathbf{v})\mathbf{n}\, dA$$
$$= \langle\frac{\partial \mathbf{u}}{\partial t}\rangle + \frac{1}{V}\int_{S} (\mathbf{u} \otimes \mathbf{v})\mathbf{n}\, dA. \tag{B.9.12}$$

B.10 Derivatives of Objective Fields

If ϕ denotes an objective *scalar* field (i.e., a real-valued field associated with inertial observers) so that [see (12.2.69)]

$$\phi^*(\mathbf{x}^*) = \phi(\mathbf{x}), \tag{B.10.1}$$

then, for any displacement $\mathbf{h}$ for O [witnessed as $\mathbf{h}^* = \mathbf{Q}_0\mathbf{h}$ by O^*: see (12.2.40)],

$$\phi^*(\mathbf{x}^* + \mathbf{h}^*) = \phi(\mathbf{x} + \mathbf{h}). \tag{B.10.2}$$

Thus (see Subsection B.4.1)

$$\phi^*(\mathbf{x}^*) + \nabla^*\phi^*(\mathbf{x}^*) .^* \mathbf{h}^* + o(\mathbf{h}^*) = \phi(\mathbf{x}) + \nabla\phi(\mathbf{x}) . \mathbf{h} + o(\mathbf{h}). \tag{B.10.3}$$

Accordingly, noting that $\|\mathbf{h}^*\| = \|\mathbf{h}\|$ (Why?),

$$\nabla^*\phi^*(\mathbf{x}^*) .^* \mathbf{Q}_0\mathbf{h} = \nabla\phi(\mathbf{x}) . \mathbf{h} + o(\mathbf{h}). \tag{B.10.4}$$

Writing $\mathbf{h} = s\hat{\mathbf{u}}$ where $\|\hat{\mathbf{u}}\| = 1$, dividing by s, and letting $s \to 0$, it follows that

$$\nabla^*\phi^*(\mathbf{x}^*) .^* \mathbf{Q}_0\hat{\mathbf{u}} = \nabla\phi(\mathbf{x}) . \hat{\mathbf{u}} \tag{B.10.5}$$

for all unit vectors in $\mathcal{V}$, and hence *all* vectors in $\mathcal{V}$ (Why?). Hence (Show this!)

$$\mathbf{Q}_0^T \nabla^*\phi^*(\mathbf{x}^*) = \nabla\phi(\mathbf{x})$$

whence [recall (A.16.4)]

$$\nabla^*\phi^*(\mathbf{x}^*) = \mathbf{Q}_0\nabla\phi(\mathbf{x}). \tag{B.10.6}$$

If $\mathbf{u}$ denotes an objective vector field, so that [see (12.2.70)]

$$\mathbf{u}^*(\mathbf{x}^*) = \mathbf{Q}_0\, \mathbf{u}(\mathbf{x}), \tag{B.10.7}$$

then

$$\mathbf{u}^*(\mathbf{x}^* + \mathbf{h}^*) = \mathbf{Q}_0\, \mathbf{u}(\mathbf{x} + \mathbf{h}), \tag{B.10.8}$$

and so (see Subsection B.4.3)

$$\mathbf{u}^*(\mathbf{x}^*) + \nabla^*\mathbf{u}^*(\mathbf{x}^*)\mathbf{h}^* + o(\mathbf{h}^*) = \mathbf{Q}_0(\mathbf{u}(\mathbf{x}) + \nabla\mathbf{u}(\mathbf{x})\mathbf{h} + o(\mathbf{h})). \tag{B.10.9}$$

Exercise B.10.1. Show from (B.10.9) and (B.10.7) that the argument employed in obtaining (B.10.5) yields

$$\nabla^*\mathbf{u}^*(\mathbf{x}^*)\mathbf{Q}_0\hat{\mathbf{u}} = \mathbf{Q}_0\, \nabla\mathbf{u}(\mathbf{x})\hat{\mathbf{u}} \tag{B.10.10}$$

for any unit vector $\hat{\mathbf{u}} \in \mathcal{V}$ and hence *all* vectors in $\mathcal{V}$. Deduce that

$$\nabla^*\mathbf{u}^*(\mathbf{x}^*) = \mathbf{Q}_0\nabla\mathbf{u}(\mathbf{x})\mathbf{Q}_0^T. \tag{B.10.11}$$

It follows from (B.7.1), on taking the trace of relation (B.10.11), that [note (A.12.14)]

$$(\mathrm{div}^*\mathbf{u}^*)(\mathbf{x}^*) = \mathrm{tr}\{\mathbf{Q}_0\nabla\mathbf{u}(\mathbf{x})\mathbf{Q}_0^T\} = \mathrm{tr}\{\mathbf{Q}_0^T\mathbf{Q}_0\nabla\mathbf{u}(\mathbf{x})\} = \mathrm{tr}\{\nabla\mathbf{u}(\mathbf{x})\}. \tag{B.10.12}$$

That is,

$$(\mathrm{div}^*\mathbf{u}^*)(\mathbf{x}^*) = (\mathrm{div}\,\mathbf{u})(\mathbf{x}). \tag{B.10.13}$$

If $\mathbf{A}$ denotes an objective linear transformation field, so that [see (12.2.71)]

$$\mathbf{A}^*(\mathbf{x}^*) = \mathbf{Q}_0\, \mathbf{A}(\mathbf{x})\mathbf{Q}_0^T, \tag{B.10.14}$$

then, for any $\mathbf{k} \in \mathcal{V}$ for O, regarded as $\mathbf{k}^* = \mathbf{Q}_0\mathbf{k}$ by O^*, and suppressing arguments $\mathbf{x}^*$ and $\mathbf{x}$,

$$(\mathrm{div}^*\{\mathbf{A}^*\}) .^* \mathbf{k}^* = \mathrm{div}^*\{(\mathbf{A}^*)^T\mathbf{k}^*\} = \mathrm{div}^*\{\mathbf{u}^*\}. \tag{B.10.15}$$

Here $\quad \mathbf{u}^* := (\mathbf{A}^*)^T\mathbf{k}^* = (\mathbf{Q}_0\mathbf{A}\mathbf{Q}_0^T)^T\mathbf{Q}_0\mathbf{k}$

$$= ((\mathbf{Q}_0^T)^T\mathbf{A}^T\mathbf{Q}_0^T)\mathbf{Q}_0\mathbf{k} = \mathbf{Q}_0(\mathbf{A}^T\mathbf{k}) =: \mathbf{Q}_0\,\mathbf{u}, \tag{B.10.16}$$

where

$$\mathbf{u} := \mathbf{A}^T\mathbf{k}. \tag{B.10.17}$$

Thus $\mathbf{u}$ is an objective vector field and so, from (B.10.13),

$$\text{div}^*\{(\mathbf{A}^*)^T\mathbf{k}^*\} = \text{div}\{\mathbf{A}^T\mathbf{k}\}, \tag{B.10.18}$$

and hence

$$(\text{div}^*\{\mathbf{A}^*\}).^*\mathbf{k}^* = (\text{div}\,\mathbf{A}).\mathbf{k}. \tag{B.10.19}$$

Accordingly, for any $\mathbf{k} \in \mathcal{V}$,

$$(\text{div}\,\mathbf{A}).\mathbf{k} = (\text{div}^*\{\mathbf{A}^*\}).^*\,\mathbf{Q}_0\,\mathbf{k} = \mathbf{Q}_0^T\text{div}^*\{\mathbf{A}^*\}.\mathbf{k}. \tag{B.10.20}$$

Thus

$$\text{div}\,\mathbf{A} = \mathbf{Q}_0^T(\text{div}^*\{\mathbf{A}^*\}), \tag{B.10.21}$$

and hence

$$\text{div}^*\{\mathbf{A}^*\} = \mathbf{Q}_0(\text{div}\,\mathbf{A}). \tag{B.10.22}$$

Summarising, *if* ϕ, $\mathbf{u}$, *and* $\mathbf{A}$ *are objective fields, then so too are* $\nabla\phi, \nabla\mathbf{u}, \text{div}\,\mathbf{u}$, *and* $\text{div}\,\mathbf{A}$.

B.11 Calculus in Phase Space $\mathbb{P}$ When Identified with $\mathbb{R}^{6N}$

B.11.1 Basic Concepts

The *displacement* from point $\mathbf{X}$ to point $\mathbf{X}'$ in $\mathbb{P}$ is

$$\mathbf{X}' - \mathbf{X} := (\mathbf{x}_1' - \mathbf{x}_1, \ldots, \mathbf{x}_N' - \mathbf{x}_N; \mathbf{p}_1' - \mathbf{p}_1, \ldots, \mathbf{p}_N' - \mathbf{p}_N), \tag{B.11.1}$$

where $\mathbf{X}$ and $\mathbf{X}'$ are the ordered lists $(\mathbf{x}_1, \ldots, \mathbf{p}_N)$ and $(\mathbf{x}_1', \ldots, \mathbf{p}_N')$, respectively [see (14.2.1)]. Accordingly, from (14.2.2) and (14.2.3),

$$\mathbf{X}' - \mathbf{X} \in \mathcal{V}^{2N}, \tag{B.11.2}$$

while individually $\mathbf{X}$ and $\mathbf{X}'$ lie in $\mathcal{E}^N \times \mathcal{V}^N$. If a Cartesian co-ordinate system is selected for $\mathcal{E}$, then the distinction between points and displacements is lost: $\mathbf{X}, \mathbf{X}'$ and $\mathbf{X}' - \mathbf{X}$ are all identified with elements of[10] $\mathbb{R}^{6N}$ [see (14.3.4)]. Of course, $\mathbb{R}^{6N}$ has the natural structure of an inner-product space, with distinguished orthonormal basis

$$\mathbf{E}_1 := (1,0,0,\ldots,0), \qquad \mathbf{E}_2 := (0,1,0,\ldots,0), \ldots, \qquad \mathbf{E}_{6N} := (0,0,\ldots,0,1). \tag{B.11.3}$$

If (here summation over repeated indices is intended, from 1 to $6N$)

$$\mathbf{U} = U_p\mathbf{E}_p \qquad \text{and} \qquad \mathbf{U}' = U_q'\mathbf{E}_q, \qquad \text{then} \qquad \mathbf{U}.\mathbf{U}' := U_pU_p'. \tag{B.11.4}$$

[10] Points are identified with their displacements from the distinguished element $(0,0,\ldots,0)$ of $\mathbb{R}^{6N}$, precisely as is the case for $\mathcal{E}$ when a Cartesian co-ordinate system is adopted.

The norm $$\|\mathbf{U}\|_{\mathbb{P}} := (\mathbf{U}.\mathbf{U})^{1/2} = (U_1^2 + U_2^2 + \cdots + U_{6N}^2)^{1/2}, \tag{B.11.5}$$

and the *distance* between $\mathbf{X}$ and $\mathbf{X}'$ is

$$d_{\mathbb{P}}(\mathbf{X},\mathbf{X}') := \|\mathbf{X}' - \mathbf{X}\|_{\mathbb{P}} = ((X_1' - X_1)^2 + \cdots + (X_{6N}' - X_{6N})^2)^{1/2}. \tag{B.11.6}$$

A generalised 'rectangular box' with 'edges' parallel to $\mathbf{E}_1,\ldots,\mathbf{E}_{6N}$ and edge 'lengths' $\Delta X_1,\ldots,\Delta X_{6N}$ (here $\Delta X_p > 0$, $p = 1,2,\ldots,6N$) is said to have a *phase-space volume*

$$\Delta V := \prod_{p=1}^{6N} \Delta X_p. \tag{B.11.7}$$

A unique alternating $6N$-linear form $\boldsymbol{\omega}_{\mathbb{P}}$ on $\mathbb{P}$ (identified with $\mathbb{R}^{6N}$) is defined by

$$\boldsymbol{\omega}_{\mathbb{P}}(\mathbf{E}_1,\mathbf{E}_2,\ldots,\mathbf{E}_{6N}) = 1. \tag{B.11.8}$$

Thus $$\boldsymbol{\omega}_{\mathbb{P}} : \mathbb{R}^{6N} \to \mathbb{R}, \tag{B.11.9}$$

where $\boldsymbol{\omega}_{\mathbb{P}}$ is linear in each of its $6N$ arguments, and interchange of any pair of arguments changes its value by a factor -1.

Exercise B.11.1. Convince yourself that $\boldsymbol{\omega}_{\mathbb{P}}$ is uniquely determined by (B.11.8). (This is a direct generalisation of ATF 1 in Appendix A.10.)

Notice that displacements along the edges of the box considered earlier are $\Delta X_p \mathbf{E}_p (p = 1,2,\ldots,6N)$ and that

$$\begin{aligned}\boldsymbol{\omega}_{\mathbb{P}}(\Delta X_1\mathbf{E}_1, \Delta X_2\mathbf{E}_2,\ldots,\Delta X_{6N}\mathbf{E}_{6N}) &= \left(\prod_{p=1}^{6N} \Delta X_p\right) \boldsymbol{\omega}_{\mathbb{P}}(\mathbf{E}_1,\mathbf{E}_2,\ldots,\mathbf{E}_{6N}) \\ &= \Delta V. \end{aligned} \tag{B.11.10}$$

For this reason $\boldsymbol{\omega}_{\mathbb{P}}$ is termed a *volume form* for $\mathbb{P}$. There is a natural generalisation of the notion of a parallelipiped and its associated volume (see Appendix A.5). The *6N-dimensional parallelipiped* defined by 'vertex' $\mathbf{X}_0 \in \mathbb{R}^{6N}$, and displacements therefrom of $\mathbf{U}_1,\ldots,\mathbf{U}_{6N} \in \mathbb{R}^{6N}$, is the set

$$P(\mathbf{X}_0;\mathbf{U}_1,\ldots,\mathbf{U}_{6N}) := \{\mathbf{X} \in \mathbb{R}^{6N} : \mathbf{X} = \mathbf{X}_0 + \lambda_p \mathbf{U}_p \text{ where } 0 \le \lambda_p \le 1\} \tag{B.11.11}$$

with volume $$V(P(\mathbf{X}_0;\mathbf{U}_1,\ldots,\mathbf{U}_{6N})) := |\boldsymbol{\omega}_{\mathbb{P}}(\mathbf{U}_1,\ldots,\mathbf{U}_{6N})|. \tag{B.11.12}$$

Exercise B.11.2. (Generalisation of ATF 3 in Appendix A.11) Show that $\boldsymbol{\omega}_{\mathbb{P}}(\mathbf{U}_1,\ldots,\mathbf{U}_{6N}) \neq 0$ if and only if $\mathbf{U}_1,\ldots,\mathbf{U}_{6N}$ are linearly independent. Note the implication for volumes of parallelipipeds.

Principal invariants of any linear transformation $\mathbf{A}$ on $\mathbb{R}^{6N}$ are defined as simple generalisations of the results of Appendix A.12 on noting the analogue of ATF 2 of Appendix A.11. If $\boldsymbol{\omega}$ denotes any non-zero $6N$-linear alternating form on $\mathbb{R}^{6N}$, then

$$\boldsymbol{\omega} = k\boldsymbol{\omega}_{\mathbb{P}} \tag{B.11.13}$$

for some unique $k \in \mathbb{R}(k \neq 0)$. In particular,

$$\omega_{\mathbf{A}}^{(1)}(\mathbf{U}_1,\ldots,\mathbf{U}_{6N}) := \omega_{\mathbb{P}}(\mathbf{A}\mathbf{U}_1,\mathbf{U}_2,\ldots,\mathbf{U}_{6N}) + \omega_{\mathbb{P}}(\mathbf{U}_1,\mathbf{A}\mathbf{U}_2,\mathbf{U}_3,\ldots,\mathbf{U}_{6N}) + \ldots + \omega_{\mathbb{P}}(\mathbf{U}_1,\ldots,\mathbf{U}_{6N-1},\mathbf{A}\mathbf{U}_{6N}) \tag{B.11.14}$$

and

$$\omega_{\mathbf{A}}^{(6N)}(\mathbf{U}_1,\mathbf{U}_2,\ldots,\mathbf{U}_{6N}) := \omega_{\mathbb{P}}(\mathbf{A}\mathbf{U}_1,\mathbf{A}\mathbf{U}_2,\ldots,\mathbf{A}\mathbf{U}_{6N}) \tag{B.11.15}$$

define $6N$-linear alternating forms on $\mathbb{R}^{6N}$. Accordingly, from (B.11.13),

$$\omega_{\mathbf{A}}^{(1)} = k^{(1)}\omega_{\mathbb{P}} \quad \text{and} \quad \omega_{\mathbf{A}}^{(6N)} = k^{(6N)}\omega_{\mathbb{P}} \tag{B.11.16}$$

for unique $k^{(1)}$ and $k^{(6N)} \in \mathbb{R}$. These are, respectively, the *trace* $\operatorname{tr}\mathbf{A}$ and *determinant* $\det\mathbf{A}$ of $\mathbf{A}$. Standard properties follow by direct generalisation of (A.12.1) and (A.12.3). In particular, $\mathbf{A}$ is characterised by the $6N \times 6N$ matrix $[A_{pq}]$, where [cf. (A.9.1)]

$$A_{pq} := \mathbf{E}_p \,.\, \mathbf{A}\mathbf{E}_q. \tag{B.11.17}$$

Further,

$$\operatorname{tr}\mathbf{A} = \sum_{p=1}^{6N} A_{pp} = A_{pp}. \tag{B.11.18}$$

B.11.2 Deformations and Differential Calculus in $\mathbb{R}^{6N}$

A *deformation* $\mathbf{d}$ on a subset $\mathbb{D}_0 \subset \mathbb{R}^{6N}$ is a bijective map

$$\mathbf{d} : \mathbb{D}_0 \to \mathbb{R}^{6N}. \tag{B.11.19}$$

A deformation of form

$$\mathbf{d}(\mathbf{X}) = \mathbf{Y}_0 + \mathbf{A}(\mathbf{X} - \mathbf{X}_0), \tag{B.11.20}$$

where $\mathbf{A}$ is an invertible linear map on $\mathbb{R}^{6N}$ and $\mathbf{X}_0, \mathbf{Y}_0 \in \mathbb{R}^{6N}$, is said to be *homogeneous*. Homogeneous deformations map (generalised) parallelipipeds into parallelipipeds, and the associated volume magnification factor is given by the modulus of the determinant of the associated linear map.

Exercise B.11.3. Show that homogeneous deformation $\mathbf{d}$ of (B.11.20) maps $P(\mathbf{X}_1;\mathbf{U}_1,\ldots,\mathbf{U}_{6N})$ into $P(\mathbf{Y}_0;\mathbf{A}(\mathbf{X}_1 - \mathbf{X}_0);\mathbf{A}\mathbf{U}_1,\ldots,\mathbf{A}\mathbf{U}_{6N})$. Noting that the volume of $P(\mathbf{X}_1;\mathbf{U}_1,\ldots,\mathbf{U}_{6N})$ is given by (B.11.12), deduce that the *volume magnification factor* associated with $\mathbf{d}$ is $|\det\mathbf{A}|$.

In visualising calculations in $\mathbb{R}^{6N}$ it is helpful to note the following simple generalisations of *line* and *plane*. Given a *direction* in $\mathbb{R}^{6N}$ defined by $\mathbf{N} \in \mathbb{R}^{6N}$ $(\mathbf{N} \neq \mathbf{0})$, the *line through* $\mathbf{X}_1 \in \mathbb{R}^{6N}$ *parallel to* $\mathbf{N}$ is

$$\operatorname{line}(\mathbf{X}_1;\mathbf{N}) := \{\mathbf{X} \in \mathbb{R}^{6N} : \mathbf{X} - \mathbf{X}_1 = \lambda\mathbf{N}, \lambda \in \mathbb{R}\}, \tag{B.11.21}$$

and the *plane through* $\mathbf{X}_1$ *orthogonal to* $\mathbf{N}$ is

$$\operatorname{plane}(\mathbf{X}_1;\mathbf{N}) := \{\mathbf{X} \in \mathbb{R}^{6N} : (\mathbf{X} - \mathbf{X}_1)\,.\,\mathbf{N} = 0\}. \tag{B.11.22}$$

Exercise B.11.4. Convince yourself that the dimension associated with line (B.11.21) (the dimension of the space of displacements 'along' the line) is 1, while plane

(B.11.22) has dimension $6N-1$ [in the sense that displacements in the plane form a $(6N-1)$-dimensional space].

The notion of derivative discussed in Appendix B.4 readily generalises to fields on open[11] subsets $\mathbb{D}$ of $\mathbb{P}$. Thus if

$$\phi : \mathbb{D} \subset \mathbb{P} \to \mathbb{R} \tag{B.11.23}$$

is a scalar field, then its derivative at $\mathbf{X} \in \mathbb{D}$ (if it exists) is that vector $\nabla_{\mathbb{P}}\phi(\mathbf{X})$ for which (here $\mathbf{H}$ is a displacement in $\mathbb{P}$)

$$\phi(\mathbf{X}+\mathbf{H}) = \phi(\mathbf{X}) + (\nabla_{\mathbb{P}}\phi(\mathbf{X})).\mathbf{H} + o(\mathbf{H}) \qquad \text{as} \quad \|\mathbf{H}\|_{\mathbb{P}} \to 0 \tag{B.11.24}$$

[cf. (B.4.9)]. With $\mathbf{H} = s\mathbf{E}_p$ it follows, on letting $s \to 0$, that

$$\nabla_{\mathbb{P}}\phi(\mathbf{X}).\mathbf{E}_p = \frac{\partial \check{\phi}}{\partial X_p}(X_1,\ldots,X_{6N}), \tag{B.11.25}$$

where
$$\check{\phi}(X_1,\ldots,X_{6N}) := \phi(\mathbf{X}). \tag{B.11.26}$$

Accordingly
$$\nabla_{\mathbb{P}}\phi(\mathbf{X}) = \check{\phi}_{,p}(X_1,\ldots,X_{6N})\mathbf{E}_p. \tag{B.11.27}$$

Exercise B.11.5. Prove (B.11.25) and (B.11.27) by generalising the discussion of Appendix B.4.

The derivative of a deformation

$$\mathbf{d} : \mathbb{D} \to \mathbb{P} \tag{B.11.28}$$

at $\mathbf{X} \in \mathbb{D}$ (if it exists) is that linear transformation $\nabla_{\mathbb{P}}\mathbf{d}(\mathbf{X})$ on $\mathbb{R}^{6N}$ for which

$$\mathbf{d}(\mathbf{X}+\mathbf{H}) = \mathbf{d}(\mathbf{X}) + (\nabla_{\mathbb{P}}\mathbf{d}(\mathbf{X}))\mathbf{H} + o(\mathbf{H}) \qquad \text{as} \quad \|\mathbf{H}\|_{\mathbb{P}} \to 0. \tag{B.11.29}$$

Exercise B.11.6. Show that

$$(\nabla_{\mathbb{P}}\mathbf{d}(\mathbf{X}))\mathbf{E}_q = \frac{\partial \check{\mathbf{d}}}{\partial X_q}(X_1,\ldots,X_{6N}), \tag{B.11.30}$$

where
$$\check{\mathbf{d}}(X_1,\ldots,X_{6N}) := \mathbf{d}(\mathbf{X}). \tag{B.11.31}$$

Writing the co-ordinate form of $\mathbf{d}$ as

$$\mathbf{d}(\mathbf{X}) =: (\check{d}_1(X_1,\ldots,X_{6N}),\ldots,\check{d}_{6N}(X_1,\ldots,X_{6N}), \tag{B.11.32}$$

show that

$$\mathbf{E}_p.(\nabla\mathbf{d}(\mathbf{X}))\mathbf{E}_q = \frac{\partial \check{d}_p}{\partial X_q}(X_1,\ldots,X_{6N}) =: \check{d}_{p,q}(X_1,\ldots,X_{6N}). \tag{B.11.33}$$

[Generalise the argument of (B.4.2)].

[11] Defined in terms of the metric structure on $\mathbb{P}$ endowed by the distance function $d_{\mathbb{P}}$ defined in (B.11.6). See Apostol [91], p. 48.

Relation (B.11.33) defines the $6N \times 6N$ matrix $[\check{d}_{p,q}]$ which represents $\nabla_{\mathbb{P}}\mathbf{d}$ [cf. (B.4.40)].

The derivative of a vector field

$$\mathbf{U} : \mathbb{D} \to \mathcal{V}^{2N} \leftrightarrow \mathbb{R}^{6N} \tag{B.11.34}$$

at $\mathbf{X} \in \mathbb{D}$ (if it exists) is that linear map $\nabla_{\mathbb{P}}\mathbf{U}(\mathbf{X})$ from $\mathcal{V}^{2N}$ into $\mathcal{V}^{2N}$ for which

$$\mathbf{U}(\mathbf{X}+\mathbf{H}) = \mathbf{U}(\mathbf{X}) + (\nabla_{\mathbb{P}}\mathbf{U}(\mathbf{X}))\mathbf{H} + o(\mathbf{H}) \qquad \text{as} \quad \|\mathbf{H}\|_{\mathbb{P}} \to 0. \tag{B.11.35}$$

Given the identification of both $\mathbb{P}$ and $\mathcal{V}^{2N}$ with $\mathbb{R}^{6N}$, this is formally the same as (B.11.29). Thus, from Exercise B.11.6,

$$(\nabla_{\mathbb{P}}\mathbf{U}(\mathbf{X}))\mathbf{E}_q = \check{\mathbf{U}}_{,q}(X_1,\ldots,X_{6N}) \tag{B.11.36}$$

and $\nabla_{\mathbb{P}}\mathbf{U}(\mathbf{X})$ has the matrix representation

$$[(\nabla\mathbf{U}(\mathbf{X}))_{pq}] = \mathbf{E}_p \,.\, (\nabla\mathbf{U}(\mathbf{X}))\mathbf{E}_q = \check{U}_{p,q}(X_1,\ldots,X_{6N}). \tag{B.11.37}$$

The divergence of a vector field $\mathbf{U}$ is

$$\mathrm{div}_{\mathbb{P}}\mathbf{U} := \mathrm{tr}\{\nabla_{\mathbb{P}}\mathbf{U}\}. \tag{B.11.38}$$

From (B.11.18) and (B.11.37)

$$\mathrm{div}_{\mathbb{P}}\,\mathbf{U} = \check{U}_{p,p}. \tag{B.11.39}$$

If $\mathbf{A}$ denotes an invertible linear transformation field on $\mathbb{D}$, then $\det\mathbf{A}$ is a non-vanishing scalar field on $\mathbb{D}$. Straightforward generalisation of (B.4.7) yields

$$(\nabla_{\mathbb{P}}(\det))(\mathbf{A}) = (\det\mathbf{A})\mathbf{A}^{-T}. \tag{B.11.40}$$

Here $\mathbf{A}^{-T}$ denotes the inverse of the transpose $\mathbf{A}^T$ of $\mathbf{A}$. That is [cf. (B.8.1)],

$$\mathbf{A}^T\mathbf{U}\,.\,\mathbf{U}' := \mathbf{U}\,.\,\mathbf{A}\mathbf{U}' \tag{B.11.41}$$

holds for all $\mathbf{U}, \mathbf{U}' \in \mathbb{R}^{6N}$. Of course [see (B.11.17))

$$(\mathbf{A}^T)_{pq} := \mathbf{E}_p \,.\, \mathbf{A}^T\mathbf{E}_q = \mathbf{A}\mathbf{E}_p \,.\, \mathbf{E}_q = A_{qp}. \tag{B.11.42}$$

The analysis of Appendix B.5 leading to Results B.5.1 and B.5.2 may formally be repeated on making the identifications [see (14.2.13) and (14.2.16)]

$$\mathbf{x} \leftrightarrow \mathbf{X}, \qquad \hat{\mathbf{x}} \leftrightarrow \hat{\mathbf{X}}, \qquad \chi_{t_0} \leftrightarrow \boldsymbol{\phi}_0, \qquad \mathbf{F} \leftrightarrow \nabla_{\mathbb{P}}\boldsymbol{\phi}_0, \qquad \text{and} \qquad \mathbf{v} \leftrightarrow \mathbf{V}. \tag{B.11.43}$$

Specifically, the analogues of these results are [cf. (B.5.23)]

$$\nabla_{\mathbb{P}}\mathbf{V} = \dot{\widehat{\nabla\mathbf{d}}}(\nabla\mathbf{d})^{-1} \tag{B.11.44}$$

and [cf. (B.5.36)]

$$\dot{\widehat{\det\nabla\mathbf{d}}} = \mathrm{tr}\{\dot{\widehat{\nabla\mathbf{d}}}(\nabla\mathbf{d})^{-1}\}\det\nabla\mathbf{d}. \tag{B.11.45}$$

Thus, from (B.11.38),

$$\dot{\widehat{\det\nabla\mathbf{d}}} = (\mathrm{div}_{\mathbb{P}}\mathbf{V})\det\nabla\mathbf{d}. \tag{B.11.46}$$

B.11.3 Integration in $\mathbb{R}^{6N}$

The discussion of Riemann integration in Appendix B.6 readily generalises to $\mathbb{R}^{6N}$ once the appropriate notion of a smooth surface is established. A smooth surface in $\mathbb{R}^{6N}$ is associated with a real-valued function f on $\mathbb{R}^{6N}$ with a continuous derivative [see (B.11.24)]. Specifically, the *smooth surface defined by f which passes through* $\mathbf{X}_0 \in \mathbb{R}^{6N}$ is

$$\mathcal{S}(f;\mathbf{X}_0) := \{\mathbf{X} \in \mathbb{R}^{6N} : f(\mathbf{X}) = f(\mathbf{X}_0) \text{ and } \nabla_{\mathbb{P}} f(\mathbf{X}) \neq \mathbf{0}\}. \tag{B.11.47}$$

The unit normal fields are

$$\mathbf{N} := \pm \frac{\nabla_{\mathbb{P}} f}{\|\nabla_{\mathbb{P}} f\|_{\mathbb{P}}}. \tag{B.11.48}$$

Continuity of $\mathbf{N}$ on $\mathcal{S}$ is ensured if $\nabla_{\mathbb{P}} f$ is a continuous non-zero function of position in $\mathbb{R}^{6N}$.

A closed and bounded region R in $\mathbb{R}^{6N}$ with piece-wise smooth boundary[12] is termed *regular.*

Exercise B.11.7. Consider how Appendix B.6 generalises from $\mathcal{E}$ to $\mathbb{R}^{6N}$. Specifically, what constitutes a 'rectangular box' in $\mathbb{R}^{6N}$? How is a regular region partitioned, and what is the 'fineness' of a partition?[13] Note the analogue of requirement (B.6.4) which is necessary to establish the volume of R. Consider the integral of a scalar field f over R by examining the direct analogues of (B.6.7) through (B.6.14), and note how vector and higher-order tensor fields are thereby defined [see (B.6.15) and (B.6.16)].

The existence of integrals of *continuous* fields over regular regions and their evaluation by repeated ($6N$-fold!) Riemann integrals, each taken over an interval or disjoint union of intervals in $\mathbb{R}$, are direct analogues of Answers 1 and 2 in Appendix B.6.

Analogues of Theorems B.6.1 and B.6.2 should be evident. In particular, the effect of a deformation $\mathbf{d}$ upon a partition of a regular region into generalised parallelipipeds is approximated locally by its linearisation [see (B.11.29), and compare (B.4.28) and (B.4.29)]

$$\mathbf{d}_{\text{lin}}(\mathbf{X}+\mathbf{H}) := \mathbf{d}(\mathbf{X}) + \nabla\mathbf{d}(\mathbf{X})\mathbf{H}. \tag{B.11.49}$$

This homogeneous deformation [see (B.11.20): here $\mathbf{Y}_0 = \mathbf{d}(\mathbf{X}), \mathbf{A} = \nabla\mathbf{d}(\mathbf{X})$, and $\mathbf{H} = (\mathbf{X} - \mathbf{X}_0)$] has associated volume magnification factor [the *Jacobian* of $\mathbf{d}$ at $\mathbf{X}$: cf. (B.5.6)]

$$J_{\mathbf{d}}(\mathbf{X}) := |\det \nabla\mathbf{d}(\mathbf{X})|. \tag{B.11.50}$$

It follows that the analogue of (B.6.32) with $\hat{R} = \mathbb{S}, f = P(.,t)$, and $\mathbf{d} = \boldsymbol{\phi}_0(.,t)$ is [see (14.2.14)]

$$\int_{\boldsymbol{\phi}_0(\mathbb{S},t)} P(\mathbf{X},t)dV_{\mathbf{X}} = \int_{\mathbb{S}} P(\boldsymbol{\phi}_0(\hat{\mathbf{X}},t),t)J_{\boldsymbol{\phi}_0}(\hat{\mathbf{X}},t)dV_{\hat{\mathbf{X}}}. \tag{B.11.51}$$

[12] That is, the boundary consists of a finite number of smooth surfaces in the sense of (B.11.47).

[13] Note that the maximum distance between points of parallelipiped $P(X_0;\mathbf{U}_1,\ldots,\mathbf{U}_N)$ in (B.11.11) is $\|\mathbf{U}_1 + \ldots + \mathbf{U}_{6N}\|_{\mathbb{P}}$.

From (14.2.14) this yields

$$\int_{\mathbb{S}} \{P_0(\hat{\mathbf{X}}) - P(\boldsymbol{\phi}_0(\hat{\mathbf{X}},t),t)J_{\boldsymbol{\phi}_0}(\hat{\mathbf{X}},t)\}dV_{\hat{\mathbf{X}}} = 0. \tag{B.11.52}$$

Accordingly, if (14.2.14) is to hold for every regular region $\mathbb{S}$ in $\mathbb{R}^{6N}$, and the integrand is continuous, then the analogue of Theorem B.6.1 implies that

$$P(\boldsymbol{\phi}_0(\hat{\mathbf{X}},t),t)J_{\boldsymbol{\phi}_0}(\hat{\mathbf{X}},t) = P_0(\hat{\mathbf{X}}). \tag{B.11.53}$$

Noting that locally $J_{\boldsymbol{\phi}_0}$ is either $+\det\nabla\boldsymbol{\phi}_0$ or $-\det\nabla\boldsymbol{\phi}_0$ and $\det\nabla\boldsymbol{\phi}_0$ is continuous, (B.11.46) with $\mathbf{d} = \nabla\boldsymbol{\phi}_0$ yields

$$\dot{J}_{\boldsymbol{\phi}_0} = (\mathrm{div}_{\mathbb{P}}\,\mathbf{V})J_{\boldsymbol{\phi}_0}. \tag{B.11.54}$$

Differentiation of (B.11.53) with respect to time thus yields

$$\left(\frac{\partial P}{\partial t} + \nabla_{\mathbb{P}} P.\mathbf{V} + P\,\mathrm{div}_{\mathbb{P}}\,\mathbf{V}\right)J_{\boldsymbol{\phi}_0} = 0. \tag{B.11.55}$$

The result (14.2.22) of Exercise 11.3.2 (with $f = P$ and $\mathbf{U} = \mathbf{V}$), together with the non-vanishing of $J_{\boldsymbol{\phi}_0}$, yield the key relation (14.2.15).

References

[1] Gurtin, M. E., 1981. *An Introduction to Continuum Mechanics.* Academic Press, New York.
[2] Truesdell, C., & Noll, W., 1965. The nonlinear field theories of mechanics. In: *Handbuch der Physik* III/1 (ed. S. Flügge). Springer-Verlag, Berlin.
[3] Chadwick, P., 1976. *Continuum Mechanics.* Allen & Unwin, London.
[4] Landau, L. D., & Lifschitz, E. M., 1959. *Fluid Mechanics.* Pergamon Press, London.
[5] Paterson, A. R., 1983. *A First Course in Fluid Dynamics.* Cambridge University Press, Cambridge, UK.
[6] Brush, S. G., 1986. *The Kind of Motion we call Heat.* North Holland, Amsterdam.
[7] Goldstein, H., Poole, C., & Safko,J., 2002. *Classical Mechanics* (3rd ed.). Addison-Wesley, San Francisco.
[8] Truesdell, C., 1977. *A First Course in Rational Continuum Mechanics*, Vol. 1. Academic Press, New York.
[9] Born, W., & Wolf, E., 1999. *Principles of Optics* (7th ed.). Cambridge University Press, Cambridge, UK.
[10] Hardy, R. J., 1982. Formulas for determining local properties in molecular-dynamics simulations: Shock waves. *J. Chem. Phys.* **76**, 622–628.
[11] Gurtin, M. E., 1972. *The linear theory of elasticity*. In: *Handbuch der Physik* VIa/2 (ed. C. Truesdell). Springer-Verlag, Berlin.
[12] Zadeh, L. A., 1965. Fuzzy sets. *Information and Control* **8**, 338–353.
[13] Bear, J., 1972. *Dynamics of Fluids in Porous Media.* Elsevier, Amsterdam.
[14] Murdoch, A. I., & Kubik, J., 1995. On the continuum modelling of porous media containing fluid: a molecular viewpoint with particular attention to scale. *Transport in Porous Media* **19**, 157–197.
[15] Murdoch, A. I., & Hassanizadeh, S. M., 2002. Macroscale balance relations for bulk, interfacial and common line systems in multiphase flows through porous media on the basis of molecular considerations. *Int. J. Multiphase Flow* **28**, 1091–1123.
[16] Noll, W., 1955. Der Herleitung der Grundgleichen der Thermomechanik der Kontinua aus der statischen Mechanik. *J. Rational Mech. Anal.* **4**. 627–646. (Translated as: Lehoucq, R., & von Lilienfeld, O. A., 2010. Derivation of the fundamental equations of continuum mechanics from statistical mechanics. *J. Elasticity*, **100**, 5–24.)
[17] Atkins, P. W., 1996. *The Elements of Physical Chemistry* (2nd ed.). Oxford University Press, Oxford, UK.
[18] Murdoch, A. I., 2007. A critique of atomistic definitions of the stress tensor. *J. Elasticity* **88**, 113–140.
[19] Carlsson, T., & Leslie, F. M., 1999. The development of theory for flow and dynamic effects for nematic liquid crystals. *Liquid Crystals* **26**, 1267–1280.
[20] Murdoch, A. I., 2003. On the microscopic interpretation of stress and couple stress. J. Elasticity **71**, 105–131.

[21] Kellogg, O. D., 1967. *Foundations of Potential Theory* (reprint of first edition of 1929). Springer-Verlag, Berlin.

[22] Israelachvili, J. N., 1974. The nature of van der Waals forces. *Contemp. Phys.* **15**, 159–177.

[23] Gurtin, M. E., Fried, E., & Anand, L., 2010. *The Mechanics and Thermodynamics of Continua*. Cambridge University Press, New York.

[24] de Gennes, P. G., 1974. *The Physics of Liquid Crystals*. Oxford University Press, Oxford, UK.

[25] Toupin, R. A., 1962. Elastic materials with couple-stress. *Arch. Rational Mech. Anal.* **11**, 385–414.

[26] Mindlin, M. D., & Tiersten, H. F., 1962. Effects of couple-stresses in linear elasticity. *Arch. Rational Mech. Anal.* **11**, 415–448.

[27] Murdoch, A. I., 1987. On the relationship between balance relations for generalised continua and molecular behaviour. *Int. J. Eng. Sci.* **25**, 883–914.

[28] Sedov, L. I., 1965. *Introduction to the Mechanics of a Continuous Medium*. Addison-Wesley, Reading, MA.

[29] Reddy, J. N., 2006. *An Introduction to Continuum Mechanics*. Cambridge University Press, Cambridge, UK.

[30] Kröner, E. (ed.), 1968. *Mechanics of Generalized Continua*. Springer-Verlag, New York.

[31] Carlson, D. E., 1972. Linear thermoelasticity. In: *Handbuch der Physik*, VIa/2 (ed. C. Truesdell). Springer-Verlag, Berlin.

[32] Ohanian, H. C., 1985. *Physics* (Vol. 1). Norton, New York.

[33] Truesdell, C., & Toupin, R. A., 1960. The classical field theories. In: *Handbuch der Physik* III/1 (Ed. S. Flügge). Springer-Verlag, Berlin.

[34] Truesdell, C., 1969. *Rational Thermodynamics*. McGraw-Hill, New York.

[35] Atkin, R. J., & Craine, R. E., 1976. Continuum theories of mixtures: basic theory and historical development. *Q. J., Mech. Appl. Math.* **XXIX**, 211–244.

[36] Bowen, R. M., 1976. *Theory of mixtures*. In: *Continuum Physics*, III (Ed. A. C., Eringen). Academic Press, New York.

[37] Gurtin, M. E., Oliver, M. L., & Williams, W. O., 1973. On balance of forces for mixtures. *Q. Appl. Math.* **30**, 527–530.

[38] Williams, W. O., 1973. On the theory of mixtures. *Arch. Rational Mech. Anal.* **51**, 239–260.

[39] Oliver, M. L., & Williams, W. O., 1975. Formulation of balance of forces in mixture theories. *Q. Appl. Math.* **33**, 81–86.

[40] Morro, A., & Murdoch, A. I., 1986. Stress, body force, and momentum balance in mixture theory. *Meccanica* **21**, 184–190.

[41] Murdoch, A. I., & Morro, A., 1987. On the continuum theory of mixtures: motivation from discrete considerations. *Int. J. Eng. Sci.* **25**, 9–25.

[42] Bedford, A., 1983. Theories of immiscible and structured mixtures. *Int. J. Eng. Sci.* **21**, 863–890.

[43] Alblas, J. B., 1976. A note on the physical foundation of the theory of multipole stresses. *Arch. Mech. Stos.* **28**, 279–298.

[44] Murdoch, A. I., 1985. A corpuscular approach to continuum mechanics. *Arch. Rational Mech. Anal.* **88**, 291–321.

[45] Noll, W., 1973. Lectures on the foundations of continuum mechanics and thermodynamics. *Arch. Rational Mech. Anal.* **52**, 62–92.

[46] MacLane, S., & Birkhoff, G., 1967. *Algebra*. Macmillan, London.

[47] Roberts, P. H., & Donnelly, R. J., 1974. Superfluid mechanics. In: *Annual Review of Fluid Mechanics* **6** (ed. van Dyke, M., Vincentini, W. G., & Wehausen, J. V.). Annual Reviews, Palo Alto, CA.

[48] Hills, R. N., & Roberts, P. H., 1977. Superfluid mechanics for a high density of vortex lines. *Arch. Rational Mech. Anal.* **66**, 43–71.

[49] Piquet, J., 2001. *Turbulent Flows* (revised 2nd printing). Springer-Verlag, Berlin.

[50] Murdoch, A. I., 2003. Objectivity in classical continuum mechanics: a rationale for discarding the 'principle of invariance under superposed rigid body motions' in favour of purely objective considerations. *Continuum Mech. Thermodyn.* **15**, 309–320.
[51] Murdoch, A. I., 2006. Some primitive concepts in continuum mechanics regarded in terms of objective space-time molecular averaging: the key rôle played by inertial observers. *J. Elasticity* **84**, 69–97.
[52] Rivlin, R. S., 1970. Red herrings and sundry unidentified fish in nonlinear continuum mechanics. In: *Inelastic Behavior of Solids* (ed. Kanninen, M. G., Adler, D., Rosenfield, A. R., Jaffee, R. I.), McGraw-Hill, New York.
[53] Müller, I., 1972. On the frame dependence of stress and heat flux. *Arch. Rational Mech. Anal.* **45**, 241–250.
[54] Murdoch, A. I., 1983. On material frame-indifference, intrinsic spin, and certain constitutive relations motivated by the kinetic theory of gases. *Arch. Rational Mech. Anal.* **83**, 185–194.
[55] Wang, C. C., 1975. On the concept of frame-indifference in continuum mechanics and in the kinetic theory of gases. *Arch. Rational Mech. Anal.* **58**, 381–393.
[56] Truesdell, C., 1976. Correction of two errors in the kinetic theory of gases which have been used to cast unfounded doubt upon the principle of material frame-indifference. *Meccanica* **11**, 196–199.
[57] Speziale, G., 1981. On frame-indifference and iterative procedures in the kinetic theory of gases. *Int. J. Eng. Sci.* **19**, 63–73.
[58] Svendsen, B., & Bertram, A., 1999. On frame-indifference and form invariance in constitutive theory. *Acta Mechanica* **132**, 195–207.
[59] Edelen, G. B., & McLennan, J. A., 1973. Material indifference: a principle or a convenience. *Int. J. Eng. Sci.* **11**, 813–817.
[60] Söderholm, L. H., 1976. The principle of material frame-indifference and material equations of gases. *Int. J. Eng. Sci.* **14**, 523–528.
[61] Woods, L. C., 1981. The bogus axioms of continuum mechanics. *Bull. Inst. Math. Applications* **17**, 98–102.
[62] Burnett, D., 1935. The distribution of molecular velocities and the mean motion in a non-uniform gas. *Proc. Lond. Math. Soc.* **40**, 382–435.
[63] Murdoch, A. I., 1982. On material frame-indifference. *Proc. R. Soc. Lond. A* **380**, 417–426.
[64] Liu, I.-S., 2003. On Euclidean objectivity and the principle of material frame-indifference. *Continuum Mech. Thermodyn.* **16**, 309–320.
[65] Liu, I.-S., 2005. Further remarks on Euclidean objectivity and the principle of material frame-indifference. *Continuum Mech. Thermodyn.* **17**, 125–133.
[66] Murdoch, A. I., 2005. On criticism of the nature of objectivity in classical continuum physics. *Continuum Mech. Thermodyn.* **17**, 135–148.
[67] Edelen, D. G. B., 1976. Nonlocal field theories. In: *Continuum Physics*, IV (ed. A. C., Eringen). Academic Press, New York.
[68] Silling, S., 2000. Reformulation of elasticity theory for discontinuities and long-range forces. *J. Mech. Phys. Solids* **48**, 175–209.
[69] Lehoucq, R. B., & Sears, M. P., 2011. Statistical mechanical foundation of the peridynamic nonlocal continuum theory: Energy and momentum laws. *Physical Review E* **84**(031112), 1–7.
[70] Silling, S. A., Epton, M., Weckner, O., & Askari, E., 2007. Peridynamic states and constitutive modelling. *J. Elasticity* **88**, 151–184.
[71] Stone, A. J., 1984. Intermolecular forces. In: *Molecular Liquids – Dynamics and Interactions* (ed. A. J., Barnes). Reidel, Dordrecht, The Netherlands.
[72] Landau, L. D., & Lifschitz, E. M., 1980. *Statistical Physics* (3rd ed. part 1). Pergamon Press, Oxford, UK.
[73] Irving, J. H., & Kirkwood, J. G., 1950. The statistical mechanical theory of transport processes. IV. The equations of hydrodynamics. *J. Chem. Phys.* **18**, 817–829.

[74] Pitteri, M., 1986. Continuum equations of balance in classical statistical mechanics. *Arch. Rational Mech. Anal.* **94**, 291–305.
[75] Admal, N. C., & Tadmor, E. B., 2010. A unified interpretation of stress in molecular systems. *J. Elasticity* **100**, 63–143.
[76] Murdoch, A. I., & Bedeaux, D., 1993. On the physical interpretation of fields in continuum mechanics. *Int. J. Eng. Sci.* **31**, 1345–1373.
[77] Murdoch, A. I., & Bedeaux, D., 1994. Continuum equations of balance via weighted averages of microscopic quantities. *Proc. R. Soc. London A* **445**, 157–179.
[78] Murdoch, A. I., & Bedeaux, D., 1996. A microscopic perspective on the physical foundations of continuum mechanics: 1. Macroscopic states, reproducibility, and macroscopic statistics, at prescribed scales of length and time. *Int. J. Eng. Sci.* **34**, 1111–1129.
[79] Murdoch, A. I., & Bedeaux, D., 1997. A microscopic perspective on the physical foundations of continuum mechanics – II. A projection operator approach to the separation of reversible and irreversible contributions to macroscopic behaviour. *Int. J. Eng. Sci.* **35**, 921–949.
[80] Zwanzig, R., 1960. Ensemble method in the theory of irreversibility. *J. Chem. Phys.* **33**, 1338–1341.
[81] Zwanzig, R., 2004. *Nonequilibrium Statistical Mechanics*. Oxford University Press, Oxford, UK.
[82] Kröner, E., 1971. *Statistical Continuum Mechanics*. C. I. S. M. E., Courses and Lectures No. 92. Springer-Verlag, Vienna.
[83] Belleni-Morante, A., 1994. *A Concise Guide to Semigroups and Evolution Equations*. World Scientific, Singapore.
[84] Jacobson, N., 1951. *Lectures in Abstract Algebra*, Vol. 1 – *Basic Concepts*. van Nostrand, Princeton, NJ.
[85] Lamb, W., Murdoch, A. I., & Stewart, J., 2001. On an operator identity central to projection operator methodology. *Physica A* **298**, 121–139.
[86] Grabert, H., 1982. *Projection Operator Techniques in Nonequilibrium Statistical Mechanics*. Springer-Verlag, Berlin.
[87] Murdoch, A. I., & Bedeaux, D., 2001. Characterisation of microstates for confined systems and associated scale-dependent continuum fields via Fourier coefficients. *J. Phys. A: Math. Gen.* **34**, 6495–6508.
[88] Halmos, P. R., 1958. *Finite-Dimensional Vector Spaces*. van Nostrand, Princeton.
[89] Goertzel, G., & Tralli, N., 1960. *Some Mathematical Methods of Physics*. McGraw-Hill, New York.
[90] Greub, W., 1978. *Multilinear Algebra*, 2nd edn. Springer-Verlag, New York.
[91] Apostol, T. M., 1957. *Mathematical Analysis*. Addison-Wesley, Reading, MA.
[92] Moeckel, G. P., 1975. Thermodynamics of an interface. *Arch. Rational Mech. Anal.* **57**, 255–280.
[93] Gurtin, M. E., & Murdoch, A. I., 1975. A continuum theory of elastic material surfaces. *Arch. Rational Mech. Anal.* **57**, 291–323.
[94] Rusanov, A. I., 1971. Recent investigations on the thickness of surface layers. In: *Progress in Surface and Membrane Science*, Vol. 4. (ed. Danielli, J. F., Rosenberg, M. D., & Codenhead, D. A.) Academic Press, New York.
[95] Rowlinson, J. S., & Widom, B., 1982. *Molecular Theory of Capillarity*. Oxford University Press, London.
[96] Murdoch, A. I., 2005. Some fundamental aspects of surface modelling. *J. Elasticity* **80**, 33–52.
[97] Ericksen, J. E., 1976. Equilibrium theory of liquid crystals. In: *Advances in Liquid Crystals*, Vol. 2. (ed. G. H., Brown). Academic Press, New York.
[98] Leslie, F. M., 1979. Theory of flow phenomena in liquid crystals. In: *Advances in Liquid Crystals*, Vol. 4. (ed. G. H., Brown). Academic Press, New York.
[99] Maugin, G. A., 1993. *Material Inhomogeneities in Elasticity*. Chapman & Hall, London.
[100] Gurtin, M. E., 2000. *Configurational Forces as a Basic Concept of Continuum Physics*. Springer-Verlag, Berlin.

[101] Coulson, C. A., 1961. *Electricity*. Oliver & Boyd, London.
[102] Rutherford, D. E., 1957. *Vector Methods (9th ed.)*. Oliver & Boyd, London.
[103] Murdoch, A. I., 2003. *Foundations of Continuum Modelling: a Microscopic Perspective with Applications. Center of Excellence for Advanced Materials and Structures.* IPPT, Polish Academy of Sciences, Warsaw.
[104] Day, W. A., 1972. *The Thermodynamics of Simple Materials with Fading Memory.* Springer-Verlag, Berlin.

Index